INTEGRAL METHODS IN SCIENCE AND ENGINEERING–90

INTEGRAL METHODS IN SCIENCE AND ENGINEERING–90

Edited by

A. Haji-Sheikh
Constantin Corduneanu
John L. Fry
Tseng Huang
Fred R. Payne

The University of Texas at Arlington

⊙ **HEMISPHERE PUBLISHING CORPORATION**
A member of the Taylor & Francis Group

New York   Washington   Philadelphia   London

INTEGRAL METHODS IN SCIENCE AND ENGINEERING–90

Copyright © 1991 by Hemisphere Publishing Corporation. All rights reserved. Printed in the United States of America. Except as permitted under the United States Copyright Act of 1976, no part of this publication may be reproduced or distributed in any form or by any means, or stored in a database or retrieval system, without the prior written permission of the publisher.

1 2 3 4 5 6 7 8 9 0   E B E B   9 8 7 6 5 4 3 2 1

A CIP catalog record for this book is available from the British Library.

Library of Congress Cataloging-in-Publication Data

Integral methods in science and engineering-90 / edited by A. Haji-Sheikh.
       p. cm.
  Includes bibliographical references and index.

   1. Engineering mathematics. I. Haji-Sheikh, A.
TA330.I58   1991
620'.0015154—dc20   91-6935
                                            CIP

ISBN 1-56032-066-4

Contents

Preface **xi**

IMSE-90 Organization **xv**

Reviewers **xvii**

STOCHASTIC ANALYSIS

Trends in Science: The Infinitely Large, the Infinitely Small and the Infinitely Complex
H. Mark **3**

The Use of Multiple Integrals in the Characterization of Renewal Counting Processes
B. D. Sivazlian **11**

Molecular Dynamics and Monte Carlo Simulations of Polymeric Materials: Chain Relaxation Dynamics
W. Brostow, R. Cook, and J. Kubát **22**

A Comparison of Some Approximate Methods of Computing Bayes Estimators
S. K. Bhattacharya, A. Pandey, and A. K. Singh **35**

Exact Distribution of the Least-Squares Estimator of the Linear-Stress Failure Model
D. A. Miller and A. K. Singh **43**

On Weak Solutions of Nonlinear Volterra Stochastic Equations
M. Lewin **52**

Modeling of Dynamic Systems by Itô-Type Systems of Stochastic Differential Equations
G. S. Ladde **63**

Itô-Type Systems of Stochastic Integro-Differential Equations
G. S. Ladde and S. Sathananthan **75**

INTEGRAL EQUATIONS

Representation Theory and Nonlinear Systems
I. W. Sandberg **92**

Integral Methods for Solving Fokker-Planck-Type Equations
V. Protopopescu **111**

Absolute Stability of Certain Delay Systems
    C. Corduneanu   **123**

On the Use of Product Integration for the Solution of Finite Part Singular Integral Equations
    B. Bertram   **132**

On Nonlinear Sturm-Liouville Boundary Value Problems at Resonance
    H. Matala-aho and S. Seikkala   **137**

Comparison of a Direct Method for Inverting Fredholm Equations of the First Kind with the Method of Regularization
    J. A. Merson and T. S. Cale   **147**

Computational Aspects of Accelerated Projection Methods for Nonlinear Integral Equations
    D. R. Dellwo, M. B. Friedman, and R. Aggarwal   **158**

CONTINUUM MECHANICS

Multiple Stable Solutions to the Dynamical Equations of an Excitable Medium
    A. T. Winfree   **172**

The Rigorous Solution of the Classical Theory of Plates
    C. Constanda   **184**

Explicit Stiffness and Error Estimator Expressions for a Family of Axisymmetric Triangular Finite Elements
    B. E. Bergmann, R. V. Nambiar, and K. L. Lawrence   **191**

Integral Method for Hygrothermal Effect of Composite Plates
    S. Nomura and J. S. Chang   **201**

Theoretical Studies of Some Neuropharmacological Effects on Directional Selectivity
    H. Öğmen   **206**

FINITE ANALYSIS

Conservation Laws and Boundary Conditions in Fluid Mechanics
    K. Oshima   **222**

Some Remarks on the Boundary Element Method for Nonlinear Problems
    K. Ruotsalainen   **228**

A Mixed Finite Element Method for Stokes Problem—Acceleration, Pressure, Formulation
    C. L. Chang and G. J. Fix   **240**

Numerical Solution of the Two-Dimensional Scalar Helmholtz Equation Using the Bymoment Method
A. Cangellaris and R. Lee **249**

Combined Finite Element/Boundary Integral Approach to the Prediction of Acoustic Signatures and Radar Cross Sections
E. T. Moyer Jr. and E. A. Schroeder **260**

Finite Analytic Method for Numerical Solution of Mathematical Models
F. Civan **270**

Quadrature and Cubature Methods for Numerical Solution of Integro-Differential Equations
F. Civan **282**

Numerical Solution of Second Order Differential Equations
W. Squire **298**

Progressive Aitken Midpoint Quadrature
W. Squire **307**

TRANSPORT PROCESSES

Over the Design of Supersonic Aircraft and Space Vehicles of Minimum Drag in Supersonic and Hypersonic Flow
A. Nastase **319**

A Field Integral Method for Solutions of the Full Potential Equation in Rectangular Grids
Z. Fang and I. Paraschivoiu **342**

A Triad of Solutions for 2-D Navier-Stokes: Global, Semi-Local and Local
M. Nair and F. R. Payne **352**

A New Integral Method for Magnetohydrodynamic (MHD) Flow Analysis
Y.-M. Lee and D. R. Wilson **360**

Bifurcation in Rayleigh-Bénard Convection with Gases in Rectangular Containers
J. R. Leith and J. G. Maveety **376**

Simulation of EM Wave Scattering from a Dense Distribution of Spherical Scatterers
Y. C. Tzeng, J. Bredow, and A. K. Fung **388**

A Comparison Between a Frequency Domain Method and the Standard Moment Method in Surface Scattering
K. S. Chen and A. K. Fung **411**

Use of First and Higher Order Influence Functions in the Duhamel's Integral
Solution of the IHCP
   B. Litkouhi and M. H. N. Naraghi   **423**

A Method for Solving Coupled Initial Boundary Value Systems of Second
Order Partial Differential Equations, Exact and Accurate Approximate
Solutions
   L. Jódar   **437**

Integral Equations with Monte Carlo Boundary Conditions: Conduction in
Electronic Devices
   A. Haji-Sheikh and S. P. Kinsey   **448**

SYSTEMS AND DYNAMICS

Predicting the Autonomous Behaviour of Simple Nonlinear Feedback
Systems
   D. P. Atherton   **460**

A Global View of Analytical Optimization Methods
   S. S. L. Chang   **476**

A Methodology for Computation of the Exact Distribution for an Acceptance
Testing Procedure for a Parallel System
   A. K. Singh and A. Singh   **482**

SUPERCOMPUTER APPLICATIONS

Scientific Supercomputing in the '90's: The Grand Challenges, the Global
Village and the Great Debate
   J. W. D. Connolly   **493**

Parallel Computing in Science
   P. C. Pattnaik   **501**

Supercomputer Study of a Particle Arch
   T. D. Kelly   **507**

Instability Approach to the Magnetic Structure of Transition Metals
   K. Schwartzman   **524**

Computational Approach to Determine Magnetic Order in High Temperature
Superconductors
   J. L. Fry, T. W. Chang, E. C. Ethridge, and P. C. Pattnaik   **531**

Large Scale Scientific Computing Initiative for Solids
   *N. E. Brener, J. Callaway, J. M. Tyler, G. Fuster, J. L. Fry,
   and E. C. Ethridge*   **539**

Integral Methods in the Quasi-particle Theory of Electronic Structure
  M. Zaider, J. L. Fry, and D. E. Orr   **546**

Self-Consistency in Embedded Green Function Calculations of Surface
Electronic Structure
  J. H. Kaiser and R. Marroum   **564**

Dynamics of Relaxation and Reconstruction at Semiconductor Surfaces
  J. Gryko and R. E. Allen   **572**

Light Particles in Fluids: Mean Field Theory versus Path-Integration
  B. N. Miller   **581**

Path Integral Monte Carlo Simulation of Ortho-Positronium in a 2-
Dimensional Lennard-Jones Fluid
  G. A. Worrell and T. Reese   **591**

Electronic Energy Levels in Condensed Media: New Techniques for
Theoreticians
  N. R. Kestner and J.-A. Yang   **600**

Index   **611**

Preface

The second international conference on global methods (specifically integral techniques), entitled "Integral Methods in Science and Engineering (IMSE-90)," was hosted by the University of Texas at Arlington on 15-18 May, 1990. Four continents and 11 countries were represented. Total attendance was approximately 90, with seven invited papers, and 66 contributed papers from which 54 were selected for publication. The invited speakers, all of whom addressed "real," i.e., nonlinear problems, were, in order of appearance:

Hans Mark, Chancellor of the University of Texas System, gave the keynote address opening the conference.

Adriana Nastase, Director, Aerodynamics Institute, RWTH-Aachen, Germany, spoke on design of vehicles for minimum drag at supersonic and hypersonic speeds.

Koichi Oshima, Institute of Space and Astronautical Sciences and President of the Japan CFD Society, discussed the conservation laws and boundary conditions in fluid mechanics.

J. W. D. Connolly, Director of the Center for Computational Sciences, University of Kentucky and past NSF-Supercomputer Director, cited scientific applications of supercomputers.

D. P. Atherton, Professor of Control Engineering, University of Sussex, England, surveyed the rich behaviors of certain "simple" nonlinear systems.

I. W. Sandberg, Regents Chair in Engineering, EE and Computer Engineering, University of Texas-Austin, discussed representation theory and nonlinear systems.

A. T. Winfree, Regents Professor, Biosciences, University of Arizona, described his numerical experiments in three dimensions for excitable media.

IMSE-90 is the second in a series of conferences emphasizing a group of methodologies and their application across the full spectrum of physical problems as opposed to the more usual focus upon one or more disciplines. The goal of IMSE-90, identical to that of IMSE-85, also sponsored by the University of Texas at Arlington, appears well-received. That goal is to promote synergistic interaction of researchers using integral methods in one or more of its manifold manifestations in their assaults upon problems of technological or intrinsic interest. Some examples are: Volterra and Fredholm equations arising in mechanics and transport processes, Baysian statistics, Itô-systems, Fokker-Planck equations, Monte Carlo, control theory, delay equations, Schroedinger equation, and other topics. Thus, the scope of applications in IMSE-90 was quite broad.

Genesis of the IMSE concept in 1982 was a suggestion by F. Payne to John Rouse, the former Dean of Engineering, and subsequent work by the first IMSE committee. Selected, peer-reviewed papers were published, as herein, under hardcover. At IMSE-85, John Fry suggested adjoint supercomputer activities, which was done at IMSE-90.

The IMSE concept can be characterized as a "parallel" one whose umbrella encompasses any disciple of study in contrast to the "serial" viewpoint of more conventional conferences exploring one or more disciplines.

The IMSE can serve as a renewal for the attendees by cross-fertilization of ideas from nominally differing disciplines due to the unifying impact of similar methodologies. This, IMSE's ultimate benefit, appears well-served by this pair of conferences. An international advisory committee, with members from four continents, was formed at IMSE-90 to promote this concept.

ACKNOWLEDGMENT OF FINANCIAL SUPPORT OF IMSE-90

Significant contributions were made in support of this event by the following University of Texas at Arlington administrators: Wendell H. Nedderman, President of the University; Bob F. Perkins, Vice-President for Research and Dean of the Graduate School; Howard J. Arnott, Dean of the College of Science; and John H. McElroy, Dean of the College of Engineering. Without the financial and moral support of these University administrators, IMSE-90 would not have been possible. In addition, good support by the home departments of the Organizing Committee played essential roles; these were Aerospace, Civil, Electrical and Mechanical Engineering together with Chemistry, Mathematics and Physics.

ACKNOWLEDGEMENT OF SUPERIOR SERVICES TO IMSE-90

William M. Jacqmein, Colonel USA (ret), efficiently coordinated a myriad of details and thereby relieved the Organizing Committee to concentrate upon the scientific, rather than managerial, aspects. Rosemary Haji-Sheikh acted as technical secretary, tracking some hundred addresses and 60 contributed papers. She initiated and/or finalized a huge amount of correspondence and prepared the conference schedule; her help, as a purely volunteer worker, eased greatly program composition and editorial tasks for the Committee. Barbara Sanderson, AE Secretary, produced hundreds of memos and letters over the period June 1988 to September 1990, and beyond. She also made hundreds of calls, mostly to get the Committee together. Judy Gomez, AE Executive Secretary, acted as financial officer and tracked a myriad of inputs and expenditures from three separate accounts. Seven departments, AE, CE, EE, ME, Chemistry, Mathematics, and Physics, not only provided Committee members and secretarial aid but also students who volunteered as unpaid aides to Session Chairmen in their conduct of IMSE-90 technical sessions.

SPECIAL THANKS

A. Haji-Sheikh, Professor of Mechanical Engineering, member of both IMSE Organizing Committees and Chairman, IMSE-90 Editorial Committee was the glue which held together the technical program. He processed literally thousands of pieces of correspondence and phone calls. In 1985 he and the Chairman split these duties; this time he had virtually all of it.

Fred R. Payne

IMSE-90 Organization

IMSE-90 Organizing Committee:

Fred R. Payne, *IMSE-90 Chairman*
John L. Fry, *Supercomputer Chairman*
A. Haji-Sheikh, *Editorial Committee Chairman*
C. Corduneanu, *Mathematics*
Adrian K. Fung, *Electrical Engineering*
Tseng Huang, *Civil Engineering*
Zoltan A. Schelly, *Chemistry*
Bo Ping Wang, *Mechanical Engineering*

IMSE-90 Editorial Committee:

B. Bertram, *Michigan Technological University, USA*
Faruk Civan, *University of Okahoma, USA*
C. Constanda, *University of Strathyclyde, Scotland*
C. C. Corduneanu, *University of Texas at Arlington, USA*
John L. Fry, *University of Texas at Arlington, USA*
A. Haji-Sheikh, *University of Texas at Arlington, USA*
G. S. Ladde, *University of Texas at Arlington, USA*
J. R. Leith, *University of New Mexico, USA*
W. von Maltzahn, *University of Texas at Arlington, USA*
A. Nastase, *RWTH-Aachen, Germany*
P. C. Pattnaik, *IBM Research Division, Yorktown Heights, NY, USA*
F. R. Payne, *University of Texas at Arlington, USA*
K. Schwartzman, *Indiana University of Pennsylvania, Indiana, USA*
A. K. Singh, *New Mexico Institute of Mining and Technology, USA*
B. P. Wang, *University of Texas at Arlington, USA*

Steering Committee Members for the 3rd IMSE Conference:

C. Constanda (Committee Chairman), *University of Strathclyde, Scotland*
H. H. Chiu, *Institute Aero/Astro, NCKU, Taiwan*
C. A. Fletcher, *University of Sydney, Australia*
V. P. Korobeinikov, *USSR Academy of Science, USSR*
A. Nastase, *Aerodynamics Institute, RWTH-Aachen, Germany*
W. J. Minkowycz, *University of Illinois at Chicago, USA*
K. Oshima, *Inst. Space and Astro Sciences, Japan*
I. Paraschivoiu, *Ecole Polytechnique, Montreal, Canada*

S. Seikkala, *University of Oulu, Finland*
H. Wu, *Computing Center, Academia Sinica, Beijing, China*
K. S. Yajnik, *National Aeronautical Laboratory, Bangalore, India*
D. Yarbrough, *Tennessee Technological University, USA*
C. Corduneanu, *University of Texas at Arlington, USA*
J. L. Fry, *University of Texas at Arlington, USA*
A. Haji-Sheikh, *University of Texas at Arlington, USA*
T. Huang, *University of Texas at Arlington, USA*
F. R. Payne, *University of Texas at Arlington, USA*

Reviewers

The IMSE-90 Organizing Committee appreciates the assistance of the following individuals in reviewing the manuscripts and some generously did multiple reviews.

B. Bertram
F. Civan
S. S. L. Chang
C. Corduneanu
A. Fung
J. L. Fry
J. Gryko
S. Gutman
A. Hjjafar
A. Haji-Sheikh
T. Huang
J. H. Kaiser
M. A. Karam
N. R. Kestner
S. P. Kinsey
A. R. Kukreti
W. T. Kyner
G. S. Ladde
K. L. Lawrence
J. R. Leith
B. Litkouhi
B. Miller
D. A. Miller
E. T. Moyer, Jr.
M. H. N. Naraghi
A. Nastase

S. Nomura
A. Pandey
I. Paraschiviou
P. C. Pattnaik
F. R. Payne
V. Protopopescu
A. Razani
J. Reynolds
M. E. Rudd
O. G. Ruehr
M. Sambandham
S. Sathananthan
E. A. Schroeder
K. Schwartzman
A. Singh
A. K. Singh
B. D. Sivazlian
W. W. von Maltzan
B. P. Wang
L. W. White
D. D. Wilson
P. C. Wollan
A. Yucel
Y.-M. Lee
M. Zaider
M. M. Zaman

STOCHASTIC ANALYSES

Summary

The Keynote address by Hans Mark, Chancellor of the University of Texas System, entitled "Trends in Science: The Infinitely Large, The Infinitely Small and The Infinitely Complex," opened the conference at an appropriate level of insight and challenge. "Large" refers to astrophysics, "small" to nuclear science, each of which is rather well in hand by advances in this century; "complex" refers to life and other nonlinear processes finally coming under effective attacks. Chancellor Mark's remarks range several centuries of scientific endeavors, including conjectures in regard to entropy, self-organization, "strange attractors,", "chaos," usw. The reader is encouraged to peruse, with care, his paper as a good introduction to the technical content and "flavor" of this volume.

Sivazlian, "The Use of Multiple Integrals in the Characterization of Renewal Counting Processes," seeks a novel and more natural approach to obtaining joint probability density functions for arrival times assuming normality and identical probability distributions; then elementary calculations lead to rigorous derivation of JPDF. A number of examples enrich the paper.

Brostow, Cook, and Kubát, "Molecular Dynamics and Monte Carlo Simulations of Polymeric Materials: Chain Relaxation Dynamics," contains a sizable survey contrasting the two approaches.

Bhattacharya, Pandey, and Singh, "A Comparison of Some Approximate Methods of Computing Bayes Estimators," described a comprehensive series of numeric tests using Tierney-Kadane approximates and Laguerre quadratures for computing conditional probability estimates. Exact Bayes estimators are computed for test samples upon SUN 3/150 microcomputer with IMSL STAT package and FORTRAN 77. Several tables compare the two approaches for various ranges.

Miller and Singh, "Exact Distribution of the Least-Squares Estimator of the Linear-Stress Failure Model," compute a number of test cases. Paper is excellent as a series of model evaluations.

Lewin, "On Weak Solutions of Nonlinear Volterra Stochastic Equations," uses recent work by Metivier and Voit on weak solutions for PDE and treats a variational problem for stochastic Volterra IE. The Skorohod representation is used and Foedo-Galerkin approximates; a good amount of analysis is necessary in her proofs and an integral Fleming-Voit equation is validated to have at least a weak solution.

Ladde's, "Modeling of Dynamic Systems by Itô-Type Systems of Stochastic Differential Equations," treats Itô-Doob systems with both deterministic and random initial value problems (DIVP, RIVP) and proves five theorems dealing with integral representation of solutions, error estimates, relative stability (under variation of parameter) and stability analyses, corollaries and four examples.

Ladde and Sathananthan, "Itô-Type Systems of Stochastic Integro-Differential Equations," complements the prior paper. Similar results are obtained under somewhat different hypotheses; three examples vary.

Additional papers on applied stochastic processes are included in other sections of this volume. For example, applications to electromagnetic wave theory, random walk, and molecular dynamics are in the sections on *Transport Processes* and *Supercomputer Applications*.

Fred R. Payne

Trends in Science: The Infinitely Large, the Infinitely Small and the Infinitely Complex

HANS MARK
The University of Texas System
601 Colorado Street
Austin, Texas 78701, USA

A few years ago, Professor Victor Weisskopf of MIT, was asked to summarize the progress of science in the 20th Century. In his response to that question, he formulated a viewpoint which I think is particularly useful. He said that the major achievements of science in this century had been to understand the "infinitely large," the "infinitely small," and that we had made progress in understanding the "infinitely complex." It is of some interest perhaps to define more clearly what Weisskopf meant by these terms.

In the case of the "infinitely large," he was speaking of cosmology. The development of accurate cosmic distance scales using Cepheid variables, that is, variable stars whose frequency of pulsation is related to their absolute brightness, allowed Edwin Hubble in the 1920's to develop an accurate picture of the structure of the universe. Using these methods, Hubble measured the distribution of galaxies in the universe and, in fact, unravelled the structure of galaxies as well. In 1917, even before Hubble started his epoch-making work, Albert Einstein had written down the field equations of general relativity and discovered that the universe must be dynamic. Einstein was, in fact, unwilling to believe the implication of his own equations and amusingly, invented the fudge factor that he called the "cosmological constant" to permit the existence of a steady state universe. Ten years later, Edwin Hubble observed the "red shift" in the light coming from distant galaxies and discovered that the further away a galaxy is, the faster it seems to be moving away from us. Einstein's dynamic universe, therefore, seemed to be confirmed by observation.

In the case of the "infinitely small," we have the identification of electrons by J. J. Thompson in the 1890's and then the development of models of the atom which culminated in the work of Ernest Rutherford, who discovered experimentally that the "planetary" model of the atom was correct. An atom consisted of a massive nucleus surrounded by electrons confined to moving in well defined and calculable orbits. Once the structure of the atom was established, it was possible to

apply the laws of quantum mechanics which had been developed earlier
by Max Planck, Albert Einstein, Niels Bohr and others. With this
powerful new tool, the structure of matter was, for all practical
purposes, revealed. Weisskopf regarded the understanding of the
infinitely large and the infinitely small as the major triumph of the
human intellect in the 20th Century.

What is, of course, most interesting is that the infinitely large and
the infinitely small are coming together. The fact that the universe is
expanding is now well accepted. This means that the expansion must have
started with some violent "big bang" which is the accepted term for the
beginning of things today. Working backward from currently observed
expansion velocities, one can calculate that the "big bang" must have
occurred about twenty billion years ago. The fact that something like
this must have happened was elegantly confirmed about thirty years ago
by the discovery of the residual black body radiation, with a temperature
of about 3°K, left by the "big bang." These things have now been well
established.

At the same time, people began to look at the structure of the nucleus
and sub-nuclear particles using particle accelerators that were capable
of ever-higher energies. Families of sub-nuclear particles were
discovered that could exist only at very high energies (or temperatures).
These particles determine the structure of the nucleus and there is now
a reasonably good empirical model called the "standard model" that
explains the behavior of the "infinitely small."

In addition, great progress has been made in understanding the four
forces that seem to govern the behavior of the universe. These forces
are gravity, the "weak" force responsible for the nuclear process called
"beta decay," the electromagnetic force, and the "strong" nuclear force
that is responsible for holding the atomic nucleus together. As the
energy (or temperature) is increased, these forces apparently become
more like each other and ultimately coalesce, probably into one
"universal" force. It has already been shown that the "weak" force and
the "electromagnetic" force coalesce into the same force at sufficiently
high energies. It is conjectured that if the energy is raised, then the
other two will follow suit.

How are the infinitely small and the infinitely large related? The
answer is, of course, that they are related through the events that must
have occurred twenty billion years ago right after the "big bang." Right
after that event, the temperature of the universe was extremely high,
corresponding to many thousands of billions of electron volts. Thus, the
particles that have been discovered in our high energy accelerators must
have existed in some kind of thermo-dynamic equilibrium or quasi-
equilibrium because the temperature was sufficiently high to permit their
existence. Also, at very high temperatures, radiation rather than matter,
governs the behavior of the universe and energy is rapidly exchanged
between radiation and matter through the process of pair production.

In order to understand the behavior of the universe at very early
times, it is necessary to understand the properties of the matter
that existed at the time. This is, of course, where the behavior
of sub-nuclear particles becomes very important, and it is for this
reason that the currently observed properties of the "infinitely large"
depend ultimately on the behavior of particles of matter that are
"infinitely small." We are learning that the behavior of matter at the

moment and immediately after the "big bang" determines how the universe looks today. In this sense, the recently launched space telescope, which is appropriately named after Edwin Hubble, and the Superconducting Super Collider particle accelerator (SSC), soon to be built in Central Texas, are studying the same thing.

The story I have just told is intellectually very satisfying and is very definitely the triumph of 20th Century science that Weisskopf said it was. There is no question in my mind that Weisskopf is correct in that judgment.

What about the "infinitely complex?" By that term, Weisskopf wanted to describe the very complex chemical processes that are permitted by the properties of certain elements. The most important of these are, of course, hydrogen, carbon, oxygen and nitrogen, on which the "infinitely complex" chemical processes that govern the behavior of living beings are based. The understanding of the nature of life itself is the most important result that the infinitely complex will reveal.

How do the atoms (hydrogen, carbon, oxygen and nitrogen) come together to form more complicated molecules? Stanley Miller and Harold Urey showed that if one mixes these simple compounds of atoms and simply adds energy in any form, be it radiation, electrical discharge, or even ultra-sound, then more complicated molecules, specifically those necessary for the creation of the polymers governing life, can be formed. The first step is, therefore, fairly easy to explain. What happens next? How do molecules such as amino acids organize themselves to become proteins and nucleic acids? Not much is known about this and we are just beginning to understand, using sophisticated computer modeling, how this might happen. How do these molecules then organize themselves into living cells? Almost nothing is known about this process which is at the very origin of life. These are some of the most important questions that can be raised in discussing the infinitely complex.

In discussing these matters with a group of distinguished mathematicians such as this one, I thought that it might be appropriate to take another view. What I would like to do is to look at the same problems of the infinitely complex in terms of the differential equations and their solutions which will be discussed later on at this meeting. I have a suspicion that our understanding of the infinitely complex will come from mathematics -- the mathematics of non-linear systems.

It is a remarkable fact that almost everything we know and can predict about the physical world depends on a small number of differential or partial differential equations for which there exist exact solutions. These are the systems of linear equations for which the principle of superposition is applicable. The harmonic oscillator is probably the most important example of this class of relationship. The basic harmonic oscillator is, of course, the pendulum and it is interesting that a great many physical phenomena can at least be approximated as harmonic oscillators that obey an equation in which the driving force is proportional to some displacement. All of quantum field theory is based essentially on the mathematics of the harmonic oscillator.

The principle of superposition is a very powerful technique for dealing with linear differential equations. It permits the treatment of the equations by expanding the arbitrary solutions as a series of functions which have a definite protocol for calculating the coefficients of that series. The best example is, of course, the Fourier series but others such as spherical harmonics are very commonly used as well. Thus, the principle of superposition allows the development of solutions that are not necessarily purely harmonic in character and therefore permits the quantitative explanation of many different phenomena. The principle of superposition as well is one of the most powerful techniques that was applied in the development of quantum mechanics.

In addition to linear differential equations, there are also certain nonlinear differential equations that are "integrable." The best example of this is, of course, the Newtonian two body problem in which a force proportional to the inverse square of the distance between two bodies is assumed to exist. These integrable but nonlinear equations have the property that there are certain "constants" of the motion, that is, integrals of the motion such as the energy (for conservative systems) and the angular momentum (for systems with central forces) that are constant no matter what the parameters of the motion are under any given circumstances. It is most fortunate that the gravitational two body problem that Issac Newton first treated turned out to be such an integrable system. If that were not the case, it would have been much longer, very probably, before what we today call "rational mechanics" was invented.

In addition to the development of exact solutions for linear equations and for integrable nonlinear equations, powerful perturbation methods were developed that could be employed to treat certain non-linearities and/or divergences from integrability. If the perturbations are small enough, then it is possible to use the exact solutions as a starting point and develop mathematical methods that permitted the precise calculation of the effect of the perturbation. All of this led to the opening of a new era in the expansion of human knowledge starting at the end of the 17th Century. The ability to solve so many problems with a small arsenal of mathematical methods gave rise to a very optimistic "world view" which in the 18th Century came to be called "Age of Reason." The essential point is that the world was felt to be predictable. The fact that it was possible to explain very complicated phenomena with some relatively simple mathematical methods created an atmosphere in which people felt that in the end, it was necessary only to understand the "clockwork" which constituted the universe and then to perform the necessary calculations to predict the future. There is no doubt at all that this "rational" approach had a very powerful effect on people's thinking about politics, economics and all facets of life. There are even some who say that the American Constitution with its almost "mechanistic" approach to election processes and other procedures is an example of Newtonian politics in action.

Having invented the mechanics of particles using the Newtonian laws of motion, people turned their attention to other branches of science as the 18th Century drew to a close. The development of a theory of heat was particularly important because of the industrial machinery, especially the steam engines, then coming into existence. The equivalence of mechanical energy and heat was established by Benjamin Thompson (Count Rumford) at the end of the 19th Century and this eventually led to what we call today the first law of thermodynamics.

It was from the first law of thermodynamics that people began to
realize that the conservation of energy, which had been recognized
already by Newton in connection with particle mechanics, was indeed a
much more general principle.

Another general principle became apparent when the properties of heat
engines were examined in greater detail. It was discovered that
mechanical work extracted from a heat source could never be extracted
with perfect efficiency. That is, all of the heat energy in the source
could never be completely converted into mechanical energy. On the other
hand, it was possible to convert mechanical energy completely into heat.
There was, in short, some fundamental "irreversibility" which seemed to
characterize the behavior of heat engines of this type. Sadi Carnot and
his successors quantified this principle which we now call the second
law of thermodynamics. Carnot defined a new thermodynamic variable
called "entropy" which defines the amount of heat that is irretrievably
lost in any effort to convert heat energy to mechanical energy. It is
important to recognize that all of these analyses of heat engines were
carried out without once appealing to the "rational mechanics" of Newton
and his followers. These two bodies of knowledge developed quite
independently of each other during the 18th Century and they together
formed the basis of what Weisskopf called the "infinitely complex."

The resolution of what exactly it was that turned out to be "complex"
was when people began to suspect the existence of atoms from a detailed
examination of chemical processes. John Dalton, Antoine Laurent
Lavoisier, Joseph Priestley and their collaborators were responsible
for these developments in the last years of the 18th Century and the
first years of the next. Toward the end of the 19th Century, the most
capable theoretical physicists began to set themselves the problem of
combining atomic theory with what was then known about thermodynamics.
The great Ludwig Boltzmann was the one who finally succeeded in
accomplishing that objective. The dream was to explain thermodynamics
in terms of Newtonian mechanics. If that were indeed possible, then it
should be easy to understand the world and, at least in principle, it
should be possible to calculate it as well. Boltzmann's brilliant
insight was that it should be possible to relate the "average" variables
that defined the working fluid in heat engines, that is, the pressure,
the temperature, the flow properties of the fluid, to the behavior
of the individual atoms of which the fluid is constituted. In short,
he felt that he could relate everyday experience with the "infinitely
complex" behavior of thousands of billions of atoms. If this could be
done, then calculations could be made that would lead to "predictable"
results just as in the case of Newtonian mechanics.

In order to accomplish his objective, Boltzmann invented what we today
call "statistical mechanics." There is no question that the triumph of
Boltzmann's physics was the "explanation" of the second law of thermodynamics in terms of statistical mechanics. He derived the "Boltzmann
Transport Equation" which turns out to be a highly non-linear integral-
differential equation (an equation which is precisely the subject of
this conference). By developing certain solutions of this equation,
Boltzmann was able to achieve his breakthrough. Boltzmann also
developed the concept of "phase space" which defines the variables
describing a system of a large number of particles that constitute
the fluid or gas that Boltzmann was describing. In the Newtonian two
body problem, for example, the phase space has only six variables,
that is, the three displacement variables describing the relative

positions of the particles and the three conjugate momenta. In the case
of Botlzmann's system, which had a very large number of atoms, say "n",
there are six "n" variables describing the system. So far I have
described things that are quite familiar. Let me now turn to something
that is not as commonly understood. In developing his theories,
Boltzmann had to assume that in order for the second law of thermodynamics to hold, what is called the "ergodic hypothesis" had to be true.
The ergodic hypothesis says that any large system described in statistical
terms eventually occupies all possible states in phase space. This means
that every possible state of the system will eventually be occupied if
enough time passes. Another way of saying this is that the system
explores all possible states in phase space. Thus, if the ergodic
hypothesis is in fact true, then Boltzmann's marriage of Newtonian
mechanics and the thermodynamics of Rumford and Carnot would be complete.

Unfortunately, even at the time that Boltzmann was completing his great
work, there were people who questioned the validity of the ergodic
hypothesis. We now know that the ergodic hypothesis is in fact not true
and we now have rigorous proof to that effect. In Boltzmann's day, people
pointed out a fundamental paradox: If the universe really is a clockwork,
then the direction of time can go either way -- that is -- the universe
would be reversible. This statement seems to contradict the second law
of thermodynamics which, in fact, gives a direction to the passage of
time. This paradox has still not really been resolved. We do know that
the Newtonian systems, that is the "integrable systems," occupy only an
infinitesimal part of the phase space described by a large number of
material bodies, or particles. For some time, the great Enrico Fermi
believed that quantum mechanics was the answer to the ergodic paradox.
He believed that any perturbation would eventually cause even a
completely deterministic Newtonian system to pass through all possible
states in phase space. Fermi believed that the indeterminacy inherent
in quantum mechanics would provide such perturbations spontaneously and
that, therefore, the paradox was resolved. It turns out that even
Fermi was wrong. Even with such perturbations, it can be shown that
the ergodic hypothesis does not hold.

If all of this is true, what can be said about statistical mechanics?
Can we really believe it? For all practical purposes, the answer is yes.
But the intellectual foundations of Boltzmann's work, as I have already
said, are still not completely understood, and this is still the case a
century after Boltzmann developed his basic theorems.

Non-linear equations such as the Boltzmann equation among many others,
have generally defied attempts to find comprehensive general solutions.
Numerical methods are, therefore, the only way in which at least a
partial approach can be developed in the general sense. The trouble
with numerical methods is, of course, that they are extremely labor
intensive. Therefore, little progress in dealing with non-linear systems
of differential equations was made until the advent of high speed
computers. While it is true that non-linear and non-integrable
equations do not have general solutions, it is possible with high
speed computers to explore a large fraction of the "solution space" of
some of these equations. By doing this, it has become possible to
discern regularities in non-linear systems that are simply not apparent
from the form of the equation itself or from the physical laws on
which that equation is based. It is in looking at many hundreds of
millions of numerical solutions of non-integrable and non-linear
equations that these irregularities become apparent. It is important

to recognize that it is not only the existence of the high speed
computer itself but also the development of graphic readout techniques
with these computers that have allowed the solution of some of the
problems I have defined (perhaps "solution" is too strong a word -- I
should really say treatment).

About twenty years ago, a number of people began looking at the behavior
of non-linear systems using advanced computer graphic techniques. Perhaps
the most important work in the early days was done by Professor Edward N.
Lorenz of MIT. Lorenz is a meteorologist and is, therefore, interested
in the equations of fluid mechanics that govern the behavior of the
atmosphere. He was interested in solving the Navier-Stokes equations
with the boundary conditions that are imposed by the input of energy,
the presence of water vapor, and a host of other factors that cause the
weather. The Navier-Stokes equations are complex non-linear differential
equations and, interestingly enough, are an approximation of the
Boltzmann equation that I mentioned earlier. If the collision integrals
that are usually written on the right side of the Boltzmann equation are
approximated by a single constant (called the viscosity) then the
Navier-Stokes equations are the result.

What Lorenz found in exploring the "solution space" of the Navier-Stokes
equation for his boundary conditions was that there is a level of "order"
that can be shown to exist only after the capability to perform vast
numbers of numerical calculations was in hand. What Lorenz showed is
that the initial conditions of the problem defined a "solution space"
for the equations that exhibited quite regular patterns. Another way of
saying this is that phase space occupied by the solutions of the system
was quite limited and exhibited a regular pattern depending on how the
calculation was started. Lorenz called these regularities in the
"solution space" of the equation the "strange attractors" around which
the solutions of the non-linear dynamical system he was considering
seemed to congregate.

The appearance of these new regularities, by the way, gives the lie to
the statement that computers are only tools or computational aids. The
fact is that brand new and genuine scientific insights can arise from
large scale computations precisely because they permit us to explore
the "solution space" of a non-linear system more completely than we have
ever been able to do before. I suppose that if we look at Johannes
Kepler's work three hundred years ago, we should not be surprised. When
Kepler painstakingly used Tycho Brahe's observations of the orbit of
Mars to derive the laws of planetary motion, he really did demonstrate
the importance of computational physics! It is, indeed, true that the
most important lessons have to be relearned every once in a while.

Why should we be interested in this computational treatment of nonlinear dynamics and in Edward Lorenz's "strange attractors?" I have no
rigorous answer to this question, but I do have some speculations that
might be interesting. The fact that the ergodic hypothesis is not true,
and the fact that non-linear systems occupy a restricted volume of phase
space that seems to exhibit certain inherent patterns, may provide
the clue to unravelling the riddle of the "infinitely complex" by
mathematical methods. This was, of course, the speculation that I made
earlier in this paper. Let me return to the discussion of the nature of
life. Earlier in this paper, I asked the fundamental question which is
why certain complex molecules seem to organize themselves in very

definite patterns. In a statistical sense, the behavior of these
molecules is, of course, governed by non-linear equations of some kind.
Is it possible that Edward Lorenz's patterns somehow govern the "self
organization" that is exhibited by some of these complex molecules?
Will we eventually be able to understand the "infinitely complex"
through an understanding of the behavior of these highly non-linear
systems? I do not know the answers to these questions, but I suspect
that there may be a very fruitful area of research by exploring them.

Just as in the past century, we have cleared up many of the mysteries of
Weisskopf's "infinitely small" and "infinitely large" so, I believe, we
will in the next century understand the "infinitely complex." To me
this is the next and most important trend in science and I believe that
those of you who are advancing the discipline of mathematics as a way of
modeling the physical world will have very much to contribute to this
important effort.

The Use of Multiple Integrals in the Characterization of Renewal Counting Processes

B. D. SIVAZLIAN
Department of Industrial and Systems Engineering
The University of Florida
Gainesville, Florida 32611, USA

ABSTRACT

It is shown that a class of multiple integrals may be used as a novel
mathematical methodology to solve problems arising in renewal counting
processes. Multiple integrals provide a natural vehicle to approach
these complex problems as one is essentially dealing with sums of
independent random variables in the context of interarrival times. From
the joint distribution function of the number of renewals and the
interarrival times, the distribution of the number of renewals is
obtained. Using a similar approach the joint distribution function of
the number of renewals and the forward (backward) recurrence time is
generated for both the ordinary and the delayed renewal process. From
these, the well-known distributions of the forward and backward
recurrence times are obtained. The corresponding limit distributions
are derived. The results obtained are indicative of the type of
statistical properties that can be generated when using this new
methodology in characterizing renewal counting processes. In addition,
the new approach offers the advantage of elegance and of a unified
theory of renewals.

INTRODUCTION

The objective of the present paper is to show that a class of multiple
integrals may be used as a novel mathematical methodology to solve
problems arising in renewal counting processes. Multiple integrals
provide a natural vehicle to approach these complex problems as one is
essentially dealing with sums of independent random variables in the
context of interarrival times. The class of multiple integrals
typically arising in these problems is of the generalized Liouville type
[7]. In this paper we use this methodology to provide a new derivation
to a number of distributions arising in a renewal process. The primary
emphasis is to demonstrate the use of multiple integrals as a method of
analysis and solution, rather than to derive the specific intended
result in the shortest number of steps. The available methods for
obtaining some of the results in the existing literature rely on event
arguments relating waiting times to number of renewals, which are
restrictive (see e.g. Cox [3]). Our approach which is based on deriving
joint distribution functions suggest a more elegant and a more unified
methodology which has been used to solve complex problems in renewal
theory such as deriving the probability law of a renewal counting

process (Sivazlian [8]), an unsolved problem so far. We first state a result in multiple integrals. Next we derive 1) the distribution of the number of renewals, 2) the joint distribution of the number of renewals and of the forward recurrence time, and 3) the joint distribution of the number of renewals and of the backward recurrence time. For a comprehensive treatment of renewal theory, the reader is referred to Cohen [2], to Daley and Vere-Jones [4], and Arrow et al [1]. The ramifications of the new methodology are expounded in Sivazlian [9].

A RESULT IN MULTIPLE INTEGRAL

<u>Theorem</u>

Define for $t > 0$, the function $g(t) \in \mathcal{C}$ (i.e. continuous) and the function $\phi_i(t) \in \mathcal{K}$ (i.e. with at most a finite number of points of discontinuity in every finite interval and such that the integral

$$\int_0^t |\phi_i(u)| du$$

has a finite value for every $t > 0$), $i=1,2,\ldots,$ n where n is a positive integer (Mikusinski [5]). Then

$$\underset{0 < t_1+t_2 + \cdots + t_n \leq t}{\int\int \cdots \int} g(t_1+t_2 + \cdots + t_n)$$

$$\phi_1(t_1)\phi_2(t_2) \cdots \phi_n(t_n) \, dt_1 \, dt_2 \cdots dt_n$$

$$= \int_0^t g(u) [\phi_1(u)*\phi_2(u) * \cdots *\phi_n(u)] \, du$$

where the integrand of the right hand single integral is a function of class $\mathcal{K}$. Here the notation * refers to the usual convolution operation. For a proof see Sivazlian [7]. Note here that the theorem on convolution of measures based on the Riesz representation theorem (see e.g. Rudin [6], p. 219) does not provide a means to reduce the above multiple integral.

THE DISTRIBUTION OF THE NUMBER OF RENEWALS

Let $\{T_i\}$, $i=1,2,\ldots,$ be the sequence of interarrival times in an ordinary renewal process, assumed to be independently and identically distributed random variables with probability density function $f(x)$, $0 < x < \infty$, and distribution function $F(x)$. Let $\{N(t), t \geq 0\}$ be the total number of renewals in $[0,t]$ where $N(0)=0$. The joint distribution of $N(t)$ and T_1, T_2, $\ldots$, $T_{N(t)}$ is:

a - For $N(t)=0$, $t \geq 0$:

$$P\{N(t)=0, T_1 > t\} = 1- F(t)$$

b - For $N(t)=n \geq 1$, $\quad 0 < t_1+t_2 + \cdots + t_n \leq t$:

$P\{N(t)=n,\ t_1 < T_1 \leq t_1+dt_1,\ t_2 < T_2 \leq t_2+dt_2,\ \ldots,$

$\quad t_n < T_n \leq t_n+dt_n,\ T_{n+1} > t-(t_1+t_2 + \cdots + t_n)\}$

$\quad = f(t_1) \cdot f(t_2) \cdots f(t_n)\ \{1- F[t-(t_1+t_2 + \cdots + t_n)]\}\ dt_1\ dt_2 \cdots dt_n$

3a. Distribution of N(t) for the Ordinary Renewal Process

Thus, the probability mass function of $N(t)$ is:

For $N(t) = 0$:

$P\{N(t)=0\} = 1-F(t)$

For $N(t)=n \geq 1$:

$$P\{N(t)=n\} = \int\int \cdots \int_{0 < t_1+t_2 + \cdots + t_n \leq t} f(t_1) \cdot f(t_2) \cdots f(t_n)$$

$$\{1 - F[t-(t_1+t_2 + \cdots + t_n)]\}\ dt_1\ dt_2 \cdots dt_n$$

$$= \int_0^t f^{*(n)}(u)\ [1-F(t-u)]\ du$$

$$= \int_0^t f^{*(n)}(u)\ du - \int_0^t f^{*(n)}(u) \cdot F(t-u)\,du$$

$$= \int_0^t f^{*(n)}(u)\ du - \int_0^t f^{*(n+1)}(u)\,du$$

$$= F^{(n)}(t) - F^{(n+1)}(t)$$

3b. Distribution of N(t) for the Modified Renewal Process

For the modified or delayed renewal process, let $f_1(\cdot)$ be the probability density function of the time till the first renewal and $F_1(\cdot)$ be its distribution function, the distribution function $F(\cdot)$ for the other interarrival times not being affected. Then:

For $\quad N(t) = 0$:

$P\{N(t) = 0\} = 1- F_1(t)$

For $\quad N(t) = n \geq 1$

$$P\{N(t) = n\} = \int\int \cdots \int_{0 < t_1 + t_2 + \cdots + t_n \le t} f_1(t_1) \cdot f(t_2) \cdots f(t_n)$$

$$\{1 - F[t - (t_1 + t_2 + \cdots + t_n)]\} \, dt_1 \, dt_2 \, \ldots \, dt_n$$

$$= \int_0^t f_1(u) * f^{*(n-1)}(u) \, [1 - F(t-u)] \, du$$

$$= \int_0^t f_1(u) * f^{*(n-1)}(u) \, du - \int_0^t f_1(u) * f^{*(n)}(u) \, du$$

For the ordinary renewal process, the renewal function (see e.g. Cox [3], p. 45) is defined as:

$$M(t) = E\{N(t)\} = \sum_{n=0}^{\infty} nP\{N(t) = n\} = \sum_{n=1}^{\infty} F^{(n)}(t)$$

and the renewal density function is defined as

$$m(t) = \frac{dM(t)}{dt} = \sum_{n=1}^{\infty} f^{*(n)}(t)$$

THE DISTRIBUTION OF THE FORWARD RECURRENCE TIME, V(t)

The forward recurrence time at time t is the time elapsed from t to the next renewal. Let for $0 \le t < \infty$, $V(t)$ be the forward recurrence time or excess time or residual life at time t. By first obtaining the joint distribution of 1) the number of renewals $N(t)$ up to time t, 2) the interarrival times $\{T_i\}$, and 3) the forward recurrence time $V(t)$, the joint distribution function of $N(t)$ and $V(t)$ is generated. The result is then used to derive the distribution of $V(t)$.

4a. Distribution of V(t) for the Ordinary Renewal Process

To obtain the joint distribution of $N(t)$, $\{T_i\}$ and $V(t)$, we consider two cases:

Case 1: $N(t) = 0$

Clearly $T_1 = t + V(t)$, and we have for $0 \le \tau < \infty$

$$P\{N(t) = 0, T_1 > t, V(t) \le \tau\} = P\{N(t) = 0, V(t) \le \tau\}$$

$$= P\{N(t) = 0, T_1 > t, T_1 - t \le \tau\}$$

$$= P\{N(t) = 0, t < T_1 \le t + \tau\}$$

$$= F(t + \tau) - F(t) \tag{1}$$

Case 2: $N(t) = n$, $\qquad n = 1, 2, \ldots$

Here $\qquad\qquad V(t) = T_1 + T_2 + \cdots + T_n + T_{n+1} - t$

Thus, for $0 \le \tau < \infty$,

$$P\{(N(t) = n, \; t_1 < T_1 \le t_1 + dt_1, \; \ldots \; , \; t_n < T_n \le t_n + dt_n,$$

$$T_1 + T_2 + \cdots + T_n + T_{n+1} > t, \; V(t) \le \tau\}$$

$$= P\{N(t) = n, \; t_1 < T_1 \le t_1 + dt_1, \; \ldots \; , \; t_n < T_n \le t_n + dt_n,$$

$$t - (t_1 + t_2 + \cdots + t_n) < T_{n+1} \le t + \tau - (t_1 + t_2 + \cdots + t_n)\}$$

Hence, using our theorem:

$$P\{N(t) = n, \; V(t) \le \tau\}$$

$$= \iint \cdots \int_{0 \, < \, t_1 + t_2 \, + \, \cdots \, + \, t_n \, \le \, t} f(t_1) \, f(t_2) \, \cdots \, f(t_n)$$

$$\{F[t + \tau - (t_1 + t_2 + \cdots + t_n)]$$

$$- F[t - (t_1 + t_2 + \cdots + t_n)]\} \, dt_1 \, dt_2 \, \ldots \, dt_n$$

$$= \int_0^t f^{*(n)}(u) \, [F(t+\tau-u) - F(t-u)] \, du \qquad n = 1, 2, \ldots$$

It immediately follows that

$$P\{V(t) \le \tau\} = F(t + \tau) - F(t) + \sum_{n=1}^{\infty} \int_0^t f^{*(n)}(u) \, [F(t + \tau - u)$$

$$- F(t - u)] \, du \tag{2}$$

Let $m(t)$ be the renewal density function. Then

$$m(t) = \sum_{n=1}^{\infty} f^{*(n)}(t) \tag{3}$$

Hence using (3) in (2):

$$P\{V(t) \leq \tau\} = F(t + \tau) - F(t) + \int_0^t m(u) \ [F(t+\tau-u) - F(t-u)] \ du, \tag{4}$$

$$0 \leq \tau < \infty$$

4b. Distribution of V(t) for the Modified Renewal Process

For the modified or delayed renewal process, let $f_1(\cdot)$ be the probability density function of the time till the first renewal and $F_1(\cdot)$ be its distribution function, the distribution function of the interarrival times for the other renewals remaining the same namely $F(\cdot)$. Then,

$$P\{V(t) \leq \tau\} = F_1(t + \tau) - F_1(t) + \int_0^t f_1(u) \ [F(t + \tau - u) - F(t - u)] \ du$$

$$+ \sum_{n=2}^{\infty} \int_0^t f_1(u) * f^{*(n-1)}(u) \ [F(t + \tau - u) - F(t - u)] \ du \tag{5}$$

4c. Limiting Behavior of V(t) as $t \to \infty$

To find the limiting behavior of $V(t)$ as $t \to \infty$, we have from (4)

$$\lim_{t \to \infty} P\{V(t) \leq \tau\} = \lim_{t \to \infty} \{F(t + \tau) - F(t)$$

$$+ \int_0^t m(u) \ [F(t + \tau - u) - F(t - u)] \ du$$

$$= \lim_{t \to \infty} \{F(t + \tau) - F(t)]$$

$$+ \lim_{t \to \infty} \int_0^t m(u) \ [1 - F(t - u)] \ du$$

$$- \lim_{t \to \infty} \int_0^t m(u) \ [1 - F(t + \tau - u)] \ du \tag{6}$$

Let $\mu = E[T]$ be the expected interarrival time. A well known result in renewal theory (see e.g. Cox [3], p.46) is

$$\lim_{t \to \infty} m(t) = \frac{1}{\mu} \tag{7}$$

We note in (6) that the value of the first limit is zero. To evaluate
the second and third limit we make use of Smith's key renewal theorem
(see e.g. Cohen [2], p. 102). The value of the second limit is unity.
Thus

$$\lim_{t\to\infty} P\{V(t) \le \tau\} = 1 - \frac{1}{\mu} \int_0^\infty [1 - F(v + \tau)] \, dv$$

$$= \frac{1}{\mu} \int_0^\tau [1 - F(w)] \, dw, \qquad 0 \le \tau < \infty \tag{8}$$

4d. Example

Consider the case when $f(x)$ has the negative exponential distribution:

$$f(x) = \lambda e^{-\lambda x} \quad , \quad F(x) = 1 - e^{-\lambda x} \quad , \quad 0 \le x < \infty$$

Then $\qquad m(x) = \lambda$

Substituting in (4)

$$P\{V(t) \le \tau\} = e^{-\lambda t} - e^{-\lambda(t+\tau)} + \int_0^t \lambda [e^{-\lambda(t-u)} - e^{-\lambda(t+\tau-u)}] \, du$$

$$= e^{-\lambda t}(1 - e^{-\lambda\tau}) + \lambda e^{-\lambda t}(1 - e^{-\lambda\tau}) \int_0^t e^{\lambda u} \, du$$

$$= 1 - e^{-\lambda\tau}$$

$$E[V(t)] = \int_0^\infty \tau \lambda \, e^{-\lambda\tau} \, d\tau = \frac{1}{\lambda} \tag{9}$$

THE DISTRIBUTION OF THE BACKWARD RECURRENCE TIME, B(t)

The backward recurrence time at time t is either the length of time t,
if no renewals have taken place, i.e. if $N(t) = 0$, or is the time
elapsed from the last renewal before t to time t, if at least one
renewal has taken place, i.e. if $N(t) \ge 1$.

5a. Distribution of B(t) for the Ordinary Renewal Process

To obtain the joint distribution of $N(t)$, $\{T_i\}$ and $B(t)$, we consider two
cases:

Case 1: $N(t) = 0$. Clearly $B(t) = t$.

   i. For $\qquad 0 \le \theta < t$

$$P\{N(t) = 0, B(t) \le \theta\} = 0 \tag{10}$$

ii. For $\theta = t$

$$P\{N(t) = 0,\ B(t) \le \theta\} = 1\text{-}F(t) \tag{11}$$

iii. For $t < \theta < \infty$

$$P\{N(t) = 0,\ B(t) \le \theta\} = 1 \tag{12}$$

Case 2: $N(t) \ge 1$. Clearly $0 \le B(t) \le t$.

Following an approach similar to Section 4, we have:

i. For $0 \le \theta < t$

Since $B(t) = t - (T_1 + T_2 + \cdots + T_n)$

$$P\{N(t) = n,\ t_1 < T_1 \le t_1 + dt_1,\ \ldots\ ,\ t_n < T_n \le t_n + dt_n$$

$$T_1 + T_2 + \cdots + T_{n+1} > t,\ B(t) \le \theta\}$$

$$= P\{N(t) = n,\ t_1 < T_1 \le t_1 + dt_1,\ \ldots\ ,\ t_n < T_n \le t_n + dt_n$$

$$t\text{-}\theta \le T_1 + T_2 + \cdots + T_n < t,\ t - (T_1 + T_2 + \cdots + T_n)$$

$$< T_{n+1} < \infty\}$$

Hence

$$P\{N(t) = n,\ B(t) \le \theta\}$$

$$= \iint \cdots \int_{t - \theta \le t_1 + t_2 + \cdots + t_n < t} f(t_1)\ f(t_2)\ \cdots\ f(_n)$$

$$\{1 - F[t - (t_1 + t_2 + \cdots + t_n)]\}\ dt_1\ dt_2\ \ldots\ dt_n$$

$$= \int_{t-\theta}^{t} f^{*(n)}\ (u)\ [1 - F(t\text{-}u)]\ du \tag{13}$$

ii. For $\theta = t$

$$P\{N(t) = n,\ t_1 < T_1 \le t_1 + dt_1,\ \ldots\ ,\ t_n < T_n \le t_n + dt_n,$$

$$T_1 + T_2 + \cdots + T_{n+1} > t,\ B(t) \le \theta\}$$

$$= P\{N(t) = n,\ t_1 < T_1 \le t_1 + dt_1,\ \ldots\ ,\ t_n < T_n \le t_n + dt_n$$

$$0 \le T_1 + T_2 + \cdots + T_n < t,\ t - (T_1 + T_2 + \cdots + T_n) < T_{n+1} < \infty\}$$

Thus, $P\{N(t) = n, \ B(t) \le \theta\}$

$$= \int\int \cdots \int_{0 \le t_1+t_2 + \cdots + t_n < t} f(t_1) \ f(t_2) \ \cdots \ f(t_n)$$

$$\{1- F[t-(t_1+t_2 + \cdots + t_n)]\} \ dt_1 \ dt_2 \ \cdots \ dt_n$$

$$= \int_0^t f^{*(n)}(u) \ [1- F(t-u)] \ du \tag{14}$$

iii. For $t < \theta < \infty$

$$P\{N(t) = n, \ t_1 < T_1 \le t_1 + dt_1, \ \cdots \ , \ t_n < T_n \le t_n + dt_n,$$

$$T_1 + T_2 + \cdots + T_{n+1} > t, \qquad B(t) \le \theta\} = 0$$

Hence

$$P\{N(t) = n, \ B(t) \le \theta\} = 0 \tag{15}$$

Combining (10) to (15), we obtain

$$P\{B(t) \le \theta\} = \sum_{n=0}^{\infty} P(N(t) = n, \ B(t) \le \theta\}$$

$$= \sum_{n=1}^{\infty} \int_{t-\theta}^t f^{*(n)}(u) \ [1- F(t-u)] \ du \qquad 0 \le \theta < t$$

$$= 1- F(t) + \sum_{n=1}^{\infty} \int_0^t f^{*(n)}(u) \ [1- F(t-u)] \ du \qquad \theta = t$$

$$= 1 \qquad\qquad\qquad\qquad t < \theta < \infty$$

Using (3) and noting that for $\theta = t$ the right-hand side reduces to 1, we have

$$P\{B(t) \le \theta\} = \int_{t-\theta}^t m(u) \ [1- F(t - u)] \ du \qquad 0 \le \theta < t$$

$$\tag{16}$$

$$= 1 \qquad\qquad\qquad\qquad t \le \theta < \infty$$

or $\quad P\{B(t) \leq \theta\} = \displaystyle\int_0^\theta m(t-v) \, [1- F(v)] \, dv \qquad\qquad 0 \leq \theta < t$

$$\qquad\qquad\qquad\qquad = 1 \qquad\qquad\qquad\qquad\qquad\qquad t \leq \theta < \infty \tag{17}$$

5b. Distribution of B(t) for the Modified Renewal Process

Again, for the modified or delayed renewal process, if $f_1(\cdot)$ is the probability density function of the time till the first renewal, and $F_1(\cdot)$ is its distribution function, then

$$P\{B(t) \leq \theta\} = \sum_{n=2}^\infty \int_{t-\theta}^t f_1(u) * f^{*(n-1)}(u) \, [1- F(t-u)] \, du \qquad 0 \leq \theta < t$$

$$\qquad\qquad\qquad = 1 \qquad\qquad\qquad\qquad\qquad\qquad t \leq \theta < \infty \tag{18}$$

5c. Limiting Behavior of B(t) as $t \to \infty$

To find the limiting behavior of $B(t)$ as $t \to \infty$, we have from (17):

$$\lim_{t\to\infty} P\{B(t) \leq \theta\} = \lim_{t\to\infty} \int_0^\theta m(t- v) \, \{1- F(v)\} \, dv \qquad 0 < \theta < \infty \tag{19}$$

Using a result from Cohen [2], (p. 112), we have

$$\lim_{t\to\infty} P\{B(t) \leq \theta\} = \frac{1}{\mu} \int_0^\theta [1- F(v)] \, dv, \qquad\qquad 0 < \theta < \infty \tag{20}$$

where $\mu = E[T]$ is the expected interarrival time.

5d. Example

Let $f(x)$ be the negative exponential function as introduced in Section 4d; we have, using (17):

$$P\{B(t) \leq \theta\} = \int_0^\theta \lambda \, e^{-\lambda v} \, dv = (1- e^{-\lambda\theta}) \qquad\qquad 0 \leq \theta < t$$

$$\qquad\qquad\qquad = 1 \qquad\qquad\qquad\qquad\qquad\qquad t \leq \theta < \infty \tag{21}$$

$$E[B(t)] = \int_0^{\Theta} \Theta \lambda e^{-\lambda\Theta} \, d\Theta + t \, e^{-\lambda t}$$

$$= \frac{1}{\lambda} (1 - e^{-\lambda t}) \tag{22}$$

REFERENCES

[1] K. J. Arrow, S. Karlin and H. Scarf, "<u>Studies in the Mathematical Theory of Inventory and Production</u>", Stanford University Press, Stanford, 1958.

[2] J. W. Cohen, <u>The Single Server Queue</u>, North-Holland Publishing Co., Amsterdam, 1969.

[3] D. R. Cox, <u>Renewal Theory</u>, John Wiley, New York, 1962.

[4] D. J. Daley and D. Vere-Jones, <u>An Introduction to the Theory of Point Processes</u>, Springer-Verlag, New York, 1988.

[5] J. Mikusiński, <u>Operational Calculus</u>, Pergamon Press, Oxford, 1959.

[6] W. Rudin, <u>Functional Analysis</u>, McGraw Hill, New York, 1973.

[7] B. D. Sivazlian, A Class of Multiple Integrals, <u>SIAM J. Math. Anal.</u>, 2, pp. 72-75, 1971.

[8] B. D. Sivazlian, "On the Joint Distribution of the Number of Renewals in a Renewal Process", <u>Stochastic Analysis and Applications</u>, 7, pp. 475-495, 1989.

[9] B. D. Sivazlian, "New Methodologies in Renewal Theory", <u>Transactions of the Seventh Army Conference on Applied Mathematics and Computing</u>, Report No. 90-1, U. S. Army Research Office, Research Triangle Park, 1990.

Molecular Dynamics and Monte Carlo Simulations of Polymeric Materials: Chain Relaxation Dynamics

WITOLD BROSTOW
Center for Materials Characterization and Department of Physics
University of North Texas, Denton, Texas 76203-5308, USA

ROBERT COOK
Lawrence Livermore National Laboratory
University of California, Livermore, California 94550, USA

JOSEF KUBÁT
Department of Polymeric Materials
Chalmers University of Technology, 412 96 Gothenburg, Sweden

1. INTRODUCTION AND SCOPE

Simulation algorithms of the Monte Carlo type (MC) have found applications in a large number of disciplines, including areas as diverse as fundamental mathematics, applied physics and engineering, molecular biology, and even the social sciences. Molecular dynamics methods (MD) have a narrower application range - essentially in materials science and engineering, physics, chemistry and molecular biology - but are often more powerful. These methods have been applied to the study of solid polymers, polymer melts, and polymer solutions. We shall discuss some general characteristics of these methods, along with their particular usefulness in the study of the relaxation properties of polymeric systems. Relaxation constitutes the most essential feature of polymeric materials; it determines whether a material will survive an applied mechanical force without fracture; it also determines the service lifetime of the materials under repetitive force applications.

Simulation of any material relies on sampling an appropriate phase space. That space is often defined by a set of values of coordinates and momenta. The set, along with a probability distribution over it, constitutes a stochastic variable. In turn, a stochastic process is a function of the stochastic variable and a second variable, in most cases of time t. Hence one may look upon our simulations as realizations of stochastic processes. A book by van Kampen [1] approaches a number of processes occurring in physical systems from this point of view.

The order of presentation is as follows. First, we discuss reasons why computer simulation of physical systems is worthwhile. Then we tell a brief history and characterize the essential features of the MC and MD methods. We then discuss some of the key properties of polymeric materials we wish to simulate. This is followed by simulation results that throw some light on the nature and behavior of polymers, and also which make possible the evaluation of the capabilities of the simulation methods.

2. SIMULATIONS OF PHYSICAL SYSTEMS: MOTIVATION

Our goal as scientists when we study a new system, whether it be the core of a neutron star, the weather patterns over North Texas, the nature of the HIV virus, or a polymeric material, is first to be able to describe the system, specify the parts and their interactions, and then to be able to make predictions about what the system will do in the future under imposed conditions. The first goal generally takes the form of building a mental model of the system. In doing so it is necessary to make decisions about what is really important to the system and what is merely window dressing. These decisions are rarely clear cut; they are often driven by our second goal, the ability to make predictions based on our model. If the model is too complex, it may represent the actual physical system very well but because of its complexity be useless as a model for predicting the system's behavior. On the other hand, if we model the system too simply, the model may be very easy to use as a basis of prediction, but the predictions are of no value, or worse are incorrect.

Particularly in physical science and engineering, the route from models to predictions is a mathematical one which we often call "theory." It generally turns out that in doing "theory" it is necessary to make mathematical approximations in order to use the model. Thus when a theorist proposes a theory to explain or predict some aspect of a physical system and the theory's predictions disagree with experiment, there are three possible causes. First, it could be that the model is missing some essential aspect of the physical system. Second, it is possible that the mathematical hardware brought to bear on the model is inadequate (or wrong) to handle the problem. Third, the experimental results may be in error, or, as is more often the case, the experimentalist was not measuring what he thought he was. Nature can be tricky.

Computer simulation allows the theorist to clarify the situation by simulating the model in a well controlled and precise way. In this way the hidden variables that lurk about in nature are eliminated. The results of these simulations are often called "computer experiments" and in a sense sit midway between the theorist and the true experimentalist. By comparing the predictions of a theory with a simulation result, the theorist can tell if the mathematical hardware is faulty and often if so, where. This is possible because the theory and computer experiment are both based on *exactly* the same model. In addition the viability of the model can often be tested by comparing the simulation results with the true experimental results, and further the parameters in the model can be varied to see how the simulation results depend upon them. This can lead to better model building and thus ultimately better theory.

Just as there are traps into which an experimentalist can fall, there are also dangers inherent to conducting simulations. Is the program really doing what the programmer <u>thinks</u> it is doing? And a different question: does the algorithm contain essential features of the system to be simulated - and nothing else? These are problems of verification and validation of simulations, discussed in some detail by Bratley, Fox and Schrage (BFS)[2]. For instance, an interesting verification procedure is <u>stress testing</u>: if we impose nonsensical conditions (temperature of the material T = - 800K, or mole fraction of a component x_A = 1.75), but we keep getting similar results as before, then something is wrong. As for validation, BFS note cases when a zealous programmer put in a feature which he <u>expected</u> to get out of his simulation. While there is no general validation recipe, BFS outline several possible procedures.

3. THE MONTE CARLO AND MOLECULAR DYNAMICS METHOD

The Monte Carlo method has an interesting history. It was
conceived first by a group of Polish mathematicians associated
with the John Casimir University in Lvov. The group included
Stanislaw Ulam and they spent much of their time in a café near
the University, discussing all kinds of topics between heaven and
earth. In 1938 or so, somebody pointed out that playing roulette
in the casino in Monte Carlo, Monaco, is really a problem in
mathematics. Somebody else suggested that those gathered should
develop a scientific procedure for breaking the bank at the
casino. Interesting attempts were made and argued about, until
World War II broke out in 1939. Of course the original objective
was never achieved. However, a by-product of one of the attempts
is the present Monte Carlo procedure. Because of the war and
dispersion of the original participants around the globe, the
first account of the method was published only in 1949 by
Metropolis and Ulam [3].

The MC algorithm has applications in many disciplines, _not_ only
in physical sciences. As an example of problems which appear
vastly different from each other but can be treated by similar
methods, propagating energy along a polymer chain with branches
is in some respects similar to propagating gossip between curious
human beings! We shall, however, confine ourselves to polymeric materials, which are of interest to a broad spectrum of
scientists and engineers.

One can construct a _phase space_ for a material, such that each
particle or element of the system is represented by six coordinates: three Cartesian coordinates of position and three of
momentum. In the MC approach one attempts to calculate equilibrium properties of a system by averaging over a collection of
states chosen in a "Monte Carlo" fashion from the appropriate
phase space. For example, if one were interested in the radius
of gyration of a polymer, a well defined experimental quantity,
usually represented by $<r^2>_g$, one might "create" 10,000 chains on
a computer in a manner that makes them representative of the
equilibrium distribution, calculate r^2_g for each and average.
One has to make the sampling representative, and numerous approaches exist for this purpose. Such a computer experiment
gives one a numerical result, essentially by doing a configurational phase space average. The method does not, however, tell
us how the chains move.

We could also evaluate $<r^2>_g$ in the following manner. We set up
a single chain on the computer and completely specify the
interaction potential between all pairs of segments, which,
through appropriate derivatives, gives the forces acting on each
atom. Neglecting for the moment the action of solvent on the
chain, if we give each of the particles an initial velocity
appropriate to the temperature of the system, we have a
completely specified set of coupled differential equations:

$$F_i = m_i a_i \tag{1}$$

where F is the force, m the mass and a the acceleration, all
pertaining to the i-th particle. There exist numerous algorithms
for numerically solving this N-body problem, where solving simply
means obtaining the time evolution of the system under study. In

Section 5 we will outline such a simple approach. If we do this
and average the configurations of a single chain over a long
enough time, so that it adequately samples the phase space
available to it, we can also determine $<r^2>_g$.

The two methods described are similar in several respects. Both
depend upon adequate sampling to obtain accurate results. In the
MC approach we average over configurations which need not be
temporally related to one another, but rather they must sample
the available phase space adequately. The second approach, is
that of molecular dynamics (MD). It was developed by Berni J.
Alder and collaborators at the Lawrence Livermore National
Laboratory; the first paper on the subject which dealt with
simple hard sphere fluids was published in 1957 by Alder and
Wainwright [4]. In MD we also sample the phase space, but the
sequential configurations are related temporally. Because of
this restriction, it often turns out that the MD approach is much
too costly in computer time if one is only interested in
equilibrium properties. The time "trajectories" of many systems
are simply too inefficient for sampling phase space.

However, the MD approach does give us something that the MC
approach does not: it shows us how the system gets from one
configuration to another. Because of this, the method is able to
look at non-equilibrium dynamic events such as flow, fracture and
diffusion. Particularly in the case of polymeric systems, this
is a very powerful feature, since the motions of polymer chains
is the key to the understanding of their properties. We shall
outline the most important of those properties in the following
section.

4.   CHAIN RELAXATION AND DESTRUCTIVE PROCESSES
       IN POLYMERIC MATERIALS

As already noted in Section 1, chain relaxation constitutes the
most essential feature of polymers. Other classes of materials
such as metals exhibit elasticity and plasticity. But because of
relaxation, polymers exhibit also viscoelasticity, the variation
of mechanical properties with time t. There are entire books on
viscoelasticity [5,6], as well as on its consequences for service
lifetime and fracture of polymeric materials [7]. In this section
we shall briefly outline two mutually complementary approaches to
relaxation.

Experiments on viscoelasticity can be performed in a variety of
regimes. One of them is stress relaxation, where the decrease of
stress σ with time, $\sigma(t)$, is determined for a specimen held at
constant deformation. A common feature of earlier theories of
the process was the assumption that movements of the elementary
flow units take place independently of each other. Instead, one
of us developed a cooperative model [8-12]. The flow units can
occupy two states, relaxed and unrelaxed (essentially an Ising-
type model). Nearest neighbor interactions are accounted for.
When a flow unit relaxes, the probability that a neighboring
element would relax changes, which produces a broad relaxation
spectrum.

Let us consider the cooperative model a little more. When the
material is strained, the flow units are raised to the upper
energy level; during the stress relaxation process they fall to
the lower level. The fall is accompanied by energy emission;

that energy can be envisaged in the form of phonons which may
induce transitions of other unrelaxed flow units. The phonons
will form clusters obeying the Bose-Einstein distribution. The
transition probability of the unrelaxed flow units will be
related to the size of such phonon clusters. We do not provide
here the details of the theory, since they can be found for
instance in [10]. But we would like to point out two results of
the analysis. First, there exists a longest relaxation time t^*.
The clustering of the phonons produces a spectrum of relaxation
times characterized by

$$t^*, \ t^*/2, \ t^*/3, .. \tag{2}$$

associated with clusters having the sizes 1, 2, 3,... .

The second interesting consequence of the cooperative model
pertains to σ vs. ln t diagrams normalized with respect to the
initial effective stress. Except for nearly horizontal tails and
sometimes also nearly horizontal initial sections, a large
central part of the curve can be well aproximated by a straight
line with the slope

$$F^t = - \ (d\sigma/d \ \ln \ t)_{max} \tag{3}$$

It can be deduced from the cooperative model that

$$F^t = c(\sigma_0 - \sigma_i) \tag{4}$$

where $\sigma_0 = \sigma(0)$ while the internal stress is defined as $\sigma_i =
\sigma(\infty)$; c is a constant which for many materials is equal to 0.1.
Eq. (3) in conjunction with (4) has been confirmed experimentally
for a variety of materials [10]; hence there is an interest in
establishing whether simulations will bring about a similar
result. We [13] are studying a system of particles on a triangular
lattice simulated with the MD method. Fig. 1 shows such a

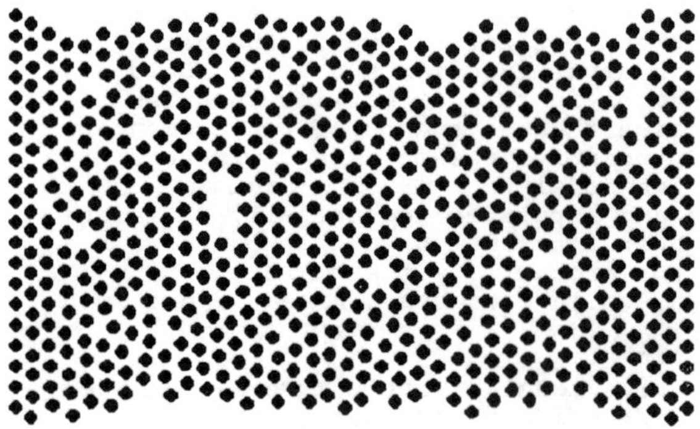

Figure 1; explanation in text.

lattice with visible cracks some time after the application of a
tensile force perpendicular to the shorter sides of the
rectangle. Fig. 2 shows an example of MD results; the descending

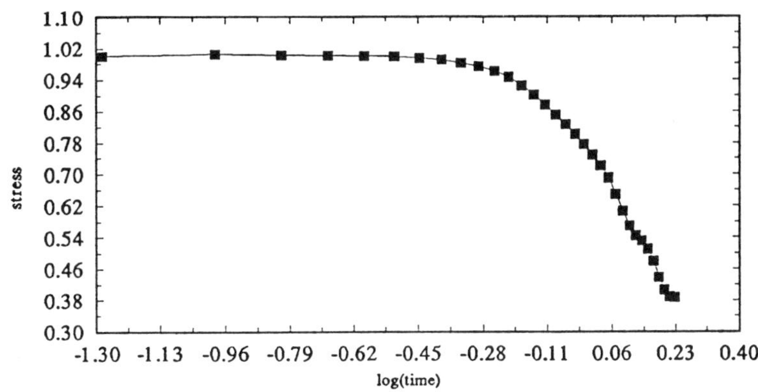

Figure 2; explained in text.

part is approximately linear - demonstrating that Eq. (4) is
obeyed. It remains to be seen whether introduction of particle
connectedness - that is converting the system into chains - will
also confirm that model.

There is a somewhat complementary approach based on the concept
of chain relaxation capability (CRC) [14-16]. CRC is defined [14] as
the amount of external energy dissipated by relaxation in a unit
of time per unit weight of the polymer. All of the
nondestructive processes contribute to CRC, including in
particular: transmission of energy across the chain producing
intensified vibrations of the segments; transmission by
entanglements and to a lesser degree by segment motions of energy
from the chain to its neighbors (a connection to the cooperative
model of relaxation discussed above clearly exists);
conformational rearrangements executed by the chains; and elastic
energy storage resulting from bond stretching and angle changes.

Important in the CRC approach is the competition between relaxa-
tional and destructive processes. Fortunately relaxation <u>has
priority</u>: only when the CRC is exceeded, do <u>destructive</u> processes
manifest themselves [14]. This approach made possible the development of a theory of impact [15], the formulation of a criterion
for elimination of rapid crack propagation [17] and the derivation
of equations for the velocity of slow crack propagation [18].

5.   THE MOLECULAR DYNAMICS METHOD APPLIED TO POLYMERIC MATERIALS

Molecular dynamics simulations of polymeric materials can be used
to test some of the relaxation ideas outlined above. In the
molecular dynamics method one first defines the configurational
energy U^c in terms of binary interactions, $u_{ab}(R_{ab})$, that is

$$U^c = \Sigma_a \Sigma_b \, u_{ab} \, (R_{ab}) \tag{5}$$

where R_{ab} is the distance between particles a and b and the summation extends over all pairs. When dealing with a polymer sample it is convenient to label the particles of each chain by two indices, one which labels the chain and the other which labels the segment on a given chain. Thus the force acting at time t on the j-th segment in i-th chain is

$$F_{ij}(t) = - \partial U^c(t)/\partial R_{ij} \qquad (6)$$

where R_{ij} is the relevant coordinate. The force defined by Eq. (6) is an internal one. When an external force such as stretching is applied, it has to be added to the force resulting from Eq. (6). At the time $t + \Delta t$ one calculates the new position of each particle from the simple difference equation:

$$R_{ij}(t + \Delta t) = R_{ij}(t) + w_{ij}(t - \Delta t/2)\Delta t + F_{ij}(t)\Delta t^2 \qquad (7)$$

where the segment velocity w_{ij} is

$$w_{ij}(t - \Delta t/2) = (R_{ij}(t)) - (R_{ij}(t - \Delta t)/\Delta t. \qquad (8)$$

Equation (7) is the key relation providing information on the dynamics of particles (in our case polymer segments). Since this is a one-component system, unit masses have been assumed without a loss of generality. The $R_{ij}(t)$ trajectories for each segment can thus be followed.

There are several other questions that must be addressed before one can apply MD methods to polymeric samples. First, we must realize that a macroscopic sample of a polymeric material has 10^{23} atoms which interact with fairly complex (if known) interaction potentials. Because of the limitations of computer storage and execution time, we are by necessity limited to at most perhaps 10,000 particles, more typically about 1000 and often much fewer. Further, even if known precisely, the interaction potentials are often very complex and thus computationally time consuming. There is often a trade-off, the more exact (complex) the potential, the smaller the system one can study. And even if one does look at a system with 10,000 atoms, how can that possibly be representative of a system of 10^{23} atoms?

Fortunately, it turns out that physics comes to our rescue at least part of the time. The difficulty with a small number of particles is generally not the smallness of the number, but the fact that in the simulation they have to be put into some kind of "box" which has as a consequence the creation of surface effects. If we have 10^{23} particles in a cube, only roughly 10^{16} of them are on the surface, a large number but only 10^{-5} percent of the total. If on the other hand we have only 10,000 particles in a cube, roughly 23 percent of them will be on the surface. This problem is avoided (or at least moderated) by the use of <u>periodic</u>

<u>boundary conditions</u>. Picture again a cubical box; however,
instead of the surfaces being boundaries, imagine that the
opposite sides of the box are actually connected, so that if a
particle leaves the bottom of the box it simultaneously enters
the top of the box. In a similar manner particles that are near
a surface of the box interact with particles that are near the
opposite surface, as if there were periodic images of the box
filling space. In this fashion the particles never see a
surface. The key to this "trick" working is that the range of
the interparticle potentials must be much shorter than the length
of the box. Otherwise, particles would interact with their own
images and an artificial long range order would be set up.
Consequences of such artifact interactions are discussed in [19].

Luckily, most physical phenomena and related interaction
potentials are relatively short range, acting only over a few
molecular diameters. There are exceptions, however, where long
range correlations play a critical role, such as in phase
transitions. Polymers offer a special challenge since single
macromolecules are often much larger than one can comfortably
simulate. One is then forced to either study shorter chains,
perhaps letting each particle in the simulation represent a large
section of chain and thus trusting in scaling laws, or to limit
oneself to properties that are local and thus molecular-weight-
independent. In doing so we are assuming that correlations exist
along a chain backbone for only a few segments. As noted above,
the exact treatment of intermolecular potentials is at best
difficult and time consuming. Historically, most MD simulations
of polymers have focused primarily on the feature that separates
polymers from other molecular forms, namely the connectivity of
atoms in a chain. Thus models as simple as a string of beads
connected by simple springs can serve as an adequate model for
chain diffusion or radius of gyration studies. Slightly more
decorated models that add bond angle potentials, bond rotational
potentials, and excluded volume forces can improve the model
substantially without excessive additional computational effort.

A second serious limitation of the MD method is that the
actual time spans simulated are frequently quite short. If, for
example, we are trying to simulate a polymer model in which we
allow bond vibrations, we must use a time step that is on the
order of 10^{-15} seconds, since the vibrations have frequencies of
about 10^{14} sec. Thus, at best, we can look at motions on time
scales down to perhaps 10^{-10} seconds in a 100,000 step simulation.
For some rapid processes this is adequate, but many of the more
important polymer properties depend upon molecular motions that
have much longer relaxation times. Thus a simulation of flow or
creep behavior, which might have a characteristic relaxation time
from microseconds to seconds or even longer, cannot use femto-
second time steps. One must adapt the model to the time scale of
interest. Thus if we are interested in flow, we have to give up
looking at individual bond vibrations in a detailed fashion.

The treatment of polymers in solution requires special
techniques. How does one deal with the solvent? In most cases
it is impractical to simulate the solvent molecules individually.
First, there are too many, and second, one immediately locks the
simulation into the femto-second (or less) time scale. If we are
interested in motions on the nano-second or longer time scale
(this includes most motions of interest in polymers) then we
really do not care what the individual solvent molecules are
doing; rather we are concerned with their collective action on
the time scale of interest. This action has two components.

First the solvent provides a viscous drag to the chain segments
of the polymer as they attempt to move through it. However, the
solvent also exchanges energy with the chain through the
collective action of the impacts of the solvent molecules against
the chain segments. These impacts are the molecular origin of
the Brownian motion that the chain undergoes due to its
interaction with the fluid medium in which it resides.
Simulations which treat particles in this fashion are referred to
as Brownian dynamics (BD). The equations of motion to be solved
have the general form

$$ma_i = F_i - \eta v_i + R_i(t) \qquad\qquad (9)$$

where F_i is the interparticle force as one would have in normal
MD simulation, ηv_i is the viscous drag on particle i which acts
to oppose its motion and $R_i(t)$ is the Brownian or Langevin term
that supplies a statistical "kick" to each particle. If the time
scales of interest are long compared to the velocity relaxations,
the acceleration term can be neglected; one has then the simple
Langevin equations to solve numerically. The form of $R_i(t)$ is
governed by certain statistical laws; the magnitude of the
"kicks" the particles receive are drawn from a Gaussian
distribution whose variance depends upon the time step size,
temperature, and solvent viscosity. Thus most simulation studies
of polymers in solution use these techniques. With less firm
theoretical footing, BD can be used to treat polymers in the melt
or in the solid phase above T_g. The requirement is that the time
scale be long compared to the local motions in the thermal bath.

Before proceeding to a discussion of the MD results we have
obtained for polymeric materials, it is worth noting some of the
important earlier results that have had an enormous impact on our
understanding and approach to computer simulation. Though there
are many examples that we could take, we focus on only two. The
first is concerned with the work of Alder [20] and others on the
importance of intermolecular potentials to MD simulations. In
considering the properties of simple fluids, they showed that
only the existence of a repulsive potential was essential to the
qualitative aspects of the simulation, and that in particular
attractive potentials had only a minor effect on the basic
physics. The importance of this results was that it demonstrated
that extremely simple models could be used to study complex
systems. Not only were the essential physics not sacrificed, but
new insights were gained. This general result has allowed us to
model polymer systems quite simply and still learn a great deal
about the underlying dynamics.

The second example is drawn from the Brownian dynamics work on
polymer chains by Helfand [21-23]. The rotational potentials of
simple polyethylene-like chains are three-fold, with two gauche
states and an energetically lower trans state. It is because the
barriers to rotation are reasonably low (~12kJ/mol) that polymers
have the unique properties that they do. Because a single
conformational change in the middle of a chain, from say gauche
to trans, would entail swinging the chain tails through solution
by a relative angle of 120°, it was thought that conformational
changes must always occur in pairs, such that the chain tails
need not be displaced. These local conformational changes were
often labeled "crankshaft" rotations and numerous possibilities

were theorized. One of the consequences of such motions,
however, is that the rotational bonds of the chain must pass over
two barriers simultaneously, leading to the prediction that an
experimentally measured activation energy for conformational
motion would have a value of twice the barrier height to
rotation. As the experimental evidence [24] became available it
became clear that this was not the case, the measured activation
energies corresponded to only single conformational transitions.
The computer simulation work of Helfand clarified this apparent
inconsistency. The simulations showed that in fact
conformational transitions did occur one at a time, and thus the
single barrier was correct. But the simulations were able to say
more. By appropriate analysis Helfand [22] was able to determine
what fraction of the transitions were correlated, in the sense
that one transition enhanced the likelihood of a neighboring
transition. Some similarity exists here with the model of Kubat
described in Section 4. These coupled transitions, which
amounted to only about 20% of the total number of transitions,
were in fact crankshaft-like. The simulations also showed that
the local stresses that occurred due to an isolated transition
could be easily taken up by small perturbations in the
neighboring bonds [23]. Our understanding of this phenomena has
reshaped the way we think about polymer motion, and is due
largely to the insights gained through computer simulation.

6.   MOLECULAR DYNAMICS RESULTS FOR POLYMERIC MATERIALS

The treatment of polymeric materials near and below the glass
transition temperature, T_g, is especially difficult because the
relaxation times are long, yet the relaxation processes
themselves may be very local. The typical stress-strain curve
for a polymeric sample generally shows an elastic region followed
by a yield in which the sample elongates without appreciable
increase in applied stress. After the yield is completed, the
modulus increases abruptly and the material usually fractures.
One molecular interpretation of this sequence of events might run
as follows. In the unstressed polymer sample below T_g, the
individual chains are in some random, entangled conformation (we
consider only non-crystalline polymers at this point) which
depends intimately upon the history of the sample. Because we
are below T_g there is relatively little molecular motion other
than local vibrations. As we begin to stress the material the
individual chains respond by stretching their bonds, bond angles,
and bond rotational angles to accommodate the deformation.
However, there is no substantial chain motion or reorganization,
and thus the deformation is purely elastic. At the yield point
there is a large deformation of the sample which can be
accommodated only by significant chain motions and reorienta-
tions. These motions have the net effect of elongating and
orienting the individual chains into very different conformations. Such a transition to new conformations must involve the
passage over barriers, which simply do not occur until the
applied stress is high enough. Once the chains have passed over
these barriers, they become locked into their new conformation,
even if the imposed stress is relaxed. Thus the deformation
passed the yield point and is not elastic. Once the chains have
been stretched and aligned to the maximum degree consistent with
their various structures and entanglements, the yield must stop;
any further deformation is made elastically at high modulus.
Material failure is the ultimate result.

The process just described has been modeled with a collection of interacting bead-spring chains [25]. Each spring has two minima representing a long and a short state (extended and coiled). Particles on adjacent chains interact via shearing-like forces that make it difficult to pull one particle past another. An important feature of the model is that it is mathematically a purely one-dimensional system. Each particle, which can be thought to represent a portion of a polymer chain, is characterized by a single coordinate. We deform the model by pulling on the end atoms and then solving the equations of motion of the system. The stress energy is constantly withdrawn from the system via damping terms so there is no build up in temperature. In essence the simulation is run at 0 K. As we pull on the sample, we monitor its length which allows us to plot a stress-strain relationship. Even this very simple model mimics polymer behavior. At small stress levels all bonds respond elastically giving a modulus that is consistent with the collective spring potential. At stresses that are sufficient to pull a spring from a short state to a long one, we see yielding associated with this barrier crossing. The dynamics and exact form of the yield region of the curve depend intimately on the interaction potential between chains.

The above simulation study models the polymeric material in a very coarse fashion. The beads may represent sections of chain and the double-welled spring potentials simply try to mimic the barriers that chain segments must feel as they attempt to move. In another study [26] we decided to examine the actual microscopic deformations that a section of chain experiences when subjected to a deformation below or near T_g. To do this we took the detailed polyethylene potential of Helfand [22] and set up a single chain of 4, 6 or 8 atoms with the central bond in a gauche conformation. We then represented the surrounding environment in a mean field way by attaching each particle to the background by a constraining spring that restricted motion in directions perpendicular to the deformation direction. The goal here was to model the constraining forces of the surrounding chains without actually including them in the simulation. The chain was then extended along the line of the end atoms. The deformations in the bond lengths, angles, and rotational angles was monitored as a function of elongation and constraining spring strength. In the absence of any constraining spring (a free chain), an orderly rotation was seen from the gauche state to the trans state with increasing elongation. Very little energy is stored in any of the bonds during such a process. However, as the constraining forces are applied, the gauche to trans conformational change becomes increasingly impeded since this relaxation motion involves significant motions of the chain atoms. For very strong constraining forces the gauche to trans relaxation is completely impeded and the chain deformation is taken up in the bond lengths and angles which can deform with much less motion perpendicular to the deformation direction. This results in significant amounts of energy being stored in these bonds and bond angles with the result that at very high constraining forces they can break before they conformationally relax. This result is consistent with the CRC ideas presented earlier.

7. SOME CONCLUDING REMARKS

Since the late 50's and the pioneering work of Alder and others,
the use of computer simulation as a research tool for both the
calculation of relevant material properties and perhaps even more
importantly for the insight provided in understanding the
microscopic origin of these properties has been an ever expanding
field of scientific activity. The early work focused on simple
solids and liquids, but with increasing computer power we now
study complex proteins. The pioneers in this field were by and
large theorists, but now its practitioners span all scientists,
and the marketing of "simulation packages" has become a multi-
million dollar business.

In the area of polymer science, MC and MD studies have given us
significant insights into the properties of individual polymer
chains, particularly in solution. The real challenges ahead lie
in the area of solid polymers and their material properties. We
have outlined a few pioneering efforts with these goals in mind.
We are confident that the use of MC and MD methods to study
polymeric materials will become an area of rapidly expanding
activity.

ACKNOWLEDGMENT

Part of this work was performed under the auspices of the U.S.
Department of Energy by the Lawrence Livermore National
Laboratory under Contract No. W-7405-ENG-48.

REFERENCES

1. van Kampen, N.G., Stochastic Processes in Physics and
    Chemistry, North Holland, Amsterdam-New York-Oxford, 1981.

2. Bratley, P., Fox, B.L. and Schrage, L.E., _A Guide to
    Simulation_, Springer, New York-Berlin-Heidelberg-Tokyo,
    1983.

3. Metropolis, N. and Ulam, S., _J. Am. Statist. Assoc._ **44**, 335,
    1949.

4. Alder, B.J. and Wainwright, T.E., _J. Chem. Phys._ **27**, 1208,
    1957.

5. Ferry, J.D., _Viscoelastic Properties of Polymers_, 3rd ed.,
    Wiley, New York, 1980.

6. Aklonis, J.J. and McKnight, W.J., _Introduction to Polymer
    Viscoelasticity_, 2nd ed., Wiley, New York, 1983.

7. _Failure of Plastics_, editors W. Brostow and R.D.
    Corneliussen, Hanser Publ., Munich-Vienna-New York, 1986.

8. Bohlin, L. and Kubát, J. and Rigdahl, M., _Colloid & Polymer
    Sci._ **265**, 803, 1987.

9. Kubát, J., _Phys. Status Solidi B_, **111**, 599, 1982.

10. Kubát, J. and Rigdahl, M., Chapter 4 in Ref. 17.

11. Ek, C.-G., Kubát, J. and Rigdahl, M., _Colloid & Polymer Sci._ _265_, 803, 1987.

12. Kubát, J., Maurer, F.H.J., Rigdahl, M. and Welander, M., _Rheol. Acta_, _28_, 147, 1989.

13. Brostow, W., Kubát, J. and Latka, M., work in progress.

14. Brostow, W., _Mater. Chem. & Phys._ _13_, 47, 1985.

15. Brostow, W., Chapter 10 in Ref. 7.

16. Brostow, W. and Macip, M.A., _Macromolecules_, _22_, 2761, 1989.

17. Brostow, W. and Müller, W.F., _Polymer_, _27_, 76, 1986.

18. Brostow, W., Fleissner, M. and Müller, W.F., _Polymer_, _31_, in press, 1990.

19. Krishnamurthy, V., Brostow, W. and Sochanski, J. S., _Mater. Chem. & Phys._ _20_, 451, 1988.

20. Einwohner, T. and Alder, B.J., _J. Chem. Phys._, _49_, 1458, 1968.

21. Helfand, E., Wasserman, Z. R., and Weber, T. A., _J. Chem. Phys._ _70_, 2016, 1979.

22. Helfand, E., Wasserman, Z.R. and Weber, T. A., _Macromolecules_ _13_, 526, 1980.

23. Skolnick, J. and Helfand, E., _J. Chem. Phys._ _72_, 5489, 1980.

24. See for example Matsuo, K., Kuhlmann, K. F., Yang, H. W.-H., Geny, F., Stockmayer, W. H. and Jones, A. A., _J. Polymer Sci. Phys._ _15_, 1347, 1977; Liao, T.-P., Okamoto, Y., Morawetz, H., _Macromolecules_ _12_, 535, 1979.

25. Cook, R., _J. Polymer Sci. Phys._, _26_, 1337, 1988.

26. Cook, R., _J. Polymer Sci. Phys._, _26_, 1349, 1988.

A Comparison of Some Approximate Methods of Computing Bayes Estimators

SAMIR K. BHATTACHARYA
Allahabad University, India

ALOK PANDEY and ASHOK K. SINGH
New Mexico Institute of Mining and Technology
Socorro, New Mexico 87801, USA

ABSTRACT

In Bayesian estimation problems, one is faced with evaluation of posterior moments. For the continuous case, this involves computing a ratio of two integrals. There are many problems in Bayesian estimation for which closed form expressions for these integrals do not exist. Several methods of approximation exist for computing the Bayes estimator in such cases. The purpose of this paper is to compare some of these approximations by using Monte Carlo simulation.

1. INTRODUCTION

One of the main results of Wald's statistical decision theory is that in any given statistical problem, one can restrict attention to Bayes or limits of Bayes rules [3] . Bayes methods also provide means of incorporating any prior information into the statistical inference problem at hand.

The computation of Bayes estimator involves determining a ratio of two integrals. For natural conjugate priors, these two integrals can be evaluated in closed form. There are many situations, however, in which it is not possible to evaluate these integrals analytically; numerical methods are needed for such problems. Tierney and Kadane [4] have used Laplace's approximation of order $0(n^{-1})$ on both numerator and denominator to obtain a $0(n^{-2})$ approximation for the ratio.

The purpose of this paper is to investigate the approximation of Tierney and Kadane, and also to compare it to a suitable numerical quadrature. We have chosen a problem in Bayesian life testing as a test problem.

2. THE TEST PROBLEM

Let $x_1 < x_2 < \ldots < x_r$ be a Type II censored sample of size r obtained by putting n items on a destructive life test; failed items are not replaced. The time to failure of each of the n items is assumed to be exponential with mean θ . Bhattacharya [2] derived the Bayes estimator of θ for various choices of $g(\theta)$, the prior pdf of θ. For the natural conjugate prior

$$g(\theta) = \frac{\exp\left(\frac{-\mu}{\theta}\right)\left(\frac{\mu}{\theta}\right)^{\nu+1}}{\mu\Gamma(\nu)} \tag{1}$$

the Bayes estimator of θ is [2]:

$$\hat{\theta}_B = \frac{\mu + T_r}{r + v - 1}, \quad v > 1.$$
(2)

where $T_r = \sum_{i=1}^{r} X_i + (n-r)X_r$.

3. THE APPROXIMATION OF TIERNEY AND KADANE

If $\underline{x}$ is a random sample form the conditional pdf $f(x|\theta)$, and θ has the prior pdf $g(\theta)$, then Bayes estimator of a function $h(\theta)$ under squared-error loss is given by [3]

$$E[h(\theta)|x] = \frac{\int h(\theta)\, \exp[L(\theta)]\, g(\theta)\, d\theta}{\int \exp[L(\theta)]\, g(\theta)\, d\theta}$$
(3)

where $L(\theta) = \ln \prod_{i=1}^{n} f(X_i|\theta)$ is the conditional log-likelihood function.

For the special case of $h(\theta) > 0$, Tierney and Kadane [4] used Laplace's method of order $0(n^{-1})$ on both numerator and denominator of (3) to obtain the following $0(n^{-2})$ approximation:

$$E[h(\theta)|X] \approx \frac{\sigma_2}{\sigma_1} \exp\{n(L_2(\hat{\theta}_2) - L_1(\hat{\theta}_1))\}$$
(4)

where

$$L_1(\theta) = \frac{(\log g(\theta) + L(\theta))}{n}$$

$$L_2(\theta) = \frac{(\log h(\theta) + \log g(\theta) + L(\theta))}{n}$$

$$\hat{\theta}_1 = mode\ of\ L_1(\theta)$$

$$\hat{\theta}_2 = mode\ of\ L_2(\theta)$$

$$\sigma_1^2 = \frac{-1}{L_1^{(2)}(\hat{\theta}_1)}$$

$$\sigma_2^2 = \frac{-1}{L_2^{(2)}(\hat{\theta}_2)} \ .$$

4. COMPARISON OF T-K METHOD WITH A NUMERICAL QUADRATURE

For the test problem described in Section 2, the Bayes estimator of θ under the squared-error loss assumption is

$$E(\theta|\underline{x}) = (\mu+T_r) \frac{\int_0^\infty e^{-w} w^{r+v-2} \, dw}{\int_0^\infty e^{-w} w^{r+v-1} \, dw}$$

We use Laguerre quadrature [1] on both numerator and denominator to obtain

$$\hat{\theta}_L = (\mu+T_r) \frac{\sum_{i=1}^{N} W_i \, x_i^{r+v-2}}{\sum_{i=1}^{N} W_i \, x_i^{r+v-1}} \tag{5}$$

where W_i are the weights and x_i's the zeros of Laguerre polynomials.

For the T-K approximation [4], we have:

$$L_1(\theta) = \frac{1}{n}\left[K - \frac{\mu+T_r}{\theta} - (r+v+1)\ln\,\theta \right]$$

$$L_2(\theta) = \frac{1}{n}\left[K - \frac{\mu+T_r}{\theta} - (r+v)\ln\theta \right]$$

Differentiating $L_i(\theta)$ with respect to θ and solving $L_i'(\theta) = 0$, we obtain

$$\hat{\theta}_1 = \frac{(\mu+T_r)}{(r+v+1)}$$

$$\hat{\theta}_2 = \frac{(\mu+T_r)}{(r+v)}$$

$$\sigma_1^2 = \frac{n(\mu+T_r)^2}{(r+v+1)^3}$$

$$\sigma_2^2 = \frac{n(\mu+T_r)^2}{(r+v)^3}$$

Substitution of these values in (4) leads to:

$$\hat{\theta}_{TK} = \frac{(\mu+T_r)(v+r)^{v+r+(3/2)}}{(v+r+1)^{v+r+(1/2)}} \, e \tag{6}$$

We have used Monte Carlo simulation to compare the approximations (5) and (6) with the exact Bayes estimator (2). The simulation experiment is briefly described below:

[1] Generate one observation U from the gamma pdf with parameters v and μ and set $\theta = 1/U$. This yields a θ-$value$ from the inverted gamma pdf (1).

[2] Generate T_r from the gamma pdf with parameters r and θ .

[3] Compute the exact Bayes estimator $\hat{\theta}$ given by (2), and the two approximation (5) and (6).

5. RESULTS OF SIMULATION

The IMSL STAT library subroutine RNGAM was used to generate gamma random variables needed for Steps 1 and 2 of the simulation experiment. All computations were done in FORTRAN77 on a SUN 3/150 computer.

The simulation experiment described in Section 4 was run for μ = 1, 10, 100, r = 10, 40(10), v = 10, 40(10). Tables 1 - 3 show the values of generated θ and T_r , and also the computed values of $\hat{\theta}_{B}$, $\hat{\theta}_{TK}$ and $\hat{\theta}_{L}$. We used N = 15 nodes for the Laguerre quadrature in (5).

6. DISCUSSION OF RESULTS

It can be seen from Tables 1 - 3 that for $r + v - 1 \leq 28$, the approximation obtained by using Laguerre quadrature has no error. This is to be expected since the Laguerre quadrature (5) has degree of precision $2(N-1) = 28$ [1]. Even for these cases, $\hat{\theta}_{TK}$ is quite close to $\hat{\theta}_{B}$. For example, when $\mu = 1$ (Table 1), $r = 10$, $v = 10$, the θ-value generated from the prior pdf (1) is $\theta = 5.71$ e-2; the value of T_r generated from the conditional Gamma(10,θ) is $T_r = 0.4607$. The exact Bayes estimator is $\theta_B = 7.69e-2$, and its approximations are $\theta_{TK} = 7.67e-2$, $\theta_L = 7.69e-2$.

For larger values of $r + v - 1$, the aproximation of Tierney and Kadane gives better results than the Laguerre quadrature. For example, when $\mu = 100$ (Table 3), $r = 40$, $v = 40$, generated $\theta = 2.94$, generated $T_r = 105.7$, and the true Bayes estimate is $\theta_B = 2.60$; the two approximations to the Bayes estimator are: $\theta_{TK} = 2.60$, $\theta_L = 4.28$.

REFERENCES

[1] Abramowitz, M. and Stegun, I., "Handbook of Mathematical Functions", Dover Publications, Inc., New York, pp. 890-923, 1972.

[2] Bhattacharya, S.K., "Bayesian Approach to Life Testing and Reliability Estimation", Jour. Amer. Stat. Assoc., Vol. 62, pp. 49-62, 1967.

[3] Ferguson, Thomas S., "Mathematical Statistics - A Decision Theoretic Approach", Academic Press, New York, pp. 86-89, 1967.

[4] Tierney, L. and Kadane, J.B., "Accurate Approximations for Posterior Moments and Marginal Densities", Jour. Amer. Stat. Assoc., Vol. 81, pp. 82-86, 1986.

[5] IMSL STAT/LIBRARY Reference Manual, Houston, Texas, 1987.

Table 1: Results of simulation experiment
for $\mu = 1$

		r = 10	r = 20	r = 30	r = 40
$\nu=10$	θ	5.71 E-2	0.119	6.81 E-2	0.126
	T_r	0.4607	2.583	2.651	4.523
	$\hat{\theta}_B$	7.69 E-2	0.124	7.45 E-2	0.113
	$\hat{\theta}_{TK}$	7.67 E-2	0.124	7.45 E-2	0.113
	$\hat{\theta}_L$	7.69 E-2	0.123	7.93 E-2	0.120
$\nu=20$	θ	4.41 E-2	6.20 E-2	6.81 E-2	5.77 E-2
	T_r	0.3775	1.591	2.651	2.474
	$\hat{\theta}_B$	4.75 E-2	6.64 E-2	7.45 E-2	5.89 E-2
	$\hat{\theta}_{TK}$	4.74 E-2	6.64 E-2	7.45 E-2	5.89 E-2
	$\hat{\theta}_L$	4.75 E-2	6.65 E-2	7.93 E-2	7.27 E-2
$\nu=30$	θ	3.48 E-2	3.06 E-2	4.35 E-2	2.72 E-2
	T_r	0.2821	0.4961	0.8390	1.003
	$\hat{\theta}_B$	3.29 E-2	3.05 E-2	3.12 E-2	2.90 E-2
	$\hat{\theta}_{TK}$	3.29 E-2	3.05 E-2	3.12 E-2	2.90 E-2
	$\hat{\theta}_L$	3.29 E-2	3.25 E-2	3.85 E-2	4.17 E-2
$\nu=40$	θ	2.13 E-2	2.18 E-2	2.58 E-2	2.16 E-2
	T_r	0.3148	0.4684	0.6196	0.9251
	$\hat{\theta}_B$	2.68 E-2	2.49 E-2	2.35 E-2	2.44 E-2
	$\hat{\theta}_{TK}$	2.68 E-2	2.49 E-2	2.35 E-2	2.44 E-2
	$\hat{\theta}_L$	2.85 E-2	3.07 E-2	3.37 E-2	4.01 E-2

Table 2: Results of simulation experiments
for $\mu = 10$

		r = 10	r = 20	r = 30	r = 40
$\nu=10$	θ	0.512	1.36	1.38	0.836
	T_r	7.687	26.67	36.15	26.66
	$\hat{\theta}_B$	0.931	1.264	1.18	0.748
	$\hat{\theta}_{TK}$	0.928	1.263	1.18	0.748
	$\hat{\theta}_L$	0.931	1.264	1.19	0.796
$\nu=20$	θ	0.425	0.373	0.458	1.01
	T_r	2.170	5.888	15.13	31.14
	$\hat{\theta}_B$	0.420	0.407	0.513	0.697
	$\hat{\theta}_{TK}$	0.420	0.407	0.513	0.697
	$\hat{\theta}_L$	0.420	0.408	0.546	0.861
$\nu=30$	θ	0.294	0.293	0.281	0.328
	T_r	3.481	5.514	8.635	15.89
	$\hat{\theta}_B$	0.346	0.317	0.316	0.375
	$\hat{\theta}_{TK}$	0.345	0.316	0.316	0.375
	$\hat{\theta}_L$	0.346	0.337	0.390	0.539
$\nu=40$	θ	0.218	0.259	0.283	0.268
	T_r	2.418	6.051	9.693	13.67
	$\hat{\theta}_B$	0.253	0.272	0.285	0.299
	$\hat{\theta}_{TK}$	0.253	0.272	0.285	0.299
	$\hat{\theta}_L$	0.270	0.336	0.410	0.493

Table 3: Results of simulation experiments
for $\mu = 100$

		r = 10	r = 20	r = 30	r = 40
$\nu=10$	θ	5.89	7.88	9.52	12.04
	T_r	78.59	177.3	348.1	540.0
	$\hat{\theta}_B$	9.40	9.56	11.49	13.06
	$\hat{\theta}_{TK}$	9.37	9.55	11.48	13.06
	$\hat{\theta}_L$	9.40	9.56	11.51	13.90
$\nu=20$	θ	6.08	3.84	5.66	5.05
	T_r	82.51	94.61	174.7	208.4
	$\hat{\theta}_B$	6.29	4.99	5.61	5.23
	$\hat{\theta}_{TK}$	6.29	4.99	5.60	5.23
	$\hat{\theta}_L$	6.29	5.00	5.97	6.45
$\nu=30$	θ	3.81	3.34	3.95	4.77
	T_r	62.45	40.40	148.1	199.3
	$\hat{\theta}_B$	4.17	2.87	4.21	4.34
	$\hat{\theta}_{TK}$	4.16	2.86	4.20	4.34
	$\hat{\theta}_L$	4.17	3.05	5.19	6.24
$\nu=40$	θ	2.74	2.45	2.87	2.94
	T_r	38.94	49.90	112.4	105.7
	$\hat{\theta}_B$	2.84	2.54	3.08	2.60
	$\hat{\theta}_{TK}$	2.83	2.54	3.08	2.60
	$\hat{\theta}_L$	3.02	3.14	4.43	4.28

Exact Distribution of the Least-Squares Estimator of the Linear Stress Failure Model

DAVE ALAN MILLER and ASHOK K. SINGH
Department of Mathematics
New Mexico Institute of Mining and Technology
Socorro, New Mexico 87801, USA

ABSTRACT

Let the failure time distribution of an item under a constant stress V_i be exponential with hazard rate $\lambda_i = BV_i$, $i = 1, 2, \ldots, k$, where B is an unknown parameter to be estimated. One model for λ_i, the maximum likelihood estimate of the hazard rate λ_i from a Type II censored sample obtained at stress V_i, is $\hat{\lambda}_i = B\hat{V}_i + \epsilon_i$, where ϵ_i is an unobservable random variable. The problem of estimation of B by the method of Least-Squares [LS] has been investigated in the literature.

The LS estimate of B is given by $b_R = \Sigma\hat{\lambda}_i/\Sigma V_i$. The distribution of the statistic b_R is needed for testing and confidence interval estimation problems involving the parameter B. It is known that the asymptotic distribution of b_R is normal, but the small sample distribution of b_R is not available.

In this paper we derive the exact distribution of the statistic b_R. To this end we first compute the moment generating function which is the Laplace transform of the pdf. This turns out to be

$$M(s) = \prod_{i=1}^{k} 2\left(\xi_i s\right)^{r_i/2} K_{r_i}\left(2\sqrt{\xi_i s}\right)/\Gamma(r_i), \tag{1}$$

where $\xi_i = r_i/(\Theta_i \Sigma V_j)$, $\Theta_i = 1/\lambda_i$, r_i is the number of failures under stress V_i, and $K_{r_i}(\bullet)$ is the modified Bessel function K of integer order r_i.

Since we are interested in computing cut-off points of the distribution, we use (1) to compute the Laplace transform of the cdf:

$$\frac{M(s)}{s} = \mathscr{L}\left\{\int_0^t f(t)\,dt\right\}.$$

INTRODUCTION

Often it is difficult if not impossible to ascertain the life
behavior under normal use stresses of certain devices (e.g. nuclear
reactor or space vehicle components) in a laboratory simulation for
reasons of time or economy. One method to circumvent these
difficulties is to subject the device being tested to different
combinations of increased stresses (e.g. temperature stress,
voltage stress, etc.) in order to gain more knowledge about the
relation of the variation of failure with different environments.
It is then desired to make inferences from the information obtained
under these simulated environments to predict the behavior of the
device under normal use-stress conditions. This is better known
as the method of accelerated life testing.

ASSUMPTIONS

In this paper the failure time distribution is assumed to be
exponential with location at the origin since all the necessary
information about the mean life with respect to the environment is
contained in the regression function [1].

Two assumptions are made about the failure behavior of the device.
These are:

i)   That considering the device as a whole the failure modes may
     be of the "single" or "multiple" variety with no distinction
     being made between modes.

ii) That the different levels of stress do not change the type of
     life-time distribution but do have an effect on the parameters
     [1].

If $f(t;\theta)$ is the probability density function of the failure time
random variable of a device under an environment defined by a
vector of stresses S where θ is a vector of parameters, then in
order to make an inference about the failure behavior under an
environment which cannot be simulated in a life test the functional
relationship between θ and S must be determined. This may be
derived empirically or physically. Once the relationship between θ
and S is known, the properties of $f(t;\theta)$ under normal use stress
may be studied [1].

GOAL OF THIS PAPER

Let the failure time distribution be in the form
$f(t;\lambda_i) = \lambda_i \exp(-\lambda_i t)$, $\lambda_i > 0$, $t \geq 0$, under a constant stress V_i with
hazard rate $\lambda_i = BV_i + e_i$, $i = 1, 2, \ldots, k$, where B is an unknown
parameter to be estimated, $\hat{\lambda}_i = 1/\hat{\theta}_i$ is the estimator of λ_i, and e_i
is an unobservable random variable which may be assumed to be
uncorrelated because of randomization procedures.

If n_i items are tested under constant stress V_i until r_i failures
have occurred noting the times t_{1i}, $t_{2i}, \ldots t_{ri}$, then with no
replacement as items fail, the maximum likelihood estimator of θ_i
is given by [1]

$$\hat{\theta}_i = \left(\sum_{j=1}^{r_i} t_{ji} + (n_i - r_i) t_{ri} \right) / r_i .$$

and the inverse Laplace transform is [2]

$$g(t) = \mathcal{L}^{-1}\{P(s)\} = \frac{1}{2\pi i} \int_{c-i\infty}^{c+i\infty} e^{st} P(s)\,ds.$$

The distribution of $\hat{\Theta}_i$ is given by [1]

$$g(\hat{\Theta}_i) = \exp(-r_i\hat{\Theta}_i/\Theta_i)(\Gamma(r_i))^{-1}(r_i/\Theta_i)^{r_i}(\hat{\Theta}_i)^{r_i-1}, \ldots \hat{\Theta}_i, r_i > 0 \tag{2}$$
$$= 0 \qquad \text{otherwise.}$$

The distribution of $\hat{\lambda}_i$ is given by [1]

$$g(\hat{\lambda}_i) = \exp(-r_i/\Theta_i\hat{\lambda}_i)(\Gamma(r_i))^{-1}(r_i/\Theta_i)^{r_i}(\hat{\lambda}_i)^{-r_i-1}, \ldots \hat{\lambda}_i, r_i > 0$$
$$= 0 \qquad \text{otherwise.}$$

If one censors at the same r_i, the least-squares estimator of B is given by [1]

$$b_R = \Sigma\hat{\lambda}_i / \Sigma V_i .$$

For r large, the variance of b_R is [1]

$$VAR(b_R) = (B^2\Sigma V_i^2)/[r(\Sigma V_i)^2] .$$

It has been shown by [1] that the distribution of b_R is asymptotically normal. It is the purpose of this paper to derive the exact distribution of b_R numerically.

THE DISTRIBUTION OF b_R

If one substitutes $\Sigma V_i = S$ and $\hat{t}_i = \hat{\lambda}_i/S$, then

$$b_{Rk} = \hat{t}_1 + \hat{t}_2 + \ldots \hat{t}_k .$$

Since the distribution of $\hat{\lambda}_i$ is known, and $\hat{t}_i$ is monotonic in the range of λ_i, the distribution of t_i can be obtained as [1]

$$g(\hat{t}_i) = \exp(-r_i/S\Theta_i\hat{t}_i)(\Gamma(r_i))^{-1}(r_i/S\Theta_i)^{r_i}(\hat{t}_i)^{-r_i-1}, \ldots \hat{t}_i \geq 0 \tag{3}$$
$$= 0 \qquad \text{otherwise.}$$

The exact small sample distribution of b_{Rk} can be obtained by the k-fold convolution of the distribution of the individual t_i's [1]. For ease of computation, if one sets $\xi_i = r_i / (\Theta_i\Sigma V_j)$ then the distribution of (3), the inverse gamma, becomes

$$g(\hat{t}_i) = \exp(-\xi_i/\hat{t}_i)(\Gamma(r_i))^{-1}\xi_i^{r_i}\hat{t}_i^{-r_i-1}, \ldots \hat{t}_i \geq 0$$
$$= 0 \qquad \text{otherwise.}$$

The k-fold convolution is cumbersome to calculate in the time domain. Since it is a well known fact that the convolution of distributions in the time domain is simply the multiplication of the Laplace transforms of the distributions in the Laplace domain, computation may be simplified by using Laplace transforms.

If g and g' are piecewise continuous for all t≥0, and g satisfies

$$\int_0^\infty | g(t) | \, dt < \infty,$$

then the Laplace transform of g is given by

$$P(s) = \mathcal{L}\{g(t)\} = \int_0^\infty e^{-st} g(t)\,dt,$$

The Laplace transform of $g(\hat{t}_i)$ is

$$P_i(s) = \mathcal{L}\{g(\hat{t}_i)\} = \int_0^\infty \frac{e^{-s\hat{t}_i} \, e^{-\xi_i/\hat{t}_i} \, \xi^{r_i}}{\Gamma(r_i) \, \hat{t}_i^{\,r_i+1}} \, d\hat{t}_i. \tag{4}$$

Using the fact that $K_{-r}(\cdot) = K_r(\cdot)$ [3], $K_r(\cdot)$ being the modified Bessel function K of integer order r, (4) may be computed yielding [4]

$$P_i(s) = 2\left(\xi_i s\right)^{r_i/2} \left(\Gamma(r_i)\right)^{-1} K_{r_i}\left(2\sqrt{\xi_i s}\right).$$

The Laplace transform of the k-fold convolution becomes

$$M(s) = \prod_{i=1}^{k} P_i(s) = \prod_{i=1}^{k} \left[\frac{2(\xi_i s)^{r_i/2}}{\Gamma(r_i)} K_{r_i}\left(2\sqrt{\xi_i s}\right) \right]. \tag{5}$$

To obtain the probability density function, one simply finds the inverse Laplace transform of (5). To find the cumulative distribution function, one may employ the fact that [5]

$$\frac{M(s)}{s} = \mathcal{L}\left[\int_0^t f(t)\,dt\right].$$

A program to obtain cut-off points of the distribution utilizing a Bessel function generating subroutine [6] (accurate to 7 digits) and the numerical Laplace inverting subroutine [7] was checked using a simplified case. The accuracy of the inverting subroutine was checked against several functions with known Laplace transforms.

With $n=10$, $r_1=1$, $t(1,1)=1$ and $V_1=1$, the simplified cdf becomes

$$G(t) = \int g(t)\,dt = e^{-0.1/t}. \tag{6}$$

For purposes of comparison, the actual and computed pdf and cdf for the above simplified case are necessarily graphed separately in Figures 1 through 4 due to the resolution of the printer. The exact computed cdf's appear in Figures 5 through 8 for two sets of values that were chosen arbitrarily for B, the Type II sampling cut-off number, and the stresses. Table I compares the calculated values of (6) with the values computed from the numerical inverting program.

NOMENCLATURE

B	:	an unknown parameter to be estimated.
b_R	:	the LS estimator of B.
b_{Rk}	:	the sum of $\underline{k}$ independent, non-identically distributed, non-normally distributed random variables.
cdf	:	cumulative distribution function.
LS	:	least-squares estimator.
pdf	:	probability density function.

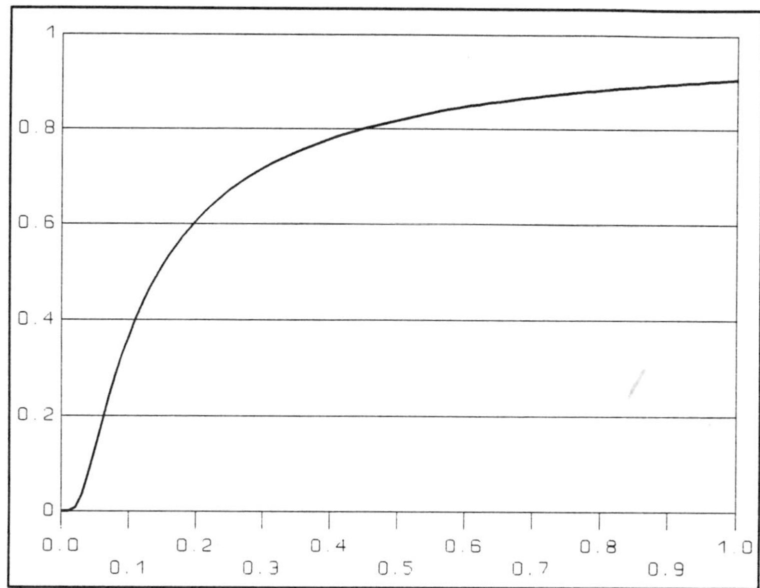

Figure 3. Graph of the actual cdf for the simplified case.

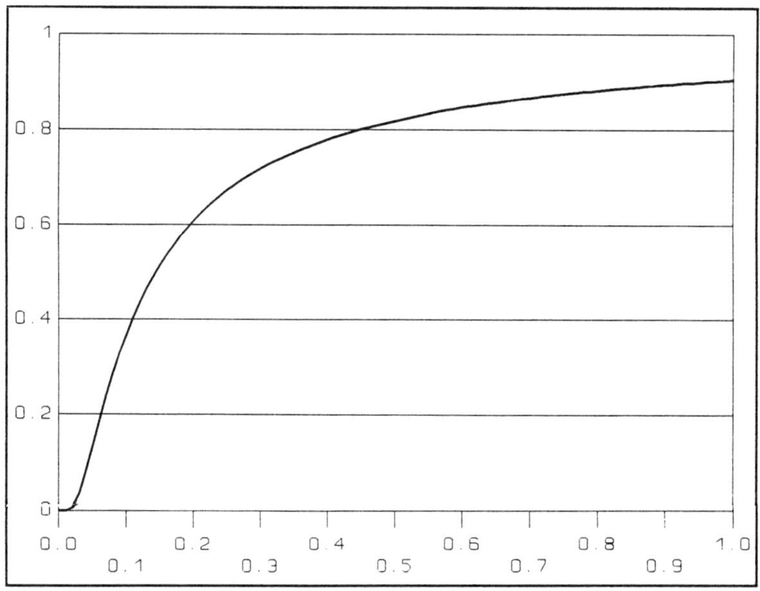

Figure 4. Graph of the computed cdf for the simplified case.

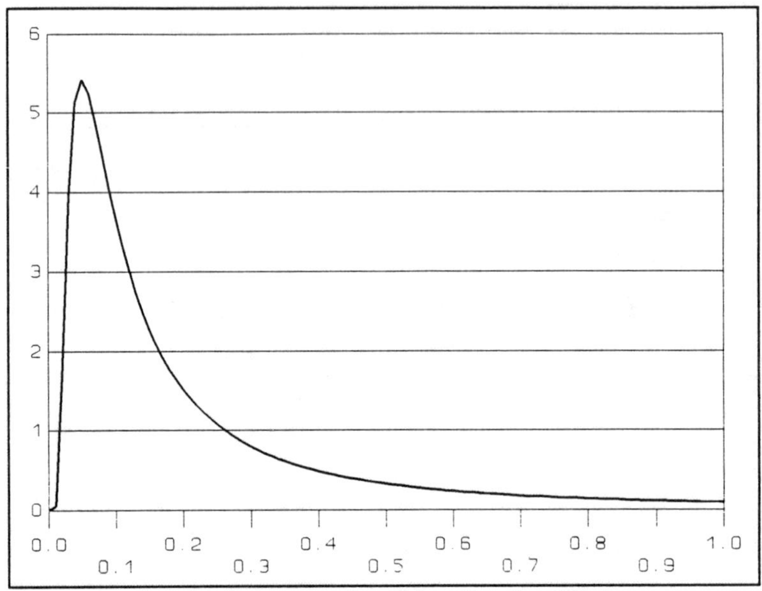

Figure 1. Graph of the actual pdf for the simplified case.

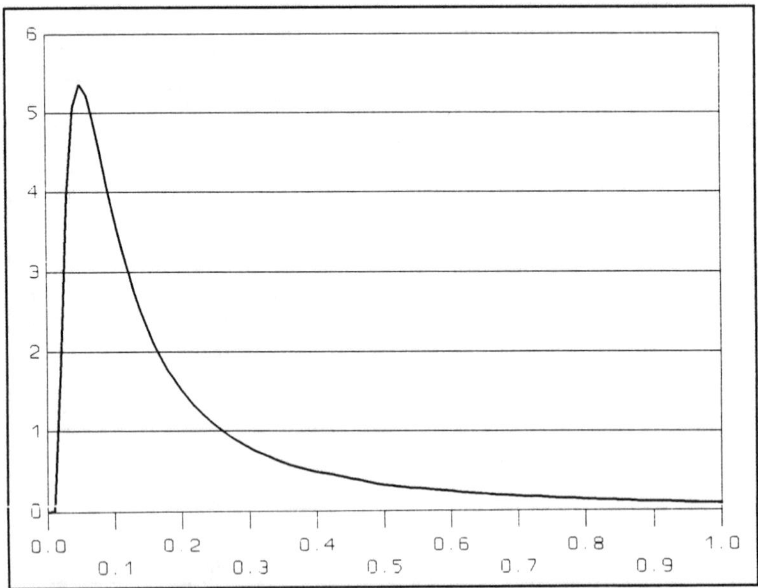

Figure 2. Graph of the computed pdf for the simplified case.

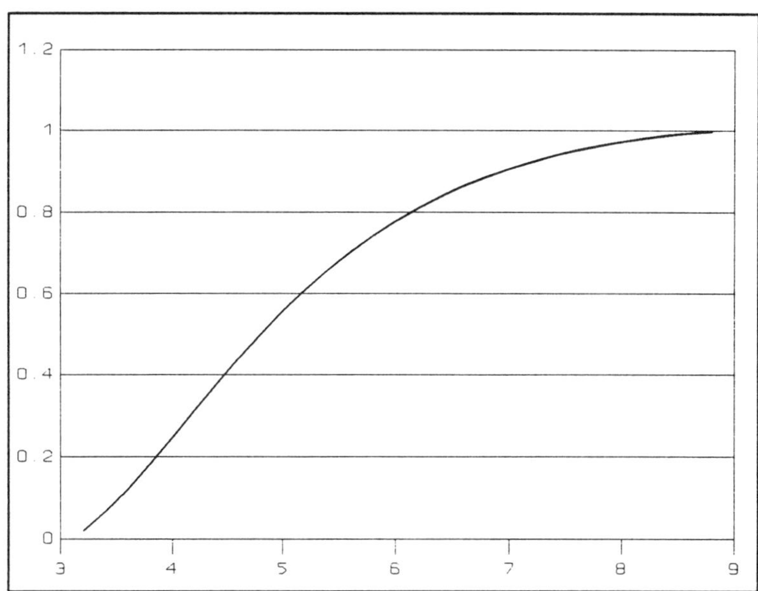

Figure 5 Exact computed cdf with B=4, k=10, r=5, and ten evenly incremented stresses from 2 to 20.

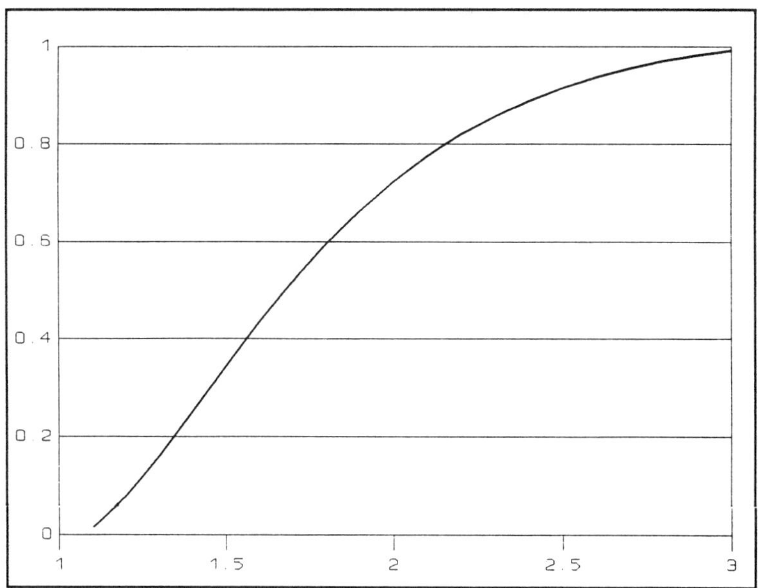

Figure 6 Exact computed cdf with B=1.5, r=7, and five stresses, 2, 4, 5, 6, and 7.

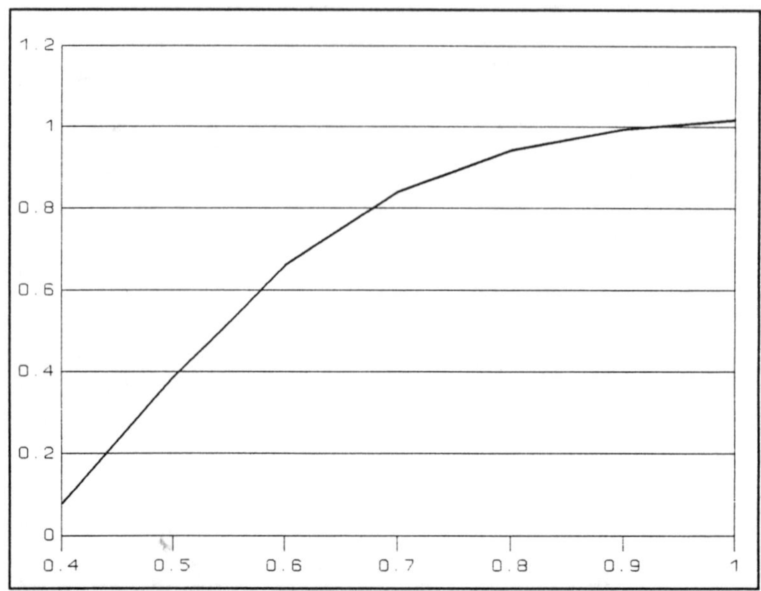

Figure 7 Exact computed cdf with B=0.5, r=10, and eight stresses evenly incremented 2 through 9.

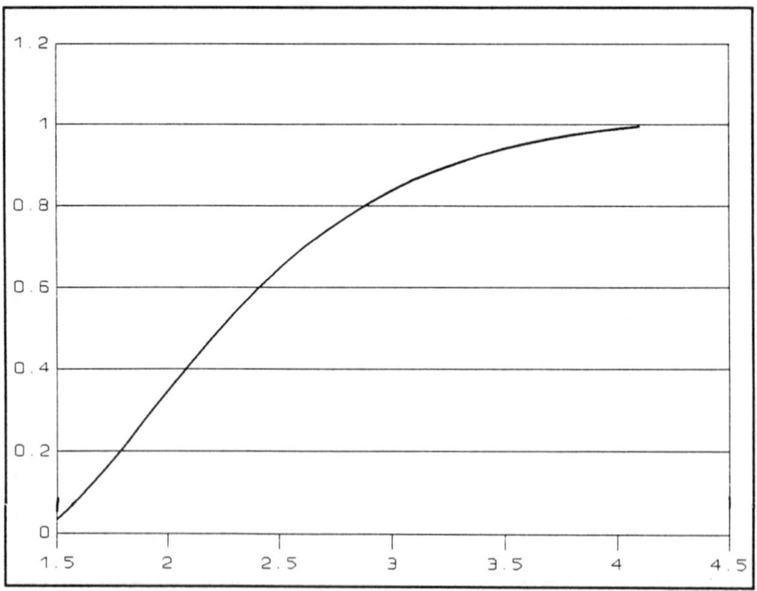

Figure 8 Exact computed cdf with B=2, r=7, and five stresses 1, 4, 5, 6, and 7.

ACKNOWLEDGEMENTS

The authors wish to thank the Research and Economic Development
Division of NMIMT for the professional and financial support, the
secretaries at NMIMT especially those at the PRRC, Humanities,
Petroleum and Mathematics departments, and Gordon R. Kane of the
Electrical Engineering Department at NMIMT for his immense
assistance with the project.

REFERENCES

[1]   Singpurwalla, N. D. "Inference from Accelerated Life Tests
      When Observations are Obtained from Censored Samples",
      Technometrics, vol. 13, No. 1, pp. 161-170, Feb. 1971.

[2]   Andrews, L. C. and Shivamoggi, B. K. Integral Transforms for
      Engineers and Applied Mathematicians, pp. 163-164, Macmillan
      Publishing Company, 1988.

[3]   Abramowitz, M. and Stegun, I. Hand Book of Mathematical
      Functions, p. 375, Dover Publications Inc., New York, 1965.

[4]   Gradshteyn, I. S. and Ryzhik, I. M. Table of Integrals,
      Series and Products, p. 340, Academic Press, 1965.

[5]   Gardner, M. F. and Barnes, J. L. Transients in Linear
      Systems, p. 271, John Wiley & Sons Inc., New York, 1958.

[6]   Press, W. H. et al. Numerical Recipes, pp. 176-179, Cambridge
      University Press, Cambridge, 1986.

[7]   Stehfest, H. "Algorithm 368: Numerical Inversion of Laplace
      Transforms [D5]", Collected Algorithms from CACM.

On Weak Solutions of Nonlinear Volterra Stochastic Equations

MARICA LEWIN
Department of Industrial Engineering and Management
Technion—Israel Institute of Technology
Haifa 32000, Israel

ABSTRACT

We treat the existence of a weak solution for the equation

$$x(t) = \int_o^t a(t-s)Ax(s)ds + \int_o^t a(t-s)Bx(s)dw(s) + f(t) \tag{1}$$

We assume V is a real, separable Hilbert space, V^* its dual, H a Hilbert space, and we have the continuous, dense injection

$$V \subset H \subset V^*$$

$a(t)$ is a positive kernel, $f(t)$ a stochastic process taking values on V, A an operator mapping V into V^*, and B a linear (not necessarily continuous) operator from H to H. w is a Wiener process in the Hilbert space H, with covenance Q. We give sufficient conditions for $a(t)$, $f(t)$, the operators A and B, such that the equation (1) possesses at least a weak solution

$$x(t) \in L_\sigma^2([0,T],V] \cap L^2[[0,T],V^*], w.p.\ 1$$

This research was supported in part by the US–Israel Binational Science Foundation (86–00285/1) and the Israel Office for Absorption of New Scientists.

1. INTRODUCTION

We consider the following equation

$$x(t) = \int_0^t a(t-s)Ax(s)\,ds + \int_0^t a(t-s)Bx(s)\,dw(s) + f(t) \tag{1.1}$$

The following general framework is considered: (Ω, F, P, E) is a filtered probability space, satisfying the usual assumptions.

1.2) V is a separable Hilbert space with its dual V^* and H is a separable Hilbert space identified with its dual.

   We assume that we have $V \subset H \subset V^*$.

1.3) A maps V into V^*.

1.4) w is a Wiener process in the Hilbert space H, with covariance Q. For any $v \in H$, $B(v)$ is a linear operator (not necessarily continuous) from H to H.

1.5) $a(t)$ is a positive kernel.

The work has as point of departure the lecture of Métivier and Viot, on weak solutions of partial differential equations, [8].

They treat the existence of the solution for the equation

$$\begin{cases} dx(t) + Ax(t)\,dt = Bx(t)\,dw(t) \\ x(t_o) = x_o \end{cases} \tag{1.2}$$

and they prove that the sought weak solution is a stochastic process whose trajectories belong to $C([0,T], H)$ w.p.1.

They solve *the martingale problem*, which for the equation 1.6) is equivalent with the existence of a weak solution, using the energy equality.

The integral equations cannot be treated using the martingale problem, because the stochastic integral $\int_0^t a(t-s)Bx(s)\,dw(s)$ is not anymore a martingale, the kernel $a(t-s)$ depending on t. So, we shall investigate the existence of a weak solution using other methods, more precisely the Theorem of representation of Skorohod (see [6]).

The stochastic Volterra equations were deeply studied when the function under the Bochner integral is defined on $[0,T] \times H$, with values in H, and with Caratheodory conditions (see [11]).

For the first time, we treat a variational problem for a random integral equation of Volterra type.

Let us recall the notions of strong and weak solutions for the problem we are concerned with.

<u>Strong solutions</u> The basic probability space (Ω, F, P) with its filtration (F_t) and a H–valued

Wiener process on this space being given, a strong solution is a H–valued stochastic process, such that

a) $x(\cdot,\omega)\in L^1_\sigma[[0,T],V]$, and $Ax(\cdot,\omega)\in L^1[[0,T],V^*]$, P.a.s.

b) $x(t)$ satisfies the equation (1.1) w.p.1. in V^*.

<u>Weak solutions</u>   By a weak solution of (1.1) we mean to find a system $(\tilde\Omega,\tilde F,\tilde P,\tilde F_t,\tilde f(t),$ $\tilde w(t),\tilde x(t))$, where

al) $(\tilde\Omega,\tilde F,\tilde P,\tilde F_t)$ is a filtered probability space.

b1) $\tilde w(t)$ is an H–valued Wiener process with covariance Q.

c1) $\tilde w(t)$ has the same laws as $w(t)$ on $C([0,T],H)$. $\tilde f(t)$ has the same law as $f(t)$ on $C([0,T],V)$.

c)   The conditions a) and b) from the Definition of strong solutions are satisfied in the new system $(\tilde\Omega,\tilde F,\tilde P,(\tilde F_t),\tilde f(t),\tilde w(t),\tilde x(t))$.

2. GENERAL ASSUMPTIONS

We shall use the following standard methods: $(\,,)$ and $|\ |$ will denote the scalar product, respectively for the norm in H.

$\|\ \|$ will denote the norm in V and $\|\ \|$ the norm in V^*, while the duality between V and V^* will be written $<\,,\,>$. w–lim will be the notation for the weak limit.

We make the following assumptions on the operators A and B and the spaces V,H,V^*.

<u>I. (Assumptions on A)</u> $A:V\to V^*$.

Ia.   *(Monotony)* For every $u,v\in V$,
   $<u-v,A(u)-A(v)>\ \geq 0$.

Ib.   *(Hemicontinuity on V if w $\lim\limits_{t\to 0} A(u+tv)=Au$, for any $u,v\in V$.*

Ic.   *Boundedness.*   There exists a constant K, such that for every $u\in V$ $\|Au\|\leq K(1+\|u\|)$.

Id.   *Coercivity.*   For any $u_n\in V$, $\|u_n\|\to\infty$, $\lim\limits_{n-\infty}\dfrac{<u_n,Au_n>}{\|u\|}=\infty$.

<u>II. Assumption on B</u>   For every u, $B(u)$ is a linear operator from H to H, such that, for every $u\in H$, the operator

IIa. $b(u)=B(u)QB^*(u)$ is defined and it is a nuclear operator in H (B^* the adjoint operator of B. Q, the covariance operator is by definition a linear continuous, compact and self–adjoint operator on H).

If w is a Wiener process with Q as a covariance operator, then $L^2(H,H,w)$ denotes the class of processes f, defined on $[0,T]\times\Omega$, having as values a linear operator from H to H, strongly nonanticipating and such that

$$E(\int_o^T \text{trace }(f(s)Qf^*(s))\,ds)<\infty$$

54

$L^2(H, H, W)$ is a Hilbert space with the norm IIb) (see [9]). B is supposed to be a continuous operator from H into $L^2(H, H, W)$, and moreover

IIb) trace $B(u)QB^*(u) \le K_1(1 + |u|^2)$.

<u>III. Assumptions on the spaces V, H, V'</u> We deal with the continuous, densely defined injections $V \to H \to V^*$, and moreover, the injection from V into H is supposed to be compact.

<u>IV.</u> $a \in L^1_{loc}(0, \infty)$ and $\operatorname{Re} \hat{a}(i\eta) \ge 0$ for every $\eta \in R'$. a is a continuous function on $[0, T]$.

<u>V.</u> $f(t, \omega)$ is a process defined on $[0, T] \times \Omega$, taking values in V continuous in t, strongly nonanticipating, such that

$$E \|f(t, \omega)\|_V^2 < \infty \quad \text{for all } t \in [0, T] \tag{V.1}$$

<u>Remark.</u> The condition a) in the definition of strong solutions (respectively d in the definition of weak solutions) makes the equation (1.1) meaningful. A is defined on V, Ax has to be integrable so x has to belong to $L^1[0, T, V]$, and the result Ax has to be in V^∞.

3. MAIN RESULT

<u>Theorem</u> *Under the hypotheses I, II, III, IV, V, the equation (1.1) has a weak solution.*

The general scheme for the existence theorem is the following:

a)   We construct a sequence x_n of "approximative solutions" whose laws P_n are probability measures on $G_\sigma = C([0, T], V^*) \cap L^2_\sigma([0, T], V)$, and the family P_n are relatively compact on the space mentioned above. L^2_σ is the space L^2 endowed with the weak topology.

b)   We apply the theorem of representation of Skorohod, so we construct a probability space $(\widetilde{\Omega}, \widetilde{F}, \widetilde{P})$ and on it a sequence of random variables taking values on G_σ, whose laws are P_n (the same laws as x_n), and $\tilde{x}_n$ converges w.p.1 to $\tilde{x}$ in G_σ.

c)   We prove that $\tilde{x}(t)$ satisfies the equation (1.1) in the new space $(\widetilde{\Omega}, \widetilde{F}, \widetilde{P})$.

<u>Proof</u> We consider the Foedo–Galerkin method of approximation used also in the case of partial differential equations by Pardoux, Métivier, Viot, etc. [8][10].

The approximations consist of:

a) e_i is an orthonormal laws in H, with $e_i \in V$, and V_n is the subspace generated by $\{ e_i : i = 1, 2, \ldots, n \}$. One defines for every $v' \in V'$.

$$\Pi_n(v') = \sum_{1 \le i \le n} < e_i, v' > e_i \tag{3.1}$$

By $\Pi_n \circ h$, we denote the projection of an element of H or V_n. Then, x_n would be a solution of

$$x_n(t) = \Pi_n \circ f(t) + \int_0^t a(t-s)\Pi_n(Ax_n(s))\,ds + \int_0^t a(t-s)\Pi_n \circ B(x_n(s))\,d(\Pi_n \circ W(s))$$

$$(3.2)$$

$$= C_n(t) + D_n(t) + Y_n(t)$$

The equation (3.2) is considered in a finite dimensional space V_n, so endowed with a unique topology. The bypothesis imposed on A, implies it is demicontinuous from V to V^*, so $\Pi_n A$ will be a continuous operator from V_n to V_n. The growth conditions Ic), IIb) will result in the existence of a solution (weak) for (3.2).

x_n, the solution of (3.2) defines a V_n–valued process, whose laws, P_n can be considered as a family of probability measures on G_σ.

If we denote by τ_1 the topology of $C([0,T], V^*)$ and by τ_2 the topology of $L_\sigma^2([0,T], V)$, G_σ will be furnished with the topology $\tau = \tau_1 \wedge \tau_2$, the supremum of both τ_1 and τ_2. (G_σ, τ) is not anymore a Polish space, but a Lusin space, because $L_\sigma^2([0,T], V)$ is a Borel set of $C([0,T], V^*)$ (see[7]).

A sufficient condition for the sequence P_n to be relatively compact on (G_σ, z) is to be relatively compact in both spaces, $C([0,T], V^*)$ and $L_\sigma^2([0,T], V)$ (see [   ]), which are Polish spaces.

For this purpose, the conditions Ic), IIb) the continuity of $a(t)$, and the fact that all the functions from (3.2) are F_t adapted, will result in

$$E\, \|x_n^{(t)}\|_V^2 \le 3(E\, \|f_n(t)\|_V^2 + \int_0^t K_1(1 + E\, \|x_n(s)\|_V^2\,ds + \int_0^t K_2(1 + E\, \|x_n\|_V^2\,ds \qquad (3.3)$$

By the Gronwall inequality (see [4]) and the assumptions on f, we obtain

$$E\, \|x_n(t)\|_V^2 \le KE\, \|f_n(t)\|_V^2 \le KE\, \|f(t)\|_V^2 \qquad (3.4)$$

Hence,

$$E \int_0^T \|x_w\|_V^2\,ds \le K \int_0^T \|f(t)\|_V^2 \le M \qquad (3.5)$$

which implies that w.p.1 x_n belongs to a bounded set from $L^2([0,T], V)$, which results in the relatively compactness of P_n on $L_\sigma^2([0,T], V)$.

For studying the relatively compactness of P_n on $C([0,T], V^*)$ we apply a theorem of characterization from ([5], Th. 7.2, page 128), so we have to prove

a)   For every $\eta > 0$ and rational $t \geq 0$, there exists a compact set $\Gamma_{\eta,t} \subset V^*$, such that

$$\inf_n P \{ x_n(t) \in \Gamma_{\eta,t} \} \geq 1 - \eta \tag{3.6}$$

b)   the tightness of P_n, that is, for every $\eta > 0$ and $t > 0$, there exists $\delta > 0$, such that

$$\sup_n P \{ W'(x_n, \delta, t) \geq \eta \} \leq \eta$$

By W' it has been denoted the modulus of continuity on $C([0,T], V^*)$. (3.3) leads immediately to (3.6), because (3.3) implies, applying some properties of the stochastic integral a stronger form of (3.4),

$$E(\sup_{0 \leq t \leq T} \| x_n(t) \|_V^2) \leq KE\left[\sup_{t \in [0,T]} \| f(t) \|_V^2 \right] \leq M$$

(3.4) implies (3.6) for $\Gamma_{\eta,t}$ a bounded set from V. But the injection $V \to H$ is compact and $H \to V^*$ continuous, so $\Gamma_{\eta,t}$ will be a compact set in V^*.

For proving a) we need a new condition, III1) *the existence of a special basis.* There exists on orthonormal basis (e_i) of H, included in V, such that the projection operators Π_n defined by (3.1) are such that

$$\sup_n \| \Pi_n \|_{L(V^*, V^*)} < \infty \tag{3.7}$$

The hypothesis imposed on the operator A implies the condition (3.7) is satisfied. (See [10]).

We continue the proof applying one of the criterion for tightness in the space of continuous functions on an interval, taking values in a Hilbert space.

So, finally, we obtain that P_n makes up a sequence relatively compact on (G_σ, z), and we deduce that $C_n(t), D_n(t), Y_n(t)$ (the first, second, respectively third term of the right hand side of (3.2) are also relatively compact in $C([0,T], V^*)$ $Y_n(t)$, the stochastic integral is relatively compact in $C([0,T], H)$.

In the Theorem of representation of Skorohod that we intend to apply now, are considered the Polish space X on which P_n are probability measure and its Borel σ–algebra $B(x)$.

But G_σ is a Lusin space, so we have to precise that its Borel σ–algebra is the trace of the Borel σ–algebra of $C[0,T], V^*)$ on (G_σ, z) (conformly to [7], this trace coincides with the trace of the Borel σ–algebra of $L_\sigma^2[0,T, V]$ on (G_σ, z)).

So, we apply now the Theorem of Skorohod which allows us to construct a probability space $(\tilde{\Omega}, \tilde{F}, \tilde{P})$ and on this space a sequence of random variables $(\tilde{x}_n, \tilde{w}_n, \tilde{C}_n, \tilde{\mathcal{D}}_n, \tilde{Y}_n)$ taking values in $G_\sigma \times C[[0,T], H] \times C([0,T], W) \times C([0,T], V^*) \times C([0,T], H)$, having the same law as $x_n, w_n, C_n, \mathcal{D}_n, Y_n)$ and moreover, converging w.p.1 to a random variable $(\tilde{x}, \tilde{w}, \tilde{C}, \tilde{\mathcal{D}}, \tilde{\mathcal{E}})$ taking values in the same space.

On the new space $(\tilde{\Omega}, \tilde{F}, \tilde{P})$, due to the coincidence of the finite dimensional distributions, $\tilde{w}_n(t)$ has the same probability law as $\Pi_n \circ w(t)$, $\tilde{f}_n(t)$, the same law as $\Pi_n \circ f(t)$, and $\tilde{x}_n(t)$ has the same distribution as $x_n(t)$. The properties remain true also for the limiting processes. The process $\tilde{w}_n(t)$ is a Wiener process with respect to the σ–algebra $F_t^{(n)}$, defined as the completion of $\sigma\{\tilde{x}_n(s), \tilde{w}_n(s), \ s \le t\}$.

Furthermore, for $n \ge 0$, and each $s \le t$, the variable $\tilde{x}_n(s)$, $F_t^{(n)}$ measurable. Since $\tilde{x}_n(t)$ is continuous, $\tilde{x}_n(t)$ is a progressively measurable process with respect to $\{F_t^{(n)}\}$.

On $(\tilde{\Omega}, \tilde{F}, \tilde{P})$ we shall consider the filtration

$$\tilde{F}_t = \bigcup_n F_t^{(n)}$$

Baffico's lemma (see [2], Lemma 2.2) allows $\tilde{x}_n(t)$ to satisfy the same equation as $x_n(t)$ in the new probability space, that is

$$\tilde{x}_n(t) = \Pi_n \circ f + \int_0^t a(t-s)\Pi_n A(\tilde{x}_n(s))\, ds + \int_0^t a(t-s)\Pi_n B(\tilde{x}_n(s))\, d(\tilde{w}_n(s)) \qquad (3.8)$$

We intend now to prove that $\tilde{x}(s)$ the limit of $\tilde{x}_n(s)$ is the sought solution of (1.1) in $(\tilde{\Omega}, \tilde{F}, \tilde{P})$.

We are dealing with $\tilde{\mathcal{D}}_n = \int_0^t a(t-s)\Pi_n A(\tilde{x}_n(s))\, ds$.

The Skorohod's theorem results in the convergence of $\tilde{\mathcal{D}}_n$ to $\tilde{\mathcal{D}}$ w.p.1, but now we have to prove that the $\tilde{\mathcal{D}}$ has the desired expression that is $\tilde{\mathcal{D}} = \int_0^t a(t-s)A\tilde{x}(s)\, ds$.

For making the life easier, we introduce the following notations:

$$\mathcal{V} = L^2([0,T], V), \quad \mathcal{H} = L^2([0,T], H), \quad \mathcal{V}^\infty = L^2([0,T], V^*)$$

$\ll \ \gg$ will be the natural pairing between $\mathcal{V}$ and $\mathcal{V}^\infty$ i.e.

$$\ll u, v \gg \ = \int_0^T <u(t), v(t)>\, dt \ , \qquad u \in V, \ v \in V^* \qquad (3.9)$$

Let us denote by $\bar{A} : v \to v^*$, the operator defined by

$$(\bar{A}u)(t) = Au(t) \ , \qquad \forall \ u \in V \quad \forall \ t > 0, \quad \text{w.p.1.} \tag{3.10}$$

The assumptions done on the operator A, imply that $\bar{A}$ too, is a monotone, bounded, hemicontinuous and coercive operator.

From Skorohod's theorem, we obtain that w.p.1 $\tilde{x}_n$ converges weakly to $\tilde{x}(s)$ in $\mathcal{V}$. Then, necessarily $\tilde{x}_n$ is strongly bounded in $\mathcal{V}$. $\bar{A}$ being a bounded operator from $\mathcal{V}$ to $\mathcal{V}''$ and $\| \Pi_n \|_{L(V^*,V^*)} < \infty$ results in the existence of a subsequence $\tilde{\Pi}_n \bar{A}\tilde{x}_n$ converging weakly in $L^2([0,T], \mathcal{V}^*)$ and we denote the weak limit by α^*.

We want to prove that $\alpha^* = \bar{A}\tilde{x}$.
Let v be a fixed element from $\mathcal{V}$, the domain of $\bar{A}$. $\bar{A}$ is monotone, so we have

$$\ll \tilde{x}_n - v, \tilde{\Pi}_n \bar{A}\tilde{x}_n - \tilde{\Pi}_n \bar{A}v \gg \ \geq 0 \tag{3.11}$$

First, we shall deduce from (3.11)

$$\ll \tilde{x} - v, \alpha^* - \bar{A}v \gg \ \geq 0 \tag{3.12}$$

The left hand side of (3.11) can be written as

$$\ll \tilde{x}_n - v, \tilde{\Pi}_n \bar{A}\tilde{x}_n - \tilde{\Pi}_n \bar{A}v \gg \ = \ \ll \tilde{x}_n, \tilde{\Pi}_n A\tilde{x}_n - \alpha^* \gg \ +$$

$$+ \ll \tilde{x}_n, \alpha^* - \bar{A}v \gg \ + \ \ll \tilde{x}_n, \bar{A}v - \tilde{\Pi}_n \bar{A}v \gg \ + \ \ll -v, \tilde{\Pi}_n \bar{A}\tilde{x}_n - \tilde{\Pi}_n \bar{A}v \gg \tag{3.13}$$

$$= a_n^1 + a_n^2 + a_n^3 + a_n^4$$

(The i–th term of the right hand side of (3.13) was denoted by e_i).

We prove first that $\displaystyle\limsup_{n \to \infty} a_1 = 0$.
Without loss of generality let us suppose that all $\tilde{x}_n$ belong to the unit sphere of $\mathcal{V}$, let it be denoted to $\mathcal{V}_o$. Then

$$a'_n \leq \sup_{v \in \mathcal{V}_o} | \ll v, \tilde{\Pi}_n A\tilde{x}_n - \alpha^* \gg | = \varphi(\tilde{\Pi}_n \bar{A}\tilde{x}_n - \alpha^*), \quad \forall \ n \in N \tag{3.14}$$

So, we define $\displaystyle\varphi(v^*) = \sup_{v \in \mathcal{V}_o} | \ll v, v^* \gg |$

In [7], Métivier proved that $\varphi(v^*)$ is a convex, lower semicontinuous functional on $\mathcal{V}^*$. φ being convex, it is proved in [3, page 51] that φ is lower semicontinuous if and only it it is lower–semicontinuous with respect to the weak topology of $\mathcal{V}^*$. If we define (see [3], page 51), *the effective domain of* φ

$$D(\varphi) = \{\, v^* \in \mathcal{V}^*, \; \varphi(v^*) < +\infty \,\} \tag{3.15}$$

in [3], it is proved that φ is continuous on the interior of $D(\varphi)$. If we consider $\mathcal{V}^*$ endowed with the weak topology, $\widetilde{\Pi}_n A\bar{x}_n - \alpha^*$ converges w^* to 0, and 0 is an interior point of $D(\varphi)$. Hence $\lim\limits_{n-\infty} \varphi(\widetilde{\Pi}_n \bar{A}x_n - \alpha^*) = 0$, and from (3.14) we conclude that,

$$\limsup_{n-\infty} a_n^1 \le 0 \; . \tag{3.16}$$

Now, $|a_n^3| \le \|\bar{x}_n\| \, \|\bar{A}v - \widetilde{\Pi}_n \bar{A}v\|$, $\bar{x}_n$ strongly bounded in v, and $\widetilde{\Pi}_n Av$ converges strongly to $\bar{A}v$ in $\mathcal{V}^*$, so $a_n^3 \to 0$ when $n - \infty$.

a_n^2 is a continuous functional on $\mathcal{V}_\sigma$ and a_n^4 a continuous functional on $\mathcal{V}_\sigma^*$ (see [11]), from which we deduce (3.12).

Using the same pattern as Barbu in [3], page 251], we prove that (3.12) happens for any $v \in \mathcal{V}$. We choose now $\tilde{v}$ of the form $\tilde{v}(s, \omega) = \bar{x} - \rho U(\omega, s)$, for $\rho > 0$, and U any bounded, measurable mapping from $\Omega \times T$, to $\mathcal{V}_\sigma$.

We obtain

$$\int_e^T \ll U(s, \omega), \alpha^* - \bar{A}(\bar{x} - \rho U(s, \omega)) \gg ds \ge 0 \tag{3.16}$$

Because of the hemicontinuity of the operator $\bar{A}$, the expression $\ll u(s), \alpha^* - \bar{A}(\bar{x} - \rho U) \gg$ tends to $\ll U(s), \alpha^* - \bar{A}\bar{x} \gg$, which is ≥ 0 too.

Because it happens for all $u(s)$ bounded and measurable, the relation can take place only if $\alpha^* = A\bar{x}$.

The operator $Lu = \int_o^t a(t-s)u(s)\,ds$ is weakly continuous on $\mathcal{V}^*$, taking into account the assumption on $a(t)$. So finally we obtain that $\widetilde{\mathcal{D}}_n$ converges weakly to $\int_o^t a(t-s)\bar{A}\bar{x}(s)\,ds$ w.p.1. in $\mathcal{V}^*$.

For the convergence of $\mathcal{E}_n$ to the stochastic integral $\int_o^t a(t-s)B(\bar{x})\,d\bar{w}(s)$ we only sketch the proof.

$\bar{x}_n(s)$ belong to a bounded set of $\mathcal{V}$, w.p.1., so taking into account that the injection from v to $\mathcal{H}$ is compact (consequence of Assumption III) we get that $\bar{x}_n(s)$ belongs to a compact set from $\mathcal{H}$. Taking into account the continuity of $\bar{B}$, defined as $\bar{B}u(s) = Bu(s)$ on $\mathcal{H}$, and the fact that $\bar{w}_n(s)$ have the same laws as the projections $\Pi_n \circ w$ of $\mathcal{H}$, following, roughly speaking the technique from [12], we obtain the result. So, finally, $\bar{x}(t)$ satisfies (1.1) on the new space $(\widetilde{\Omega}, \bar{F}, \bar{P}, \bar{F}_t)$ with the Wiener process $\bar{w}(t)$. We can show that (1.1) is satisfied also in H, not only in V^*.

4. EXAMPLE.

We consider the following equation (integral form of the Fleming–Viot equation.

O is a bounded domain in R^d.

$$\begin{cases} u(t,x) = f(t,x) + \int_o^t a(t-s)\Big[\Lambda\, u(s,x) + \rho h(u(x))\, ds \,+ \\[2em] + \int_o^t a(t-s)\,[u(x)(1-u(x))] + {}^{1\!/2}\, dw(s,x) \end{cases} \tag{4.1}$$

$$\frac{\partial u}{\partial n}(t,x) = 0 \qquad x \in \partial 0\ . \tag{4.2}$$

The following assumptions are supposed:

a) $\rho(t,x) \in C[[0,T], H_o^1(0)]$

b) $w(t)$ is a Wiener process with nuclear covariance in $H = L^2(0)$

c) $h : R \to R$ is a bounded, measurable function.

The solutions of (4.1) will be interpreted as mappings from $[0,T]$ in a convenient Sobolev space of distributions, for example as functions from $[0,T]$ in $H^{-1}(0)$, which is the dual of $H_o^1(0)$.

We have the continuous, densely defined injection

$$H_o^1(0) \subset L^2(0) \subset H^{-1}(0)\ , \tag{4.3}$$

and moreover, the injection from $H_o(0)$ in $L^2(0)$ is compact.

The operator $\Lambda : H_o^1 \to H^{-1}(0)$ is defined by

$$< v, \Lambda u > \; = \int_{R^d} \sum_{k=1}^{d} D_k v(x) D_k(u(x))\, dx \tag{4.4}$$

for any $v \in H_o^1$.

The operator A will be

$$A(u) = \Lambda(u) + \rho(x) f(u) \quad u \in H_o^1(0)$$

B is a linear operator from $(L^2 0)$ to $L(L^2 0), L^2(0))$ defined for every u, as

$$B(u)h = \sqrt{u(1-u)+h} \qquad h \in L^2(0)$$

We see that all the hypothesis I, II, III are satisfied so (4.1) has at least a weak.

REFERENCES

(1) Aizicovici, S., *On a Class of Functional Differential Equations*, Rendicoute di Matematica (3), vol. 8, Serie VI, 1975.

(2) Baffico, R., *Una Estensione del Teorema di Esistenza di Skorohod*, Bollettino U.M.I., (5), 16–B, 1979.

(3) Baru, V., *Nonlinear Semigroups and Differential Equations in Banach Spaces*, Editura Academici Bucaresti Romania, Moordhoff International Publishing, Leyden, The Netherlands, 1976.

(4) Corduneaunu, C., *Principles of Differential and Integral Equations*, Chelsea Publ. Co., New–York, 1977.

(5) Ethier, S., and Kurtz, Th., *Markov Processes*, John Wiley & Sons, New–York, 1986.

(6) Krilov, *Controlled Diffusion Processes*, Springer–Verlag, New–York, 1980.

(7) Métivier, M., *Stochastic Partial Differential Equations*, Scuola Superiora di Pisa, 1989.

(8) Métivier, M., and Viot, M., *On Weak Solutions of Stochastic Partial Differential Equations*, Lectures Notes in Mathematics, 1322, 1988,.

(9) Métivier, M., *Semimartingales*, Springer–Verlag, 1982.

(10) Pardoux, *Equations aux Dérivées Partielles Stochastiques Nonlinéaires Monotónes*, Etudes des Solutions Fortes de Type Ito Thèse, Université de Paris Suel, Orsay, 1975.

(11) Yoshida, K., *Functional Analysis*, Springer–Verlag, 1980.

(12) Tudor, C., *On Volterra Stochastic Equations*, Bollettino U.M.I., (6) 5–A, 1986.

Modeling of Dynamic Systems by Itô-Type Systems of Stochastic Differential Equations

G. S. LADDE
Department of Mathematics
The University of Texas at Arlington
Arlington, Texas 76019, USA

ABSTRACT.

By developing an integral representation of a function of
solution process of a system of Itô-type stochastic
differential equations, the error estimate on absolute p-th
moment deviation of a solution process with the solution of
the mean of the corresponding deterministic system of
differential equations is analyzed. The obtained results
would provide a partial solution to one of the fundamental
problems in modeling of dynamic systems, namely, to what
extent incorporating randomness in the system causes the
change in behavior of the system relative to its
deterministic version. As a byproduct of this study, the
p-th moment relative stability and stability properties are
investigated.

INTRODUCTION.

The stochastic versus deterministic issue in the modeling
of dynamic systems is a subject of great interest.
Problems of this nature with regard to the random
polynomials [3, 6], stochastic boundary value problems [1,
2], stochastic initial value problems [5, 7] with regard to
random parameters have been investigated. The present work
deals with the Itô-type stochastic systems of differential
equations.

INTEGRAL REPRESENTATION.

Let us consider a system of stochastic differential
equations of Itô-Doob type

The research reported herein was supported by the U.S.
Army Research Office Grant No. DAAL03-89-G-0107.

"

$$dy = F(t,y)dt + \sigma(t,y)dz(t), \quad y(t_0,\omega) = y_0(\omega) \tag{1}$$

and corresponding deterministic initial value problem (DIVP)

$$m' = F(t,m), \quad m(t_0) = m_0 = E[y_0(\omega)] \tag{2}$$

which is obtained by ignoring the 'random disturbances in the system described by (1). In our presentation, we utilized the following random initial value problem (RIVP)

$$x' = F(t,x), \quad x(t_0,\omega) = x_0(\omega). \tag{3}$$

In (1), (2), and (3), y, m, and $x \in R^n$; x_0, $y_0 \in R[\Omega, R^n]$; E stands for an expected value of a random variable; $\Omega \equiv (\Omega, F, P)$ is a complete probability space; $z(t) \in R[\Omega, R^k]$ is a normalized Wiener process and it is independent of x_0, y_0. $F \in C[R_+ \times R^n, R^n]$, and $\sigma \in C[R_+ \times R^n, R^{nk}]$. We assume that F and σ satisfy desired regularity conditions so that (1) has a solution process.

The following result gives an integral representation of a function of a solution process of (1) with respect to the solution process of (3) through $(t_0, y_0(\omega))$.

<u>Theorem 1</u>: Assume that F and σ in (1) satisfy desired regularity conditions to assure the existence of solution process $y(t,\omega)$ of (1) for $t \geq t_0$. Assume that $F(t,x)$ is twice continuously differentiable with respect to x for fixed $t \geq t_0$, and $V \in C[R_+ \times R^n, R^m]$, V_t, V_x and V_{xx} exist and continuous for $(t, x) \in R_+ \times R^n$. Then

$$V(t,y(t,\omega)) = V(t_0,x(t,\omega))$$

$$+ \int_{t_0}^{t} V_x(s,x(t,s,y(s,\omega)))\Phi(t,s,y(s,\omega))\sigma(s,y(s,\omega))dz(s)$$

$$+ \int_{t_0}^{t} [V_s(s,x(t,s,y(s,\omega)))$$

$$+ \frac{1}{2}V_x(s,x(t,s,y(s,\omega)))tr\left(\frac{\partial^2 x}{\partial x_0 \partial x_0}(t,s,y(s,\omega))b(s,y(s,\omega))\right)$$

$$+ \frac{1}{2}tr(V_{xx}(s,x(t,s,y(s,\omega)))c(t,s,y(s,\omega)))]ds, \tag{4}$$

where $c(t,s,y) = \Phi(t,s,y)b(s,y)\Phi^T(t,s,y),$

$b(s,y)=\sigma(s,y)\sigma^T(s,y)$, and $x(t,\omega)=x(t,t_0,y_0(\omega))$ is the solution process of (3).

<u>Proof</u>. Let $y(t,\omega)=y(t,t_0,y_0(\omega))$ and $x(t,\omega)=x(t,t_0,y_0(\omega))$ be solution processes of (1) and (3) through $(t_0,y_0(\omega))$, respectively. Under the assumption of F, it is known that the solution $x(t,t_0,x_0)$ of (3) is continuously differentiable with respect to t_0 and twice continuously differentiable with respect to x_0. As a result, applying Itô's formula [4] to $V(s,\ x(t,s,y(s,\omega)))$, we have

$$d_sV(s,\ x(t,s,y(s,\omega)))\ =\ V_s(s,\ x(t,s,y(s,\omega)))ds$$
$$+V_x(s,x(t,s,y(s,\omega)))d_sx(t,s,y(s,\omega))$$
$$+\tfrac{1}{2}tr(V_{xx}(s,x(t,s,y(s,\omega)))d_sx(t,s,y(s,\omega))d_sx^T(t,s,y(s,\omega))).$$

From this and from the fact

$$d_sx(t,s,y(s,\omega))\ =\ \Phi(t,s,y(s,\omega))\sigma(s,y(s,\omega))dz(s)$$
$$+\tfrac{1}{2}tr\Big(\frac{\partial^2x}{\partial x_0\partial x_0}(t,s,y(s,\omega))b(s,y(s,\omega))\Big)ds,\tag{5}$$

we obtain

$$d_sV(s,x(t,s,y(s,\omega)))$$
$$=V_x(s,x(t,s,y(s,\omega)))\Phi(t,s,y(s,\omega))\sigma(s,y(s,\omega))dz(s)$$
$$+[V_s(s,x(t,s,y(s,\omega)))$$
$$+\tfrac{1}{2}V_x(s,x(t,s,y(s,\omega)))tr\Big(\frac{\partial^2x}{\partial x_0\partial x_0}(t,s,y(s,\omega))b(s,y(s,\omega))\Big)$$
$$+\tfrac{1}{2}tr\Big(V_{xx}(s,x(t,s,y(s,\omega)))\Phi(t,s,y(s,\omega))b(s,y(s,\ \omega))$$
$$\times\ \Phi^T(t,s,y(s,\omega))\Big)]ds.$$

Integrating this form t_0 to t, the desired result (4) follows. The following result shows the scope of Theorem 1.

<u>Corollary 1</u>. Let the assumptions of Theorem 1 be satisfied with $n=m$ and $V(t,\ x)=x$. Then

$$y(t,\omega)\ =\ x(t,\omega)\ +\int_{t_0}^{t}\Phi(t,s,y(s,\omega))\sigma(s,y(s,\omega))dz(s)$$

$$\tag{6}$$

$$+\ \frac{1}{2}\sum_{i,j=1}^{n}\int_{t_0}^{t}\frac{\partial^2x}{\partial x_{0i}\partial x_{0j}}(t,s,y(s,\omega))b_{ij}(x,y(s,\omega))ds$$

65

where $b(s,y)=(b_{ij}(s,y))$ and it is defined in Theorem 1.

We note that this corollary is an analogue of the variation of constants formula for deterministic nonlinear systems due to Alekseev [9]. Observe that the requirement of twice continuous differentiability of F with respect to x is due to stochastic nature of the perturbations.

The following result provides an integral representation of a function of a deviation of solution process (1) with the solution process of either (2) or (3).

<u>Theorem 2</u>. Let the hypotheses of Theorem 1 be satisfied. Then

$$V(t,y(t,\omega)-\overline{x}(t)) = V(t_0, x(t,\omega)-\overline{x}(t))$$

$$+\int_{t_0}^{t} V_x(s,x(t,s,y(s,\omega))$$

$$- x(t,s,\overline{x}(s)))\Phi(t,s,y(s,\omega))\sigma(s,y(s,\omega))dz(s)$$

$$+\int_{t_0}^{t} [\; V_s(s,x(t,s,y(s,\omega))-x(t,s,\overline{x}(s)))$$

$$+\tfrac{1}{2}V_x(s,x(t,s,y(s,\omega))$$

$$-x(t,s,\overline{x}(s)))\mathrm{tr}\Big(\frac{\partial^2 x}{\partial x_0\partial x_0}(t,s,y(s,\omega))\sigma(s,y(s,\omega))\Big)$$

$$+ \tfrac{1}{2}\mathrm{tr}(V_{xx}(s,x(t,s,y(s,\omega))-x(t,s,\overline{x}(s)))c(t,s,y(s,\omega)))]ds \quad (7)$$

where b and c are as defined in Theorem 1 and $\overline{x}(s)=x(s,t_0,z_0)$ is the solution of (2) or (3) depending on the choice of z_0.

<u>Proof</u>. The proof of this theorem can be formulated from the proof of Theorem 1. We leave the details to the reader.

<u>Example 1</u>. Let us consider

$$dy=\hat{A}(t)ydt + B(t)xdz \quad (8)$$

where $\hat{A}(t)$ and $B(t)$ are $n \times n$ continuous matrix functions, $x\in R^n$, $z(t)\in R$ is a normalized Wiener process. For $V(t,x) = \|x\|^2$, the relations (4) and (7) reduce to

$$\|y(t,\omega)\|^2 = \|x(t,\omega)\|^2 + 2 \int_{t_0}^{t} y^T(s,\omega)\Phi^T(t,s)\Phi(t,s)B(s)y(s,\omega)dz(s)$$

$$+ \int_{t_0}^{t} t_r(\Phi(t,s)B(s)y(s,\omega)y^T(s,\omega)B^T(s)\Phi^T(t,s))ds, \qquad (9)$$

and

$$\|y(t,\omega)-\bar{x}(t)\|^2 = \|x(t,\omega)-\bar{x}(t)\|^2$$

$$+ \int_{t_0}^{t} 2(y(s,\omega)-\bar{x}(s))^T\Phi^T(t,s)\Phi(t,s)B(s)y(s,\omega)dz(s)$$

$$+ \int_{t_0}^{t} t_r(\Phi(t,s)B(s)y(s,\omega)y^T(s,\omega)B^T(s)\Phi^T(t,s))ds, \qquad (10)$$

respectively, where $\bar{x}(t)$ is a solution process of

$$x' = \hat{A}(t)x, \quad \bar{x}(t_0,\omega) = z_0 \qquad (11)$$

where z_0 is either $y_0(\omega)$ or m_0.

ERROR ESTIMATES.

We state error estimate results in the context of Itô-type system of differential equations (1).

<u>Theorem 3</u>. Let the assumption of Theorem 2 be satisfied. Further assume that

(i) $\quad b(\|x\|^P) \leq \sum_{i=1}^{m} |V_i(t,\ x)| \leq a(\|x\|^P)$,

(ii) $\quad \sum_{i=1}^{m} |L_sV_i(s,x(t,s,y)-x(t,s,z))| \leq \lambda_1(s)C(\|y-z\|^P) + \lambda_2(s,\|z\|)$

and

(iii) $E\left[V_x(s,x(t,s,y(s,\omega))-x(t,s,\bar{x}(s)))\Phi(t,s,y(s,\omega))\sigma(s,y(s,\omega))\right]$

and the continuous time derivative of a $(E\left[\|x(t,\omega)-\bar{x}(t)\|^P\right])$

exists for $t \geq t_0$, where $a \in C\mathfrak{K}$ and its derivative belongs to

$C[R_+,R_+]$, $b \in \mathcal{VK}$, $C \in C\mathfrak{K}$, and $L_sV_i(s,x(t,s,y)-x(t,s,z))$ is the

i-th component of

$$L_s V(s,x(t,s,y)-x(t,s,z)) = V_s(s,x(t,s,y)-x(t,s,z))$$

$$+ \ V_x(s,x(t,s,y)-x(t,s,z))tr(\ \frac{\partial^2 x}{\partial x_0 \partial x_0}(t,s,y)b(s,y)))$$

$$+ \ \tfrac{1}{2}\ tr(V_{xx}(s,x(t,s,y)-x(t,s,z))c(t,s,y)), \tag{12}$$

$p \geq 1$, $\lambda_1 \in C[\ R_+,R_+]$ and $\lambda_2 \in C[R_+ \times R_+,\ R[\Omega,\ R_+]]$. Let us define $H(s)$, $\frac{dH}{ds}=1/h(s)$, $h(s)=C(b^{-1}(s))$ and assume that $H \in \mathcal{K}$. Then

$$E\left[\|y(t,\omega)-\bar{x}(t))\|^p\right] \leq H^{-1}(\int_{t_0}^{t} \lambda_1(s)ds$$

$$+H(a(E\left[\|y_0(\omega)-z_0\|^p\right])+\int_{t_0}^{t} \beta(s)ds)) \tag{13}$$

for $t \geq t_0$, where $y(t,\omega)$ and $x(t,\omega)$ are the solution processes of (1) and (3) through $(t_0,y_0(\omega))$, respectively, and $\bar{x}(t)=x(t,t_0,z_0)$ is the solution process of either (2) or (3) depending on the choice of z_0; $\beta(s)$ is the absolute value of the sum of $E\left[\lambda_2(s,\ \|\bar{x}(s)\|)\right]$ and the time derivative of $a(E\left[\|x(t,\ \omega)-\bar{x}(t)\|^p\right])$.

<u>Proof</u>. Let $y(t,\omega)$, $x(t,\omega)$ and $\bar{x}(t)$ be solution processes as described in the theorem. Using Theorem 2, hypothesis (iii) and taking expected value, we have

$$\sum_{i=1}^{m} E\left[V_i(t,y(t,\omega)-\bar{x}(t))\right] = \sum_{i=1}^{m} E\left[V_i(t_0,x(t,\omega)-\bar{x}(t))\right]$$

$$+ \ E\left[\int_{t_0}^{t}\sum_{i=1}^{m}L_s V_i(s,x(t,s,y(s,\omega))-x(t,s,\bar{x}(s)))\right]ds.$$

This together with hypotheses (i), (ii) and the nature of functions, one obtains

$$b(E\left[\|y(t,\omega)-\bar{x}(t))\|^{P}\right]) \leq a(E\left[\|x(t,\omega)-\bar{x}(t)\|^{P}\right])$$

$$+ \int_{t_0}^{t} (\lambda_1(s)C(E\left[\|y(s,\omega)-\bar{x}(s)\|^{P}\right])+ E\left[\lambda_2(s,\|\bar{x}(s)\|)\right])ds. \qquad (14)$$

Set

$$R(t)= \int_{t_0}^{t} (C(E\left[\|y(s,\omega)-\bar{x}(s)\|^{P}\right])\ \lambda_1(s)\ +\ E\left[\lambda_2(s,\ \|\bar{x}(s)\|)\right])ds.$$

Therefore

$$R'(t) =C(E\left[\|y(t,\omega)-\bar{x}(t)\|^{P}\right])\lambda_1(t)+E\left[\lambda_2(t,\|\bar{x}(t)\|)\right],\quad R(t_0)=0. \qquad (15)$$

From (14) we get

$$b(E\left[\|y(t,\omega)-\bar{x}(t)\|^{P}\right])\leq a(E\left[\|x(t,\omega)-\bar{x}(t)\|^{P}\right])\ +\ R(t),\ t\geq t_0.$$

This together with (15) after certain computation, yields

$$R'(t)\leq h(a(E\left[\|x(t,\omega)-\bar{x}(t)\|^{P}\right])+R(t))\lambda_1(t)+E\left[\lambda_2(t,\|\bar{x}(t)\|)\right],\ t\geq t_0$$

where   h   is as defined in the theorem.   By setting

$$u(t)=a(E\left[\|x(t,\ \omega)-\bar{x}(t))\|^{P}\right])\ +\ R(t)\ \text{with}$$

$$u(t_0)=a(E\left[\|y_0(\omega)-z_0\|^{P}\right]),\beta(t)=|E\left[\lambda_2(t,\|\bar{x}(t)\|)\right]$$

$$+\frac{d}{dt}\ (a(E\left[\|x(t,\omega)-\bar{x}(t)\|^{P}\right])))\,|,$$

and imitating the proof of Theorem 2.1 [8], the conclusion of theorem follows, immediately. This completes the proof.

The following example illustrates the scope of Theorem 3.

Example 2.   Let us consider Example 1.   Here $V(t,x)=\|x\|^2$, $a(s)=b(s)=s$, p=2.   To verify the hypotheses of Theorem 3, we assume that

$$\|\Phi(t,\ t_0)\|\leq\ K\ \exp\ \left[-\alpha(t-t_0)\right]\quad \text{for}\quad t\geq t_0. \qquad (16)$$

From this it is obvious that

$$L_sV(s,x(t,s,y)-x(t,s,z))\leq K^2\exp[-2\alpha t](e^{2\alpha s}\|y-z\|^2+e^{2\alpha s}\|z\|^2)\lambda(s).$$

This together with relation (10) in Example 1, we get

$$E\left[V(t,y(t,\omega)-\bar{x}(t))\right]\leq E\left[V(t_0,x(t,\omega)-\bar{x}(t))\right]$$

$$+ K^2\exp\ [-2\alpha t]\int_{t_0}^{t} e^{2\alpha s}E(V(s,y(s,\omega)-\bar{x}(s))+V(s,\bar{x}(s)))\lambda(s)ds$$

for $t \geq t_0$. This implies

$$\exp[2\alpha t]E\bigl[V(t,y(t,\omega)-\bar{x}(t))\bigr]\leq\exp\ [2\alpha t]E\bigl[V(t_0,x(t,\omega)-\bar{x}(t))\bigr]$$

$$+K^2\int_{t_0}^{t}\lambda(s)\,(e^{2\alpha s}E\bigl[V(s,y(s,\omega)-\bar{x}(s))\bigr]$$

$$+e^{2\alpha s}E\bigl[V(s,\bar{x}(s))\bigr])\,ds.$$

By setting

$$R(t)=K^2\int_{t_0}^{t}(e^{2\alpha s}E\bigl[V(s,y(s,\omega)-\bar{x}(s))\bigr]+e^{2\alpha s}E\bigl[V(s,\bar{x}(s))\bigr])\lambda(s)\,ds$$

and following the proof of the Theorem 2.1 [8], one obtains

$$E\bigl[\|y(t,\omega)-\bar{x}(t)\|^2\bigr]\leq K^2\exp\bigl[-2\alpha(t-t_0)\bigr](E\bigl[\|y_0(\omega)-z_0\|^2(K^2+e^{2\alpha t_0})+\|z_0\|^2\bigr])$$

$$\times\int_{t_0}^{t}\lambda(s)\exp\left[\int_{s}^{t}\lambda(u)\,du\right]ds.$$

RELATIVE STABILITY.

The sufficient conditions for relative stability of Itô type systems of differential equations in the context of method of variation of parameters are given in the next result.

<u>Theorem 4</u>. Let the assumption of Theorem 2 be satisfied. Further assume that

(i) $F(t,\ 0)\equiv 0$ for $t\in R_+$;

(ii) $b(\|x\|^P)\leq\sum_{i=1}^{m}|V_i(t,\ )|\leq a(\|x\|^P)$

   for $(t,x)\in R_+\times R^n$, $b\in\mathcal{VK}$, $a\in\mathcal{CK}$, $p\geq 1$;

(iii) $\sum_{i=1}^{m}|V_i(s,x(t,s,y)-x(t,s,z))|$

$$\leq\eta(t-s)\left[\lambda_1(s)\sum_{i=1}^{m}|V_i(s,y-z)|+\lambda_2(s)\sum_{i=1}^{m}|V_i(s,z)|\right]$$

   for $t_0\leq s\leq t,\ \|y\text{-}z\|^P<\rho,\ \|z\|^P<\rho$ some $\rho>0$, λ_i are locally integrable functions and $\eta\in\ell$;

(iv) $\displaystyle\sum_{i=1}^{m}\left|V_i(t_0,y(t,\omega))-\overline{x}(t))\right|\leq\alpha(\|y_0(\omega)-z_0\|^P)\beta(t-t_0)$, $t\geq t_0$,

whenever $E\left[\|y_0-z_0\|^P\right] < \rho$ and $E\left[\|z_0\|^P\right] < \rho$, where $\alpha\in C\mathcal{K}$, $\beta\in\mathcal{L}$;

(v) there exists a positive number K such that

$\eta(t-s)\beta(s-t_0)\leq K\beta(t-t_0)$ where η and β are defined as above.

Then

(1) the boundedness of $\displaystyle\beta(t-t_0)\exp\left[K\int_{t_0}^{t}\lambda_1(s)ds\right]$ and

$$\beta(t-t_0)\int_{t_0}^{t}\lambda_2(s)\exp\left[K\int_{s}^{t}\lambda_1(\nu)d\nu\right]ds$$

imply the p-th moment relative stability (RM_1) of (1) and (2);

(2) $\displaystyle\lim_{t\to\infty}\left[\beta(t-t_0)\exp\left[K\int_{t_0}^{t}\lambda_1(s)ds\right]\right]=0$ and

$$\lim_{t\to\infty}\left[\beta(t-t_0)\int_{t_0}^{t}\lambda_2(s,\omega)\exp\left[\int_{s}^{t}\lambda_1(s)ds\right]\right]=0$$

imply the p-th moment relative asymptotic stability (RM_2) of (1) and (2).

<u>Proof</u>. The proof of the theorem can be reconstructed by following the standard arguments of Theorems [4, 9]. The details are left as an exercise.

Finally, we represent an example to illustrate the importance of Theorem 4.

<u>Example 3</u>. Let us consider Example 2. From the discussion of this example, it is obvious that hypotheses (i)$-$(v) of Theorem 4 are satisfied with

$$V(t,x)=\|x\|^2, \eta(t-s)=\beta(t-s)=K^2\exp\left[-2\alpha(t-s)\right], \quad \lambda_1=\lambda_2=\lambda, \quad p=2,$$

and a, b and α in the theorem are identity functions.

If we assume that $\displaystyle\exp\left[-2\alpha(t-t_0)+\int_{t_0}^{t}\lambda(s)ds\right]$ is bounded and

$$\lim_{t\to\infty}\left[\frac{1}{(t-t_0)}\int_{t_0}^{t}\lambda(s)ds\right]<2\alpha, \qquad \text{then (8) and (11) have (RM_1)}$$

and (RM$_2$) properties, respectively.

STABILITY ANALYSIS.

The following theorem gives a sufficient conditions for the p-th mean stability of the equilibrium solution process of (1).

<u>Theorem 5</u>.   Let the hypotheses of Theorem 1 be satisfied. Assume that $V,x(t,\omega),\Phi(t,s,y(s,\omega))$, $\sigma(s,y(s,\omega))$, and $F(t,y)$ satisfy the following conditions

(i) $b(\|x\|^P)\leq\sum_{i=1}^{m}|V_i(t,x)|\leq a(\|x\|^P)$ for all $(t,x)\in R_+\times R^n$ where

   $p\geq1$, $a\in\mathcal{VS}$ and $a\in\mathcal{CS}$;

(ii) $F(t,0)\equiv$ and $\sigma(t,0)\equiv 0$ for $t\in R_+$;

(iii) $E\big[V_x(s,x(t,s,y(s,\omega))\Phi(t,s,y(s,\omega))\sigma(s,y(s,\omega))\big]$

   exists for $t\geq t_0$;

(iv) $\sum_{i=1}^{m}|L_sV_i(s,x(t,s,y))|\leq\lambda(s)\eta(t-s)\sum_{i=1}^{m}|V_i(s,y)|$ for

   $t_0\leq s\leq t$ and $\|y\|^P\leq\rho$, where $L_sV_i(s,x(t,s,y))$ is the

   i-th component of

$$L_sV(t,x(t,s,y))=V_s(s,x(t,s,y))$$
$$+ V_x(s,x(t,s,y))tr(\tfrac{1}{2}\frac{\partial^2 x}{\partial x_0\partial x_0}(t,s,y)b(s,y))$$
$$+ tr(\tfrac{1}{2} V_{xx}(s,x(t,s,y))c(t,s,y)),$$
$$c(t,s,y)=\Phi(t,s,y)b(s,y)\Phi^T(t,s,y),$$
$$b(s,y)=\sigma(s,y)\sigma^T(s,y);$$

   $x(t,s,y)$ is the solution process of (3) through

   (s,y), $\rho > 0$ and $\lambda\in C[R_+,R_+]\cap L^1(R_+,R_+)$;

(v) $E\left[\sum_{i=1}^{m}V_i(t_0,x(t,\omega))\right]\leq\beta(t-t_0)\alpha(E[\|y_0(\omega)\|^P])$,

   whenever $E[\|y_0(\omega)\|^P])\leq\rho$

   for some $\rho > 0$, where $\alpha\in\mathcal{CS}$ and η, $\beta\in\mathcal{L}$ and satisfy

$$\lim_{t\to\infty}\left[\int_{t_0}^{t}\eta(t-s)\beta(s-t_0)ds\right] = 0.$$

Then the trivial solution process of (1) is asymptotically
stable in the p-th mean.

<u>Proof</u>. The proof of the theorem can be formulated based on
the proof of Theorems 1 and 4. The details are left to the
reader.

<u>Example 4</u>. Let us consider Example 2. As in Example 3, by
taking $V(t,x)=\|x\|^2$, we obtain
$$L_sV(s,x(t,s,y)) = tr(\Phi(t,s)B(s)yy^TB^T(s)\Phi(t,s)).$$

From (16) $L_sV(s,x(t,s,y))$ satisfies the following relation

$$L_sV(s,x(t,s,y)) \leq \lambda(s)\exp\left[-2\alpha(t-s)\right]V(s,y) \tag{17}$$

for all $y\in R^n$ and $t_0\leq s\leq t$, where $\lambda\in C[R_+,R_+]$. We assume
the λ in (17) and α in (16)

satisfy the following condition

$$\lim_{t\to\infty}\left[\frac{1}{(t-t_0)}\int_{t_0}^t \lambda(s)ds\right]<2\alpha. \tag{18}$$

From the definition of V, p=2, (16), (17) and (18), all
the hypotheses of Theorem 5 are satisfied. Thus, the
trivial solution of (8) is asymptotically stable in the
second moment.

REFERENCES.

1.   Chandra, J., Ladde, G. S. and Lakshmikantham, V., On
      the fundamental theory of nonlinear second order
      stochastic boundary value problems, *Stoch. Anal. Appl.*, Vol.
      1(1983), pp. 1-19.
2.   Chandra, J., Ladde, G. S. and Lakshmikantham, V.,
      Stochastic analysis of compressible gas lubricated
      slider bearing problem, *SIAM J. on Appl. Math.*, Vol. 43
      (1983), pp. 1174-1186.
3.   Christensen, M. J. and Bharucha-Reid, A.T., Stability
      of the roots of random algebraic polynomial, *Comm. Simu.
      Comp. B.*, Vol. 9 (1980), pp. 179-192.
4.   Ladde, G. S. and Lakshmikantham, V., *Random Differential
      Inequalities*, Academic Press, New York, 1980.

5.   Ladde, G. S., Lakshmikantham, V. and Sambandham, M.,
     Comparison theorem and error estimates of stochastic
     differential systems, *Stoch. Anal. Appl.*, Vol 3 (1985), pp.
     23-62.
6.   Ladde, G. S. and Sambandham, M., Stochastic versus
     deterministic, *Math. Comp. in Simulation*, Vol. 24 (1982), pp. 507-514.
7.   Ladde, G. S. and Sambandham, M., Error estimates of
     solution and mean of solutions of stochastic
     differential system, *J. Math. Phys.*, Vol. 24 (1983),
     pp. 815-822.
8.   Lakshmikantham, V., A variation of constants formula
     and Bellman-Gronwall-Reid Inequalities, *J. Math. Anal. Appl.*,
     Vol. 41 (1973), pp. 199-204.
9.   Lakshmikantham, V. and Leela, S., *Differential and Integral
     Inequalities, Vol. I,* Academic Press, New York, 1969.

Itô-type Systems of Stochastic Integro-Differential Equations

G. S. LADDE
Department of Mathematics
The University of Texas at Arlington
Arlington, Texas 76019, USA

S. SATHANANTHAN
Division of Science and Mathematics
Jarvis Christian College
Hawkins, Texas 75765, USA

ABSTRACT.

By obtaining an integral representation of a function of
solution process of a system of stochastic integro-
differential equations, the p-th moment stability property
of the steady state of the system is studied. Furthermore,
some results are developed to investigate the behavior of
the system relative to its deterministic version.

INTRODUCTION.

The Mathematical modelling of several real world problems
leads to systems of differential and integro-differential
equations that involve some kind of randomness due to
uncertainities. They arise in Biological, Physical, Social
Sciences and various other disciplines [1,3,5,6,8]

In this paper, we attempt to obtain an integral
representation of a function of solution processes of an
Itô-type systems of stochastic integro-differential
equation (SIDE for short) in the context of the function of
solution processes of the corresponding deterministic
initial valued problem (DIVP) and the random initial valued
problem (RIVP). Further we investigate the p-th moment
stability properties of the steady states of the SIDE
relative to the stability properties of the steady states
of DIVP and RIVP. Finally the error estimates on p-th
moment deviation of a solution process with respect to the
solution process of DIVP or RIVP are obtained. These
results would provide a partial solution to one of the
fundamental problems in modelling of dynamic systems,
namely to what extent incorporating randomness in the

Research partially supported by U. S. Army Research Grant
DAAL03-90-G-0180

system causes the change in behavior of the system relative
to its deterministic version. Here, the method of study
is based on the integral representation of solution process
of SIDE.

PRELIMINARIES.

Consider a system of stochastic integro-differential
equations of Itô-Doob type,

$$dy=[\hat{A}(t)y+\int_{t_0}^{t}\hat{K}(t,s)y(s)ds]dt+B(t)ydz$$

$$+[\int_{t_0}^{t}\sigma(t,s)y(s)dw(s)]dt,y(t_0,\omega)=y_0(\omega) \qquad (1)$$

and corresponding deterministic initial value problem
(DIVP)

$$m'=\hat{A}(t)m +\int_{t_0}^{t}\hat{K}(t,s)m(s)ds, \quad m(t_0)=m_0 \qquad (2)$$

which is obtained by ignoring the random disturbances in
the system described by (1). In our presentation we
utilized the following random initial value problem (RIVP)

$$x'=\hat{A}(t)x+\int_{t_0}^{t}\hat{K}(t,s)x(s)ds, \qquad x(t_0,\omega) =x_0(\omega) \ . \qquad (3)$$

In (1), (2), and (3), y,m, and $x \in R^n, x_0, y_0 \in R[\ \Omega, R^n]$; E
stands for the expected value of a random variable;
$\Omega \equiv (\Omega,F,P)$ is a complete probability space; $z(t)$, $w(t)$ are
mutually independent normalized Wiener processes and are
independent of x_0, y_0; $\hat{A}$ is continuous nxn matrix function
and $\hat{K} \in C[R^+ \times R^+, R^n]$. We assume that $\hat{K}$, B and σ satisfy the
desired regularity conditions so that (1) has a solution
process.

VARIATION OF CONSTANTS METHOD.

The variation of constants method [3,4,6] is an important
technique in studying the qualitative behavior of
stochastic integro-differential equations. In this section
we investigate a generalized variation of constants formula
for Itô-type integro-differential equations.

<u>Theorem 1.</u> Let $V \in C[R_+ \times R^n, R^m]$, V_x, V_t and V_{xx} exist and
continuous for $(t,x) \in R_+ \times R^n$. Then,

$$V(t,y(t,\omega))=V(t_0,x(t,\omega))+\int_{t_0}^{t} V_s(s,x(t,s,y(s,\omega)))ds$$

$$+\int_{t_0}^{t} V_x(s,x(t,s,y(s,\omega)))R(t,s)B(s)y(s,\omega)dz(s)$$

$$+\int_{t_0}^{t}\int_{s}^{t}[V_x(\tau,x(t,\tau,y(\tau,\omega)))$$

$$-V_x(s,x(t,s,y(s,\omega)))]R(t,\tau)\hat{K}(\tau,s)y(s,\omega)\ d\tau ds$$

$$+\int_{t_0}^{t}\int_{s}^{t}V_x(\tau,x(t,\tau,y(\tau,\omega)))R(t,\tau)\sigma(\tau,s)y(s,\omega)d\tau dw(s)$$

$$+\ \tfrac{1}{2}\int_{t_0}^{t}\operatorname{tr}[V_{xx}(s,x(t,s,y(s,\omega)))[R(t,s)B(s)y(s,\omega)]$$

$$\times\ [R(t,s)B(s)y(s,\omega)]^{T}]ds. \tag{4}$$

<u>Proof:</u>   Let $x(t,s,y(s,\omega))$ be the solution process of (3) through $(s,y(s,\omega))$ where $y(s,\omega)=y(s,t_0,y_0(\omega))$ is the solution process of (1) through $(t_0,y_0(\omega))$.   The solution $x(t,t_0,x_0)$ of (3) is twice continuously differentiable with respect to x_0. As a result,   applying Itô's formula to $V(s,x(t,s,y(s,\omega)))$,   we have,

$$d_sV(s,x(t,s,y(s,\omega)))=V_s(s,x(t,s,y(s,\omega)))ds$$

$$+V_x(s,x(t,s,y(s,\omega)))d_sx(t,s,y(s,\omega))$$

$$+\tfrac{1}{2}\operatorname{tr}[V_{xx}(s,x(t,s,y(s,\omega)))$$

$$\times\ d_sx(t,s,y(s,\omega))d_sx^{T}(t,s,y(s,\omega))] \tag{5}$$

From this and from the fact,

$$d_sx(t,s,y(s,\omega))=R_s(t,s)y(s,\omega)ds+R(t,s)dy(s,\omega)$$

$$=R_s(t,s)y(s,\omega)ds+R(t,s)[[\hat{A}(s)y(s,\omega)$$

$$+\int_{t_0}^{s}\hat{K}(s,\tau)y(\tau)d\tau]ds+B(s)y(s)dz(s)+[\int_{t_0}^{s}\sigma(s,\tau)y(\tau)dw(\tau)]ds]$$

and

$$d_sx(t,s,y(s,\omega))d_sx^{T}(t,s,y(s,\omega)$$

$$=[R(t,s)B(s)y(s,\omega)][R(t,s)B(s)y(s,\omega)]^{T}ds.$$

We obtain

$$d_s V(s, x(t, s, y(s, \omega)))$$

$$= V_s(s, x(t, s, y(s, \omega))) ds + V_x(s, x(t, s, y(s, \omega)))$$

$$\times [R_s(t, s) y(s, \omega) ds + R(t, s) \hat{A}(s) y(s, \omega) ds]$$

$$+ V_x(s, x(t, s, y(s, \omega))) R(t, s) [\int_{t_0}^{s} \hat{K}(s, \tau) y(\tau) d\tau ds$$

$$+ B(s) y(s) dz(s) + \int_{t_0}^{s} \sigma(s, \tau) y(\tau) dw(\tau) ds]$$

$$+ \frac{1}{2} tr [V_{xx}(s, x(t, s, y(s, \omega))) [R(t, s) B(s) y(s, \omega)]$$

$$\times [R(t, s) B(s) y(s, \omega)]^T] ds. \tag{6}$$

Integrating from t_0 to t,

$$V(t, y(t, \omega)) - V(t_0, x(t, \omega)) = \int_{t_0}^{t} V_s(s, x(t, s, y(s, \omega))) ds$$

$$+ \int_{t_0}^{t} V_x(s, x(t, s, y(s, \omega))) [R_s(t, s) + R(t, s) \hat{A}(s)] y(s, \omega) ds$$

$$+ \int_{t_0}^{t} V_x(s, x(t, s, y(s, \omega))) R(t, s) [\int_{t_0}^{s} \hat{K}(s, \tau) y(\tau, \omega) d\tau] ds$$

$$+ \int_{t_0}^{t} V_x(s, x(t, s, y(s, \omega))) B(s) y(s, \omega) dz(s)$$

$$+ \frac{1}{2} \int_{t_0}^{t} tr [V_{xx}(s, x(t, s, y(s, \omega)))$$

$$\times [R(t, s) B(s) y(s, \omega)] [R(t, s) B(s) y(s, \omega)]^T] ds$$

$$+ \int_{t_0}^{t} V_x(s, x(t, s, y(s, \omega))) R(t, s) \int_{t_0}^{s} \sigma(s, \tau) y(\tau, \omega) dw(\tau) ds. \tag{7}$$

Using the Fubini's theorem of Itô type integral the last
term can be rewritten as

$$\int_{t_0}^{t}\int_{t_0}^{s} V_x(s,x(t,s,y(s,\omega)))R(t,s)\sigma(s,\tau)y(\tau,\omega)dw(\tau)ds$$

$$=\int_{t_0}^{t}\int_{s}^{t} V_x(\tau,x(t,\tau,y(\tau,\omega)))R(t,\tau)\sigma(\tau,s)y(s,\omega)d\tau dw(s).$$

From this, (7) and the application of Fubini's theorem, the desired result (4) follows.

<u>Corollary 1</u>. Let $V(t,x)=x$. Then, (4) reduces to

$$y(t,\omega)=x(t,\omega)+\int_{t_0}^{t} R(t,s)B(s)y(s)dz(s)$$

$$+\int_{t_0}^{t}\int_{s}^{t} R(t,\tau)\sigma(\tau,s)y(s,\omega)d\tau dw(s). \tag{8}$$

We note this corollary is an analog of the variation of constants formula for deterministic linear integro-differential equations [2].

<u>Corollary 2</u>. Let $V(t,x)=\|x\|^2$. Then, (4) reduces to

$$\|y(t,\omega)\|^2=\|x(t,\omega)\|^2+\int_{t_0}^{t} 2y^T(s,\omega)R^T(t,s)R(t,s)B(s)y(s,\omega)dz(s)$$

$$+\int_{t_0}^{t} tr[[R(t,s)B(s)y(s,\omega)][R(t,s)B(s)y(s,\omega)]^T]ds$$

$$+\int_{t_0}^{t}\int_{s}^{t} 2y^T(\tau,\omega)R^T(t,\tau)R(t,\tau)\sigma(\tau,s)y(s,\omega)d\tau dw(s)$$

$$+\int_{t_0}^{t}\int_{s}^{t} 2[y^T(\tau,\omega)R^T(t,\tau)-y^T(s,\omega)R^T(t,s)]R(t,\tau)\hat{K}(\tau,s)y(s,\omega)d\tau ds. \tag{9}$$

<u>Theorem 2</u>. Suppose all the hypotheses of Theorem 1 hold. Then,

$$V(t,y(t,\omega)-\bar{x}(t,\omega))$$
$$=V(t_0,x(t,\omega)-\bar{x}(t))+\int_{t_0}^{t} V_s(s,x(t,s,y(s,\omega))-\bar{x}(t,s,\bar{x}(s)))ds$$
$$+\int_{t_0}^{t} V_x(s,x(t,s,y(s,\omega))-\bar{x}(t,s,\bar{x}(s)))R(t,s)B(s)y(s,\omega)dz(s)$$

$$+\int_{t_0}^{t}\int_{s}^{t} V_x(\tau,x(t,\tau,y(\tau,\omega))-\bar{x}(t,\tau,\bar{x}(\tau)))$$

$$\times\ R(t,\tau)\sigma(\tau,s)y(s,\omega)d\tau dw(s)+\int_{t_0}^{t}\int_{s}^{t}[V_x(\tau,x(t,\tau,y(\tau,\omega))$$

$$-\bar{x}(t,\tau,\bar{x}(\tau)))-V_x(s,x(t,s,y(s,\omega))$$

$$-\bar{x}(t,s,\bar{x}(s)))]R(t,\tau)\hat{K}(\tau,s)\,[y(s,\omega)-\bar{x}(s)]d\tau ds$$

$$+\frac{1}{2}\int_{t_0}^{t}tr[V_{xx}(s,x(t,s,y(s,\omega))-\bar{x}(t,s,\bar{x}(s)))$$

$$\times\ [R(t,s)B(s)y(s,\omega)]\,[R(t,s)B(s)y(s,\omega)]^T]ds, \tag{10}$$

where $y(t,\omega)$ is a solution process of (1) and $\bar{x}(t)$ is a solution process of either (2) or (3) depending on the initial data.

<u>Proof</u> : Consider

$$d_sV(s,x(t,s,\ y(s,\omega))-\bar{x}(t,s,\bar{x}(s)))$$
$$=\ V_s(s,x(t,s,y(s,\omega))-\bar{x}(t,s,\bar{x}(s)))ds$$
$$+\ V_x(s,x(t,s,y(s,\omega))-\bar{x}(t,s,\bar{x}(s)))$$
$$\times\ d_s[x(t,s,y(s,\omega))-\bar{x}(t,s,\bar{x}(s))]$$
$$+\ \frac{1}{2}\ tr[V_{xx}(s,x(t,s,y(s,\omega))-\bar{x}(t,s,\bar{x}(s)))$$
$$\times\ d_s[x(t,s,y(s,\omega))-\bar{x}(t,s,\bar{x}(s))]$$
$$\times\ d_s[x^T(t,s,y(s,\omega))-\bar{x}^T(t,s,\bar{x}(s))]].$$

Using this and following the steps of Theorem 1, the theorem can be easily proved.

<u>Example 1</u>.   Consider the Itô-type integro-differential equation

$$dy=[\tfrac{1}{2}\ y-\int_{t_0}^{t}9e^{-7(t-s)}y(s)ds]dt+B(t)ydz(t)$$

$$+[\int_{t_0}^{t}\sigma(t,s)y(s)dw(s)]dt,\quad y(t_0,\omega)\ =\ y_0(\omega), \tag{11}$$

where $y\in R$, z and w are one dimensional normalized independent Wiener processes.

Consider the corresponding unperturbed equation,

$$x'=\tfrac{1}{2}\ x-\int_{t_0}^{t}9e^{-7(t-s)}x(s)ds,\quad x(t_0,\omega)=x_0(\omega)=y_0(\omega). \tag{12}$$

Here, $x(t,t_0,x_0)=R(t,t_0)x_0$, where $R(t,t_0)$ is given by

$$R(t,t_0)=\tfrac{4}{3}e^{-(t-t_0)}-\tfrac{1}{3}e^{-\frac{11}{2}(t-t_0)}$$

Therefore,

$$\Phi(t,t_0,x_0)=\frac{\partial}{\partial x_0}\,x(t,t_0,x_0)$$
$$=R(t,t_0)=\tfrac{4}{3}\,e^{-(t-t_0)}-\tfrac{1}{3}e^{-\frac{11}{2}(t-t_0)}.$$

Let $V(t,x)=x$. Applying Theorem 1, we get

$$y(t,\omega)=x(t,\omega)+\int_{t_0}^{t}\left[\tfrac{4}{3}e^{-(t-s)}-\tfrac{1}{3}e^{-\frac{11}{2}(t-s)}\right]B(s)y(s)dz(s)$$

$$+\int_{t_0}^{t}\int_{s}^{t}\left[\tfrac{4}{3}e^{-(t-\tau)}-\tfrac{1}{3}e^{-\frac{11}{2}(t-\tau)}\right]\sigma(\tau,s)y(s,\omega)d\tau dw(s).$$

STABILITY ANALYSIS.

In this section p-th moment stability properties [4] of the trivial solution process of (1) are studied.

<u>Theorem 3.</u> Let the hypotheses of Theorem 1 be satisfied. Assume that the following conditions holds:

(i) $b(\|x\|^P)\leq\sum_{i=1}^{m}|V_i(t,x)|\leq a(\|x\|^P)$ for all $(t,x)\in R_+\times R^n$

where $p\geq 1$, $b\in\mathcal{VK}$ and $a\in\mathcal{CK}$;

(ii) $E[V_x(s,x(t,s,y))R(t,s)B(s)y(s)$

$$+\int_{s}^{t}V_x(\tau,x(t,\tau,y(\tau,\omega)))R(t,\tau)\sigma(\tau,s)y(s,\omega)d\tau]$$

exists for $t\geq t_0$;

(iii) $\sum_{i=1}^{m}|L_sV_i(s,x(t,s,y))|\leq\lambda_1(t-s)\sum_{i=1}^{m}|V_i(s,y)|$

$$+\int_{t_0}^{s}\lambda_2(s)\bar{K}(t,\tau)\sum_{i=1}^{m}|V_i(\tau,y(\tau))|d\tau$$

for $t_0\leq s\leq t$, and $\|y\|^P\leq\rho$, where L_sV_i is the i-th component of

$L_s V(s,x(t,s,y)) = V_s(s,x(t,s,y))$

$$+V_x(s,x(t,s,y))\int_{t_0}^{s} R(t,s)\hat{K}(s,\tau)y(\tau)d\tau$$

$$-V_x(s,x(t,s,y))\int_{s}^{t} R(t,\tau)\hat{K}(\tau,s)y(s,\omega)d\tau$$

$$+ \tfrac{1}{2}\mathrm{tr}[V_{xx}(s,x(t,s,y(s,\omega)))][R(t,s)B(s)y(s,\omega)]$$

$$\times [R(t,s)B(s)y(s,\omega)]^T]$$

and $x(t,s,y)$ is the solution process of (3) through (s,y), $\rho>0$ and $A\in C[R_+,R_+]\cap L^1[R_+,R_+]$,

$$A(t,t_0)=\lambda_1(t-t_0)+\int_{t_0}^{t}\lambda_2(t)\bar{K}(t,\tau)d\tau+\int_{t_0}^{t}\int_{t_0}^{s}\lambda_2(s)\frac{\partial\bar{K}}{\partial t}(t,\tau)d\tau ds$$

and $\frac{\partial\bar{K}}{\partial t}(t,\tau)$ is non-negative;

(iv) $E[\sum_{i=1}^{m}V_i(t_0,x(t,\omega))]\leq\alpha(E[\|y_0(\omega)\|^P])$, whenever $E[\ \|y_0\|^P]\leq\rho$

for some $\rho > 0$,

where $\alpha\in C\mathcal{K}$. Then the trivial solution process of (1) is stable in the p-th moment.

<u>Proof</u>:   Let $y(t,\omega)$ be a solution process of (1) and let $x(t,s,y(s,\omega))$ and $x(t,\omega)=x(t,t_0,y_0(\omega))$ be the solution process of (3) through $(s,y(s,\omega))$ and $(t_0,y_0(\omega))$, respectively, for $t_0\leq s\leq t$ and $t_0\in R_+$. From (7) and assumption (ii), we obtain

$$E[|V_i(t,\ y(t,\omega))|]\leq E[|V_i(t_0,x(t,\omega))|]$$

$$+\int_{t_0}^{t}E[|L_sV_i(s,x(t,s,\ y(s,\omega)))|]\ ds.$$

This together with hypotheses (iii) and (iv), yields

$$m(t)\leq\alpha(E[\|y_0(\omega)\|^P])+\int_{t_0}^{t}\lambda_1(t-s)m(s)+\int_{t_0}^{t}\int_{t_0}^{s}\lambda_2(s)\bar{K}(t,\tau)m(\tau)d\tau ds \quad (13)$$

as long as $E[\|y(s,\omega)\|^P]\leq\rho$,   where

$$m(t) = \sum_{i=1}^{m} E[|V_i(t,y(t,\omega))|] .$$

By applying Corollary 1. [7], we get

$$m(t) \leq \alpha(E[\|y_0(\omega)\|^P]) \exp\left[\int_{t_0}^{t} A(s,t_0) \, ds\right],$$

as long as $E[\|y(s,\omega)\|^P] \leq \rho$ for $t_0 \leq s \leq t$, where

$$A(s,t_0) = \lambda_1(s-t_0) + \int_{t_0}^{s} \lambda_2(s)\bar{K}(s,\tau)d\tau + \int_{t_0}^{s} \int_{t_0}^{\xi} \lambda_2(\xi)\bar{K}_t(s,\tau)d\tau d\xi .$$

This implies

$$\sum_{i=1}^{m} E[|V_i(t,y(t,\omega))|] \leq \alpha(E[\|y_0(\omega)\|^P]) \exp\left[\int_{t_0}^{t} A(s,t_0)ds\right], \tag{14}$$

as long as $E[\|y(s,\omega)\|^P] \leq \rho$ for $t_0 \leq s \leq t$. First we show that the above inequality is valid for all $t \geq t_0$. For this purpose, we choose $y_0(\omega)$ such that

$$K\alpha(E[\|y_0(\omega)\|^P]) < b(\rho), \tag{15}$$

where $K \geq \exp\left[\int_{t_0}^{t} A(s,t_0)ds\right]$. We claim that $E[\|y(t,\omega)\|^P] < \rho$ for all $t \geq t_0$ whenever (15) holds.

Assume this claim is false, then there exists $\bar{t}$ and $y_0(\omega)$ such that $\bar{t} > t_0$, $E[\|y_0(\omega)\|^P] < \rho$, $E[\|y(t,\omega)\|^P] < \rho$ for $t_0 \leq t < \bar{t}$ and $E[\|y(\bar{t},\omega)\|^P] = \rho$. From this (14) is valid for $t \in [t_0,\bar{t}]$ and hence using hypotheses (i) and (15), we get

$$b(\rho) = b(E[\|y(\bar{t},\omega)\|^P]) \leq E[b(\|y(\bar{t},\omega)\|^P] \leq \sum_{i=1}^{m} E[|V_i(\bar{t},y(\bar{t},\omega))|] < b(\rho)$$

This contradiction establishes the impossibility of the existence of such a $\bar{t}$. Hence (14) is valid for $t \geq t_0$. Finally, the stability of the trivial solution process of (1) can be concluded by following the standard argument [4] in the stability theory.
This completes the proof.

The next theorem provides sufficient conditions for the p-th moment asymptotic stability of the trivial solution of (1).

$\underline{\text{Theorem 4}}$. Assume that the hypotheses of Theorem 3 hold except that (iii) and (iv) are replaced by

$$(\text{iii})' \quad \sum_{i=1}^{m} |L_s V_i(s,x(t,s,y))| \leq \lambda_1(s)\eta_1(t-s) \sum_{i=1}^{m} |V_i(s,y)|$$

$$+ \lambda_2(s)\eta_2(t-s) \sum_{i=1}^{m} \int_{t_0}^{s} \eta_3(s-\tau) |V_i(\tau,y(\tau))| \, d\tau$$

for $t_0 \leq s \leq t$,

$$(\text{iv})' \quad E\left[\sum_{i=1}^{m} V_i(t_0,x(t,\omega))\right] \leq \alpha(E[\|y_0(\omega)\|^P]) \beta(t-t_0), \quad t \geq t_0,$$

provided $E[\|y_0(\omega)\|^P] \leq \rho$ where α is as defined in (iv);

$\eta_1, \eta_2, \beta \in \mathcal{L}$, and satisfies

$$\eta_1(t-s)\beta(s-t_0) \leq K\beta(t-t_0),$$

$$\eta_2(t-s)\eta_3(\tau-s) \leq K\eta_1(t-s)\lambda(t-s) \quad \text{where} \quad \lambda \in \mathcal{L}^1, \quad \lambda' \text{ exists} \quad \text{and}$$

$$\lim_{t \to \infty} \left[\beta(t-t_0) \exp\left[\int_{t_0}^{t} A(s,t_0) \, ds \right] \right] = 0 \tag{16}$$

where $A(s,t_0) = K\lambda_1(s) + K\int_{t_0}^{s} [\lambda_2(s)\lambda(s-\xi) + \int_{\xi}^{s} \lambda_2(\tau)\lambda'(s-\xi) \, d\tau] \, d\xi$. Then

the trivial solution process is asymptotically stable.

$\underline{\text{Proof}}$. This theorem can be obtained by following the proof of Theorem 3.

$\underline{\text{Example 2}}$ Let us consider the Itô-type stochastic integro-differential equation

$$dy = \left[-5y + \int_{t_0}^{t} 3e^{-7(t-s)} y(s) \, ds\right] dt + B(t) y \, dz(t)$$

$$+ \left[\int_{t_0}^{t} \sigma(t,s) y(s) \, dw(s)\right] dt, \quad y(t_0,\omega) = y_0(\omega) \tag{17}$$

where $y_0 \in R$, $z(t)$, $w(t)$ are independent one dimensional normalised Wiener processes.

The mean equation corresponding to (17) is

$$x' = -5x + \int_{t_0}^{t} 3e^{-7(t-s)} x(s) \, ds, \quad x(t_0,\omega) = y_0(\omega) \tag{18}$$

we have

$$x(t,t_0,y_0(\omega))=R(t,t_0)y_0(\omega)=\left[\tfrac{1}{4}e^{-8(t-t_0)}+\tfrac{3}{4}e^{-4(t-t_0)}\right]y_0(\omega).$$

Take $V(t,x)=e^{\beta t}|x|^2,\beta>0.$

$$L_sV(s,x(t,s,y))$$

$$=V_s(s,x(t,s,y))+V_x(s,x(t,s,y))\int_{t_0}^{s}R(t,s)\hat{K}(s,\tau)y(\tau)d\tau$$

$$-V_x(s,x(t,s,y))\int_{s}^{t}R(t,\tau)\hat{K}(\tau,s)d\tau+\tfrac{1}{2}\mathrm{tr}[V_{xx}(s,x(t,s,y))$$

$$\times\ [R(t,s)B(s)y(s,\omega)]\,[R(t,s)B(s)y(s,\omega)]^T]$$

$$=\beta\ e^{\beta s}|x(t,s,y)|^2+2e^{\beta s}x(t,s,y)\,[\int_{t_0}^{s}\hat{K}(s,\tau)y(\tau)R(t,s)d\tau$$

$$-\int_{s}^{t}R(t,\tau)\hat{K}(\tau,s)d\tau]+\ e^{\beta s}[R(t,s)B(s)y(s,\omega]^2$$

$$\leq\beta e^{\beta s}R^2(t,s)y^2(s)+\ 2e^{\beta s}R^2(t,s)y(s)\int_{t_0}^{s}\hat{K}(s,\tau)y(\tau)d\tau$$

$$+e^{\beta s}[R(t,s)B(s)y(s,\omega)]^2$$

$$\leq\beta\ e^{\beta s}R^2(t,s)y^2(s)$$

$$+\ \int_{t_0}^{s}e^{\beta s}R^2(t,s)\hat{K}(s,\tau)\,[y^2(s)+y^2(\tau)]d\tau$$

$$+e^{\beta s}[R(t,s)B(s)y(s,\omega)]^2$$

$$\leq\ C_1\lambda_1(t-s)V(s,y)+\int_{t_0}^{s}C_2\lambda_2(s)\ \bar{K}(t,\tau)V(\tau,y(\tau))d\tau.$$

where $\lambda_1(t-s)=\exp[-8(t-s)]$, $\lambda_2(s)=\exp[(1+\beta)s]$,
$\bar{K}(t,\tau)=\exp[-8t+7\tau-\beta\tau]$, C_1,C_2 are constants and
$\lambda_1\in C[R_+,R_+]\cap L^1[R_+,R_+]$. Now $|x(t,\omega)|^2=\ R^2(t,t_0)|y_0(\omega)|^2$ and
$E[|x(t,\omega)|^2]=E[|y_0(\omega)|^2R^2(t,t_0)]$

$$\leq E[|y_0(\omega)|^2]^{\frac{1}{2}}(E[|y_0(\omega)|^2R^4(t,t_0)])^{\frac{1}{2}}$$

$$\leq(E[|y_0(\omega)|^2])^{\frac{1}{2}}(\rho R^4(t,t_0))^{\frac{1}{2}}$$

whenever $E[|y_0(\omega)|^2]\leq\rho$ for some $\rho>0$. Here $\alpha(|y_0|^2)=(|y_0^2(\omega)|)^{\frac{1}{2}}$

and $p=2$, $\alpha \in C\mathcal{K}$. Therefore (17) satisfies all the
hypotheses of Theorem 3.

Thus,
$$E[|y(t,\omega)|^2] \leq (E[|y_0(\omega)|^2])^{\frac{1}{2}} + \int_{t_0}^{t} A(s,t_0)E[|y(s,\omega)|^2]\,ds,$$
which implies that
$$E[|y(t,\omega)|^2] \leq (E[|y_0(\omega)|^2])^{\frac{1}{2}}\exp[\int_{t_0}^{t} A(s,t_0)\,ds],\text{where } A(s,t_0) \text{ is}$$
given by

$$A(s,t_0) = C_1\exp[-8(s-t_0)] - \frac{C_2}{7-\beta}\exp[-(7+\beta)(s-t_0)]$$

$$+\frac{C_2}{7-\beta}\exp[-8(s-t_0)] + \frac{8C_2}{(7-\beta)(1+\beta)}\exp[-(7-\beta)(s-t_0)].$$

From this we can conclude that the trivial solution of
(17) is stable in the mean square.

ERROR ESTIMATES.

We state error estimate results in the context of Itô-type
integro-differential equations (1). We shall employ the
method of variation of constants to derive the error
estimates on the p-th moment deviation of a solution
process of (1) with the solution process of either (2) or
(3).

<u>Theorem 5.</u>   Let the assumptions of Theorem 2 be
satisfied.   Further assume that

(i) $b(\|x\|^P) \leq \sum_{i=1}^{m} |V_i(t,x)| \leq a(\|x\|^P)$,

(ii) $\sum_{i=1}^{m} |L_sV_i(s,x(t,s,y)-x(t,s,z))|$
$$\leq \lambda_1(t-s)C(\|y-z\|^P) + \lambda_2(t-s)C(\|z\|)^P$$

$$+\int_{t_0}^{s} \lambda_3(t-s)\hat{K}(t,\tau)C(\|y(\tau)-z(\tau)\|^P)\,d\tau$$
(iii) $E[V_x(s,x(t,s,y(s))-x(t,s,\bar{x}(s)))R(t,s)B(s)y(s)$

$$+\int_{s}^{t} V_x(\tau,x(t,\tau,y(\tau))-x(t,\tau,\bar{x}(\tau)))R(t,\tau)\sigma(\tau,s)y(s)\,d\tau]$$

and the continuous time derivative of $a(E[\|x(t,\omega)-\bar{x}(t)\|^P])$
exists for $t \geq t_0$, where $a \in C\mathcal{K}$, and its derivative belongs
to $C[R_+,R_+]$, $b \in V\mathcal{K}$, $C \in C\mathcal{K}$, and $L_sV(s,x(t,s,y)-x(t,s,z))$ is
the i-th component of

$$L_S V(s,x(t,s,y)-x(t,s,z))$$
$$=V_s(s,x(t,s,y)-x(t,s,z))$$
$$-\int_s^t [V_x(s,x(t,s,y(s,\omega))$$
$$-x(t,s,\bar{x}(s)))R(t,\tau)\hat{K}(\tau,s)[y(s,\omega)-\bar{x}(s)]d\tau$$
$$+\int_{t_0}^s [V_x(s,x(t,s,y(s,\omega))-x(t,s,\bar{x}(s)))R(t,s)\hat{K}(s,\tau)[y(\tau)$$
$$-\bar{x}(\tau)]d\tau+\tfrac{1}{2}tr[V_{xx}(s,x(t,s,y(s,\omega))-x(t,s,z))$$
$$\times [R(t,s)B(s)y(s,\omega)][R(t,s)B(s)y(s,\omega)]^T] \tag{19}$$

$p\geq 1$, $\lambda_1,\lambda_2,\lambda_3 \in C[R_+,R_+]\cap L^1[R^+,R^+]$. Let us define
$H(s)$, $\frac{d}{ds}H=\frac{1}{h(s)}$, $h(s)=C(b^{-1}(s))$ and assume that $H\in\mathcal{H}$.
Then,

$$E[\|y(t,\omega)-\bar{x}(t)\|^P]\leq b^{-1}[r(t,\omega)] \tag{20}$$

for $t \geq t_0$, where $y(t,\omega)$ and $x(t,\omega)$ are the solution
process of (1) and (3) through $(t_0,y_0(\omega))$, and
$\bar{x}(t)=x(t,t_0,z_0)$ is the solution process of (2) or (3)
depending on the choice of z_0: $\beta(s)$ is the absolute value of
the sum of $E[\lambda_2(t-s)\|\bar{x}(s)\|^P]$ and the time derivative of a
$(E[\|x(t,\omega)-\bar{x}(t)\|^P])$; $r(t,\omega)$ is the maximal solution process
of the integro-differential equation

$$m'(t)=\beta(t)+\lambda_1(0)h(m(t))+$$
$$+\int_{t_0}^t\left[\lambda_1'(t-s)+\lambda_3(0)\hat{K}(t,s)+\int_s^t\lambda_3'(t-\tau)\hat{K}(\tau,s)d\tau\right]h(m(s))ds.$$

<u>Proof</u>: Let $y(t,\omega)$, $x(t,\omega)$ and $\bar{x}(t)$ be solution processes
as described in the theorem. Using Theorem 2, hypotheses
(iii) and taking expected value and following the proof of
Theorem 3, the proof of the theorem can be reformulated.
Further details are left to the reader.

We remark that this result provides the effects of
randomness relative the corresponding deterministic
systems.

<u>Example 3</u>. Let us consider the problem (17). Take
$V(t,x)=\|x\|^2$, $a(s)=b(s)=2$.

Then,

$$L_S \; V(s,x(t,s,y)-x(t,s,z))$$

$$=V_s(s,x(t,s,y)-x(t,s,z))$$

$$-\int_s^t [V_x(s,x(t,s,y(s,\omega))-x(t,s,\bar{x}(s)))R(t,\tau)$$

$$\times \; \hat{K}(\tau,s)\,[y(s,\omega)-\bar{x}(s)]\,d\tau$$

$$+\tfrac{1}{2}\;tr[V_{xx}(s,x(t,s,y(s,\omega))-x(t,s,z))$$

$$\times \; [R(t,s)B(s)y(s,\omega)]\,[R(t,s)B(s)y(s,\omega)]^T$$

$$+\int_{t_0}^s [V_x(s,x(t,s,y(s,\omega))-x(t,s,\bar{x}(s)))$$

$$\times \; R(t,s)\hat{K}(s,\tau)\,[y(\tau)-\bar{x}(\tau)]\,d\tau$$

$$\leq \int_{t_0}^s (x^T(t,s,y(s,\omega))-x^T(t,s,\bar{x}(s)))R(t,s)\hat{K}(s,\tau)\,(y(\tau,\omega)$$

$$-\bar{x}(\tau))\,d\tau+[R(t,s)B(s)y(s,\omega)]\,[R(t,s)B(s)y(s,\omega)]^T$$

$$\leq \left[\,C_1\|y-z\|^2+C_2\|z\|^2+\int_{t_0}^s C_3\hat{K}(s,\tau)\|y(\tau)-\bar{x}(\tau)\|^2d\tau\,\right]\exp[-8(t-s)]$$

where $\lambda_1(t-s)=C_1\exp[-8(t-s)]$, $\quad\lambda_2(t-s)=C_2\exp[-8(t-s)]$,
$\lambda_3(t-s)=C_3\exp[-8(t-s)]$ and $C(s)=s$. Here all the
hypotheses of Theorem 5 are satisfied and hence
$E[\|y(t,\omega)-\bar{x}(t)\|^2]\leq r(t)$, $\quad t\geq t_0$.

Where $r(t)$ is the solution process of the linear integro-
differential equation

$$m'(t)=\beta(t)+C_1m(t)+\int_{t_0}^t C_3\hat{K}(t,s)m(s)\,ds$$

and

$$\beta(t)=|C_2E[\|\bar{x}(t)\|^2])-\int_{t_0}^t C_2\exp[-8(t-s)]E[\|\bar{x}(s)\|^2])\,ds$$

$$+\tfrac{d}{dt}(E[\|x(t,\omega)-\bar{x}(t)\|^2]))|.$$

References:

1. Bharucha-Reid, A . T., Random Integral Equations, Academic Press, New York, 1972.
2. Grosman, S . I and Miller, R . K., Perturbation Theory for Volterra Integro-Differential systems Journal of Differential equations 8, 457-474, 1970.
3. Ladde, G . S., Variational Comparision Theorem and Perturbations of Nonlinear systems, Proc. of AMS, Vol. 52, pp. 181-187, 1975.
4. Ladde, G. S. and Laksmikantham, V., Random Differential Inequalities, Academic Press, New York 1980.
5. Ladde, G. S. and Sambandham, M., Error estimate of solutions and means of solutions of stochastic differential systems, Jour. of Math. Phys., Vol. 24 1983.
6. Ladde, G. S. and Sambandham, M., Variation of Constant formula and error estimates to stochastic difference equations, J. Math. and Phys. Sci., Vol. 22(1988), pp. 557-584.
7. Laksmikantham, V., Leela, S. and Martynyuk, A. A., Stability Analysis of Nonlinear Systems, Marcel Dekker, New York and Basel, 1989.
8. Tsokos, C. P. and Padgett, W. J., Random Integral Equations with applications to Life sciences and Engineering, Academic Press, New York, 1974.

INTEGRAL EQUATIONS

Summary

The papers in this section have a common feature, consisting of the presence of integral operators/equations in the treatment of various problems. Five of these papers have a predominantly theoretical character:

The invited paper by Irwin W. Sandberg, Regents Chair in EE and Computer Science Engineering at the University of Texas-Austin deals with the representation theory and nonlinear systems.

Protopopescu uses integral methods to solve Fokker-Planck-type equations. Corduneanu discusses absolute stability of certain nonlinear feedback systems involving delays. Bertram uses product integration to solve singular integral equations. A paper investigating Sturm-Liouville problems at resonance is presented by Matala-aho and Seikkala.

Two papers are primarily concerned with computational aspects: Merson and Cale consider the regularization method for inverting Fredholm operators. Dellwo, Friedman, and Aggarwal discuss acceleration of projection methods for solving nonlinear integral equations.

The role of integral operators/equations in the investigation of many problems occurring in science and engineering has known a tremendous increase during the last three decades. A multitude of applied fields is generating new and complicated problems for the theory of integral operators/equations. Since in many cases information is expected for finite interval of time or bounded regions of the space, the integral operators/equations happen to be the ideal tool for global approaches (as opposed to the local approach, better described by differential equations).

The contributions included in this section illustrate to various degrees and with various specific procedures, the indispensable role acquired by the integral operators/equations in contemporary research.

C. Corduneanu

Representation Theory and Nonlinear Systems

IRWIN W. SANDBERG
Department of Electrical and Computer Engineering
University of Texas at Austin
Austin, Texas 78712, USA

ABSTRACT

One of the main results is a proposition to the effect that under some typically mild conditions finite sums of the form

$$\sum_\ell \kappa_\ell \, \sigma \left[\sum_m \eta_{\ell m} Q_m(\cdot) + \rho_\ell \right]$$

are dense in an important sense in the set of shift-invariant approximately-finite-memory maps $G(\cdot)$ that take a certain type of subset U of R into R, where R is the set of real-valued functions defined on $\mathcal{R}^n$ or $\mathcal{Z}^n$. Here the $Q_m(\cdot)$ are linear, σ is any element of a certain set of nonlinear maps from $\mathcal{R}$ to $\mathcal{R}$, and the κ_ℓ, ρ_ℓ, and $\eta_{\ell m}$ are real constants. Approximate representations comprising only affine elements and lattice nonlinearities are also presented.

I. INTRODUCTION

One of the main results given in this paper, as an application of Theorem 1 in Section II, is a proof that under some typically mild conditions finite sums of the form

$$\sum_\ell \kappa_\ell \, \sigma \left[\sum_m \eta_{\ell m} Q_m(\cdot) + \rho_\ell \right]$$

are dense in an important sense in the set of shift-invariant approximately-finite-memory maps $G(\cdot)$ that take a certain type of subset U of R into R, where R is the set of real-valued functions defined on $\mathcal{R}^n$ or $\mathcal{Z}^n$ (the set of real n-vectors or integer-valued n-vectors, respectively). Here the $Q_m(\cdot)$ are linear, σ is any element of a certain set of nonlinear maps from $\mathcal{R}$ to $\mathcal{R}$, and the κ_ℓ, ρ_ℓ, and $\eta_{\ell m}$ are real constants. Corresponding results are given for related cases in which R is replaced by the family of real-valued functions defined on $\mathcal{R}_+$ or $\mathcal{Z}_+$ (i.e. on $[0, \infty)$ or $\{0, 1, \ldots\}$, respectively), and each $G(\cdot)$ as well as the $Q_m(\cdot)$ are causal. Results of this kind are of interest in connection with studies of the capabilities of neural networks.

From another perspective, and roughly speaking, the results show that the members $G(\cdot)$ of a large class of nonlinear input-output maps are not essentially more complicated than the maps of finite parallel combinations of nonlinear systems that are simple in the sense that they consist of just a linear element followed by a nonlinearity.

It is shown also (see Section 2.9) that maps $G(\cdot)$ of the type addressed can be approximated arbitrarily well, in the sense of uniform approximation over a certain subset S of U, by the maps of certain systems that contain only affine elements (i.e. elements that are represented by a linear operator and a constant term) and a finite number of lattice nonlinearities (which take the pointwise maximum or minimum of pairs of functions). An immediate corollary is that these $G(\cdot)$'s can be approximated arbitrarily well by the maps of systems that contain only affine elements and a finite number of absolute-value nonlinearities (or only affine elements and a finite number of ideal-diode nonlinearities). This has applications concerning signal processing.

The most closely related background material is contained in [1] and the references cited there. In [1] attention is focused on causal time-invariant nonlinear maps that take one set of functions on the half-line $\mathcal{R}_+$ into another. A theorem is proved which shows that, under typically reasonable conditions, an input-output map can be approximated arbitrarily well in a meaningful sense by a finite Volterra-series. The approximately-finite-memory hypotheses used in Sections II and III of this paper have their origins in [1].

Referring to cases in which the domain of G is a set of functions on $\mathcal{R}^n$ or $\mathcal{Z}^n$, while Volterra-series representations are not the main concern in this paper, we note that Theorem 1 in Section II establishes that arbitrarily good Volterra-series approximations exist for maps of the type we consider (see Section IV and the remarks concerning polynomial maps in Section 2.8). The existence of such approximations was shown earlier (in a somewhat different setting) only for the $n = 1$ causal map case [2].

A theorem due to Stone plays a key role in this paper, but the Stone-Weierstrass theorem is not used. Also, as will become clear we consider maps G that have a more general domain than indicated above. With regard to the organization of the paper, general approximation results are given in Sections II and III, and specific applications are considered in Section IV.

In the oral version of this paper exact integral representations [3,4] were also described for an important class of input-output operators. These representations generalize the familiar convolution representations for linear systems, and exist for nonlinear systems governed by, for example, ordinary differential equations that satisfy certain typically mild conditions. The reader is referred to [3,4] for the details.

II. APPROXIMATION THEOREM

2.1 Preliminaries

Throughout Section II, V denotes a linear space with zero element θ, and X stands for the set of all V-valued functions on K, where K is $\mathcal{R}^n$ or $\mathcal{Z}^n$ with n an arbitrary positive integer. We are interested mainly in the case in which $V = \mathcal{R}^p$ for some p.

Let $\{c_a : a \in A\}$ be a family of nonempty subsets of K, and for $\alpha \in K$ and $a \in A$, let $W_{\alpha,a} : X \to X$ be defined by

$$(W_{\alpha,a}x)(\beta) \;=\; x(\beta), \quad (\beta - \alpha) \in c_a$$
$$\phantom{(W_{\alpha,a}x)(\beta)} \;=\; \theta, \quad\;\;\; (\beta - \alpha) \notin c_a .$$

For example, K, A, and c_a, respectively, might be $\mathcal{R}^n$, $(0, \infty)$, and the hypercube $\{\gamma \in \mathcal{R}^n : |\gamma_j| \leq a \; \forall j\}$. In this case, $W_{\alpha,a}x$ is just a "windowed version" of x, with the (n-dimensional) window centered at α and having width $2a$.

Let S (the set of input signals over which our approximations will be seen to hold) be a nonempty subset of X, and for each $a \in A$, let $S_a := \{s|_{c_a} : s \in S\}$, where $s|_{c_a}$ stands for the restriction of s to c_a.

For each $\beta \in K$, define $T_\beta : X \to X$ to be the (shift) operator given by

$$(T_\beta x)(\alpha) = x(\alpha - \beta)$$

for $\alpha \in K$.

Let U (the domain of the maps G that we consider) be a subset of X closed under the T_β such that

$$U \supseteq S \cup \{\cup_a W_{0,a}(S)\} \cup \{\cup_a W_{0,a}(U)\} .$$

2.2 Hypotheses Concerning S

Assume that

(i) S is closed under the T_β.

(ii)   For each a, there is associated with $U_a := \{u|_{c_a} : u \in U\}$ a topology and with this topology S_a is a relatively compact subset of U_a.

2.3 Additional Preliminaries

A map M from U to the set R of real-valued functions on K is *shift invariant* if $(MT_\beta u)(\alpha) = (Mu)(\alpha - \beta)$ for α and β in K and $u \in U$. A map $F : U \to R^k$ (k a positive integer) is *shift invariant* if each of its k component maps is shift invariant.

We say that $M : U \to R$ belongs to $\mathcal{A}(S)$ if given any $\varepsilon > 0$, there is an $a \in A$ such that

$$|(Ms)(\alpha) - (MW_{\alpha,a}s)(\alpha)| \; < \; \varepsilon, \quad \alpha \in K$$

for all $s \in S$.

Assuming that the sets c_a are bounded, $M \in \mathcal{A}(S)$ has the natural interpretation that M has "approximately finite memory" on S in the sense indicated. As a simple example, if K, A, and c_a are as indicated in the earlier example, if S and U are sets of bounded (Lebesgue) measurable functions from $\mathcal{R}^n$ to $\mathcal{R}$, and if $M : U \to R$ is given by

$$(Mu)(\alpha) = \int_{\mathcal{R}^n} h(\alpha - \beta)u(\beta)d\beta$$

with $h : \mathcal{R}^n \to \mathcal{R}$ measurable and such that

$$\int_{\mathcal{R}^n} |h(\beta)|d\beta < \infty ,$$

then it easily follows that $M \in \mathcal{A}(S)$.

For each a, let $c\ell(S_a)$ denote the closure of S_a and let $\mathcal{I}_a$ be the "insertion map" from U_a to U given by

$$\begin{aligned}
(\mathcal{I}_a v)(\alpha) &= v(\alpha), &&\alpha \in c_a \\
&= \theta, &&\alpha \notin c_a .
\end{aligned}$$

2.4 Hypotheses Concerning G

Assume that $G : U \to R$ (R is defined in Section 2.3) is shift invariant, belongs to $\mathcal{A}(S)$, and is such that for each a the functional $G(\mathcal{I}_a \cdot)(\theta) : c\ell(S_a) \to \mathcal{R}$ is continuous.

2.5 Fundamental Sets and Lattice Maps

By a *fundamental set* (or a *generating set*) $\{F_\lambda : \lambda \in \Lambda\}$ of maps from U to R we mean that the F_λ are shift invariant, that each $F_\lambda(\mathcal{I}_a \cdot)(\theta) : c\ell(S_a) \to \mathcal{R}$ is continuous, and that for any $a \in A$ and any pair of points v and w in $c\ell(S_a)$ there is a $\lambda \in \Lambda$ such that

$$F_\lambda(\mathcal{I}_a v)(\theta) = G(\mathcal{I}_a v)(\theta)$$

and

$$F_\lambda(\mathcal{I}_a w)(\theta) = G(\mathcal{I}_a v)(\theta) .$$

The *lattice operations* $a \vee b$ and $a \wedge b$ on pairs of real numbers a and b are defined by $a \vee b = \max(a, b)$ and $a \wedge b = \min(a, b)$. We say that a map L from $\mathcal{R}^k$ to $\mathcal{R}$, where k is a positive integer, is a *lattice map* if Lx for $x \in \mathcal{R}^k$ is generated from the elements $x_1, \ldots, x_k$ of x by a finite number of lattice operations that do not depend on x. (The operations of addition and of multiplication by scalars are not allowed.)

2.6 The Main Theorem

Our main result is the following:

<u>**Theorem 1.**</u> Let $G : U \to R$ be as described, and let $\{F_\lambda : \lambda \in \Lambda\}$ be a corresponding fundamental set. Then for any $\varepsilon > 0$, there are an $a \in A$, a positive integer k, elements $F_{\lambda_j}, \ldots, F_{\lambda_k}$ of $\{F_\lambda : \lambda \in \Lambda\}$ and a lattice map $L : \mathcal{R}^k \to \mathcal{R}$ such that

$$|(Gs)(\alpha) - L[(FW_{\alpha,a}s)(\alpha)]| < \varepsilon, \quad \alpha \in K \tag{1}$$

for all $s \in S$, where $F : U \to R^k$ is given by

$$(Fu)_j = F_{\lambda_j}(u), \quad u \in U, \quad j = 1, \ldots, k.$$

Also, the map $Q : U \to R^k$ defined by $(Qu)(\alpha) = (FW_{\alpha,a}u)(\alpha)$ for $u \in U$ and $\alpha \in K$ is shift invariant, and, with $\|\cdot\|$ any norm on $\mathcal{R}^k$, one has

$$\sup\{\|(FW_{\alpha,a}s)(\alpha)\| : s \in S, \alpha \in K\} < \infty. \tag{2}$$

2.7 Proof

Given $\varepsilon > 0$, choose $a \in A$ such that

$$|(Gs)(\alpha) - (GW_{\alpha,a}s)(\alpha)| < \tfrac{1}{2}\varepsilon, \quad \alpha \in K \tag{3}$$

for $s \in S$.

Let the *lattice operations* $x \vee y$ and $x \wedge y$ on pairs of real-valued functions x and y on S_a be defined by

$$(x \vee y)(\alpha) \;=\; \max[x(\alpha), \, y(\alpha)]$$
$$(x \wedge y)(\alpha) \;=\; \min[x(\alpha), \, y(\alpha)]$$

for $\alpha \in K$.

We will use the following lemma which is an immediate corollary of [5, Theorem 1] (recall that $c\ell(S_a)$ is compact).

<u>**Lemma 1.**</u> Let $\mathcal{X}$ be the family of continuous real-valued functions on $c\ell(S_a)$, $\mathcal{X}_0$ an arbitrary subfamily of $\mathcal{X}$, and $U(\mathcal{X}_0)$ the family of all functions generated from $\mathcal{X}_0$ by the lattice operations and uniform passage to the limit. Suppose that $f \in \mathcal{X}$ has the property that for any v and w in $c\ell(S_a)$ there exists a $g \in \mathcal{X}_0$ such that $g(v) = f(v)$ and $g(w) = f(w)$. Then $f \in U(\mathcal{X}_0)$.

By Lemma 1 (see also [5, pp. 33-34]) with $f = G(\mathcal{I}_a \cdot)(\theta)$ and $\mathcal{X}_0 = \{F_\lambda(\mathcal{I}_a \cdot)(\theta) : \lambda \in \Lambda\}$, there are a positive integer k, elements $F_{\lambda_1}, \ldots, F_{\lambda_k}$ of $\{F_\lambda : \lambda \in \Lambda\}$, and a lattice map $L : \mathcal{R}^k \to \mathcal{R}$ such that

$$|G(\mathcal{I}_a v)(\theta) - L(F_a v)| < \tfrac{1}{2}\varepsilon \tag{4}$$

for $v \in S_a$, where $F_a : S_a \to \mathcal{R}^k$ is given by

$$(F_a)_j = F_{\lambda_j}(\mathcal{I}_a \cdot)(\theta), \quad j = 1, \ldots, k.$$

We now use the fact that (4) can be written as

$$|G(\mathcal{I}_a v)(\theta) - L[(F\mathcal{I}_a v)(\theta)]| < \tfrac{1}{2}\varepsilon,$$

where F is as defined in the theorem. Choose any $\alpha \in K$ and $s \in S$, and let $v \in S_a$ be given by

$$v(\beta) = (T_{-\alpha}s)(\beta) , \quad \beta \in c_a .$$

Then $\mathcal{I}_a v = W_{\theta,a}T_{-\alpha}s$, and by the observation that $W_{\alpha,a} = T_\alpha W_{\theta,a}T_{-\alpha}$ on X, and the shift invariance of G and F, we have

$$(G\mathcal{I}_a v)(\theta) = (GW_{\alpha,a}s)(\alpha)$$

and

$$(F\mathcal{I}_a v)(\theta) = (FW_{\alpha,a}s)(\alpha) ,$$

so that

$$|(GW_{\alpha,a}s)(\alpha) - L[(FW_{\alpha,a}s)(\alpha)]| < \tfrac{1}{2}\varepsilon .$$

This, together with (3), gives (1).

The following shows the shift invariance of Q.

Lemma 2. If $H : U \to R$ is shift invariant and $J : U \to R$ is defined by $(Ju)(\alpha) = (HW_{\alpha,a}u)(\alpha)$ for $\alpha \in K$ and $u \in U$, then J is shift invariant.

<u>Proof of Lemma 2</u>

We use the observation that $W_{\alpha,a}T_\beta = T_\beta W_{(\alpha-\beta),a}$ on X for α and β in K. Thus, for any $u \in U$ and any α and β,

$$\begin{aligned}
(JT_\beta u)(\alpha) &= (HW_{\alpha,a}T_\beta u)(\alpha) = (HT_\beta W_{(\alpha-\beta),a}u)(\alpha) \\
&= (HW_{(\alpha-\beta),a}u)(\alpha - \beta) = (Ju)(\alpha - \beta) ,
\end{aligned}$$

proving the lemma.

With regard to the boundedness of $\{\|(FW_{\alpha,a}s)(\alpha)\| : s \in S, \alpha \in K\}$, let s and α be given. By the shift-invariance of Q, we have $FW_{\alpha,a}s)(\alpha) = F(W_{0,a}T_{-\alpha}s)(\theta)$. Since $W_{0,a}T_{-\alpha}s = \mathcal{I}_a v$ for some $v \in S_a$, and since $cl(S_a)$ is compact and the components of $F(\mathcal{I}_a \cdot)(\theta)$ are continuous, we have the claimed boundedness.

2.8 Comments

There are important cases in which S and U coincide (e.g. see Section 4.1).

Let C denote the set of all continuous maps from $\mathcal{R}^k$ to $\mathcal{R}$, and let D stand for any subset of the set of maps from $\mathcal{R}^k$ to $\mathcal{R}$ that is *dense* in C on compact sets, in the (usual) sense that given $\varepsilon > 0$, $f \in C$, and a compact $E \subset \mathcal{R}^k$, there is a $g \in D$ such that $|f(\beta) - g(\beta)| < \varepsilon$ for $\beta \in E$ (e.g. one may take D to be the subset of polynomial maps). Since lattice maps are continuous, by (2) we see that G can be approximated in the sense of (1) with L drawn from D.

2.9 Preview of Consequences of the Theorem

With σ a continuous map from $\mathcal{R}$ to R, let N_σ denote the set of all maps of $\mathcal{R}^k$ into $\mathcal{R}$ of the form

$$\sum_\ell \kappa_\ell \, \sigma(\eta_\ell \cdot + \gamma_\ell)$$

in which the sum is finite, the κ_ℓ and γ_ℓ are elements of $\mathcal{R}$ and the η_ℓ are row k-vectors. (Here we view $\mathcal{R}^k$ as a set of *column* k-vectors.) It is known [6,7] that D of Section 2.8 may be taken to be N_σ for a large class of maps σ, including the "sigmoidal" σ's defined by the condition that $\sigma(x) \to 1$ as $x \to \infty$ and $\sigma(x) \to 0$ as $x \to -\infty$.

Therefore, with any such σ, maps from U to R of the form

$$\sum_\ell \kappa_\ell \, \sigma(\eta_\ell Q(\cdot) + \gamma_\ell) \tag{5}$$

in which $Q : U \to R^k$ is a shift-invariant map of the type described in the theorem, are dense over S (in the sense of the theorem) in the set of maps $G(\cdot)$ addressed. We will see in Section IV that often for a given $G(\cdot)$ a fundamental set can be chosen so that the F_λ are the restrictions to U of affine maps from some linear space U_1 to R, and then (after redefining the γ_ℓ in an obvious way) Q in (5) can be assumed to be linear and from U_1 to R^k. This has an interesting neural-network interpretation: Refering to a very large class of approximately-finite-memory shift-invariant input-output maps $G(\cdot)$ that take $U \subset X$ into R, where U is a certain extension of the subdomain S of $G(\cdot)$ of interest, dynamic feedforward networks with one hidden nonlinear layer are universal approximators, in the sense of uniform approximation on S.

In a paper by this writer and J. Park ("Universal Approximation Using Radial-Basis-Function Networks," in preparation), it is proved that another choice of D is the family of maps from $\mathcal{R}^k$ to $\mathcal{R}$ of the form

$$\sum_\ell \kappa_\ell \, M\left(\frac{\cdot - z_\ell}{\nu}\right)$$

where the sum is finite, ν and the κ_ℓ are real numbers with $\nu > 0$, the z_ℓ belong to $\mathcal{R}^k$, and M is any map from $\mathcal{R}^k$ to $\mathcal{R}$ that is continuous, bounded, and Lebesgue integrable with $\int_{\mathcal{R}^k} M(\alpha)d\alpha \neq 0$. Thus, maps from U to R of the form

$$\sum_\ell \kappa_\ell \, M\left(\frac{Q(\cdot) - z_\ell}{\nu}\right)$$

are also dense over S in the sense indicated above.

With regard to additional implications of the theorem (under the supposition that the F_λ can be taken to be affine), we see that systems governed by $G(\cdot)$'s of the type addressed can be approximated arbitrarily well on S by certain systems that contain only affine elements $(Q)_1, \ldots, (Q)_k$ and lattice nonlinearities (which take the pointwise maximum or minimum of pairs of functions). And this implies that our approximating systems can be viewed to contain, for example, only affine elements and absolute-value nonlinearities, or only affine elements and ideal-diode nonlinearities, since $\max(a, b) = \frac{1}{2}[a + b + |a - b|]$, $\min(a, b) = \frac{1}{2}[a + b - |a - b|]$, and

$$|a| = f_\xi(\xi^{-1}a) + f_\xi(-\xi^{-1}a)$$

for each pair of real numbers a and b, in which $f_\xi : \mathcal{R} \to \mathcal{R}$ for any $\xi > 0$ is given by $f_\xi(x) = 0$ for $x < 0$ and $f_\xi(x) = \xi x$ for $x \geq 0$.

Corresponding results hold for causal maps $G(\cdot)$ that operate on V-valued functions defined on $\mathcal{R}_+$ or $\mathcal{Z}_+$. This is discussed in Section III.

III. RELATED HALF-LINE RESULTS

As in Section II, V denotes a linear space with zero element θ. Here X denotes the set of all V-valued functions on $K = \mathcal{R}_+$ or $\mathcal{Z}_+$, and K^+ stands for $\{a \in K : a > 0\}$.

For each $\beta \in K$, let T_β, T^β, and P_β denote the maps of X into itself defined by

$$
\begin{aligned}
(T_\beta x)(t) &= \theta, \ t < \beta \\
&= x(t - \beta), \ \ t \geq \beta \\
(T^\beta x)(t) &= x(t + \beta), \ \ t \in K \\
(P_\beta x)(t) &= x(t), \ \ t \leq \beta \\
&= \theta, \ \ t > \beta .
\end{aligned}
$$

Let $W_{t,a} : X \to X$ $(t \in K, a \in K^+)$ be defined throughout Section III by

$$
\begin{aligned}
(W_{t,a}x)(\alpha) &= x(\alpha), \ \ t - a \leq \alpha \leq t \\
&= \theta, \ \ \text{otherwise} .
\end{aligned}
$$

Let S be a nonempty subset of X, and for each $a \in K^+$ let $S_a = \{s|_{c_a} : s \in S\}$ where here c_a denotes $[0, a] \cap K$ and $s|_{c_a}$ stands for the restriction of s to c_a.

Let U be a subset of X closed under the T_β, T^β, and P_β such that $S \subseteq U$.

3.1 Assumptions on S and Additional Definitions

Assume that

(i) S is closed under the T^β.

(ii) For each $a \in K^+$, there is associated with $U_a := \{u|_{c_a} : u \in U\}$ a topology and with this topology S_a is a relatively compact subset of U_a.

We say that a map $M : U \to R$ (R is the set of $\mathcal{R}$-valued maps on K) is *time invariant* if for each β and u, $(MT_\beta u)(t) = 0$ for $t < \beta$ and $(MT_\beta u)(t) = (Mu)(t - \beta)$ for $t \geq \beta$. A map $F : U \to R^k$ is *time invariant* if each of its k component maps is time invariant.

By $M : U \to R$ *causal* we mean that for each $a \in K$: x and y in U with $P_a x = P_a y \Rightarrow (Mx)(t) = (My)(t)$, $t \in c_a$. A map from U to R^k is *causal* if each of its k component maps is *causal*.

As (essentially) before, $M : U \to R$ belongs to $\mathcal{A}(S)$ if given any $\varepsilon > 0$ there is an $a \in K^+$ such that

$$|(Ms)(t) - (MW_{t,a}s)(t)| < \varepsilon, \quad t \in K$$

for all $s \in S$.

It is not difficult to verify that an example of a map $M : U \to R$ that belongs to $\mathcal{A}(S)$ with $K = \mathcal{Z}_+$ and S and U sets of bounded functions from $\mathcal{Z}_+$ to $\mathcal{R}$ is given by

$$(Mu)(t) = \sum_{\ell=0}^{t} h_{t-\ell} u_\ell, \quad t \in K$$

where $u_\ell = u(\ell)$ and $h_{(\cdot)}$ is $\mathcal{R}$-valued and satisfies $\sum_{\ell=0}^{\infty} |h_\ell| < \infty$.

For any $a \in K^+$ and any causal map M from U to R, M_a denotes the functional defined on U_a by

$$M_a v = (My)(a), \quad v \in U_a$$

where y is any element of U such that $y(t) = v(t)$ for $t \in c_a$.

3.2 Hypotheses on G

Let $G : U \to R$ be causal, time-invariant, belong to $\mathcal{A}(S)$, and be such that $G_a : U_a \to \mathcal{R}$ is continuous on $c\ell(S_a)$ for $a \in K^+$.

3.3 Fundamental Sets and Theorem 2

In Section III by a fundamental set $\{F_\lambda : \lambda \, \varepsilon \, \Lambda\}$ of maps from U to R we mean that $F_\lambda(\cdot) = c_\lambda + M_\lambda(\cdot)$ for each λ, with $c_\lambda \in \mathcal{R}$ and M_λ causal and time invariant, that each $M_{\lambda a}$ is continuous on $c\ell(S_a)$, and that for any $a \in K^+$ and v and w in $c\ell(S_a)$ there is a $\lambda \in \Lambda$ such that

$$F_{\lambda a}(v) = G_a(v) \quad \text{and} \quad F_{\lambda a}(w) = G_a(w) \, .$$

We will refer to c_λ as the *constant part* of F_λ.

By $L : \mathcal{R}^k \to \mathcal{R}$ a *lattice map* we mean the same thing as in Section II. In the following theorem, proved in Appendix A, we use K_a to denote $\{t \in K : t \geq a\}$.

<u>Theorem 2</u>. Let $G : U \to R$ be as described in Section 3.2, and let $\{F_\lambda : \lambda \in \Lambda\}$ be a corresponding fundamental set. Then for any $\varepsilon > 0$ there are an $a \in K^+$, a positive integer k, elements $F_{\lambda_1}, \ldots, F_{\lambda_k}$ of $\{F_\lambda : \lambda \in \Lambda\}$ and a lattice map $L : \mathcal{R}^k \to \mathcal{R}$ such that

$$|(Gs)(t) - L[(FW_{t,a}s)(t)]| < \varepsilon , \quad t \in K_a \tag{6}$$

for all $s \in S$, where $F : U \to R^k$ is given by

$$(Fu)_j = F_{\lambda_j}(u), \quad u \in U, \quad j = 1, \ldots, k \ .$$

Also, with V the k-vector of constant parts of the F_{λ_j}, the map $Q : U \to R^k$ defined by $(Qu)(t) = (FW_{t,a}u)(t) - V$ for $u \in U$ is causal and time invariant, and with $\| \cdot \|$ any norm on $\mathcal{R}^k$ one has

$$\sup\{\|(FW_{t,a}s)(t)\| : s \in S, \quad t \in K_a\} < \infty \ .$$

In addition, the conclusion above holds with K_a replaced with K provided that S is closed under the P_β and T_β.

3.4 Comments

The theorem is proved in Appendix B.

In light of the similarities of Theorems 1 and 2 it is clear that comments along the lines of those in Section 2.9 apply here too concerning the denseness of maps of the form (5) and approximations by systems containing only affine elements and lattice nonlinearities, etc. Of course here the concepts of denseness and approximation are interpreted in a related context that allows what might be called "eventual-uniform" approximation.

In the next section we give a related result concerning approximations on all of K for maps G that possess a certain composite representation.

3.5 Approximations for $G = AB$

Let V, θ, K, K^+, X, T_β, T^β, P_β, c_a, and $W_{t,a}$ be as described at the beginning of Section III and let V_1 be a second linear space.

Let S and S_1 be nonempty sets of maps from K to V and from K to V_1, respectively.

3.5.1 Conditions on A, B, S, and S_1

Define T_β in the same way also on X with V replaced with V_1, and assume that S and S_1 are closed under the T_β, and that S is closed under the T^β and P_β.

Here G is assumed to map S into R, and to have the composite representation $G = AB$ where B takes S into S_1 and A maps S_1 to R.

Assume also that A and B are causal and time invariant and that G belongs to $\mathcal{A}(S)$, in the sense of Section 3.1.

Here each $S_{1a} := \{x|_{c_a} : x \in S_1\}$ is taken to be a topological space with $S_a := \{x|_{c_a} : x \in B(S)\}$ a relatively compact subset, and each A_a defined in terms of A in the next section is assumed to be continuous on $c\ell(S_a)$.

3.5.2 <u>Fundamental Sets and Theorem 3</u>

Following the development in Section 3.1, for any causal map $M : S_1 \to R$ and any $a \in K^+$, let M_a stand for the functional on S_{1a} defined by $M_a(v) = (My)(a)$ where y is any element of S_1 for which $y(t) = v(t)$ for $t \in c_a$.

Here by a fundamental set $\{F_\lambda : \lambda \in \Lambda\}$ of maps from S_1 to R we mean that $F_\lambda(\cdot) = c_\lambda + M_\lambda(\cdot)$, $\lambda \in \Lambda$ with $c_\lambda \in \mathcal{R}$ and M_λ causal and time invariant, that each $M_{\lambda a}$ is continuous on $c\ell(S_a)$, and that given a together with v and w in $c\ell(S_a)$ there is a $\lambda \in A$ such that

$$F_{\lambda a}(v) = A_a(v) \quad \text{and} \quad F_{\lambda a}(w) = A_a(w) \ .$$

Arguments similar to those used in the proof of Theorem 2 establish the following (see Appendix B).

<u>Theorem 3</u>. Let A, B, and G be as indicated, and suppose that $\{F_\lambda : \lambda \in \Lambda\}$ is an associated fundamental set. Then given $\varepsilon > 0$ there are $a \in K^+$, elements $F_{\lambda_1}, \ldots,$ F_{λ_k} of $\{F_\lambda : \lambda \in \Lambda\}$, and a lattice map $L : \mathcal{R}^k$ to $\mathcal{R}$ such that

$$|(Gs)(t) - L[(FBW_{t,a}s)(t)]| < \varepsilon , \quad t \in K$$

for every $s \in S$, in which $F : S_1 \to R^k$ is defined by

$$(Fu)_j = F_{\lambda_j}(u), \quad u \in S_1$$

for all j. Also, with V the k-vector of constant parts of the F_{λ_j}, the map $Q : S \to R^k$ given by $(Qs)(t) = (FBW_{t,a}s)(t) - V$ is causal and time invariant, and with $\| \cdot \|$ any norm on $\mathcal{R}^k$ we have

$$\sup\{\|FBW_{t,a}s)(t)\| : s \in S, \quad t \in K\} < \infty \ .$$

3.5.3 <u>Comment</u>

It is observed in Section 4.2.3 that there is an important large class of maps G for which the hypotheses of Theorem 3 are met.

IV. SPECIFIC CASES

4.1 Discrete-Time Systems

Referring to Theorem 1, let $V = \mathcal{R}^p$ (viewed as a set of *column* vectors), and let $K = \mathcal{Z}^n$, $A = \{0, 1, \ldots\}$, and $c_a = \{\gamma \in \mathcal{Z}^n : |\gamma_j| \leq a \ \forall j\}$. Take $S = U$ to be the set of functions from K to any compact subset E of $\mathcal{R}^p$ with $\theta \in E$. View each U_a as a metric space with metric ρ_a given by

$$\rho_a(v_1, v_2) = \max_{\alpha \in c_a} \|v_1(\alpha) - v_2(\alpha)\| , \tag{7}$$

where $\| \cdot \|$ denotes a norm on $\mathcal{R}^p$. The conditions on S of Section 2.2 are met with each S_a compact (for a proof of the compactness see Appendix C).

Let G satisfy the conditions of Section 2.4. (Note that the conditions are reasonable.) We are going to find that an associated fundamental set $\{F_\lambda : \lambda \in \Lambda\}$ can be taken to be the family of restrictions to U of maps H described by

$$(Hx)(\alpha) = c + \sum_{\beta \in \mathcal{Z}^n} h(\alpha - \beta)x(\beta), \quad \alpha \in K \tag{8}$$

with $c \in \mathcal{R}$ and h a map of bounded support from $\mathcal{Z}^n$ to the set of real $1 \times p$ matrices. Consider this possibility.

The shift-invariance condition concerning the F_λ as well as the continuity requirement on the $F_\lambda(\mathcal{I}_a \cdot)(\theta)$ are obviously met.

Now suppose that $a \in A$ and v and w in S_a are given. Let γ_v and γ_w be the corresponding values of $G(\mathcal{I}_a \cdot)(\theta)$. It is a simple matter to verify that we have $(H\mathcal{I}_a v)(\theta) = \gamma_v$ and $(H\mathcal{I}_a w)(\theta) = \gamma_w$ for h and c in (8) satisfying

$$h(\beta) = r[v(-\beta) - w(-\beta)]^{tr}, \quad \beta \in c_a$$

and

$$c = \gamma_v - \sum_{\beta \in c_a} h(-\beta)v(\beta) \, ,$$

where $r \in \mathcal{R}$ is such that

$$r \sum_{\beta \in c_a} \|v(\beta) - w(\beta)\|_2^{\,2} = \gamma_v - \gamma_w$$

("tr" stands for transpose and $\| \cdot \|_2$ denotes the Euclidean norm). This shows that the hypotheses of Theorem 1 are met.

It is not difficult to show the existence of additional fundamental sets for the case addressed above, but we will not consider this here. Also, it is not difficult to see that we could have proceeded similarly with, for example, S the set of maps from K to any nonempty *bounded* set E', and U the set of maps from K to any closed subset E'' of $\mathcal{R}^p$ with $\theta \in E''$ and $E' \subseteq E''$. Of course in this case the S_a might not be compact in the U_a.

Finally, for the case in which $n = 1$ and G is causal in the usual sense, a natural choice of c_a is $[-a, 0] \cap K$. This choice leads to similar results, but with the elements of the analogous constructed fundamental set causal.

4.1.1 The $\mathcal{Z}_+$ Case

Now consider Theorem 2 and the case in which $V = \mathcal{R}^p$ and $K = \mathcal{Z}_+$. Assume that $S = U$ is the set of E-valued functions described earlier in this section, but with $K = \mathcal{Z}_+$. Let c_a and S_a be as indicated at the beginning of Section III, and as before regard the U_a as metric spaces with the metrics defined by (7). The conditions on S of Section 3.1 are clearly met. Assume that G satisfies the conditions of Section 3.2.

Here an associated fundamental set can be taken to be the family of restrictions to U of all maps H described by

$$(Hx)(t) = c + \sum_{\ell=0}^{t} h(t - \ell)x(\ell), \quad t \in \mathcal{Z}_+$$

with $c \in \mathcal{R}$ and with h real $1 \times p$ matrix valued and of bounded support. To see this, suppose a and v and w in $cl(S_a)$ are given. With γ_v and γ_w the corresponding values of $G_a(v)$ and $G_a(w)$, and with h and c satisfying

$$h(\ell) = r[v(a - \ell) - w(a - \ell)]^{tr}, \quad \ell \in c_a$$

and

$$c = \gamma_v - \sum_{\ell=0}^{a} h(a - \ell)v(\ell) \,,$$

where r is such that

$$r \sum_{\ell=0}^{a} \|v(\ell) - w(\ell)\|_2^2 = \gamma_v - \gamma_w \,,$$

one finds that $H_a(v) = \gamma_v$ and $H_a(w) = \gamma_w$. Therefore, the conclusion of Theorem 2 holds under the conditions described above, and we have uniform approximation on all of K.

The specific approximation result just established can be obtained also from Theorem 3 by considering the case in which $S_1 = S$ and B is the identity map on S.

4.2 Continuous-Time Systems

4.2.1 <u>Preliminaries</u>

Let ξ and ϕ, respectively, be any positive number and any function from the positive numbers into itself.

In this section, $\| \cdot \|$ stands for a norm on $\mathcal{R}^p$ or $\mathcal{R}^n$, as appropriate. We use $L(K)$ and $L_a(K)$ to denote the set of Lebesgue-measurable $\mathcal{R}^p$-valued functions ℓ on K and c_a, respectively, such that $\sup_\alpha \|\ell(\alpha)\| \leq \xi$. We view $L_a(K)$ as a metric space with metric ρ_a given by $\rho_a(u, v) = \sup_\alpha \|u(\alpha) - v(\alpha)\|$. Here c_a is the hypercube in the example of Section II if $K = \mathcal{R}^n$ or the interval of Section III if $K = \mathcal{R}_+$.

By $E(K)$ and $E_a(K)$ we mean any nonempty set of functions y from K and c_a, respectively, to $\mathcal{R}^p$ such that

(i) $\sup_\alpha \|y(\alpha)\| \leq \xi$, and

(ii)   for each $\varepsilon > 0$, one has $\|y(\beta) - y(\alpha)\| < \varepsilon$ whenever $\|\beta - \alpha\| < \phi(\varepsilon)$. (Of course this condition is met for $\phi(\varepsilon) = \kappa\varepsilon$ if $\|y(\beta) - y(\alpha)\| \leq \kappa\|\beta - \alpha\|$.)

By the Ascoli-Arzela Theorem [8, p. 85]*, $E_a(K)$ is a relatively compact subset of $L_a(K)$.

We will use the following two propositions:

---

*Using the fact that there is a positive constant ζ such that $\zeta\Sigma|u_j| \leq \|u\|$ for $u \in \mathcal{R}^p$, one sees that given any sequence $y_1, y_2, \ldots$ in $E_a(K)$ and any $j = 1, \ldots, p$ the set $\{y_{nj}\}$ of jth components is uniformly bounded and equicontinuous.

Proposition 1. Let c_a be as in the hypercube example of Section I, let v and w be continuous maps from c_a to $\mathcal{R}^p$, and let γ_v and γ_w be real numbers. Then there exists a $c \in \mathcal{R}$ and a continuous map q from c_a to the real $1 \times p$ matrices such that

$$c + \int_{\beta \, \in \, c_a} q(-\beta)v(\beta)d\beta \;=\; \gamma_v$$

$$c + \int_{\beta \, \in \, c_a} q(-\beta)w(\beta)d\beta \;=\; \gamma_w \; .$$

Proposition 2. Let $c_a = [0, a]$ for some $a \in (0, \infty)$, let v and w be continuous maps from c_a to $\mathcal{R}^p$, and let γ_v and γ_w be real numbers. Then there exists $c \in \mathcal{R}$ and a continuous q from c_a to the real $1 \times p$ matrices such that

$$c + \int_{\beta \, \in \, c_a} q(a - \beta)v(\beta)d\beta \;=\; \gamma_v$$

$$c + \int_{\beta \, \in \, c_a} q(a - \beta)w(\beta)d\beta \;=\; \gamma_w \; .$$

Proofs of the propositions can be obtained by modifying in an obvious way the related remarks in Section 4.1 concerning discrete-time cases.

In our discussion below of applications of Theorems 1–3, assume that $V = V_1 = \mathcal{R}^p$.

4.2.2 $K = \mathcal{R}^n$

Referring to Theorem 1, let $K = \mathcal{R}^n$, $A = (0, \infty)$, $c_a =$ the hypercube mentioned above, $S = E(\mathcal{R}^n)$, and $U = L(\mathcal{R}^n)$.

Let $\{F_\lambda : \lambda \in \Lambda\}$ be the set of all $F_\lambda : U \to R$ that have the representation

$$F_\lambda(x)(\alpha) = c + \int_{\beta \, \in \, \mathcal{R}^n} h(\alpha - \beta)x(\beta)d\beta$$

for some $c \in \mathcal{R}$ and some map h from $\mathcal{R}^n$ into the real $1 \times p$ matrices such that h is continuous on c_a for some a and zero otherwise. Then, using Proposition 1, it is easy to see that the hypotheses of the theorem are met when U_a is identified with $L_a(\mathcal{R}^n)$,[†] and of course when G meets the conditions of Section 2.4. Notice that these conditions are very reasonable. (The continuity condition on $G(\mathcal{I}_a \cdot)(\theta)$ is obviously satisfied if $G(\mathcal{I}_a \cdot)(\theta)$ is continuous on U_a.)

The remarks just before Section 4.1.1 concerning $n = 1$ and G causal apply here too.

4.2.3 $K = \mathcal{R}_+$

It is clear that a similar conclusion is obtained from Theorem 2 with $K = \mathcal{R}_+$, $S = E(\mathcal{R}_+)$, $U = L(\mathcal{R}_+)$, and the F_λ represented by

$$(F_\lambda x)(t) = c + \int_0^t h(t - \beta)x(\beta)d\beta \tag{9}$$

[†]In particular, notice that then each $c\ell(S_a)$ is a set of *continuous* functions.

in which h is continuous on $[0, a]$ for some $a > 0$ and zero otherwise (see Proposition 2). However, in this case we obtain an approximation for $(Gs)(t)$ that is uniform in t only for t sufficiently large. (Notice that the closure conditions on S of the theorem are not met here.)

Now consider Theorem 3. Let ξ_0 be a positive constant. Take $S_1 = L(\mathcal{R}_+)$, and suppose that $B(S)$ is a set that meets the conditions on $E(\mathcal{R}_+)$ in which S is the set of Lebesgue-measurable maps s from $\mathcal{R}_+$ to $\mathcal{R}^p$ such that $\|s(t)\| \leq \xi_0, t \geq 0$. Then the requirements on S_1 and S of Section 3.5.1 are met, and assuming that the conditions on A, B, and G are also satisfied, we see that the hypotheses of the theorem are met with $\{F_\lambda : \lambda \in \Lambda\}$ the set of all $F_\lambda : S_1 \to R$ that have a representation (9) (see Proposition 2).

This shows that the conclusion of Theorem 3 holds for a large class of causal time-invariant maps $G : S \to R$ which admit an A, B factorization with B a convolution.[‡]

4.2.4 <u>Further Comments on Theorem 2</u>

We now return to our discussion of Theorem 2 to illustrate that in a different setting a uniform approximation on *all* of $[0, \infty)$ can be obtained.

For the sake of simplicity take $p = 1$ and $\|\cdot\| = |\cdot|$. Let U denote the set of essentially bounded functions from $\mathcal{R}_+$ to $\mathcal{R}$ such that $\operatorname{ess\,sup}_t |u(t)| \leq \xi$. Assume that $(Gu)(t) = (Gw)(t), t \geq 0$ whenever u and w agree almost everywhere with respect to Lebesgue measure (so that G may be regaded as defined on a set of equivalence classes of elements of U).

Choose $\eta \geq 1$ and view each U_a as a metric space with

$$\rho_a(u, v) = \left(\int_0^a |u(\tau) - v(\tau)|^\eta d\tau \right)^{\frac{1}{\eta}} . \tag{10}$$

Let

$$S = E(\mathcal{R}_+) \cup_\beta T_\beta[E(\mathcal{R}_+)] \cup_{\alpha,\beta} T_\beta P_\alpha[E(\mathcal{R}_+)] . \tag{11}$$

Notice that S is closed under the T^β, T_β, and P_β.

It is proved in Appendix D that S_a is a relatively compact subset of U_a for each a. Therefore, subject to the conditions of Section 3.5.1 on G, and by a slight modification of Proposition 2, the hypotheses of Theorem 2 are met with F_λ of the form (8) with h continuous on some bounded subinterval of $[0, \infty)$ and zero otherwise.

In this case the required continuity of the G_a is with respect to the metric given in (10). Results along these lines, in which the elements of S need not be continuous, can be obtained also in the context of Theorem 1.

ACKNOWLEDGMENT

This research was supported in part by the National Science Foundation under Grant MIP-8915335.

[‡]See [1].

REFERENCES

1. I. W. Sandberg, "Nonlinear Input-Output Maps and Approximate Representations," *AT&T Technical Journal*, Vol. 64, No. 8, October 1985, pp. 1967–1983.

2. S. Boyd, "Volterra Series: Engineering Fundamentals," dissertation, University of California, Berkeley, 1985.

3. I. W. Sandberg, "g-Representations and Differential Equations," *Multidimensional Systems and Signal Processing*, in press.

4. D. Ball and I. W. Sandberg, "g- and h-Representations for Nonlinear Maps," *Journal of Mathematical Analysis and Applications*, in press.

5. M. H. Stone, "A Generalized Weierstrauss Approximation Theorem," in *Studies in Modern Analysis, Vol. 1*, R. C. Buck, ed., Englewood Cliffs: Prentice-Hall, 1962.

6. G. Cybenko, "Approximation by Superpositions of a Sigmoidal Function," *Mathematics of Control, Signals, and Systems*, Vol. 2, 1989, pp. 304–314.

7. K. Hornik, M. Stinchcombe, and H. White, "Multilayer Feedforward Networks are Universal Approximators," *Neural Networks*, Vol. 2, 1989, pp. 359–366.

8. K. Yosida, *Functional Analysis*, New York: Academic Press, 1965.

9. L. A. Liusternik and V. J. Sobolev, *Elements of Functional Analysis*, New York: Frederick Ungar, 1961.

APPENDIX A

Proof of Theorem 2

Given $\varepsilon > 0$, select $a \in K^+$ such that

$$|(Gs)(t) - (GW_{t,a}s)(t)| < \tfrac{1}{2}\varepsilon, \quad t \in K \tag{12}$$

for $s \in S$.

By Lemma 1 (in Section 2.7 and which holds with $c\ell(S_a)$ as defined in Section III) with f the restriction of G_a to $c\ell(S_a)$ and $\mathcal{X}_0 = \{F_{\lambda a}|_{c\ell(S_a)} : \lambda \in \Lambda\}$, there are a positive integer k, elements $F_{\lambda_1}, \ldots, F_{\lambda_k}$ of $\{F_\lambda : \lambda \in \Lambda\}$, and a lattice map $L : \mathcal{R}^k \to \mathcal{R}$ such that

$$|G_a(v) - L(F^{(a)}v)| < \tfrac{1}{2}\varepsilon \tag{13}$$

for $v \in S_a$, where $F^{(a)} : U_a \to \mathcal{R}^k$ is given by

$$(F^{(a)})_j = F_{\lambda_j a}, \quad j = 1, \ldots, k .$$

Choose any $t \in K$ and $s \in S$. Suppose first that $t \geq a$. Let $v \in S_a$ be given by

$$v(\alpha) = (T^{(t-a)}W_{t,a}s)(\alpha), \quad \alpha \in c_a . \tag{14}$$

Using the observation that $W_{t,a}s = T_{(t-a)}T^{(t-a)}W_{t,a}s$ and the time invariance and causality of G, one has

$$(GW_{t,a}s)(t) = (GT_{(t-a)}T^{(t-a)}W_{t,a}s)(t) = (GT^{(t-a)}W_{t,a}s)(a) = G_a(v) \ .$$

Similarly,

$$(FW_{t,a}s)(t) = F^{(a)}(v)$$

where F is as described in the theorem. By (13),

$$|(GW_{t,a}s)(t) - L(FW_{t,a}s)(t)| < \tfrac{1}{2}\,\varepsilon \ ,$$

which together with (12) gives

$$|(Gs)(t) - L[(FW_{t,a}s)(t)]| < \varepsilon \ .$$

The causality and time invariance of Q follow from Lemma 2, below, whose proof is essentially the same as the proof of the corresponding lemma in [1].[§]

Lemma 2. If $H : U \to R$ is causal and time invariant and $J : U \to R$ is defined by $(Ju)(t) = (HW_{t,a}u)(t)$ for $t \in K$ and $u \in U$, then J is causal and time invariant.

Continuing, given $s \in S$ and $t \in K_a$, we have

$$(FW_{t,a}s)(t) = F^{(a)}(v)$$

where v is the element of S_a described in (14). Thus, by the compactness of $c\ell(S_a)$ and the continuity on $c\ell(S_a)$ of the components of $F^{(a)}$ it is clear that

$$\sup\{\|(FW_{t,a}s)(t)\| : s \in S, \ \ t \in K_a\} < \infty \ . \tag{15}$$

Now suppose that S is closed under the P_β and T_β. For $s \in S$ and $t < a$, and using the causality and time invariance of G once more,

$$(GW_{t,a}s)(t) = (Gs)(t) = (GP_ts)(t) = (GT_{(a-t)}P_ts)(a) = G_a(w) \ ,$$

where $w \in S_a$ is defined by $w(\alpha) = (T_{(a-t)}P_ts)(\alpha)$, $\alpha \in c_a$. In a similar manner,

$$(FW_{t,a}s)(t) = F^{(a)}(w) \ ,$$

showing by (13) that in this case

$$|(Gs)(t) - L[(FW_{a,t}s)(t)]| < \tfrac{1}{2}\,\varepsilon \ ,$$

and that (by the argument given above)(15) holds with K_a replaced with K. This completes the proof.

[§]In the proof of time invariance, one uses the fact that $W_{t,a}T_\beta = T_\beta W_{(t-\beta),a}$ on U.

APPENDIX B

Comments on the Proof of Theorem 3

The argument is similar to the proof of Theorem 2:

Given $\varepsilon > 0$ choose a so that (12) is met.

By Lemma 1 with $f = A_a$ restricted to $c\ell(S_a)$ and $\mathcal{X}_0 = \{F_{\lambda a}|_{c\ell(S_a)} : \lambda \in \Lambda\}$, there are $F_{\lambda_1}, \ldots, F_{\lambda_k}$ and an L such that

$$|A_a(v) - L(F^{(a)}v)| < \tfrac{1}{2}\varepsilon, \quad v \in S_a$$

where $(F^{(a)})_j = F_{\lambda_j a}$ for each j.

Let $u \in S$. Then v defined by $v(t) = (Bu)(t)$, $t \in c_a$ belongs to S_a and one has $A_a(v) = (ABu)(a)$ and $F^{(a)}v = (FBu)(a)$. Thus,

$$|(Gu)(a) - L[(FBu)(a)]| < \tfrac{1}{2}\varepsilon, \quad u \in S .$$

Let $t \in K$ and $s \in S$ be given. Suppose first that $t \geq a$. Let $u \in S$ be given by

$$u(\alpha) = (T^{(t-a)}W_{t,a}s)(\alpha), \quad \alpha \in K .$$

By the time invariance of G,

$$(GW_{t,a}s)(t) = G(T_{t-a}T^{(t-a)}W_{t,a}s)(t) = (Gu)(a) ,$$

and similarly $(FBW_{t,a}s)(t) = (FBu)(a)$, showing that

$$|(GW_{t,a}s)(t) - L[(FBW_{t,a}s)(t)]| < \tfrac{1}{2}\varepsilon . \tag{16}$$

Now assume that $t < a$. Using the time invariance and causality of G,

$$(GW_{t,a}s)(t) = (Gs)(t) = (GP_t s)(t) = G(T_{(a-t)}P_t s)(a) .$$

Similarly, $(FBW_{t,a}s)(t) = FB(T_{(a-t)}P_t s)(a)$. Since $T_{(a-t)}P_t s \in S$, we see that (16) holds also for $t < a$.

The Remainder of the proof is essentially the same as the proof of Theorem 2 and is therefore omitted.

APPENDIX C

The Compactness of the S_a of Section 4.1

Let $\{w_n\}$ be an infinite sequence in S_a of Section 4.1 for some a (with $K = \mathcal{R}^n$ or $\mathcal{R}_+$). By the compactness of E, it easily follows that there is a subsequence $\{w_{\phi_n}\}$ of $\{w_n\}$ and an element w of S_a such that $w_{\phi_n}(\alpha) \to w(\alpha)$ in $(\mathcal{R}^p, \|\cdot\|)$ for each $\alpha \in c_a$. Since c_a is a *finite* set, $\|w_{\phi_n}(\alpha) - w(\alpha)\| \to 0$ as $n \to \infty$ uniformly in α. This shows that S_a is sequentially compact, and thus compact.

APPENDIX D

Relative Compactness of the S_a of Section 4.2.4

Let a be given. Since $S_a \subset U_a$ and U_a is a closed subset of $L_\eta(0, a)$ with the metric (10), it suffices to show that S_a is relatively compact in $L_\eta(0, a)$.

Clearly,

$$\int_0^a |e(t)|^\eta dt \;\le\; \xi^\eta \,, \quad e \in S_a \,. \tag{17}$$

Given $\varepsilon > 0$, let

$$\sigma = \max\{\varepsilon^\eta/6\xi^\eta \,,\; \phi[\varepsilon/(3a)^{1/\eta}]\} \,.$$

Then for $|\delta| < \sigma$ and any $e \in S_a$, one has

$$\int_0^a |e(t) - e(t+\delta)|^\eta dt \;\le\; a \cdot \varepsilon^\eta/3a + 2\xi^\eta \cdot \varepsilon^\eta/6\xi^\eta = \tfrac{2}{3}\,\varepsilon^\eta \,, \tag{18}$$

where $e(t+\delta) = 0$ for $(t+\delta) \notin [0, a]$. By (17) and (18) and a criterion [9, p. 44] due to Riesz, S_a is relatively compact in $L_\eta(0, a)$.

Integral Methods for Solving Fokker-Planck-Type Equations

V. PROTOPOPESCU
Engineering Physics and Mathematics Division
Oak Ridge National Laboratory
Oak Ridge, Tennessee 37831-6363, USA

ABSTRACT

Fokker-Planck-type equations occur quite often in different domains of physics and applied mathematics as various realizations of a generic degenerate parabolic equation. Even in the simplest situations, the analysis of the general Fokker-Planck equation is difficult and has been mostly confined to the linear case, where partial results have been obtained in showing existence, uniqueness, regularity, and completeness of eigenfunctions. In the present paper, we present a canonical integral approach that solves, in principle, the most general linear or nonlinear Fokker-Planck-type equations. The method is formal in the sense that it does not provide *per se* the means to prove existence and uniqueness of the solution in an abstract setting. The formalism is based on the Green's functions and their natural extensions to nonlinear systems and allows one to compute the solution (assumed to exist uniquely), by using a canonical iterative scheme. We present several applications of the integral approach in connection with previously developed methods and results.

INTRODUCTION

This paper is devoted to the presentation of an integral approach for solving Fokker-Planck-type equations.[1] The general Fokker-Planck equation is a degenerate parabolic equation of the form

$$u_t + \sum_i (b_i u)_{x_i} - \sum_{i,j} (a_{ij} u_{v_j})_{v_i} + \sum_i (a_i u)_{v_i} - au - f = 0 \qquad (1.1)$$

where the probability distribution function $u \in \mathbf{R}_+$ depends on position $x \in \Omega_1 \subseteq \mathbf{R}^m$, velocity $v \in \Omega_2 \subseteq \mathbf{R}^n$, $m \le n$, (a_{ij}) is a strictly positive definite matrix, and time $t > 0$. To specify the solutions of Eq. (1.1), additional restrictions have to be imposed in the form of: (i) initial condition at $t = 0$, and (ii) boundary conditions at the boundaries $\partial\Omega_1$ and $\partial\Omega_2$ of the domains Ω_1 and Ω_2, respectively.

Initial-boundary value problems involving Fokker-Planck-type equations occur frequently in reaction-diffusion systems, ecology and biology, electron scattering, kinetic theory, aerosol dynamics, and various stochastic phenomena. The

111

corresponding problems may be linear or nonlinear, stationary or time-dependent, scalar or vector (multi-component). In the linear case, the functions b_i, a_{ij}, a_i, a, and f are independent of u. In the stationary case, none of the functions appearing in the equation depend on time. In the vector version, the distribution u has more than one component, $u \in (\mathbf{R}_+)^k$.

Several examples are listed below for $n = 1, m = 1, k = 1$.

1. Burgers' equation:

$$u_t + uu_x - u_{xx} = 0 ; \tag{1.2}$$

2. Homogeneous Fokker-Planck equation:

$$u_t + vu_v - u_{vv} = 0 ; \tag{1.3}$$

3. Nonhomogeneous time-dependent Fokker-Planck equations:

$$u_t + vu_x + vu_v - u_{vv} = 0 , \tag{1.4}$$

$$u_t + vu_x - u_{vv} = 0 ; \tag{1.5}$$

4. Nonhomogeneous stationary Fokker-Planck equations:

$$vu_x + vu_v - u_{vv} - au = 0 , \tag{1.6}$$

$$vu_x - u_{vv} = 0 . \tag{1.7}$$

Even in the linear case, the rigorous analysis of the Fokker-Planck equation is difficult and far from being complete. Partial results have been obtained in showing existence, uniqueness, regularity, and completeness of eigenfunctions, by using a host of standard and/or problem-tailored methods.[2-18]

In this paper, we present some aspects of the integral approach used in solving the Fokker-Planck equation. The first two sections are devoted to a canonical integral formalism that can be applied, in principle, to all the cases embodied in Eq. (1.1). The method is formal in the sense that it does not provide *per se* the means to prove existence and uniqueness of the solution in an abstract setting. These issues have to be handled separately, in a problem-determined context. The final section of the paper exemplifies some typical situations. Each example uses a different variant of the integral approach thus showing the versatility and potential of the method.

GENERAL INTEGRAL FORMALISM

The linear or nonlinear system can be represented in abstract form as

$$N(u) + \delta B(u) = f + \delta g \tag{2.1}$$

where $N(u)$ describes the linear or nonlinear equation, $B(u)$ represents linear or nonlinear boundary conditions, f is the volume source, and g represents the boundary sources (including initial conditions). The source terms f and g include all inhomogeneities, so we can consider, without loss of generality, that $N(0) = 0$ and $B(0) = 0$. The δ distributions multiplying the boundary terms in (2.1) allow a formally unified abstract treatment of both evolution equations and initial/boundary conditions. These δ distributions are associated with the direct boundary space of the problem,[19] being uniquely specified for each well-posed specific problem under consideration.

The system (2.1) represents an equation for the function u belonging to a linear functional space $\mathcal{L}(\Omega)$ endowed with an inner product denoted by $<,>$. For simplicity, we shall assume $\mathcal{L}(\Omega) = \mathcal{L}^*(\Omega)$; throughout this work, Ω denotes the set (including the time domain for time-dependent problems) that defines the phase space for Eq. (2.1). Since we include the boundaries in the formal treatment, Ω is a closed set, containing the boundaries of the phase space underlying the problem.

We assume that the first Gâteaux derivatives of the operators appearing on the left-hand side of Eq. (2.1) exist, and they are defined[20] by

$$[N'(u) + \delta B'(u)]h \overset{\text{def}}{=} \{(d/d\epsilon)[N(u + \epsilon h) + \delta B(u + \epsilon h)]\}_{\epsilon=0} . \tag{2.2}$$

In the linear case, the Gâteaux derivative is independent of u; in general, (i.e., in the nonlinear case), these operators depend nonlinearly on u but act linearly on the vector h. Thus, the operator adjoint to $N'(u) + \delta B'(u)$ is defined via the usual linear duality:

$$< [N'(u) + \delta B'(u)]h, v >=< h, [N'^*(u) + \delta^* B'^*(u)]v > , \tag{2.3}$$

where v is an arbitrary function from the dual space $\mathcal{L}^*(\Omega) = \mathcal{L}(\Omega)$. In Eq. (2.3), $N'^*(u)$ is the formal adjoint of $N'(u)$, and $B'^*(u)$ includes all surface terms required in defining the actual adjoint. Note in (2.3) that the operator δ^* is not the same δ distribution as in (2.1) or (2.2), but is a distribution associated with the adjoint boundary space of the problem.[19] To highlight this distinction, we use the symbolical notation δ^*. The exact significance and the differences between the δ and δ^* symbols will become clearer in the sections devoted to applications.

Following the theory developed in Refs. 21 and 22, we define the operators

$$L(u)h \overset{\text{def}}{=} \int_0^1 N'(\epsilon u)h \, d\epsilon \qquad \beta(u)h \overset{\text{def}}{=} \int_0^1 B'(\epsilon u)h \, d\epsilon \tag{2.4}$$

and

$$L^*(u)v \stackrel{\text{def}}{=} \int_0^1 [N'(\epsilon u)]^* v \, d\epsilon \qquad \beta^*(u)v \stackrel{\text{def}}{=} \int_0^1 [B'(\epsilon u)]^* v \, d\epsilon \,, \qquad (2.5)$$

that still act linearly on h and v, respectively, while retaining a nonlinear parametric dependence on u. Like the variational operators, the integrated operators $L(u)$, $\beta(u)$, $L^*(u)$, $\beta^*(u)$, are still related by the linear duality relation

$$L^*(u) + \delta^* \beta^*(u) = [L(u) + \delta\beta(u)]^* \,. \qquad (2.6)$$

We note also the important relationship satisfied by $L(u)$ and $\beta(u)$:

$$[L(u) + \delta\beta(u)]u = N(u) + \delta B(u) \,. \qquad (2.7)$$

In the linear case, N and L coincide and do not depend on u. Equation (2.7) underscores the important role played by the integrated operators L, β: in contradistinction to the variational operators N', β', it is the pair of integrated operators $\{L(u), \beta(u)\}$ that replicates exactly the original nonlinear system (2.1) when applied to u. We define then the backward (retarded) and forward (advanced) propagators, G_u and G_u^*, as the inverses of the operators $L(u) + \delta\beta(u)$ and $L^*(u) + \delta^*\beta^*(u)$, respectively:

$$[L(u) + \delta\beta(u)]G_u = 1 \qquad (2.8)$$

and

$$[L^*(u) + \delta^*\beta^*(u)]G_u^* = 1 \qquad (2.9)$$

where 1 denotes the unit operator. In the linear case, when L and β do not depend on u, the propagators are the ordinary Green's functions of the problem. Equations (2.8) and (2.9) can be written in terms of formal integral kernels as

$$[L(u(x)) + \delta\beta(u(x))]G(u(x); x, x') = \delta(x - x') \qquad (2.10)$$

and

$$[L^*(u(x)) + \delta^*\beta^*(u(x))]G^*(u(x); x, x'') = \delta(x - x'') \qquad (2.11)$$

where x is a shorthand notation for the generic variable in the phase space domain Ω (including its boundaries).

Since the operators $L(u) + \delta\beta(u)$ and $L^*(u) + \delta^*\beta^*(u)$ act linearly on the respective propagators, the relationships between the propagators and the expression for the solution u in terms of these propagators can be derived, as previously noted, in the same spirit as for the usual Green's functions formalism in linear theory.[23,24] Thus, forming the inner products of (2.10) and (2.11) with $G^*(u(x); x, x'')$ and $G(u(x); x, x')$, respectively, leads to the reciprocity relation[21,22]

$$G^*(u(x); x, x') = G(u(x'); x', x) , \qquad (2.12)$$

which, in the linear case, reduces to the ordinary reciprocity relation between Green's functions.

The solution u of the original nonlinear system (2.1) is obtained in terms of the forward propagator G_u^* as follows by using Eqs. (2.5) and (2.7):

$$
\begin{aligned}
u &= <u, \delta> - <N(u) + \delta B(u), G_u^*> + <f + \delta g, G_u^*> \\
&= <u, [L^*(u) + \delta^* \beta^*(u)] G_u^*> - <[L(u) + \delta \beta(u)]u, G_u^*> \\
&\quad + <f + \delta g, g_u^*> = <f + \delta g, G_u^*> = <G_u, f + \delta g> .
\end{aligned}
\qquad (2.13)
$$

In terms of integral kernels (2.13) reads

$$
\begin{aligned}
u(x) &= \int_\Omega G(u(x); x, x')[f(x') + \delta g(x')] dx' \\
&= \int_\Omega G^*(u(x'); x', x)[f(x') + \delta g(x')] dx' ,
\end{aligned}
\qquad (2.14)
$$

that, once the propagators are expressed in terms of u, can be viewed as a nonlinear integral equation for u. In the linear case, G and G^* do not depend on u, and (2.13) and (2.14) simply express the solution as the propagation of the sources via the ordinary Green's functions.

INTEGRAL EQUATIONS FOR PROPAGATORS

By applying the formalism developed in the previous section, the original nonlinear problem has been reduced to a linear problem of finding G^* as the solution of Eq. (2.11), where u is assumed to be a known function. In the following, we shall sketch a general canonical method of carrying out such an inversion.

Consider (2.10) for a known vector u^0 and (2.11) for the actual solution $\tilde{u}$ of the original system (2.1), i.e.,

$$[L(u^0) + \delta\beta(u^0)] G_{u^0} = \delta(x - x') \qquad (3.1)$$

and

$$[L^*(\tilde{u}) + \delta^* \beta^*(\tilde{u})] G_{\tilde{u}}^* = \delta(x - x'') . \qquad (3.2)$$

Forming the inner product of (3.1) and (3.2) with $G_{\tilde{u}}^*$ and G_{u^0}, respectively, yields

$$<[L(u^0) + \delta\beta(u^0)] G_{u^0}, G_{\tilde{u}}^*> = G_{\tilde{u}}^* \qquad (3.3)$$

and

$$< G_{u^0}, [L^*(\tilde{u}) + \delta^* \beta^*(\tilde{u})] G_{\tilde{u}}^* > \, = G_{u^0} \, . \tag{3.4}$$

Using the duality relationship (2.6) and subtracting (3.4) from (3.3) leads to

$$G_{\tilde{u}}^* = G_{u^0} + \; < G_{u^0}, [L^*(u^0) + \delta^* \beta^*(u^0) - L^*(\tilde{u}) - \delta^* \beta^*(\tilde{u})] G_{\tilde{u}}^* > \, . \tag{3.5}$$

Equation (3.5) is a closed-form nonlinear integrodifferential equation satisfied by the forward propagator $G_{\tilde{u}}^*$. A similar equation is satisfied by the backward propagator $G_{\tilde{u}}$.

Note also that (3.5) is *exact* and its nonlinear character occurs, not from closure approximations, but reflects *exactly* the nonlinearities of the original system (2.1). Moreover, it gives a practical recipe for finding the propagator via an iteration scheme.

Because it retains the full nonlinear information contained in the original problem (2.1), Eq. (3.5) may be rather cumbersome. Moreover, it yields only the propagator, so the solution to the original system must subsequently be computed from the convolution expressions given by (2.13) or (2.14).

An alternative, and sometimes more efficient approach, is to obtain an integrodifferential equation, similar to (3.5), for the solution $\tilde{u}$ itself. For this purpose, we note that since the function u^0 is known, Eq. (3.1) can, in principle, be solved to obtain G_{u^0} and thus $G_{u^0}^*$ as the inverse of the *linear* operators $L(u^0) + \delta \beta(u^0)$ and $L^*(u^0) + \delta^* \beta^*(u^0)$, respectively. Actually, the whole idea of this method is to choose u^0 in such a way that the propagators G_{u^0} and $G_{u^0}^*$ are readily available. Then, the solution $\tilde{u}$ can be obtained by using (3.1), (3.2), and (2.1), the linearity of the operators L^* and β^*, and performing the following sequence of operations:

$$
\begin{aligned}
\tilde{u} = \; & < \tilde{u}, [L^*(u^0) + \delta^* \beta^*(u^0)] G_{u^0}^* > \\
= \; & < \tilde{u}, [L^*(\tilde{u}) + \delta^* \beta^*(\tilde{u})] G_{u^0}^* > \\
& + \; < \tilde{u}, [L^*(u^0) + \delta^* \beta^*(u^0) - L^*(\tilde{u}) - \delta^* \beta^*(\tilde{u})] G_{u^0}^* > \\
= \; & < [L(\tilde{u}) + \delta \beta(\tilde{u})] \tilde{u}, G_{u^0}^* > \\
& + \; < \tilde{u}, [L^*(u^0) + \delta^* \beta^*(u^0) - L^*(\tilde{u}) - \delta^* \beta^*(\tilde{u})] G_{u^0}^* > \\
= \; & < f + \delta g, G_{u^0}^* > \\
& + \; < \tilde{u}, [L^*(u^0) + \delta^* \beta^*(u^0) - L^*(\tilde{u}) - \delta^* \beta^*(\tilde{u})] G_{u^0}^* > \, .
\end{aligned} \tag{3.6}
$$

In the linear case, L^* and β^* do not depend on the unknown solution $\tilde{u}$ and thus (3.6) reduces to (2.14) with G^* independent of u, and solving (2.1) reduces to calculating the ordinary Green's function of the problem. In special situations, this can be done by using auxiliary problems[18] or eigenfunction expansions[16] (see the examples in the next section).

In the general situation, the linear case may turn out to be more complicated from an iterative viewpoint than the nonlinear case. This is due to the fact that G^* is the actual Green's function of the problem and may be rather difficult to obtain. In this case, one computes first G^* via a perturbative scheme along the following lines:

If $L^* + \delta^*\beta^*$ is not readily invertible, we start with a simpler operator $L_0^* + \delta^*\beta_0^*$ that *is* readily invertible, namely $[L_0^* + \delta^*\beta_0^*]^{-1} = G_0^*$. Then, we write $L^* + \delta^*\beta^*$ as $L_0^* + \delta^*\beta_0^* + L_1^* + \delta^*\beta_1^*$ and compute $G^*(u^0)$ via the usual perturbative series

$$G^* = G_0^* - G_0^*(1 + (L_1^* + \delta^*\beta_1^*)G_0^*)^{-1}(L_1^* + \delta^*\beta_1^*)G_0^* .$$

In the case of the linear Fokker-Planck equation, rigorous results for this scheme have been obtained by Weber,[4] Il'in,[5] and Eidel'man.[19]

ILLUSTRATIONS OF THE INTEGRAL APPROACH

A Nonlinear Equation: Applying the Propagator Method

In order to illustrate some basic aspects of the canonical formalism described in the previous section, we consider the equation

$$N(u) = u_{xx} - 2uu_x + 2u_x = 0 \qquad\qquad x \in \mathbf{R}_+ \qquad\qquad (4.1)$$

with the boundary conditions

$$u(0) = 0$$
$$u_x(0) = 1 . \qquad\qquad\qquad (4.2)$$

The solution is readily obtained in the form

$$u(x) = \frac{x}{x+1} . \qquad\qquad (4.3)$$

We want to obtain this solution by applying the integral formalism based on propagators. It is easy to see that

$$L(u)G = G_{xx} - uG_x - u_xG + 2G_x$$
$$L^*(u)G^* = G_{xx}^* + uG_x^* - 2G_x^* . \qquad\qquad (4.4)$$

By choosing $u^0 = 2$, we solve the equation

$$L^*(2)(G_{u^0}^*)_{xx} = (G_{u^0}^*)_{xx} = \delta(x - x') ; \qquad G_{u^0}^*(x, x') = 0 \qquad x > x'$$

and get

$$G_{u^0}^*(x, x') = \begin{cases} x' - x & x' \geq x \\ 0 & x' < x \end{cases} \qquad G_{u^0}^*(x', x) = \begin{cases} x - x' & x \geq x' \\ 0 & x < x' \end{cases}$$

$$(G_{u^0}^*)_{x'}(x', x) = -\theta(x - x')$$

where $\theta(x)$ is the Heaviside function, $\theta(x) = 1, x \geq 0, \theta(x) = 0, x < 0$. Thus, (3.6) becomes

$$
\begin{aligned}
u(x) = &\int_0^x \left(f(x') + \delta(x') \cdot 0 + \delta'(x') \cdot 1 \right) G^*_{u0}(x', x) dx' - \\
&- \int_0^x u(x')(u(x') - 2)(-\theta(x - x')) dx' = \\
&- \int_0^x \delta(x')(-\theta(x - x')) dx' + \int_0^x u(x')(u(x') - 2) dx' = \\
&1 + \int_0^x u(x')(u(x') - 2) dx' \ .
\end{aligned}
\tag{4.5}
$$

One verifies that this integral equation is solved by $\frac{x}{x+1}$.

Deriving the Green's Function from an Auxiliary Problem

As mentioned before when the auxiliary propagator G^*_{u0} cannot be computed, like in the previous section, one can apply a perturbative formalism for finding it. While having the advantage of being applicable to both nonlinear and linear problems, the perturbative scheme may turn out to be cumbersome, and alternative methods are desirable. One such method is described in the following example. Let us solve the linear one-dimensional boundary value problem

$$
Au \stackrel{\text{def}}{=} vu_x - u_{vv} = 0 \qquad v \in \mathbf{R}
\tag{4.6}
$$

with boundary conditions

$$
u(0, v) = g(v) \qquad v \in \mathbf{R}_+ \ .
\tag{4.7}
$$

Unlike the Green's function of the usual stationary Fokker-Planck equation, Eq. (1.6), the Green's function of the problem (4.6) cannot be found explicitly in terms of eigenfunctions by the method of Ref. 15. This is because the essential spectrum of the differential operator $u \to -u_{vv}$ is $\mathbf{R}_+$, and is not separated from zero, as required by the method in Ref. 15. Gorkov[10] was able to obtain an explicit solution of (4.6) by using a singular integral equation, but that approach is rather complicated.

An alternative method for deriving the Green's function for problem (4.6) has been proposed by Eidelman[25,26] and we apply it here to solve (4.6). Consider the forward Green's function $\mathcal{G}^*$ for the time-dependent problem

$$
\mathcal{G}^*_t + A^*\mathcal{G}^* = \mathcal{G}^*_t - v\mathcal{G}^*_x - \mathcal{G}^*_{xx} = 1 \ , \qquad x \in \mathbf{R}, \qquad v \in \mathbf{R} \ ,
\tag{4.8}
$$

whose solution with kernel

$$
\mathcal{G}^*(x, v, t, x', v', 0) =
$$

$$
\begin{cases}
\dfrac{\sqrt{3}}{2\pi}\dfrac{1}{t^2}\exp\left\{-\dfrac{(v-v')^2}{4t}-\dfrac{3}{t^3}\left(x-x'+\dfrac{v+v'}{2}t\right)^2\right\} & t>0 \\[3ex]
0 & t<0
\end{cases}
\tag{4.9}
$$

was found by Kolmogoroff.[2]

The crucial observation is that the forward Green's function G^* whose kernel is defined as

$$
G^*(x,v,x',v')\overset{\text{def}}{=}\int_0^\infty \mathcal{G}^*(x,v,t,x',v',0)dt=
$$

$$
\frac{\sqrt{3}}{2\pi}\int_0^\infty \frac{dt}{t^2}\exp\left\{-\frac{(x-x')}{4t}-\frac{3}{t^3}\left(x-x'+\frac{v+v'}{2}t\right)^2\right\}
\tag{4.10}
$$

satisfies the equation $A^*G^*=1$ and therefore solves the problem (4.6). Indeed, by applying formula (2.14) and letting $x\to 0_+$, one obtains an integral equation for the unknown part of the distribution $u(0,v)$, $v\in \mathbf{R}_-$:

$$
u(0,v)-\int_{-\infty}^0 v'G^*(0,v',0,v)u(0,v')dv'
$$

$$
=\int_0^\infty v'G^*(0,v',0,v)g(v')dv' \qquad v\in\mathbf{R}_- .
\tag{4.11}
$$

Once $u(0,v)$, $v\in\mathbf{R}_-$, the solution at $x>0$ is obtained from (2.14)

$$
u(x,v)=\int_{-\infty}^\infty v'G^*(0,v',x,v)u(0,v')dv' .
\tag{4.12}
$$

This method does not depend on the dimensions of the problem. Since for the three-dimensional time-dependent problem the Green function is also known,[3] this solves otherwise untreatable stationary three-dimensional problems. The application of this procedure to spherical geometry is reported in Ref. 18.

Integral Equations from Eigenfunction Expansions

In this section, we shall solve directly the integral equation for the solution itself, therefore avoiding the computation of the Green's function. We shall apply the formula (3.6) to the linear stationary one-dimensional problem (1.6) with $a\geq\epsilon>0$ to show how to take advantage of the eigenfunction expansions when they exist.[12,16]

Adapting the results obtained in Ref. 12 for $a=0$ to the present case ($a>0$), we write the solution of the kinetic half-space problem in the form

$$
u(0,v)=\sum_{n>0} c_n\varphi_n(v) ,
\tag{4.13}
$$

$$
u(0,v)=g(v) , \qquad\qquad v\in\mathbf{R}_+
\tag{4.14}
$$

where we take into account[12] that: (i) the spectrum is discrete, (ii) zero does not belong to it, and (iii) $\varphi_n(v)$ are normalized, $\int_{-\infty}^{\infty} \varphi_n \varphi_m w\, dv = \delta_{nm}$ (w is the full range weight). The coefficients c_n can be computed by using full range formulas,[12,14−16]

$$
\begin{aligned}
c_n &= \int_{-\infty}^{\infty} u(0,v)\varphi_n(v)w\, dv \\
&= \int_{0}^{\infty} g(v)\varphi_n(v)w\, dv - \int_{0}^{\infty} u(0,-v)\varphi_n(-v)w\, dv \ .
\end{aligned}
\tag{4.15}
$$

By using these coefficients in Eq. (4.13) and projecting the solution on positive and negative velocities, we obtain two (equivalent) integral equations for the unknown part of the distribution at the boundary, $u(0,v), v \in \mathbf{R}_-$, namely:

$$
\begin{aligned}
g(v) = \sum_{n>0} \varphi_n(v) &\left\{ \int_0^{\infty} g(v')\varphi_n(v')w\, dv' \right. \\
&\left. - \int_0^{\infty} u(0,-v')\varphi_n(-v')w\, dv' \right\} \quad v \in \mathbf{R}_+
\end{aligned}
\tag{4.16}
$$

$$
\begin{aligned}
u(0,v) = \sum_{n>0} \varphi_n(v) &\left\{ \int_0^{\infty} g(v')\varphi_n(v')w\, dv' \right. \\
&\left. - \int_0^{\infty} u(0,-v')\varphi_n(-v')w\, dv' \right\} \quad v \in \mathbf{R}_-
\end{aligned}
\tag{4.17}
$$

These equations are the ones obtained by Mayya[14] for the Fokker-Planck equation and by Klaus et al. in the abstract formulation of the Fokker-Planck-type equations.[15]

To solve these equations, one applies the standard techniques, i.e., multiply (4.16) by $\varphi_m(v)$, $m > 0$, and integrate over the full range to get

$$
\begin{aligned}
c_m &= \int_0^{\infty} g(v')\varphi_m(v')w\, dv' \\
&- \sum_{n>0} \int_0^{\infty} c_n\varphi_n(-v')\varphi_m(-v')w\, dv' \qquad m > 0 \ .
\end{aligned}
\tag{4.18}
$$

An equivalent system is obtained by multiplying (4.16) by $\varphi_m(-v)$, $m < 0$, to get

$$
\begin{aligned}
0 &= \int_0^{\infty} g(v')\varphi_m(v')w\, dv' \\
&- \sum_{n>0} \int_0^{\infty} c_n\varphi_n(-v')\varphi_m(v')w\, dv' \qquad m < 0 \ ,
\end{aligned}
\tag{4.19}
$$

from where one gets c_n via truncation. Once c_n are known, the distribution at the boundary is given by (4.13), $v \in \mathbf{R}$ and the full solution is obtained via a semigroup reconstruction.[12,16]

CONCLUSIONS

We have presented several aspects of an integral approach for solving Fokker-Planck type equations. The general formalism yields integral equations for the Green's functions (linear case) or propagators (nonlinear case) associated with the original problem. In general, these integral equations have to be solved iteratively, but compact solutions can be obtained sometimes by specific artifacts.

ACKNOWLEDGMENTS

This work was sponsored by the Department of Energy under contract #DE-AC05-84OR21400 with Martin Marietta Energy Systems, Inc.

REFERENCES

1. H. Risken, *The Fokker-Planck Equation, Methods of Solution and Applications*, Springer-Verlag, Berlin, 1984.

2. A. Kolmogoroff, "Aleatoric Movements," *Ann. Math.* **35**, 116-117 (1934) (in German).

3. S. Chandrasekhar, "Stochastic Problems in Physics and Astronomy," *Rev. Mod. Phys.* **15**, 1-89 (1943).

4. M. Weber, "The Fundamental Solution of a Degenerate Partial Differential Equation of Parabolic Type," *Trans. Amer. Math. Soc.* **71**, 24-37 (1951).

5. A. M. Il'in, "On a Class of Ultraparabolic Equations," *Dokl. Akad. Nauk. SSSR* **159**, 1214-1217 (1964) (in Russian).

6. J. J. Kohn, L. Nirenberg, "Degenerate Elliptic-Parabolic Equations of Second Order," *Commun. Pure Appl. Math.* **20**, 797-872 (1967).

7. C. D. Pagani, "Study of Some Questions Concerning the Generalized Fokker-Planck Equation," *Bull. Un. Mat. Ital.* **A3**, 961-986 (1970) (in Italian).

8. O. A. Oleĭnik, E. V. Radkevič, *Second Order Equations with Nonnegative Characteristic Form*, American Mathematical Society, Providence, RI, Plenum Press, New York, 1973.

9. C. D. Pagani, "On an Initial Boundary Value Problem for the Equation $w_t = w_{xx} - xw_y$," *Ann. Scuola Norm. Sup. Pisa* **22**, 219-263 (1975).

10. Iu. P. Gorkov, "Formula for the Solution of a Boundary-Value Problem for the Brownian Motion Steady-State Equation," *Dokl. Akad. Nauk. SSSR* **223**, 525-528 (1975) (in Russian).

11. R. Beals, "On the Abstract Treatment of Some Forward-Backward Problems of Transport and Scattering," *J. Funct. Anal.* **34**, 1-20 (1979).

12. R. Beals, V. Protopopescu, "Half Range Completeness for the Fokker-Planck Equation," *J. Stat. Phys.* **32**, 565-584 (1983).

13. D. C. Sahni, "An Exact Solution of Fokker-Planck Equation and Brownian Coagulation in the Transition Regime," *J. Colloid Interface Sci.* **91**, 418-429 (1983).

14. Y. S. Mayya, "Brownian Motion Near a Plane Absorber: Non-Analiticity of the Boundary Layer Solutions to the Fokker-Planck Equation," *J. Chem. Phys.* **82** 2033-2039 (1985).

15. M. Klaus, C. V. M. van der Mee, V. Protopopescu, "Half Range Solutions of Indefinite Sturm-Liouville Problems," *J. Funct. Anal.* **70**, 254-288 (1987).

16. W. Greenberg, C. van der Mee, V. Protopopescu, *Boundary Value Problems in Abstract Kinetic Theory*, Birkhaüser Verlag, Basel, 1987.

17. C. Cosner, S. M. Lenhart, V. Protopopescu, "Transport Equations with Second-Order Differential Collision Operators," *SIAM J. Math. Anal.* **19**, 797-813 (1988).

18. A. Boutet de Monvel, P. Diţă, "Brownian Motion Near an Absorbing Sphere," *J. Stat. Phys.*, to be published.

19. E. A. Coddington, N. Levinson, *Theory of Ordinary Differential Equations*, McGraw-Hill, New York, 1955.

20. M. Z. Nashed, "Differentiability and Related Properties of Nonlinear Operators. Some Aspects of the Role of Differentials in Nonlinear Functional Analysis," in *Nonlinear Functional Analysis and Applications*, L. B. Rall Ed., Academic Press, New York, 1970.

21. D. G. Cacuci, R. B. Perez, V. Protopopescu, "Duals and Propagators: A Canonical Formalism for Nonlinear Equations," *J. Math. Phys.* **29**, 353-361 (1988).

22. D. G. Cacuci, V. Protopopescu, "Propagators for Nonlinear Systems," *J. Phys. A* **22**, 2399-2414 (1989).

23. A. G. Butkovskij, *Green's Functions and Transfer Functions Handbook*, Ellis Harwood, Chichester, 1982.

24. G. F. Roach, *Green's Functions*, Cambridge University Press, Cambridge, 1982.

25. S. D. Eidel'man, *Parabolic Systems*, North-Holland, Amsterdam, 1969.

26. E. E. Levi, "On Completely Elliptic Linear Partial Differential Equations," *Rend. Circ. Matem. Palermo* **24**, 275-317 (1907) (in Italian).

Absolute Stability of Certain Delay Systems

C. CORDUNEANU
Department of Mathematics
University of Texas at Arlington
Arlington, Texas 76019, USA

ABSTRACT.

This paper deals with absolute stability problems for functional differential equations of Volterra type, such as $\dot{x}(t)=(Ax)(t)+b\varphi(\sigma(t))$, $\sigma=<c,x>$. The operator A is a general Volterra (causal) operator, linearly acting on a convenient function space. Criteria of absolute stability are obtained, under the main assumption of time-invariance of A.

The problem of absolute stability of control systems has been investigated during the last half century by many authors. Without any claim for completeness, we will mention here some of the books published on this subject, or containing a partial coverage of these matters: M. A. Aizerman and F. R. Gantmakher [1], C. Corduneanu [2], Ch. A. Desoer and M. Vidyasagar [5], A. H. Gelig et al. [6], S. Lefschetz [8], A. L. Lurie and V. N. Postnikov [9], R. K. Miller [10], K. S. Narendra and J. H. Taylor [11], V. M. Popov [13].

Initially, the Liapunov's method has been used, with emphasis on ordinary differential systems of the form

$$\dot{x}=Ax+b\varphi(\sigma), \quad \sigma=< c, x > , \tag{1}$$

or of a form related to (1).

Various generalizations have been discussed, including the case of delay systems such as

$$\dot{x}(t)=Ax(t)+Bx(t-\tau)+b\varphi(\sigma), \quad \sigma=< c, x > , \tag{2}$$

or the case of integro-differential systems

$$\dot{x}(t) = Ax(t) + \int_0^t B(t-s)x(s)ds + b\varphi(\sigma), \quad \sigma = <c, x> . \tag{3}$$

The absolute stability problem can be also formulated for integral equations of the form

$$\sigma(t) = f(t) + \int_0^t k(t-s)\varphi(\sigma(s))ds, \tag{4}$$

in which σ is scalar or vector-valued.

Roughly speaking, the absolute stability problem for (4) consists in providing conditions on the data, such that $\lim \sigma(t) = 0$ as $t \to \infty$ for the solution $\sigma(t)$ of (4), with φ belonging to a certain class of functions, usually defined by the inequalities

$$0 < \sigma\varphi(\sigma) \leq \lambda\sigma^2, \quad \sigma \in R, \quad \sigma \neq 0, \tag{5}$$

where the parameter λ is such that $0 < \lambda \leq \infty$.

It is known (see, for instance [2]) that the absolute stability problem for systems like (1), (2) or (3) can be reduced to the similar problem for the integral equation (4). In case of the system (1), under initial conditions

$$x(0) = x^0 \in R^n, \tag{6}$$

the variation of parameters formula leads immediately to the integral equation

$$\sigma(t) = <c, \ e^{At}x^0> + \int_0^t <c, \ e^{A(t-s)}b> \varphi(\sigma)s)ds, \tag{7}$$

which is obviously of the form (4).

This paper is concerned with systems of the form

$$\dot{x}(t) = (Ax)(t) + b\varphi(\sigma(t)), \quad \sigma = <c, x>, \tag{8}$$

in which A stands for an abstract Volterra operator subject to various conditions to be specified below.

A noteworthy particular case for A is provided by

$$(Ax)(t) = \sum_{j=0}^{\infty} A_j x(t-t_j) + \int_0^t B(t-s)x(s)ds, \tag{9}$$

under suitable conditions for the matrices A_j, $j \geq 0$, and $B(t)$, and assuming $t_j \geq 0$, $j \geq 0$, which means that A is a delay operator. It is obvious that (9) contains as special cases those encountered above: (1), (2), (3).

An interesting fact is the possibility of reducing the
general situation (8), to the case of an integral equation
of the form (4). Of course, this takes place under
convenient assumptions on the abstract Volterra operator A.
We also notice the fact that an initial functional
condition has to be attached in case we deal with operators
like those in (2) or (9):

$$x(t)=h(t).\tag{10}$$

Condition (10) must be verified on the whole negative real
axis in case of infinite delays (for instance, when the
sequence t_j is unbounded in (9)), or only on a finite
interval in case of equations (2).

Let us go back now to the system (8), under the initial
condition (6). If we assume, for instance, that

$$A:\ L^2_{loc}(R_+,\ R^n)\to L^2_{loc}(R_+,\ R^n)\tag{11}$$

is of Volterra type [4], and continuous, then there exists
a resolvent kernel $X(t,\ s)$, $0\leq s\leq t < \infty$, such that the
following representation formula (variation of parameters)
holds true:

$$x(t)=X(t,\ 0)x^0+\int_0^t X(t,s)b\varphi(\sigma(s))ds.\tag{12}$$

Substituting this $x(t)$ in the second equation in (8), we
obtain the integral equation in σ

$$\sigma(t)=<c,\ X(t,\ 0)x^0>+\int_0^t <c,\ X(t,s)b>\varphi(\sigma(s))ds,\tag{13}$$

which is similar to (4), but it is not of convolution type.
This last feature is desirable because only in such cases
can we take full advantage of the transform method.

Of course, it is possible to provide results on the
stability of equation (13), i.e., to formulate conditions
under which the solution of (13) tends to zero at infinity.
Such results are available in the literature and we send
the reader to our books [2], [4]. See also [5], [6], [7],
[8], [10], [11].

A more fruitful approach could be used if (13) is of
convolution type, i.e., $X(t,\ s)=k(t-s)$, for some
map $k:\ R_+\to\mathcal{L}(R^n,\ R^n)=$ the set of all n by n matrices
with real entries. If such a case does occur, the equation
(12) takes the form (4), and frequency domain criteria of
stability can be obtained.

The hypothesis on A which will assure that

$$X(t,s)=k(t-s),\ 0\leq s\leq t<\infty, \tag{14}$$

is that of "time-invariance". In other words, using engineering language, identical inputs must produce identical outputs, regardless of the time the system is working. Mathematically, this property can be expressed as $Ax_h=(Ax)_h$, where $x_h(t)=x(t+h)$, $h\in R$. It is appropriate to mention here that the space $L^2_{loc}(R_+,\ R^n)$ is regarded as the subspace of $L^2_{loc}(R,\ R^n)$ whose elements vanish a.e. for $t\leq 0$.

If we consider now the linear system

$$\dot{x}(t)=(Ax)(t)+f(t),\ \text{a.e.}\quad \text{on}\quad R_+, \tag{15}$$

under initial conditions

$$x(0)=\theta\in R^n, \tag{16}$$

then the variation of parameters formula gives

$$x(t)=\int_0^t X(t,\ s)f(s)ds. \tag{17}$$

It is now assumed that $f\in L^2_{loc}(R_+,\ R^n)$.

We want to show that the integral operator defined by (17) is time-invariant, if we assume the operator A in (15) to be time-invariant, i.e., such that $Ax_h=(Ax)_h$, for any $h\in R$.

In order to reach this conclusion, we notice that changing t in $t-h$, the equation (15) becomes with $y(t)=x(t-h)$:

$$\dot{y}(t)=(Ay)(t)+f(t-h), \tag{18}$$

taking into account the fact that A is time-invariant. Under intial condition $y(0)=\theta$, (18) produces the solution $y(t)=x(t-h)$, or $y(t+h)=x(t)$. If we now consider the formula (17), as well as the similar formula for $y(t)$, then the condition $y(t+h)=x(t)$ means the time-invariance of the integral operator (17). This implies that the kernel $X(t,s)$ must be a difference kernel, i.e., a relationship of the form (14) holds true. It follows that $X(t,s)\equiv X(t-s, 0)$ for $0\leq s\leq t<\infty$, and we agree to denote by $X(t)$ the kernel of the operator (17). Hence, (17) becomes

$$x(t)=\int_0^t X(t-s)f(s)ds. \tag{19}$$

Using (19), and noticing that $X(t)$ is nothing else but the matrix solution of $\dot{x}(t)=(Ax)(t)$ with $X(0)=I=$ the unit matrix, we can write (13) in the form

$$\sigma(t) = \,<c,\ X(t)x^0> + \int_0^t <c,\ X(t-s)b> \varphi(\sigma(s))ds. \qquad (20)$$

The integral equation (20) has the form (4), in which

$$f(t) = \,<c,\ X(t)x^0>, \qquad (21)$$

$$k(t) = \,<c,\ X(t)b> \qquad (22)$$

In order to deal with equation (20) and obtain frequency domain stability criteria, we have to make ourselves sure that the kernel k, given by (22), is integrable on R_+. Only in this case, or in related situations such as the square integrability of k, can we expect successfully using the transform method.

In dealing with special cases for the operator A, such as (9), we have shown [3] that under conditions

$$\sum_{j=0}^{\infty} ||A_j|| < +\infty, \qquad ||B(t)|| \in L^1(R_+,\ R), \qquad (23)$$

the operator A is continuous on the space $L^2(R_+,\ R^n)$, and the asymptotic stability of the zero solution of the linear system $\dot{x}(t)=(Ax)(t)$ is the consequence of the property

$$\|X(t)\| \in L^1(R_+,\ R) \qquad (24)$$

of the fundamental matrix. Property (24) is equivalent to a frequency-domain condition of stability, which will be indicated below.

Of course, condition (24) is a severe restriction on the fundamental matrix. But as seen in [3], in the special case (9) for the operator A, such a condition must be imposed if we want to secure the absolute stability of the system (8). Actually, in that special case for A, condition (24) is basically equivalent to the asymptotic stability of the zero solution of the homogeneous system associated to (15).

Now, we will return to the system (8), in which A stands for an abstract Volterra operator on $L^2_{loc}(R_+,\ R^n)$, b and c are constant vectors, while $\varphi(\sigma)$ is a scalar function satisfying an inequality of the form (5).

<u>Theorem 1</u>. Let us assume the following conditions hold

true in regard to the system (8):

1) The operator A is continuous on $L^2_{loc}(R_+, R^n)$, nonanticipative, and satisfies the time-invariance condition

$$(Ax)_h = Ax_h, \quad x \in L^2_{loc}(R_+, R^n); \tag{25}$$

2) The (fundamental) matrix $X(t)$ occuring in (19) verifies the condition (24), i.e., it is integrable on R_+, and $L^1(R_+, R^n)$ is an invariant subspace for A:

$$AL^1(R_+, R^n) \subset L^1(R_+, R^n); \tag{26}$$

3) The function $\varphi : R \to R$ is continuous, $\sigma\varphi(\sigma) > 0$ for $\sigma \neq 0$, and there exists $M > 0$ such that

$$|\varphi(\sigma)| \leq M[1 + \Phi(\sigma)], \quad \sigma \in R, \tag{27}$$

where
$$\phi(\sigma) = \int_0^\sigma \varphi(u)\,du \to \infty \quad \text{as} \quad |\sigma| \to \infty. \tag{28}$$

4) There exists $q \geq 0$, such that

$$Re\left\{(1 + i\omega q) < c, \tilde{X}(i\omega)b>\right\} \leq 0, \quad \omega \in R_+, \tag{29}$$

with
$$\tilde{X}(i\omega) = \int_0^\infty e^{-i\omega t} X(t)\,dt. \tag{30}$$

Then the system (8) is absolutely stable, i.e., it is globally asymptotically stable for each function φ satisfying the condition 3).

The proof of Theorem 1 is based on a result of J. A. Nohel and D. Shea [12], concerning the integral equation (4). As seen above, the system (8), under initial condition (6), can be reduced to an integral equation of the form (4), in which $f(t)$ and $k(t)$ are given by (21) and (22) respectively.

The conditions of Theorem 1 assure the properties $f \in L^1(R_+, R)$ and $k, k' \in L^1(R_+, R)$. The fact that $k' \in L^1(R_+, R)$ is a consequence of condition 2), requiring the invariance of the space $L^1(R_+, R^n)$ with respect to the operator A. We have to keep in mind that each column of $X(t)$ is a solution of the homogeneous system $\dot{x}(t) = (Ax)(t), \ t \in R_+$.

Of course, the result of J. A. Nohel and D. Shea [12]

for the equation (4), or (20), implies the global asymptotic stability of the zero solution of the system (8). Indeed, once it has been established that $\sigma(t) \to 0$ as $t \to \infty$, the variation of parameters formula (12), in which $X(t, s)$ has to be replaced by $X(t-s)$, leads to the conclusion $x(t) \to 0$ as $t \to \infty$, for any initial data $x^0 \in R^n$. First, let us notice that (24) implies also $|X(t)| \to 0$ as $t \to \infty$, because of condition (26). Second, the convolution term in (12) also tends to zero at infinity due to the fact that $|X(t)| \in L^1(R_+, R)$ and $\varphi(\sigma(t)) \to 0$ as $t \to \infty$.

This ends the proof of Theorem 1.

A similar result can be obtained for the system

$$\begin{cases} \dot{x}(t) = (Ax)(t) + b\varphi(\sigma(t)), \\ \dot{\xi}(t) = \varphi(\sigma(t)) \\ \sigma(t) = <c, \ x(t)> - \rho\xi(t), \quad \rho > 0. \end{cases} \tag{31}$$

Using again the variation of parameters formula for the first equation in (31), under assumption that A is a Volterra operator, continuous on $L^2_{loc}(R_+, R^n)$, one obtains

$$x(t) = X(t)x^0 + \int_0^t X(t-s)b\varphi(\sigma(s))ds. \tag{32}$$

Integrating both sides of the second equation in (31), from 0 to $t > 0$, and assuming $\xi(0) = \xi^0 \in R$, the last equation in (31) becomes after substitution of $x(t)$ and $\xi(t)$:

$$\sigma(t) = <c, \ X(t)x^0> - \rho\xi^0 \tag{33}$$

$$+ \int_0^t \{<c, \ X(t-s)b> - \rho\}\varphi(\sigma(s))ds.$$

The integral equation (33) for $\sigma(t)$ is again of the form (4). However, different assumptions are now natural in regard to the free term $f(t)$ and the kernel $k(t)$. We shall apply, as an illustration, a result given in our book [2] (Theorem 3.1).

<u>Theorem 2.</u> Assume that the following conditions hold true in regard to the system (33):

1) The same as in Theorem 1.

2) The same as in Theorem 1, with the extra assumption that $X(t)$ is twice differentiable on R_+, and

$$|X''(t)| \in L^1(R_+, R); \tag{34}$$

3) $\varphi : R \to R$ is continuous, and $\sigma\varphi(\sigma) > 0$ for $\sigma \neq 0$;

4) There exists $q \geq 0$, such that for $\omega \neq 0$

$$\text{Re}\ \left\{(1+i\omega q)\ \left[< c,\ \tilde{X}(i\omega)b> - \rho(i\omega)^{-1}\right]\right\} \leq 0,\qquad(35)$$

where $\rho > 0$.

Then the system (33) is absolutely stable.

The proof can be conducted on the same lines as in case of Theorem 1. The integral equation (33) is obviously of the form (4), in which

$$f(t) = <c,\ X(t)x^0> - \rho\xi^0,\qquad(36)$$

$$k(t) = < c,\ X(t)\ b > - \rho.\qquad(37)$$

The assumptions made in the statement of Theorem 2 imply those appearing in Theorem 3.1 of [2]. Consequently, one obtains $\sigma(t) \to 0$ as $t \to \infty$. Formula (32) implies $x(t) \to 0$ as $t \to \infty$ for all $x^0 \in R^n$. Finally, the last equation (31) shows that $\xi(t) \to 0$ as $t \to \infty$.

Taking into account that the properties stated above are valid for any φ satisfying the condition 3) in Theorem 2, we conclude that the system (31) is absolutely stable.

There are many possibilities in generalizing the results stated in Theorems 1 and 2. In our paper [3], systems of the form $\dot{x}(t) = (Ax)(t) + b*\varphi(\sigma(t))$ are investigated, in which $*$ stands for the convolution product. The case $b(t) = \sum_{j=0}^{\infty} b_j \delta(t-t_j) + \beta(t)$ has been investigated, where b_j, $j \geq 0$, are n-vectors such that $\sum_{j=0}^{\infty} |b_j| < +\infty$, and $\beta \in L^1(R_+,\ R^n)$. Frequency domain conditions like (35) have been obtained in the special case of operators A given by (9).

REFERENCES

1. M. A. Aizerman and R. R. Gantmakher: "Absolute Stability of Regulator Systems":Holden-Day Inc., San Francisco, 1963.
2. C. Corduneanu: "Integral Equations and Stability of Feedback Systems", Academic Press, New York, 1973.
3. C. Corduneanu: Recent contributions to the theory of differential systems with infinite delay. Institut de Math. Pure Appl., Université Catholique de Louvain, Rapport No. 95, Vander, Louvain, 1976.
4. C. Corduneanu: "Integral Equations and Applications", Cambridge University Press, 1990.
5. Ch. A. Desoer, M. Vidyasagar: "Feedback Systems: Input-Output Properties", Academic Press, New York, 1975.

6. A. H. Gelig, G. A. Leonov and V. A. Yakubovic: "The stability of Nonlinear Systems with a Nonunique Equilibrium State" (Russian), Nauka, Moscow, 1978.
7. G. Gripenberg, S. O. Londen and O.Staffans: "Volterra Integral and Functional Equations", Cambridge Univ. Press, 1990.
8. S. Lefschetz: "Stability of Nonlinear Control Systems", Academic Press, New York, 1965.
9. A. I. Lurie and V. N Postnikov: "Some Nonlinear problems in the Theory of Automatic Regulation" (Russian), GITTL, Moscow, 1951.
10. R. K. Miller: "Nonlinear Volterra Integral Equations". W. A. Benjamin, Menlo Park, CA, 1971.
11. K. S. Narendra and J. H. Taylor: "Frequency Domain Criteria for Absolute Stability". Academic Press, New York, 1973.
12. J. A. Nohel and D. F. Shea: Frequency Domain Methods for Volterra Equations. Advances in Mathematics, Vol. 22 (1976), 278-304.
13. V. M. Popov: "Hyperstability of Control Systems" Springer, Berlin, 1973.

On the Use of Product Integration for the Solution of Finite Part Singular Integral Equations

B. BERTRAM
Department of Mathematical Sciences
Michigan Technological University
Houghton, Michigan 49931, USA

Abstract. This paper discusses the use of product integration for computing the solution of hypersingular integral equations of the form

$$f(x) = g(x) + \fint_a^b \frac{k(x,t)f(t)}{(t-x)^\alpha}\, dt \tag{1}$$

where $\alpha > 1$, k, f and g are smooth functions on (a,b), and the integral is interpreted in the Hadamard or finite part sense. Several test solutions are computed (particularly for $\alpha = 2$). The technique is then successfully applied numerically to a fracture mechanics problem of an infinite strip containing a crack perpendicular to its boundaries.

1. Introduction. In [1], Kaya and Erdogan use a weighted residual method to produce a numerical solution to such singular equations as (1). We instead consider a product integration approach. The smooth part of the integrand ($k(x,t)f(t)$) is interpolated by a polynomial interpolant and the resultant weights are computed analytically. Nyström's method is used to produce a linear system resulting in an approximate solution for f .

Although the integrals occuring in the calculation of the weights are divergent, they can be successfully evaluated by interpretation in the Hadamard or finite part sense [1,2].

2. Finite Part Integrals. To consider the numerical evaluation of finite part integrals, we make the following definition [1]:

Let

$$I = \fint_a^b \frac{f(t)}{(t-x)^2}\, dt. \tag{2}$$

Then

$$I \equiv \frac{d}{dx} \int_a^b \frac{f(t)}{t-x}\, dt, \tag{3}$$

where the integral in (3) is interpreted in the Cauchy Principal Value sense. For example, if $f(t) \equiv 1$, we have

$$\fint_a^b \frac{1}{(t-x)^2} \, dt = -\frac{1}{b-x} - \frac{1}{x-a}, \tag{4}$$

for $a < x < b$. Hence $\fint_{-1}^1 \frac{1}{t^2} \, dt = -2$ in the finite part sense!

3. Product Integration. To solve (1) numerically, we employ a product integration technique (see, for instance, [3]). Let $a = t_1 , t_2 , t_3 , ..., t_{n+1} = b$ be a partition of [a,b] into n parts. On $[t_k , t_{k+2}]$, we interpolate $k(x , t) f(t)$ using a quadratic polynomial. The integral is then approximated by a sum of the form

$$\sum_{k=1}^{n+1} w_k(x) \, k(x, t_k) \, f(t_k). \tag{5}$$

The weights $w_k(x)$ in (5), are obtained by doing moment integrals on each pair of subintervals (using the finite part interpretation) and then collecting coefficients of each $k(x, t_k) f(t_k)$. Using Nyström's method, we let $x = t_k , k = 2...n$ and get an $n-1$ by $n-1$ linear system which can then be solved to obtain approximate values for f(x) at $x = t_2 , ..., t_n$.

4. Some Examples. To examine this method we consider some simple examples. (These examples were run on a Sun 3/50 system.)

Example 1: $\fint_0^1 \frac{1}{(t-x)^2} \, dt = \frac{-1}{1-x} - \frac{1}{x}$.

With $n = 4$, we have the following results (where I_a is the approximate value of the integral and I is the exact value of the interval at the point t:

t	I_a	I
.25	-5.3333333333333	-5.3333333333333
.50	-4.0000000000000	-4.0000000000000
.75	-5.3333333333333	-5.3333333333333

We see that machine accuracy is achieved with only four intervals.

Example 2: $\fint_0^1 \frac{t(1-t)}{(t-x)^2} \, dt = (1-x) \log\left(\frac{1-x}{x}\right) - 1 - x \left(\frac{1}{1-x} + \frac{1}{x}\right)$.

With $n = 4$, we have the following results:

t	I_a	I
.25	-1.45069385566594	-1.45069385566594
.50	-2.0000000000000	-2.0000000000000
.75	-1.45069385566594	-1.45069385566594

Once again, machine accuracy is obtained using only four intervals. (It should be noted that the interpolation is exact for both Examples 1 and 2.)

Example 3: $f(x) = 1 + \dfrac{1}{1-x} + \dfrac{1}{x} + \displaystyle\oint_0^1 \dfrac{f(t)}{(t-x)^2}\, dt$.

With $n = 4$, we have the following results (where f_a denotes the approximation to f at t:

t	f_a	f
.25	1.0000000000000	1.0000000000000
.50	1.0000000000000	1.0000000000000
.75	1.0000000000000	1.0000000000000

With this example, we have achieved machine accuracy for the solution of a simple integral equation with only four intervals.

Example 4: $-\pi = \displaystyle\oint_0^1 \dfrac{f(t)}{(t-x)^2}\, dt$.

In this case, $f(t) = \sqrt{t(1-t)}$. With $n = 32$, we have the following results:

t	f_a	f
.125	0.33183302220668	0.33071891388307
.250	0.43403351594746	0.43301270189222
.375	0.48488639735798	0.48412291827593
.500	0.50070135027352	0.50000000000000
.625	0.48488639735798	0.48412291827593
.750	0.43403351594746	0.41339864235384
.875	0.33183302220668	0.33071891388307

There is, in general, agreement to three places for the points chosen. For both Examples 4 and 5, this method is better toward the interior of the interval.

Example 5: $f(x) = \pi + \sqrt{x(1-x)} + \displaystyle\oint_0^1 \dfrac{f(t)}{(t-x)^2}\, dt$

As in the preceding example, $f(t) = \sqrt{t(1-t)}$. With $n = 32$, we have the following results:

t	f_a	f
.125	0.33172497458163	0.33071891388307
.250	0.43393524410284	0.4330127018922
.375	0.48478486378099	0.48412291827593
.500	0.50059895851614	0.50000000000000
.625	0.48478486378100	0.48412291827593
.750	0.43393524410285	0.41339864235384
.875	0.33172497458163	0.33071891388307

These results compare with those in Example 4; there is agreement to three places.

It should be noted that Examples 4 and 5 were computed in two ways: one which ignored a value for the endpoints of the unknown function and one for which such values were computed by a limiting procedure. This issue arises since f is generally assumed to be continuous on [a,b] while g may be continuous only on (a,b). The numerical computation of the integral itself involves the use of the endpoints, but these are not used in solution of the integral equation. As expected, use of reasonable end-point values was superior.

5. A Problem from fracture mechanics. We now apply our method to a fracture mechanics problem of studying an internal crack perpendicular to the boundaries of an infinite strip. The integral equation formulation of this problem [1] is as follows:

$$\oint_a^b \frac{V(t)}{(t-x)^2}\, dt + \int_a^b V(t)\, K(t,x)\, dt = -\pi\, c\, p(x) \tag{6}$$

where c is a constant and $a < x < b$. We wish to approximate the normalized stress intensity factors at either end ($k_1(a)$, $k_1(b)$) which can be calculated as

$$k_1(a) \equiv \frac{1}{c} \lim_{t \to a} \frac{V(t)}{\sqrt{2(t-a)}} \tag{7}$$

for $t > a$, and

$$k_1(b) \equiv \frac{1}{c} \lim_{t \to b} \frac{V(t)}{\sqrt{2(b-t)}} \tag{8}$$

for $t < b$.

As a test case, we let $p(x) \equiv p_0$ (a constant, i.e. uniform loading) and consider an internal crack in the half plane so that $a > 0$ and

$$K(t,x) = \frac{-1}{(t+x)^2} + \frac{12x}{(t+x)^3} - \frac{12x^2}{(t+x)^4}. \tag{9}$$

We consider this test case over four different intervals: [0.5,10.5], [1,3], [1,2], and [3,5].

Kaya-Erdogan Results		
$\dfrac{b+a}{b-a}$	$\dfrac{k_1(a)}{p_0\sqrt{(b-a)/2}}$	$\dfrac{k_1(b)}{p_0\sqrt{(b-a)/2}}$
1.1	1.7587	1.2108
2.0	1.0913	1.0539
3.0	1.0345	1.0246
4.0	1.0182	1.0141

Product Integration Results		
$\dfrac{b+a}{b-a}$	$\dfrac{k_1(a)}{p_0\sqrt{(b-a)/2}}$	$\dfrac{k_1(b)}{p_0\sqrt{(b-a)/2}}$
1.1	1.7681	1.2121
2.0	1.0913	1.0539
3.0	1.0270	1.0174
4.0	1.0182	1.0141

We see that the product integration results agree with those of Kaya and Erdogan to at least three places. (Note: the limit necessary to approximate the normalized stress intensity factors was produced numerically by examination of the case when n=256, observing the smooth data and then employing a linear extrapolation.)

6. Conclusions. We conclude that the product integration method for solution of hypersingular integral equations can be successfully implemented and can be used as an alternative method to the weighted residual method of Kaya and Erdogan. While our method does not (at this point) appear to be more computationally efficient, we believe that it is easier to implement.

Preliminary experiments for $\alpha > 2$ using product integration with both quadratic and linear interpolation leads to the conjecture that polynomial interpolation of degree at least α is required. Indeed, using linear interpolation clearly fails for $\alpha = 2$. On the other hand, linear interpolation with $\alpha = 1$ (for fixed singularities) has been proven to converge [4].

Future research will include error and convergence analysis as well as efforts to make this technique more efficient.

Acknowledgements. The author would like to thank Otto Ruehr and Debra Adler for many helpful ideas and conversations.

References.

[1] Kaya, A.C. and R. Erdogan, On the Solution of Integral Equations with Strongly Singular Kernels, *Quarterly of Applied Mathematics*, Vol. XLV, No. 1, pp. 105-122, 1987.

[2] Hadamard,J., *Lectures on Cauchy's problem in linear partial differential equations*, pp. 133-141, Dover, 1952.

[3] Atkinson, K.E., *A Survey of Numerical Methods for the Solution of Fredholm Integral Equations of the Second Kind*, pp. 106-122, SIAM, Philadelphia, 1976.

[4] Bertram, B., On the product integration method for solving singular integral equations in scattering theory, *Journal of Computational and Applied Mathematics*, Vol. 25, pp. 79-92, 1989.

On Nonlinear Sturm-Liouville Boundary Value Problems at Resonance

HELLEVI MATALA-AHO and SEPPO SEIKKALA
University of Oulu
90570 Oulu, Finland

Abstract

A method for studying two point boundary value problems at resonance is introduced. Compared to the widely used methods based on Leray-Schauder principle our method is more suitable for numerical solution.

1. INTRODUCTION

We shall study the nonlinear Sturm - Liouville boundary value problem

$$\begin{cases} Lu(x) & = f(x, \mathcal{A}u(x), \mathcal{B}u'(x)), \quad a.e.\,x \in (0,1), \\ B_i u & = d_i, \quad i = 1, 2, \end{cases} \tag{1.1}$$

where L is the linear second order self-adjoint differential operator $Lu = (pu')' + qu$ with $p \in C^1[0,1], q \in C[0,1]$ and the boundary conditions are either of mixed type

$$B_1 u \equiv a_0 u(0) + b_0 u(1) = d_0,$$
$$B_2 u \equiv a_1 u'(0) + b_1 u'(1) = d_1, \tag{1.2}$$

or of separated type

$$B_1 u \equiv a_0 u(0) + b_0 u'(0) = d_0,$$
$$B_2 u \equiv a_1 u(1) + b_1 u'(1) = d_1. \tag{1.3}$$

Here $|a_0| + |b_0| > 0$ and $|a_1| + |b_1| > 0$. In case of (1.2) we require that $a_0 a_1 p(1) = b_0 b_1 p(0)$. The problem is assumed to be regular i.e. p is either positive or negative on $[0,1]$. Moreover, we assume that $\mathcal{A}$ and $\mathcal{B}$ map $L_2(0,1)$ into itself and that f is a real valued function on $[0,1] \times \mathbf{R} \times \mathbf{R}$ such that the mapping $t \to f(t, \mathcal{A}u(t), \mathcal{B}u'(t))$ is square integrable on $[0,1]$ for every $u \in H^2(0,1)$ where $H^2(0,1)$ is the Sobolev space $H^2(0,1) = \{u \in L_2(0,1) | u' \in L_2(0,1), u'' \in L_2(0,1)\}$.

The problem will be of resonance type i.e. the completely homogeneous problem $Lu = 0$, $B_i u = 0$, $i = 1, 2$, has nontrivial solutions. This implies that (1.1) can not be transformed directly to an equivalent integral equation using Green's functions. Hence, instead of (1.1) we first shall study a parametrized problem

$$\left\{ \begin{array}{l} Lu(x) = f(x, \mathcal{A}u(x), \mathcal{B}u'(x)) - \delta_1 \phi_{01}(x) - \delta_2 \phi_{02}(x), \ a.e. x \in (0,1), \\ B_i u = d_i, \\ \int\limits_0^1 u(x)\phi_{0i}(x)dx = \lambda_i, \quad i = 1,2, \end{array} \right. \tag{1.4}$$

where λ_1, λ_2 are the real parameters and ϕ_{01}, ϕ_{02} are the spanning orthonormal functions of the linear space H_0 of solutions to the completely homogeneous problem. If H_0 is one dimensional we have only ϕ_{01} and choose $\delta_2 = \lambda_2 = 0, \phi_{02} \equiv 0$ in (1.4). The constants δ_1 and δ_2 depend on the solution u and hence on λ_1 and λ_2. The original problem (1.1) with boundary conditions (1.2) or (1.3) has a solution if there exist λ_1 and λ_2 such that $\delta_1 = \delta_2 = 0$. Problem (1.4) is transformed into an equivalent integral equation

$$u(x) = \lambda_1 \phi_{01}(x) + \lambda_2 \phi_{02}(x) + c(x) + \int\limits_0^1 k(x,y) f(y, \mathcal{A}u(y), \mathcal{B}u'(y))dy, \tag{1.5}$$

where k is a modified Green's function and c is the unique solution of the boundary value problem (2.7) given below. Existence results for (1.4) and for the original problem (1.1) are given and one example is solved numerically. The method used in this paper is a generalization of that in [1] where the special case $Lu = u''$ was considered and instead of the linear combination $\lambda_1 \phi_{01} + \lambda_2 \phi_{02}$ we always had $\lambda \phi_0$ with $\phi_0 \equiv 1$. The general forms of $Lu = (pu')' + qu$ and of the boundary conditions in (1.1) allow the treatment of numerous physical problems of resonance type, e.g. the forced vibration of a string with fixed end points under the applied pressure whose frequency equals the natural frequency of the system.

2. THE PARAMETRIZED PROBLEM AND THE EQUIVALENT INTEGRAL EQUATION

We shall prove the equivalence of the boundary value problem (1.4) to the integral equation (1.5) by first considering homogeneous boundary conditions. For the sake of simplicity we assume that dim $H_0 = 2$ and only give the analogous results in case dim $H_0 = 1$.

For notational convenience we use the Dirac's delta function and refer to ([2]) for the existence of a modified Green's function $k(x,y)$ which, as a function of x, is the unique solution of the boundary value problem

$$\left\{ \begin{array}{l} Lk(x,y) = \delta(x-y) - \phi_{01}(x)\phi_{01}(y) - \phi_{02}(x)\phi_{02}(y), \\ B_1 k(x,y) = B_2 k(x,y) = 0, \\ \int\limits_0^1 k(x,y)\phi_{0i}(x)dx \equiv 0, \quad i = 1,2. \end{array} \right. \tag{2.1}$$

Let $g \in L_2(0,1)$. By a solution of the boundary value problem

$$\left\{ \begin{array}{l} Lu(x) = g(x) - \delta_1 \phi_{01}(x) - \delta_2 \phi_{02}(x), \ a.e. x \in (0,1), \\ B_1 u = B_2 u = 0, \\ \int\limits_0^1 u(x)\phi_{0i}(x)dx = \lambda_i, \quad i = 1,2, \end{array} \right. \tag{2.2}$$

we mean a function $u \in H^2(0,1)$ which satisfies (2.2). From the formal selfadjointness of L and from the nature of boundary conditions (in case of (1.2) the condition $a_0 a_1 p(1) = b_0 b_1 p(0)$ is needed here) it follows that

$$p(1)[u'(1)v(1) - v'(1)u(1)] - p(0)[u'(0)v(0) - v'(0)u(0)] = 0 \qquad (2.3.a)$$

and hence

$$(Lu, v) = (u, Lv) \qquad (2.3.b)$$

$[(u,v)$ is the usual inner product of $L_2(0,1), (u,v) = \int_0^1 u(x)v(x)dx]$ for $u, v \in H^2(0,1)$ satisfying the homogeneous boundary conditions. From this and from the orthonormality of the functions ϕ_{01} and ϕ_{02} it follows that necessarily

$$\delta_i = \int_0^1 \phi_{0i}(x)g(x)\,dx, \quad i = 1, 2, \qquad (2.4)$$

since $(Lu, \phi_{0i}) = (u, L\phi_{0i}) = 0$.

LEMMA 2.1. *If δ_1 and δ_2 are given by (2.4) then the boundary value problem (2.2) has a solution u which is given by*

$$u(x) = \lambda_1 \phi_{01}(x) + \lambda_2 \phi_{02}(x) + \int_0^1 k(x,y)g(y)dy. \qquad (2.5)$$

PROOF: If u is given by (2.5) then from the conditions (2.1) for the modified Green's function k it follows that u satisfies (2.2).
On the other hand, if u is a solution of (2.2) we will now show that u satisfies (2.5). From the differential equations of (2.1) and (2.2) it follows that

$$k(x,y)Lu(x) - u(x)Lk(x,y) = k(x,y)g(x) - \delta_1 k(x,y)\phi_{01}(x)$$
$$- \delta_2 k(x,y)\phi_{02}(x) - u(x)\delta(x-y) + \phi_{01}(y)u(x)\phi_{01}(x) + \phi_{02}(y)u(x)\phi_{02}(x). \qquad (2.6)$$

By using the Lagrange identity for L,

$$k(x,y)Lu(x) - u(x)Lk(x,y) = \frac{d}{dx}[p(x)k(x,y)u'(x) - p(x)k_x(x,y)u(x)],$$

and taking into account (2.3.a) and

$$\int_0^1 k(x,y)\phi_{0i}(x)dx = 0, \int_0^1 u(x)\phi_{0i}(x)dx = \lambda_i, \quad i = 1, 2,$$

we get from (2.6), by integrating,

$$0 = \int_0^1 k(x,y)g(x)dx - u(y) + \lambda_1 \phi_{01}(y) + \lambda_2 \phi_{02}(y),$$

from which Eq. (2.5) follows by the symmetry of k. $\blacksquare$

LEMMA 2.2. *For each pair $(d_1, d_2) \in \mathbf{R}^2$ there exists a unique triplet $(u, c_1, c_2) \in H^2(0,1) \times \mathbf{R}^2$ for which*

$$
\begin{cases}
Lu(x) = c_1 \phi_{01}(x) + c_2 \phi_{02}(x), \\
B_1 u = d_1, B_2 u = d_2 \\
\int_0^1 u(x)\phi_{0i}(x)dx = 0, \quad i = 1, 2.
\end{cases}
\tag{2.7}
$$

PROOF: Suppose first that $d_1 = d_2 = 0$. Then, by Lemma 2.1, $(u, c_1, c_2) = (0, 0, 0)$ is the only solution of (2.7), since now $\lambda_1 = \lambda_2 = 0$ and $g \equiv 0$ in Eq. (2.5).

Let $d_1, d_2 \in \mathbf{R}$. In a standard way it can be shown that there exist nontrivial functions u_1 and u_2 satisfying (2.7) (possibly with different constants c_1 and c_2) except that only the boundary condition $B_1 u_1 = 0$ holds for u_1 and $B_2 u_2 = 0$ for u_2. This implies, by the first part of this proof, that $B_2 u_1 \neq 0$ and $B_1 u_2 \neq 0$. Then

$$
u(x) = \frac{d_2}{B_2 u_1} u_1(x) + \frac{d_1}{B_1 u_2} u_2(x)
\tag{2.8}
$$

satisfies (2.7) for some constants $c_1 = \gamma_1$ and $c_2 = \gamma_2$. But if $\bar{u}$ also is a solution of (2.7) with $c_1 = \bar{\gamma}_1$ and $c_2 = \bar{\gamma}_2$, then $u - \bar{u}$ is a solution of (2.7) with $d_1 = d_2 = 0$. Hence, again by the first part of the proof, $u = \bar{u}, \gamma_1 = \bar{\gamma}_1$ and $\gamma_2 = \bar{\gamma}_2$. This completes the proof. ∎

THEOREM 2.1. *The boundary value problem (1.4),*

$$
\begin{cases}
Lu(x) = f(x, \mathcal{A}u(x), \mathcal{B}u'(x)) - \delta_1 \phi_{01}(x) - \delta_2 \phi_{02}(x), a.e.x \in (0,1), \\
B_1 u = d_1, B_2 u = d_2, \\
\int_0^1 u(x)\phi_{0i}(x)dx = \lambda_i, \quad i = 1, 2,
\end{cases}
$$

is equivalent to the integral equation (1.5),

$$
u(x) = \lambda_1 \phi_{01}(x) + \lambda_2 \phi_{02}(x) + c(x) + \int_0^1 k(x, y)f(y, \mathcal{A}u(y), \mathcal{B}u'(y))dy,
$$

where c is the unique solution of (2.7). Moreover, if u is a solution of (1.4) then the constants δ_1 and δ_2 are given by

$$
\delta_i = c_i + \int_0^1 \phi_{0i}(x)f(x, \mathcal{A}u(x), \mathcal{B}u'(x))dx, \quad i = 1, 2,
\tag{2.9}
$$

where c_1 and c_2 are the unique constants of (2.7).

PROOF: Let u be a solution of (1.4) and denote $g(x) = f(x, \mathcal{A}u(x), \mathcal{B}u'(x))$. Then $w = u - c$ satisfies (2.2) and hence, by Lemma 2.1,

$$
w(x) = u(x) - c(x) = \lambda_1 \phi_{01}(x) + \lambda_2 \phi_{02}(x) + \int_0^1 k(x, y)g(y)dy,
$$

i.e. u is a solution of the integral equation (1.5). If, on the other hand, u is a solution of (1.5) then obviously u is a solution of (1.4).

Since $w = u - c$ is a solution of (2.2) with δ_i replaced by $\delta_i - c_i$, Eq. (2.9) follows from (2.4). ∎

EXAMPLE 2.1. Let $Lu = u'' + 4\pi^2 u$, $B_1 u = u(0) - u(1)$ and $B_2 u = u'(0) - u'(1)$. Now the completely homogeneous problem (periodic boundary value problem)

$$\begin{cases} u''(x) + 4\pi^2 u(x) = 0, & 0 < x < 1, \\ u(0) = u(1) \\ u'(0) = u'(1) \end{cases}$$

has nontrivial solutions and the solution set H_0 is spanned by the orthonormal functions

$$\phi_{01}(x) = \sqrt{2} \cos 2\pi x, \qquad \phi_{02}(x) = \sqrt{2} \sin 2\pi x.$$

The solution of (2.1), i.e. the modified Green's function $k(x,y)$, is given by

$$k(x,y) = \begin{cases} 2(l_1 \cos 2\pi x + l_2 \sin 2\pi x + k_0), & x < y \\ 2(l_3 \cos 2\pi x + l_4 \sin 2\pi x + k_0), & x \geq y, \end{cases} \tag{2.10}$$

where

$$l_1 = -\frac{1}{16\pi^2} \cos 2\pi y - \frac{y - \frac{1}{2}}{4\pi} \sin 2\pi y,$$

$$l_2 = l_4 - \frac{1}{4\pi} \cos 2\pi y,$$

$$l_3 = l_1 - \frac{1}{4\pi} \sin 2\pi y,$$

$$l_4 = -\frac{1}{4\pi^2} \sin 2\pi y \cos^2 2\pi y,$$

and

$$k_0 = \frac{x}{4\pi}(\cos 2\pi x \sin 2\pi y - \cos 2\pi y \sin 2\pi x),$$

and the function $c(x)$, solution of (2.7) corresponding to the nonhomogeneous boundary conditions, is

$$c(x) = \left(\frac{1}{2}d_1 - \frac{1}{8\pi^2}d_2\right) \cos 2\pi x + \frac{1}{4\pi}(d_2 - d_1) \sin 2\pi x$$

$$- d_1 x \cos 2\pi x - \frac{1}{2\pi}d_2 x \sin 2\pi x. \tag{2.11}$$

The unique constants c_1 and c_2 of (2.7) are given by

$$c_1 = -\sqrt{2}d_2, \qquad c_2 = 2\sqrt{2}\pi d_1.$$

Analogously to the above we can prove

THEOREM 2.2. Let $\dim H_0 = 1$ and ϕ_{01} be the spanning normed function of H_0 and let $\phi_{02} \equiv 0$. Then the boundary value problem (1.4) is equivalent to the integral equation (1.5), where $(c, c_1) \in H^2(0,1) \times \mathbf{R}$ is the unique solution of (2.7) and $k(x,y)$ is the corresponding modified Green's function satisfying (2.1). Moreover, (2.9) holds true, with $\delta_2 = c_2 = 0$.

EXAMPLE 2.2.. Let $Lu = u'' + \pi^2 u$, $B_1 u = u(0)$ and $B_2 u = u(1)$. The solution set H_0 to the completely homogeneous Dirichlet's problem

$$u''(x) + \pi^2 u(x) = 0$$
$$u(0) = 0, u(1) = 0,$$

is spanned by $\phi_{01}(x) = \sqrt{2}\sin\pi x$. The modified Green's function is now

$$k(x,y) = \begin{cases} \frac{1}{\pi}[(x\cos\pi x - \frac{1}{2\pi}\sin\pi x)\sin\pi y - (1-y)\cos\pi y\sin\pi x], & x < y, \\ \frac{1}{\pi}[(y\cos\pi y - \frac{1}{2\pi}\sin\pi y)\sin\pi x - (1-x)\cos\pi x\sin\pi y], & x \geq y, \end{cases} \tag{2.12}$$

the solution $c(x)$ of (2.7) (with $\phi_{02} \equiv 0$)

$$c(x) = d_1\cos\pi x - \frac{1}{2\pi}(d_1 + d_2)\sin\pi x - (d_1 + d_2)x\cos\pi x \tag{2.13}$$

and the constant c_1 equals $2\pi(d_1 + d_2)$.

The integral equation (1.5) may be studied in different ways, e.g. using fixed point methods. We shall apply the contraction mapping principle and to this end we introduce the linear operators

$$K_0 u(x) = \int_0^1 k(x,y)u(y)dy$$

and

$$K_1 u(x) = \int_0^1 k_x(x,y)u(y)dy$$

with k given by (2.1), and henceforth assume that the Lipschitz conditions

$$|f(x,u,v) - f(x,\bar{u},\bar{v})| \leq M|u - \bar{u}| + N|v - \bar{v}|, u,\bar{u},v,\bar{v} \in \mathbf{R}, x \in [0,1], \tag{2.14}$$

$$\|\mathcal{A}u - \mathcal{A}\bar{u}\| \leq A_0\|u - \bar{u}\|, u,\bar{u} \in L_2(0,1), \tag{2.15}$$

and

$$\|\mathcal{B}v - \mathcal{B}\bar{v}\| \leq B_0\|v - \bar{v}\|, v,\bar{v} \in L_2(0,1), \tag{2.16}$$

hold for the function f and for the operators $\mathcal{A}, \mathcal{B} : L_2(0,1) \to L_2(0,1)$. The norm $\|\|$ denotes both the L_2-norm and the L_2-operator norm.

THEOREM 2.3. *If*

$$\|K_0\|^2 M^2 A_0^2 + \|K_1\|^2 N^2 B_0^2 < \frac{1}{2}, \tag{2.17}$$

then the integral equation (1.5), and hence the boundary value problem (1.4), has a unique solution for any $\lambda_1, \lambda_2 \in \mathbf{R}$.

PROOF: We shall prove that the operator $T : H^1(0,1) \to H^1(0,1) = \{u \in L_2(0,1)|u' \in L_2(0,1)\}$ defined by

$$Tu(x) = \lambda_1\phi_{01}(x) + \lambda_2\phi_{02}(x) + c(x) + Ku(x), \quad u \in H^1(0,1),$$

where

$$Ku(x) = \int_0^1 k(x,y) f(y, \mathcal{A}u(y), \mathcal{B}u'(y)) dy,$$

is a contraction on $H^1(0,1)$. Now for $u, \bar{u} \in H^1(0,1)$ it follows from (2.14) - (2.16) that

$$
\begin{aligned}
\|Ku - K\bar{u}\|^2 &= \|K_0 f(\cdot, \mathcal{A}u(\cdot), \mathcal{B}u'(\cdot)) - K_0 f(\cdot, \mathcal{A}\bar{u}(\cdot), \mathcal{B}\bar{u}'(\cdot))\|^2 \\
&\leq \|K_0\|^2 \|f(\cdot, \mathcal{A}u(\cdot), \mathcal{B}u'(\cdot)) - f(\cdot, \mathcal{A}\bar{u}(\cdot), \mathcal{B}\bar{u}'(\cdot))\|^2 \\
&\leq 2\|K_0\|^2 (M^2 \|\mathcal{A}u - \mathcal{A}\bar{u}\|^2 + N^2 \|\mathcal{B}u' - \mathcal{B}\bar{u}'\|^2) \\
&\leq 2\|K_0\|^2 (M^2 A_0^2 \|u - \bar{u}\|^2 + N^2 B_0^2 \|u' - \bar{u}'\|^2).
\end{aligned}
\tag{2.18}
$$

Since $(K_0 u)' = K_1 u$ we get similarly

$$\|(Ku)' - (K\bar{u})'\|^2 \leq 2\|K_1\|^2 (M^2 A_0^2 \|u - \bar{u}\|^2 + N^2 B_0^2 \|u' - \bar{u}'\|^2). \tag{2.19}$$

By choosing the norm

$$\|u\|_e = (M^2 A_0^2 \|u\|^2 + N^2 B_0^2 \|u'\|^2)^{1/2}$$

in $H^1(0,1)$ we obtain from (2.18) and (2.19)

$$
\begin{aligned}
\|Tu - T\bar{u}\|_e &= \|Ku - K\bar{u}\|_e \\
&\leq \sqrt{2}(\|K_0\|^2 M^2 A_0^2 + \|K_1\|^2 N^2 B_0^2)^{1/2} \|u - \bar{u}\|_e
\end{aligned}
$$

and hence, by (2.17), T is a contraction in $H^1(0,1)$. Thus Eq. (1.5) has a unique solution. $\blacksquare$

COROLLARY 2.1. *The boundary value problem*

$$
\begin{cases}
u''(x) + 4\pi^2 u(x) = f(x, u(x), u'(x)) - \delta_1 \cos 2\pi x - \delta_2 \sin 2\pi x \\
u(0) - u(1) = d_1 \\
u'(0) - u'(1) = d_2 \\
\int_0^1 u(x) \cos 2\pi x \, dx = \lambda_1, \int_0^1 u(x) \sin 2\pi x \, dx = \lambda_2
\end{cases}
\tag{2.20}
$$

has a unique solution for any $\lambda_1, \lambda_2 \in \mathbf{R}$, *if*

$$\frac{1}{16\pi^4} M^2 + \frac{1}{9\pi^2} N^2 < \frac{1}{2}. \tag{2.21}$$

PROOF: Using the eigenfunction expansion of the compact self adjoint operator K_0 it can be shown that

$$\|K_0\| = \frac{1}{4\pi^2} \quad \text{and} \quad \|K_1\| = \frac{1}{3\pi}.$$

Since $A_0 = B_0 = 1$, inequality (2.17) holds true by (2.21) and the conclusion follows by Theorem 2.3. $\blacksquare$

If the right hand side of (1.4) does not depend on the derivative u', i.e. $f(x, u, v)$ is independent of v, then by the proof of Theorem 2.3 it is obvious that inequality (2.17) can be replaced by the sharper inequality

$$\|K_0\| M A_0 < 1, \tag{2.22}$$

which yields optimal upper bounds for the Lipschitz constant M in (2.14) (now $N = 0$). For example, we have the following one.

COROLLARY 2.2. *The boundary value problem*

$$\begin{cases} u''(x) + 4\pi^2 u(x) = f(x, u(x)) - \delta_1 \cos 2\pi x - \delta_2 \sin 2\pi x \\ u(0) - u(1) = d_1 \\ u'(0) - u'(1) = d_2 \\ \int_0^1 u(x) \cos 2\pi x \, dx = \lambda_1, \ \int_0^1 u(x) \sin 2\pi x \, dx = \lambda_2 \end{cases} \quad (2.23)$$

has a unique solution for any $\lambda_1, \lambda_2 \in \mathbf{R}$, *if* $M < 4\pi^2$. *The upper bound* $4\pi^2$ *for* M *is optimal.*

The optimality of $4\pi^2$ in Corollary 2.2 can be seen by taking $\delta_1 = \delta_2 = d_1 = d_2 = \lambda_1 = \lambda_2 = 0$ and $f(x, u) = 4\pi^2 u$ in (2.23). In this case problem (2.23) no longer has a unique solution but infinitely many solutions $u(x) \equiv c, c \in \mathbf{R}$.
If H_0 is one dimensional we easily obtain a result analogous to Theorem 2.3 but we will only take an optimal result similar to Corollary 2.2.

COROLLARY 2.3. *The boundary value problem*

$$\begin{cases} u''(x) + \pi^2 u(x) = f(x, u(x)) - \delta_1 \sin \pi x \\ u(0) = d_1, u(1) = d_2 \\ \int_0^1 u(x) \sin \pi x \, dx = \lambda_1 \end{cases} \quad (2.24)$$

has a unique solution for any λ_1, *if* $M < 3\pi^2$. *This upper bound for* M *is optimal.*

PROOF: The upper bound $3\pi^2$ for M follows from the fact that now the norm of the compact self adjoint operator K_0 is $\|K_0\| = \frac{1}{3\pi^2}$ which is the maximum absolute value of the eiqenvalues of K_0. The optimality of $3\pi^2$ can be seen by choosing in (2.24) $\delta_1 = d_1 = d_2 = \lambda_1 = 0$ and $f(x, u) = -3\pi^2 u$ in which case problem (2.24) has infinitely many solutions $u(x) = c \sin 2\pi x, c \in \mathbf{R}$. ∎

3. SOLUTION OF THE ORIGINAL PROBLEM

With the assumptions of Theorem 2.3 or Corollaries 2.1 and 2.2, for example, there exists the unique solution $u = u_{\lambda_1, \lambda_2}$ of the parametrized problem and hence Eq. (2.9) defines functions $\delta_1 = \delta_1(\lambda_1, \lambda_2)$ and $\delta_2 = \delta_2(\lambda_1, \lambda_2)$ which can be shown to be Lipschitz continuous on λ_1 and λ_2. If we can find constants λ_1 and λ_2 such that $\delta_1(\lambda_1, \lambda_2) = \delta_2(\lambda_1, \lambda_2) = 0$ then the corresponding solution u_{λ_1, λ_2} of (1.4) is a solution of the original problem (1.1). We will here give only a numerical example in the above case where $\dim H_0 = 2$ and after that in case $\dim H_0 = 1$ we will give some sufficient conditions under which there exists a λ_1 such that $\delta_1(\lambda_1) = 0$.

EXAMPLE 3.1. *The function* $u(x) = \arctan(\frac{1}{2} \sin 2\pi x)$ *is a solution to the periodic boundary value problem*

$$\begin{cases} u''(x) + 4\pi^2 u(x) = f(x, u(x)), & 0 < x < 1 \\ u(0) = u(1), & u'(0) = u'(1) \end{cases} \quad (3.1)$$

where $f(x,u) = -2\pi^2 \sin 2\pi x \cdot (1+\tan^2 u)^{-1} - \pi^2 \sin 2\pi x \cdot \cos^2 2\pi x(1+\frac{1}{4}\sin^2 2\pi x)^{-2} + 4\pi^2 \arctan(\frac{1}{2}\sin 2\pi x)$. *Since*

$$\left|\frac{\partial f(x,u)}{\partial u}\right| \leq 2\pi^2$$

we can choose the Lipschitz constant $M = 2\pi^2$ in (2.14). Then Corollary 2.2 also quarantees the existence of a solution to (2.23). By numerical calculation using Picard's iterations it is found that

$$\delta_1(0.0, 0.3215) = 0.0, \quad \delta_2(0.0, 0.3215) = 0.000027$$

and the maximum error of the numerical solution is 0.00017. The true values of λ_1 and λ_2 for the actual solution are $\lambda_1 = 0.0$ and $\lambda_2 = 0.33385$. We used 40 subintervals in the trapezoidal rule for numerical integration and the absolute stopping criterion with $\varepsilon = 5 \cdot 10^{-5}$ for iterations.

PROPOSITION 3.1. *Let the assumptions of Corollary 2.3 hold and assume moreover that for a $g \in L_1(0,1)$,*

$$|f(x,u)| \leq g(x), \quad 0 \leq x \leq 1, u \in \mathbf{R},$$

and that the limits $f(x, \pm\infty) = \lim_{u \to \pm\infty} f(x,u)$ exist and satisfy

$$\int_0^1 f(x, \infty) \sin \pi x \, dx \cdot \int_0^1 f(x, -\infty) \sin \pi x \, dx < 0. \tag{3.2}$$

Then the boundary value problem

$$\begin{cases} u''(x) + \pi^2 u(x) = f(x, u(x)), & 0 < x < 1, \\ u(0) = 0, \quad u(1) = 0 \end{cases}$$

has a solution.

PROOF: By Corollary 2.3 for any $\lambda \in \mathbf{R}$ there exists a unique solution u_λ to problem (2.24),

$$u_\lambda(x) = \lambda\sqrt{2}\sin \pi x + \int_0^1 k(x,y)f(y, u_\lambda(y))dy, \quad 0 \leq x \leq 1,$$

where k is given by (2.12). Since $u_\lambda(x) \to \pm\infty, 0 < x < 1$, if $\lambda \to \pm\infty$, then by Lebesgue's convergence theorem

$$\delta(\lambda) = \int_0^1 \sqrt{2}\sin \pi x \cdot f(x, u_\lambda(x))dx \to \int_0^1 \sqrt{2}\sin \pi x \cdot f(x, \pm\infty)dx, \lambda \to \pm\infty.$$

Hence, by (3.2) and by the continuity of $\delta(\lambda)$, there exists a λ_0 such that $\delta(\lambda_0) = 0$. ∎

REFERENCES

1. Saranen, J. and Seikkala, S. *On second order two point boundary value problems, Proc. Int. Conf. on Theory and Applications of Differential Equations*, Ohio University, 1988, 355–361.
2. Stakgold, I. *Boundary Value Problems of Mathematical Physics*, vol. I, The MacMillan Company, New York, 1967.

Comparison of a Direct Method for Inverting Fredholm Equations of the First Kind with the Method of Regularization

J. A. MERSON and T. S. CALE
Chemical, Bio, & Materials Engineering Department
and
Center for Energy Systems Research
Arizona State University
Tempe, Arizona 85287-6006, USA

ABSTRACT

The inversion of "ill-posed" Fredholm integrals of the first kind is
often required in science and engineering applications. In such cases,
small uncertainties in the data can lead to large variability in the
solution. Solution properties are incorporated, explicitly or
implicitly, into the numerical inversion of ill-posed integral equations
in order to produce meaningful results. The explicit incorporation is
done through the use of a "solution property function" (SPF). A
"direct" method is described where the SPF is minimized so that the sum
of the squared errors between experimental and back-calculated data is
less than a given magnitude. This direct method is compared, from an
engineering point of view, to the more common method of regularization.
The example problem considered is the determination of a magnetization
profile from voltage data obtained using an AC permeameter. While both
methods can yield satisfactory solutions, the direct method formulation
offers greater flexibility in the choice of SPFs. The role of the sum
of squared errors in the analysis is also discussed.

INTRODUCTION

Integral equations find many applications in science and engineering.
One widespread use is in the determination of physical properties from
experimental information obtained using linear, integral detectors.
Fredholm equations of the first kind are a special class of these
integral equations and are of the form

$$H(y) = \int_a^b K(y,w)\ P(w)\ dw \tag{1}$$

where y and w may, for example, represent position or time, P(w) is the
physical property detected, H(y) is the measured instrument response and
the kernel of the integral K(y,w) is the sensitivity of the instrument
to the physical property being detected. Each measurement represents a
transformation, usually a smearing, of the desired physical property
[1,2]. If unknown, the property P(w) can in principle be determined
from a series of measurements H(y) by inverting Equation 1.

The numerical inversion of Equation 1, needed for the determination of
the physical property of interest, can be "ill-posed". Such problems
are said to be ill-posed because small changes in H(y) lead to large

changes in the calculated physical properties [3]. Thus, small
uncertainty levels in the data can lead to unrealistic solutions
representing P(w). Heuristically, this usually occurs because the
kernel (detector sensitivity) varies slower in w than the desired
physical property over the length of the integration in Equation 1.
Solutions obtained by naively inverting Equation 1 for these ill-posed
problems are usually unrealistic and of little use.

Several methods have been used successfully to invert ill-posed Fredholm
equations of the first kind to determine physical properties from
experimental measurements. Typically, information about the solution,
contained in a "solution property function" (SPF), is included in the
problem formulation. Examples of this include non-negative least
squares [3-5], regularization [6-9], and projection [10]. Other methods
are discussed by Delves and Mohamed [9] and Sacher and Morrison [5].
Munn and Smith [4] measured the NMR spin lattice relaxations of fluid
contained in porous solids to determine pore size distributions.
Because pore sizes must be positive, the method of non-negative least
squares was used for the analysis of the experimental data. Chemburkar
et al. [8] relied on the method of regularization, with a smoothness
condition for an SPF, in a study of the mathematical basis for
determining temperature profiles in geothermal reservoirs from
measurements of a reacting chemical tracer. Likewise, Britten *et al.*
[7] used regularization with a smoothness condition to estimate surface
site energy distributions from adsorption isotherms determined from
temperature programmed desorption experiments. Luo [10] measured
"folded" nuclear radiation spectra to determine actual radiation spectra
where the distortion of the actual spectra was caused by the integral
nature of the detector. Cale and Merson [6] determined the local cross
sectional average solid temperature profile in a fixed bed catalytic
reactor during chemical reaction from measurements of voltage induced in
the coils of an AC permeameter by the superparamagnetic catalyst sample.
To invert the governing Fredholm equation, they relied on regularization
with a condition of minimum variance about an experimentally determined
average temperature.

The major purpose of this paper is to compare two methods for inverting
ill-posed Fredholm equations of the first kind; the method of
regularization and a nonlinearly constrained optimization formulation.
The latter is labeled the "direct method" for convenience. The
regularization method is an unconstrained optimization problem which is
solved by Gauss-Jordan elimination. Trial calculations are performed
using a test (user defined) magnetization profile which might arise
during the determination of axial temperature profiles [6]. The purpose
of the comparison in this paper is to compare the two formulations from
an engineering view point (*i.e.* with respect to ease of calculation and
application). No comparison of efficiencies of the algorithms is made.
An integral part of the comparison is the use of several different SPFs.
While it is common to use minimum variance or some other smoothness
property of the solution as an SPF, we focus on SPFs which are, or could
be, available experimentally. The use of trial calculations to improve
experimental designs is also discussed.

EXPERIMENTAL DESCRIPTION

In this laboratory, we use an AC permeameter (ACP) to make low field
magnetic measurements of supported nickel catalyst samples [6,11-13].

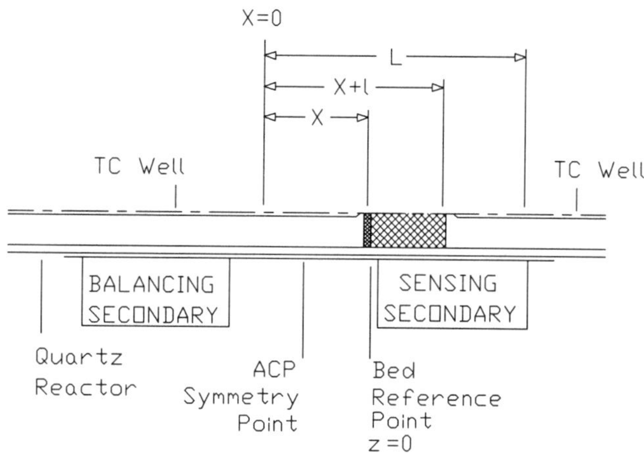

FIGURE 1. Schematic showing ACP coil and related nomenclature.

The experimental procedure for determining the axial magnetization profile in the sample is detailed in the literature [6] and is discussed briefly here in order to demonstrate the origin of the governing Fredholm equation of the first kind. The ACP consists of two secondary coils wound in opposition surrounded by a common primary coil [13]. A schematic of the ACP is shown in Figure 1 to illustrate some of the nomenclature used in the analysis (also see NOMENCLATURE). The primary coil is excited by 20 V RMS at 47 Hz using a constant voltage supply and the net voltage induced in the secondaries is measured.

The carefully matched secondary coils are centered in the primary and the net ACP output voltage is small in the absence of sample. When a superparamagnetic sample is placed in one secondary, the voltage generated due to the mutual inductance of this coil with the primary coil increases. With the sample centered between the two secondary coils, the net output voltage is zero. This position is defined as X=0.0. The voltage generated in the ACP by the sample is measured every 1.0 mm as the sample is pulled stepwise from this null point through one of the secondary coils [6]. It can be shown that this voltage, with the sample reference position at X_j in the ACP secondary coils, can be described by

$$V(X_j) = \int_0^{\ell} S'(X_j+z)\ I(z)\ A_0\ dz = \int_0^{\ell} S(X_j+z)\ M(z)\ dz \qquad (2)$$

where $I(z)$ is the average cross sectional magnetization (moment/volume) as a function of position z in the sample. A_0 is the nominal cross-sectional area of the secondary coils. The voltage generated due to the presence of magnetic material at position X+z in the ACP coils is $S'(X+z)$, which is referred to as the sensitivity of the ACP coils and has units of μV/moment. For convenience, we let $S(X+z) = S'(X+z)\ A_0 I_0$, where I_0 is a reference magnetization. The experimentally determined coil sensitivity $S(X+z)$ and the voltage measurements $V(X)$ represent the known values in Equation 2. We determine the change in relative magnetization $M(z)=I(z)/I_0$ by inverting Equation 2, using Simpson's rule to approximate the integral. In turn, this is converted to a

temperature profile [6]. The inversion is ill-posed, largely because
the distance between the centers of the secondary coils is 11 cm, while
the length of the catalyst sample is typically 1 cm.

FORMULATION OF TEST CALCULATIONS

Test Magnetization Profile

The test magnetization profile used for all calculations to compare the
two methods is representative of results obtained in previous work [6].
The magnetization profile is nonlinear and is represented by a fifth
order polynomial function

$$M(z) = \sum_{k=0}^{k=5} b_k z^k \quad .$$
(3)

The derivative of the test profile is zero at the outlet of the catalyst
bed (z=0.0), because this is expected for these systems. Forty voltage
"data" points are generated to simulate experimental results using the
test magnetization profile and a coil sensitivity, $S(X+z)$, which is
characteristic of our experimental system. The voltages range between -
14 and 2300 μV for X values between 0.0 and 3.9 cm . Random noise is
then added to the voltage "data" at a level consistent with our previous
experimental results. For the calculations of this paper, the sum of
squared differences between the noiseless and modified data used is 358
μV^2. Using these voltage data, relative magnetization profiles are
back-calculated by both the method of regularization and the direct
method.

The sum of the squared differences between the generated data and back-
calculated voltages is a vital part of both formulations. The
difference between the generated and back-calculated voltage at each
position X_j of the sample in the ACP is

$$G_j(\mathbf{b}) = V(X_j) - \sum_{k=0}^{k=5} b_k \int_0^{\ell} z^k S(X_j+z) \, dz$$
(4)

where the vector $\mathbf{b}$ contains the coefficients of the polynomial. $\mathbf{G}^T\mathbf{G}$,
the sum of squared errors, is represented by $\|\mathbf{G(b)}\|$. If each G is set
to zero and a naive inversion is attempted using n+1 data points, the
resulting magnetization profile is a wild function of z. Essentially,
this is because the approach does not allow for any lack of fit due to
uncertainty in the data. Since the precision of the data is finite, the
minimum realistic value of $\|\mathbf{G(b)}\|$ is some finite and perhaps predictable
value. Solutions in which $\|\mathbf{G(b)}\|$ is significantly smaller than the sum
of the squared errors introduced into the data (358 μV^2) cannot be
expected to be valid, in general. A least squares minimization of
$\|\mathbf{G(b)}\|$ allows a finite lack of fit, while still involving the solution
of a linear system of equations. However, this formulation yields
unrealistic profiles for the problem under consideration, because the
resulting sum of squared errors is 333 μV^2. This value serves as a
lower bound on the values of $\|\mathbf{G(b)}\|$ for reasonable answers. Values of
$\|\mathbf{G(b)}\|$ associated with reasonable answers lie between 333 μV^2 and 358
μV^2.

Regularization Method

The method of regularization [9], as used in this paper, is the unconstrained minimization of a modified least squares objective function

$$\underset{\mathbf{b}}{\text{minimize}} \quad \|\mathbf{G}(\mathbf{b})\| + \alpha\, F(\mathbf{b}) \tag{5}$$

where α is the regularization parameter and $F(\mathbf{b})$ represents the SPF. This unconstrained problem involves the solution of linear equations, as in linear least squares, provided that the SPF is chosen so that Equation 5 is a quadratic function of the unknowns. As in the least squares formulation, the problem reduces to the solution of n+1 linear equations in the n+1 unknown polynomial coefficients, once a value for α is selected. The use of the SPF allows $\|\mathbf{G}(\mathbf{b})\|$ to be larger and meaningful solutions become possible. The primary difficulty is the choice of the regularization parameter α, or the effective weight of the SPF. If the magnitude of α is too large, then the solution is dominated by the properties of the SPF. If the magnitude of α is too small then the SPF has little effect and the solution varies wildly because $\|\mathbf{G}(\mathbf{b})\|$ is too small. Typically, extensive trial calculations similar to those used in this work are done to determine a good choice for the value of α, which depends on the uncertainty in the experimental data. As a final validation of the choice of α, the solution arrived at through this process is used as a test function and the trial calculations are repeated.

Direct Method

Delves and Mohamed [9] suggest that the inversion of ill-posed Fredholm equations of the first kind can be formulated in a more direct and revealing manner. For the problem at hand, the direct method can be formulated as

$$\underset{\mathbf{b}}{\text{minimize}}\ F(\mathbf{b}) \tag{6}$$

$$\text{subject to} \quad \|\mathbf{G}(\mathbf{b})\| \le \epsilon \tag{6a}$$

where the value of ϵ depends on the precision of the data. Contributions due to the inadequacy of the proposed polynomial representation of the magnetization profile are also included in ϵ. To serve the purposes of this paper, that contribution is zero since the solution can be exactly represented by a fifth order polynomial. The solution of the nonlinearly constrained minimization problem of Equations 6 is more tedious numerically than the solution to the linear least squares problem which results from Equation 5, when using the usual SPFs [9]. This difference in complexity has led to the preferential use of the method of regularization. In this paper, we use the technique of successive quadratic programming (SQP) to solve Equations 6. The general SQP solution scheme for this nonlinearly constrained optimization has been described in the literature [14], and is implemented using the IMSL subroutine NCONF [15] by rewriting Equations 6 as

$$\underset{\mathbf{b}}{\text{min}}\ \ F(\mathbf{b}) \tag{7}$$

subject to $\epsilon - \|G(\mathbf{b})\| \geq 0$ (7a)

and $\mathbf{b}_\ell \leq \mathbf{b} \leq \mathbf{b}_u$ (7b)

where $\mathbf{b}_\ell$ and $\mathbf{b}_u$ are vectors containing the lower and upper bounds of the polynomial coefficients, b_k. All functions are assumed to be continuously differentiable. Theoretical details about the solution procedure and algorithm can be found in the literature [14-18].

Solution Property Functions

The average relative magnetization of the sample is routinely determined in our experimental procedure; therefore, it seems reasonable to incorporate that information through an appropriate SPF:

$$F(\mathbf{b}) = (M_{ave} - \int_0^\ell M(z)\,dz / \int_0^\ell dz)^2 \ . \tag{8}$$

The physically based zero slope property of the profile at $z=0$ can also be incorporated into each problem formulation as

$$F(\mathbf{b}) = b_1^2 \ . \tag{9}$$

The temperatures at both ends of the bed can be determined using thermocouples; therefore, these can be included as desired properties of the solution as

$$F(\mathbf{b}) = (0.96 - b_0)^2 + (1.003 - \sum_{k=0}^{k=5} b_k \,(0.9)^k)^2 \tag{10}$$

where 0.960 and 1.003 are the values of the relative magnetization at the ends of the 0.9 cm bed which would correspond to measured temperatures.

RESULTS

The test magnetization profile we seek to reproduce is shown as curve A of Figure 2. We consider the three SPFs individually and in combination. When more than one solution property function is used, they are combined into a single SPF. Additional SPFs are relatively easy to incorporate into the direct method using NCONF, since they can be simply added as constraints. Conversely, inequality constraints are not really feasible for the method of regularization, and equality constraints cannot be rigidly enforced. To compare the two methods, the SPFs are all written as quadratic functions of the unknown elements of $\mathbf{b}$ to be compatible with the method of regularization. No constraints beyond those listed in Equations 7 are included in the direct method. In addition, the penalty function approach is more satisfying for experimentally based SPFs, since experimental information cannot be considered exact. When using combined SPFs, they are given equal weight in both methods. Thus, each method has a single parameter, α or ϵ. As mentioned, we focus on experimentally available or physically based SPFs. The minimum variance SPF serves as a reference for comparison, since we have successfully used it in previous work on the same kind of problem [6]. Curve B shows the "best" profile computed by the method of regularization using this SPF. This reasonably satisfactory profile is obtained by varying α, and $\|G(\mathbf{b})\|$ is 346 μV^2. The solution is sensitive to the value of α. At lower values of $\|G(\mathbf{b})\|$, the solution varies

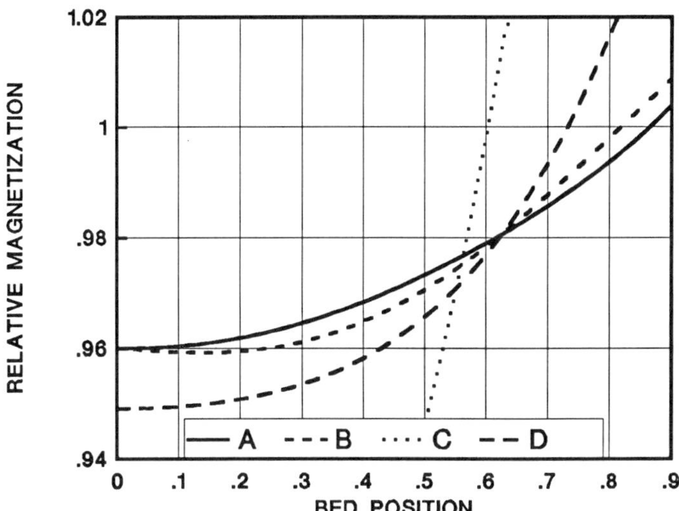

FIGURE 2. See text for description of curves.

wildly with z. At $\|G(b)\|$=358 μV^2, the profile is inverted; *i.e.*, the magnetization decreases with increasing z. "Experimentally", there is no reason to choose a given value of α in the range of α's which provide reasonable profiles. This SPF could provide quantitative profile determinations, if extensive trial calculations are performed. Using the same SPF, the direct method yields a flat profile at the value of the average magnetization.

The magnetization profile calculated by the direct method for the SPF of Equation 8 is shown as curve C in Figure 2. The value of $\|G(b)\|$ for this solution is 344 μV^2. It is not a oscillatory function of z; however, it is not satisfactory. The magnetization changes from 0.848 to 1.304, instead of from 0.960 to 1.003. The minimum value of $\|G(b)\|$ obtained by the direct method is 344 μV^2 with this SPF. For $\|G(b)\|$ = 358 μV^2, the solution is inverted. The minimum value of $\|G(b)\|$ obtained by the method of regularization is 341 μV^2. No profiles are shown for the method of regularization, because they vary wildly with z, even for values of $\|G(b)\|$ greater than 358 μV^2. The computed average magnetization matches the test value for all calculations. Thus, the average magnetization does not provide enough information to compute reasonable profiles.

Curve D of Figure 2 shows the profile computed by the direct method using both Equations 8 and 9 as the SPF. $\|G(b)\|$ for this solution is 343 μV^2. While it is physically reasonable, the minimum variance SPF can give a better profile when using the method of regularization. No corresponding profile computed by the method of regularization is plotted in Figure 2, since the method does not provide a realistic solution for any reasonable value of α. The combined SPF is better for the case of the direct method; however, the solution is sensitive to the value of ϵ used. Extensive trial calculations would be required to obtain quantitative profiles.

Curve B of Figure 3 shows the profile computed by the method of regularization for the SPF of Equation 10; *i.e.*, assuming that the magnetization is known at the ends of the bed. As in the previous cases, it varies wildly. $\|G(b)\|$ is 341 μV^2 for this profile, which is the largest value obtained for any α. Curve D of Figure 3 shows the profile computed by the method of regularization when both Equations 9 and 10 are used as the SPF. The value of $\|G(b)\|$ is 342 μV^2, which is also the maximum value obtained for any reasonable α. The profile is not physically meaningful. Curves C and E of Figure 3 are solutions computed by the direct method using Equation 10 and the combination of Equation 10 and Equation 9, respectively. The value of $\|G(b)\|$ is 358 μV^2 for each of these curves and the solution is not sensitive to the choice of ϵ, for values above 344 μV^2. No solutions are obtained for lower values of ϵ. The value of 358 μV^2 is used since it is the value computed from the generated data and may be estimated from actual data.

Using the combination of all three SPFs, Equations 8 through 10, profiles are computed using both methods. In the method of regularization, the profiles remain oscillatory for any value of α. The profile computed using the direct method is sensitive to the choice of ϵ and does not offer an improvement to the solutions shown as curve C or E in Figure 3. The sensitivity to ϵ is reintroduced by the inclusion of Equation 8 in the combined SPF and is characteristic of the solutions obtained when using only Equation 8 as the SPF.

For the test problem considered, the direct method can yield better results than the method of regularization when physically and experimentally based SPFs are used. Knowledge of the solution values at the endpoints is particularly advantageous, since the solution is not sensitive to the parameter of the direct method (ϵ). The method of

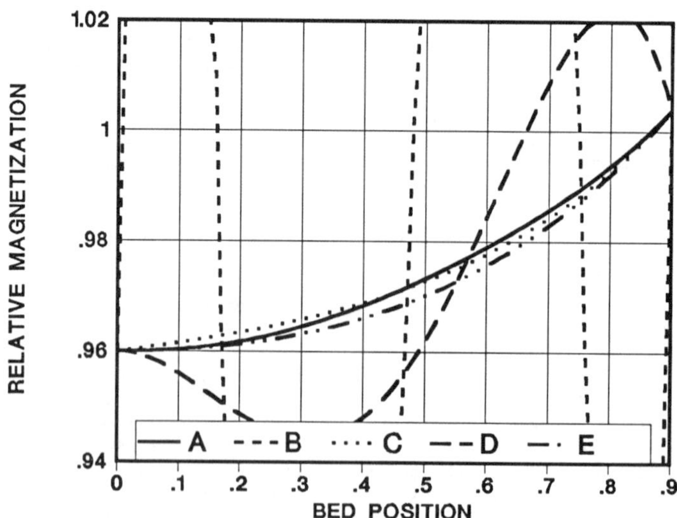

FIGURE 3. See text for description of curves.

regularization, as commonly used, cannot take full advantage of such
information. In situations where the solution obtained is not sensitive
to ϵ, the relatively complicated computations involved in the direct
method are justified since fewer trial calculations are needed to ensure
a reliable inversion. In addition, it is instructive to work directly
with the sum of squared errors, as in the direct method, though there is
a one to one relationship between it and the regularization parameter α
[9].

The direct method has one other distinct engineering advantage over the
method of regularization. Using SQP, multiple inequality and equality
constraints of the solution can be incorporated into the problem. The
latter is also possible in the method of regularization; however, they
should be quadratic to maintain the numerical simplicity of the linear
least squares formulation. This is not a requirement with the direct
method and qualitative information such as a derivative being positive
is easily incorporated into the solution. Thus, properties of the
solution that are known from an engineering point of view can be
included in the solution procedure.

Trial calculations provide insight for the design of experiments. Using
knowledge about shapes of instrument sensitivities and expected physical
properties of interest, experimental configurations can be designed or
suggested from these trial calculations. For example, the results shown
in Figure 3 point out the importance of measuring the temperatures at
the inlet and outlet of the catalyst bed in our experiments. Thus, the
equipment should be designed to provide these values. As another
example, consider the ACP which we describe briefly. If the coils were
redesigned so that the spatial scale of the sensitivity variation was
reduced from 10 cm to 1 cm, the inversion becomes much easier for the
same uncertainty in the data. If at all possible, the integral nature
of the physical property detector should be eliminated. If redesign of
the experiment is not feasible, trial calculations still provide useful
insight into the sensitivity of the measurement to noise in the data and
validity of possible physical or experimental solution property
functions.

CONCLUSIONS

Physically or experimentally available SPFs are desirable in the
determination of physical properties when the inversion of an ill-posed
Fredholm integral equation of the first kind is required. For the
example problem considered, knowing the values at the endpoints is
particularly advantageous. The direct method uses this experimental
information to give more reliable magnetization profiles than the method
of regularization. In addition, the parameter of the direct method is
the upper limit on the sum of squared errors, an integral part of the
problem. In the absence of physical or experimental SPFs, the method of
regularization using a minimum variance about the mean gives reasonably
satisfactory profiles. Extensive trial calculations are necessary to
provide a good value for the regularization parameter α. When either
method is used, trial calculations should be done before the
experimental design is finalized.

ACKNOWLEDGEMENTS

We gratefully acknowledge support of this project by the National
Science Foundation.

NOMENCLATURE

Roman Symbols

A	nominal cross sectional area (cm^2)
b	vector containing the polynomial coefficients
F	solution property function
G	vector of differences between generated and back-calculated data (μV)
H	measurement from linear integral detector
I	magnetization or moment density (arbitrary units)
K	kernel of Fredholm integral equation
ℓ	sample length (cm)
L	range of X of which data are generated (cm)
M	relative magnetization or moment density (dimensionless)
n	order of polynomial representing the magnetization profile
P	physical property being measured
S	response of the entire ACP to a unit moment at X+z (μV/moment)
V	ACP output voltage due to sample (μV)
w	independent variable in P(w)
X	position of sample reference point (z=0) (cm)
y	independent variable in H(y)
z	position in sample relative to sample reference point (cm)

Greek Symbols

α	regularization parameter
ϵ	direct method parameter

Subscripts

j	indices for each data point in the voltage profile
k	indices for polynomial terms of magnetization profile
ℓ	lower bound of b in SQP
u	upper bound of b in SQP
0	nominal or reference value

REFERENCES

1. Collins, R.E, *Mathematical Methods for Physicists and Engineers*.
    Reinhold, New York, 1968.

2. Hildebrand, F.B, *Methods of Applied Mathematics*. Prentice-Hall,
    Englewood Cliffs, New Jersey, 1965.

3. Butler, J.P., Reeds, J.A., and Dawson, S.V., Estimating solutions
    of first kind integral equations with nonnegative constraints and
    optimal smoothing. *SIAM J. Numer. Anal.*, vol. 18, no. 3, pp. 381,
    1981.

4.   Munn, K. and Smith, D.M., A NMR technique for the analysis of pore structure: numerical inversion of relaxation measurements. *J. Colloid. Inter. Sci.*, vol. 119, no. 1, 1987.

5.   Sacher, R.S. and Morrison, I.D, An improved CAEDMON program for the adsorption isotherms of heterogeneous substrates. *J. Colloid. Inter. Sci.*, vol. 70, no. 1, pp. 153, 1979.

6.   Cale, T.S. and Merson, J.A., Magnetic determination of axial catalyst temperature profiles. *AIChE J.*, vol. 35, no. 9, pp. 1428, 1989.

7.   Britten, J.A., Travis, B.J., and Brown, L.F., Calculation of surface site-energy distributions from adsorption isotherms and temperature-programmed desorption data. *AIChE Symp. Ser. No. 230*, vol. 79, no. 7, 1983.

8.   Chemburkar, R., Brown, L., Travis, B., and Robinson, B., Numerical determination of temperature profiles in flowing systems from conversions of chemically reacting tracers. paper 168d presented at AIChE National Meeting, Nov. 28 to Dec. 2, Washington D.C. 1988.

9.   Delves, L.M. and Mohamed, J.L., *Computational Methods for Integral Equations*. Cambridge University Press, Cambridge, 1985.

10. Luo, Z., A numerical method for solving the Fredholm integral equation of the first kind and its application to restore the folded radiation spectrum. *Nuc. Instrum. and Meth. Phys. Res. A*, vol. 255, pp. 152, 1987.

11. Cale, T.S., Nickel crystallite thermometry during ethane hydrogenolysis. *J.Catal.*, vol. 90, no. 1, 40, 1984.

12. Cale, T.S. and Ludlow, D.K., Magnetic crystallite thermometry. *J. Catal.*, vol. 86, no. 2, pp. 450, 1984.

13. Cale, T. S. and Ludlow, D. K., Application of ac permeametry to catalytic crystallite thermometry. *Anal. Inst.*, vol. 13, no. 2, pp. 183, 1984.

14. Schittkowski, K., NLPQL: A Fortran subroutine solving constrained nonlinear programming problems. *Annals of Operations Research*, vol. 5, pp. 485, 1985.

15. IMSL, Math/Library *Fortran Subroutines for Mathematical Applications*. Houston Texas, Version 1.0, 1987.

16. Gill, P.E., Murray, W., Sanders, M.A., and Wright, M.H., Model building and practical aspects of nonlinear programming. *Computational Mathematical Programing*, NATO ASI Series, vol. 15, pp. 209, 1985.

17. Powell, M.J.D., A fast algorithm for nonlinearly constrained optimization calculations. *Numerical Analysis*, ed. G.A. Watson, Lecture Notes in Mathematics, vol. 620, pp. 144, 1978.

18. Stoer, J., Principles of sequential quadratic programming methods for solving nonlinear programs. *Computational Mathematical Programming*, NATO ASI Series, vol. 15, pp. 165, 1985.

Computational Aspects of Accelerated Projection Methods for Nonlinear Integral Equations

DAVID R. DELLWO
Department of Mathematics
United States Merchant Marine Academy
Kings Point, New York 11024, USA

MORTON B. FRIEDMAN and RAJESH AGGARWAL
Department of Civil Engineering and Engineering Mechanics
Colombia University
New York, New York 10027, USA

ABSTRACT

An acceleration procedure is described that significantly improves the convergence rate associated with projection methods for constructing isolated fixed points of nonlinear operators. The technique produces a sequence of identically ranked finite dimensional systems whose solutions define a sequence of successively more accurate approximations to the projected solution and consequently to the fixed point. The focus of this paper is on the computational aspects of the acceleration procedure applied to integral equations. Numerical solutions are constructed for representative cases and it is shown that acceleration is particularly effective when implemented on a parallel computer.

1. INTRODUCTION

Projection methods are a class of conceptually simple approximate procedures for constructing isolated fixed points of nonlinear operators. In particular, they are widely used in the numerical treatment of integral equations [2-4, 6-8]. These techniques employ a bounded finite rank projection P to represent the solution of

$$y = F(y) \tag{1.1}$$

in the form

$$y = Py + (I-P)y \tag{1.2}$$

Here y is an element of Banach space B with norm $||\bullet||$, F is a completely continuous mapping from a closed subset of B into B, and P is chosen so that Py is a good approximation to y.

In general, the projected solution Py can be determined only approximately, as a solution of a set of nonlinear algebraic equations. For instance, the conventional Galerkin approximate $\bar{y}$ to Py is obtained by solving the nonlinear finite-dimensional system

$$\bar{y}-PF(\bar{y}) \tag{1.3}$$

see [2-4, 6-8]. The magnitude of the error $||y - y|| \le ||Py -y|| + ||(I - P)y||$ depends on the choice of elements and dimension of the range space of P. In order to improve the accuracy for a particular type of projection, it is necessary to increase the rank N of the projection. Consequently the solution of (1.3) by Newton's method, for example, requires repeated inversions of N x N updated Jacobian matrices. The work of solving (1.3) grows as N^3 times the number of iterations required to attain a given accuracy.

The present study focuses on the computational aspects of a new approach to projection methods developed by the authors [6,7] in which the fixed point is resolved to any desired accuracy without modifying the structure or increasing the rank of P. Since N is fixed and not necessarily large, the computational work associated with matrix inversion is significantly reduced compared to the conventional approach which requires N to be large for high accuracy.

Accelerated projection methods are discussed briefly in the next section. Detailed theoretical support for both accelerated projection and iterated projection schemes can be found in [7] for linear problems and in [6] for nonlinear problems. Here, the computational attributes of a power series approach to acceleration are illustrated in section 3, where a quadratic two point boundary value problem is analyzed by serial programming methods. The highly efficient parallel implementation of accelerated methods is discussed in section 4, where a linear equation arising in the theory of polymer physics is analyzed as an example.

2. ACCELERATED PROJECTION METHODS

The approach taken in [6] exploits the fact that the components Py and Qy = (I-P)y of the decomposition (1.2) satisfy a system

$$Py = PF(Py + Qy) \tag{2.1a}$$

$$Qy = QF(Py + Qy) \tag{2.1b}$$

in which equation (2.1b), for the infinite dimensional solution component Qy involves a contraction operator QF. It follows that (2.1b) has a unique solution for each Py. If this solution is substituted for Qy, then (2.1a) provides an exact finite dimensional system for Py. Thus, in principle, (2.1) and (1.2) can be employed to construct the fixed point from its finite dimensional component, Py.

In practice (2.1b) is solved approximately for Qy in terms of Py and the solution is used in (2.1a) to obtain an approximate equation for Py. The Galerkin scheme (1.3), for instance, can be derived from (2.1a) using Qy = 0 to approximate the solution of (2.1b). Moreover, a sequence of successive refinements to the Galerkin estimate of Qy can be used in (2.1a) to define a sequence of projection schemes whose solutions provide a sequence of higher order approximations to Py. The approximations to Py and Qy are combined using (1.2) to define a sequence of higher order approximations, $y_n(N)$, $n \geq 1$, to the fixed point. Here n refers to the order of the acceleration scheme and $y_n(N)$ to an approximate fixed point derived from the nth approximation of Qy for a given projection of rank N.

It is shown in [6] that

$$\lim_{n \to \infty} ||y-y_n(N)|| = 0$$

for a sufficiently large but fixed N and also that

$$\lim_{N \to \infty} ||y-y_n(N)|| = 0 \tag{2.2}$$

for each order n. Most important, conditions are obtained in [6] which guarantee that the convergence rates associated with (2.2) increase with n. The sequence of accelerated projection methods exhibit progressively higher algebraic degree but all are of the same dimension N. The first order method (n = 1) is essentially equivalent to the Galerkin technique. The second and higher order schemes involve iterates of F, and possibly Frechet derivatives of F. For an integral operator, it may be necessary to employ numerical integration in their evaluation.

Various construction techniques are available for developing approximate solutions to (2.1b). Alternate methods of approximating Qy yield, from (2.1a), different sequences of finite algebraic systems for the numerical estimation of Py. A power series approach to the solution of (2.1b) is summarized below. The interested reader is referred to [6] for a complete discussion of this method.

The power series technique embeds (2.1b) in the family of equations

$$w(z;\delta) = \delta QF(z + w(z;\delta)), \quad 0 \le \delta \le 1 \tag{2.3}$$

obtained by substituting the symbols z and w for Py and Qy respectively. The relation $Qy = w(Py;1)$ provides the connection to the desired solution of (2.1b). The solution of (2.3) is expanded in the form

$$w(z;\delta) = \sum_{k=0}^{\infty} w_k(z)\,\delta^k \tag{2.4}$$

to obtain a sequence

$$\phi_n(Py) = \sum_{k=0}^{n-1} w_k(Py), \quad n \ge 1 \tag{2.5}$$

of approximations to Qy. The coefficients in (2.4) and consequently the partial sums (2.5) can be evaluated by substituting (2.4) in (2.3) and matching like powers of δ. The first three sums are

$$\phi_1\,(Py) = 0$$
$$\phi_2\,(Py) = QF(Py)$$
$$\phi_3\,(Py) = QF(Py) + QF'(Py)QF(Py)$$

where F' denotes the Frechet derivative of F.

The first order accelerated projection scheme

$$z_1 = PF(z_1) \tag{2.6a}$$

$$y_1 = z_1 \tag{2.6b}$$

is obtained from (1.2) and (2.1) when y, Py and Qy are replaced with y_1, z_1 and $\phi_1 = 0$, respectively. System (2.6) is the Galerkin scheme (1.3). The second order acceleration method is

$$z_2 = PF(z_2 + QF(z_2)) \tag{2.7a}$$

$$y_2 = z_2 + QF(z_2) \tag{2.7b}$$

and the third order method is

$$z_3 = PF(z_3 + QF(z_3) + QF'(z_3)\,QF(z_3)) \tag{2.8a}$$

$$y_3 = z_3 + QF(z_3) + QF'(z_3)QF(z_3). \tag{2.8b}$$

The accuracy of these techniques is discussed in [6].

3. A MODEL BOUNDARY VALUE PROBLEM

Accelerated projection methods (2.6), (2.7) and (2.8) are employed in this section to construct the solutions of

$$y(x) - \lambda \int_0^1 G(x, \tau) y^2(\tau) \, d\tau = 1 \tag{3.1}$$

on $0 \le x \le 1$. Here λ is a real parameter, and

$$G(x, \tau) = \begin{cases} \tau(1-x) & \tau < x \\ x(1-\tau) & \tau \ge x \end{cases}$$

Equation (3.1) is equivalent to the two point boundary value problem.

$$y'' + \lambda y^2 = 0; \quad y(0) = y(1) = 1.$$

In [6], an acceleration technique is employed to determine the critical value of λ at which (3.1) has a singular solution. Here it will be shown how acceleration can be used to construct accurate solutions for noncritical λ's.

The present numerical analysis of (3.1) employs a piecewise constant interpolator

$$Py(x) = \sum_{i=1}^{N} y(x_i^*) e_i(x), \quad N \ge 1 \tag{3.2}$$

supported by a uniform grid: $0 = x_0 < x_1 \cdots < x_N = 1$, $x_i = i/N$, to define the projection. The midpoint and the characteristic function of the interval $(x_{i-1}, x_i]$ are denoted by x_i and $e_i(x) = 1(0)$ for $x_{i-1} < x \le x_i$ (otherwise).

The Galerkin method (2.6a)

$$a_i = 1 + \lambda \sum_{j=1}^{N} \alpha_{ij}^{(0)} a_j^2, \quad 1 \le i \le N \tag{3.3a}$$

determines the coefficients of the first order interpolant

$$y_1(x) = \sum_{i=1}^{N} a_i e_i(x) \tag{3.3b}$$

to the solution of (3.1). The matrix of coefficients in (3.3a) is defined by

$$\alpha_{ij}^{(p)} = T(x_j^*, \frac{1}{N}, p, x_i^*) \qquad\qquad (3.4)$$

with

$$T(m, h, p, x) = (h/2)^{p+1} \int_{-1}^{1} G(x, m + \frac{h}{2}\tau) \tau^p d\tau$$

The estimate

$$\max_{0 \leq x \leq 1} |y(x) - y_1(x)| = 0(N^{-1})$$

for the first order convergence rate is obtained from the results reported in [6].

Likewise, system (2.7a)

$$b_i = 1 + \lambda \sum_{j=1}^{N} \alpha_{ij}^{(0)} b_j^2 + \lambda^2 \sum_{j=1}^{N} \{2\alpha_{ij}^{(1)} b_j K_j(\overline{b}) - \alpha_{ij}^{(2)} b_j^3\}$$
$$+ \lambda^3 \sum_{j=1}^{N} \{\alpha_{ij}^{(2)} K_j^2(\overline{b}) - \alpha_{ij}^{(3)} b_j^2 K_j(\overline{b}) + \frac{1}{4}\alpha_{ij}^{(4)} b_j^4\} \qquad (3.5a)$$

with $1 \leq i \leq N$ determines the coefficients in the second order interpolant

$$y_2(x) = \sum_{i=1}^{N} (b_i + \lambda \sum_{j=1}^{N} [T(x_j^*, \frac{1}{N}, 0, x) - \alpha_{ij}^{(0)}] b_j^2) e_i(x) \qquad (3.5b)$$

to the solution of (3.1). Here b is the vector with components b_i,

$$K_i(\overline{b}) = \frac{1}{N}\sum_{k=i}^{N} b_k^2 - \frac{1}{2N} b_i^2 - \frac{1}{N}\sum_{k=1}^{N} x_k^* b_k^2, 1 \leq i \leq N$$

and the convergence rate $0(N^{-2})$ follows from the error estimates reported in [6]. The coefficients of the third order interpolant

$$y_3(x) = \sum_{i=1}^{N} (c_i + \lambda \sum_{j=1}^{N} [T(x_j^*, \frac{1}{N}, 0, x) - \alpha_{ij}^{(0)}] c_j^2$$
$$+ 2\lambda^2 \sum_{j=1}^{N} [T(x_j^*, \frac{1}{N}, 1, x) - \alpha_{ij}^{(1)}] c_j K_j(\overline{c})$$
$$- \lambda^2 \sum_{j=1}^{N} [T(x_j^*, \frac{1}{N}, 2, x) - \alpha_{ij}^{(2)}] c_j^3) e_i(x) \qquad (3.6a)$$

defined by (2.8b) satisfy

$$c_i - 1 + \lambda \sum_{j=1}^{N} \alpha_{ij}^{(0)} c_j^2 + \lambda^2 \sum_{j=1}^{N} \{2\alpha_{ij}^{(1)} c_j K_j(\overline{c}) - \alpha_{ij}^{(2)} c_j^3\}$$

$$+ \lambda^3 \sum_{j=1}^{N} \{4\alpha_{ij}^{(4)} c_j \overline{K}_j(\overline{c}) + \alpha_{ij}^{(2)} K_j(\overline{c}) - \frac{5}{3} \alpha_{ij}^{(3)} c_j^2 K_j(\overline{c}) + \frac{5}{12} \alpha_{ij}^{(4)} c_j^4\}$$

$$+ \lambda^4 \sum_{j=1}^{N} \{4\alpha_{ij}^{(2)} K_j(\overline{c}) \overline{K}_j(\overline{c}) - 2\alpha_{ij}^{(3)} c_j^2 \overline{K}_j(\overline{c}) - \frac{2}{3} \alpha_{ij}^{(4)} c_j K_j^2(\overline{c})$$

$$+ \frac{1}{2} \alpha_{ij}^{(5)} c_j^3 K_j(\overline{c}) - \frac{1}{12} \alpha_{ij}^{(6)} c_j^5\}$$

$$+ \lambda^5 \sum_{j=1}^{N} \{4\alpha_{ij}^{(2)} \overline{K}_j^{(2)}(\overline{c}) - \frac{4}{3} \alpha_{ij}^{(4)} c_j \overline{K}_j(\overline{c}) K_j(\overline{c}) + \frac{1}{3} \alpha_{ij}^{(5)} c_j^3 \overline{K}_j(\overline{c})\}$$

$$+ \frac{1}{9} \alpha_{ij}^{(6)} c_j^2 K_j^2(\overline{c}) - \frac{1}{18} \alpha_{ij}^{(7)} c_j^4 K_j(\overline{c}) + \frac{1}{144} \alpha_{ij}^{(8)} c_j^{(6)}\} \qquad (3.6b)$$

with $1 \leq i \leq N$ and

$$\overline{K}_i(\overline{c}) - \frac{1}{24N^3} \sum_{j=1}^{N} x_j^* c_j^3 - \frac{1}{12N^3} \sum_{j=1}^{N} c_j K_j(\overline{c}) + \frac{1}{8N^2} c_i K_i(\overline{c})$$

$$+ \frac{1}{48N^3} c_i^3 - \frac{1}{24N^3} \sum_{j=i}^{N} c_j^3$$

where c is a vector with components c_i. The third order convergence rate is $0(N^{-3})$

The quadratures in (3.4) were evaluated exactly and systems
(3.3a), (3.5a), (3.6b) solved by Newton's method using serial
programming techniques. For illustration, λ was chosen to be 1.5.
Figure 1 is a sketch of the solutions to (3.1) for $\lambda = 1.5$ obtained from the second order interpolant (3.5b) with N = 64.

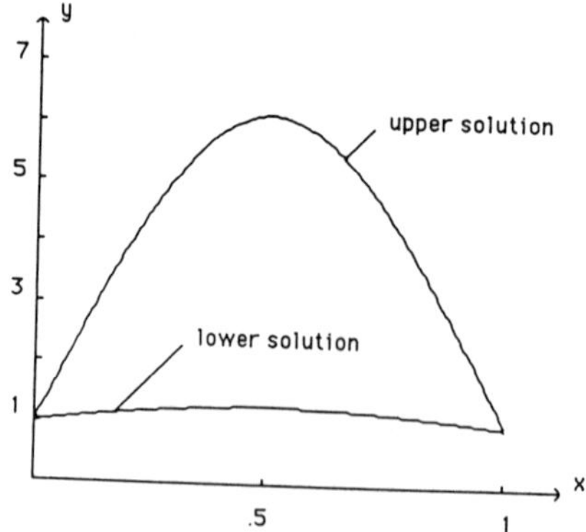

FIGURE 1. The solutions of (3.1) with $\lambda = 1.5$. As λ increases,
numerical evidence suggests that $\lambda = 2.42060$ (see [6]) corresponds
to a "fold," a merging of solutions.

Estimates for the maximum errors $||y\text{-}y_n||$, $n = 1,2,3$ associated with (3.3b), (3.5b), (3.6a) employing 4,8,16 and 32 nodes are shown in Tables 1 and 2 for $\lambda = 1.5$. These estimates were obtained by evaluating

$$\max_{1\leq i \leq 100} |y^*(i/100) - y_n(i/100)|, \quad n = 1,2,3$$

with $y^*(x)$ given by the 64 node second order result (3.5b).

	N=4	N=8	N=16	N=32
1st order	1.9E-0	9.6E-1	4.8E-1	2.4E-1
2nd order	6.2E-2	8.0E-3	9.4E-4	9.9E-5
3rd order	4.2E-2	6.4E-3	8.8E-4	1.1E-4

TABLE 1. Maximum Error Associated with the Calculation of the Upper Solution to (3.1) with $\lambda = 1.5$ by the Accelerated Projection Methods (3.2), (3.4) and (3.5).

Inspection of the tables clearly indicates the Galerkin method (3.3a) converges at a slower rate than the second and third order schemes. The orders of magnitude of reduction in the error produced by the higher order schemes confirms the power of the acceleration method. Surprisingly, the tables suggest the second order method (3.5) converges at the same rate, $O(N^{-3})$, as the third order scheme (3.6) for this calculation. It is reasonable to conjecture that the unexpected accuracy of (3.5) is due to the midpoint choice for interpolation. However, no attempt has been made to obtain the refined estimate required to establish this rigorously.

	N=4	N=8	N=16	N=32
1st order	1.2E-1	6.0E-2	3.3E-2	1.7E-2
2nd order	1.1E-3	1.4E-4	1.7E-5	1.9E-6
3rd order	1.5E-4	2.1E-5	2.7E-6	3.1E-7

TABLE 2. Maximum Error Associate with the Calculation of the Lower Solution to (3.1) with $\lambda = 1.5$ by the Accelerated Projection Methods (3.2), (3.4) and (3.5).

4. PARALLEL VARIANTS OF ACCELERATED PROJECTION METHODS

The focus of discussion in this section is the parallel implementation of accelerated projection methods. However, for nonlinear problems, this approach is still under development [1]. In order to demonstrate the inherently parallel nature of the acceleration method, a weakly singular integral equation is chosen as an example.

$$y(x) = x^2 - 1/2 \int_{-1}^{1} |x-\tau|^{-1/2} y(\tau) \, d\tau, \quad |x| \leq 1, \tag{4.1}$$

This equation which appears in the theory of intrinsic viscosity
[9], is analyzed by the acceleration method developed by the
authors in [7] for linear systems. The approach is equivalent to
a Picard solution of (2.1b) in the general nonlinear setting of
[6]. The first, second and third order projection schemes are
discussed here. They employ the piecewise constant interpolant
(3.2) on a uniform mesh: $-1 = x_0 < x_1 < \cdots < x_N = 1$, $-1 + 2i/N$.
The midpoint and characteristic function of the interval $(x_{i-1}, x_i]$
are denoted by $\overset{*}{x_i}$ and $e_i(x)$ respectively.

The discrete first order system is

$$u_1 - PKu_1 = Pg + PKQg \tag{4.2}$$

where $g(x) = x^2$, $u_1(x)$ approximates Py in (3.2) and K denotes the
integral operator

$$Ky(x) = -\frac{1}{2} \int_{-1}^{1} |x-t|^{-1/2} y(t) \, dt.$$

The first order solution is $y_1 = u_1 + Qg$. The discrete second
order system is

$$u_2 - PK(I + QK)u_2 = Pg + PK(I + QK)Qg \tag{4.3}$$

where $u_2(x)$ is the approximate interpolant and
$y_2 = (I + QK)(u_2 + Qg)$ is the approximate fixed point. The third
order scheme is

$$u_3 - PK(I + QK + QKQK)u_3 = Pg + PK(I + QK + QKQK)Qg$$

$$y_3 = (I + QK + QKQK)(u_3 + Qg) \tag{4.4}$$

Systems (4.2), (4.3) and (4.4) were analyzed by serial means in
[7] to estimate the solution of (4.1) at the end points $x = \pm 1$
where the effects of the singular kernel is most pronounced. The
detailed matrix development of these schemes can be found there.

Parallel computation and assembly of the terms in (4.2), (4.3) and
(4.4) is split over p processor as described in [1]. The required
inversions (i.e. solving the systems) are performed by a single
processor since set-up time dominates the solution process for
each of the small systems. As higher and higher order schemes are
considered, the inversion step is a rapidly shrinking proportion
of the execution time. Although the algorithm employed is appropriate for both shared memory and message passing machines, it was
implemented on the Balance 21000 Computer Systems [5]. The
architecture of that machine features 32-bit CPUs operating
independently but sharing a common central memory through a system
bus. Each CPU has 8K of local RAM and 8K of cashe RAM.

The objective in using parallel processing is to speed up the
elapsed time of an application. The speedup of p processors

$$S_p(n,N) = \frac{time\ using\ one\ processor}{time\ using\ p\ processors} \qquad (4.5)$$

is a widely used measure of performance of a parallel algorithm.
The fact that the speedup depends on the order n and the rank N of
the projection method is explicitly indicated in (4.5).

The speedup associated with the assembly of terms in (4.2), (4.3),
(4.4) is sketched in Fig. 2 for N=50. Figure 3 contains a similar
sketch of the speedup but for N=100. End point errors (at x=1)
associated with each of the schemes are also indicated on the
figures.

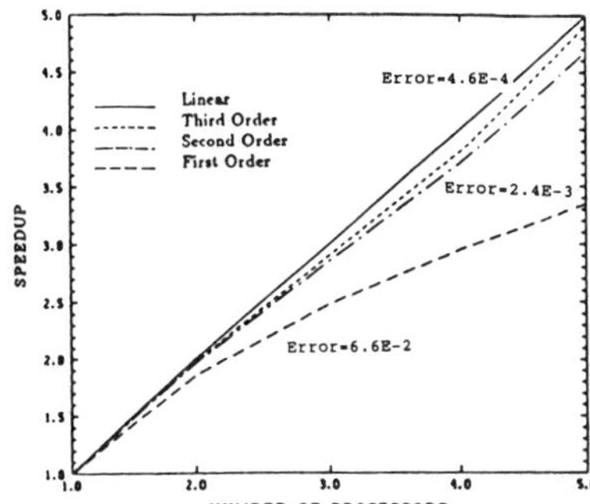

FIGURE 2. The assembly speedup for systems (4.2), (4.3) and (4.4)
with N=50.

Systems (4.2), (4.3), (4.4) are each of size N x N. Thus parallel
assembly is expected to be more efficient for larger N. This is
confirmed by Figs. 2 and 3. Most importantly, it is apparent from
the graphs that the speedup improves with increasing order n. the
speedup of the parallel 3rd order scheme is more nearly linear
than the speedup of the parallel 2nd order scheme, which in turn
is more nearly linear than the speedup of the parallel 1st order
scheme. The improved performance is a consequence of the observation, (see [1]) that the rate of growth of the total computation
with increasing order greatly exceeds the rate of growth of
communication required by the computation.

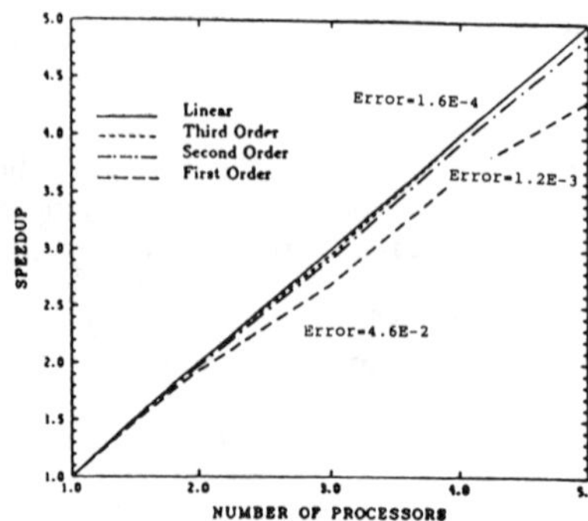

FIGURE 3. The assembly speedup for systems (4.2), (4.3) and (4.4) with N=100.

These results confirm that at least in the linear case, accelerated methods are particularly effective in a parallel environment. If the efficiency of the parallel implementation is measured by the ratio of the speedup S_p to the number of processors p, then the third order method clearly provides better than 90% efficiency as seen by examining Figs. 2 and 3. It is possible to achieve even greater efficiencies by increasing N/p because more uniform load balancing occurs as the amount of work per processor increases.

The object of accelerations is to obtain an accurate solution utilizing small systems N. It is evident from the results that the more accuracy desired, the more efficient becomes the parallel implementation.

REFERENCES

[1]   Aggarwal, R., Dellwo, D.R. and Friedman, M.B., *Parallel Algorithms for Accelerated Projection Methods*, (forthcoming)

[2]   Atkinson, K.E., *A Survey of Numerical Methods for the Solution of Fredholm Integral Equations of the Second Kind*, Society for Industrial and Applied Mathematics, Philadelphia, 1976.

[3]   Atkinson, K.E. and Potra, F.A., *Projection and Iterated Projection Methods for Nonlinear Integral Equations*, SIAM J. Numer. Anal., 24 (1987), pp. 1352-1373.

[4]   Baker, C.T.H., *The Numerical Treatment of Integral Equations*, Oxford University Press, Oxford, 1978.

[5]   Balance Technical Summary, November 1987.

[6]   Dellwo, D.R. and Friedman, M.B., *Accelerated Projection and Iterated Projection Methods with Applications to Nonlinear Integral Equations*, SIAM J. Numer. Anal., (forthcoming).

[7]   Friedman, M.B. and Dellwo, D.R., *Accelerated Projection Methods*, J.Comput. Phys., 45 (1982), pp. 108-126.

[8]   Kantorovich, L.V. and Akilov, G.P., *Functional Analysis in Normed Spaces*, Pergamon Press, New York, 1964.

[9]   Kirkwood, J.G. and Riseman, J., *Chem. Phys.*, 16 (1948), pp. 565-573.

CONTINUUM MECHANICS

Summary

The papers in this section cover a broad range of topics in continuum mechanics.

An invited paper "Multiple Stable Solutions to the Dynamical Equations of An Excitable Medium," by A. T. Winfree, Regents Professor at the University of Arizona, presents particle-like stable solutions of a certain class of partial differential equations, describing a 3-dimensional excitable medium. Such media play fundamental roles in phenomena of interest to many scientists.

In the first paper, Constanda discussed the rigorous mathematical treatment of the boundary problems of the classical Kirchhoff plate theory, via the boundary integral equation method. A scheme for the approximate computation of the solution was introduced and a numerical example was given.

Bergmann, Nambiar and Lawrence developed a closed form, hierarchic triangular element stiffness matrix for the axisymmetric class of elasticity problems. In addition, a closed form local error estimate in the energy norm using simple nodal stress averaging as an estimate of the exact solution was presented.

Nomura and Chang presented an integral method approach to hygrothermal behavior of composite laminates by using symbolic algebra software. Galerkin functions satisfying given homogeneous boundary conditions were derived analytically using symbolic software and manipulated to generate the elements in the matrix algebraic equations. An example for the thermal buckling of composites was given.

The last paper in this section was presented by Öğmen in which he introduced an analytical model to study the effects of neuropharmacological agents in a neural network and investigated a directionally selective motion detection model for the fly visual system. The results were compared with the experimental data obtained by Schmid-Bülthoff.

Tseng Huang

Multiple Stable Solutions to the Dynamical Equations of an Excitable Medium

A. T. WINFREE
326 Biological Sciences West
Department of Ecology and Evolutionary Biology
University of Arizona
Tucson, Arizona 85718, USA

INTRODUCTION

This paper displays some recently discovered particle-like stable solutions of a certain class of partial differential equations. These equations describe a 3-dimensional 'excitable medium'. Such media play fundamental roles in phenomena that interest physical chemists and physiologists. These stable objects were discovered while trying to understand a kind of turbulence called 'fibrillation' in an excitable medium called 'heart muscle'. Similar objects occur in a chemically excitable medium called the 'Belousov-Zhabotinsky reagent'. They are probably endemic to all sorts of excitable media. They were discovered numerically, using supercomputers to solve the partial differential equations of reaction and diffusion in a motionless 3-dimensional simply-connected continuum.

EXCITABILITY

The general idea of excitability, in a 0-dimensional or spatially uniform sense, is that a small but not too-small stimulus, applied to such a dynamical system while it is in a certain range of states ('excitable' states), may provoke it to execute a relatively large and rapid excursion. In an excitable **medium**, the local dynamics are excitable and neighboring regions are so coupled that excitation in one region can provide its neighbors with the kind of stimulus required to provoke excitation there. A solitary pulse of asymptotically stable shape and speed then propagates by this mechanism. Molecular diffusion in a chemical system, for example, can mediate such coupling.

A chain of dominoes is a discrete 1-dimensional excitable medium, in the sense that if one domino is perturbed far enough off equilibrium to fall over, and if its neighbors are near enough and not too firmly anchored at their bases, then this impulse will propagate at an asymptotically constant speed. A prairie of dry grass is a nearly continuous 2-dimensional excitable medium in the sense that if a sufficient nucleus of fire ignites it, then the flame front spreads at a speed characteristic of the medium. A keg of gunpowder is a 3-dimensional excitable medium in a similar sense.

In the domino and gunpowder examples a 1-shot transition propagates: the medium moves from one locally-attracting equilibrium to another and sticks there. Or, as in the grassfire example, there may *be no* attracting second equilibrium: the grass grows back spontaneously, eventually to dry again. This is a **regenerative** excitable medium, in which the unique spatially uniform steady-state (that consists of new grass growing in a layer of last-season's dry grass tinder) is a global attractor, but if the system is locally heated enough (that is to say, if it is perturbed far enough in certain direction in state space from that asymptotic quiescence), then before returning it first makes a much larger excursion in the same direction. This can trigger adjacent sites, if sites are spatially coupled in an appropriate way. Nerve membrane and heart muscle membrane are much like prairie grass in this respect. So are certain kinds of gaseous and liquid chemical solutions. In chemical and electrophysiological cases the coupling mechanism can be accurately described mathematically by the Laplacian operator, representing flux of molecular species' in the chemical examples,or the gradient of electric current in the neurobiological examples.

One simple idealization of excitability in nerve membranes was provided during the middle of this century by K.F. Bonhoeffer in Germany, Richard FitzHugh in the United States, and Jin-Ichi Nagumo in Japan. It is written as a partial differential equation in Equation (1):

$$
\begin{aligned}
(1) \qquad \partial u/\partial t &= (u - v - u^3/3)\,/\varepsilon + \mathbf{D}\,\nabla^2 u \qquad &\text{'propagator'} \\
\partial v/\partial t &= \varepsilon\,(u - \gamma v + \beta) \qquad &\text{'controller'}
\end{aligned}
$$

At each point in a spatial continuum, two local state variables, denoted u and v, change continuously in time. The local kinetics are described by a second order ordinary differential equation. The kinetics adopted are quite simple: except for one term, rates of change are linear in these two variables. Quantity u, the 'excitation variable' or 'propagator variable', is degraded at a rate proportional to v, the 'recovery variable' or 'controller variable', and is autocatalytically regenerated in proportion to u. Adding degradation at rate $u^3/3$, the first equation's nullcline becomes a cubic, intersecting the recovery variable's linear nullcline at one to three points, depending on parameters. The recovery variable, v, grows at a rate proportional to u plus a constant, while also undergoing first-order decay. Intercept β and slope γ may be chosen to create an intersection between the linear and cubic nullclines (equilibria) on the phase portrait. If the local state is perturbed from an attracting equilibrium by displacing u across the middle branch of the cubic nullcline (thus crossing a threshold), then it departs farther before returning toward equilibrium. Parameter ε governs the ratio of rates of excitation and regeneration. If ε is small, then u-motion is much quicker than v-motion: from equilibrium u rapidly increases until it crosses the upper branch of the cubic, then glides along it until 'falling off' to the lower branch, which escorts it back to the original equilibrium. The 'ε' of Equation (1) is the '$\sqrt{\varepsilon}$' of Tyson and Keener (1988).

We make an 'excitable **medium**' by spatially coupling a continuum of such sites, usually by means of the second spatial derivative of u. The local time rate of change of u now consists of local excitable kinetics plus a term representing diffusion of u from adjacent sites.

Omitting term $u^3/3$, equation (1) would be a linear partial differential equation. Its analytical solution would relax boringly and promptly from any initial distribution of u and v toward uniform quiescence. But with the nonlinear term intact, the local dynamic is excitable and we encounter the possibility of interesting behavior: an action potential may propagate like a shock front. This is an 'excitable medium'.

We now confront a problem analogous to classical problems in physics: a 3-dimensional continuum has been described in idealized terms by a local partial differential equation, and we wish to make use of this formulation to discover the medium's physically significant behaviors. In this case, unlike the analogous vector-field problems in a Navier-Stokes medium or Maxwell medium, we deal only with a field of real numbers. Also there is but a single nonlinear term involved. The problem takes the form of a general reaction-diffusion system:

$$(2) \qquad \partial \underline{u}(\underline{x},t)/\partial t = \underline{K}(\underline{u}(\underline{x},t)) + \mathbf{D}\, \nabla^2 \underline{u}(\underline{x},t)$$

where $\underline{u} \; \varepsilon \; \mathcal{R}^n$ of state space, $\underline{x} \; \varepsilon \; \mathcal{R}^3$ of physical space, and $\mathbf{D} \; \varepsilon \; \mathcal{R}^{n \times n}$ is a square matrix of real-valued diffusion coefficients (usually a diagonal matrix, often with only D_{11} non-zero).

ORGANIZING CENTERS

By definition, an excitable medium supports a pulse-like plane wave or a train thereof. A train of equispaced traveling pulses is equivalent to a single pulse circulating on a 1-dimensional ring. There is a 1-parameter family of solutions, depending on wave spacing or ring perimeter. One might expect similar circulation on a thin enough planar annulus or within a thin enough solid torus. This circulating solution persists as the annulus or torus fattens. Surprisingly, it persists even when the hole in the annulus or torus is finally closed. As the hole closes the 1-parameter family of solutions converges to a unique vortex-like solution called a 'rotor'. From this rotating vortex a spiral-shaped wave radiates. In 2 dimensions these vortices have a discrete spectrum of distinct (period,wavelength) combinations determined by the medium's parameters. They also differ in rotating either clockwise or anti-clockwise. In 3-dimensions there are vortex **rings** of various sizes and corresponding periods, but there are no more left-right enantiomers to distinguish. There are also fancier vortex rings, involving topologically distinct knots and links and their enantiomers.

Without such 'organizing centers', as they are called, a medium with globally attracting equilibrium promptly reverts to uniform quiescence, or if stimulated, to more-or-less uniform response to each stimulus with intervals of quiescence between stimuli. This is the normal situation in human heart muscle, for example. But heart muscle, like any other excitable medium, is also susceptible to periodic modes of self-organization. The study of periodic self-organized activity in excitable media is largely the study of organizing centers made of vortex filaments. Such organizing centers may play a role in short-period cardiac arrhythmias, e.g. those leading to ventricular fibrillation and sudden death (Winfree 1988, 1989, 1990bcde).

NUMERICAL SOLUTIONS

When it comes to quantitative questions, topology is no help. How small can an organizing center be? What is the characteristic period of vortex rotation? Do they move only at 'subsonic' speeds, << wave propagation speed? Do they collide and reconnect? Are any organizing centers stable? To answer such questions one must solve the equations in some analytical approximation or attempt numerical experiments. So far no-one has been able to do it analytically, so we resort to numerical experiments.

Two-Dimensional Behavior

Numerically solving this kind of equation is not difficult in two dimensions. Vortices were first obtained from such equations numerically in the early 1970's. Once started, the vortex is tremendously resistant to attempts at erasure, for reasons that have much to do with topology. Its **dynamics,** however, cannot be inferred from mere arguments of continuity: computations and laboratory experiments are required.

Compound rotary motion. In some versions (and not others) of the Belousov-Zhabotinsky chemically excitable medium it was noticed that the vortex center spontaneously moves. As long as 15-20 years ago paths as simple as a repeated 3-petal flower were recorded, but irregular paths were equally common, so this behavior was called 'meander'. What aspects of this motion derive from time-varying non-uniformities in the medium, and what aspects are inherent in equations of form (2) in a uniform 2-dimensional context? The question was posed, but for a decade not addressed. It was also proposed that meander is an essentially 3-dimensional phenomenon, stemming from the **twist** imposed on vortex lines by a gradient of chemical parameters away from the liquid's air interface (Agladze et al. 1988). At the 2-dimensional stage of debugging the first 3-dimensional supercomputer simulations of excitable media, Nandapurkar and Winfree (1985 unpublished) were surprised to observe regular flower-like motion of vortex centers in a variety of simple models. Jahnke et al. (1989) pursued this phenomenon through systematic variation of parameters in the Oregonator model of the Belousov-Zhabotinsky medium. They found a locus in parameter space at which 'meander' bifurcates by the appearance of a **second** discrete frequency close to the original frequency of rigid rotation. Beyond this border, the vortex center moves along a path that can be described as a superposition of two rotations. Close to the bifurcation border, both can be accurately described as uniform-speed circles, but the loops are less symmetric in more fully developed meander. Much the same motion, dubbed 'epicycloidal', was been discovered in numerical experiments on more abstract models by Zykov (1986, 1987). These were repeated, extended, and redescribed by Lugosi (1989). Turning to laboratory experiments, the same phenomena were then sought in the Belousov-Zhabotinsky reagent, and found in surprising detail (Jahnke et al. 1989). By changing recipe parameters (acidity and bromate concentration), the amplitudes and frequencies of the two components could be varied. Accidental non-uniformities in the medium played an important modifying role but when they were finally eliminated, floral meander persisted stably during dozens of vortex rotations (but not for as long as 100 rotations because the medium gradually depletes essential reactants, with eventually severe impact on parameters of wave propagation.) Meander was thus found computationally, and then in the laboratory, to originate from nothing more than the interplay of reaction and diffusion in uniform, isotropic, 2-dimensional continua. Meron (1991) has discovered a bifurcation in spiral-wave solutions to the reaction-diffusion equations of excitable media, which may correspond to the onset of floral meander.

The discovery that meander in uniform 2-dimensional media is deterministic and not chaotic, though still not understood in terms of dynamical systems theory, nonetheless subverted the dogma of 20 years standing that the rotor far from any boundaries in a uniform excitable medium simply spins in place, motionless in the absence of an intrinsically preferred direction such as might be provided by a parameter gradient or anisotropy in the medium.

Distinct rotor types in a given medium. A long-standing dogma holds that a given excitable medium like equation (2) has a unique rotor solution: whatever initial conditions may be contrived, they create some field of propagating waves that eventually radiate away, plus some residual assortment of left- and right-handed rotors. All these rotors radiate waves of exactly the same period and wavelength (unless some are close enough together to interact, or equivalently, half that far from a no-flux boundary; the necessary closeness is less than twice the wavelength/π, the nominal rotor diameter). The period, wavelength, and possible motion of the rotor are universally acknowledged to be unique functions of the medium's parameters, independent of initial conditions (with the recent pseudo-exception noted above, that the phase of a meander pattern, like the phase of a rigid rotation and the geographical location of the rotor, is set by initial conditions). Every computation for 20 years and every laboratory experiment in a uniform medium has had this result. But one could doubt, at least in principle, at least in media with non-monotonic dispersion curves. The dispersion curve plots the speed of a periodic wavetrain against the wavelength (in one representation; other familiar representations plot frequency vs. wavenumber, period vs. wavelength, etc.) Suppose the dispersion curve had zig-zag segments, or even isolas, corresponding to alternative pulse profiles and wave speeds at a given wavelength. There are abundant examples, e.g. the piecewise-linear kinetics first used to demonstrate the stability of computational rotors (Winfree 1974ab, 1978); in this particular case the wavelengths at which stable pulse trains may have discretely different speeds happens **not** to be close to the familiar and apparently unique rotor's wavelength, but this instance suggests the possibility that whatever principle selects the rotor's location on the dispersion curve, it might select more than one location. The consequences of non-monotonicity in dispersion curves have been elegantly explored by Karfunkel and Seelig (1977), Elphick et al. (1986), and Meron (1989), and probably have a bearing on the following observations.

Suppose my original guess were correct, that speed * wavelength $\cong 8\pi^2 D$ in the wavetrain emanating from any rotor (Winfree 1973; Jahnke et al. 1989) or that the theory of Keener (1986) and Tyson and Keener (1988) for a spiral wave circulating around a hole could be extrapolated to arbitrarily small holes (with the result that speed * wavelength $\cong 6\pi D$), or more generally, that the wavetrain radiated from the rotor must lie not only on the dispersion curve but also on some intersecting 'critical curve' of generally hyperbolic appearance on the (speed, wavelength) plane. Then could there not be multiple intersections? Last winter I went looking for some, hill-climbing in the 3-parameter space of equation (1) to enhance non-monotonicity in the dispersion curve. Such a case was found in the medium in which stable vortex rings were first demonstrated, viz. equation (1) with $\varepsilon = 0.30$, $\beta = 0.7$, $\gamma = 0.5$ (Skaggs et al. 1988, Courtemanche et al. 1990). In this medium there are two alternative rotors, the originally known rotor with wavelength 21 (period 11, wavespeed 1.9, and a narrow gap in which the recovered medium lingers near its excitable equilibrium awaiting the next front) and another with wavelength 29 (period 17, wavespeed 1.7, and a wide excitable gap but otherwise essentially the same pulse shape). This simple counter-example (illustrated in Winfree 1991ab) excludes the comfortable idea that the behavior of vortex filaments (their lateral motions per unit time or per rotation period as functions of local curvature, for example) must be uniquely determined by the intrinsic properties of the medium. Watching the dispersion curve change, one does indeed find the spiral wavelength and period always clinging near the dispersion curve for that ε, and the values observed do indeed trace a monotone decreasing critical curve parametrized by ε --- but it looks little like a hyperbola and in any case lies far outside the foregoing estimates. There seems still much to learn before rotors are quantitatively understood in a mathematical way.

These explorations concerned 2-dimensional vortices, otherwise interpreted as trivially 3-dimensional vortex lines without curvature. A uniformly curved (circular) filament is the same in every radial 2-dimensional cross-section, so only one need be computed. The equation to be solved is the same except for a transformation of coordinates to reduce 3 dimensions to 2 by rewriting the Laplacian operator then noting that azimuthal derivatives vanish identically, by the imposed symmetry. (For examples see Nandapurkar and Winfree (1989) and Winfree and Jahnke (1989)). Such vortex rings typically drift along their symmetry axis, meanwhile shrinking, at a rate that depends on the radius. But those dependences are utterly different for the vortex rings based on the period-11 rotor and the period-17 rotor: the former shrinks to a preferred stable radius, while the latter expands, apparently without limit, at a rate that increases the area of the circle's disk at a fixed rate.

Three-Dimensional Behavior

Computation in three dimensions without resort to special symmetry constraints --- in order to study curved, helical, linked, knotted vortex filaments --- is not more complicated but is much more time-consuming. If at least 50 x 50 gridpoints suffice in two of those dimensions, then at least 50 times that many are needed to represent a box of the same size in three dimensions, and a box about twice this large in at least two directions is needed for the vortex filament to bend into a closed ring. Thus we need $10^{5\text{-}6}$ gridpoints, each recording the local u and v (and additional variables in the more complex dynamics required for realistic representation of the chemical kinetics or membrane excitability). In each rotation using explicit Euler method with fixed space grid and time step, $10^{2\text{-}3}$ incrementations are needed, depending on the fineness of spatial resolution required for tolerably accurate simulation of a continuum. To follow the evolution of the singular filament we need to continue the computation through $10^{1\text{-}2}$ such rotations. Thus we need $10^{8\text{-}11}$ updates of u and v per to this reaction-diffusion equation.

This is what we have done, and the rest of this paper concerns only the **results** of such computations. The simple methods implemented on Cyber 205, ETA-10, and Cray XMP supercomputers are thoroughly discussed by P. Nandapurkar in Zykov's book "Simulation of Excitable Media" (1988). The general organization of the FORTRAN integrator utility is described in Lugosi and Winfree (1988). Utilities for geometric analysis of the results are outlined in Winfree and Guilford (1988), Winfree (1990a), and Henze et al. (1990).

<u>Extracting results from numerical experiments.</u> Having a look at the results is less easy and has been the main focus of our concern during the past 2 years. The results at any instant consist of an array of about a million concentrations. Most of them represent waves propagating through the bulk of the medium. These waves are little different from those found in 1-dimensional pulse propagation, except near the vortex filament, where they may be so sharply curved that their propagation speed is much affected. The novel part of an organizing center is thus the immediate neighborhood of the vortex filament, their source. There are several ways to extract it numerically, throwing away the larger volume occupied merely by outbound waves of uniform spacing and negligible curvature. For example we may watch every gridpoint during one period of vortex rotation, retaining only those which failed to excite. Or we may, very nearly equivalently, at each instant retain only those gridpoints whose u and v (or whatever) concentrations lie **not** along the usual trajectory of excitation and recovery, but deep in the interior of that loop in state space. Or we may locate the inner edge of the spiral wave rotating around the filament by finding the instantaneous maximum of the vector cross product of u and v gradients.

Having done this we then sample the local geometry at each of 100 equispaced stations along the vortex filament. Then the u(x,y,z) and v(x,y,z) arrays are updated and a similar snapshot is obtained again. Each arc between stations on the original filament is found to have moved. That vector displacement, divided by the time between snapshots, is resolved into perpendicular components: V_n along the local Frenet frame's normal vector (in the direction of the filament's curvature, k) and V_b along the binormal vector. (The Frenet frame is the orthogonal triplet of unit vectors tangent to the filament, perpendicular to it in the direction of the center of its local curvature, and perpendicular to those two). Both components of velocity are plotted against quantities describing the local geometry, in hopes of discerning a simple law of motion. What are the quantities describing local geometry? There are only two (plus perhaps their derivatives): the curvature of the filament and the twist rate of its chemical gradients, as functions of arclength along the filament (Winfree and Guilford, 1988; Keener, 1989; Winfree,

1990a; Henze et al. 1990). Both can be determined automatically by proven numerical procedures.

Examples of Stable Organizing Centers

<u>Vortex rings.</u> The simplest organizing center is a vortex ring. Its filament is a perfect circle of some (generally changing) radius, r(t). In the case of equal diffusion, the filament was analytically expected (Panfilov and Pertsov, 1984) and computationally found to shrink at a rate equal to the product of its curvature, $k = 1/r$, times the diffusion coefficient:

$$dr/dt = Dk = D/r,$$

or in time units of one rotation period, τ_0, and one spiral wavelength, λ_0,

$$\tau = t/\tau_0, \quad \rho = r/\lambda_0,$$

(3) $\quad d\rho/d\tau = [D/(\lambda^2_0/\tau_0)] /\rho$

This is convenient since, according to the estimates given above, a rotor's $(\lambda^2_0/\tau_0) \approx \alpha D$, where α ranges from about 20 to 100; thus

$$d\rho/d\tau \approx (0.01 \text{ to } 0.05) /\rho, \text{ and}$$

$$\rho^2 - \rho_0^2 \approx (2 \text{ to } 10\%) (\tau - \tau_0), \text{ independent of reaction and diffusion specifics.}$$

If diffusion coefficients are **not** equal, shrinkage need not continue all the way to extinction. In some excitable media, shrinkage stops just before the hole inside the ring has been squeezed shut. Then we retain a stable vortex ring, which glides through the medium like a particle with momentum. This particle-like solution was discovered by Skaggs et al. (1988), and examined in more detail by Courtemanche et al. (1990), using equation (1) with parameters limited to a narrow range around $\varepsilon = 0.3$, $\beta = 0.7$, $\gamma = 0.5$, outside which the great stability of this solution is lost. (This same range supports the multiple alternative rotors mentioned above.)

Using equation (1) as well as some other piecewise linear kinetic schemes, another kind of stability for vortex ring solutions was discovered by Nandapurkar and Winfree (1989): if the ring spontaneously expands (as it does in a broad range of parameters) then it eventually encounters the no-flux boundary conditions. It then converts to one of the alternative rotors and proceeds to shrink; e.g., equation (1)'s period-17 solution discussed above hits the wall and switches to the smaller period-11 rotor, and the ring then shrinks back to its preferred stable radius. Or it may be absorbed into the wall, after which the medium reverts to uniform quiescence. Or it may be repelled, thus standing off from the walls at a distance of 1-2 vortex core radii as defined above (a cylinder of perimeter equal to the wavelength of the periodic wavetrain emitted from this rotating source, i.e. the propagation speed times the period of rotation.) In parameter ranges admitting such repulsion, the confined ring typically rolls up the walls until it encounters a ceiling, then rolls along it, contracting, until it either becomes too small (ring radius comparable to core radius) and abruptly vanishes, or resists further compression and increase of curvature, and stabilizes at a finite radius.

By symmetry, this situation is geometrically identical to the apposition of such a vortex ring against a mirror-image ring with **or without** a no-flux boundary in the mirror plane. But is it dynamically identical, i.e., is this mirror-pairing stable when the two rings are not actual mirror images? Might there be stable organizing centers composed of such parallel-paired rings, in addition to the linked-ring organizing centers foreseen mathematically (Winfree and Strogatz 1984) and then discovered computationally (Nandapurkar and Winfree 1987)? The general answer is unknown, but in one numerical experiment (Courtemanche et al. 1990) we found the pairing to be unstable: vortex cores react against one another like slippery grapes. (Linked rings also appear to be unstable, generalizing from the few examples observed in Nandapurkar and Winfree (1987)).

Shrinking, expanding or seemingly stable vortex rings (depending on recipe parameters) have also been computed using a chemical kinetic model of the Belousov-Zhabotinsky chemical reaction. Early observations of curved 3-dimensional filaments in this reagent showed that curved segments contract toward their centers of curvature, leading to the demise of all rings (Winfree 1974). In computations the rate adheres nicely to a linear dependence on curvature of the filament with coefficient not far from the diffusion coefficient, as expected from simple analytical approximations; the corresponding laboratory measurements also remain nicely linear until the tiny vortex ring can no longer be measured during its abrupt collapse (Jahnke et al. 1988; Henze et al. 1989; Winfree and Jahnke 1989). While shrinking, the ring also glides along its symmetry axis at a speed that increases with curvature. (In the special case of equal-diffusion media, this rate is theoretically identically zero; the counter-example in Jahnke et al. (1989) was probably a misinterpretation, non-zero drift being induced by interaction with a nearby no-flux boundary.)

<u>Helices.</u> The next-simplest organizing center starts from an initially uncurved vortex filament given uniform twist. In the initial condition every horizontal x,z plane contains the same 2-dimensional vortex computed from the chemical kinetic equations of the Belousov-Zhabotinsky reaction. Identical planes are stacked up like bills on a spindle, but each is slightly further rotated about the common vertical y axis, accumulating a full $360°$ turn between floor and ceiling. Floor and ceiling are computationally identified (periodic boundary conditions). The wave radiating from this axis may be viewed as a skewered stack of rotationally-staggered spirals. By repeating this experiment in boxes of different heights the effect of twist rate (shortening the rotation period, for example) can be determined. It was discovered numerically, violating every anticipation, that if and only if the twist rate exceeds a certain threshold, the filament springs into helical form. It then grows at a rate that can be measured until either stabilizing at finite radius or hitting the no-flux walls, where its expansion may or may not be arrested, depending on the kinetics chosen (Henze et al. 1990). The purpose of these calculations was to observe the dependence of filament motion (resolved into components V_n toward the local center of curvature, and V_b perpendicular to that and to the local tangent vector) on local curvature and twist. One discovery hoped for was that one derivation of this dependence in the limit of small twist and $(small)^2$ curvature (Keener 1988) would be found to remain reasonably accurate in the domain of feasible experiments with filaments of finite twist and curvature, far beyond the strict limits of the singular-perturbation argument. These numerical experiments instead demonstrated that, for a non-meandering vortex filament:

> Within the practical range of curvature, k, and twist, w, V_b depends roughly linearly on $\sqrt{k^2+(w-w_0)^2}$, where w_0 is the initial twist of the uncurved filament;

Within the computationally practical range of w at k=0 , V_n remains 0 up to a threshold of w_0 then becomes abruptly quite negative, and varies with k and w as the helix grows, in a way that often leaves it still quite negative even when k and w have both become small;

With the computational resources at our disposal, we have not been able to approach small enough k and w to exhibit the laws of motion originally foreseen.

These phenomena came as a surprise and are still not understood, but they have resisted our vigorous and prolonged attempt to interpret them as numerical artifacts. It now seems possible to describe them by appropriate addition of supplementary terms and corresponding adjustable parameters (Keener and Tyson 1990) to the linear laws of motion originally foreseen (Keener 1988). This correction works only in the range of k and w within which the coils of the helix do not come too close together (Keener and Tyson (1991). If adjacent coils of the helix significantly interact across distances $\leq$ d then laws of motion based exclusively on local geometry must be expected to fail in helices so compact that every part of the filament passes that close to some other part. Keener and Tyson (1991) find that with only a few new terms and parameters they can fit at least the V_n (not the V_b) data of Jahnke et al. (1990) on the assumption that d is as large as one wavelength. Prompted by the suggestion of Keener and Tyson, a numerical experiment was conducted, confronting an Oregonator rotor with a **co**-rotating replicas above and below, just as in a helix, by using periodic boundary conditions. The rotor indeed began to move at d $\approx$ one wavelength; as separation decreased to 3/4 wavelength, the imposed vortex motion increased to 5% of wave propagation speed, comparable to the previously observed motions of the helical filament. The magnitude of the motions induced is comparable to the magnitude of V_n departures from theory, but whether the changes are in the right direction and quantitatively correct, and whether they account for the peculiar V_b observations as well, all remain to be seen. It also remains to be seen how much such interactions are altered by the rotor's deformation in the pertinent 3-dimensional context, i.e., in a curved and twisted filament.

These phenomena have not yet been observed, nor even sought, in the laboratory, though it should be relatively easy to do so.

<u>Knots.</u> The foregoing examples were (a) uniformly curved, without twist or torsion, and (b) uniformly curved and twisted, with uniform torsion. The last example in this brief bestiary is a knotted vortex ring, in which curvature, twist, and torsion continuously vary along the filament with a three-fold repeat. This also sometimes proves to be a stable configuration. Much as in the stable vortex rings, no arc of such a filament is much more (or much less) than a core diameter away from some other arc: the entire knot fits within a sphere only one wavelength in radius. This compactness may play an essential role in the knot's stability: each segment of the filament is moving in a way that we have not yet described entirely in terms of local geometry, perhaps because interactions with adjacent segments of vortex filament also play a role. They certainly do when the knot breaks, as it does in some excitable media. Two filaments ('red' and 'green') snagging on one another sometimes obstruct one another's motion or break and reconnect to form red-green hybrids. Through such breakage, knots transmute to other topological forms (Winfree and Strogatz 1984). For example when a trefoil knot touches itself the vortex reconnects to become two linked rings that eventually also collide and fuse; this final fused ring gradually flattens and ends as a planar circle with no twist, which shrinks and eventually vanishes (Winfree 1990a). Such numerically observed transmutations dramatically underscore the discovery that filaments separated by a core diameter or less cannot be regarded as independent for purposes of inferring laws of motion from the local geometry of either alone.

The anatomy of a stable knot is revealed using software prepared in C for the Silicon Graphics IRIS. Detailed quantitative analysis of its differential geometry, using the Pascal toolkit mentioned above, reveals remarkable complexity.

CONCLUSION

Three-dimensional vorticity in excitable media is a tractable subject from the viewpoints of analysis, computation, and chemical experiment. Excitable media are as pervasive in the universe around us and inside us too, as are the Maxwell and Navier-Stokes media that began to command attention a century ago. The importance of organizing center dynamics in **excitable** media is that their geometry is so unfamiliar that they will not be seen until that geometry is **made** familiar, a game that has only started in the past few years.

ACKNOWLEDGEMENTS

This work has been generously supported for a long time by the National Science Foundation (Chemical Dynamics and Applied Mathematics). I thank John Tyson for some corrections. Without my graduate students and postdocs at the University of Arizona, Marc Courtemanche, William Guilford, Chris Henze, Pramod Nandapurkar, and Michael Wolfson, much less would have happened. We thank the University of Arizona Mathematics Department for access to their Silicon Graphics IRIS.

REFERENCES

Agladze, K.I., Panfilov, A.V., and Rudenko, A.N., Non-stationary Rotation of Spiral Waves: Three-dimensional Effects, *Physica,* vol. 29D, pp. 409-415, 1988.

Casten, R.G. , Cohen, H. , and Lagerstrom, P.A. Perturbation Analysis of and Approximation to the Hodgkin-Huxley Theory. Quart. Appl. Math. 32 , pp. 365-402, 1975.

Courtemanche, M., Skaggs, W.E., and Winfree, A.T., Stable Three-dimensional Action Potential Circulation in the FitzHugh-Nagumo Model, *Physica*, vol. 41D, pp. 173-182, 1990.

Courtemanche, M. and Winfree, A.T., Two-dimensional Rotating Depolarization Waves in a Modified Beeler-Reuter Model of Cardiac Cell Activity, *Science at the John von Neumann National Supercomputer Center*, vol. 3, pp. 79-86. ed. G. Cook. Consortium for Scientific Computing, Princeton, 1990a.

Courtemanche, M. and Winfree, A.T., A Two-Dimensional Model of Electrical Waves in the Heart , Pixel vol. 1(3) , pp. 24-31, 1990b.

Elphick, C., Meron, E., and Spiegel, E.A., Spatiotemporal Complexity in Travelling Patterns, *Phys. Rev. Lett.* vol. 63 pp. 496-499, 1986.

Henze, C., Courtemanche, M., Jahnke, W., and Winfree, A.T., Modeling Chemical Vortex Dynamics in Excitable Media, *Science at the John von Neumann Supercomputer National Supercomputer Center*, vol. 2 (1988 Annual Report), pp. 79-86, ed. G. Cook. Consortium for Scientific Computing, Princeton, N.J., 1989.

Henze, C., Lugosi, E., and Winfree, A.T., Helical Organizing Centers in Excitable Media. *Canad.J.Physics* , in press for September 1990.

Jahnke,W., Henze, C., and Winfree, A.T., Chemical Vortex Dynamics in Three-dimensional Excitable Media, *Nature* vol. 336 , pp. 662-665, 1988.

Jahnke, W., Skaggs, W.E., and Winfree, A.T., Chemical Vortex Dynamics in the Belousov-Zhabotinsky Reaction and in the 2-Variable Oregonator Model, *J. Chem. Phys.* vol. 93, pp. 740-749,1989.

Karfunkel, H.R. and Kahlert, C., Excitable Chemical Reaction Systems II. Several Pulses on the Ring Fiber, *J.Math. Biology* vol.4, pp. 183-185, 1977.

Keener, J.P., A Geometrical Theory for Spiral Waves in Excitable Media, *SIAM J. Appl. Math.* 46, pp. 1039-1056, 1986.

Keener, J.P., The Dynamics of Three Dimensional Scroll Waves in Excitable Media, *Physica* vol. 31D(2), pp. 69-276, 1989.

Keener, J.P. and Tyson, J.J., Helical and Circular Scroll Wave Filaments, *Physica D*, in press 1990.

Keener, J.P. and Tyson, J.J., The Dynamics of Helical Scroll Waves in Excitable Media *Physica D,* submitted 1990.

Lugosi, E., Analysis of Meandering in Zykov Kinetics, *Physica* , vol 40D, pp. 331-337, 1989.

Lugosi, E. and Winfree, A.T., Simulation of Wave-Propagation in Three Dimensions using Fortran on the Cyber 205, *J. Comput.Chem.* vol. 9, pp. 689-701,1989 .

Meron, E., Nonlocal Effects in Spiral Waves. *Phys. Rev. Lett.* , vol. 63, pp. 684-687, 1989.

Meron, E., The Role of Curvature and Wavefront Interactions in Spiral Wave Dynamics, Physica D, in press, 1991.

Nandapurkar, P.J., Computation of Three Dimensional Waves in Supercomputers, Appendix II in: V.S. Zykov, <u>Simulation of Wave Processes in Excitable Media</u> (translation) Manchester Univ. Press 1988.

Nandapurkar, P.J. and Winfree, A.T., A Computational Study of Twisted Linked Scroll Waves in Excitable Media, *Physica* vol. 29D, pp. 9-83, 1987.

Nandapurkar, P.J. and Winfree, A.T., Dynamical Stability of Untwisted Scroll Rings in Excitable Media, *Physica* vol. 35D, pp. 277-288, 1989.

Panfilov, A.V. and Pertsov, A.M., Vortex Ring in Three-Dimensional Active Medium Described by Reaction Diffusion Equation, *Dokl. Akad. Nauk. USSR.* 274(6), pp. 1500-1503,1984.

Skaggs, W.E., Lugosi, E., and Winfree, A.T., Stable Vortex Rings of Excitation in Neuroelectric Media, *IEEE Trans.Cir.Sys.* vol. 35(7), pp. 784-787,1989. There is a typo in the equation: parameters 0.7 and 0.5 are interchanged.

Tyson,J .J. and Keener, J.P., Singular Perturbation Theory of Traveling Waves in Excitable Media. *Physica* vol. 32D, pp. 327-361, 1988.

Winfree, A.T., Spatial and Temporal Organization in the Zhabotinsky Reaction, *Adv. Med. Phys.* vol. 16, pp. 115-136, 1973.

Winfree, A.T., Two Kinds of Wave in an Oscillating Chemical Solution, *Faraday Symp. Chem. Soc.* vol. 9, pp. 38-46, 1974.

Winfree, A.T., When Time Breaks Down: The 3-dimensional Dynamics of Electrochemical Waves and Cardiac Arrhythmias, Princeton University Press, 1987.

Winfree, A.T., Understanding the Onset of Fibrillation in the Heart Muscle: Two-dimensional Vortices in Healthy Myocardium, *Science at the John von Neumann National Supercomputer Center* vol. 1, pp. 125-130, ed. G. Cook, Consortium for Scientific Computing, Princeton, 1988.

Winfree, A.T., Electrical Instability in Cardiac Muscle: Phase Singularities and Rotors, *J.Theor.Biol.* vol. 183, pp. 353-405, 1989.

Winfree, A.T., Stable Particle-like Solutions to the Nonlinear Wave Equations of Excitable Media, *SIAM Review* vol. 32, pp. 1-53, 1990a.

Winfree, A.T., Estimating the Ventricular Fibrillation Threshold, In: Theory of Heart, ed. L. Glass and P. Hunter, Springer-Verlag, 1990b.

Winfree, A.T., Vortex Action Potentials in Normal Ventricular Muscle. In: Mathematical Approaches to Cardiac Arrhythmias, ed. J. Jalife, *Ann. N.Y.A.S.*, vol. 591, pp. 190-207,1990c.

Winfree, A.T., Rotors in Normal Ventricular Myocardium. Einthoven Lecture, Netherlands Royal Academy of Sciences, in press 1990d.

Winfree, A.T., Ventricular Reentry in Three Dimensions, In: Cardiac Electrophysiology, from Cell to Bedside, ed D.P. Zipes and J. Jalife, W.B. Saunders Co., pp. 224-234, 1990e.

Winfree, A.T., Alternative Stable Rotors in an Excitable Medium, *Physica D*, in press 1991a.

Winfree, A.T., Discrete Spectrum of Rotor Periods in an Excitable Medium, Phys. Lett. A, in press, 1990.

Winfree, A.T. and Guilford, W., The Dynamics of Organizing Centers: Numerical Experiments in Differential Geometry, In: Biomathematics and Related Computational Problems, pp. 697-716, ed: Riccardi,L.M., Kluwer Academic Publishers 1988.

Winfree, A.T. and Jahnke, W., Three-dimensional Scroll Ring Dynamics in the Belousov-Zhabotinsky Reagent and in the 2-variable Oregonator Model, *J.Phys.Chem.* vol. 93, pp. 2823-2832, 1989.

Winfree, A.T. and Strogatz, S.H., Singular Filaments Organize Chemical Waves in Three Dimensions: 4: Wave Taxonomy, *Physica* vol. 13D, pp. 221-233, 1984.

Zykov, V.S., Simulation of Wave Processes in Excitable Media (in Russian), Nauka, Moscow, 1984; also in 1988 English translation by Manchester University Press.

Zykov, V.S., Cycloidal Circulation of Spiral Waves in an Excitable Medium, *Biofizika* vol. 31, pp. 862-865, 1986.

Zykov, V.S., Kinematics of Nonstationary Circulation of Spiral Waves in an Excitable Medium, *Biofizika* vol. 32, pp. 337-340, 1987.

The Rigorous Solution of the Classical Theory of Plates

CHRISTIAN CONSTANDA
Department of Mathematics
University of Strathclyde
Glasgow, Scotland, UK

Abstract

The boundary integral equation method is used to reduce the Dirichlet boundary value problem for the biharmonic equation to a system of second kind singular integral equations of index zero. A numerical approximation scheme is also indicated, and an example is given.

Introduction

The classical theory of bending of thin plates proposed by Kirchhoff [1] arose from the necessity to find a mathematical model which, while considerably simpler than the full three-dimensional equations of elasticity, still provides sufficiently accurate information on the solution. In this paper we intend to discuss certain mathematical aspects of the Dirichlet boundary value problem, arising in the application of the boundary integral equation method. This technique has not been used fully so far in conjunction with the classical model because of difficulties connected with the bi-Laplacian and the representation of its solutions in terms of potential-type functions. We also indicate a scheme for the approximate computation of the solution, and give a numerical example.

Preliminary results

Let S be a finite domain in $\mathbf{R}^2$ bounded by a smooth curve ∂S, ν the unit outward normal to ∂S, s the arc parameter on ∂S, $w(x)$ the deflection of a point $x = (x_1, x_2) \in S$, p the resultant load on the top and bottom faces of the plate, D the modulus of rigidity of the material, and Δ the two-dimensional Laplacian. Mathematically, the classical model of bending of thin elastic plates consists in solving the equation

$$\Delta\Delta w = \frac{p}{D} \quad \text{in } S, \tag{1}$$

together with an appropriate set of boundary conditions. Thus [2], for a clamped-edge plate we have

$$w = 0, \quad \frac{\partial w}{\partial \nu} = 0 \quad \text{on } \partial S, \tag{2}$$

for a simply supported plate

$$w = 0, \quad \frac{\partial^2 w}{\partial \nu^2} + \sigma \left(\frac{\partial^2 w}{\partial s^2} + \kappa \frac{\partial w}{\partial \nu} \right) = 0 \quad \text{on } \partial S,$$

and for a plate with an unloaded edge

$$(1 - \sigma) \frac{\partial^2}{\partial s \partial \nu} \left(\frac{\partial w}{\partial s} \right) + \frac{\partial}{\partial \nu} (\Delta w) = 0,$$

$$\frac{\partial^2 w}{\partial \nu^2} + \sigma \left(\frac{\partial^2 w}{\partial s^2} + \kappa \frac{\partial w}{\partial \nu} \right) = 0 \quad \text{on } \partial S, \tag{3}$$

where σ is Poisson's ratio and κ the curvature of ∂S.

These problems have received much attention in the past, various authors attempting to solve them by means of the boundary integral equation method (see, for example, [3] and [4]). However, their efforts have not been entirely successful, in the sense that they resorted to numerical approximations before a rigorous investigation of the ensuing equations could be completed to show that such approximations were warranted. The reason for this lies in the nature of the procedures adopted.

More specifically, in [3] the solution of (1), (2) is sought in the form

$$w(x) = |x|^2 \varphi(x) + \psi(x),$$

where $|x|^2 = x_1^2 + x_2^2$ and φ and ψ are harmonic functions represented by single layer potentials. This leads to a pair of coupled integral equations on ∂S for the unknown densities of the potentials, which, in view of their complexity, are not discussed analytically. It should be pointed out that the situation is not improved if φ and ψ are sought as a single layer potential and a double layer potential, respectively, since in this case the problem reduces to a system of integro-differential equations on ∂S.

In [4] the technique is based on Green's representation of the solution of the biharmonic equation, for $x \in S$, in the form

$$-w(x) = \int_{\partial S} \left[E(x, y) \frac{\partial}{\partial \nu} \Delta w(y) - \left(\frac{\partial}{\partial \nu(y)} E(x, y) \right) \Delta w(y) \right.$$

$$\left. + (\Delta(y) E(x, y)) \frac{\partial}{\partial \nu} w(y) - \left(\frac{\partial}{\partial \nu(y)} \Delta(y) E(x, y) \right) w(y) \right] ds(y), \tag{4}$$

where

$$E(x, y) = \frac{1}{8\pi} |x - y|^2 \ln |x - y|$$

is a fundamental solution for the $\Delta\Delta$ operator. A replacement of (3) in (4) and integration by parts leads to a formula for $w(x)$, $x \in S$, which is then differentiated in the direction of $\nu(x^0)$, $x^0 \in \partial S$. As $x \to x^0$, the expressions of $w(x)$ and $\partial w / \partial \nu$ give rise to a pair a coupled singular integral equations for these quantities on ∂S. An alternative formulation is also proposed, involving the limiting values of $\partial w / \partial \nu$ and $\partial w / \partial s$ on the boundary. However, the analytic treatment of the corresponding equations in both cases is abandoned, since the systems in question are degenerate.

This degeneracy may be due to the fact that additional equations obtained by differentiation from the same relation have a tendency to force compatibility conditions on the solution, which a regular system does not require and cannot guarantee to satisfy.

To avoid this drawback, the method needs to be re-examined and suitably modified. We exemplify one possible such scheme in the case of a clamped-edge plate.

Analytic solution

Let ∂S be a C^4-curve. Since (1) can be made homogeneous by means of a particular solution (for example, provided by the domain potential of density $p/(2D)$ in the case where p is sufficiently smooth), we aim to find a function $u \in C^4(S) \cap C^3(\bar{S})$ such that

$$\Delta\Delta u = 0 \quad \text{in } S, \tag{5}$$

$$u = f, \quad \frac{\partial u}{\partial \nu} = g \quad \text{on } \partial S, \tag{6}$$

where $f \in C^{3,\alpha}(\partial S)$ and $g \in C^{2,\alpha}(\partial S)$, $0 < \alpha < 1$, are given functions. Replacing (6) in (4), we find that for all $x \in S$

$$8\pi u(x) = -\int_{\partial S} |x - y|^2 \ln|x - y| \frac{\partial}{\partial \nu} \Delta u(y)\, ds(y)$$

$$+ \int_{\partial S} \left(\frac{\partial}{\partial \nu(y)} |x - y|^2 \ln|x - y| \right) \Delta u(y)\, ds(y) + 4\tilde{f}(x) - 4\tilde{g}(x), \tag{7}$$

where

$$\tilde{f}(x) = \int_{\partial S} \left(\frac{\partial}{\partial \nu(y)} \ln|x - y| \right) f(y)\, ds(y),$$

$$\tilde{g}(x) = \int_{\partial S} (\ln|x - y| + 1) g(y)\, ds(y). \tag{8}$$

By the regularity properties of harmonic potentials [5], $\tilde{f}, \tilde{g} \in C^4(S) \cap C^{3,\alpha}(\bar{S})$. Then, applying the Δ operator on both sides of (7), we arrive at

$$\Delta u(x) = \frac{1}{2\pi} \int_{\partial S} \left(\frac{\partial}{\partial \nu(y)} \ln|x - y| \right) \Delta u(y)\, ds(y)$$

$$- \frac{1}{2\pi} \int_{\partial S} (\ln|x - y| + 1) \frac{\partial}{\partial \nu} \Delta u(y)\, ds(y). \tag{9}$$

We choose the direction of the unit tangent vector τ to ∂S so that $\{\nu, \tau\}$ has positive orientation. If we take into account the behaviour of the single and double layer potentials near the boundary [6], we find that

$$\frac{\partial}{\partial s}\Delta u(x) + \frac{1}{\pi}\int\limits_{\partial S}\left(\frac{\partial}{\partial \nu(x)}\ln|x-y|\right)\frac{\partial}{\partial s}\Delta u(y)\,ds(y)$$

$$+ \frac{1}{\pi}\int\limits_{\partial S}\left(\frac{\partial}{\partial s(x)}\ln|x-y|\right)\frac{\partial}{\partial \nu}\Delta u(y)\,ds(y) = 0, \quad x \in \partial S, \qquad (10)$$

where the second integral on the left-hand side is understood in the sense of principal value.

We do not use (9) again to derive a similar relation for $\partial\Delta u(x)/\partial\nu$, since, as remarked above, the resulting system would be degenerate. Instead, we use Green's formula

$$\int\limits_{S}(v\Delta\Delta u - u\Delta\Delta v)\,da = \int\limits_{\partial S}\left(v\frac{\partial\Delta u}{\partial\nu} - \frac{\partial v}{\partial\nu}\Delta u + \Delta v\frac{\partial u}{\partial\nu} - \frac{\partial\Delta v}{\partial\nu}u\right)ds \qquad (11)$$

in the standard way, with u satisfying (5) and v replaced by

$$G(x, y) = \frac{1}{8\pi}|x-y|^2\theta(x-y) = \frac{1}{8\pi}|x-y|^2\tan^{-1}\left(\frac{x_2-y_2}{x_1-y_1}\right),$$

to find that for all $x \in S$

$$\int\limits_{\partial S}\left[G(x, y)\frac{\partial}{\partial\nu}\Delta u(y) - \left(\frac{\partial}{\partial\nu(y)}G(x, y)\right)\Delta u(y)\right.$$

$$\left. + (\Delta(y)G(x, y))\frac{\partial}{\partial\nu}u(y) - \left(\frac{\partial}{\partial\nu(y)}\Delta(y)G(x, y)\right)u(y)\right]ds(y) = 0. \qquad (12)$$

Using (6) and the fact that

$$\Delta(x)G(x, y) = \Delta(y)G(x, y) = 4\theta(x-y), \quad x \neq y,$$

and bearing in mind that $\theta(x-y)$ is the harmonic conjugate of $\ln|x-y|$, which means that

$$\frac{\partial}{\partial s(y)}\theta(x-y) = \frac{\partial}{\partial\nu(y)}\ln|x-y|,$$

$$\frac{\partial}{\partial\nu(y)}\theta(x-y) = -\frac{\partial}{\partial s(y)}\ln|x-y|, \qquad (13)$$

we see that (12) yields

$$-\int\limits_{\partial S}|x-y|^2\theta(x-y)\frac{\partial}{\partial\nu}\Delta u(y)\,ds(y)$$

$$+ \int\limits_{\partial S}\left(\frac{\partial}{\partial\nu(y)}|x-y|^2\theta(x-y)\right)\Delta u(y)\,ds(y) + 4\hat{f}(x) - 4\hat{g}(x) = 0, \quad x \in S, \quad (14)$$

where

$$\hat{f}(x) = -\int\limits_{\partial S} \left(\frac{\partial}{\partial s(y)} \ln|x - y|\right) f(y)\, ds(y),$$

$$\hat{g}(x) = \int\limits_{\partial S} \theta(x - y) g(y)\, ds(y).$$

As before, $\hat{f}, \hat{g} \in C^4(S) \cap C^{3,\alpha}(\bar{S})$.

Applying the Δ operator in (14) and taking (13) into account, we deduce that

$$-\int\limits_{\partial S} \left(\frac{\partial}{\partial s(y)} \ln|x - y|\right) \Delta u(y)\, ds(y) - \int\limits_{\partial S} \theta(x - y) \frac{\partial}{\partial \nu} \Delta u(y)\, ds(y) = 0, \quad x \in S.$$

If we now proceed as in the case of (10), we find that, in view of the behaviour of the kernels of the above integrals near the boundary [6],

$$\frac{\partial}{\partial \nu} \Delta u(x) + \frac{1}{\pi} \int\limits_{\partial S} \left(\frac{\partial}{\partial s(x)} \ln|x - y|\right) \frac{\partial}{\partial s} \Delta u(y)\, ds(y)$$

$$- \frac{1}{\pi} \int\limits_{\partial S} \left(\frac{\partial}{\partial \nu(x)} \ln|x - y|\right) \frac{\partial}{\partial \nu} \Delta u(y)\, ds(y) = 0, \quad x \in \partial S, \qquad (15)$$

where the first integral on the left-hand side is understood in the sense of principal value.

Thus, setting

$$\varphi_1 = \frac{\partial}{\partial s} \Delta u, \quad \varphi_2 = \frac{\partial}{\partial \nu} \Delta u \quad \text{on } \partial S,$$

we conclude that φ_1 and φ_2 satisfy the system of singular integral equations

$$(I - K)\varphi = 0, \tag{16}$$

where I is the (2×2)-matrix identity operator, K the (2×2)-matrix singular integral operator with kernel

$$k(x, y) = \frac{1}{\pi} \begin{pmatrix} -\dfrac{\partial}{\partial \nu(x)} \ln|x - y| & -\dfrac{\partial}{\partial s(x)} \ln|x - y| \\[2ex] -\dfrac{\partial}{\partial s(x)} \ln|x - y| & \dfrac{\partial}{\partial \nu(x)} \ln|x - y| \end{pmatrix},$$

and $\varphi = (\varphi_1, \varphi_2)^{\mathrm{T}}$.

The system (16) can be written in the form

$$\varphi(x) + \frac{1}{\pi} \begin{pmatrix} 0 & 1 \\ 1 & 0 \end{pmatrix} \int\limits_{\partial S} \left(\frac{\partial}{\partial s(y)} \ln|x - y|\right) \varphi(y)\, ds(y) + (L\varphi)(x) = 0, \quad x \in \partial S, \tag{17}$$

where L is an integral operator with a proper γ-singular kernel on ∂S, $0 < \gamma < 1$ [6]. It is easily seen that the index of (17) is zero, consequently, Fredholm's theorems are applicable to this system, which can now be investigated by means of classical methods.

The uniqueness of u is shown immediately by means of Green's formula (11).

Numerical approximation

To avoid having to handle singular integral equations, we propose to approximate the solution of (5), (6) by using a perturbation method. More precisely, we set

$$u_\alpha = -\frac{\partial u}{\partial x_\alpha}, \quad u_3 = u, \quad \alpha = 1, 2, \tag{18}$$

and consider the boundary value problem [7]

$$\frac{1}{2}\varepsilon(2u_{1,11} + u_{1,22} + u_{2,12}) - u_1 - u_{3,1} = 0,$$
$$\frac{1}{2}\varepsilon(u_{1,12} + u_{2,11} + 2u_{2,22}) - u_2 - u_{3,2} = 0, \tag{19}$$
$$u_{1,1} + u_{2,2} + \Delta u_3 = 0 \quad \text{in } S,$$

$$u_i = f_i, \quad i = 1, 2, 3, \quad \text{on } \partial S, \tag{20}$$

where $(\ldots)_{,\alpha} = \partial(\ldots)/\partial x_\alpha$, $\alpha = 1, 2$, and f_1 and f_2 are computed from (6) and (18).

This is an interior Dirichlet problem for a version of the system governing the bending of plates with transverse shear deformation, which was extensively studied in [6]. Also, in [8] it was shown that the difference between the solution of (5), (6) and that of (19), (20) is $O(\varepsilon^{1/2})$, and that $u_\alpha + u_{3,\alpha} = O(\varepsilon)$, $\alpha = 1, 2$.

Approximate values of u_1, u_2 and u_3 can be computed, for instance, by means of the generalized Fourier series method [6].

As an example, consider a circular plate of radius R, with a clamped edge and a load distributed uniformly over its top face. Then $p = \text{const}$ in (1), which admits the particular solution

$$w_0 = \frac{p}{48D}(x_1^4 + x_2^4), \quad x \in S.$$

Consequently, the boundary conditions for the biharmonic function $u = w - w_0$ are

$$u(x) = -\frac{p}{48D}(x_1^4 + x_2^4),$$
$$\frac{\partial}{\partial\nu}u(x) = -\frac{p}{12RD}(x_1^4 + x_2^4), \quad x \in \partial S,$$

or, alternatively,

$$u_{,\alpha}(x) = -\frac{p}{12D}x_\alpha^3, \quad \alpha = 1, 2, \quad x \in \partial S.$$

Taking $R = 1$ and $p/(12D) = 1$, for simplicity, using the circle concentric with ∂S and of radius 1.1 as the auxiliary curve [6], and restricting ourselves to the first 51 terms in the series, we obtain the following results for various points between the centre of the circle and ∂S:

	point	$(0,0)$	$(0.1,0)$	$(0.5,0)$	$(0.9,0)$

(i) $\varepsilon = 0.01$

	$(0,0)$	$(0.1,0)$	$(0.5,0)$	$(0.9,0)$
w_{approx}	0.1684	0.1631	0.0885	0.0055
relative error	0.1018	0.1125	0.1607	0.1926

(ii) $\varepsilon = 0.001$

	$(0,0)$	$(0.1,0)$	$(0.5,0)$	$(0.9,0)$
w_{approx}	0.1851	0.1814	0.1035	0.0066
relative error	0.0127	0.0131	0.0185	0.0314
w_{exact}	0.1875	0.1838	0.1055	0.0068

Here w_{exact} is the exact value of w satisfying (5), (6), w_{approx} the numerical approximation of the corresponding solution u of (19), (20), and the relative error is equal to $|w_{\text{exact}} - w_{\text{approx}}|/|w_{\text{exact}}|$. This error represents the combined effect of two approximations: that of w by u, and that of u by the result of the computational algorithm. It can be seen that, although the error worsens as we approach the boundary, the overall approximation remains very good if ε is sufficiently small.

References

1. Kirchhoff, G., Über das Gleichgewicht und die Bewegung einer elastischen Scheibe, *J. Reine Angew. Math.* **40** (1850), 51–58.

2. Love, A.E.H., *A Treatise on the Mathematical Theory of Elasticity*, 4th ed., Dover Publications, New York, 1944.

3. Jaswon, M.A. and Maiti, M., An Integral Equation Formulation of Plate Bending Problems, *J. Engng Math.* **2** (1968), 83–93.

4. Hansen, E.B., Numerical Solution of Integro-differential and Singular Integral Equations for Plate Bending Problems, *J. of Elasticity* **6**(1976), 39–56.

5. Günter, N.M., *Die Potentialtheorie und ihre Anwendungen auf Grundaufgaben der mathematischen Physik*, Teubner, Leipzig, 1957.

6. Constanda, C., *A Mathematical Analysis of Bending of Plates with Transverse Shear Deformation*, Longman Scientific & Technical, Harlow, Essex, 1990.

7. Zienkiewicz, O.C., *The Finite Element Method*, 3rd ed., McGraw-Hill, New York, 1977.

8. Destuynder, P., Méthode d'Éléments Finis pour le Modèle des Plaques en Flexion de Naghdi-Reissner, *R.A.I.R.O. Analyse Numérique* **15**(1981), 201–230.

Explicit Stiffness and Error Estimator Expressions for a Family of Axisymmetric Triangular Finite Elements

BRAD E. BERGMANN, RAJIV V. NAMBIAR,
and KENT L. LAWRENCE
Mechanical Engineering Department
The University of Texas at Arlington
Arlington, Texas 76019-0023, USA

INTRODUCTION

Efficiency in computer computation becomes paramount in the finite element analysis of increasingly large and complex problems. Consequently closed form integration in lieu of numerical integration becomes attractive from a computer applications standpoint. In this paper we develop closed form expressions of the element stiffness matrix and simple error estimator for the axisymmetric class of elasticity problems [1].

The type of element discussed here is the hierarchic, triangular element based upon the family of element shape functions of arbitrary order as described by Peano [2].

Our work is modeled after that of Subramanian and Bose [3] who derived explicit expressions for closed form stiffness matrices of non-hierarchic, triangular elements applied to both the plane and the axisymmetric case.

Recently, much work has been devoted to the development of various error estimation methods in finite element analysis. Zienkiewicz and Zhu [4] have proposed an *A Posteriori* error analysis in which an initial analysis is performed and its results are used to form an estimate of the error. The initial mesh is refined using the error estimation as an indicator. Further analysis and error estimation iterations are performed until a desired solution accuracy is achieved.

Simple yet effective local error estimators allow for easy implementation into computer code while providing accurate indicators for subsequent adaptive mesh refining. Presented herein is a closed form local error estimate in the energy norm using simple nodal stress averaging as an estimate of the exact solution.

ELEMENT STIFFNESS MATRIX

In the development of the element stiffness matrix in this work, an approximate method in integration was employed. This approach introduces a centroidal radius $\bar{r}$ of an element to simplify subsequent integration [5]. Use of the centroidal radius of an element yields an approximate stiffness matrix and is designated the approximate integration method (AIM).

For the axisymmetric problem, a triangular element in the r-z plane represents a toroidal volume in three dimensions as shown in Figure 1. By employing area or natural coordinates [5], any point within an element can be expressed in terms of the nodal coordinates as

$$r = L_1 r_1 + L_2 r_2 + L_3 r_3 \tag{1}$$

$$z = L_1 z_1 + L_2 z_2 + L_3 z_3 \tag{2}$$

The centroidal radius of an element is defined as

$$\bar{r} = \frac{r_1 + r_2 + r_3}{3} \tag{3}$$

Also useful is the natural coordinate identity.

$$L_1 + L_2 + L_3 = 1 \tag{4}$$

Consider L_1 and L_2 as the two independent variables.

Displacements within an element can be expressed as functions of the nodal displacements in the r and z directions (δu and δw, respectively).

$$\begin{bmatrix} u \\ w \end{bmatrix} = \begin{bmatrix} N & 0 \\ 0 & N \end{bmatrix} \begin{bmatrix} \delta u \\ \delta w \end{bmatrix} \tag{5}$$

For a CST element

$$N = \begin{bmatrix} L_1 & L_2 & L_3 \end{bmatrix} \tag{6}$$

For a LST element

$$N = \begin{bmatrix} L_1 & L_2 & L_3 & 2L_1 L_2 & 2L_2 L_3 & 2L_3 L_1 \end{bmatrix} \tag{7}$$

Strains are related to displacements by

$$\begin{bmatrix} \varepsilon_r \\ \varepsilon_z \\ \gamma_{rz} \\ \varepsilon_\theta \end{bmatrix} = \begin{bmatrix} \partial u / \partial r \\ \partial w / \partial z \\ \partial u / \partial z + \partial w / \partial r \\ u / r \end{bmatrix} \tag{8}$$

Using (1), (2), (4), and (5), chain differentiating according to (8), and rearranging gives

$$\varepsilon = \mathbf{P}\,\mathbf{R}\begin{bmatrix} \delta u \\ \delta w \end{bmatrix} \tag{9}$$

where

$$\mathbf{P} = \frac{1}{J}\begin{bmatrix} 0 & z_{23} & z_{31} & 0 & 0 \\ 0 & 0 & 0 & r_{32} & r_{13} \\ 0 & r_{32} & r_{13} & z_{23} & z_{31} \\ J/\bar{r} & 0 & 0 & 0 & 0 \end{bmatrix} \tag{10}$$

also

$$z_{23} = z_2 - z_3, \quad r_{32} = r_3 - r_2 , \ldots \text{ etc.} \tag{11}$$

$$J = r_{13}\, z_{23} - r_{32}\, z_{31} \tag{12}$$

and

$$\mathbf{R} = \begin{bmatrix} N & 0 \\ N_{,L_1} & 0 \\ N_{,L_2} & 0 \\ 0 & N_{,L_1} \\ 0 & N_{,L_2} \end{bmatrix} \tag{13}$$

Here $N_{,L_1}$ and $N_{,L_2}$ denotes differentiating N with respect to L_1, and L_2, respectively. See [6].

The strain energy in an element is defined as

$$U = \frac{1}{2}\int_{\Omega} \varepsilon^{T}\,\mathbf{D}\,\varepsilon\, d\Omega \tag{14}$$

where $\mathbf{D}$ is the appropriate material elasticity matrix. In the axisymmetric problem, the differential volume can be represented as

$$d\Omega = 2\,\pi\,\bar{r}\, dr\, dz \tag{15}$$

converting to natural coordinates and inserting (10) gives

$$U = \pi\,\bar{r}\,J \int_0^1 \int_0^{1-L_2} [\delta u \ \delta w]\,\mathbf{R}^{T}\,\mathbf{G}\,\mathbf{R}\begin{bmatrix} \delta u \\ \delta w \end{bmatrix} dL_1\, dL_2 \tag{16}$$

where

$$\mathbf{G} = \mathbf{P}^{T}\,\mathbf{D}\,\mathbf{P} \tag{17}$$

The matrix **G** is a 5 x 5 symmetric matrix whose terms are dependent only on material properties and element nodal coordinates. Element strain energy can also be represented as

$$U = \frac{1}{2} [\delta u \; \delta w] \, \mathbf{K} \begin{bmatrix} \delta u \\ \delta w \end{bmatrix} \tag{18}$$

By comparing (17) and (19), it is evident that the element stiffness matrix **K** is given by

$$\mathbf{K} = 2\,\pi\,\bar{r}\,J \int_0^1 \int_0^{1-L_2} \mathbf{R}^T \, \mathbf{G} \, \mathbf{R} \, dL_1 \, dL_2 \tag{19}$$

Expanding $\mathbf{R}^T \mathbf{G} \mathbf{R}$ reveals that six distinct integrations involving shape functions and their derivatives must be carried out for the approximate integration methods [8]. These are:

$$A = \int_0^1 \int_0^{1-L_2} \mathbf{N}^T_{,L_1} \, \mathbf{N}_{,L_1} \, dL_1 \, dL_2$$

$$B = \int_0^1 \int_0^{1-L_2} \mathbf{N}^T_{,L_1} \, \mathbf{N}_{,L_2} \, dL_1 \, dL_2$$

$$C = \int_0^1 \int_0^{1-L_2} \mathbf{N}^T_{,L_2} \, \mathbf{N}_{,L_2} \, dL_1 \, dL_2$$

$$W = \int_0^1 \int_0^{1-L_2} \mathbf{N}^T \, \mathbf{N} \, dL_1 \, dL_2 \tag{20}$$

$$X = \int_0^1 \int_0^{1-L_2} \mathbf{N}^T \, \mathbf{N}_{,L_1} \, dL_1 \, dL_2$$

$$Y = \int_0^1 \int_0^{1-L_2} \mathbf{N}^T \, \mathbf{N}_{,L_2} \, dL_1 \, dL_2$$

The above integrations can be performed using the formula

$$\int_0^1 \int_0^{1-L_2} L_1^p \, L_2^q \, dL_1 \, dL_2 = \frac{p!\,q!}{(p+q+2)!} \tag{21}$$

The matrices **A, B, C, W, X, Y** implicitly encompass the entire family of hierarchic shape functions **N** and each p-level shape function represents a sub-matrix in the next higher p-level matrix. For instance, the above six matrices evaluated for a QST element would also contain CST and LST sub-matrices.

The element stiffness matrix can now be represented in a closed form as

$$\mathbf{K} = 2\,\pi\,\bar{r}\,J \begin{bmatrix} XX & XY \\ XY^T & YY \end{bmatrix} \qquad (22)$$

where XX, XY, YY matrices are expressed by

$$XX = G_{11}\,W + G_{12}(X + X^T) + G_{13}(Y + Y^T) + G_{22}A + G_{23}(B + B^T) + G_{33}\,C$$

$$XY = G_{14}X + G_{15}Y + G_{24}A + G_{25}B + G_{34}B^T + G_{35}C \qquad (23)$$

$$YY = G_{44}A + G_{45}(B + B^T) + G_{55}C$$

Thus, the closed form representation of the element stiffness matrix for axisymmetric problems may be readily incorporated in computer code since the matrices **A, B, C, W, X,** and **Y** are composed of constant terms that can be stored internally as data. Only $\bar{r}$, J, and terms in the G matrix will vary from element to element.

ELEMENT ERROR ESTIMATOR

Error in stresses appears to be a logical basis in which to develop the error estimator for axisymmetric problems.   The error can be defined as the difference between the exact stresses and the finite element stresses

$$e_\sigma = \sigma - \hat{\sigma} \qquad (24)$$

where σ is the exact stresses and $\hat{\sigma}$ is the finite element stresses. Zienkiewicz and Zhu [4] have proposed using an error energy norm as the basis of the error estimator.

$$\|e_i\|^2 = \int_\Omega \left(\sigma - \hat{\sigma}\right)^T C \left(\sigma - \hat{\sigma}\right) d\Omega \qquad (25)$$

where C is the material compliance matrix.

To proceed, an estimate of the exact stresses σ must be made. This "exact" stress estimate will be denoted subsequently by σ^*. Numerous methods of estimating σ^* have been proposed [5]. The simple method of nodal stress averaging coupled with the interpolation of these stresses over the individual element domain utilizing the shape functions presented by Byrd [7] is employed here. Stress continuity is ensured across interelement boundaries by this method. However, the stress distribution may not necessarily be smooth from element to element. Inserting the averaged stress estimate σ^* and expanding the energy norm yields.

$$\|e_i\|^2 = \int_\Omega \sigma^{*T} C \sigma^* \, d\Omega - 2 \int_\Omega \sigma^{*T} C \hat{\sigma} \, d\Omega + \int_\Omega \hat{\sigma}^T C \hat{\sigma} \, d\Omega \tag{26}$$

Integrating each of the three terms in (31) provides explicit matrices similar to the matrices derived for the element stiffness matrix.

Following the error estimation procedure set forth by Zienkeiwicz and Zhu [4], the total error energy norm in a finite element mesh is

$$\|e\|^2 = \Sigma \|e_i\|^2 \tag{27}$$

while the total strain energy norm is given by

$$\|\hat{u}\|^2 = \Sigma \|\hat{u}_i\|^2 \tag{28}$$

where the summation is taken over the total number of elements in a mesh.

An approximation of the exact strain energy norm is obtained by correcting the finite element strain energy norm as follows

$$\|u\|^2 = \|\hat{u}\|^2 + \|e\|^2 \tag{29}$$

Thus, the total or global error η in a finite element mesh can be expressed by

$$\eta = \frac{\|e\|}{\|u\|} \tag{30}$$

Moreover, if a specified accuracy $\bar{\eta}$ is desired in a finite element analysis, a mean permissible error $\bar{e}_m$ per element can be determined from

$$\bar{e}_m = \bar{\eta} \sqrt{\frac{\|u\|^2}{m}} \tag{31}$$

where m is the total number of elements. An element subdivision indicator ξ_i for each element can be calculated using

$$\xi_i = \frac{\|e_i\|}{\bar{e}_m} \tag{32}$$

The element subdivision indicator estimates the extent to which a particular element should be subdivided to achieve the specified accuracy.

A useful ratio identified as an effectively index can be computed to ascertain the effectivity of the total error when the exact total error is known. The effectivity index is defined as

$$\theta = \frac{\|e\|}{\|e\|_E} \tag{33}$$

where $\|e\|_E$ is the exact error. An insight into the accuracy of the error estimator is gained when applying the effectivity index to finite element analyses of classical problems where the exact stress distribution is known.

RESULTS

In order to evaluate the effectiveness of the closed form element stiffness matrix and error estimator discussed above, several problems were solved. The results for a stepped cylinder with pressure on its upper surface are presented here.

The element subdivision indicators ξ_i, are depicted within each element while the degrees of freedom (DOF), error estimate, exact error, and effectivity index are given in tabular form above each mesh. Both uniform and adaptive mesh refinement procedures (Figures 2 and 3, respectively) were carried out to achieve a specified accuracy of five percent.

A stress singularity exists at the re-entrant corner. Uniformly refining an initial 70 element, 320 DOF mesh met the desired accuracy in two iterations at an error estimate of 4.82 percent with a 280 element, 1200 DOF mesh. Adaptive refinement shown in Figure 4 also satisfied the accuracy criterion in two steps at an error estimate of 4.32 percent using a 90 element, 408 DOF mesh. Figure 4 reveals the greater efficiently realized in DOF when adaptively refining the mesh for this problem. Note the relatively large error indicators in those elements near the singularity as result of the large stress gradient. This observation suggests that adaptive refinement guided by the element error indicators is particularly advantageous in those problems containing a region of stress concentration.

CONCLUSIONS AND RECOMMENDATIONS

Closed form representations of the element stiffness matrix and error estimator for axisymmetric problems are intended to provide computational speed and efficiency without compromising solution accuracy. The approximate integration method, AIM, was implemented in the element stiffness matrix and error estimator derivation to accomplish this goal, i.e. six closed form matrices of constant terms were developed. The purpose of the element error estimator is to facilitate adaptive mesh refinement. Adaptive refinement in the stepped cylinder was shown to be much more efficient than uniform refinement. Ideally, an adaptive mesh refinement procedure will equal or surpass the desired solution accuracy in one iteration. As seen in the problems considered here, two steps were required. This can be attributed to the fact that the error estimator is an estimate only.

REFERENCES

1. R. W. Clough and Y.Rashid, "Finite Element Analysis of Axi-symmetric Solids," *Proc. ASCE 91*, EM.1, 71-85, (1965).
2. A. Peano, "Hierarchies of Conforming Finite Elements for Plane Elasticity and Plate Bending," *Computers & Math. with Appls.* 2, No. 3/4, 211-24, (1976).
3. G. Subramanian and C. J. Bose, "Convenient Generation of Stiffness Matrices for the Family of Plane Triangular Elements," *Computers and Structures* 15, No. 1, 85-89, (1982).
4. O. C. Zienkiewicz and J. Z. Zhu, "A Simple Error Estimator and Adaptive Procedure for Practical Engineering Analysis," *Int. J. Num. Meth. Eng.* 24, 337-57, (1987).
5. O. C. Zienkiewicz and R. L. Taylor, *The Finite Element Method*, 4th Edition, McGraw-Hill Publishing Company, London, 1989.
6. R. V. Nambiar, "Closed Form Expressions for Hierarchic Triangular and Tetrahedral Finite Elements," Doctoral Dissertation, The University of Texas at Arlington, 1989.
7. D. E. Byrd, "Identification and Elimination of Errors in Finite Element Analysis," Doctoral Dissertation, University of Colorado, 1988.
8. Brad E. Bergmann, "Closed Form Stiffness Matrix and Error Estimator for Hierarchic Traingular Elements Applied to Axisymmetric, Elasticity Problems," Masters Report, Department of Mechanical Engineering, Univ. of Texas at Arlington, Dec. 1989.

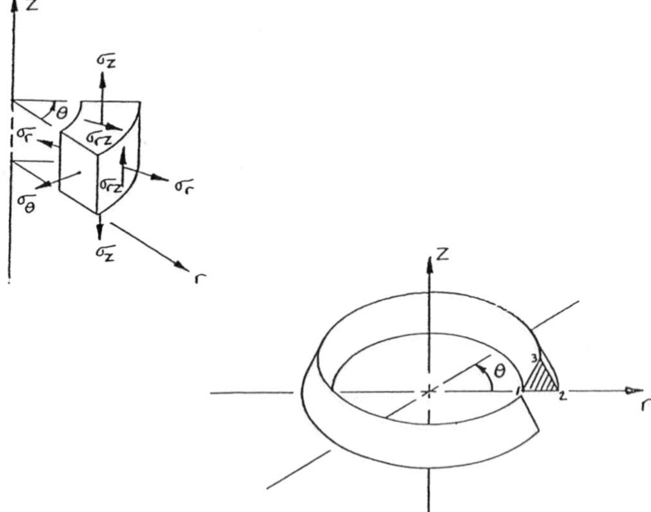

Figure 1.   Toroidal volume of a triangular element in an axisymmetric problem.

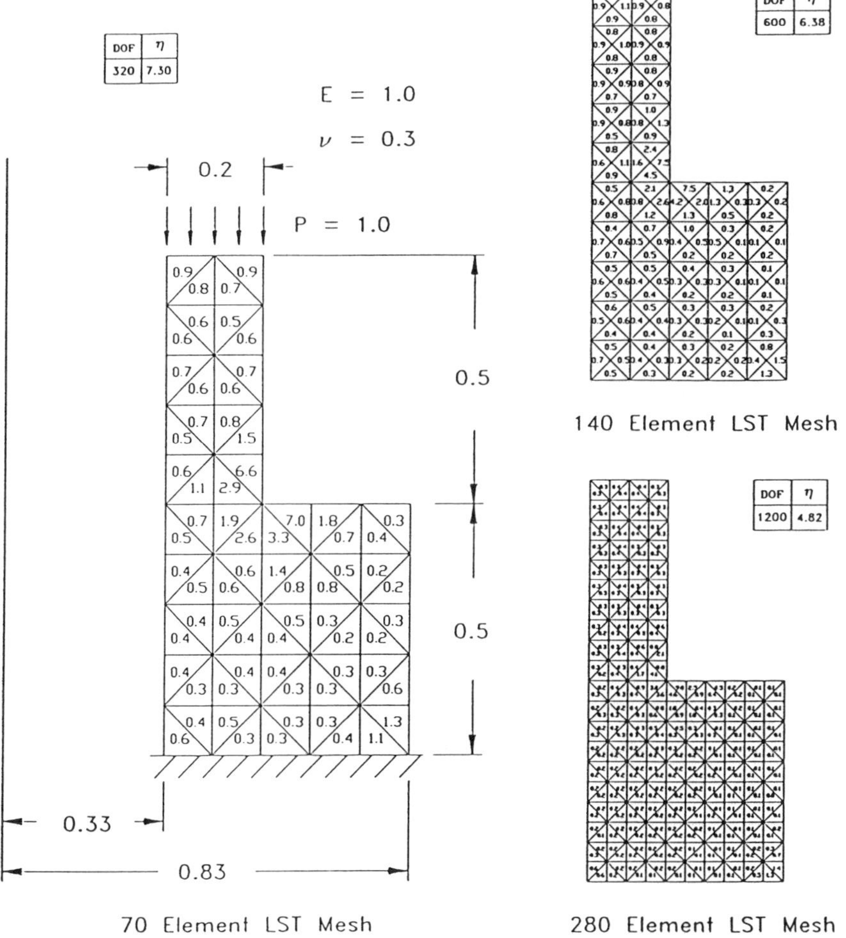

Figure 2. Uniform mesh refinement of a stepped cylinder modeled using LST elements.

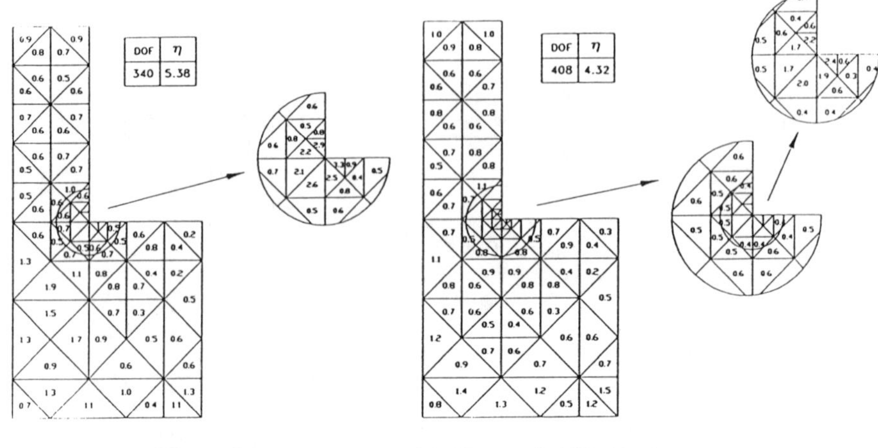

Figure 3.   Adaptive mesh refinement of a stepped cylinder modeled using LST elements.

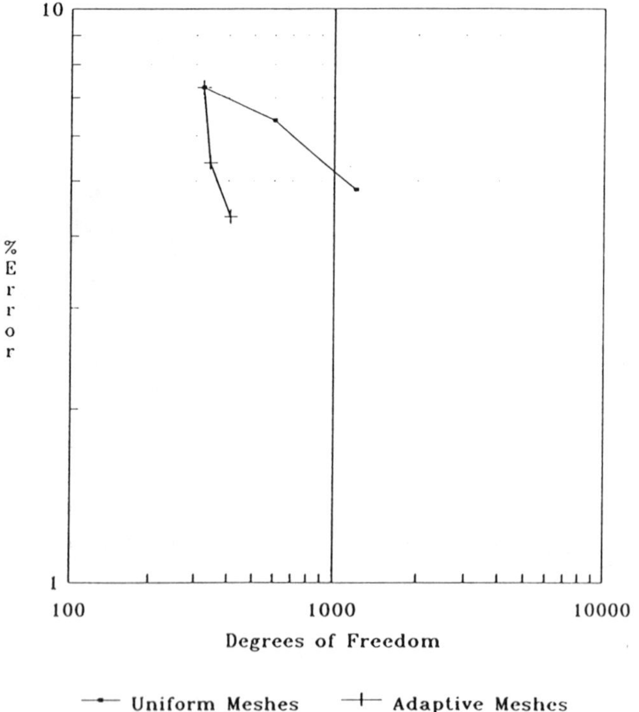

Figure 4.   Comparison of the estimated errors versus DOF for uniform and adaptive meshes of a stepped cylinder modeled using LST elements.

200

Integral Method for Hygrothermal Effect of Composite Plates

SEIICHI NOMURA
Department of Mechanical Engineering
University of Texas at Arlington
Arlington, Texas 76019-023, USA

J. S. CHANG
Institute of Applied Mechanics
National Taiwan University
Taipei, Taiwan, ROC

ABSTRACT

This paper presents an integral method approach to hygrothermal behavior of composite laminates by taking advantage of symbolic algebra software. Recent development of symbols software packages such as *REDUCE* and *Mathematica* revitalized the classical Galerkin method that is applicable for a wide range of typical physical and engineering problems. Galerkin functions can be derived analytically using symbolic software that satisfy given homogeneous boundary conditions. Those Galerkin functions can be further manipulated by symbolic algebra to generate necessary matrix elements for algebraic equations. It is generally known that trial functions that satisfy given boundary conditions yield excellent numerical accuracy. As an example of the proposed method, thermal buckling of composites is taken and the result shows good agreement with known solutions.

INTRODUCTION

The recent development of symbolic algebra software has attracted attention among engineers as it brings a new perspective to the conventional analytical methods [1]. With a proper combination of symbolic software and conventional numerical software, it is now possible to treat a wide range of physical and engineering problems in which purely numerical methods were the only available approaches before.

Symbolic algebra software can perform expansion, integration and differentiation of polynomials and transcendental functions analytically. It can also handle vector and tensor manipulations that would otherwise be formidable by human hands. Some of widely available packages include *REDUCE [2]*, *Mathematica [3]*, *Maple*, *Derive* and *Macsyma* and others (See [1] for overview).

In this paper, an integral method is presented to demonstrate the usefulness of symbolic algebra software for the analysis of thermal buckling of composite plates (For other types of application, see [4-5]). The lateral displacement for (two-dimensional) composite plates is governed by a fourth order partial differential equation. This equation with associated boundary conditions is usually difficult to solve analytically and the finite element method has been widely used instead. In the proposed approach, the solution is assumed as a linear combination of polynomial Galerkin functions each of which satisfies the given homogeneous boundary conditions. The coefficients of such polynomials can be determined analytically by symbolic algebra software. Symbolic algebra software is also used to generate matrix elements in the Galerkin procedure. Without the help of symbolic algebra software, manual calculation is next to impossible.

It has been known that if the Galerkin functions satisfy given boundary conditions (which is not always necessarily the case), the accuracy of the solution is significantly improved. This is demonstrated in this paper for the buckling of plates compared with the exact known solution. Moreover, the present approach is suitable for parametric study as it retains all the relevant parameters.

ANALYSIS

Consider a two-dimensional anisotropic composite plate subject to uniform temperature and moisture distribution. For simplicity, classical laminate theory is adopted here although the extension to higher order theories is possible. The governing equation for the lateral displacement, w, is expressed in tensorial form as

$$D_{ijkl}\, w_{,ijkl} + (N_{ij}^{H}\, w_{,i})_{,j} = - M_{ij,ij}^{H} \tag{1}$$

where partial derivative with respect to x_i is denoted by $,_i$ and the repeated index implies summation over x_1 and x_2 [6]. Each quantity in equation (1) is defined as

$$D_{ijkl} = \int_{-\frac{h}{2}}^{\frac{h}{2}} z^2\, C_{ijkl}(z)\, dz \tag{2}$$

$$N_{ij} = \int_{-\frac{h}{2}}^{\frac{h}{2}} \sigma_{ij}(z)\, dz \tag{3}$$

$$N_{ij}^{H} = \int_{-\frac{h}{2}}^{\frac{h}{2}} C_{ijkl}(z)\, (\alpha_{kl}(z)\, \Delta T + \beta_{kl}(z)\, \Delta m)\, dz \tag{4}$$

$$M_{ij}^{H} = \int_{-\frac{h}{2}}^{\frac{h}{2}} z\, C_{ijkl}(z)\, (\alpha_{kl}(z)\, \Delta T + \beta_{kl}(z)\, \Delta m)\, dz \tag{5}$$

where $C_{ijkl}(z)$ is the (anisotropic) elastic modulus of each lamina, α_{kl} is the thermal expansion coefficient, β_{kl} is the moisture expansion coefficient and h is the total thickness of the plate. The quantities of ΔT and Δm are uniform temperature rise and moisture rise, respectively. It should be noted that M_{ij}^{H} of equation (5) vanishes if 1) the laminate stacking is symmetrical and 2) the temperature and moisture distribution is uniform.

Three type of boundary conditions are expressed as

$$\text{Simply supported: } w=0,\ M_n=0 \tag{6}$$

$$\text{Clamped: } w=0,\ \frac{\partial w}{\partial n}=0 \tag{7}$$

$$\text{Free: } M_\nu = 0, \quad \frac{\partial M_s}{\partial s} + M_{ij,j}\, n_{ij} = 0 \tag{8}$$

where n_i is the normal to the plate boundary, M_{ij} is the bending moment and M_s and M_s are normal and tangential components of M_{ij} along the composite plate boundary, respectively. It should be noted that the bending moment, M_{ij} and the lateral deflection, w, are related as

$$M_{ij} = -D_{ijkl}\, w_{,kl} \tag{9}$$

SOLUTION PROCEDURE

Temperature and moisture rise causes buckling of a composite plate with clamped boundaries. The Galerkin procedure can convert equation (1) to a corresponding algebraic eigenvalue problem suitable for symbolic manipulation. The solution to equation (1) is assumed to be a linear combination of base functions as

$$w(x,y) = \sum_{\alpha=1}^{n} c_\alpha\, \psi_\alpha(x,y) \tag{10}$$

where $\psi_\alpha(x,y)$'s are base functions that satisfy a given homogeneous boundary condition. Following the standard Galerkin procedure, equation (10) is substituted into equation (1), multiplied by $\psi_\beta(x,y)$ and integrated over the entire plate. Under the conditions that 1) the laminate is symmetrical and 2) the temperature and the moisture are uniform, this procedure yields the following matrix eigenvalue problem:

$$A\,c = N_{ij}^H\, B^{ij}\, c \tag{11}$$

where

$$(a_{\alpha\beta}) = \int\!\!\int D_{ijkl}\, \psi_{\alpha,ij}\, \psi_{\beta,kl}\, dS \tag{12}$$

and

$$(b_{\alpha\beta}^{ij}) = \int\!\!\int \psi_{\alpha,i}\, \psi_{\beta,j}\, dS \tag{13}$$

where the integral range is over the entire plate surface.

The base function $\psi_\alpha(x,y)$ in equation (10) is assumed to be constructed from a series of polynomials as

$$\psi_\alpha(x,y) = \sum_{i=1}^{n} h_i^\alpha\, x^{L_i}\, y^{M_i} \tag{14}$$

where L_i and M_i are ordered integers. The coefficients, h_i^α's, can be determined so as to satisfy the given boundary conditions. The minimum order of polynomials, $\psi_\alpha(x,y)$, is chosen in such a way that the number of unknowns (h_i^α's) exceeds the number of independent equations derived from the boundary condition (rank).

Symbolic algebra software (*Mathematica*) was used to generate h_i^α's from the boundary conditions which yield a set of simultaneous (underdetermined) equations. The solution to these underdetermined equations is a null space of the matrix (h_i^α). *Mathematica* was also used to evaluate equations (12) and (13) by expressing each component of the matrices, $(a_{\alpha\beta})$ and $(b_{\alpha\beta}^{ij})$ as functions of h_i^α's to be fed into an eigenvalue solver routine, thus avoiding time-consuming numerical integrals. It should be noted that all the variables retain their symbolic form until the last moment for substitution.

EXAMPLE AND CONCLUSION

Justification of the above developed procedure can be done by directly comparing numerical results with analytical solutions for a particular problem. A good example is found in buckling of a square plate due to compressional forces on all the sides. The governing equation is

$$\nabla^2\nabla^2 w + \lambda\nabla^2 w = 0 \tag{15}$$

where

$$\lambda = \frac{N^T}{D}, \quad D = \frac{Eh^3}{12(1-\nu^2)} \tag{16}$$

$$N^T = \frac{E}{1-\nu} \int_{-\frac{h}{2}}^{\frac{h}{2}} (\alpha \triangle T + \beta \triangle m) \, dz \tag{17}$$

The value of initial buckling with the clamped boundary condition on all the sides can be found by the Weinstein's method [6]. The following table summarizes the comparison with the current approach.

TABLE 1. Eigenvalues for different number of terms

n	λ
8	5.47
9	5.47
10	5.31
Analytical Solution	5.31

n: order of polynomials on x and y

Boundary conditions:

$$x=0, \ 1 : w=0, \ \frac{\partial w}{\partial x}=0$$

$$y=0, 1 : w=0, \ \frac{\partial w}{\partial y}=0$$

It is seen from the above table that the first eigenvalue, λ, by the present approach converges to the analytical solution by taking the maximum tenth order polynomials on both x and y. The six independent polynomials for the case of order 10 that satisfy the given boundary conditions are derived by symbolic algebra as :

$$w_1 = 3x^2y^2 - 4x^3y^2 + x^6y^2 - 6x^2y^3 + 8x^3y^3 - 2x^6y^3 + 3x^2y^4 - 4x^3y^4 + x^6y^4$$

$$w_2 = 4x^2y^2 - 6x^3y^2 + 2x^5y^2 - 6x^2y^3 + 9x^3y^3 - 3x^5y^3 + 2x^2y^5 - 3x^3y^5 + x^5y^5$$

$$w_3 = 3x^2y^2 - 6x^3y^2 + 3x^4y^2 - 4x^2y^3 + 8x^3y^3 - 4x^4y^3 + x^2y^6 - 2x^3y^6 + x^4y^6$$

$$w_4 = 2x^2y^2 - 3x^3y^2 + x^5y^2 - 4x^2y^3 + 6x^3y^3 - 2x^5y^3 + 2x^2y^4 - 3x^3y^4 + x^5y^4$$

$$w_5 = 2x^2y^2 - 4x^3y^2 + 2x^4y^2 - 3x^2y^3 + 6x^3y^3 - 3x^4y^3 + x^2y^5 - 2x^3y^5 + x^4y^5$$

$$w_6 = x^2y^2 - 2x^3y^2 + x^4y^2 - 2x^2y^3 + 4x^3y^3 - 2x^4y^3 + x^2y^4 - 2x^3y^4 + x^4y^4 \tag{18}$$

Those polynomials are fed into equations (12) and (13) to arrive at the matrix eigenvalue problem. It should be noted that by a proper programming in symbolic algebra this lengthy and tedious process can be completely automated without human interface.

Finally, the application of present method is not limited to the classical laminate theory. Adoption of higher order theories will yield a non-linear eigenvalue problem which can be handled routinely by symbolic algebra. Results will be reported subsequently.

ACKNOWLEDGMENT

One of the authors (S.N.) acknowledges the financial support by the Texas Higher Education Coordinating Board.

REFERENCES

1. Nomura, S. and Haji-Sheikh, A., Heat Transfer in Arbitrary Shaped Duct with Constant Wall Temperature, in *Symbolic Computation Fluid Mechanics and Heat Transfer*, pp.75-80, ed.by H. H. Bau, Herbert and M. M. Yovanovich, ASME, 1988.

2. Hearn, A., *REDUCE 3: User's Manual*, Rand Corporation, 1985.

3. Wolfram, S., *Mathematica*, Addison Wesley, 1989.

4. Nomura, S., Integral Method for Composite Materials, *Proceedings of the Seventh International Conference on Composite Materials*, Volume 3, pp.100-106, Pergamon Press, ed. by Wu Yunshu, Gu Zhenlong and Wu Renjie, Beijing, China, 1989.

5. Nomura, S. and Wang, B.P., Free Vibration of Plate by Integral Method, *Computers & Structures*, Vol.32, No.1, pp.245-247 1989.

6. Timoshenko, S. P. *Theory of Elastic Stability*, McGraw-Hill, 1961.

Theoretical Studies of Some Neuropharmacological Effects on Directional Selectivity

H. ÖĞMEN
Department of Electrical Engineering
University of Houston
Houston, Texas 77204-4793, USA

ABSTRACT

Recent efforts in the understanding of motion detection and directional selectivity include electrophysiological studies using single photoreceptor stimulations and a combination of electrophysiology and neuropharmacology. Results of the former have been interpreted in favor of facilitatory models while results of the latter have been interpreted in favor of inhibitory models. This chapter addresses this conflicting data interpretation problem by studying neuropharmacological predictions of a neural network model which was proposed in Öğmen & Gagné (1990b) for the motion sensitive units of the fly visual system.

INTRODUCTION

Recently, there has been a remarkable convergence in the functional properties of models proposed for directionally selective motion perception in various species ranging from invertebrates to humans. The core of these models is based on a model originally developed to explain *functional* properties of visually guided behavior of insects. By considering the nervous system as a black box and by using a Volterra series representation, based on experimental data, a correlator type model was proposed for the fly visual system (Reichardt 1961). Recently, variants of this model have been extended to the human visual system (van Doorn & Koenderink, 1983; van Santen & Sperling, 1984, 1985; Wilson, 1985). However, although the Reichardt correlator was successful in predicting many aspects of optomotor behavior, it does not specify the neural architecture controlling this behavior and its predictions can not be extended to a large number of real-time dynamic environments. Many experimental work has been devoted to the understanding of the neural architecture underlying directional selectivity (DS). Some authors considered "minimal networks " and classified plausible models into *facilitatory* and *inhibitory* categories as illustrated in Figure 1. In both cases one needs a single (nonlinear) cell and two spatially distinct input channels to achieve directionally selective motion detection. In the facilitatory model (Figure 1a), both channels make excitatory connections with the elementary motion detection cell (EMD). The output of one channel is delayed by Δt. If an input signal perturbs this channel first and then moves to the next one and if the time lapse between the

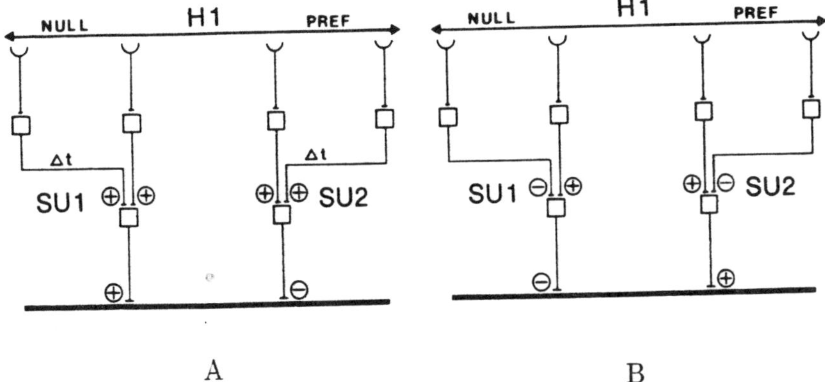

Figure 1: Excitatory (a) and inhibitory (b) motion detection models. Null and preferred directions of the H1 neuron are indicated at the top of the figure. SU1 and SU2 represent hypothetical subunits of the model. Signals flow from the top to the bottom of the figure. Adapted from Schmid & Bülthoff (1988).

excitation of channels (which is determined by the spatial separation of sampling channels and the velocity of the stimulus) is close to Δt, then the EMD will register temporally overlapping excitatory postsynaptic potentials (EPSP). The threshold of the EMD is chosen such that only overlapping EPSPs can yield an output signal. Then, a motion in this direction (preferred direction) within a given range of velocities will be detected by the EMD. If the stimulus moves in the opposite direction, EPSPs will not overlap and the cell will stay silent. The EMD of an inhibitory model (Figure 1b) receives an inhibitory signal from one channel and an excitatory signal from the second. The inhibitory signal is assumed to last longer than the excitatory one. If a stimulus excites the excitatory channel first, then the inhibitory one, since the inhibition arrives after the excitation, the cell will respond (preferred direction). In the opposite direction (null direction), the stimulus excites the inhibitory channel first and then the excitatory channel. The inhibitory signal arriving before and lasting longer than the excitatory signal, the cell will remain silent. Finally, the outputs of two mirror image like EMDs are combined at another cell in an opponent fashion.

Riehle and Franceschini used a special technique on the fly to stimulate only two photoreceptors making connections with two neighboring "channels". (Riehle & Franceschini, 1983). By using such a minimal stimulation they were able to observe the behavior of the minimal configuration able to detect motion and their results favored a facilitatory model such as the model of Figure 1a rather than the inhibitory model of Figure 1b.

Another line of experiments combined neuropharmacology and electrophysiology to study the role of inhibition in motion detection in the insect visual system (Schmid & Bülthoff, 1988). An antagonist of the inhibitory neurotransmitter GABA, picrotoxinin, was used to block inhibitory interactions conveyed by GABA. By recording from a wide-field directionally selective cell in the lobula plate (the *H1* cell, the same cell recorded by Riehle-Franceschini) they observed changes in the directional selectivity of the network.

Injection of GABA (200 nl) caused a decrease in the spike rate for both resting activity and responses to the motion in the preferred direction. The response for the null direction stayed at zero. The cell recovered slowly and complete recovery was reached in about 35 minutes after injection.

Injection of picrotoxinin at small amounts (0.5-1 nl) to lobula or medulla caused a general increase in spike activity. The cell begun to respond to motion in the null direction but the amplitude of the responses was higher for the preferred direction than the null direction, thus DS was preserved although the cell responded to motion in both directions. For medium amounts (1-3 nl) basically similar effects were observed. However, in the case where picrotoxinin was injected to lobula the null response became equal to the preferred one shortly after injection (1-2 minutes). Thus, H1 lost its directional selectivity. The abolishment of DS lasted about 1 to 2 minutes. At high amounts (3-8 nl) the detected spike activity increased for 2 minutes and then completely disappeared for the following 5 minutes. Then, spike activity started to increase again. The null response became higher than the preferred response at 8 minutes after injection for the following 4-5 minutes. Thus, directional selectivity of H1 was inverted during this period. Complete recovery was obtained 60 minutes after injection.

Based on the simple models of Figure 1 these authors concluded that their results favored an inhibitory scheme rather than a facilitatory one. Furthermore, they modified the inhibitory model by adding extra EMDs so that stimulation of a single channel activates two EMDs with opposite preferred directions thereby eliminating the sensitivity of the model to single channel inputs in order to make it compatible with Riehle-Franceschini data. However such a modification does not make the inhibitory model fully compatible with Riehle-Franceschini data because its predictions are inconsistent with responses obtained for overlapping stimuli. Moreover, the *inversion* of DS poses a serious problem to both of these simplified schemes. Although it is not perfectly clear whether the inversion of directional selectivity is due to the properties of motion detection network or to the action of the picrotoxinin on the integrating cell *H1*, this point deserves attention because none of the simplified models of Figure 1 can explain this result. Thus, if this result is really due to the properties of the neural network rather than being an experimental artifact both of the simplified models can be rejected.

In this paper, this apparent conflict between recent experimental data is addressed by modeling effects of neuropharmacological agents in neural networks and by applying this formalism to a directionally selective motion detection model proposed for the fly visual system (Öğmen & Gagné 1988, 1990b). Since a good agreement has already been found between predictions of this model and Riehle-Franceschini data (Öğmen & Gagné 1990b), we focus herein on the pharmacological properties of the model and compare its predictions with Schmid-Bülthoff data as an attempt to resolve the "conflict". For a review of invertebrate neuropharmacology the reader is referred to Leake & Walker (1980) and Hardie (1989).

A DIRECTIONALLY SELECTIVE MOTION DETECTION MODEL

Figure 2 illustrates the neural model for directionally selective motion detection (Öğmen & Gagné 1988, 1990b).

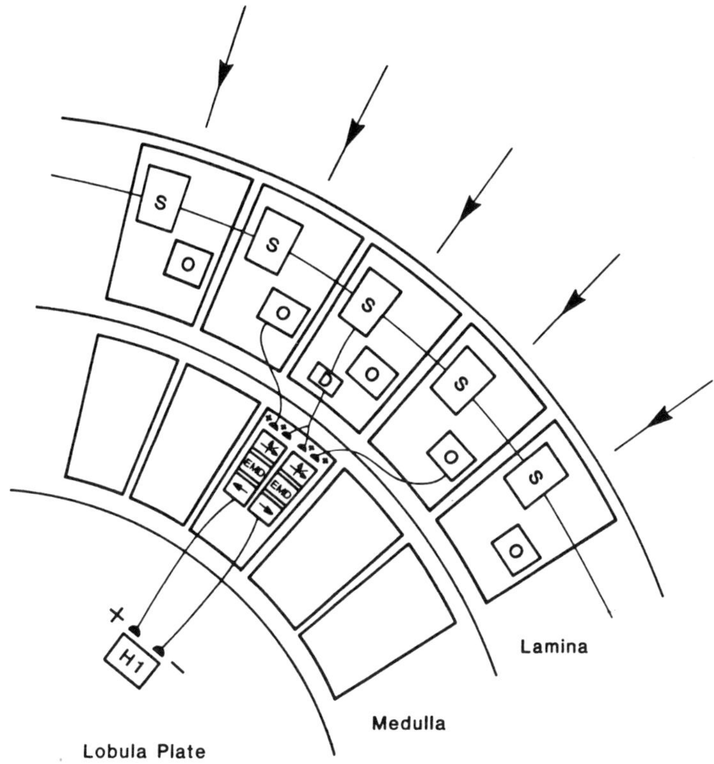

Figure 2: Neural model for directionally selective motion detection in the visual system of the fly. From Öğmen & Gagné (1990b).

The input signals are first processed by sustained (boxes marked S) and transient channels (boxes marked O). The output of the sustained unit is passed through a cell with slow dynamics (box marked D). This first stage of processing realizes spatial and temporal adaptations. Unlike many versions of Reichardt model, these "filters" are *nonlinear* and *time-variant*. Corresponding neural correlates are the spiking units recorded by Arnett (1972) and conjectured to be *L4* and *L5* neurons of lamina (Laughlin 1984, Shaw 1981, 1984). Analysis of models for these units and comparisons with electrophysiological data can be found in Öğmen & Gagné (1990a). The output of neighboring sustained and transient units are combined nonlinearly at the elementary motion detector cell (EMD). The mechanism yielding directional selectivity is similar to the excitatory model of Figure 1a: Temporally overlapping EPSPs exceed the threshold while nonoverlapping EPSPs remain subthreshold. However, the early processing differs in a crucial way from the simple models of Figure 1 due to nonlinear intensity dependence, nonlinear signal rectification, sustained/transient channel interactions, and adaptivity. These properties are achieved by use of extensive inhibition. Thus, although inhibition is absent in elementary motion detection stage of the facilitatory model, it plays an important role in our model and the purpose of this study is to establish whether the pharmacological findings can be explained based on pharmacological alterations of these inhibitory interactions. Finally, EMDs responding to regressive (progressive) motion make excitatory (inhibitory) connections to the wide-field cell *H1*. For a detailed exposition of this model and comparisons with Riehle-Franceschini data the reader is referred to Öğmen & Gagné (1990b). The mathematical description of the model is provided in Appendix A and in order to illustrate model equations and motivate the pharmacological modifications we will briefly introduce the equation describing the dynamics of the sustained cell (Box S; note that symbols S and D in Figure 2 and in the following equation denote different things):

$$\frac{dx_i}{dt} = -Ax_i + (B - x_i)S^+ - (D + x_i)S^- \tag{1}$$

This shunting equation (Grossberg 1973) is a membrane equation describing the time domain behavior of the membrane potential x_i of *ith* sustained unit. Its rate of change is controlled by three terms: The term $(B - x_i)S_i^+$ represents the effect of depolarizing inputs. S^+ is the total transmitter gated excitatory input to the *ith* cell, and B is the upper saturation level ($B > 0$) of x_i. The term $-(D + x_i)S^-$ accounts for the effect of hyperpolarizing inputs with S^- representing the total transmitter gated inhibitory input to the *ith* cell, and $-D$ ($D > 0$) is the lower saturation level of x_i. Finally the term $-Ax_i$ represents the effect of passive channels ($A > 0$).

MATHEMATICAL CHARACTERIZATION OF PHARMACOLOGICAL EFFECTS

Two major theories have been proposed for kinetics of the physiological responses induced by drugs. Both postulate that there exists reversible reactions between drug and receptor molecules and that the law of mass action applies. However, the occupation theory (Stephenson 1956; Ariens 1966) links the physiological response to the proportion of receptors occupied by the drug while the rate theory (Paton 1961) attributes the response to the *process* of occupation. Thus, according to the rate theory, the response to the drug is proportional to the rate of drug-receptor association. However, both theories predict equivalent equilibrium response-dose relationships and the response is a *strictly increasing* function of the dose. In addition to these local drug-receptor interactions, the diffusion of the drug in the tissue should be characterized by taking into account diffusion barriers and tissue properties. However, given the difficulties associated with the characterization of these local properties and the lack of precise anatomical corroboration of sub elements in the model, a macroscopic, lumped description of drug action becomes desirable. We will use a simple formalism which can be reduced to microscopic interactions by using Michaelis-Menten kinetic model and mass action laws. We introduce a coefficient Φ which depends both on the amount of substance injected to the brain and on time. This coefficient gates the interactions carried out by the pharmacologically targeted neurotransmitter. In this case the targeted neurotransmitter is inhibitory and accordingly, equation (1) becomes:

$$\frac{dx_i}{dt} = -Ax_i + (B - x_i)S^+ - \Phi(D + x_i)S^- \tag{2}$$

Thus for this study we will consider modified version of equations in Appendix A so that they include the coefficient Φ gating the inhibitory interactions using the targeted transmitter. We will simply assume that the neuropharmacological effect is a strictly increasing function of the injected dose and decays slowly with time (possibly due to a combination of diffusion and inactivation processes). We will neglect the delay between drug injection and the first noticeable physiological response. Thus, the following differential equation can give a simple description for the pharmacological manipulation:

$$\frac{d\Phi}{dt} = \xi(1 - \Phi) + \mathcal{G}^+ - \Phi\mathcal{G}^- \tag{3}$$

where Φ is the pharmacological coefficient, $\mathcal{G}^+$ and $\mathcal{G}^-$ represent respectively the dose of injected agonist and antagonist of the neurotransmitter and ξ is the macroscopic passive decay constant. Hence, Φ has the following interpretation: Normal conditions: $\Phi = 1$; injection of the antagonist: $0 \leq \Phi < 1$; and injection of agonist: $\Phi > 1$. *Injection of GABA* is characterized by putting $\mathcal{G}^- = 0$ and $\mathcal{G}^+(t) = \mathcal{G}_0^+$ for $0 \leq t \leq t_0$ and 0 otherwise. Then, solving equation (3) and approximating x_i at its equilibrium value one finds:

$$x_i = \frac{BS^+ - D\Phi(t)S^-}{A + S^+ + \Phi(t)S^-} \tag{4}$$

$$\Phi(t) = \begin{cases} 1 & \text{for } t < 0 \\ \Phi_1(t) = \frac{\rho}{\lambda}(1 - e^{-\lambda t}) + e^{-\lambda t} & \text{for } 0 \leq t \leq t_0 \\ 1 - (1 - \Phi_1(t_0))e^{-\xi(t - t_0)} & \text{for } t > t_0 \end{cases} \tag{5}$$

where $\rho = \xi + \mathcal{G}_0^+$ and $\lambda = \xi$. According to equation (5) during the injection of GABA, Φ increases monotonically and once the injection is stopped it decays monotonically with a time constant ξ.

Injection of picrotoxinin which is an antagonist of GABA is modeled similarly: Let $\mathcal{G}^+ = 0$ and $\mathcal{G}^-(t) = \mathcal{G}_0^-$ for $0 \leq t \leq t_0$ and 0 otherwise. Then equation (5) will hold with $\rho = \xi$ and $\lambda = \xi + \mathcal{G}_0^-$. After the injection of picrotoxinin, Φ decreases towards its asymptotic value $\frac{\xi}{\xi + \mathcal{G}_0}$ with a time constant $\xi + \mathcal{G}_0$, and once the injection is stopped, the system recovers with a time constant ξ.

SOME PREDICTIONS OF THE MODEL

In this section we study some predictions of the motion detection model coupled with pharmacological equations. Formal results and intuitive explanations are presented and the reader is referred to Appendices A and B, and Öğmen & Gagné (1990b) for mathematical details and proofs.

For $g(a)$, the input-output nonlinearity of neurons, a positive semi-definite sigmoidal function admitting the upperbound U is used: $g(a) = 0$ for $a \leq 0$, g is faster-than-linear in the interval $[0, \Lambda_1]$, linear in $[\Lambda_1, \Lambda_2]$, and slower-than-linear in $[\Lambda_2, \infty[$. We will call the interval $\mathcal{R}_{sub} =] - \infty, 0]$ the *subthreshold region*; and $\mathcal{R}_{sup} =]0, \infty[$ the *suprathreshold region*. The suprathreshold region comprises an *amplification region* $\mathcal{R}_a =]0, \Lambda_2]$ and a *saturation region* $\mathcal{R}_s =]\Lambda_2, \infty[$. g is assumed to be strictly increasing in the suprathreshold region. But since the intracellular activities are bounded by B, in order to make all portions of the input-output curve effective, B is chosen such that when the membrane potential tends to its upper limit B, the output tends to its upper limit U.

In order to relate the time behavior of the pharmacological coefficient Φ to the DS property of the motion detecting network, one needs to relate variations of Φ to variations of cell variables, and variations of cell variables to the DS property of the network. The following lemmas provide these relationships:

Lemma 1 x, y, *and* x_{lob} *are strictly decreasing functions of* Φ.

Lemma 2 *Let* $h(a) = g(a - \Gamma_{emd})$. *DS is preserved iff*

$$D_h(g(x - \Gamma_x), g(y - \Gamma_y)) \tag{6}$$

is positive; DS is abolished iff it is zero; and DS is inverted iff it is negative.

where D_h is the deficiency of h (i.e. $D_h(a, b) = h(a + b) - [h(a) + h(b)]$).

Lemma 3 (Necessary conditions) *If DS is inverted then* $g(x - \Gamma_x) > \Gamma_{emd}$ *and* $g(y - \Gamma_y) > \Gamma_{emd}$.

Lemma 1 shows how to relate unambiguously the time behavior of the pharmacological coefficient to that of activities of sustained, on-off, and *H1* neurons (respectively x, y, and x_{lob}; subscripts are dropped for simplicity, c.f. Appendix B) while Lemma 2 gives the conditions that relate variations in x and y to the DS property.

By Lemma 1 and Equation (5), *injection of GABA* suppresses the activities of early units and recovery occurs slowly. The suppression of early units leads to a suppression of activities of EMDs since even in the case of overlapping EPSPs the threshold cannot be exceeded. Therefore, a general decrease in the spike activity of *H1* and a recovery with time constant ξ is predicted.

By Lemma 1 and Equation (5) with the substitutions corresponding to the *injection of picrotoxinin*, there will be a general increase in the activities of early units as well as lobula units receiving opponent inputs. The implications of these increases are highly dependent on the nonlinear properties of g. Administration of sufficient amounts of antagonistic drugs drives intracellular potentials of these cells into the suprathreshold region of g, thus results into spontaneous activity (or increase in the spontaneous activity if it already exists). Accordingly, EMDs start to respond to motion in both preferred and null directions, since the pooled EPSPs from early units can reach the suprathreshold region even without temporal overlap. Changes in the DS property of the wide-field cells (e.g. *H1*) are established by the following propositions:

Proposition 1 (Small amounts) *If* $(g(x - \Gamma_x) + g(y - \Gamma_y)) \in \mathcal{R}_a$ *then DS is preserved.*

Proposition 2 (Large amounts-existence) *Assume that conditions of Lemma 3 hold, and that $g(x - \Gamma_x) \to U$, $g(y - \Gamma_y) \to U$ (large amounts). If $U > \Lambda_2 + \Gamma_{emd}$ then DS is inverted.*

Corollary 1 (Abolishment) *There exists an intermediate amount of drug for which DS is abolished.*

These results are based upon the fact that for motion in the preferred (null) direction overlapping (nonoverlapping) EPSPs are transformed by g and signaled down to *H1*. Thus, as stated by Lemma 2, one needs to compare the transformation of overlapping and nonoverlapping signals. If the overlapped activities are in the amplification region, then the same is true for nonoverlapping activities (or they are subthreshold). And since in that region g is faster-than-linear or linear, overlapped activity, thus the preferred direction, prevails upon nonoverlapped activities (null direction), i.e. DS is preserved. When the amount of drug is increased such that activities are in the saturation region (g slower-than-linear) then the preferred direction starts to lose its advantage coming from the temporal overlap. When these signals are balanced on the saturation curve DS is abolished and a further increase in the amount of drug inverts DS. The time course of these changes as well as recovery is dictated by Equation (5).

CONCLUDING REMARKS

Although many experimental data indicate that lamina units are a good candidate for our sustained and on-off units (Laughlin 1984, Shaw 1981, 1984), there is no direct demonstration of this anatomical mapping. Similarly, anatomical location of EMDs is unknown. In order to answer questions of similar nature, Schmid & Bülthoff conducted experiments where the activity of *H1* and the electroretinogram (ERG) potentials in lamina were recorded simultaneously. At small amounts of drug injection no drug correlated changes in the ERG were observed. This seems to indicate that since even at small amounts important changes in the DS occurs, lamina units do not play an important role in the DS property. However, these findings are in principle insufficient to reach that conclusion since the ERG in the lamina is essentially caused by large monopolar cells L1 and L2 and not the spiking cells recorded by Arnett which are the candidates for our early units (Coombe 1986). Moreover, at high amounts of drug injection the ERG was modified and this can suggest that similar effects may exist for small amounts but are too weak to be detected as a prominent ERG component yet strong enough to alter DS properties.

Despite these local difficulties, neuropharmacology provides a powerful tool for probing various properties of neural *networks*. At the network level, one can use a macroscopic characterization of neuropharmacological effects and study network equations coupled with a pharmacological model. Our main motivation in this study was to formally analyze pharmacological properties of a neural network model proposed for motion detection in the visual system of the fly in order to seek a resolution to paradoxical data. From the theoretical point of view, such a resolution is offered. Experimentally, local circuit analysis can be performed by

213

more precise techniques and these results can be compared with model predictions by reducing the neuropharmacological equations to microlevel interactions through application of Michaelis-Menten equations and the law of mass action. Combined interpretation of pharmacological and focused stimulation experiments rejects both simplified models and suggests that motion detection should not be confined to a single synaptic interaction but should be analyzed at the network level. A very interesting experimental approach would be to combine these two techniques for the same preparation.

APPENDIX A

Mathematical description of the model

Sustained units

The dynamics of the ith sustained unit is characterized by the following equations:

$$\frac{dx_i}{dt} = -Ax_i + (B - x_i)[I + J_i]z_i - (D + x_i)\sum_{k \neq i}[I + J_k]z_k \tag{7}$$

$$\frac{dz_i}{dt} = \alpha(\beta - z_i) - (I + J_i)z_i \tag{8}$$

where J_i and I are respectively specific and arousal inputs. z_i denotes the amount of available transmitter and A, B, D, α, β are positive constants.

On-off units

$$\frac{dx_{on,i}}{dt} = -Ax_{on,i} + (B - x_{on,i})[I + J_i]z_{on,i} - (D + x_{on,i})Iz_{off,i} \tag{9}$$

$$\frac{dx_{off,i}}{dt} = -Ax_{off,i} + (B - x_{off,i})Iz_{off,i} - (D + x_{on,i})[I + J_i]z_{on,i} \tag{10}$$

$$\frac{dz_{on,i}(t)}{dt} = \alpha(\beta - z_{on,i}(t)) - \gamma(I + J_i)z_{on,i}(t) \tag{11}$$

$$\frac{dz_{off,i}(t)}{dt} = \alpha(\beta - z_{off,i}(t)) - \gamma I z_{off,i}(t) \tag{12}$$

$$\frac{dy_i}{dt} = -Ay_i + (B - y_i)(Fg(x_{on,i} - \Gamma_{on,i}) + Gg(x_{off,i} - \Gamma_{off,i})) \tag{13}$$

where g is a nonlinear function, F, G are constants, and Γ is the signal threshold. We call this structure an *augmented gated dipole*.

EMDs

Let us denote by $x_{emd,i,j}(t)$ the activity of the EMD at position i receiving inputs from the sustained unit at i and on-off unit at $i + j$, i.e. with a sampling base equal to j. The dynamics of this EMD is described by:

$$\frac{dx_{emd,i,j}}{dt} = -Ax_{emd,i,j} + (B - x_{emd,i,j})(f(g(x_i - \Gamma_i)) + g(y_{i+j} - \Gamma_{y,i+j})) \tag{14}$$

and for the EMD with opposite preferred direction:

$$\frac{dx_{\overline{emd},i,j}}{dt} = -Ax_{\overline{emd},i,j} + (B - x_{\overline{emd},i,j})(f(g(x_i - \Gamma_i)) + g(y_{i-j} - \Gamma_{y,i-j})) \tag{15}$$

where $f(.)$ represents the transformation by slow interneurons.

Wide-field cells

The following equation is proposed for the activity x_{lob} of the lobula cells:

$$\frac{dx_{lob}}{dt} = -Ax_{lob} + (B - x_{lob})\sum_{i,j} H_{i,j}g(x_{emd,i,j} - \Gamma_{emd,i,j}) - (D + x_{lob})\sum_{i,j} J_{i,j}g(x_{\overline{emd},i,j} -]$$

where the coefficients $H_{i,j}$ and $J_{i,j}$ account for the connection strengths. These connection strengths determine the receptive field profiles of lobula plate cells. Consequently, the general expression (16) can be specialized to various classes of cells found in the lobula plate including the *H1* cell, by a proper choice of $H_{i,j}$ and $J_{i,j}$.

APPENDIX B

Proof of Lemma 2: In order to establish the DS property one needs to compare x_{H1} for preferred and null direction stimulations. Given the mirror image symmetry of EMDs, the excitatory (inhibitory) input for the preferred direction will be equal to the inhibitory (excitatory) input for the null direction. For the measured averaged steady-state response, the corresponding lobula plate equation is given by:

$$x_{H1} = \frac{BK_e - \Phi DK_i}{A + K_e + \Phi K_i} \tag{17}$$

where $K_e = \sum_{i,j} H_{i,j}g(x_{emd,i,j} - \Gamma_{emd,i,j})$, $K_i = \sum_{i,j} J_{i,j}g(x_{\overline{emd},i,j} - \Gamma_{\overline{emd},i,j})$ and the DS property is determined by the sign of

$$\frac{BK_e - \Phi DK_i}{A + K_e + \Phi K_i} - \frac{BK_i - \Phi DK_e}{A + K_i + \Phi K_e} \tag{18}$$

which in turn is determined by the sign of $K_e - K_i$. Moreover, for a spatially homogeneous stimulus pattern (i.e. a stimulus pattern like the one used in Schmid-Bülthoff experiments) the summations can be replaced by a single typical EMD output. Assume for simplicity that $H_{i,j} = J_{i,j}$. It then suffices to study the sign of one summand in the numerator. Dropping the indices and approximating $g(x_{emd} - \Gamma_{emd})$ and $g(x_{\overline{emd}} - \Gamma_{emd})$ respectively by $g(g(x - \Gamma_x) + g(y - \Gamma_y) - \Gamma_{emd})$ (overlapping EPSPs) and $g(g(x - \Gamma_x) - \Gamma_{emd}) + g(g(y - \Gamma_y) - \Gamma_{emd})$ (nonoverlapping EPSP signaled by two different EMDs) for motion in the preferred direction and respectively $g(g(x - \Gamma_x) - \Gamma_{emd} + g(g(y - \Gamma_y) - \Gamma_{emd}$ and $g(g(x - \Gamma_x) + g(y - \Gamma_y) - \Gamma_{emd})$ for motion in the null direction one obtains (6). Note that inhibition at the *H1* level is not critical for the DS property.

Proof of Lemma 3: Since g is monotone increasing, by applying Lemma 2, it is necessary to have both $g(x - \Gamma_x)$ and $g(y - \Gamma_y)$ in the suprathreshold region.

Proof of Proposition 1: If $(g(x - \Gamma_x) + g(y - \Gamma_y)) \in \mathcal{R}_a$ then $(g(x - \Gamma_x) - \Gamma_{emd}) \in \mathcal{R}_a \cup \mathcal{R}_{sub}$ and $(g(y - \Gamma_y) - \Gamma_{emd}) \in \mathcal{R}_a \cup \mathcal{R}_{sub}$. Since g is strictly increasing, faster-than-linear or linear in $\mathcal{R}_a$, preservation of DS follows.

Proof of Proposition 2: $lim_{g(x-\Gamma_x)\to U; g(y-\Gamma_y)\to U} D_h(g(x - \Gamma_x), g(y - \Gamma_y)) = h(2U) - 2h(U)$. But if $U > \Lambda_2 + \Gamma_{emd}$ then $h(a)$ is slower-than-linear for $a > U$, thus the deficiency is negative and by Lemma 2 DS is inverted.

Corollary 1 follows from continuity of D_h.

References

[1] Ariens, E. J. Receptor theory and structure-action relationships. In: N. J. Harper, & A. B. Simmonds (Eds.), *Advances in drug research*, London: Academic Press, 1966.

[2] Arnett, D. W. Spatial and temporal integration properties of units in first optic ganglion of dipterans. *Journal of Neurophysiology*, vol. 35, pp. 429-444, 1972.

[3] Coombe, P. E. The large monopolar cells L1 and L2 are responsible for ERG transients in *Drosophila*. *J. Comp. Physiol. A*, vol. 159, pp. 655-665, 1986.

[4] van Doorn, A. J., & Koenderink, J. J. The structure of the human motion detection system. *IEEE Transactions on System, Man, and Cybernetics*, vol. 13, pp. 916-922, 1983.

[5] Franceschini, N. Early processing of colour and motion in a mosaic visual system. *Neuroscience Research*, vol. Suppl. 2, pp. S17-S49, 1985.

[6] Grossberg, S. Contour enhancement, short term memory, and constancies in reverberating neural networks. *Studies in Applied Math.*, vol. 52, pp. 217-257, 1973.

[7] Hardie, R. C. Neurotransmitters in compound eyes. In: D. G. Stavenga, & R. C. Hardie (Eds.), *Facets of vision*, Berlin: Springer, 1989.

[8] Laughlin, S. The Roles of parallel channels in early visual processing by the anthropod compound eye. In: M. A. Ali (Ed.), *Photoreception and vision in invertebrates*, (pp. 457-481). New York: Plenum Press, 1984.

[9] Leake, L. D., & Walker, LR. J. *Invertebrate Neuropharmacology*, New York: Wiley, 1980.

[10] Öğmen, H., & Gagné, S. Short-range motion detection in the insect visual system. *Neural Networks*, vol. 1, suppl. 1, (Proceedings of the First International Neural Network Society Conference)', 519, 1988.

[11] Öğmen, H., & Gagné, S. Neural models for sustained and on-off units of insect lamina. *Biological Cybernetics*, vol. 63, pp. 51-60, 1990a.

[12] Öğmen, H., & Gagné, S. Neural network architectures for motion perception and elementary motion detection in the fly visual system. *Neural Networks*, (1990b, in press).

[13] Paton, W. D. M. A theory of drug action based on the rate of drug-receptor combination. *Proc. Roy. Soc. B*, vol. 154, pp. 21-69, 1961.

[14] Reichardt, W. Autocorrelation, a principle for evaluation of sensory information by the central nervous system. In W. A. Rosenblith (Ed.), *Principles of sensory communications* (pp. 303-317). New York: Wiley, 1961.

[15] Riehle, A., & Franceschini, N. Motion detection in flies: Parametric control over on-off pathways. *Experimental Brain Research*, vol. 54, pp. 390-394, 1984.

[16] van Santen, J. P. H., & Sperling, G. Temporal covariance model of human motion perception. *Journal of the Optical Society of America A*, vol. 1, pp. 451-473, 1984.

[17] van Santen, J. P. H., & Sperling, G. Elaborated Reichardt detectors. *Journal of the Optical Society of America A*, vol. 2, pp. 300-321, 1985.

[18] Schmid, A., & Bülthoff, H. Using neuropharmacology to distinguish between excitatory and inhibitory movement detection mechanisms in the fly *calliphora erythrocephala*. *Biological Cybernetics*, vol. 59, pp. 71-80, 1988.

[19] Shaw, S. R. Anatomy and physiology of identified non-spiking cells in the photoreceptor-lamina complex of the compound eye of insects, especially diptera. In: A. Roberts, & B. M. H. Bush (Eds.) *Neurones without Impulses*, pp. 61-116, Cambridge University Press: Cambridge, 1981.

[20] Shaw, S. R. Early visual processing in insects. *J. exp. Biol.*, vol. 112, pp. 225-251, 1984.

[21] Stephenson, R. P. A modification of the receptor theory. *Br. J. Pharmac. Chemother.*, vol. 11, pp. 379-386, 1956.

[22] Wilson, H. R. A model for direction selectivity in threshold motion perception. *Biological Cybernetics*, vol. 51, pp. 213-222, 1985.

FINITE ANALYSIS

Summary

Koichi Oshima, President of the Japan CFD Society and founder of the series of ISFD, International Symposia on CFD (Nagoya 1985, Sydney 1987, and Nagoya 1989) gave the invited lecture in this area, "Conservation Laws and Boundary Conditions in Fluid Mechanics." He addressed the problem of non-uniqueness, due to unknown bound circulation, of two-dimensional, steady incompressible viscous flow. In steady flows, the usual application of the Kutta-Joukovskiy condition suffices to fix the circulation. For unsteady flows, his discrete vortex or panel model applied slightly downstream of the tailing edge works well. Details of the modeling of vortex shedding are given as well as numerics, water channel and wind tunnel experiments of a number of elliptic cylinder airfoils. Reynolds numbers ranged from 2,000 to 120,000. Conclusions include flow field and structure factors and that the mechanism whereby vorticity is generated in multi-energetic flows with large wake regions is still unknown.

Ruotsalainen's "Some Remarks on the Boundary Element Method for Nonlinear Problems," treats the Galerkin BEM approximation to nonlinear IE. Direct and indirect methods are applied to monotonic nonlinearities with some growth constraints. Extension of uniqueness proofs are given with additional regularity assumption on the nonlinearity. L-p order estimates are provided. The theory of a-proper maps enables him to show quasioptimality of Galerkin solutions.

Chang and Fix, "A Mixed Finite Element Method for Stokes Problem— Acceleration, Pressure, Formulation," develops a new solver by converting Stokes ("Creeping Flow" of a Newtonian fluid) PDE into a first-order system via a least squares approach. They show that this leads to optimal convergence without the restrictions required in Galerkin-type formulations. Convergence proofs and rates are given.

Cangellaris and Lee, "Numerical Solution of the Two-Dimensional Scalar Helmholtz Equation Using the Bymoment Method," remark that the major obstacle to FEM application in unbounded regions in that differential-based methods are always formulated as BVP[really, ABVP]; finite computers must artificially truncate the computational domain. They briefly review several attempts to treat this difficulty: hybrid FEM (which forfeit the banded matrix property so dear to numericists), surface integral equations or SIE, "unimoment" methods replacing SIE by eigenvalue expansions which is rather complex, and, lastly, their "bymoment" method as a new approach to E-M field scaling in unbounded regions. The latter decouples the two regions by a scatterer-conforming surface and surface tangential fields expanded in coefficients to be determined and interior solutions preserving the unimoment scheme. This paper epitomizes IMSE philosophy by its novel and creative use of "global" or "integral" techniques to solve problems of either technological or intrinsic interest. Several computations are described with good agreement with other methods.

Moyer and Schroeder, "Combined Finite Element/Boundary Integral Approach to the Prediction of Acoustic Signatures and Radar Cross-Sections," treat Helmholtz in three-dimensional radiation and scaling problems. The unbounded domain is approximated via mapped infinite elements whose shape is fixed by free-space Green's function, yielding accurate solution for the near-field. A boundary integral generates the far-field. Acoustic results for hard spheres and a range of wavenumbers and E-M field results for circular cylinder are given. A numerical instability in the calculations of the infinite element matrices is discussed. These last two papers nicely compliment each other.

The following four papers serve as a mini-book; they deal with various approaches to the numeric problem of evaluating definite integrals.

Civan, "Finite Analytic Method for Numerical Solution of Mathematical Models," reviews the uses of analytic solutions over small elements to develop numeric schemes; application is made to two examples, the hyperbolic, nonlinear PDE of Bellman and Adomian in their 1985 PDE book and the test problem of Civan and Sliepcevich, IJNME, 1983 of generalized elliptic type. Civan's motivation is the plethora of nonlinear problems and the fact that usual methods, FEM and FDM, create accuracy problems for two basic reasons; 1) low-order accuracy to avoid complex schemes and to reduce computer effort (costs) and 2) discretization of differing terms in the DE's tend to be nonuniform. Hence, difficulties arise in numerical dispersion, oscillations, grid orientation unless severe grid expansion is made. He feels the finite analytic method, Chen and Li (1980, 1984) is not widely recognized as efficiently addressing these problems; this uses local analytics to construct higher-order numerics and allows relatively larger grids over conventional methods. Cited pre-coding manipulations can be tedious but a major tool, the method of Frobenius for nonlinearities, appears a prime candidate for symbolic manipulators such as MACSYMA, REDUCE 3, ... The Bellman-Adomian problem is solved for three distinct initial functions. A number of general and specific suggestions/ recommendations are given, e. g., for Neumann and Dirichlet boundary conditions.

Civan, "Quadrature and Cubature Methods for Numerical Solution of Integro-Differential Equations," uses, as a vehicle, the Buckley-Leverett problem of water flooding of naturally fractured oil reservoirs. "Quadrature" denotes numerical evaluation of definite integrals of a single variable; "cubature" denotes the correspondent for two or more variables. He gives numerous examples of both numerics and rules for finding the weights of interpolating polynomials. A major conclusion is "In general, . . . , cubature (solutions) are of better quality and free of grid-orientation effects." Paper is a detailed expansion of parts of Engels' book, "Numerical Quadrature and Cubature," Academic Press, 1980 and thus valuable to practitioners who preform numerical integrations.

Squire, "Numerical Solution of Second-Order Differential Equations," remarks on the common practice of reducing systems of high-order DE to systems of first order DE (Runge-Kutta, etc.) but that in some cases a convenience obtains for codes directly solving certain frequently arising forms. He details several methods for second-order DE, a common type in physics and engineering. A single equation is treated and examples given but he notes that extension to systems of second-order DE is straightforward. Specifically, he considers only solvers of fourth-order accuracy (RK4 is common). The Run-Kutta-Nystrom, Devogelaire and modifications and several other methods are computed for various test cases.

Paper is a valuable addition to the numeric literature by its relaxation of common, basic postulates and for its numerical guidance.

Squire, "Progressive Aitken Mid Point Quadrature," continues his work comparing Romberg (trapezoidal rule with Richardson extrapolation—as used by Euclid, c 500 BC!) to Aitken extrapolation which he finds superior. This uses three times the modes of each prior evaluation, hence, is "progressive." He cautions that the convergence characteristics of the two methods differ; Aitken has no theorem on rates of convergence along diagonals, rows and columns. A FORTRAN 77 code is provided and exercised for a number of detailed cases. He concludes with some practical hints to the calculator.

Fred R. Payne

Conservation Laws and Boundary Conditions
in Fluid Mechanics

KOICHI OSHIMA
Institute of Space and Astronomy Science
Yoshinodai 3, Sagamihara 229, Japan

ABSTRACT

Here we consider a two-dimensional, steady, incompressible, viscous flow around a
two-dimensional airfoil placed in a uniform flow. The flow fields described by the
Navier-Stokes equations in this case are not unique, but can have arbitrary values of
the bound circulation. Even the non-slip condition applied along the airfoil surface
cannot specify this value. For steady flow, in which there exists no free vorticity
within the flow field, the Kuttar-Joukovskiy condition has been usually applied to
determine this value. Although this condition is purely phenomenological, its
applicability has been well established experimentally as well as analytically.

For unsteady flows, such as those caused by an oscillating or rotating airfoil in a
uniform flow, the difficulty remains, because the shedded vorticity does exist and,
then, their strengths have to be defined. There have been proposed various ways to
define this flow situation, and our numerical analyses based on discrete vortex
method as well as on the full Navier-Stokes solution using FEM have proved that the
so-called global Kutta condition, which applies the usual Kutta condition at a point
slightly downstream from the trailing edge, gives the most plausible result.

On the other hand, it has been found numerically as well as experimentally that the
vortex street shedded behind an oscillating airfoil grows and rearranges into other
vortex street with different Strouhal numbers. This eventual Strouhal number
depends only on the mean blockage ratio of the oscillating airfoil. Although the newly
shedded vortices are solely determined by the local flow condition around the trailing
edge, the global flow condition depends only time averaged flow condition. The global
flow pattern is determined solely by the global conservation laws, in this case the
momentum conservation, regardless the local flow pattern which is determined by the
local boundary condition. This local condition can be chosen from various
phenomenological reasoning, such as the Kutta condition, or the no-slip condition
and its influence is limited only within the local flow.

222

1. INTRODUCTION

Behind a wing oscillating in a uniform flow, there exists vortical region which usually forms vortex street. A vehicle which steadily flies with a large incidence angle in air usually accompanies a large separation vortex which sheds vortices intermittently. In these flows, the vorticity generated on the body plays a dominant role. and the flow fields around these wing or body are unsteady. Numerical methods solving such unsteady or separated flows are not established yet even using supercomputers.

The vortical flow is a new category of flow, introduced by Oshima & Oshima [1], which is based on the idea of three classifications of fluid flows; the zero Reynolds number flow, the boundary layer and outside potential flows, and the infinite Reynolds number flow or the inviscid flow. The vorticity has the effects of generating the vorticity along body surface and of diffusing the vorticity already created. In the vortical flow corresponding to the infinite Reynolds number flow, the vorticity which is generated on the body surface due to the viscosity flows down without diffusion nor decay. The theoretical formulation of the infinite Reynolds number flow was accomplished by Vorus [2]. He constructed the integrable formulation of the two-dimensional steady Navier-Stokes equations using vorticity and velocity and fulfilling the boundary condition at the body surface and the infinity. The convection terms are linearized by the uniform velocity and the flow around a flat plate, which includes separation zone at a large angle of attack, was analyzed. This results suggest that the potential flow theory combined with generation and transport of the vorticity, such as discrete vortex method or panel method which accounts for the wake, can be applied to the separated flow.

The vortical flow includes many interesting and important phenomena. The various connecting phenomena observed both in nature and in human work have been systematically described by Lugt [3]. In flowing fluid, both of two- and three-dimensional solid bodies in motion of oscillation or rotation generate vortices with various shapes and intensities from their moving surfaces and shed them into the flow field. In the two-dimensional flow field, the mechanism of these vortex generation and the influences to the solid bodies have been studied extensively [4]. In the three-dimensional flow, even for the stationary bodies, the vortices flowing down along the body surface interact with the body, and changes their characters. This phenomenon also has been investigated numerically as well as experimentally by many authors [5][6]. However, flows around three-dimensional, moving bodies have not yet made clear enough on the fluid dynamical point of view, owing to its complexity of three-dimensional interaction of vortices. Although, in connection with unsteady spin of recovering rocket body, several works have been reported [7][8].

Mathematical background of the vortical flow is briefly discussed in § 2, in which the existence theorem of the solutions of the Navier-Stokes equations and the Euler equations are discussed and the mathematical foundation of discrete vortex method is given. It is pointed out that the nature of two-dimensional and three-dimensional flow is completely different. As a matter of fact, the two-dimensional flow is merely a mathematical illusion with no relation with physical reality. In § 3, the flow fields around the rotating cylinder with various cross sections were measured using a water channel and a wind tunnel. The model was set to rotate around the center axis perpendicular to its generator and the rotating axis was fixed perpendicularly to the on-coming flow. Dynamic forces such as drag, lift, moment and torque were measured using strain gauge balance and all the data were taken time-sequentially and analyzed using micro-computer system. Also the flow field was visualized for each cylinder. The incident angle was measured from the on-coming flow direction.

2. MATHEMATICAL BACKGROUND

2.1 Existence Theorem of the Solutions

Two-dimensional weak global solutions of the Euler equations as well as the Navier-Stokes equations starting with smooth initial conditions have been proved to exist (Leray-Hoph). Three-dimensional weak global solution of the Euler equations does not exist. Three-dimensional weak global solution of the Navier-Stokes equations does exist in the case of small Reynolds number flows. For high Reynolds number flow, its existence is most possible, rigorous proof has not reported yet, though.

In two-dimensional flow, based on the continuity equation, that is the divergent free condition of the velocity field, the streamfunction formulation is possible and the velocity field is determined from the vorticity distribution only. The vorticity is determined from the divergence-free condition of itself which is derived from the divergent-free condition of the velocity field. Thus the velocity field of two-dimensional flow is determined without the momentum equation, which describes time evolution of vorticity. This is the reason of difficulty to numerically solve the viscous flow problems, that is, the flow field is degenerated in two-dimensional constraint. In two-dimensional flow, the vorticity has single component perpendicular to the flow field, and is fixed to the material. That is, the material elements are convected by the velocity field and carry its own vorticity. This is the base of discrete vortex method. However, in three-dimensional flows, the material elements do not carry its own vorticity, and discrete vortex method is suffered much difficulty for extension to the three-dimensional cases. This is particularly true, when generation of vorticity on the solid boundary is to be discussed.

2.2 Discrete Vortex Method

Discrete vortex method has been applied to solve the autorotation problem in two-dimensions, which is based on the potential flow theory combined with a model of vortex generation and has demonstrated to be suitable for analyzing the incompressible high Reynolds number flow past a bluff body. Discrete vortex method was firstly applied to study the instability of the vortex sheet in 1930', in which a vortex sheet is replaced by a group of vortex filaments. Later, roll up of vortices from an aircraft wing was calculated by this method. However, these have resulted in unstable flows as increase of the number of vortex filaments, due to strong interactions among the vortex filaments, because the induced velocity by a vortex filament is proportional to the reciprocal of the distance between the filaments. After that, it has been recognized that some relaxation process is necessary in order to proceed stable computation. Chorin introduced finite size core of vortex filaments. The merging of vortex filaments within a close distance was assumed for concentrated vortex region.

The conformal mapping is used for simple bodies. The flow field around a physical body is transformed onto a unit circle and the computation is carried out in the mapped plane. In case of complicated bodies, the boundary line of the body is replaced by a number of vortex filaments so as to satisfy the specified boundary condition, in which the sub-vortex technique was utilized in order to weaken the singularity of the vortices composing the boundary of the body for stable computation. The various generation models of the vortex filaments have been applied and the flow fields around various shaped bodies have been solved.

Three-dimensional interaction of vortices with finite core has been investigated numerically using the Rosenhead-Moore' approximation with a success. As the computational scheme in this approximation can not deal with the structure of the vortex core, a bundle of vortex elements is employed to represent a single physical vortex tube. The validity of this method has been confirmed by comparing the analytical results for a single vortex ring. Comparison with the experiments showed also good agreement.

2.3 Modeling of Vortex Shedding

In order to determine the strengths of the newly generated vortices at each time step, the no-net-slip conditions is used. Since the tangential component of the velocity difference between the body surface and the fluid is zero. To save the computation time, only two sections on the body surface are usually chosen. In case of a flat plate, two discrete vortices are supposed to be generated at the leading and trailing edge, at which the Kutta condition is applied. That is, the no slip condition is adopted along the infinitesimal elements at the both edges. In case of a circular cylinder, the body surface is divided into two parts of the upper and lower surfaces and the no-net-slip condition was applied along each half surface. In case of an elliptic cylinder, the no-net-slip condition is applied along the partial section around the leading and trailing edges, at each of which a discrete vortex is generated. The extent of each section is determined so that the solution of the potential flow about the rotating ellipse in still fluid approximately satisfies the no-net-slip condition along the rest sections. This angle is uniquely determined by the thickness ratio of the ellipse, 23° for 15 % thickness and 90° for 50 % thickness ratios.

2.4 Numerical Simulation of Autorotating Airfoil

Numerical simulation of forced rotation of an elliptic cylinder with 15 % thickness ratio driven at a constant angular velocity was done in 7 cases of the reduced angular velocities S of 0.42 to 0.035, where $S = \omega c / 2\pi U$ and ω is the angular velocity, c is the chord length and U is the incoming flow velocity. The computation time is 5 to 10 minutes by using FACOM M-380 computing system. On the time sequences of the instantaneous streamlines on which the flow comes from left to right and the elliptic cylinder rotates clockwise, it is seen that one positive or clockwise rotating vortex is shed from the retreating edge of the elliptic cylinder, and another negative or counter-clockwise rotating vortex is released from the advancing edge of the wing, during each half revolution of the body. Then, the vortices shed from the body makes a Karman type vortex street and flow downward due to the downwash caused by the high-speed rotation. In the case of smaller angular velocity with S less than 0.14, however, the positive vortices dominate and show upwash. The reason of this is supposed to be the vortex interactions near the wing. Based on these simulation results, the criterion for sustained autorotation was found.

3. EXPERIMENTS

3.1 Water Channel Experiments

The water channel of the Institute of Space and Astronautical Science has been used, which has the test section of $50 \times 50 \, cm^2$, and the flow velocity is controlled in the range of 10 to 90 cm/s by on-off control of two parallel pumping stations and also adjusting the water depth. This channel is also used as a towing tank with traveling speed of up to 12 cm/sec and traveling span of 4 m.

For two-dimensional experiments, the brass or aluminum model with the chord length of 3 cm and the aspect ratio of 6 was mounted in the water channel. The corresponding Reynolds number based on the chord length of the model are from 9,000 to 21,000. Their thickness ratio and the non-dimensional moments of inertia to water varies from 15 % to 50 % and 0.7 to 2.6, respectively. The models were set in the autorotating state, however, some tests were carried out in the forced rotating state with the constant angular velocity driven by a stepping motor for the comparison of the flow pattern with the autorotating state.

For three-dimensional experiments, five types of cylinder models with various cross sections were tested. All of them have the same generator length of 20 cm and the same maximum thickness of 2 cm. The model was set in the center of the channel test section and rotated by a stepping motor at constant angular velocities of 25.7 and 55.4 degree/s. The flow field around the model is visualized by milk tracer and tuft methods and the instantaneous flow patterns were taken by a motor driven camera, synchronized with several angular positions of the model. The Reynolds number based on the model thickness is 2,000 to 18,000.

3.2 Wind Tunnel experiments

The variable pressure wind tunnel used is of the Institute of Space and Astronautical Science and the open jet type with the circular test section of 1.6 m diameter. The experiments were carried out at flow velocities of 6 to 45 m/s. The turbulence level of this wind tunnel is 0.6 % at flow speed less than 30 m/s and 1 % at flow speed beyond it.

In two-dimensional experiments, an acrylic cylinder with the chord length of 15 cm, the aspect ratio of 3, the thickness ration of 15 %, and the non-dimensional moment of inertia of 20 to air flow was mounted by a ball bearing system in the wind tunnel and the acrylic end plates with the 20 cm in diameter were provided on both sides of the model. The corresponding Reynolds numbers based on the chord length of the model are from 60,000 to 45,000. A pulse motor and a torque meter are installed in tandem for the forced rotation test. The model were set either in the free rotating state or in the forced rotating state with a constant angular velocity driven by a stepping motor.

In three-dimensional experiments, the models used have the size five times as large as of the water channel models. Dynamic forces are measured by a strain gauge balance system provided with micro-computer aided data processing system. The flow field is visualized by tuft method, in which the wool tufts are fixed to the model surface at every 4 cm grid point, and multi-exposure photograph at every 30 degree angular position is taken using synchronized strobo flash light. The Reynolds number of this wind tunnel experiment is 70,000 to 120,000.

CONCLUSIONS

Vorticity generation mechanism in multienergetic flows, such as separation point accompanying large wake zone, is unknown yet.

Far field structure does not depend on local character, such as the vortex shedding mechanism near the separation point. The so-called global Kutta condition is applicable.

Flow field character can be defined only by the boundary condition with the same scale, just like in turbulence modeling or meteorological applications.

<u>REFERENCES</u>

1. OSHIMA, Y. & OSHIMA, K.: Vortical flow behind an oscillating air foil, 15th ICTAM, North Holland, p357 (1980)
2. VORUS, W.S.: A theory of flow separation, J.Fluid Mech. vol.132 pp.163-183 (1983)
3. IZUTSU, N.: Numerical and experimental study of dynamics of two- dimensional body, ISAS Rept 619 (1986)
4. LUGT, H.L.: Autorotation, Ann. Rev. Fluid Mech. vol.15 p123 (1983)
5. WANG, K.C.: Separation patterns of boundary layer over an inclined body of revolution, AIAA J vol.10 p.1044 (1972)
6. ISHII, Y. & OSHIMA, Y.: Study of three-dimensional separation on a spheroid at incidence, ISAS Rept SP4, p43 (1986)
7. KUBOTA, H., ARAI, I. & MATSUZAKA, M.: Flat spin of slender bodies at high angles of attack, J. Spacecraft vol.20 p108 (1983)
8. ERICSSON,L.E. & REDDING,J.P.: Fluid dynamics of unsteady separated flow, Prog.Aerospace Sci. vol.23 p.1 (1986)

Some Remarks on the Boundary Element Method for Nonlinear Problems

KEIJO RUOTSALAINEN
University of Oulu, Faculty of Technology
Section of Mathematics
SF-90570, Oulu, Finland

Abstract

In this paper we analyse the Galerkin boundary element method applied to nonlinear boundary integral equations. We formulate nonlinear boundary value problems as boundary integral equations. Both the direct and indirect methods are applied. Assuming the nonlinearities to be of monotone type and to possess some restrictive growth conditions we prove the existence and uniqueness of the solution. Furnishing the theory of a-proper mappings we derive the convergence of the Galerkin method. Assuming more regularity on the nonlinearity we finally prove the optimal order error estimates in L^p-spaces.

1. INTRODUCTION

In this paper we consider the numerical approximation of some nonlinear boundary value problems. We analyse the Galerkin boundary element method for finding an approximate solution. In this study it is assumed that the differential equation is linear, but the boundary values are subjected to nonlinear boundary conditions. These kind of equations find many applications in the theory of elasticity, in steady state heat transfer problems having variable heat-conductivity on the boundary [4, 11].

The problems to be considered are as follows:

PROBLEM 1. *Find Φ in some approriate function space such that*

$$\begin{cases} \Delta\Phi = 0, & in\ \Omega \\ \Phi|_{\Gamma_1} = f, & on\ \Gamma_1 \\ \partial_n\Phi|_{\Gamma_2} = G(\Phi|_{\Gamma_2}), & on\ \Gamma_2. \end{cases} \tag{1}$$

Here and f is the known boundary data, and $G(\cdot)$ is the given nonlinearity, which is supposed to be of monotone type. The exact form of the nonlinearity will be defined in later sections, where the numerical approximation scheme will be examined more closely. The symbol ∂_n stands for the outer normal derivative on the boundary.

First we convert the boundary value problem to an equivalent boundary integral equation by means of the Green representation formula. Namely, every harmonic function can be represented by its Cauchy data:

$$\Phi(x) = \frac{1}{2\pi} \int_\Gamma \Phi(y)\partial_n \log|x-y|\,ds_y - \frac{1}{2\pi}\int_\Gamma \partial_n \Phi(y)\log|x-y|\,ds_y. \qquad (2)$$

Now using the jump relations of the boundary potentials one derives a (2×2)-system of equations [16: chap. 2]:

$$\begin{pmatrix} V_{11} & -K_{12} - V_{12}\circ G(\cdot) \\ -V_{21} & I + K_{22} + V_{22}\circ G(\cdot) \end{pmatrix}\begin{pmatrix} u \\ v \end{pmatrix} = \begin{pmatrix} f + K_{11}f \\ -K_{21}f \end{pmatrix}. \qquad (3)$$

The functions u and v denotes the a priori unknown part of the Cauchy data, i.e. $u = \partial_n\Phi|_{\Gamma_1}$ and $v = \Phi|_{\Gamma_2}$. The boundary operators in (3) are realized for smooth functions by setting

$$K_{ij}v(x) = \frac{1}{\pi}\int_{\Gamma_j} v(y)\partial_n(y)\log|x-y|\,ds_y, \quad x \in \Gamma_i;$$

$$V_{ij}u(x) = -\frac{1}{\pi}\int_{\Gamma_j} u(y)\log|x-y|\,ds_y, \quad x \in \Gamma_i. \qquad (4)$$

In applications also the indirect formulation of the problem is very common. Here we simplify the problem by dropping the first boundary condition in (1). In other words we assume that the unknown function Φ in (1) is subjected only to the nonlinear Neumann condition. We are to solve the boundary value problem

PROBLEM 2. *Find Φ such that*

$$\begin{cases} \Delta\Phi = 0, & \text{in } \Omega \\ -\partial_n\Phi = G(\Phi) - f, & \text{on } \Gamma. \end{cases} \qquad (5)$$

In the indirect method we make the **ansatz**: The solution of the potential problem (5) can be represented by the monopole potential

$$\Phi(x) = -\frac{1}{2\pi}\int_\Gamma u(y)\log|x-y|\,ds_y. \qquad (6)$$

Using the boundary conditions of (1) and the properties of the normal derivative of the monopole potential we may conclude [17] that such a function Φ is a solution of (1) provided u satisfies the nonlinear boundary integral equation

$$(\frac{1}{2}I - K^*)u + G(Vu) = f. \qquad (7)$$

Here the operator K^* is the spatial adjoint of the double layer operator K, and V is the Symm's integral operator. They are defined by setting

$$K^*u(x) = \frac{1}{2\pi}\int_\Gamma u(y)\partial_{n(x)}\log|x-y|\,ds_y,$$

$$Vu(x) = -\frac{1}{2\pi}\int_\Gamma u(y)\log|x-y|\,ds_y. \qquad (8)$$

The paper is organised as follows. In the first section we introduce the necessary functions spaces. Also the approximate spaces are defined and their basic properties are recalled. There we give also the formal approximate schemes that are then studied in forthcoming sections. The second chapter is devoted for the study of the mixed problem. We derive the optimal order convergences of the approximate solution. Then in the final section we focus our attention to the numerical analysis of the equation (7). There we allow strong nonlinearities whereas in the former case we just allow at most linear growth for the function G.

2. THE PRELIMINARIES

2.1 The boundary spaces. Let us first fix notations for the function spaces on the boundary curve Γ. Here we assume that Ω is a bounded open domain in $\mathbf{R}^2$ with a smooth boundary Γ. In other words Γ has a regular parameter representation $x : \mathbf{R} \to \Gamma$ with nonvanishing Jacobian, i.e. $|\frac{dx}{dt}| \neq 0$ and the representation is a C^∞-mapping. To avoid difficulties related to the mapping properties of the Symm's integral operator we require that the transfinite diameter $\mathrm{cap}(\Gamma) \neq 1$.

By $L^p(\Gamma)$ we denote the conventional space of L^p-integrable functions with respect to the Lebesque measure on Γ. This space is endowed with usual L^p-norm, $1 \leq p \leq \infty$,

$$\|u\|_{L^p(\Gamma)} = \{ \int_\Gamma |u(x)|^p \, ds_x \}^{\frac{1}{p}}.$$

The dual spaces are defined with respect to the L^2-inner product. As it is already well-known fact $L^q(\Gamma)$ is the dual space of $L^p(\Gamma)$, if q is the conjugate exponent of p: $\frac{1}{p} + \frac{1}{q} = 1$.

For our analysis we need also function spaces containing functions which are more differentiable. On the boundary the Sobolev-Slobodetckii space is denoted by $W^{t,p}(\Gamma)$. For these spaces holds as well that $W^{-t,q}(\Gamma)$ is the dual space of $W^{t,p}(\Gamma)$ with respect to the L^2-inner product. Especially, when $p = 2$ one frequently uses the notation $H^t(\Gamma)$ for the Sobolev space $W^{t,2}(\Gamma)$. For the basic properties of these spaces we quote to the reference [1].

Moreover, for the proper treatment of the mixed boundary value problem, we define the Sobolev space $H^s(\Gamma_j)$, $s \geq 0$ [13, 23]:

$$H^s(\Gamma_j) = \{ f \in L^2(\Gamma_j) \mid \exists \tilde{f} \in H^s(\Gamma), \tilde{f}|_{\Gamma_j} = f \},$$

with the norm defined by

$$\|f\|_{H^s(\Gamma_j)} = \inf \|\tilde{f}\|_{H^s(\Gamma)}.$$

Furthermore, we need the Sobolev space $\tilde{H}^s(\Gamma_j)$. It is the subspace of $H^s(\Gamma)$. It will contain the $L^2(\Gamma_j)$-functions whose nullextensions f^* over all Γ are contained in $H^s(\Gamma)$. As above we introduce by duality [10: Thm. 2.5.1]

$$\tilde{H}^{-s}(\Gamma_j) = [H^s(\Gamma_j)]^*;$$
$$H^{-s}(\Gamma_j) = [\tilde{H}^s(\Gamma_j)]^*, \ s > 0$$

with the norms [13: p. 79]

$$\|f\|_{\tilde{H}^{-s}(\Gamma_j)} = \sup_{\|\phi\|_{H^s(\Gamma_j)}\leq 1} |(f,\phi)_{L^2(\Gamma_j)}|$$

$$\|f\|_{\tilde{H}^{-s}(\Gamma_j)} = \|f^*\|_{H^{-s}(\Gamma)}.$$

On the boundary parts Γ_1 and Γ_2 we introduce partitions Ξ_1 and Ξ_2, respectively. The distribution of grid points is completely arbitrary. By means of the parameter representation we identify the functions on the boundary parts with functions on the interval of the real line. Thus for the approximation we transplant the spline spaces on the interval to the boundary curve via the parameter representation. We denote these approximation spaces by $S_{n,1}^d$ and $S_{n,2}^d$. They contain $(d-1)$-times continuously differentiable functions that are polynomials of degree d on each subinterval. The approximation properties are now well-known [2, 22]:

Approximation property. For every $u \in H^s(\Gamma_j)$, $s \leq d+1$ there exists $\psi \in S_{n,j}^d$ such that

$$\|u - \psi\|_{H^t(\Gamma_j)} \leq c\,n^{t-s}\|u\|_{H^s(\Gamma_j)},$$

where $t \leq s$, $t < d + \frac{1}{2}$. The constant c depends only s,t and d; but not on the mesh size.

We remind the reader that the above approximation property holds also with respect to the Sobolev norm $\|\cdot\|_{H^s(\Gamma)}$. But besides these conventional L^2-estimates we shall require the approximation properties also with respect to the L^p-topology. In [7] the uniform boundedness of the L^2-projection in L^p-spaces is derived by means of the Riesz-Thorin theorem. As a consequence the L^2-projection possesses the following error estimates for a function $u \in W^{s,p}(\Gamma)$, $s < d+1$,

$$\|u - P_n u\|_{W^{t,p}(\Gamma)} \leq c\,n^{t-s}\|u\|_{W^{s,p}(\Gamma)},$$

where $-d - 1 \leq t \leq \min\{s,0\}$.

2.2 The approximation procedure. We use the spaces $S_{n,1}^d$ and $S_{n,2}^d$ to approximate the solutions u and v of the equation (3). The problem, formally, is to find the coefficients α_i and β_i, $i = 1,\cdots,n$, such that

$$u_n = \Sigma_{i=0}^n \alpha_i \psi_i$$
$$v_n = \Sigma_{i=1}^n \beta_i \psi_i$$

approximate u and v as good as possible.

We fix the coefficients α_i and β_i using the Galerkin method. Let us denote by $\mathcal{P}_n : L^2(\Gamma_1) \times L^2(\Gamma_2) \to S_{n,1}^d \times S_{n,2}^d$ the orthogonal projection. The Galerkin scheme reads now as to find $\mathcal{U}_n = (u_n, v_n)$ such that

$$\mathcal{P}_n \mathcal{A}(\mathcal{U}_n) = \mathcal{P}_n \mathcal{A}(\mathcal{U}), \tag{9}$$

where $\mathcal{U} = (u,v)$ is the solution of equation (3) and $\mathcal{A}$ is the matrix operator defined in (3).

If $P_n : L^2(\Gamma) \to S_n^d$ is the L^2-projection, the Galerkin equations corresponding to the integral equation (7) can be written as

$$\frac{1}{2}u_n - P_n K^* u_n + P_n G(V u_n) = P_n f. \tag{10}$$

In both cases the Galerkin equations leads to a nonlinear system of equations.

Our problem is to prove the existence of the Galerkin solutions to both of the finite dimensional equations (9) and (10). This will be done in the next chapter.

3. THE CONVERGENCE OF THE METHOD

In deriving the error estimates for the boundary element approximations we will need some basic properties of the a-proper mappings. To begin with we start with the definition of a-proper mapping.

DEFINITION 3.1. *Let $T : X \to Y$ be an operator, where X and Y are reflexive Banach spaces. We say that T is **a-proper** with respect to a projectionally complete scheme $\{X_n, P_n, Y_n, Q_n\}$ provided:*
(1) $T_n = Q_n T : X_n \to Y_n$ is continuous for every $n \in \mathbf{N}$;
(2) If $\{v_{n(j)} \mid v_{n(j)} \in X_{n(j)}\}$ is a bounded sequence with

$$T_{n(j)} v_{n(j)} \to b, \; j \to \infty$$

strongly in Y, then there exists $v \in X$ and a subsequence $\{v_{n_{j(k)}}\}$ such that $v_{n_{j(k)}} \to v$ strongly in X and $Tv = b$.

In what follows we shall apply the following theorem [15: Thm. 4.3G] or [8: Thm. 21.3, pp. 263-264]:

THEOREM 3.2. *Let $T : X \to Y$ be a-proper and $u_0 \in X$ the solution of the equation*

$$Tu = f.$$

If T is also Fréchet-differentiable such that the Fréchet- derivative $DT(u_0)$ is injective and a-proper, then there exists $n_0 \in \mathbf{N}$ such that for each $n \geq n_0$ the finite dimensional equation

$$T_n(x_n) = Q_n f$$

has a solution and the sequence of the approximate solutions tend to u_0.

3.1 Approximation of the mixed problem. Let us first define the exact form of the nonlinearity to be considered. We assume that the nonlinear operator G is the Nemitskyi-operator corresponding to the Caratheodory function $g(\cdot, \cdot) : \Gamma_2 \times \mathbf{R} \to \mathbf{R}$. Besides to the usual Caratheodory conditions [14]

(i). $g(\cdot, u) : \Gamma_2 \to \mathbf{R}$ is measurable for all $u \in \mathbf{R}$;

(ii). $g(x, \cdot) : \mathbf{R} \to \mathbf{R}$ is continuous for almost all $x \in \mathbf{R}$;

we require that

(iii). $\frac{\partial}{\partial u}g(\cdot,\cdot)$ is also a Caratheodory function such that

$$0 < \gamma \le \frac{\partial}{\partial u}g(x,u) \le \gamma^{-1} < \infty$$

Assuming that these conditions are valid then for the nonlinear mapping

$$\mathcal{A}_0 + \mathcal{G} = \begin{pmatrix} V_{11} & -K_{12} \\ V_{21} & I + K_{22} \end{pmatrix} + \begin{pmatrix} 0 & -V_{12} \circ G(\cdot) \\ 0 & V_{22} \circ G(\cdot) \end{pmatrix} \tag{11}$$

has the following mapping properties [16: chap. 3]:

PROPOSITION 3.3. *The non-linear mapping* $\mathcal{A} = \mathcal{A}_0 + \mathcal{G} : \tilde{H}^{-\frac{1}{2}}(\Gamma_1) \times L^2(\Gamma_2) \to H^{\frac{1}{2}}(\Gamma_1) \times L^2(\Gamma_2)$ *is a-proper with respect to the projectionally complete scheme* $\{\mathcal{P}_n, S_{n,1}^d \times S_{n,2}^d\}$.

PROPOSITION 3.4. *The mapping* $\mathcal{A}$ *is Fréchet-differentiable with the derivative*

$$\mathcal{A}'(\mathcal{U})\begin{pmatrix} u \\ v \end{pmatrix} = \begin{pmatrix} V_{11} & -K_{12} \\ V_{21} & I + K_{22} \end{pmatrix}\begin{pmatrix} u \\ v \end{pmatrix} + \begin{pmatrix} 0 & -V_{12} \circ G'(\mathcal{U}) \\ 0 & V_{22} \circ G'(\mathcal{U}) \end{pmatrix}\begin{pmatrix} u \\ v \end{pmatrix}, \tag{12}$$

for all $\mathcal{U} \in \tilde{H}^{-\frac{1}{2}}(\Gamma_1) \times L^2(\Gamma_2)$ *and* $(u,v) \in \tilde{H}^{-\frac{1}{2}}(\Gamma_1) \times L^2(\Gamma_2)$.

The linear integral operator $\mathcal{A}'(\mathcal{U})$ corresponds the mixed Dirichlet-Robin boundary value problem:

$$\begin{cases} \Delta \Psi & = 0 \; ; \\ \Psi|_{\Gamma_1} & = g \; ; \\ -\partial_n \Psi|_{\Gamma_2} & = G'(\mathcal{U})\,\Psi|_{\Gamma_2}, \end{cases} \tag{13}$$

where $g \in H^{\frac{1}{2}}(\Gamma_1)$.

Since we assumed that the Caratheodory function $\frac{\partial}{\partial n}g(x,u)$ is positive and bounded the operator $G'(u) : L^2(\Gamma_2) \to L^2(\Gamma_2)$ is linear, bounded and strongly monotone. Hence the linear boundary value problem is uniquely solvable [3, 5, 12]. This in turn implies, due the equivalence of the boundary integral and variational formulation, that the linear integral operator $\mathcal{A}'(\mathcal{U})$ is injective (see [1]).

Clearly, the perturbation $\mathcal{G}'(\mathcal{U}) : L^2(\Gamma_2) \to L^2(\Gamma_2)$ is compact. The integral operator $\mathcal{A}_0$ being strongly elliptic [6: Thm. 2.19] we obtain that the Fréchet-derivative $\mathcal{A}'(\mathcal{U})$ is a Fredholm operator with index zero. In fact, there exists a compact linear operator $C_1 : \tilde{H}^{-\frac{1}{2}}(\Gamma_1) \times L^2(\Gamma_2) \to H^{\frac{1}{2}}(\Gamma_1) \times L^2(\Gamma_2)$ such that $\mathcal{A}'(\mathcal{U}) - C_1$ is positive definite.

Thus we have proved

PROPOSITION 3.5. *The Fréchet-derivative* $\mathcal{A}'(\mathcal{U})$ *is a-proper with respect to the projectionally complete scheme defined above and injective.*

To use Theorem 3.2 we have to verify that for given $f \in H^{\frac{1}{2}}(\Gamma_1)$ the system of integral equations admit a unique solution. To do this we remind that the nonlinear mixed boundary value problem is uniquely solvable, since the nonlinear boundary operator is strongly monotone and the differential operator is strongly elliptic. From the Cauchy data of the solution of the boundary value problem we obtain the solution of (3). The uniqueness follows now essentially from the injectivity of the linear integral operator $\mathcal{A}_0$ [6] (for the details see [16: Thm. 2.2]).

Now we are in position to state our main theorem

THEOREM 3.6. *Let $\mathcal{U}$ be the solution of the nonlinear system of boundary integral equations*

$$\mathcal{A}_0(\mathcal{U}) + \mathcal{G}(\mathcal{U}) = \mathcal{F}, \tag{14}$$

where $\mathcal{F} \in H^{\frac{1}{2}}(\Gamma_1) \times H^{\frac{1}{2}}(\Gamma_2)$. Then there exists $n_0 \in \mathbf{N}$ such that for all $n \geq n_0$ the Galerkin equation

$$\mathcal{P}_n\mathcal{A}(\mathcal{U}_n) = \mathcal{P}_n\mathcal{F} \tag{15}$$

admits a solution and $\mathcal{U}_n \to \mathcal{U}$ as $n \to \infty$. Moreover, there exist a constant $\rho > 0$ independent of $\mathcal{U}$ and $\mathcal{U}_n$ such that

$$\|\mathcal{U} - \mathcal{U}_n\|_{\tilde{H}^{-\frac{1}{2}}(\Gamma_1) \times L^2(\Gamma_2)} \leq \rho \|\mathcal{U} - \mathcal{P}_n\mathcal{U}_0\|_{\tilde{H}^{-\frac{1}{2}}(\Gamma_1) \times L^2(\Gamma_2)}. \tag{16}$$

The existence of the Galerkin solutions is obtained by the general theory of a-proper mappings [15: Thm. 4.3G]. Then using the Gårding inequality we are able to prove the quasioptimality of the Galerkin solutions, since the nonlinear integral operator is of Gårding type and its linearization is strongly elliptic [16: chap. 4].

We may now utilize standard results in approximation theory to deduce the order of convergence of the Galerkin approximation. This will be done in the forthcoming paper.

3.2 Approximation of the strongly nonlinear problem. To this end the Galerkin boundary element method has been studied just for a mildly nonlinear boundary integral equation, which was obtained by the direct formulation of the nonlinear boundary value problem [16, 17, 18, 19, 20]. Here we shall allow strong nonlinearities, which permits us to study the numerical approximation of the black-body radiation with the u^4-nonlinearity.

Using the indirect approach we obtain the nonlinear boundary integral equation

$$(\frac{1}{2}I - K^*)u + G(Vu) = f. \tag{17}$$

The nonlinear mapping G is the Nemitskyi operator corresponding to the Caratheodory function $g(\cdot, \cdot) : \Gamma \times \mathbf{R} \to \mathbf{R}$. We suppose that the Caratheodory function $g(\cdot, \cdot)$ satisfies the following conditions (i)-(vi):

(i). $g(\cdot, u) : \Gamma \to \mathbf{R}$ is measurable for all $u \in \mathbf{R}$;

(ii). $g(x, \cdot) : \mathbf{R} \to \mathbf{R}$ is continuous for almost all $x \in \Gamma$.

(iii). There exists constants $a_1 > 0, a_2 \geq 0$ independent of u such that for some $p \geq 2$

$$|g(u)| \leq a_1|u|^{p-1} + a_2;$$

$$g(u)\, u \geq b_1|u|^{p-1} + b_2;$$

(iv). $g(\cdot)$ is strictly increasing function on $\mathbf{R}$;

(v). There exists constants $K_1 > 0$ and $K_2 \geq 0$ such that

$$|g(u) - g(v)| \leq |u - v|\{K_1 + K_2(|u|^{p-2} + |v|^{p-2})\}$$

(vi). The Nemitskyi operator corresponding to the Caratheodory function

$$\frac{\partial}{\partial u}g(x, u) : \Gamma \times \mathbf{R} \to \mathbf{R}$$

is bounded and continuous from $L^p(\Gamma)$ to $L^{\frac{p}{p-2}}(\Gamma)$.

Under these additional assumption the nonlinear integral operator possesses the following mapping properties.

THEOREM 3.7. *Let us assume that the Caratheodory-function* $g(\cdot) : \Gamma \times \mathbf{R} \to \mathbf{R}$ *fulfills the assumptions (i)-(vi). Then the nonlinear operator*

$$A(u) = (\frac{1}{2}I - K^*)u + G(Vu) \tag{18}$$

is bounded, continuously Fréchet-differentiable in $L^q(\Gamma)$. *Furthermore, for every* $f \in L^q(\Gamma)$ *the equation* $A(u) = f$ *admits a unique solution.*

PROOF: See [17: Theorem 4 and 7].

By definition one easily verifies that the Fréchet-derivative $DA(u)$ is defined by setting

$$DA(u)\, h = (\frac{1}{2}I - K^*)h + DG(Vu)\, Vh \tag{19}$$

and it is bounded as an operator from $L^q(\Gamma)$ into itself.

The following results are crucial for the convergence of the approximation scheme.

THEOREM 3.8. *The nonlinear operator* $A(\cdot) : L^q(\Gamma) \to L^q(\Gamma)$ *is a-proper with respect to the projectionally complete scheme* $\{P_n, S_n^d\}$.

PROOF: The statement follows from the general theory of a-proper mappings [15: prop. 1.1D], since the perturbation $-K^* + G(V(\cdot))$ is compact by the mapping properties of the single and double layer potentials.

THEOREM 3.9. *The bounded linear operator*

$$DA(u) : L^q(\Gamma) \to L^q(\Gamma)$$

is a Fredholm operator with index zero: $ind(DA(u)) = 0$. Furthermore, it is also injective.

PROOF: The first part of the statement is obvious from the general theory of Fredholmn operators [21: chap. IV and V]. Namely, the bounded linear operators $K^* : L^q(\Gamma) \to L^q(\Gamma)$ and $DG(Vu)V : L^q(\Gamma) \to L^q(\Gamma)$ are compact by the mapping properties of the single and double layer operators [9, 17]. Therefore $DA(u)$ is a Fredholm operator with index zero.

To prove the injectivity we shall show that $DA(u) : L^q(\Gamma) \to L^q(\Gamma)$ is strictly V-monotone. As in Theorem 5 [17] we use the semicoercivity of the linear operator $V(\frac{1}{2}I - K^*)$. There we noticed that for every $w \in L^q(\Gamma)$ we can find a harmonic function $\Psi \in W^{1,2}(\Omega)$ such that

$$(V(\frac{1}{2}I - K^*)w, w)_{L^2(\Gamma)} \geq \|\nabla\Psi\|^2_{L^2(\Omega)}.$$

The assumption (iv) implies that $\frac{\partial}{\partial u}g(x, u) > 0$ for almost all $x \in \Gamma$. Thus we get

$$(DA(u)w, Vw)_{L^2(\Gamma)} = \|\Psi\|^2_{L^2(\Omega)} + \int_\Gamma \frac{\partial}{\partial u}g(x, Vu(x))|Vw(x)|^2 \, ds_x > 0$$

for every $w \neq 0$, which proves our statement.

Now by [15: Thm. 4.3G] or [8: Thm. 21.3] we finally obtain

THEOREM 3.10. *For given $f \in L^q(\Gamma)$ there exists $n_0 > 0$ such that for every $n \geq n_0$ the Galerkin equations*

$$P_n A(u_n) = P_n f \tag{20}$$

admits a unique solution, and u_n converges towards the unique solution of the equation $A(u) = f$.

To this end we have not said a word about the accuracy of the approximate solution obtained by the Galerkin method. Next we shall use the Fredholm property to derive the asymptotic error estimates in $L^q(\Gamma)$-norm (compare with Thm. 4.4 in [16]). We begin by quoting the following result from [17]

LEMMA 3.11. *For the Galerkin solutions holds the following asymptotic estimates*

$$\|P_n u - u_n\|_{L^q(\Gamma)} \leq C(u)\{\|K^*(u - P_n u)\|_{L^q(\Gamma)} + \|V(u - P_n u)\|_{L^q(\Gamma)}\}. \tag{21}$$

The previous lemma may be applied to establish the asymptotic error estimates in various Sobolev norms.

THEOREM 3.12. *Let $u \in L^q(\Gamma)$ be the unique solution of the nonlinear integral equation $A(u) = f$, and $u_n \in S_n^d(\Xi)$ the Galerkin solution corresponding the mesh parameter $h = \frac{1}{n}$. Then there holds the asymptotic error estimates*

$$\|u - u_n\|_{L^q(\Gamma)} \leq c(\|u\|_{L^q(\Gamma)})\|u - P_n u\|_{L^q(\Gamma)} \tag{22}$$

and

$$\|u - u_n\|_{W^{-1,q}(\Gamma)} \le c_1(\|u\|_{L^q(\Gamma)}\|u - P_n u\|_{L^q(\Gamma)}. \tag{23}$$

PROOF: The first part of the statement is obvious. By the triangle inequality we obtain

$$\|u - u_n\|_{L^q(\Gamma)} \le \|u - P_n u\|_{L^q(\Gamma)} + \|u_n - P_n u\|_{L^q(\Gamma)}$$

For the second term on the right hand side we can use estimate obtained in the previous lemma. If we choose $n \ge n_0$, we achieve

$$\begin{aligned}
\|u - u_n\|_{L^q(\Gamma)} \le\ & \|(I - P_n)u\|_{L^q(\Gamma)} \\
& + c(\|u\|_{L^q(\Gamma)})\{\|K^*(u - P_n u)\|_{L^q(\Gamma)} + \|V(u - P_n u)\|_{L^q(\Gamma)}\}.
\end{aligned}$$

Because the operators $V : L^q(\Gamma) \to L^q(\Gamma)$ and $K : L^q(\Gamma) \to L^q(\Gamma)$ are bounded [17: Thm. 1], we get

$$\|u - u_n\|_{L^q(\Gamma)} \le \{1 + c(\|u\|_{L^q(\Gamma)})(\|K^*\| + \|V\|)\}\|u - P_n u\|_{L^q(\Gamma)}.$$

The estimate (23) follows by utilizing the mapping properties of the operators V and K^*. Since we assumed that the boundary curve Γ is smooth there holds [17]: for every $w \in W^{-1,q}(\Gamma)$

$$\|Vw\|_{L^q(\Gamma)} \le c\|w\|_{W^{-1,q}(\Gamma)}$$
$$\|K^*w\|_{L^q(\Gamma)} \le c\,\|w\|_{W^{-1,q}(\Gamma)}.$$

With these estimates and Lemma 3.11 one easily deduces

$$\|u_n - P_n u\|_{W^{-1,q}(\Gamma)} \le c(\|u\|_{L^q(\Gamma)})\{\|K^*\| + \|V\|\}\|u - P_n u\|_{W^{-1,q}(\Gamma)}.$$

But this estimate already proves the inequality (23), because by the triangle inequality we have

$$\|u - u_n\|_{W^{-1,q}(\Gamma)} \le \|u - P_n u\|_{W^{-1,q}(\Gamma)} + \|u_n - P_n u\|_{W^{-1,q}(\Gamma)}.$$

Theorem 3.12 yields immediately the asymptotic convergence rates provided we know the regularity of the solution u

COROLLARY 3.13. *Let us assume that the solution $u \in W^{s,q}(\Gamma)$. Then there holds the error estimates*

$$\|u - u_n\|_{W^{t,q}(\Gamma)} \le c(\|u\|_{L^q(\Gamma)})h^{s-t}\|u\|_{W^{-s,q}(\Gamma)},$$

for every $-1 \le t \le 0 \le s \le d+1$.

PROOF: By Theorem 3.15 the estimates follows simply by the mapping properties of spline functions.

Finally we want note that these theoretical results agree quite well with the numerical computations made in [19: ex. 2]. There the piecewise constants are used to approximate the solution. The experimental order was exactly the theoretical order obtained here. The numerical computations were carried out with the $|u|u^3$-nonlinearity.

References

[1] R.A. Adams, "Sobolev spaces," Academic Press, New York, 1975.

[2] I. Babuška and A.K. Aziz, *Survey lectures on the mathematical foundation of the finite element methods*, in "The Mathematical Foundation of the Finite Element Method with Applications to Partial Differential Equations," Academic Press, New York, 1972, pp. 3–359.

[3] V. Barbu, "Nonlinear semigroups and differential equations in Banach spaces," Noordhoff International Publishing, Leyden, 1978.

[4] R. Bialecki and A.J. Novak, *Boundary value problems in heat conduction with nonlinear material and nonlinear boundary conditions*, Appl. Math. Mod. **5** (1981), 417–421.

[5] H. Brézis, *Problemés unilatereaux*, J. Math. Pures et Appliquées **51** (1972), 1–168.

[6] M. Costabel and E. Stephan, *Boundary integral equations for mixed boundary value problem in polygonal domains and Galerkin approximation*, in "Mathematical models and methods in mechanics," Polish Scientific Publ., 1985.

[7] C. de Boor, *A bound on the L^∞-norm of L^2-approximation by splines in terms of global mesh ratio*, Math. Comp. **30** (1976), 765–771.

[8] K. Deimling, "Nonlinear Functional Analysis," Springer-Verlag, Heidelberg, 1985.

[9] P. Eggermont and J. Saranen, *L^p estimates of boundary integral equations for some nonlinear boundary value problems*, manusscript (to appear).

[10] L. Hörmander, "Linear Partial Differential Operators," Berlin-Heidelberg-New York, 1963.

[11] D.B. Ingham, P.J. Manzoor and M. Manzoor, *Boundary integral equation solution of non-linear plane potential problems*, IMA. J. Numer. Anal. **1** (1981), 415–426.

[12] B. Kawohl, Thesis, "Über nichtlineare gemischte Randwertprobleme für elliptische Differentialgleichungen zweiter Ordnung auf Gebieten mit Ecken," Technische Hochschule Darmstadt, 1978.

[13] J.L. Lions and E. Magenes, "Non-homogeneous boundary value problems and applications vol. 1," Springer-Verlag, Berlin-Heidelberg-New York, 1972.

[14] D. Pascali and S. Sburlan, "Nonlinear mappings of Monotone Type," Siijthoff and Noordhoff International Publ., Bucarest, 1978.

[15] W.V. Petryshyn, *On the approximation-solvability of equations involvinga-proper and pseudo a-proper mappings*, Bull. Am. Math. Soc. **81** (1975), 223–312.

[16] K. Ruotsalainen, *On the the boundary element method for a mixed non-linear boundary value problem*, Applicaple Analysis (to appear).

[17] K. Ruotsalainen, *On the boundary element method for strongly nonlinear problems*, manusscript (to appear).

[18] K. Ruotsalainen and J. Saranen, *On the collocation method for a nonlinear boundary integral equation*, J. Comp. Appl. Math. **28** (1989), 339–348.

[19] K. Ruotsalainen and W.L. Wendland, *On the boundary element method for some nonlinear boundary value problems*, Numer. Math. **53** (1988), 299–314.

[20] J. Saranen, *Projection methods for a class of Hammerstein equations*, SIAM J. Numer. Anal. (to appear).

[21] M. Schechter, "Principles of Functional Analysis," Academic Press, New York, 1971.

[22] L. Schumaker, "Spline Functions: Basic Theory," John Wiley, New York, 1981.

[23] W.L. Wendland, E. Stephan and G.C. Hsiao, *On the integral equation method for the mixed boundary value problem of the Laplacian*, Math. Meth. in Appl. Sci. **1** (1979), 265–321.

Keywords. boundary element method, nonlinear, boundary value problems, Galerkin and related methods, monotone operators, a-proper mappings

1980 *Mathematics subject classifications*: 65R20, 65N20, 45L10, 47A10

A Mixed Finite Element Method for Stokes Problem—Acceleration, Pressure, Formulation

CHING LUNG CHANG and GEORGE J. FIX
Department of Mathematics
University of Texas at Arlington
Arlington, Texas 76019-0408, USA

ABSTRACT

Previous finite element methods [6, 9, 11] proposed for Stokes equations have the disadvantage of element geometry restricted low accuracy for a fixed number of degrees of freedom. This paper develops a new method where the Stokes equation is treated as a first order linear system. Using a least squares method, we obtain optimal asymptotic convergence in appropriate Sobolev spaces. For example, in two dimensions, if the domain is subdivided into regular triangles and if piecewise linear functions in H^1 space are chosen to be trial functions for both velocity and pressure, then we have the following convergence rate.

$$||p - p^h||_0 + ||\underline{u} - \underline{u}^h||_0 = O(h^2)$$

for sufficiently smooth pressures p and velocities $\underline{u}$.

1. INTRODUCTION

The Stokes equation governing linear incompressible flow of Newtonian fluids at a Reynolds number ν can be written as:

$$\begin{cases} -\nu\Delta\underline{u} + \text{grad } p = \underline{f} & \text{in } \Omega \\ \text{div } \underline{u} = 0 & \text{in } \Omega \\ \underline{u} = 0 & \text{on } \Gamma \end{cases} \qquad (1.1)$$

where $\underline{u}$, p, denote the fluid velocity and kinematic pressure. The flow region Ω is restricted to be simply connected bounded domain in $\mathbb{R}^2$ with a Lipschitz continuous boundary Γ. We let $\underline{u} = (u_1, u_2)^T$ and body force $\underline{f} = (f_1, f_2)^T$.

Previous work (see e.g., [6], [8], and [9] dealt with Galerkin based methods which required substantial conditions on the finite element spaces in order to achieve optimal convergence.

In this paper, (1.1) is converted into a first order linear system, and a least squares method for this system. Such a formulation will be shown to lead to optimal convergence without the restrictions needed in Galerkin based formulation.

As a first step, we introduce new variables $\underline{\Phi} = (\phi_1,\ \phi_2,\ \phi_3,\ \phi_4)^T$ where

$$\frac{\partial u_1}{\partial x} = -\frac{\partial u_2}{\partial y} = \phi_1$$

$$\frac{\partial u_1}{\partial y} = \phi_2$$

$$\frac{\partial u_2}{\partial x} = \phi_3 \qquad\qquad\qquad\qquad (1.2)$$

$$p = \phi_4$$

Then, (1.1) can be rewritten as:

$$-\nu\left(\frac{\partial \phi_1}{\partial x} + \frac{\partial \phi_2}{\partial y}\right) + \frac{\partial \phi_4}{\partial x} = f_1$$

$$-\nu\left(\frac{\partial \phi_3}{\partial x} + \frac{\partial \phi_1}{\partial y}\right) + \frac{\partial \phi_4}{\partial y} = f_2$$

$$\qquad\qquad\qquad \text{in } \Omega \qquad\qquad (1.3)$$

$$\frac{\partial \phi_2}{\partial x} - \frac{\partial \phi_1}{\partial y} = 0$$

$$\frac{\partial \phi_1}{\partial x} + \frac{\partial \phi_3}{\partial y} = 0$$

Next, we require that u_1, u_2 be constant on Γ, i.e., $(\phi_1, \phi_2) \wedge \underline{n} = 0$ and $(\phi_3, -\phi_1) \wedge n = 0$, where $\underline{n} = (n_1, n_2)^T$ is the outer unit normal vector to the boundary Γ and $\wedge$ denotes the wedge product in $\mathbb{R}^2$. This gives

$$\begin{cases} L\underline{\Phi} = A\underline{\Phi}_x + B\underline{\Phi}_y = \underline{F} & \text{in } \Omega \\[2ex] R\underline{\Phi} = \qquad\qquad = 0 & \text{on } \Gamma , \end{cases} \qquad (1.4)$$

where $\underline{F} = (f_1, f_2, 0, 0)^T$

$$A = \begin{bmatrix} -\nu & 0 & 0 & 1 \\ 0 & 0 & -\nu & 0 \\ 0 & 1 & 0 & 0 \\ 1 & 0 & 0 & 0 \end{bmatrix}, \quad B = \begin{bmatrix} 0 & -\nu & 0 & 0 \\ \nu & 0 & 0 & 1 \\ -1 & 0 & 0 & 0 \\ 0 & 0 & 1 & 0 \end{bmatrix}$$

and

$$R = \begin{bmatrix} -n_2 & n_1 & 0 & 0 \\ -n_1 & 0 & -n_2 & 0 \end{bmatrix}$$

It is easy to check that if $(\underline{u}, p)$ solves (1.1) it will solve (1.3). Conversely, suppose Φ solves (1.4). Then we formally find the $\underline{u}$ by solving the following div-curl system [15]:

$$\begin{cases} \text{curl } \underline{u} = \phi_3 - \phi_2 & \text{in } \Omega \\ \text{div } \underline{u} = 0 & \text{in } \Omega \\ \underline{u} \cdot \underline{n} = 0 & \text{on } \Gamma \end{cases} \tag{1.5}$$

In this regard the following result ([6], p. 22) is useful.

<u>Theorem 1.1</u>. A <u>function</u> $\underline{v} \in [L^2(\Omega)]^2$ <u>satisfies</u> div $\underline{v} = 0$ <u>and</u>

$$\int_{\Gamma_i} v \cdot n = 0 \qquad \underline{\text{for}} \ 0 \le i \le P \underset{i \le p}{\cup} \Gamma_i = \Gamma$$

<u>if</u> <u>and</u> <u>only</u> <u>if</u> <u>there</u> <u>exists</u> <u>a</u> <u>function</u> $\phi \in H^1(\Omega)$ <u>such</u> <u>that</u> $\underline{v} = \text{curl } \phi$. To apply this result we note that from the third and fourth equations of (1.3), we have

$$\text{div} \begin{pmatrix} \phi_2 \\ -\phi_1 \end{pmatrix} = 0 \text{ and div} \begin{pmatrix} \phi_1 \\ \phi_3 \end{pmatrix} = 0 \qquad \text{in } \Omega.$$

The boundary condition $R\underline{v} = 0$ shows

$$\begin{pmatrix} \phi_2 \\ -\phi_1 \end{pmatrix} \cdot \underline{n} = 0 \text{ and } \begin{pmatrix} \phi_1 \\ \phi_3 \end{pmatrix} \cdot \underline{n} = 0$$

So, there exist U_1 and U_2 such that

$$\begin{pmatrix} \phi_2 \\ -\phi_1 \end{pmatrix} = \text{curl } U_1 \text{ and } \begin{pmatrix} \phi_1 \\ \phi_3 \end{pmatrix} = \text{curl } U_2,$$

and this gives us the desired velocity field.

2. NOTATIONS AND SOME FUNCTION SPACES

Throughout this paper, we assume that $\Omega \in \mathbb{R}^2$ is a simply connected, bounded open set with a Lipschitz continuous positive oriented boundary Γ. We introduce here the standard notations with Sobolev spaces and their associated norms [4, 6]. We let

$$H^s(\Omega) = \{v \in L^2(\Omega); \partial^\alpha v \in L^2(\Omega), \forall |\alpha| \le s\} \tag{2.1}$$

which is a Banach space for the norm

$$||U||_s^2 = \sum_{|\alpha| \le s} \int_\Omega |\partial^\alpha U|^2 \tag{2.2}$$

For subset $\Omega^h \subset \Omega$, we may also define the Sobolev space $H^s(\Omega^h)$ and the associated norm $||\cdot||_{s,\Omega^h}$.

We define $D(\Omega)$ to be the linear space of functions infinitely differentiable and with compact support on Ω.

$H_0^s(\Omega)$ is the closure of $D(\Omega)$ for the norm $||\cdot||_s$. We denote by $H^{-s}(\Omega)$ the dual space of $H_0^s(\Omega)$ normed by

$$||u||_{-s} = \sup\frac{|(u,v)|}{||v||_s} \quad \text{over } v \in H_0^s(\Omega), v \not\equiv 0 \tag{2.3}$$

We shall also use the following function space

$$V^s = \{\underline{\psi} \in [H^s(\Omega)]^4 ; \, R\underline{\psi} = 0 \text{ on } \Gamma\} \tag{2.4}$$

Let τ_h denote a family of regular triangulation [4] of the polygonal domain Ω^h into triangle Ω_k, $k = 1, 2, \cdots, T$, where h is the maximum diameter of these triangles. Let $P_r(\Omega_k)$ denote the space of polynomials of degree less than or equal to r defined over Ω_k and let

$$S_r^h = \{\underline{\psi}^h \in [H^1(\Omega)]^4 ; \, \psi_1^h, \psi_2^h, \psi_3^h, \psi_4^h \in P_r(\Omega_k), \, k = 1, \cdots, T\} \tag{2.5}$$

where ψ_i^h, $i = 1, 2, 3, 4$ are components of $\underline{\psi}^h$, and let

$$V_r^h = S_r^h \cap V^1 \tag{2.6}$$

Note that V_r^h satisfies the following approximation properties for $t = 0, 1$ and for every $\underline{\psi}^h \in V_r^h$:

$$||\underline{\psi} - \underline{\psi}^h||_t \leq C_{s,t} h^{r-t} ||\underline{\psi}||_s \tag{2.7}$$

Here $C_{s,t}$ is a positive constant independent of $\underline{\psi}$, $\underline{\psi}^h$ and h [4].

Let $\gamma_0 U$ denote the boundary value of U on Γ, we define $H^{s-\frac{1}{2}}(\Omega)$ as the image of $H^s(\Omega)$ by the transformation γ_0, equipped with the norm:

$$||f||_{s-\frac{1}{2}, \Gamma} = \inf ||v||_{s, \Omega} \quad \text{over } v \in H^s(\Omega), \, \gamma_0 V = f \tag{2.8}$$

and define $L_0^2(\Omega) = \{v \in L^2(\Omega); \, (v, 1) = 0\} \tag{2.9}$

3. THE LEAST SQUARES FORMULATION

Let us define a quadratic functional

$$J(\underline{\Psi}) = ||L\underline{\psi} - F||_0^2 \qquad \text{for } \underline{\psi} \in V^1 \tag{3.1}$$

It is easily seen that a solution of (1.4) minimizes (3.1). Consider the problem for which we minimize the $J(\underline{\psi})$ over the space $V_r^h \subset V^1$. Observe that $\underline{\Phi}$ minimizes (3.1), over V^1, if and only if

$$A(\underline{\Phi}, \underline{\psi}) = G(\underline{\psi}) \text{ for all } \underline{\psi} \in V^1 \tag{3.2}$$

where
$$\begin{cases} A(\underline{\Phi}, \underline{\psi}) = \int_\Omega L\underline{\Phi} \cdot L\underline{\psi} \\ \quad G(\underline{\psi}) = \int_\Omega F \cdot L\underline{\psi} \end{cases} \tag{3.3}$$

$\underline{\Phi}^h$ minimizes (3.1) over V_r^h, if and only if

$$A(\underline{\Phi}^h, \underline{\psi}^h) = G(\underline{\psi}^h) \qquad \text{for all } \underline{\psi}^h \in V_r^h \subset V^1 \tag{3.4}$$

Once a basis is chosen for V_r^h, (3.4) reduces to a symmetric algebraic system.

3.1 A Priori Estimate

It is easy to verify the operator L defined by (1.4) is strongly elliptic in the sense of Petrovski [12], since $\det(\xi A + \eta B) = -\nu(\xi^2 + \eta^2) \neq 0$ for any non-zero real pair (ξ, η).

It is also easy to verify the boundary operator R satisfies the Lopatinski condition [12]. Indeed, we conclude that (L, R) is regular elliptic, by [1, 12]. This means it is Fredholm and has finite number nullity $N \geq 0$. Thus if we add to (L, R) N linearly independent linear relations

$$\Lambda_i \underline{\Phi} = C_i \qquad i = 1, 2, \cdots, N$$

with given $C_i \in \mathbb{R}$ and linear functionals Λ_i in such a way that the extended problem has a unique solution.

For the problem (1.4) we have $N = 1$ and

$$\Lambda_1 \underline{\Phi} = 0$$

where

$$\Lambda_1 \underline{\Phi} = \int_\Gamma \phi_4 \tag{3.5}$$

This gives the following result [1, 5, 8, 12]

<u>Lemma 3.1</u>. <u>For</u> (L, R) <u>in</u> (1.4) <u>and</u> Λ_1 <u>in</u> (3.5) <u>there is a priori estimate for all</u>

$\underline{\psi} \in [H^s(\Omega)]^4 \cup [H^1(\Omega)]^4$ <u>of the following form</u>:

$$C^{-1} ||\underline{\psi}||_s \leq ||L\underline{\psi}||_{s-1} + ||R\underline{\psi}||_{s-\frac{1}{2}, \Gamma} + |\Lambda_1\underline{\psi}| \leq C||\underline{\psi}||_s. \tag{3.6}$$

<u>where</u> $s \geq 0$, $C > 0$ <u>and</u> C <u>is independent of</u> $\underline{\psi}$.

If we take $\underline{\psi} \in V_r^h$, $\Lambda_1 \underline{\psi}$ is imposed to be zero, then we have

$$C^{-1} ||\underline{\psi}||_s \leq ||L\underline{\psi}||_{s-1} \leq C||\underline{\psi}||_s \qquad s \geq 0 \tag{3.7}$$

3.2 H^1-Norm Convergence and Error Estimates

Now, if $\underline{\Phi} \in V^1$ satisfies (1.4), then $J(\underline{\Phi}) = 0$, so ϕ solves (3.2). We define the approximate solution $\underline{\Phi}^h \in V_r^h$ of (1.4) to satisfy

$$J(\underline{\Phi}^h) \leq J(\underline{\Psi}^h) \text{ for all } \underline{\Psi}^h \in V_r^h, \tag{3.8}$$

This is equivalent to (3.4). Since $V_r^h \subset V^1$, we have
$$A(\underline{\Phi} - \underline{\Phi}^h, \ \underline{\Psi}^h) = 0 \qquad \text{for all } \underline{\Psi}^h \in V_r^h \tag{3.9}$$

The coercivity of $A(\cdot, \ \cdot)$ in V^1 follows directly from the (3.7) as the case $s = 1$. Applying the Lax-Milgram theorem [13] and the approximation property (2.7), we have the following result:

Theorem 3.2.   <u>The</u> <u>problem</u> <u>of</u> <u>minimizing</u> (3.1) <u>over</u> <u>space</u> V_r^h <u>has</u> <u>a</u> <u>unique</u> <u>solution</u> Φ^h <u>for</u> <u>any</u> <u>regular</u> <u>triangulation</u> τ_h <u>with</u> h > 0.   <u>Furthermore, the</u> <u>matrix</u> <u>corresponding</u> <u>to</u> (3.4) <u>is</u> <u>symmetric</u> <u>and</u> <u>positive</u> <u>definite.</u>   <u>Moreover, there</u> <u>exists</u> <u>a</u> <u>constant</u> C > 0 <u>which</u> <u>is</u> <u>independent</u> <u>of</u> h, <u>such</u> <u>that</u>

$$||\underline{\Phi} - \underline{\Phi}^h||_1 \le Ch^{s-1}||\underline{\Phi}||_s \tag{3.10}$$

with $1 \le s \le r + 1$ if $\underline{\Phi} \in [H^s(\Omega)]^4 \cap V^1$.

3.3 L^2-Norm Convergence and Error Estimates

In this section, we use a duality argument due to Nitsche [10] in which the negative norm in the a priori estimate of (3.7) with s = 0 plays an important role.

Let $\underline{e} = \underline{\Phi} - \underline{\Phi}^h$.   Since $\underline{e} \in V^1$, by (3.7) we have

$$||\underline{e}||_0 \le C||L\underline{e}||_{-1} \tag{3.11}$$

Returning to the boundary value problem (1.4), we note that it may not be solvable for all given $\underline{F} \in [H^1(\Omega)]^4$. Let N^* denote the codimensions of the range of L in the space $[H^1(\Omega)]^4$. Since (L, R) is regular elliptic, it is Fredholm. The adjoint form of (L, R) satisfies an a priori estimate [12], and the associated nullity N^* is finite. In fact N^* is the number of the compatibility conditions required to insure the existence of a solution of (1.4).

It follows that there are N^* linear independent orthonormal function vectors $\underline{I}_k \in [H^1(\Omega)]^4$, $k = 1, \cdots, N^*$, such that

$$(\underline{I}_k, \ \underline{I}_\ell) = \delta_{k\ell}$$

$$(L\underline{\psi}, \ \underline{I}_k) = 0 \qquad \text{for any } \underline{\psi} \in V^1$$

Then, for any given $\underline{G} \in [H^1(\Omega)]^4$, there exists a unique $\underline{\Phi} \in V^1$ and $\alpha_i \in \mathbb{R}$, $i = 1, \cdots, N^*$, such that

$$L\underline{\Phi} + \sum_{i=1}^{N^*} \alpha_i \ \underline{I}_i = \underline{G}.$$

So, we have

$$|(L\underline{e}, \ \underline{G})| = |(L\underline{e}, \ L\underline{\Phi} + \sum_{i=1}^{N^*} \alpha_i \ \underline{I}_i)|$$

$$= |(L\underline{e}, \ L\underline{\Phi}|$$

$$= |(L\underline{e}, \ L(\underline{\Phi} - \underline{\Phi}^h))| \qquad \text{for any } \underline{\Phi}^h \in V_r^h$$

$$\leq \ ||L\underline{e}||_0 \ \cdot \ ||L(\underline{\Phi} - \underline{\Phi}^h||_0$$

$$\leq \ C||\underline{e}||_1 \ \cdot \ ||\underline{\Phi} - \underline{\Phi}^h||_1$$

$$\leq \ Ch||\underline{e}||_1 \ \cdot \ ||\underline{\Phi}||_2$$

$$\leq \ Ch||\underline{e}||_1 \ \cdot \ ||L\underline{\Phi}||_1$$

$$\leq \ Ch||\underline{e}|| \ \cdot \ ||L\underline{\Phi} + \sum_{i=1}^{N^*}\alpha_i \ \underline{I}_i||_1$$

$$\leq \ Ch||\underline{e}||_1 \ \cdot \ ||\underline{G}||_1$$

This gives

$$||L\underline{e}||_{-1} \ \leq \ Ch||\underline{e}||_1$$

We summarize these results in the following.

<u>Theorem 3.3</u>   <u>Under the same conditions as in Theorem 3.2, we have</u>

$$||\underline{e}||_0 \ \leq \ C||L\underline{e}||_{-1} \ \leq \ Ch||\underline{e}||_1 \ . \tag{3.12}$$

Specifically, these error estimates apply to the pressure P, and take the form

$$\begin{cases} ||P-P^h||_1 \ \leq \ Ch^{s-1} \\[2mm] ||P-P^h||_0 \ \leq \ Ch^s \end{cases}$$

where $C > 0$ is independent of h and $1 \leq s \leq r + 1$.

4. CALCULATION OF VELOCITY AND THE ASSOCIATED ERROR ESTIMATE

In first step, after minimizing (3.1) over space V_r^h we obtained a vector function $\underline{\Phi}^h$ which is the approximate to $\underline{\Phi}$. To complete the algorithm we determine an approximation $\underline{u}^h$ to the velocity $\underline{u}$.

4.1 Implementation

Consider the boundary value problem:

$$L'\underline{u} \ = \begin{pmatrix} \text{curl } \underline{u} \\[2mm] \text{div } \underline{u} \end{pmatrix} = \begin{pmatrix} \phi_3^h - \phi_2^h \\[2mm] 0 \end{pmatrix} \text{in } \Omega \tag{4.1}$$

$$R'\underline{u} = \underline{u} \cdot \underline{n} \ = \ 0 \qquad\qquad \text{on } \Gamma$$

We define the space

$$U^s \ = \ \{\underline{v}\in[H^s(\Omega)]^2; \ R'\underline{v} \ = \ 0 \text{ on } \Gamma\}$$

$$U_r^h \ = \ \{\underline{v}\in U^1; \ v_1, \ v_2\in P_r(\Omega_k) \qquad\qquad k = 1, \ 2, \ \cdots, \ T\}$$

Consider the minimization of

$$J'(\underline{u}) \;=\; \left|\left| L'\underline{u} - \begin{pmatrix} \phi_3^h - \phi_2^h \\ 0 \end{pmatrix} \right|\right|_0^2 \tag{4.2}$$

over U_r^h. If $\underline{u}^h \in U_r^h$; i.e.,

$$J'(\underline{u}^h) \;\leq\; J'(\underline{v}^h) \qquad\qquad \text{for all } \underline{v}^h \in U_r^h.$$

This gives

$$A'(\underline{u}^h,\ \underline{v}^h) = G'(\underline{v}^h) \qquad\qquad \text{for all } \underline{v}^h \in U_r^h \tag{4.3}$$

where $A'(\underline{u},\ \ \underline{v}) = \displaystyle\int_\Omega L'\underline{u} \cdot L'\ \underline{v}$ and $G'(\underline{v}) = \displaystyle\int_\Omega (\phi_3^h - \phi_2^h)v_1$.

If $\underline{\hat{u}}$ is the solution of (4.1), then it can be proved [7] that the following holds

$$||\underline{\hat{u}} - \underline{u}^h||_1 \leq Ch^{s-1}, \quad ||\underline{\hat{u}} - \underline{u}^h||_0 \leq Ch^s, \tag{4.4}$$

where $C > 0$ is independent of h and $1 \leq s \leq r + 1$.

From Section 1 it follows that the solution $\underline{u}$ for (1.1) will solve

$$\begin{cases} L'\underline{u} = \begin{bmatrix} \phi_3 - \phi_2 \\ 0 \end{bmatrix} \\ R'\underline{u} = 0 \end{cases} \tag{4.5}$$

Thus

$$\begin{cases} L'(\underline{u} - \underline{\hat{u}}) = \begin{bmatrix} (\phi_3 - \phi_3^h) - (\phi_2 \quad \phi_2^h) \\ 0 \end{bmatrix} \\ R'(\underline{u} - \underline{\hat{u}}) = 0 \end{cases} \tag{4.6}$$

The a priori estimate for (4.3) gives the following result [7, 12]:

$$||\underline{u} - \underline{\hat{u}}||_1 \leq C(||\phi_3 - \phi_3^h||_0 + ||\phi_2 - \phi_2^h||_0) \tag{4.7}$$

Combining (3.10), (3.12), (4.4) and (4.6) we have the following error estimates for velocity:

$$||\underline{u} - \underline{u}^h||_1 \leq Ch^{s-1}$$

and

$$||\underline{u} - \underline{u}^h||_0 \leq Ch^s$$

C is independent of h and $1 \leq s \leq r + 1$.
Under the same conditions as Theorem 3.2 we have

$$||\underline{u} - \underline{u}^h||_1 \leq Ch^{s-1}, \qquad\qquad ||\underline{u} - \underline{u}^h||_0 \leq Ch^s$$

$$||\underline{p} - \underline{p}^h||_1 \leq Ch^{s-1}, \qquad\qquad ||\underline{p} - \underline{p}^h||_0 \leq Ch^s$$

for $1 \leq s \leq r + 1$, where the constant $C < 0$ is independent of h.

REFERENCES

1. Agmon, S., Douglis, A. and Nirenberg, L., "Estimates Near the Boundary for Solution of Elliptic Partial Differential Equations Satisfying General Boundary Conditions (2)," *Comm. Pure Appl. Math.*, 17, pp. 35-92, 1964.

2. Babuska, I., The Finite Element Method with Lagrangian Multipliers, *Num. Math.*, 20, pp. 179-192, 1973.

3. Brezzi, F., "On Existence, Uniqueness and Approximation of Saddle Point Problems Arising from Lagrange Multipliers." R.A.I.R.O., serie rouge, R2, pp. 129-151, 1974.

4. Ciarlet, P. G., *The Finite Element for Elliptic Problems*, North-Holland, Amsterdam, 1978.

5. Dikanskij, A. S., "Conjugate Problems of Elliptic Differential and Pseudo-Differential Boundary Value Problems in a Bounded Domain," *Math. USSR Sbornik*, 20, pp. 67-83, 1973.

6. Girault, V. and Raviat, P. A., "Finite Element Approximation of the Navier-Stokes Equations," *Lecture Notes in Mathematics*, 749, Springer-Verlag, New York, 1979.

7. Neittaanmaki, P. and Saranen, J., Jyvaskyla, "Finite Element Approximation of Vector Fields Given by Curl and Divergence," *Math. Mech. in the Appl. Sci.*, 3, pp. 328-335, 1981.

8. Schechter, M., "On L^p Estimates and Regularity II," *Math. Scand.*, 13, pp. 47-69, 1963.

9. Teman, R., *Navier-Stokes Equations*, North-Holland, 1977.

10. Strang, G. and Fix, G. J., *An analysis of the Finite Element Method*, Prentice Hall, Englewood Cliffs, 1973.

11. Thomasset, F., *Implementation of Finite Element Methods for Navier-Stokes Equations*, Springer-Verlag, New York, 1981.

12. Wendland, W. L., *Elliptic Systems in the Plane*, Pitman Publishing Ltd., London, 1979.

13. Yosida, K., *Functional Analysis*, Springer-Verlag, Berlin, 1965.

14. Fix, G. J., Gunzburger, M. D., and Nicolaides, R. A., "Least Squares Finite Element Methods," *Comp. and Math. with Appl.*, 6, pp. 265-278, 1979.

15. Fix, G. J., and Rose, M. E., "A Comparative Study of Finite Element and Finite Difference Methods for the Cauchy-Riemann Equations," *SIAM Jour. of Numer. Anal.*, Vol. 22, No. 2, pp. 250-261, 1985.

Numerical Solution of the Two-Dimensional Scalar Helmholtz Equation Using the Bymoment Method

ANDREAS CANGELLARIS and ROBERT LEE
Department of Electrical and Computer Engineering
University of Arizona
Tucson, Arizona 85721, USA

1. INTRODUCTION

The major obstacle in the application of the finite-element method (FEM) to the solution of Helmholtz equation in unbounded regions is that differential equation-based methods are always formulated as boundary-value problems. From a computational point of view, this implies that an artificial boundary is required for the truncation of the computational domain. However, such a boundary should be transparent to the solution in the sense that an appropriate mechanism should be used to couple the solution inside the truncation boundary to the solution outside, ensuring the proper continuity of the solution along the boundary as well as its proper behavior at infinity. Such a method, in the context of electromagnetic field problems, was first introduced by Silvester and Hsieh [1], and McDonald and Wexler [2]. In their formulation, a surface integral equation for the fields in the exterior region was imposed as a boundary constraint on the finite-element solution in the interior region. This led to the so-called hybrid finite element methods (HFEM). Since then several modifications and/or extensions of the method have appeared [3]-[5]. A disadvantage the HFEM have is that the resulting matrices have nonuniform block submatrix structures that are not easily amenable to banded matrix algorithms. This is due to the "global" nature of the surface integral equation used for the truncation of the computational domain.

Another approach for rigorously coupling the interior and exterior solutions was the unimoment method introduced by Mei [6] and applied extensively to electromagnetic scattering problems in two and three dimensions [7]-[10]. The unimoment method, which can be thought of as an HFEM with the surface integral equation replaced by an eigenfunction expansion, uses a separable surface for the truncation boundary to decouple the solution in the interior region where the scatterer is present, from the solution in the exterior. Then the fields in the interior are found by employing the FEM to solve a standard Dirichlet boundary-value problem, while appropriate eigenfunction expansions for the specific coordinate system are used to represent the fields in the exterior region. Finally, the interior and exterior problems are coupled by imposing the continuity of the tangential electric and magnetic fields on the separable surface. The use of a separable surface seemed to be an inherent limitation of the unimoment method since the finite-element solution within a circular/spherical surface becomes rather inefficient for elongated scatterers that occupy only a small portion of the enclosed volume. A new method was then developed [11], which combined the finite-element solution of the interior region with the surface integral equation used in the extended boundary condition method [12] to circumvent the need for a separable boundary. However, this hybrid method was found to have convergence difficulties when dealing with very elongated objects. More recently, the field-feedback formulation [13] was shown to overcome some of the disadvantages of the aforementioned methods.

The *bymoment method*, which is proposed here as a new approach to the solution of electromagnetic scattering problems in unbounded regions, was presented recently in relation to two-dimensional electromagnetic scattering [14]. In this presentation, we validated the method by comparing the bymoment solution to known eigenfunction series solutions. The bymoment method decouples the exterior and interior regions by using a surface that conforms to the scatterer. On this surface, the tangential electric and/or magnetic fields are expressed in terms of a set of appropriate expansion functions whose coefficients are to be determined. The interior solution is then generated in exactly the same manner as the unimoment method, and is expressed in terms of the unknown coefficients in the expansion of the tangential fields on the enclosing surface. Finally, an elegant use of scalar Green's theorem over a volume enclosed by a surface just inside the finite-element mesh and the surface at infinity allows the coupling of the interior solution to the exterior region and the determination of the unknown coefficients. In a sense, the bymoment method tries to combine the attractive features of all the aforementioned methods. More specifically, the restriction of the separable geometry in the unimoment method is removed, and the finite-element discretization is limited to the region occupied by the scatterer. Furthermore, the partially banded-partially full matrices resulting from the application of the HFEM are avoided.

Let us now consider the bymoment method relative to integral equation techniques. The resulting matrix equation generated from the integral equation method is the same size as the finite element matrix, but instead of being banded, the matrix is full. Thus, for penetrable cylinders, the bymoment method is much more efficient compared to integral equation techniques. However, the major advantage of the integral equation methods is that they automatically account for the boundary conditions in the exterior region. Therefore, they may be more efficient than the bymoment method for solving problems involving cylinders which are perfectly conducting.

In what follows, the bymoment method is presented for the solution of the scalar Helmholtz equation in two dimensions, with specific application to the analysis of electromagnetic wave scattering by infinitely long cylinders of arbitrary shape and composition, embedded in an infinite homogeneous medium. The formulation and solution of the interior Dirichlet boundary value problem is discussed first. Then, the coupling of the interior solution to the exterior region and the solution for the unknown field on the surface truncating the finite-element grid is presented. We continue with some discussion on the choice of the expansion and testing functions and their influence on the accuracy of the solution. Finally, numerical results for the problem of plane wave scattering by cylindrical geometries are given, along with comparisons with results obtained from the literature and generated by other means.

2. THE BYMOMENT METHOD

2.1 Finite-Element Solution

Consider an infinitely long cylinder of arbitrary cross section embedded in a homogeneous medium. Without loss of generality, the medium is assumed to be free space with electric permittivity $\epsilon = \epsilon_0$ and magnetic permeability $\mu = \mu_0$. Since the proposed finite-element formulation assumes that the material properties are constant within each element of the corresponding mesh, only cylinders with piece-wise constant material properties are considered. The coordinate system is chosen such that the $z-$axis is parallel to the axis of the cylinder. It is assumed that the incident field has no z variation so that the problem reduces to a two-dimensional one. For the purposes of this discussion, the transverse magnetic case TM_z where the magnetic field vector is parallel to the $xy-$plane and the electric field has only its $z-$ component (E_z, H_x, H_y) is presented. The transverse electric case (H_z, E_x, E_y) can easily be obtained using a similar formulation.

The geometry is shown in Figure 1. The line ∂S represents the mesh boundary around the cylinder. Notice that the boundary conforms to the cylinder shape. Notice that there is **at least one** layer of elements between the cylinder boundary and the boundary of the enclosing mesh. The artificial line $\partial S'$ is wholly enclosed by the mesh boundary, and is chosen to pass through the interior of all the elements which border the boundary of the finite-element mesh. The surface within ∂S is designated to be S_1. The region exterior to the mesh is denoted by S_2. The electric field in S_1 satisfies the Helmholtz equation,

$$(\nabla^2 + k^2)E_z(x, y) = 0 \tag{1}$$

with the assumption that no sources are present. The wavenumber is given by $k = k_0\sqrt{\mu_r \epsilon_r^*}$ where $k_0 = \omega\sqrt{\mu_0 \epsilon_0}$ is the wavenumber for free space, ω is the angular frequency of the time-harmonic excitation, μ_r is the relative magnetic permeability for the material within S_1, and the relative dielectric constant ϵ_r^* incorporates the conduction loss and is given by $\epsilon_r^* = \epsilon_r - j\sigma/\omega\epsilon_0$. The $e^{j\omega t}$ time variation is suppressed.

In order to consider the finite-element problem, assume that the solution is known on the boundary ∂S. Then the solution inside S_1 may be found by solving a standard finite-element problem with Dirichlet boundary conditions. For this case, the method of weighted residuals [15] is used. This guarantees that the Helmholtz equation in (1) is true in the weak sense. The weak form of (1) is written as follows,

$$\int\int \frac{1}{\mu_r} \left[(\nabla^2 E_z + k^2 E_z)\psi_i \right] dS = 0 \tag{2}$$

where the weighting functions ψ_i $(i = 1, 2, \ldots)$ constitute a set of first order differentiable scalar functions. Using Green's first identity, expression (3) becomes

$$\int\int \frac{1}{\mu_r} \left[(\nabla E_z \cdot \nabla\psi_i - k^2 E_z \psi_i \right] dS - \int \left[\left(\frac{1}{\mu_r}\hat{n} \cdot \nabla E_z \right) \psi_i \right] dl = 0. \tag{3}$$

The electric field E_z is expanded in a set of orthogonal basis functions and (3) is evaluated over each element in the mesh. Notice that the continuity of $\hat{n} \cdot \nabla E_z/\mu_r$ at the interfaces between adjacent elements nullifies the contributions from the line integral along these interelement boundaries. This leaves only the line integral along the outer boundary ∂S. However, since we are dealing with the Dirichlet problem, the weighting functions ψ_i are chosen to be zero on ∂S; hence, the line integral along ∂S is zero and the problem can be solved by evaluating a surface integral over each of the elements. The resulting matrix is then solved using an algorithm which exploits the banded character of the matrix.

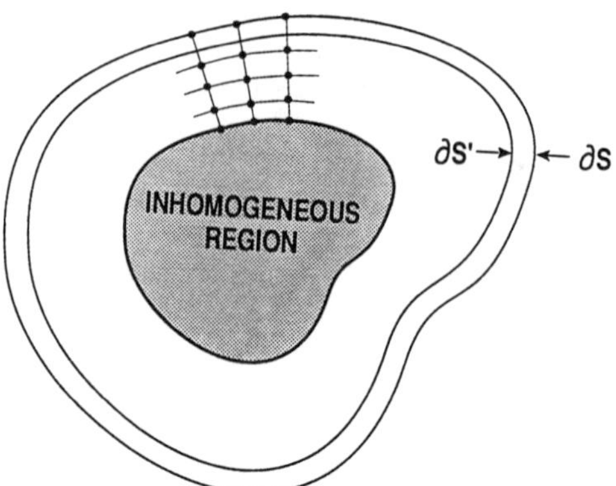

FIGURE 1. Geometry: arbitrary cylinder in free space.

2.2 Finite-Element Basis Functions

To find the electric field in S_1 it was assumed that E_z was known on ∂S. Actually, E_z on ∂S is not known apriori; hence, the interior solution cannot be obtained directly. Instead, a set of numerical finite-element solutions is generated first inside S_1 by prescribing specific boundary conditions for E_z on ∂S. These solutions form a set of *finite-element basis functions*. In order to form a complete set of finite-element basis functions, the corresponding boundary conditions are chosen such that they form a complete set of linearly independent functions on the boundary ∂S. This requirement is plainly evident from the uniqueness theorem [16].

A new variable t is defined which varies from 0 to d where d is the length of ∂S. Then the variation E_z along ∂S is expressed in terms of t and denoted $E_S(t)$. $E_S(t)$ can be expanded in terms of a complete set of linearly independent functions $\Psi_n(t)$ as follows

$$E_S(t) = \sum_{n=1}^{N} \alpha_n \Psi_n(t), \tag{4}$$

where α_n are unknown coefficients (since E_S is still unknown) and N is the number of basis functions required to represent E_S. Using $\Psi_n(t)$ $(n = 1, 2, \ldots)$ as the set of boundary conditions on ∂S, the corresponding set of finite-element basis functions is formed and denoted by $\Lambda_n(x, y)$. Therefore, the electric field inside S_1 can be written in terms of the unknown coefficients α_n as

$$E_z(x, y) = \sum_{n=1}^{N} \alpha_n \Lambda_n(x, y), \quad (x, y) \in S_1. \tag{5}$$

In the case where N is infinite, the sum needs to be truncated at some finite value. The number of terms which are used is dependent upon convergence requirements and the choice of Ψ_n. From (5) it becomes apparent that in order to complete the solution the coefficients α_n need to be found. This is done by coupling the interior solution to the properties of the exterior region as explained next.

2.3 Determination of the Coefficients

For the geometry of Figure 1, let us denote the surface enclosed by $\partial S'$ by S'_1. Let us also define S'_2 to be the surface which encompasses the area of S_2 and the area between ∂S and $\partial S'$. The electric field in S_2' can be written in terms of the incident E_z^{inc} and the scattered field E_z^{sc} as

$$E_z(x,y) = E_z^{inc}(x,y) + E_z^{sc}(x,y), \quad (x,y) \in S_2'. \tag{6}$$

The scattered field satisfies the Helmholtz equation,

$$\left(\nabla^2 + k_0^2\right) E_z^{sc}(x,y) = 0, \quad (x,y) \in S_2' \tag{7}$$

and the Sommerfeld radiation condition at infinity. Let Φ_i $(i = 1,2,\ldots)$ be a set of linearly independent *testing functions* which are chosen to satisfy the Helmholtz equation over the surface S_2' and the radiation condition at infinity. Application of the scalar Green's theorem for E_z^{sc} and any one of the Φ_i over the surface S_2' gives

$$\int_{\partial S'} \left[\Phi_i \frac{\partial E_z^{sc}}{\partial n} - E_z^{sc} \frac{\partial \Phi_i}{\partial n} \right] dl = 0. \tag{8}$$

Substituting (6) into (8) and then using (5) for E_z, one gets

$$\sum_{n=1}^{N} \alpha_n \int_{\partial S'} \left[\Phi_i \frac{\partial \Lambda_n}{\partial n} - \Lambda_n \frac{\partial \Phi_i}{\partial n} \right] dl = \int_{\partial S'} \left[\Phi_i \frac{\partial E_z^{inc}}{\partial n} - E_z^{inc} \frac{\partial \Phi_i}{\partial n} \right] dl. \tag{9}$$

Since both E_z^{inc} and Φ_i are known analytically, analytical expressions for both $\partial E_z^{inc}/\partial n$ and $\partial \Phi_i/\partial n$ can be derived. Also, the numerical values for both Λ_n and $\partial \Lambda_n/\partial n$ along $\partial S'$ are obtained from the finite-element solutions.

The only unknowns left in (9) are the coefficients α_n $(n = 1,2,\ldots,N)$. Using the first N functions Φ_i, an $N \times N$ matrix equation is formed for the determination of these coefficients. This matrix equation can be written as

$$\begin{pmatrix} Z_{11} & Z_{12} & \cdots & Z_{1N} \\ Z_{21} & Z_{22} & \cdots & Z_{2N} \\ \vdots & \vdots & \ddots & \vdots \\ Z_{N1} & Z_{N2} & \cdots & Z_{NN} \end{pmatrix} \begin{pmatrix} \alpha_1 \\ \alpha_2 \\ \vdots \\ \alpha_N \end{pmatrix} = \begin{pmatrix} T_1 \\ T_2 \\ \vdots \\ T_N \end{pmatrix} \tag{10}$$

where Z_{in} and T_i are given by,

$$Z_{in} = \int_{\partial S'} \left[\Phi_i \frac{\partial \Lambda_n}{\partial n} - \Lambda_n \frac{\partial \Phi_i}{\partial n} \right] dl \tag{11}$$

$$T_i = \int_{\partial S'} \left[\Phi_i \frac{\partial E_z^{inc}}{\partial n} - E_z^{inc} \frac{\partial \Phi_i}{\partial n} \right] dl. \tag{12}$$

3. NUMERICAL CONSIDERATIONS

From the discussion in the previous section, two key features of the bymoment method are identified. First, finite-element grids conforming to the geometry under investigation are used, thus limiting the size of the computational domain that needs to be discretized. Second, the generation of the finite element solution in the interior regions is totally decoupled from the exterior, thus avoiding the partially banded partially full matrix resulting in HFEM. This is accomplished at the expense of generating the full matrix in (10) that needs to be inverted for the calculation of the coefficients α_i, $i = 1, 2, \ldots, N$. However, the dimension of this matrix is small, as it will become apparent from the following discussion. Finally, it is pointed out that the banded FEM matrix for the interior solution and the square matrix in (10) are independent of the excitation. Hence, they need to be inverted only once for a given geometry and then be used repeatedly for various excitations.

As discussed in section 2.2, the tangential electric field along the truncating boundary is expanded in terms of a set of linearly independent functions Ψ_n. Obviously, both entire-domain and subdomain functions can be used. The major advantage of using entire-domain basis functions is that, in most cases, a smaller N than in the subdomain case is required to obtain the same accuracy. A small value for N is important since N is the dimension of the full matrix in equation (10). Another advantage of using entire-domain basis functions is the flexibility in choosing their number. To be more specific, the number of basis functions N is determined by the criteria of convergence that the user specifies. In cases where accuracy is not as important as speed, a smaller value for N can be considered. In any case, in choosing N one should keep in mind that simple physical reasoning reveals that the appropriate value of N for a given convergence criteria is dependent on the geometry under consideration and the wavelength of the incident radiation. Finally, since ∂S is a closed line, continuity of Ψ_n along ∂S imposes the following condition on Ψ_n

$$\Psi_n(0) = \Psi_n(d). \tag{13}$$

An obvious choice for Ψ_n is the following set of sinusoids

$$\Psi_n(t) = \begin{cases} \cos \frac{(n-1)}{2} \theta(t), & n \text{ odd} \\ \sin \frac{n}{2} \theta(t), & n \text{ even} \end{cases} \tag{14}$$

where $t \in [0, d]$ maps to $\theta(t) \in [0, 2\pi]$. Of course, any other set of periodic functions satisfying (13) will do. At this point, it is important to observe that, because of the oscillatory behavior of the periodic functions, the value of N is limited by the finite-element grid resolution along the boundary ∂S. For example, consider the set of functions in (14) used along the boundary ∂S with M finite-element nodes and, hence, M segments. If it is assumed that at least four segments are needed to approximate one period of the sine function, then the number of basis functions N should be less or equal to $M/2$.

The major advantage of the subdomain basis functions is the simplicity of their implementation into the finite element code. As an example, a very simple basis function is the triangle function, with support over two of the boundary elements, maximum value of one at the boundary node common to the two elements and a minimum value of zero at the adjacent boundary nodes. The variation of the basis function from the common node to the two adjacent nodes is linear. Since the behavior of this triangle function is the same as the basis functions used in the finite element problem, the corresponding finite element basis function Λ_n is generated by setting the value of one of the nodes on the boundary to 1 and setting the other boundary nodes to 0. It is apparent then that the number of basis functions N is fixed and equal to the number of boundary nodes. This is a disadvantage when dealing with meshes with a large number of nodes on the boundary. For such cases, an alternative choice of subdomain basis functions seems to be triangle functions with support over more than two boundary elements. For example, use of a triangle function with support over four elements would halve the number of basis functions

N.

There are numerous choices which can be made for the set of testing functions Φ_n used in the determination of the unknown coefficients in the expansion of the tangential electric or magnetic field on the truncation boundary. As discussed in section 2.3, for a function to qualify as a testing function it must satisfy Helmholtz equation over the surface S_2' and the radiation condition at infinity. A way to narrow the choices is by requiring the testing functions to have closed form expressions. One obvious choice is the set of cylindrical harmonic functions

$$\Phi_n = H_n^{(2)}(k_0\rho) \begin{Bmatrix} \cos n\phi \\ \sin n\phi \end{Bmatrix} \quad n = 1, 2, \ldots. \tag{15}$$

Another suitable testing function is the free-space scalar Green's function with the line source located inside S_1'. The two-dimensional Green's function is given by

$$g(\vec{\rho}, \vec{\rho}_s) = H_0^{(2)}\left(k_0|\vec{\rho} - \vec{\rho}_s|\right). \tag{16}$$

Let $\vec{\rho}_n$, $n = 1, 2, \ldots, N$ be a set of points inside S_1'. By substituting $\vec{\rho}_n$ for $\vec{\rho}_s$ in (16), a set of linearly independent functions is generated which can be used for Φ_n.

4. NUMERICAL RESULTS

In this section, results are presented from the application of the bymoment method in the analysis of plane wave scattering by perfectly conducting and material cylinders. The preprocessor FASTQ [17] was used for the finite element grid generation and the optimization of the nodal ordering in the grid. The incident field used in all cases was a plane wave. The free-space Green's functions of (16) were used as testing functions.

First, we considered a circular homogeneous dielectric cylinder of radius 0.3λ embedded in air (free-space) where λ is the wavelength of the incident field in free space. The relative dielectric constant of the cylinder was chosen to be $\epsilon_r = 4$, and a TM_z polarized incident plane wave is assumed to be propagating in the x direction. In Figure 2, we compared the bymoment solution to the eigenfunction series solution for the magnitude of the magnetic field as a function of x along the line $y = 0$ interior to the cylinder. We also plotted $|H_z|$ as a function of y along the line $x = 0$ in the same figure. The series solutions are denoted by (Ser), and the bymoment solutions are denoted by (FEM). We see that the agreement between the two solutions is excellent.

Next, we considered a circular layered dielectric cylinder embedded in air. The cross section of the cylinder is shown in the insert of Figure 3. The incident plane wave is propagating along the positive x−axis. The radii of the layers were chosen to be $r_1 = 0.15\lambda$, $r_2 = 0.2\lambda$, $r_3 = 0.25\lambda$, $r_4 = 0.3\lambda$, while the relative dielectric constants were $\epsilon_{r1} = 8$, $\epsilon_{r2} = 6$, $\epsilon_{r3} = 4$, and $\epsilon_{r4} = 2$. Figure 3 shows the echo width results for the above geometry for both TE_z (H_z, E_x, E_y) and TM_z (E_z, H_x, H_y) excitation. For the TM_z case, the echo width is defined by

$$L_e = \lim_{\rho \to \infty} \left[2\pi\rho \frac{|E_z^{sc}|^2}{|E_z^{inc}|^2} \right], \tag{17}$$

while for the TE_z case,

$$L_e = \lim_{\rho \to \infty} \left[2\pi\rho \frac{|H_z^{sc}|^2}{|H_z^{inc}|^2} \right].$$

(18)

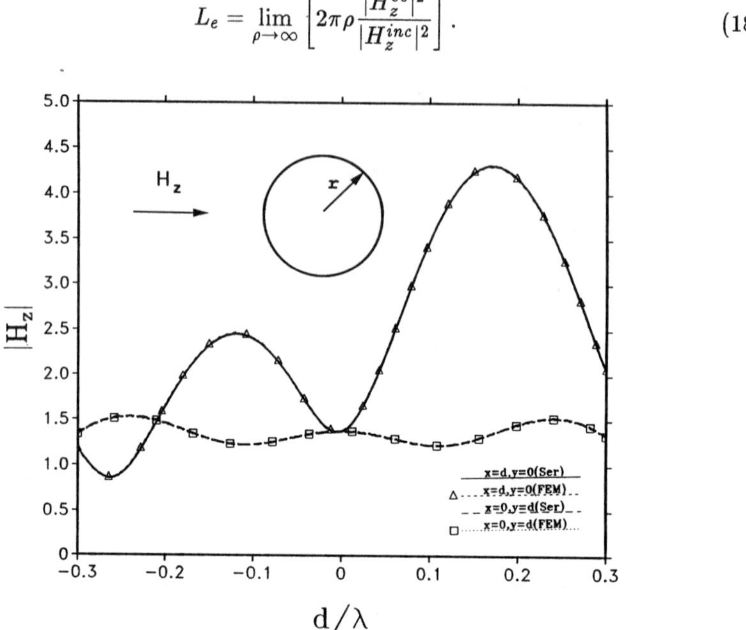

FIGURE 2. Plot of the magnitude of the magnetic field for a homogeneous dielectric circular cylinder in free space under TE_z plane wave excitation.

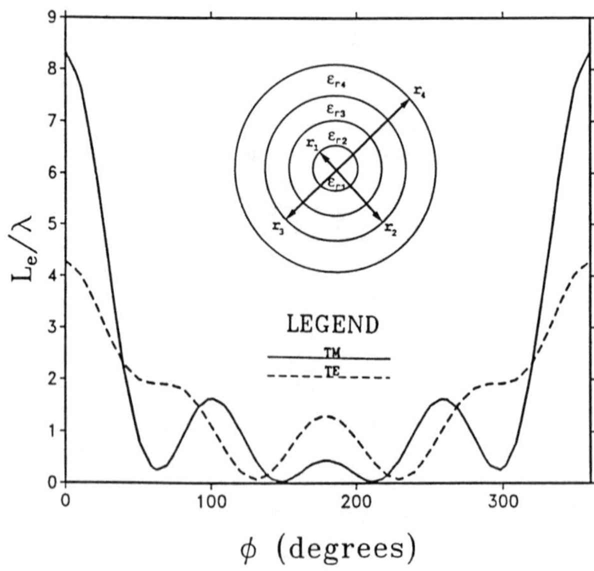

FIGURE 3. Echo width plots for a layered dielectric cylinder in free-space under TE_z and TM_z plane wave excitation.

Finally, we considered a perfectly conducting cylinder of rectangular cross-section in free-space, illuminated by a plane wave with its electric field polarized along the axis of the cylinder ($TM_z polarization$) and incident normally to the wider side of the rectanlge. The wider side was taken to be 0.25λ, while the other side was 0.1λ. The finite-element grid used for this problem is shown in Figure 4. It is well-known [18] that in the vicinity of sharp conducting edges the component of the induced current density parallel to the axis of the edge exhibits a singular behavior. In Figure 5, the normalized induced current density on the illuminated side of the cylinder is shown as computed by the bymoment method using two different grids. The first grid, corresponding to the curve labeled "FEM-1", extends only one element away from the wider sides of the cylinder. The second grid, corresponding to the curve labeled "FEM-3", extends three elements away from the wider sides of the cylinder and is the one shown in Figure 4. The bymoment method results for the two cases were compared with results obtained from an integral equation solution of the problem using the method of moments and shown by the solid curve in Figure 5 [19]. Notice that, because of the symmetry of the geometry and the excitation, the induced current density is shown for only half of the cylinder side. The significant error in the results "FEM-1" is due to the inaccuracy in the numerical derivatives of the finite-element basis functions ($\partial \Lambda_n / \partial n$) in the vicinity of the 90^0 edges. Indeed, for the "FEM-1" case, these derivatives need to be computed since they are required for the testing procedure of equation (9). However, from [18] it is known that these derivatives have a singular behavior in the vicinity of the conducting edge, which is not properly depicted by the smooth finite-element interpolation functions. While interpolation functions with built-in singularities are expected to give better results [20], we observe that the results "FEM-3" obtained used smooth interpolation functions and the grid of Figure 3 are in very good agreement with the integral equation results.

5. SUMMARY

The bymoment method has been presented for the solution of the scalar Helmholtz equation in two dimensions, with specific applications in the analysis of electromagnetic wave scattering by infinite material cylinders in an unbounded homogeneous

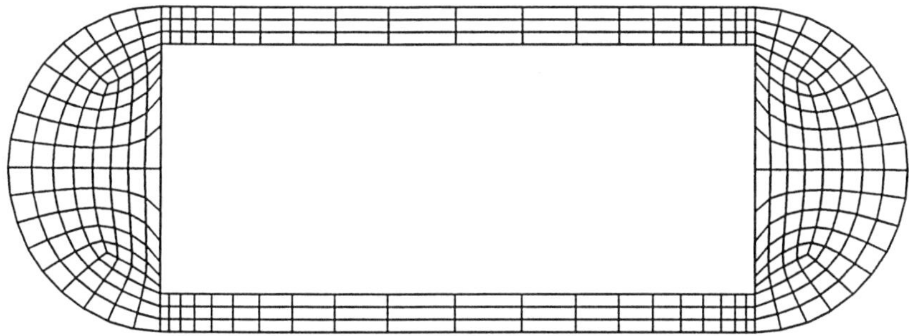

FIGURE 4. Finite-element grid for the perfectly conducting rectangular cylinder.

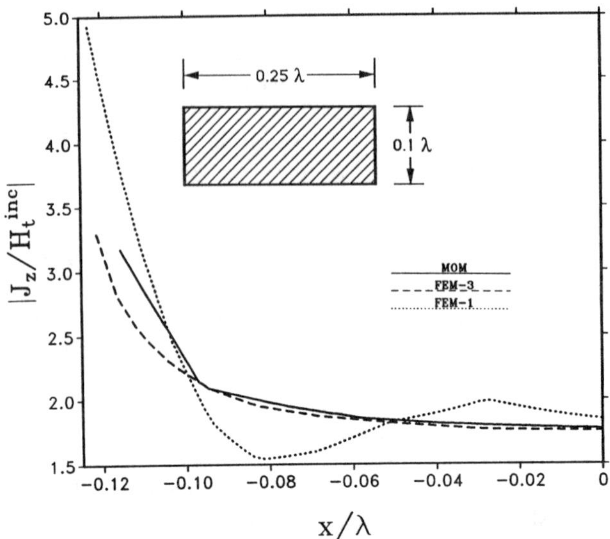

FIGURE 5. Magnitude of the normalized surface current density along the illuminated side of a perfectly conducting rectangular cylinder (TM_z case).

medium. A surface conforming to the shape of the cylinder is used to divide the computational domain into an interior region, where the scatterer exists, and an exterior unbounded region free from inhomogeneities. By using the tangential field on this surface as the key unknown, the method allows the decoupling of the solution in the interior from the solution in the exterior unbounded region. This unknown tangential field is expanded in a series of known basis functions with unknown coefficients. The various numerical issues associated with the choice of these basis functions were examined. The solution in the interior is expressed in terms of the unknown coefficients in the above expansion by solving the interior Dirichlet problem for each one of the basis functions used in the expansion. Finally, a linear system of equations for these unknown coefficients is generated making use of Green's theorem in the region exterior to the scatterer, in conjuction with the scattered field and a set of linearly independent functions which satisfy the homogeneous Helmholtz equation in the exterior region and the Sommerfeld radiation condition at infinity. Numerical results were presented from the application of the method to scattering by dielectric and perfectly conducting cylinders, and good agreement was observed between the bymoment method results and those obtained from an integral equation solution.

6. REFERENCES

1. Silvester, P., and Hsieh, M.S., "Finite-element solution of 2-dimensional exterior field problems," *Proc. Inst. Elec. Eng.*, vol. 118, pp. 1743-1747, Dec. 1971.
2. McDonald, B.H., and Wexler, A., "Finite-element solution of unbounded field problems," *IEEE Trans. Microwave Theory Tech.*, vol. MTT-20, pp. 841-847, Dec. 1972.

3. Lynch, D.R., Paulsen, K.D., and Strohbehn, J.W., "Hybrid finite element method for unbounded electromagnetic problems in hyperthermia," *Int. J. Num. Meth. Eng.*, vol. 23, pp. 1915-1937, 1986.

4. Paulsen, K.D., Lynch, D.R., and Strohbehn, J.W., "Three-dimensional finite, boundary, and hybrid element solutions of the Maxwell equations for lossy dielectric media," *IEEE Trans. Microwave Theory Tech.*, vol. MTT-36, pp. 682-693, April 1988.

5. Jin, J.M., and Liepa, V.V., "Application of hybrid finite element method to electromagnetic scattering from coated cylinders," *IEEE Trans. Antennas and Prop.*, vol. AP-36, pp. 55-70, Jan. 1988.

6. Mei, K.K., "Unimoment method for solving antenna and scattering problems," *IEEE Trans. Antennas and Prop.*, vol. AP-22, pp. 760-766, Nov. 1974.

7. Chang, S.-K., and Mei, K.K., "Application of the unimoment method to electromagnetic scattering of dielectric cylinders," *IEEE Trans. Antennas and Prop.*, vol. AP-24, pp. 35-42, March 1976.

8. Tremain, D.E., and Mei, K.K., "Application of the unimoment method to scattering from periodic dielectric structures," *J. Opt. Soc. Am.*, vol. 68, pp. 775-783, June 1978.

9. Morgan, M.A., and Mei, K.K., "Finite element computation of scattering by inhomogeneous penetrable bodies of revolution," *IEEE Trans. Antennas and Prop.*, vol. AP-27, pp. 202-214, March 1979.

10. Morgan, M.A., "Generalized coupled azimuthal potentials for electromagnetic fields in inhomogeneous media," *IEEE Trans. Antennas and Prop.*, vol. AP-36, pp. 1735-1743, Dec. 1988.

11. Morgan, M.A., Chen, C.H., Hill, S.C., and Barber, P.W., "Finite element - boundary integral formulation for electromagnetic scattering," *Wave Motion*, vol. 6, pp. 91-103, Jan. 1984.

12. Waterman, P.C., "Matrix formulation of electromagnetic scattering," *Proc. IEEE*, vol. 53, pp. 805-812, 1965.

13. Morgan, M.A., and Welch, B.E., "The field feedback formulation for electromagnetic scattering computations," *IEEE Trans. Antennas and Prop.*, vol. AP-34, pp.1377-1382, Dec. 1986.

14. Cangellaris, A.C., and Lee, R., "A finite element method for solving electromagnetic scattering problems," *1989 IEEE AP-S International Symposium Digest*, pp. 1116-1119, San Jose, CA, June 1989.

15. Huebner, K.H., *The Finite Element Method for Engineers*, pp. 106-119, John Wiley and Sons, New York, 1975.

16. Harrington, R.F., *Time-Harmonic Electromagnetic Fields*, pp. 100-103, McGraw-Hill, New York, 1961.

17. Blacker, T.D., "FASTQ User Manual, Version 1.2," Sandia Report, SAND88-1326, July 1988.

18. Meixner, J., "The behaviour of electromagnetic fields at edges," *IEEE Trans. Antennas and Prop.*, vol. AP-20, no. 4, July 1972.

19. Butler, C.H., Xu, X.-B., and Glisson, A.W., "Current induced on a conducting cylinder located near the planar interface between two semi-infinite half-spaces," *IEEE Trans. Antennas and Prop.*, vol. AP-33, no. 4, pp. 616-624, June 1985.

20. Akin, J.E., *Application and Implementation of Finite Element Methods*, pp. 122-125, Academic Press, Inc., Florida, 1984.

Combined Finite Element/Boundary Integral Approach to the Prediction of Acoustic Signatures and Radar Cross Sections

E. THOMAS MOYER, JR.
School of Engineering and Applied Science
The George Washington University
Washington, DC 20052, USA

ERWING A. SCHROEDER
Applied Mathematics Division
David Taylor Research Center
Bethesda, Maryland 20084, USA

ABSTRACT

A finite element approach for the solution of the Helmholtz equation in three dimensions is used to solve radiation and scattering problems. The infinite domain is approximated using mapped infinite elements with shape functions incorporating the free-space Green's function. This approach provides an accurate solution for the near-field response. The far field is generated using a boundary integral approach. Acoustic results are presented for a sound-hard sphere for $\kappa a = 2.0$ to $\kappa a = 25.0$. Electromagnetic results are presented for a circular cylinder. Near-field solutions and far-field bistatic cross sections are examined. Agreement with analytic results is excellent for most problems. A numerical instability in the calculation of the infinite element matrices is discussed.

INTRODUCTION

The Helmholtz partial differential equation governs many time harmonic field problems in physics and engineering. Acoustic and electromagnetic problems are prominent applications. Popular numerical techniques for the solution of the Helmholtz equation are the boundary element method, the finite element method, and the finite difference method (which is closely related to the finite volume method). For problems involving radiation and scattering in free space, the problem domain is essentially infinite.

The boundary element method models only the scattering surface of reradiating bodies. The Helmholtz integral equation is then solved. This method produces a full, complex, and asymmetric linear system of equations which is singular at internal resonances of the scatterer geometry. These difficulties have limited its applicability to lower frequencies. The method has been employed for κa less than 50 [1,2,3].

The finite element and finite difference methods model the full field by solving the partial differential equations in the region. Both methods produce systems of linear algebraic equations which are banded, symmetric, and complex, also they require significantly more degrees of freedom than the boundary element method. Since the physical domain is effectively unbounded, an outer boundary must be established sufficiently far from the scatterer on which an approximate boundary condition can be applied.

If the outer boundary could be applied close enough to the scatterer, either the finite element or the finite difference method would be the method of choice for solving scattering problems because the system matrices would be banded and symmetric. This paper presents a finite element approach in which infinite elements are used to allow truncation of the domain only a few elements from the scatterer, and which shows significant potential for scattering applications. The infinite elements are an extension of finite elements that include an infinite geometric mapping and shape functions incorporating the free-space Green's function. A unique numerical quadrature is employed for the calculation of the element matrices. At present the quadrature converges only at moderate frequencies, limiting the method to problems with κa less than 100.

The Helmholtz integral equation with a boundary element approach is used to generate far-field results. The integrand is calculated using near-field finite element results on a surface which encloses the target and is inside the computational domain. The solution at any point in the infinite domain can be generated by using that point as the field point in the Helmholtz integral equation.

A plane acoustic wave scattering from a sound hard sphere, and a plane TE polarized electromagnetic wave scattering from a perfectly conducting circular cylinder are studied. Both near-field solutions and bistatic radar cross sections are presented.

FINITE ELEMENT FORMULATION

The Helmholtz partial differential equation for time harmonic wave problems in source free media is written as

$$\nabla^2 \psi + \kappa^2 \psi = 0 \tag{1}$$

where κ is the wave number and ψ is the field variable. The essential and natural boundary conditions are

$$\psi = \psi^* \quad \text{On } S_1 \qquad \text{and} \qquad \frac{\partial \psi}{\partial n} = \frac{\partial \psi^*}{\partial n} \quad \text{On } S_2 \tag{2}$$

where ψ^* and $\dfrac{\partial \psi^*}{\partial n}$ are the prescribed boundary conditions. The weighted residual formulation leads to the integral equation

$$\int_V \left[\nabla W \nabla \psi - \kappa^2 W \psi \right] dV = \int_{S_2} W \frac{\partial \psi^*}{\partial n} \, dS \tag{3}$$

Using the standard finite element approach [4], a system of linear equations is obtained in terms of the nodal values of the field variable ψ

$$\mathbf{K} \, \Psi = \mathbf{F} \tag{4}$$

where $\mathbf{F}$ arises from natural boundary conditions and the stiffness matrix for an individual element can be written in terms of the element shape functions as

$$K_{ij} = \int_{V_e} \left[\frac{\partial N_i}{\partial X_k} \frac{\partial N_j}{\partial X_k} - \kappa^2 \, N_i \, N_j \right] dV \tag{5}$$

with N_i the finite element shape functions and V_e the element volume. This approach, with appropriate boundary conditions, produces the general solution for problems governed by the Helmholtz equation.

For the problems studied in this paper, a standard formulation of quadratic isoparametric elements was used [4]. A mesh density of four elements per wavelength will usually produce accurate solutions. In two dimensions these elements are curvilinear quadrilaterals and in three dimensions they are curvilinear hexahedra. The quadratic geometric variation allows for the modeling of most geometries without introducing boundary slope discontinuities which can give rise to nonphysical reflections.

OUTER BOUNDARY CONDITIONS AND INFINITE ELEMENTS

In finite element solutions for problems of radiation or scattering, the extent of infinite exterior domains are limited, usually by an approximation of the Sommerfeld radiation condition. A second order approximation can be obtained using linear spring and dashpot structural elements with appropriate constitutive parameters [5]. This method appears to be adequate if the outer boundary is between two and four wavelengths from the scatterer. Most structural finite element codes, using scalar analogies with the second order radiation condition, can solve Helmholtz equation problems [6]. Many authors have discussed higher order approximations [7,8], but

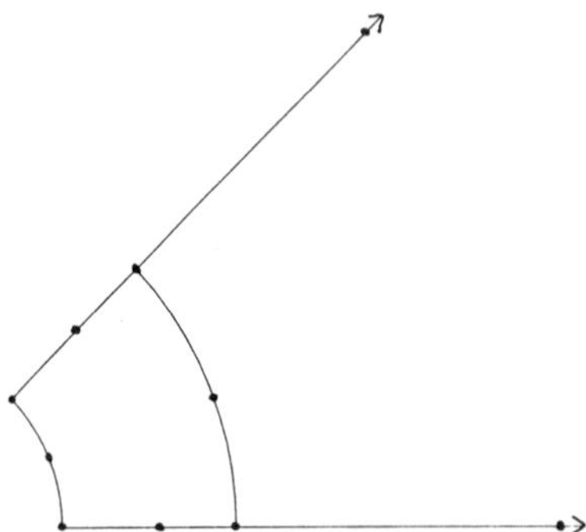

FIGURE 1. Finite and infinite elements

these approximations are difficult to employ with standard finite elements because they require derivatives of higher than first order.

The use of infinite elements appears to be a promising approach for approximating the infinite exterior region in scattering problems. Recently many publications have reported on applications of infinite elements to a variety of problems. For exterior scattering problems, an approach presented by Zienkiewicz and others [9] is most promising. This method uses a standard finite element which is mapped into an infinite region in the physical domain. The finite face of the element is then applied to the near-field modeling as shown schematically for two dimensions in Figure 1. The extrapolation functions for the element include the correct wave component for radiating waves. The geometric and extrapolation mappings are:

$$r = \frac{A}{(1-t)} \tag{6}$$

$$N_i(p,s,t) = M_i(p,s,t)\cdot\exp\left[i\kappa\frac{A}{(1-t)}\right]\left[-i\kappa\frac{A}{2}\right] \tag{7}$$

where $M_i(p,s,t)$ are the standard quadratic shape functions, A is a constant related to the location of the centroid of the element face and p,s are the standard mapped coordinates related to the two finite spatial dimensions. The extrapolation functions decay radially with leading order $1/r$ in three dimensions and with $1/\sqrt{r}$ for two-dimensions (where r is measured from the centroid of the physical domain).

The infinite elements allow for a general quadratic variation over the finite face of the element and for decay in a three-term series expansion. The leading term in the shape function is the three-dimensional free-space Green's function for the Helmholtz equation. With this approach, the infinite elements can be placed quite close to the scattering object. If infinite elements can be located near the scatterer, the number of degrees of freedom is of the same order as the boundary element approach, but produce banded and symmetric matrices. In the worst case, the order of bandwidth is $\sqrt{N}$ (where N is the number of surface nodes).

Using infinite elements requires evaluating integrals of the form

$$I = \int_{-1}^{1} \frac{f(t)}{(1-t)^2} \exp\left[\frac{2i\kappa A}{(1-t)}\right] dt \tag{8}$$

This work implements an interpolatory quadrature approach used in [9]. The approach approximates the integral in Equation (8) as

$$I = \sum_{i=1}^{i=n} W_i f(t_i) \tag{9}$$

with weight factors

$$W_j = \sum_{i=1}^{i=n} \int_{-1}^{1} X_{ij}^{-1} \frac{t^{i-1}}{(1-t)^2} \exp\left[\frac{2i\kappa A}{(1-t)}\right] dt \qquad (10)$$

where t_i are the chosen quadrature points. As in [9], the quadrature points are chosen at equally spaced intervals in the range $-1 < t < 1$ ($t = -1$ and $t = 1$ are invalid choices because the weight factors cannot be evaluated for these points). The factor X_{ij}^{-1} arises from the inversion of the interpolatory quadrature matrix and a recursion relation can be used to evaluate all terms. The final computation is a series evaluation involving simple functions and the sine and cosine integrals. In this work the IMSL library was employed to evaluate the sine and cosine integrals [10]. The procedure requires inversion of an ill-conditioned matrix and this limits the possible number of quadrature points to fewer than ten for double precision computations.

The numerical quadrature is the limiting factor in application of the infinite elements. As κA in the problem increases, the numerical quadrature becomes unstable. For example, consider the function $f(t) = 1$. The integral in Equation (8) is exact with one integration point. For arbitrary order, however, the sum of all weights should also give the exact value. Table 1 shows the predicted integral value for various orders of integration and values of κA.

TABLE 1: Quadrature performance for integral evaluation

κA	N=1	N=6	N=8
100	(2.532E-03,4.312E-03)	(2.532E-03,4.312E-03)	(2.532E-03,4.311E-03)
1000	(-4.134E-04,2.812E-04)	(-4.134E-04,2.812E-04)	(-4.137E-04,3.052E-04)
1500	(3.313E-04,-3.676E-05)	(3.312E-04,-3.677E-05)	(3.281E-04,-1.221E-04)

For κA less than 100, eight-point integration gives reasonable results, however as κA increases, this integration gives increasingly poorer results. For $\kappa A = 1500$, the six-point integration loses precision in the third decimal place. When the weight factors are closely examined, even for κA on the order of 50, fewer than nine significant digits are accurate for six integration points and fewer than seven significant digits are accurate for eight integration points. The lack of precision in the integration (and the breakdown for large arguments) is the current difficulty with infinite elements. If a frequency dependent quadrature can be developed, the infinite element will be shown to be a robust methodology for approximating infinite domains.

FAR-FIELD GENERATION USING SURFACE INTEGRALS

To compute the scattering cross section, field values must be obtained for points far from the scattering object. The finite element solution provides the field in a bounded region containing the object. To compute far-field values, a surface integral derived from Green's theorem is used. The domain for Green's theorem is the exterior of a surface containing no sources of scattered radiation, that is, the scattering object must be enclosed by the surface. For the far-field to be computed directly, values of the field and its normal derivatives on the surface must be available, that is, the surface must be contained in the region modeled with finite elements. The integral is evaluated by two-dimensional Gaussian quadrature. To perform the integration, the surface, which is the boundary of the exterior region, is divided into boundary elements that are similar in formulation to finite elements. The values of the field and its derivative are provided at nodes on the edge of each the element and then interpolated to Gauss points inside the element. In general, field values at the nodes of the boundary elements will be interpolated from values computed at nearby finite element nodes, and the normal derivatives will be obtained using a finite difference approximation. With the values and normal derivatives of the field given on the surface S, the surface integral for field values at far-field points is:

$$\psi(\mathbf{r}) = \int_S [\psi(\mathbf{r})\mathbf{n}\cdot\nabla' g(\mathbf{r})^2 - g(\mathbf{r})\mathbf{n}\cdot\nabla'\psi(\mathbf{r}')]ds' \qquad (11)$$

In these integrals, the position vector $\mathbf{r}'$ is the integration variable on S, the position vector $\mathbf{r}$ locates the far-field point, the distance $r = |\mathbf{r}' - \mathbf{r}|$, the free-space Greens function is $g(\mathbf{r}) = \dfrac{e^{-ikr}}{4\pi r}$, and the unit vector $\hat{\mathbf{n}}'$ is normal to S at $\mathbf{r}'$ [1].

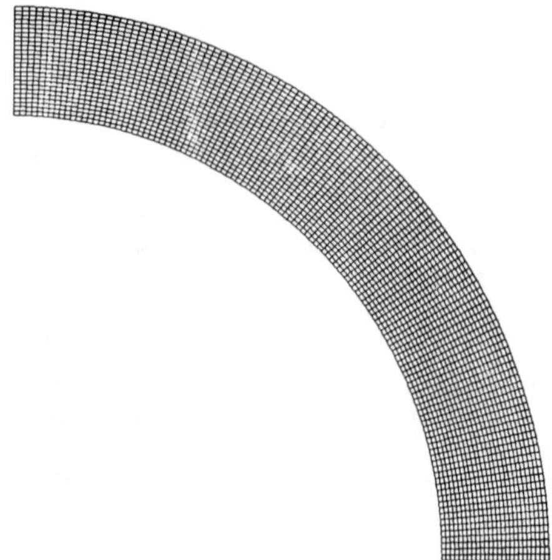

FIGURE 2. Finite element mesh for perfectly conducting cylinder

If the finite element solution exploits symmetries of the object and the fields, it will model only a part of the finite region. Depending on the symmetries used by the finite element solution, the integral will be evaluated repeatedly over the modeled part of the surface; each time different factors of 1.0 or -1.0 are applied to coordinate and field values to adapt to the various symmetry regions. The integration will be repeated one, two, four, or eight times, depending on whether the finite element mesh represents the whole, a half, a quadrant, or an octant of the finite region.

EXAMPLE PROBLEMS

The first example problem is a Transverse Electric (TE) field scattering from a perfectly conducting infinite cylinder with a radius of ten wavelengths. The incident field is assumed to be a plane wave. The problem involves only Dirichlet boundary conditions on the cylindrical surface. A series solution can be computed in terms of Bessel Functions [11]. The mesh employed is shown in Figure 2. The outer boundary was truncated by applying the second order radiation condition using springs and dashpots, and the structural code, NASTRAN, was used for the analysis [12]. Figure 3. compares the near-field solution magnitude and phase on a

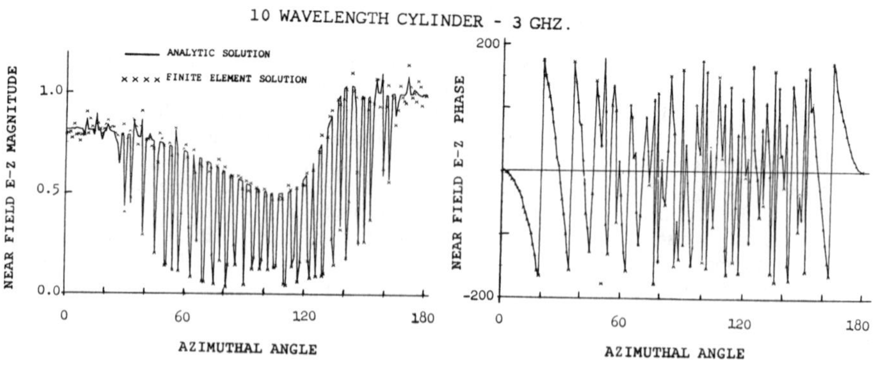

FIGURE 3. Near field solution for perfectly conducting circular cylinder

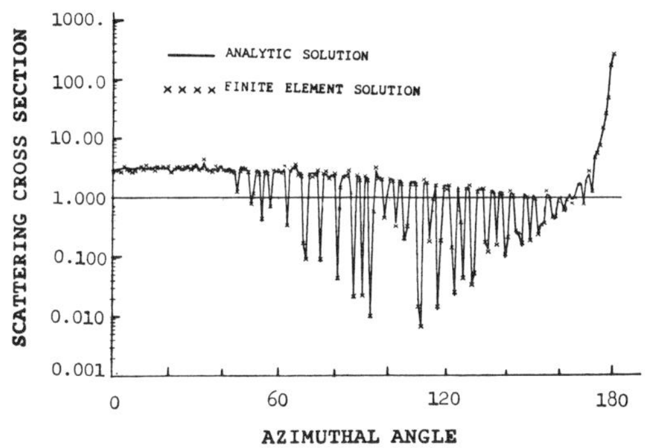

FIGURE 4. Bistatic cross section for a perfectly conducting circular cylinder

typical ring of elements (one wavelength away from the cylinder surface) with the series solution, and shows excellent agreement. These results (with the normal derivative) were integrated to the far field using the surface integral approach discussed previously. The scattering cross section is calculated from these results. Figure 4. shows excellent agreement between the predicted scattering cross section and the series solution. These results indicate that the local finite element approach with a surface integration to the far-field is a viable numerical approach for calculation of scattering cross sections.

To investigate the use of infinite elements for problems of radiation and scattering, the sound hard sphere was chosen as the example problem. A series solution is available for an incident plane wave on the surface of the sphere with Neumann boundary conditions [11]. Figure 5. shows the surface mesh employed for these examples. At four elements per wavelength, the mesh will be sufficient for problems up to $\kappa R = 25$ where R is the radius of the sphere. The full model is generated by extruding the mesh in the radial direction a certain number of elements and applying infinite elements to the outer surface. The simplest approximation will involve applying infinite elements directly on the surface mesh and employing no conventional finite elements. Figure 6. compares the predicted surface pressure amplitude and phase with the series solution when only infinite elements are applied to the surface. No conventional finite elements are used and excellent agreement is obtained. For low frequency problems with κR less than 10, the infinite elements alone produce an accurate solution. As the frequency of the incident wave

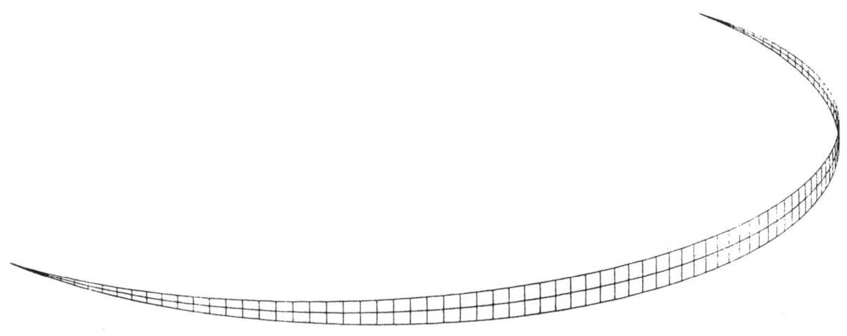

FIGURE 5. Surface mesh for the sound hard sphere

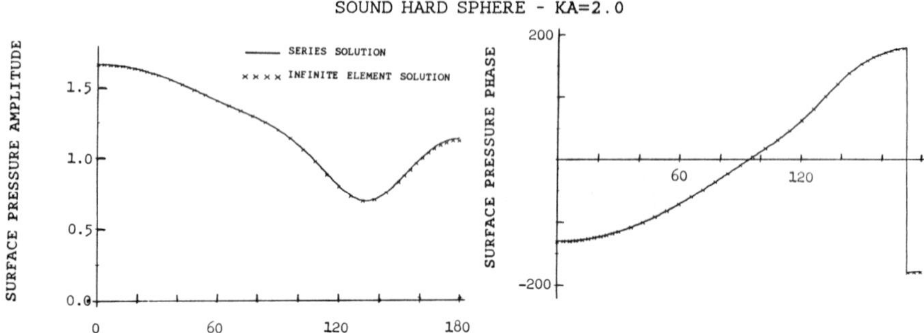

FIGURE 6. Surface pressure for $\kappa R = 2$ - infinite elements only

is increased, the infinite element solution losses accuracy. Figure 7. compares the surface pressure magnitude and phase with the series solution at $\kappa R = 10$, and shows that near the trailing edge of the sphere, the solutions do not match. The characteristics of the solutions are the same, but the peaks are overestimated. If a row of extruded finite elements is added, the situation improves significantly. Figure 8. shows excellent agreement between the series solution and the predicted surface pressure using a layer of finite elements between the scattering surface and the infinite elements. As the frequency is increased, the solution predicted using infinite elements alone degrades considerably. Figure 9. shows that the solution at $\kappa R = 25$ using only infinite elements is quite poor compared with the series solution. Figure 10. shows significant improvement is obtained by adding a layer of finite elements under the infinite elements, and that while agreement is not exact, the predictions are reasonable.

At the highest κR investigated, two factors contribute to reported inaccuracies: model breakdown as the four-element per wavelength limit is reached and instability in the numerical quadrature. For large-body radiation and scattering problems, more modeling will be required and for large shapes, the infinite element, while theoretically a viable approach, will require a more robust integration scheme.

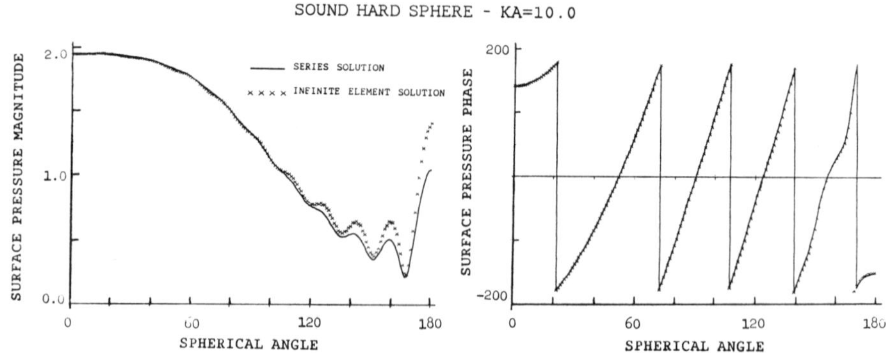

FIGURE 7. Surface pressure for $\kappa R = 10$ - infinite elements only

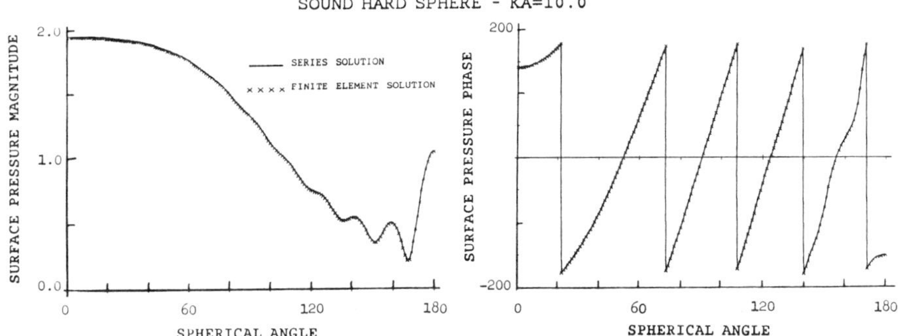

FIGURE 8. Surface pressure for $\kappa R = 10$ - finite and infinite elements

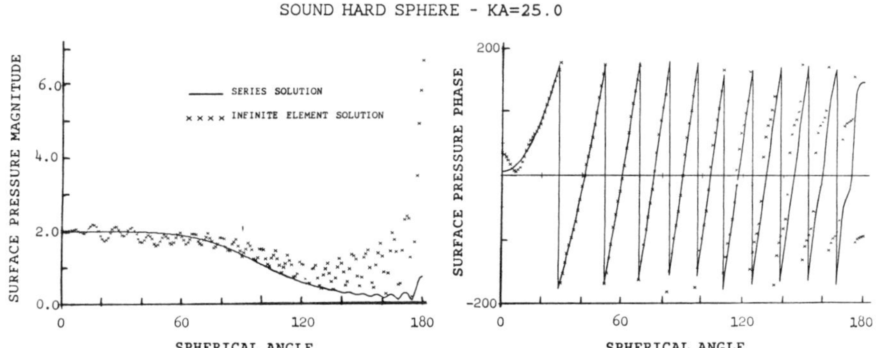

FIGURE 9. Surface pressure for $\kappa R = 25$ - infinite elements only

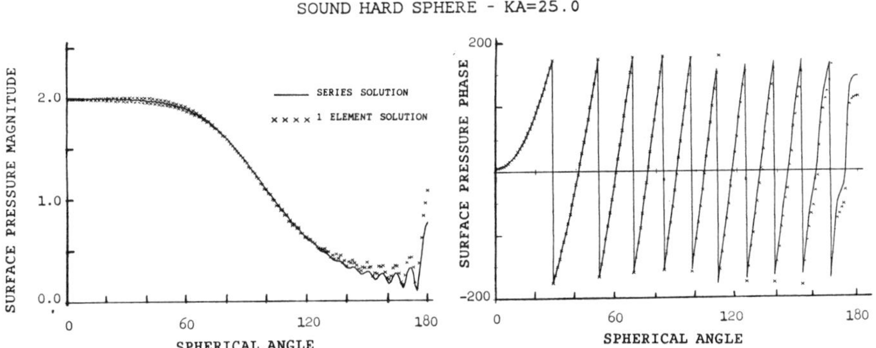

FIGURE 10. Surface pressure for $\kappa R = 25$ - finite and infinite elements

267

CONCLUSIONS

A combined finite element, infinite element approach has been presented and employed for Helmholtz problems involving radiation and scattering. The use of infinite elements proved to be a good approach for representing infinite regions for nondimensional frequencies up to $\kappa a = 100$. Applications for higher frequencies will require a more robust integration scheme for the infinite element stiffness matrices. For low frequency problems, κR less than 10, reasonable accuracy can be achieved using only infinite elements on the surface of the scatterer. As the frequency increases, at least one layer of finite elements is needed between the scattering surface and the infinite elements. For all problems studied in this paper, only one layer of finite elements is needed to obtain acceptable accuracy. For more general shapes this requirement will need further study.

A surface-integral generation scheme for far-field parameters has been presented and demonstrated for two-dimensional applications. This approach has been demonstrated previously in three dimensions [13]. This approach produces excellent far-field results when accurate near-field values are available, and the far-field results predicted are always at least as accurate as the integrated near-field values.

The approach presented shows promise for radiation and scattering problems and requires computation times on the order of N^2 for N surface mesh points. If robust iterative sparse solvers are developed from the current active research in this area, the method would be of order N since finite element matrices are naturally sparse. For problems of radiation and scattering from convex bodies with one layer of finite elements between the scattering surface and the infinite elements, typical finite element matrices have about 100 N nonzero elements. More research is needed to improve the integration scheme for the infinite elements and to investigate more general shapes. Coupled with a robust sparse solver, the approach presented is a promising move toward an order N method for the prediction of acoustic and non-acoustic signatures.

ACKNOWLEDGEMENT

This work was supported in FY-90 by the David Taylor Research Center's (DTRC) Independent Research (IR) program, sponsored by the Office of the Chief of Naval Research (OCNR), Office of Naval Research (ONR), and administered by the Research Director, DTRC 0112, under Program Element 0601152N, Task Area ZR 014 02 01, and DTRC Work Unit 1844-147, DN 500052.

REFERENCES

1. A.J. Poggio and E.K. Miller, "Solutions of Three-dimensional Scattering Problems" *Computer Techniques for Electromagnetics*, R. Mittra, ed, pp. 159-264, Pergamon Press, Oxford, 1973.

2. R.F. Harrington, *Field Computation By Moment Methods*, New, York, Macmillan, 1968.

3. H.A. Schenck, "Improved Integral Formulation for Acoustic Radiation Problems," *J. Acoust. Soc. of Amer.*, Vol. 44, pp. 41-58, 1968.

4. O.C. Zienkiewicz, *The Finite Element Method*, London, McGraw Hill, 1977.

5. E.T. Moyer Jr. and E. Schroeder, "Finite Element Modeling of Electromagnetic Fields and Waves Using NASTRAN", *Seventeenth NASTRAN Users' Colloquium*, NASA CP-3029, National Aeronautics and Space Administration, Washington, DC, pp. 214-246, 1989.

6. G.C. Everstine, "Structural Analogies for Scalar Field Problems," *Int. J. Num. Meth. in Engrg.*, Vol 17, pp. 471-476, 1981.

7. B. Enquist and A. Majda, "Absorbing Boundary Conditions for the Numerical Simulation of Waves", *Mathematics of Computation*, Vol. 31., pp. 629-651, 1977.

8. A. Bayliss and E. Turkel, "Radiation Boundary Conditions for Wave-Like Equations", *Commun. on Pure and Applied Math.*, Vol. 33, pp. 707-725, 1980.

9. O.C. Zienkiewicz, K. Bando, P. Bettess, C. Emson and T.C. Chiam, "Mapped Infinite Elements for Exterior Wave Problems", *Int. J. Num. Meth. in Engrg.*, Vol. 21, pp. 1229–1251, 1985.

10. The IMSL Library Reference Manual, IMSL Corporation, Houston, Texas, 1988.

11. J.J. Bowman, T.B.A. Senior and P.L.E. Uslenghi, *Electromagnetic and Acoustic Scattering by Simple Shapes,* Hemisphere Publishing Co. ,New York, 1987.

12. *COSMIC NASTRAN User's Manual,* Computer Software Management and Information Center, University of Georgia, Athens, Ga. 1986.

13. E.T. Moyer Jr. and E.A. Schroeder, "A Finite Element/Boundary Integral Equation Approach to the Accurate Prediction of Radar Cross Section", *Proc. of the Applied Computational Electromagnetic Society (ACES) Conference,* Monterey, pp. 321-328, 1990.

Finite Analytic Method for Numerical Solution of Mathematical Models

FARUK CIVAN
School of Petroleum and Geological Engineering
The University of Oklahoma
Energy Center, T301
Norman, Oklahoma 73019, USA

ABSTRACT

The finite analytic numerical method using power series based local
analytic solutions is described and its application and accuracy are
demonstrated.

INTRODUCTION

Mathematical models of natural phenomena and engineering processes
frequently involve nonlinear type differential equations. When
analytical solution methods become impossible or impractical, numerical
solutions are sought. Commonly used numerical solution methods are
based on some variations of local algebraic representations of
differential equations. These algebraic equations are compiled over the
solution domain, the initial and/or boundary conditions are incorporated
and then, they are solved simultaneously to yield an approximate
solution of the model equations.

Applications of conventional methods, such as finite differences and
finite elements, create certain problems which affect the accuracy of
numerical solutions. There are two basic reasons. First, usually, the
low order accurate formulations are preferred to avoid complicated
numerical schemes and to minimize the computational effort. Second, the
discretization of various terms in differential equations are carried
out independently of the other terms in the equation. Therefore,
various difficulties, including numerical dispersion and oscillation
problems and the grid orientation effects are encountered unless the
spatial and temporal steps used for discretization are reduced in a
manner to satisfy the pertinent stability criteria. While such
numerical remedies are convenient, the computational effort requirement
increases greatly.

One potential method which reduces the limitations of the conventional
methods significantly is the finite analytic method introduced by Chen
and Li.[1]   This method has been applied to various problems
successfully by several researchers including Chen et al.[1,2], Hwang et
al.[3], Manohar and Stephenson[4], Smith and Severin[5] and Iqbal and
Civan[6]. In spite of its apparent advantages, however, this method has
not been widely recognized judging from the recent literature.

270

Basically, the finite analytic method utilizes local analytic solutions of equations over small elements to develop higher order accurate numerical schemes. In other words, the finite analytic numerical schemes are derived directly from the model equations. Consequently, the numerical solutions generated by this method are less susceptible to the limitations of the conventional finite differences and finite elements methods. This method allows for use of relatively larger grid sizes and time steps without the inherent difficulties of the conventional methods. Thus, the computational time and memory requirements are reduced significantly. The basic difficulty of this method is the derivation of the local analytic solutions which may be a cumbersome task in some cases.

In this paper, systematic derivation of finite analytic numerical schemes using power series solutions is described and its application is illustrated via two simple examples.

THE FINITE ANALYTIC METHOD

As the first step, the solution domain is divided into small elements by discretizing the temporal and spatial variables similar to conventional finite differences and finite elements methods.

The second step is to derive an approximate local analytic solution. Most analytical solution methods are applicable to linear equations. Therefore, nonlinear equations must be transformed to linear forms using appropriate methods such as those given by Civan and Sliepcevich[7]. Otherwise, special solution techniques such as the Frobenius method[8] are required. However, in some cases, it is more practical to approximate the nonlinear equations by linearized forms.

The methods of obtaining linearized forms vary by the researchers and effect the accuracy of the resulting numerical schemes. In a simpler method which resulted with satisfactory results a nonlinear term can be expressed as a product of linear and nonlinear parts and the nonlinear term can be replaced with a representative average value over a small local element. For example, Smith and Severin[5] linearized the radiative heat transfer equation

$$d^2f/dx^2 = (4K\sigma f^4/k) - (1/k)dq_a/dx \qquad (1)$$

by assuming constant values of $(4K\sigma f^3/k)$ and $(1/k)dq_a/dx$ over a finite interval to obtain the following linearized equation

$$d^2f/dx^2 = \overline{(4K\sigma f^3/k)}f - \overline{(1/k)dq_a/dx} \qquad (2)$$

Sundaram and Wankat [9] linearized the following equation

$$\partial f/\partial t = f^{1/2}\partial^2 f/\partial x^2 \qquad (3)$$

by assigning $f^{1/2}$ a constant value chosen as the geometric mean of the two boundary values. Chen and Chen[2] linearized the Navier-Stokes equation,

$$\frac{\partial^2 f}{\partial x^2} + \frac{\partial^2 f}{\partial y^2} = R\, \frac{\partial f}{\partial t} + u\, \frac{\partial f}{\partial x} + v\, \frac{\partial f}{\partial y} + F \qquad (4)$$

in which $f\varepsilon(u,v)$, by separating u and v as

$$u = u' + U \qquad (5)$$

$$v = v' + V \qquad (6)$$

in which U and V denote some local representative values assigned at the mid point of a finite interval and u' and v' denote the deviations of these values from the actual values u and v or the corrections. As a result of linearization analytic solutions of Eqs. (2), (3) and (4) can be obtained by standard methods designed to work for linearized equations. However, such solutions often assume complicated mathematical forms.

In many cases, mathematically simpler solutions can be derived by means of the Frobenius method with or without linearization. Examples are given by Manohar and Stephenson [4] and by El-Halafawy and Eissa [8]. The Frobenius method is based on the assumption of a power series solution. For a multivariate function, a generalized power series in the following form can be written.

$$f(x_1,x_2,x_3,\ldots,x_m) = \sum_{r_m=0}^{k-1} \ldots \sum_{r_3=0}^{r_4} \sum_{r_2=0}^{r_3} \sum_{r_1=0}^{r_2} a_{r_1 r_2 r_3 \ldots r_m}$$

$$x_1^{r_1} x_2^{r_2 - r_1} x_3^{r_3 - r_2} \ldots x_m^{r_m - r_{m-1}} + e_k \qquad (7)$$

In some special cases extended power series may be required. A substitution of Eq. (7) into a differential equation yields a recurrence formula for the coefficients. Hence, the number of coefficients in the power series can be reduced by means of the recurrence formula and the resulting local analytic solution can be expressed in the following general form according to Manohar and Stephenson [4]:

$$f(x_1,x_2,\ldots,x_m) = \sum_{i=0}^{n} b_i B_i (x_1,x_2,\ldots,x_m) + e_k \quad ; \quad n<k-1 \qquad (8)$$

Thus, for the same order of accuracy, Eq.(8) contains less coefficients than Eq.(7); i.e. $n<k-1$. Frequently, the application of the Frobenius method requires tedious algebraic manipulations to arrive at a recurrence formula, especially for multivariable and nonlinear problems. A review of the methods for performing operations on multivariate power series is given by Forbes [10].

The algebraic representations for Neumann or mixed boundary conditions are obtained by replacing the function derivatives by the following expression obtained by differentiating Eq.(8) following Manohar and Stephenson[4]

$$\partial^n f(x_1, x_2, \ldots, x_m)/\partial x_j{}^n = \sum_{i=0}^{m} b_i \partial^n B_i(x_1, x_2, \ldots, x_m)/\partial x_j{}^n \; ;$$

$$j = 1, 2, \ldots, m \qquad (9)$$

To develop a finite analytic numerical scheme, a computational molecule having m+1 nodes is selected on the grid system. Any of m+1 nodes is chosen as a reference point and the coordinates of the remaining points are expressed with respect to this reference point. Then, Eq. (8) is applied at all of (m+1) points. The first m equations are solved simultaneously for the coefficients, $b_i : 1, 2, \ldots, m$, in terms of the nodal function values which are, then, substituted into the $(m+1)^{th}$ equation to obtain an algebraic equation relating the nodal function values of the computational molecule.

For the treatment of Neumann boundary condition Manohar and Stephenson [4] propose a convenient formulation by incorporating the Neumann boundary condition directly into the derivation of the finite analytic scheme near the Neumann boundary. This approach utilizes Eq. (8) for the interior points and Eq. (9) for the Neumann boundary points for calculation of the polynomial coefficients $b_i : i = 1, 2, \ldots, m$ for Eq. (8).

The algebraic equations obtained as described above are solved simultaneously over the solution grid to obtain the numerical solution of the problem. The stability and truncation error analysis of the finite analytic schemes are given by Iyengar and Manohar [13] and Manohar et al. [14].

APPLICATIONS

In this section, the application and accuracy of the finite analytic method is demonstrated by two examples.

<u>First Problem</u>

The first problem deals with the solution of the following hyperbolic nonlinear problem given by Bellman and Adomian [11]

$$\frac{\partial f}{\partial t} + u(f) \frac{\partial f}{\partial x} = 0 \quad , \quad 0 < x \leq 1 \quad , \quad 0 \leq t \leq 1 \qquad (10)$$

subject to the initial function

$$f = g(x) \quad , \quad 0 < x \leq 1 \quad , \quad t = 0 \qquad (11)$$

The following analytical solution is considered for the numerical experiment:

$$f = x/(t - 10) \text{ for } g(x) = -0.1 \, x \text{ and } u(f) = -f \qquad (12)$$

To obtain a local algebraic representation, first, linearize Eq.10 by replacing u(f) with a local value, U, whose value is determined by Eq.(21) given later. Then, a Frobenius power series solution is assumed as

$$f(x,t) = \sum_{\nu=0}^{k-1} \sum_{\mu=0}^{\nu} a_{\mu,\nu} x^{\nu-\mu} t^{\mu} + e_k \tag{13}$$

in which $a_{\mu,\nu}$ are some polynomial coefficients and e_k denotes the error term. A substitution of Eq.(13) into Eq.(10) yields the following recurrence relationship between the coefficients

$$a_{\mu+1,\nu} = - U(\nu-\mu) a_{\mu,\nu} / (\mu+1) \quad : \mu=0,1,\ldots,\nu \text{ and } \nu=0,1,\ldots,k-1 \tag{14}$$

Hence, using Eq. (14), Eq. (13) simplifies to

$$f(x,t) = \sum_{\nu=0}^{k-1} a_{0,\nu} (x-Ut)^{\nu} + e_k \tag{15}$$

Note that for the same order of magnitude Eq.(15) contains less number of polynomial coefficients than Eq.(13).

To develop a finite analytic numerical scheme consider Eq. (15), for instance, for k=3 as the local analytic solution

$$\tilde{f} = a_{0,0} + a_{0,1}(x - Ut) + a_{0,2}(x - Ut)^2 + e_3 \tag{16}$$

Simultaneously, consider the grid element shown in Fig. 1. Let $n=1,2,\ldots$ and $i=1,2,\ldots,N$ denote the discrete temporal and spatial steps, respectively. Assuming the location of the point (i,n+1) as reference with its coordinates as (0,0) and applying Eq.(16) at the points (i-1,n), (i-1,n+1), (i,n) and (i,n+1) yields the equations given in Table 1. Solving Eqs. (I.2-4) in Table I for $a_{0,0}$, $a_{0,1}$ and $a_{0,2}$ in terms of the discrete function values and substituting $a_{0,0}$ in Eq. (I.1) in Table I yields the following four-point formula:

$$f_i^{n+1} = f_{i-1}^n + \left[\frac{-h + Uk}{h + Uk} \right] (f_{i-1}^{n+1} - f_i^n) \quad ;$$

$$i=2,3,\ldots,N \text{ and } n=2,3,\ldots, \tag{17}$$

The initial condition is

$$f_i^1 = g(x_i) \quad ; \quad i=1,2,\ldots,N \tag{18}$$

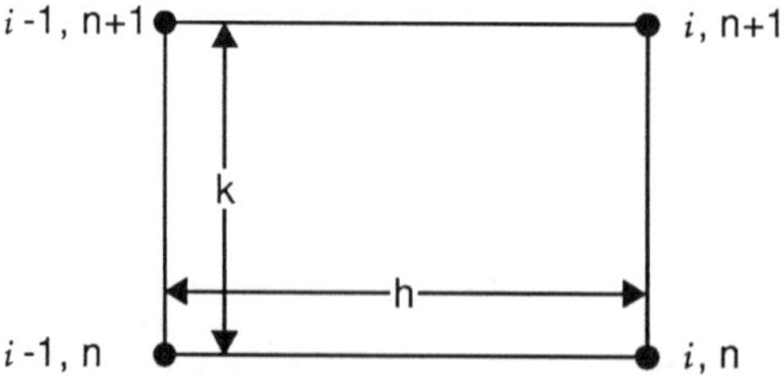

FIGURE 1. Computational molecule
for the first problem

TABLE I - Nodal equations for the first problem.

$$f_i^{n+1} = a_{0,0} \tag{I.1}$$

$$f_i^n = a_{0,0} + a_{0,1}Uk + a_{0,2}(Uk)^2 \tag{I.2}$$

$$f_{i-1}^n = a_{0,0} - a_{0,1}(h - Uk) + a_{0,2}(h - Uk)^2 \tag{I.3}$$

$$f_{i-1}^{n+1} = a_{0,0} - a_{0,1}h + a_{0,2}h^2 \tag{I.4}$$

To test Eq. (17) the problem was considered as an initial and boundary value problem by defining a boundary condition at x=0 via Eq. (12), i.e.

$$f_1^n = Eq. (12) \quad ; \quad n=2,3,\ldots, \tag{19}$$

Eq.(17) was derived previously by Iqbal and Civan[6]. However, in their Eq. (22) the parameter U was approximated as a weighted average according to

$$U = wu(f_{i-1}^{n+1}) + (1-w)u(f_i^{n+1}) \tag{20}$$

in which w denotes a weighting coefficient. In the present study Eq. (21), which gave better results than Eq. (20), has been used.

$$U = u(f_{i-1}^{n+1}) \tag{21}$$

The relative errors were in the order of 10^{-2} using h=0.01 and k=0.001. This solution is less accurate than the explicit finite difference solution which resulted in 10^{-4} relative error.

Second Problem

Consider the following test problem used by Civan and Sliepcevich[12]

$$\partial^2 f/\partial X^2 + \beta^2 \partial^2 f/\partial Y^2 = 0 \quad ; \quad 0<X \text{ and } Y<1 \tag{22}$$

subject to

$$f = 0 \quad ; \quad 0 \leq X \leq 1 \quad , \quad Y=0 \tag{23}$$

$$f = \sin(\pi X/2) \quad ; \quad 0 \leq X \leq 1 \quad , \quad Y=1 \tag{24}$$

$$f = 0 \quad ; \quad X = 0, \ 0 \leq Y \leq 1 \tag{25}$$

$$\partial f/\partial X = 0 \quad ; \quad X = 1 \ , \ 0 \leq Y \leq 1 \tag{26}$$

The exact analytical solution is given by

$$f = \sinh[\pi Y/(2\beta)]\sin(\pi X/2)/\sinh[\pi/(2\beta)] \tag{27}$$

Assume a general analytic solution in the form of a Frobenius power series as

$$f(x,y) = \sum_{\nu=0}^{k-1} \sum_{\mu=0}^{\nu} a_{\mu,\nu} x^{\nu-\mu} y^{\mu} + e_k \tag{28}$$

in which $a_{\mu,\nu}$ denote some polynomial coefficients, $x=X$, $y=Y/\beta$, and e_k is an error term. A substitution of Eq. (28) into Eq. (22) yields the following recurrence relationship between the coefficients

$$a_{\mu+2,\nu} = -(\nu-\mu)(\nu-\mu-1) a_{\mu,\nu}/[(\mu+1)(\mu+2)] \; ;$$

$$\mu=0,1,\ldots\nu \text{ and } \nu=0,1,\ldots,k-1 \tag{29}$$

Thus, by means of Eq. (29), Eq. (28) simplifies to

$$f(x,y) = a_{0,0} + a_{0,1}x + a_{1,1}y + a_{0,2}(x^2 - y^2) + a_{1,2}xy$$

$$+ a_{0,3}(x^3 - 3xy^2) + a_{1,3}(x^2y - y^3/3) + a_{1,4}(x^3y - xy^3)$$

$$+ a_{0,4}(x^4 - 6x^2y^2 + y^4) + e^5 \tag{30}$$

Note that Eq. (30) contains six polynomial coefficients less than Eq. (28) for the same order of accuracy.

To develop a finite analytic numerical scheme consider the computational molecule shown in Fig. 2 for the interior points. Assume the point (i,j) as the reference point with coordinates $(0,0)$ and the discrete points in the X and Y-directions are spaced uniformly by K and H intervals, respectively. Let N and M be the number of discrete points in the X and Y-directions, respectively. Then, the grid point intervals in terms of x and y become k=K and h=H/β, respectively. Next, drop the error term, e_5, and apply Eq. (30) at the nine discrete points shown in Fig. 2. Then, the resulting set of nine linear equations, Eqs. (II.1, 2a, 3-5, 6a, 7a, 8, 9), shown in Table II, are solved simultaneously for the polynomial coefficients in terms of the discrete function values at the nine points. These coefficients are substituted into Eq. (30) at point (i,j) to obtain the following nine-point formula:

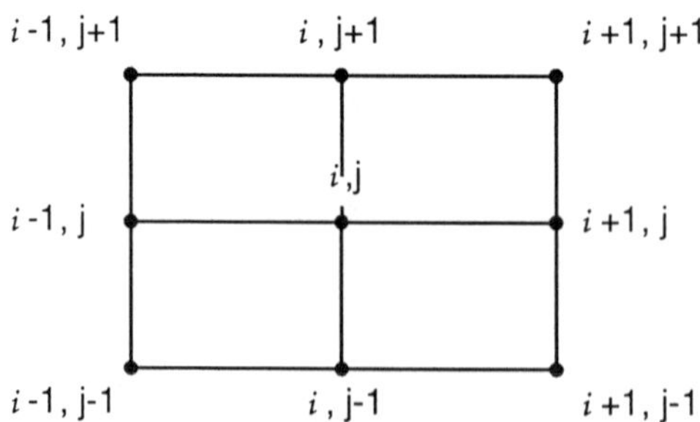

FIGURE 2. Computational molecule for the second problem

Table II. Nodal equations for the second problem.

$$u_{i,j} = a_{0,0} \tag{II.1}$$

$$u_{i+1,j} = a_{0,0} + a_{0,1}k + a_{0,2}k^2 + a_{0,3}k^3 + a_{1,4}k^4 \tag{II.2a}$$

$$\partial u_{i+1,j}/\partial x = a_{0,1} + 2a_{0,2}k + 3a_{0,3}k^2 + 4a_{1,4}k^3 = 0 \tag{II.2b}$$

$$u_{i,j+1} = a_{0,0} + a_{1,1}h - a_{0,2}h^2 - a_{1,3}h^3 + a_{1,4}h^4 \tag{II.3}$$

$$u_{i-1,j} = a_{0,0} - a_{0,1}k + a_{0,2}k^2 - a_{0,3}k^3 + a_{1,4}k^4 \tag{II.4}$$

$$u_{i,j-1} = a_{0,0} - a_{1,1}h - a_{0,2}h^2 + a_{1,3}h^3 + a_{1,4}h^4 \tag{II.5}$$

$$
\begin{aligned}
u_{i+1,j} = {}& a_{0,0} + a_{0,1}k - a_{1,1}h + a_{0,2}(k^2 - h^2) - a_{1,2}kh \\
& + a_{0,3}(k^3 - 3kh^2) + a_{1,3}(-3k^2h + h^3) + a_{1,4}(k^4 - 6k^2h^2 \\
& + h^4) + a_{0,4}(-k^3h + kh^3)
\end{aligned}
\tag{II.6a}
$$

$$
\begin{aligned}
\partial u_{i+1,j}/\partial x = {}& a_{0,1} + 2a_{0,2}k - a_{1,2}h + a_{0,3}(3k^2 - 3h^2) - 6a_{1,3}kh \\
& + a_{1,4}(4k^3 - 12kh^2) + a_{0,4}(-3k^2h + h^3) = 0
\end{aligned}
\tag{II.6b}
$$

$$
\begin{aligned}
u_{i+1,j+1} = {}& a_{0,0} + a_{0,1}k + a_{1,1}h + a_{0,2}(k^2 - h^2) + a_{1,2}kh \\
& + a_{0,3}(k^3 - 3kh^2) + a_{1,3}(3k^2h - h^3) \\
& + a_{1,4}(k^4 - 6k^2h^2 + h^4) + a_{0,4}(k^3h - kh^3)
\end{aligned}
\tag{II.7a}
$$

$$
\begin{aligned}
\partial u_{i+1,j+1}/\partial x = {}& a_{0,1} + 2a_{0,2}k + a_{1,2}h + a_{0,3}(3k^2 - 3h^2) + 6a_{1,3}kh \\
& + a_{1,4}(4k^3 - 12kh^2) + a_{0,4}(3k^2h - h^3) = 0
\end{aligned}
\tag{II.7b}
$$

$$
\begin{aligned}
u_{i-1,j+1} = {}& a_{0,0} - a_{0,1}k + a_{1,1}h + a_{0,2}(k^2 - h^2) - a_{1,2}kh \\
& + a_{0,3}(-k^3 + 3kh^2) + a_{1,3}(3k^2h - h^3) \\
& + a_{1,4}(k^4 - 6k^2h^2 + h^4) + a_{0,4}(-k^3h + kh^3)
\end{aligned}
\tag{II.8}
$$

$$
\begin{aligned}
u_{i-1,j-1} = {}& a_{0,0} - a_{0,1}k - a_{1,1}h + a_{0,2}(k^2 - h^2) + a_{1,2}kh \\
& + a_{0,3}(-k^3 + 3kh^2) + a_{1,3}(-3k^2h + h^3) \\
& + a_{1,4}(k^4 - 6k^2h^2 + h^4) + a_{0,4}(k^3h - kh^3)
\end{aligned}
\tag{II.9}
$$

$$f_{i,j} = \left[\frac{3h^2}{5(h^2 + k^2)} - \frac{1}{10} \right] (f_{i+1,j} + f_{i-1,j})$$

$$+ \left[\frac{3k^2}{5(h^2 + k^2)} - \frac{1}{10} \right] (f_{i,j+1} + f_{i,j-1})$$

$$+ \frac{1}{20} (f_{i+1,j+1} + f_{i+1,j-1} + f_{i-1,j-1} + f_{i-1,j+1}) \quad ;$$

$$i=2,3,\ldots,(N-1) \text{ and } j = 2,3,\ldots,(M-1) \tag{31}$$

Note that for k=h and k=3h, Eq.(31) simplifies to Eqs. 5 and 7 of Manohar and Stephenson[4], respectively. Eq.(31) has a truncation error of the order (k^5+h^5).

For an algebraic representation of the Neumann boundary condition given by Eq.(26); first take a derivative of Eq.(30) with respect to x to obtain

$$\partial f/\partial x = a_{0,1} + 2a_{0,2}x + a_{1,2}y + a_{0,3}(3x^2 - 3y^2) + 2a_{1,3}xy$$

$$+ a_{1,4}(3x^2y - y^3) + a_{0,4}(4x^3 - 12xy^2) + \partial e_5/\partial x \tag{32}$$

Then, apply Eq.(32) at point (i+1,j) to obtain

$$(\partial f/\partial x)_{i+1,j} = a_{0,1} + 2ka_{0,2} + 3k^2a_{0,3} + 4k^3a_{0,4} + e_4 \tag{33}$$

As suggested by Manohar and Stephenson [4], Eq. (33) can be used to incorporate the Neumann condition directly into the derivation of the finite analytic scheme. As a result a special form of Eq. (31) can be derived for the interior points next to the boundary points involving the Neumann boundary condition. This is accomplished by expressing the Neumann condition in terms of Eq. (33) and then applying it at the nodes involving this condition while using Eq. (30) directly at the interior domain points. Hence, Eqs. (II.1, 2b, 3-5, 6b, 7b, 8, 9) listed in Table II are used to obtain the expressions for the polynomial coefficients in Eq. (30) in terms of the nodal point values. Using these expressions the following finite analytic scheme is obtained:

$$f_{i,j} = \frac{1}{5h^2 + 19k^2} \left[\frac{19k^2 - h^2}{2} (f_{i,j+1} + f_{i,j-1}) + (5h^2 + k^2)f_{i-1,j} \right.$$

$$\left. + \frac{h^2 - k^2}{2} (f_{i-1,j+1} + f_{i-1,j-1}) \right] : i=N-1 \text{ and } j = 2,3,\ldots,M-1$$

$$\ldots (34)$$

Note that Eq. (34) does not contain the nodal point values of $f_{i+1,j+1}$, $f_{i+1,j}$, and $f_{i+1,j-1}$ which involve the Neumann boundary. Eq. (34) simplifies to Eq. (19) of Manohar and Stephenson [4] for k=h.

Once the numerical solutions at the interior points have been obtained
by solving Eqs. (31) and (34), simultaneously, the numerical values at
the nodes involving the Neumann boundary condition can be calculated
using Eq. (II.2a) in Table II. Again, substituting the expressions of
the polynomial coefficients obtained for Eq. (34) Eq. (II.2a) yields the
following equation for the boundary points.

$$f_{i+1,j} = \left[\frac{4}{3} - \frac{2k^2}{9h^2} \right] f_{i,j} + \left[-\frac{1}{3} + \frac{2k^2}{9h^2} \right] f_{i-1,j}$$

$$+ \frac{k^2}{9h^2} (f_{i,j+1} + f_{i,j-1} - f_{i-1,j+1} - f_{i-1,j-1})$$

$$: \; i=N-1 \text{ and } j=2,3,\ldots,M-1 \tag{35}$$

For k=h, Eq. (35) simplifies to Eq. (22) of Manohar and Stephenson [4].

The numerical solutions at the interior grid points were obtained first
by solving Eqs. (31) and (34) iteratively using the Gauss-Seidel method.
Then, the numerical solutions for the grid points along the Neumann
boundary were calculated directly from Eq. (35). In the Gauss-Seidel
method the discrete function values $f_{i-1,j}$, $f_{i-1,j-1}$, $f_{i,j-1}$ and
$f_{i+1,j-1}$ have been assigned the updated values just calculated during an
iteration step while the previously calculated values have been used for
the remaining discrete function values. Numerical calculations have
been carried out using a uniform grid system for N=M=7. A local error
defined as the difference between the analytical and the numerical
solutions was calculated at the center of the domain for various β
values. In Fig. 3, the results obtained from the finite analytic method
are compared with those reported by Ramadhyani and Patankar[15]. This
comparison reveals that the finite analytic method results in
significantly less errors than the Galerkin, control volume, and the 5-
point finite difference methods.

CONCLUSIONS

The finite analytic method provides convenient means of deriving higher
order difference methods using fewer discrete point values. Direct
implementation of Neumann and mixed type boundary conditions in this
method helps achieve accurate solutions. When local analytical
solutions are derived from linearized equations the method of
linearization effects the accuracy of the numerical results.

ACKNOWLEDGMENT

The support from the School of Petroleum and Geological Engineering and
the College of Engineering Computing facilities at the University of
Oklahoma is appreciated.

REFERENCES

1. Chen, C.J., and Li, P., The Finite Analytic Method for Steady-State
and Unsteady Heat Transfer Problems, 19th ASME/AIChE U.S. Nat. Heat
Trans. Conf., Orlando, FL, ASME Paper No. 80-HT-86, July 27-30, 1980.

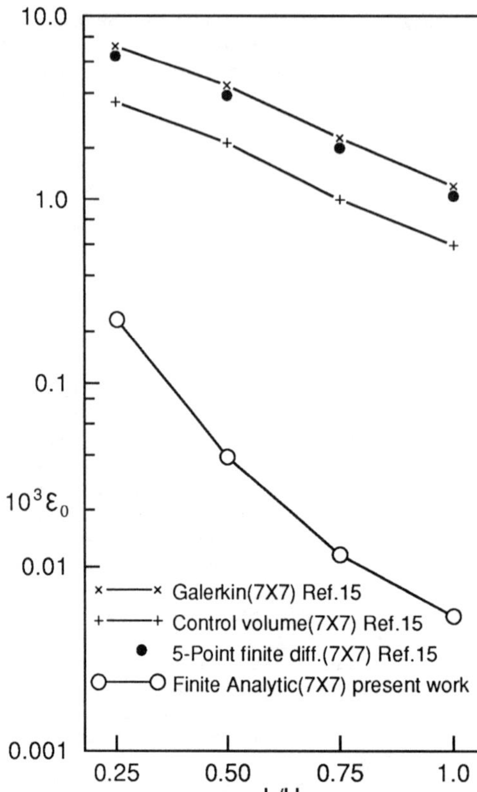

FIGURE 3. The center-point error for
the second problem

2. Chen, C.J., and Chen, H.C., Finite Analytic Numerical Method for
Unsteady Two-dimensional Navier-Stokes Equation, J. Comp. Phys., Vol.
53, No. 2, p. 209-226, Feb. 1984.

3. Hwang, J.C., et. al., Finite Analytic Numerical Solution for Two-
Dimensional Groundwater Solute Transport, Water Res. Research, Vol. 21,
No. 9, p. 1354-1360, Sept. 1985.

4. Manohar, R., and Stephenson, J.W., Optimal Finite Analytic Methods,
J. Heat Transfer, Vol. 104, p. 432-437, 1982.

5. Smith, T.F. and Severin, S.S., Development of the Finite Analytic
with Radiation Method, Trans. ASME J. Heat Transfer, Vol. 107, pp. 735-
737, August 1985.

6. Iqbal, G.M. and Civan, F. Reservoir Simulation Based on the Finite
Analytic Method, SPE 15625, Proc. 61st Annual Technical Conference and
Exhibition of the Society of Petroleum Engineers, New Orleans, LA,
October 5-8, 1986, 8 pages.

7. Civan, F. and Sliepcevich, C.M., Convenient Formulations for Convection/Diffusion Transport, Chem. Engr. Sci., Vol. 40, No. 10, pp. 1973-1974, 1985.

8. El-Halafawy, F.Z. and Eissa, M., Software for the Frobenius Method for the solution of nonlinear differential equations, Appl. Math. Modelling, Vol. 11, pp. 229-232, 1987.

9. Sundaram, N. and Wankat, P.C., Pressure Drop Effects in the Pressurization and Blowdown Steps of Pressure Swing Adsorption, Chemical Engineering Science, Vol. 43, No. 1, pp. 123-129, 1988.

10. Forbes, G.W., Truncation and Manipulation of Multivariate Power Series, J. Computational and Applied Mathematics, Vol. 15, pp. 27-36, 1986.

11. Bellman, R. and Adomian, G., Partial Differential Equations - New Methods for Their Treatment and Solution, p. 133-134, D. Reidel Publishing Company, Boston, 1985.

12. Civan, F. and Sliepcevich, C.M., Solution of the Poisson Equation by Differential Quadrature, International J. for Numerical Methods in Engineering, Vol. 19, pp. 711-724, 1983.

13. Iyengar, S.R.K., and Manohar, R., High Order Difference Methods for Heat Equation in Polar Cylindrical Coordinates, J. Computational Physics, Vol. 77, No. 2 pp. 425-438, 1988.

14. Manohar, R., Iyengar, S.R.K., and Krishnaiah, U.A., High Order Difference Methods for Linear Variable Coefficient Parabolic Equation, J. Computational Physics, Vol. 77, No. 2, pp. 513-523, 1988.

15. Ramadhyani, S. and Patankar, S.V., Solution of the Poisson Equation: Comparison of the Galerkin and Control-Volume Methods, Int. J. Num. Meth. Engng., Vol. 15, pp. 1395-1402, 1980.

Quadrature and Cubature Methods for Numerical Solution of Integro-Differential Equations

FARUK CIVAN
School of Petroleum and Geological Engineering
University of Oklahoma
Energy Center, T301
Norman, Oklahoma 73019, USA

ABSTRACT

Numerical solution of integro-differential equations by the quadrature
and cubature methods is illustrated by solving the Buckley-Leverett
problem for water flooding of naturally fractured reservoirs bearing
oil.

INTRODUCTION

Integro-differential equations are encountered frequently in the
modeling of processes involving source/sink, delay, and memory effects
or are obtained as a result of mathematical manipulation of modeling
equations. In this paper, the quadrature and cubature rules are used as
practical means of developing algebraic representations of integro-
differential equations for which numerical solution methods are well
developed.

The origin of the quadrature and cubature rules is the Gaussian
quadrature used as an algebraic rule for numerical integration. The
derivation, application, and error estimation of integral quadrature and
cubature are presented in detail by Engels[1]. Bellman et al.[2]
suggested the differential quadrature method for rapid solution of
nonlinear differential equations. Civan and Sliepcevich[3] generalized
the differential quadrature method for multi-dimensional problems.
Civan[4] proposed the differential cubature method as an accurate
alternative method to the differential quadrature method when dealing
with multi-dimensional differential equations. These methods provide
practical means of deriving higher order accurate numerical
approximation schemes for integrals and differentials. They are based on
fitting a polynomial to discrete function values and therefore, they are
subject to the limitations of the polynomial fit. In general, as the
order of polynomial increases the accuracy of numerical solutions first
increases, goes through a maximum, and then decreases due to the ill-
conditioning of the coefficient matrix of the resulting system of
algebraic equations. The best performance appears to be within the
orders of seven through eleven using equally spaced grid points.

The solution of integro-differential equations having one independent
variable by the quadrature method has been demonstrated by Civan and

Sliepcevich[5]. In this paper the solutions of the Buckley-Leverett
problem which is a two variable integro-differential equation are
presented using the quadrature and cubature approximations and compared
with an analytical solution.

QUADRATURE AND CUBATURE

The details of the generalized quadrature and cubature rules for
numerical solution of differential, integral, and integro-differential
type equations are presented by Civan[4]. Briefly, quadrature is an
algebraic rule expressing a linear transformation of a continuous
function as a weighted linear sum of discrete function values in one
variable. Thus, the quadrature approximation at the ith discrete point
is given by

$$L\{f(x)\}_i = \overset{\sim}{\sum_{j=1}^{n}} w_{ij}f(x_j) \tag{1}$$

In Eq.(1) L denotes a linear operator, such as an integration or
differentiation applied to a function f(x), where x is the independent
variable, i is the index for the discrete points shown in Fig. 1.a, n is
the total number of discrete points considered, and w_{ij} are the
weighting coefficients obtained by the solution of the following moment
equations [Engels, 1]

$$\sum_{j=1}^{n} x_j^v w_{ij} = L\{x^v\}_i \quad ; \quad v=0,1,\ldots,n-1 \quad \text{and} \quad i=1,2,\ldots,n \tag{2}$$

The coefficients (x^v) of Eq.(2) which are the monomials of polynomials
(see Table 1) compose a Vandermonde matrix. Thus, a unique solution of
Eq.(2) can be obtained analytically as described by Hamming[6] or
numerically using the method by Björk and Preyra[7]. The quadrature
error is given by [Civan, 4]

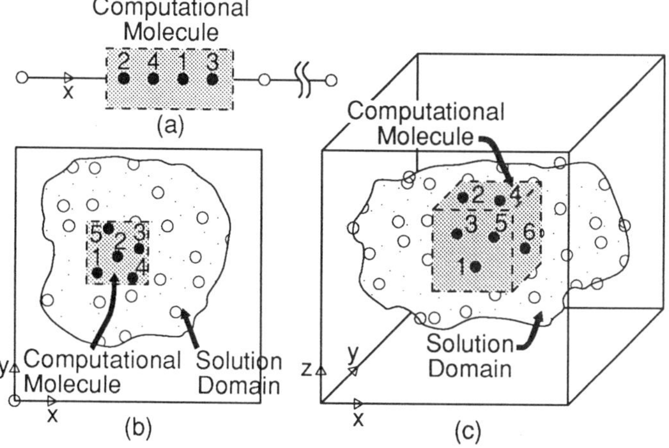

FIGURE 1. Arbitrary one-dimensional indexing of grid points
   in (a) one, (b) two, and (c) three dimensional domains

TABLE 1. Typical monomials for one, two, and three variable functions.

Number of Variables	Type of Monomials	Number of Monomials
1	v = 0, 1, 2, 3, • • • • • • , (k-1) 1 X X^2 X^3 • • • • • • X^{k-1}	m = v+1
2	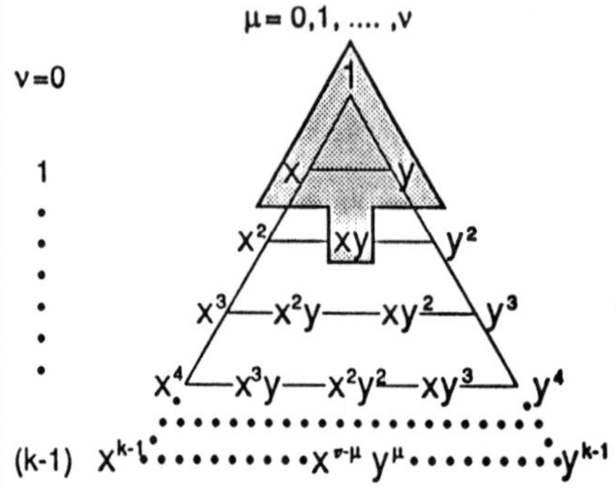	m = (v+1)(v+2)/2 1 3 6 10 15 k(k+1)/2
3	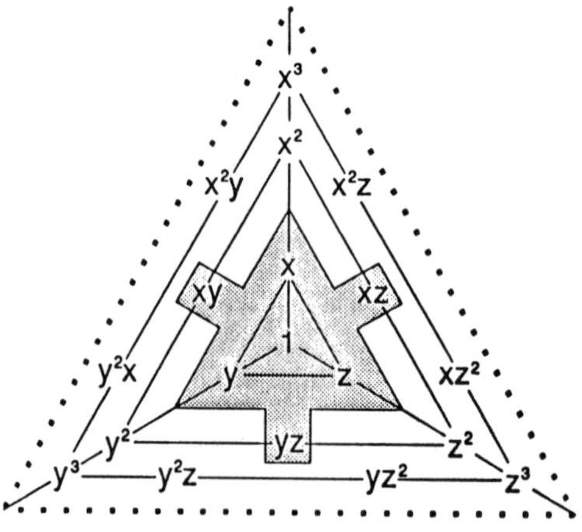	$m = \frac{k(k+1)}{2} + (k-1)^2$

$$(E_n f)_i = [(L\{x^n\})_i - \sum_{j=1}^{n} w_{ij} x_j^n] \frac{f^{(n)}(\tilde{x})}{n!} \tag{3}$$

where x is a value of x in the domain of [-1,1] selected according to the mean-value theorem.

Although the quadrature rule is applicable to multi-dimensional problems as shown by Bellman et al.[2] and Civan and Sliepcevich[3], it works best for one-dimensional problems. For multi-dimensional problems, the cubature rule yields more accurate results. For two dimensional problems the cubature rule is defined by Engels[1] as

$$L\{f(x,y)\}_i \; \tilde{=} \; \sum_{j=1}^{n} w_{ij} f(x_j, y_j) \tag{4}$$

where i denotes the one dimensional indexing of arbitrarily sequenced grid points for the two dimensional solution domain sketched in Fig. 1.b. The cubature weights, w_{ij}, are determined by solving the moment equations given by

$$\sum_{j=1}^{n} x_j^{\nu-\mu} y_j^{\mu} w_{ij} = L\{x^{\nu-\mu} y^{\mu}\}_i \quad ; \quad \mu=0,1,\ldots,\nu; \; \nu=0,1,\ldots,k-1;$$

$$\text{and } i=1,2,\ldots,n \tag{5}$$

To obtain a unique solution of Eq.(5) n-monomials, $x^{\nu-\mu} y^{\mu}$, are selected from the set of monomials, $x^{\nu-\mu} y^{\mu}$: $\mu=0,1,\ldots,\nu$ and $\nu=0,1,\ldots,k-1$, shown in Table 1.

For three dimensional problems the cubature rule is given by Civan[4] as

$$L\{f(x,y,z)\}_i \; \tilde{=} \; \sum_{j=1}^{n} w_{ij} f(x_j, y_j, z_j) \tag{6}$$

where i denotes the one dimensional indexing of arbitrarily sequenced grid points for the three dimensional solution domain sketched in Fig. 1.c. The cubature coefficients, w_{ij}, are calculated by solving the following moment equations.

$$\sum_{j=1}^{n} x_j^{\nu-\mu} y_j^{\mu-\lambda} z_j^{\lambda} w_{ij} = L\{x^{\nu-\mu} y^{\mu-\lambda} z^{\lambda}\}_i \; : \; \lambda=0,1,\ldots,\mu; \; \mu=0,1,\ldots,\nu;$$

$$\nu=0,1,\ldots,k-1; \text{ and } i=1,2,\ldots,n \tag{7}$$

Again, a unique solution of Eq.(7) can be obtained using selected monomials, $x^{\nu-\mu} y^{\mu-\lambda} z^{\lambda}$, shown in Table 1.

The error estimation expressions for the cubature rules are complicated and are not practical. For example, see Engels[1] for the error estimation for numerical integration of two dimensional problems.

WEIGHTING COEFFICIENTS

In this section, the determination of the quadrature and cubature weighting coefficients are illustrated by simple examples.

<u>Quadrature Weights</u>

As a first example, let us consider a computational molecule consisting of three grid points spaced equally by a distance of $\Delta x = h$ as shown in

Fig. 2a and determine the weights for a first order derivative and a single integral of a one variable function. Assuming the mid point as the reference, the coordinates of these three points are given by

$$x_1=-h, \quad x_2=0, \quad x_3=h \tag{8a,b,c}$$

On the other hand, the moment equations given by Eq.(2) can be written in matrix form as

$$
\begin{matrix} v=0 \\ 1 \\ \cdot \\ \cdot \\ (n-1) \end{matrix}
\quad j=1\ 2\ \dots\ n
$$

$$
\begin{bmatrix} & & \\ & x_j^{\,v} & \\ & & \end{bmatrix}
\begin{bmatrix} w_{i1} \\ w_{i2} \\ \cdot \\ \cdot \\ w_{in} \end{bmatrix}
=
\begin{bmatrix} \\ L\{x^v\}_i \\ \\ \end{bmatrix}
\tag{9}
$$

For a differential quadrature approximating a first order derivative we have

$$L=d/dx \quad \text{and} \quad L\{x^v\} = vx^{v-1} \tag{10a,b}$$

Therefore, substituting Eq.(10) and using the monomials 1, x, x^2 shown in the shaded area of Table 1, Eq.(9) becomes

$$
\begin{bmatrix} 1 & 1 & 1 \\ x_1 & x_2 & x_3 \\ x_1^2 & x_2^2 & x_3^2 \end{bmatrix}
\begin{bmatrix} w_{i1} \\ w_{i2} \\ w_{i3} \end{bmatrix}
=
\begin{bmatrix} 0 \\ 1 \\ 2x_i \end{bmatrix}
\quad ; \quad i = 1,2,3
\tag{11}
$$

Substituting Eq.(8) and solving Eq.(11) yields the following quadrature weights at point x_1 as

$$[w_{11}w_{12}w_{13}]^T = (1/2h)\,[-3 \quad 4 \quad -1]^T \tag{12}$$

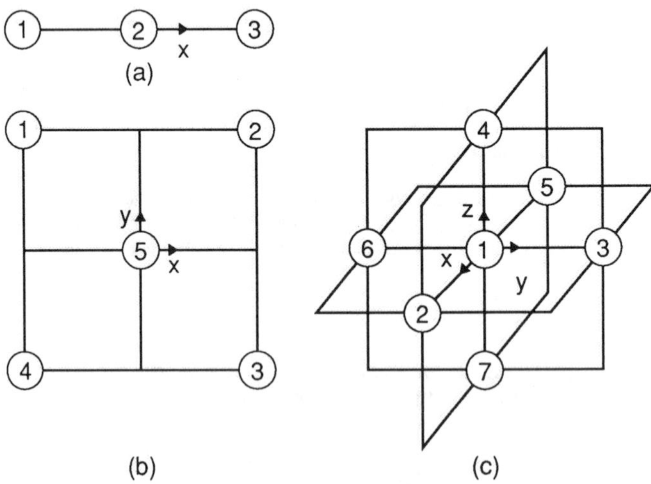

FIGURE 2. Typical computational molecules for (a) one, (b) two, and (c) three variable functions

and substituting Eq.(10a) and (12) into Eq.(1) yields the following well-known forward difference formula

$$df(x_1)/dx = [-3f(x_1) + 4f(x_2) - f(x_3)]/(2h) \qquad (13)$$

The central and backward difference formulae can be derived similarly. For an integral quadrature approximating a single integral over, for example, an interval of $(-h,x)$

$$L = \int_{-h}^{x} dx \quad \text{and} \quad L\{x^v\} = [x^{v+1} - (-h)^{v+1}]/(v+1) \qquad (14a,b)$$

The weighting coefficients can be derived by the similar procedure.

<u>Cubature Weights</u>

The second example deals with the cubature weights for a two-dimensional function. For this case, consider, as an example, the computational molecule shown in Fig. 2b having equally spaced points in both x and y directions by $\Delta x = \Delta y = h$. If the center point is selected as the reference point; then

$$x_1 = -h, \; y_1 = h; \; x_2 = h, \; y_2 = h; \; x_3 = h, \; y_3 = -h; \; x_4 = -h, \; y_4 = -h$$

$$\ldots (15a,b,c,d)$$

On the other hand, Eq.(5) can be written as

$$j=1 \; 2 \; \ldots n$$

One row for
each set of
v and μ of
selected
monomials

$$\begin{bmatrix} x_j{}^{v-\mu}y_j \end{bmatrix} \begin{bmatrix} w_{i1} \\ w_{i2} \\ . \\ . \\ w_{in} \end{bmatrix} = \begin{bmatrix} L\{x^{v-\mu}y^{\mu}\}_i \end{bmatrix} \qquad (16)$$

Since four nodal points are considered only four monomials must be selected for a unique solution of Eq.(16). This requires fitting a k-1=2nd order polynomial. The complete number of monomials is $k(k+1)/2 = 3(4)/2 = 6$. Thus, $k(k+1)/2-n = 6-4 = 2$ monomials of 2nd order need to be deleted. However, any arbitrary choice, such as deleting (x^2,xy) or (xy,y^2) does not lead to successful solutions for the weighting coefficients. In the present case, deleting (x^2,y^2) yields a successful solution. Thus, consider the monominals $1, x, y, xy$ as shown inside the shaded area for two variable functions in Table 1. If the differential cubature weights for the first order derivative at the center point O is required, then

$$L = \partial/\partial x \quad \text{and} \quad L\{x^{v-\mu}y^{\mu}\} = (v-\mu)x^{v-\mu-1}y^{\mu} \qquad (17a,b)$$

and Eq.(16) becomes

$$\begin{matrix} v=0, \; \mu=0 \\ v=1, \; \mu=0 \\ v=1, \; \mu=1 \\ v=2, \; \mu=1 \end{matrix} \begin{bmatrix} 1 & 1 & 1 & 1 \\ x_1 & x_2 & x_3 & x_4 \\ y_1 & y_2 & y_3 & y_4 \\ x_1y_1 & x_2y_2 & x_3y_3 & x_4y_4 \end{bmatrix} \begin{bmatrix} w_{i1} \\ w_{i2} \\ w_{i3} \\ w_{i4} \end{bmatrix} = \begin{bmatrix} 0 \\ 1 \\ 0 \\ y_i \end{bmatrix} \qquad (18)$$

The solution of Eq.(18) at point i=5($x_5=0$, $y_5=0$) and using Eq.(15) yields the following weighting coefficients.

$$[w_{51}w_{52}w_{53}w_{54}]^T = (1/4h)[-1 \ 1 \ 1 \ -1]^T \tag{19}$$

A substitution of Eq. (17a) and Eq.(19) into Eq.(4) yields the following differential cubature approximation

$$\partial f(x_5=0,y_5=0)/\partial x = [-f(x_1,y_1) + f(x_2,y_2) + f(x_3,y_3) - f(x_4,y_4)]/(4h)$$

$$\ldots(20)$$

Eq.(20) is a well known central difference approximation formula.

Next, assume that the integral cubature weighting coefficients are required, as an example, for

$$L = \int_{-h}^{x} \int_{-h}^{y} dydx \tag{21a}$$

Then,

$$L\{x^{\nu-\mu}y^\mu\} = \int_{-h}^{x} \int_{-h}^{y} x^{\nu-\mu}y^\mu dydx = [x^{\nu-\mu+1} - (-h)^{\nu-\mu+1}]$$

$$[y^{\mu+1} - (-h)^{\mu+1}]/[(\nu-\mu+1)(\mu+1)] \tag{21b}$$

and the procedure similar to differential cubature yields the coefficients and the integral cubature formula.

The final example deals with three variable cubature weights. Consider the computational molecule shown Fig. 2c. If all the grid points are equally spaced in x, y, and z-directions so that $\Delta x=\Delta y=\Delta z=h$ and point 1 is selected as the reference, then

$$x_1 = 0, \ y_1 = 0, \ z_1 = 0; \ x_2 = h, \ y_2 = 0, \ z_2 = 0; \ \ldots;$$

$$x_7 = 0, \ y_7 = 0, \ z_7 = -h \tag{22a,b,\ldots}$$

Since there are seven nodal points, we select seven monomials: 1, x, y, z, xy, xz, yz; as shown in Table 1. Eq.(7) can be written in matrix form as

$$
\begin{array}{l}
\text{One row for} \\
\text{each set of} \\
\nu, \ \mu \ \text{and} \ \lambda \\
\text{of selected} \\
\text{monomials}
\end{array}
\begin{bmatrix} & j=1 \ 2 \ \ldots.n \\ \\ x_j^{\nu-\mu}y_j^{\mu-\lambda}z_j^{\lambda} \\ \\ \end{bmatrix}
\begin{bmatrix} w_{i1} \\ w_{i2} \\ . \\ . \\ . \\ w_{in} \end{bmatrix}
=
\begin{bmatrix} L\{x^{\nu-\mu}y^{\mu-\lambda}z^{\lambda}\}_i \end{bmatrix} \tag{23}
$$

For the first order derivative with respect to x

$$L = \partial/\partial x \quad \text{and} \quad L\{x^{\nu-\mu}y^{\mu-\lambda}z^{\lambda}\} = (\nu-\mu)x^{\nu-\mu-1}y^{\mu-\lambda}z^{\lambda} \tag{24a,b}$$

Hence, for the monomials selected for the problem, Eq.(23) becomes

$$
\begin{array}{l}
\nu=0,\ \mu=0,\ \lambda=0 \\
\nu=1,\ \mu=0,\ \lambda=0 \\
\nu=1,\ \mu=1,\ \lambda=0 \\
\nu=1,\ \mu=1,\ \lambda=1 \\
\nu=2,\ \mu=1,\ \lambda=0 \\
\nu=2,\ \mu=1,\ \lambda=1 \\
\nu=2,\ \mu=2,\ \lambda=1
\end{array}
\begin{bmatrix}
1 & 1 & \ldots & 1 \\
x_1 & x_2 & \ldots & x_7 \\
y_1 & y_2 & \ldots & y_7 \\
z_1 & z_2 & \ldots & z_7 \\
x_1y_1 & x_2y_2 & \ldots & x_7y_7 \\
x_1z_1 & x_2z_2 & \ldots & x_7z_7 \\
y_1z_1 & y_2z_2 & \ldots & y_7z_7
\end{bmatrix}
\begin{bmatrix}
w_{i1} \\ w_{i2} \\ \cdot \\ \cdot \\ \cdot \\ \cdot \\ w_{i7}
\end{bmatrix}
=
\begin{bmatrix}
0 \\ 1 \\ 0 \\ 0 \\ y_i \\ z_i \\ 0
\end{bmatrix}
\qquad (25)
$$

A solution of Eq.(25) at i=1 and using Eq.(22) yields the values of the weighting coefficients which are then substituted into Eq.(6) to yield a differential cubature formula as follows

$$
\partial f(x_1,y_1,z_1)/\partial x = [-6f(x_1,y_1,z_1) + f(x_2,y_2,z_2) + \ldots + f(x_7,y_7,z_7)]/h
$$

$$
\ldots(26)
$$

Integral quadrature formulae can be derived similarly.

APPLICATION

The following integro-differential equation given by deSwaan[8] for the Buckley-Leverett flow in oil bearing fractured reservoirs was selected for the present example because its solution is of continuing interest. The problem is to solve the following equation:

$$
- u \frac{\partial f}{\partial x} = \frac{\partial s}{\partial t} + \frac{R_\infty \lambda}{\phi_f} \int_0^t e^{-\lambda(t-\tau)} \frac{\partial s}{\partial \tau} d\tau \quad ; \quad 0<x\le L \ , \ t>0 \qquad (27)
$$

subject to the following initial and boundary conditions, respectively

$$
s = s_1 \quad ; \quad 0 \le x \le L \ , \ t = 0 \qquad (28)
$$

$$
s = s_2 \quad ; \quad x = 0 \ , \ t > 0 \qquad (29)
$$

In Eqs (27)-(29) u, R_∞ , λ, ϕ_f, L, s_1 and s_2 are some given values and f is a function of s.

To facilitate comparison with the analytical solution given by Kazemi et al.[9] the case for u=168 ft^{3}day^{-1}, L=1,000 ft, λ=0.1 day^{-1}, R_∞=0.08, ϕ_f=0.001, f=s, s_1=0, and s_2=1 is considered for numerical solution. Using these data, Eqs. (27)-(29) can be expressed in the following dimensionless form:

$$
-\alpha \frac{\partial S}{\partial X} = \frac{\partial S}{\partial T} - \int_0^T S dT + \beta S \ ; \ 0<X\le 1 \ , \ T>0 \qquad (30)
$$

$$
S = 0 \ ; \ 0\le X\le 1 \ , \ T=0 \qquad (31)
$$

$$
S = e^{T/\gamma} \ ; \ X=0 \ , \ T>0 \qquad (32)
$$

In Eqs. (30)-(32) the dimensionless quantities are defined as follows:

$$X = x/L \tag{33}$$

$$S = se^{\lambda t} \tag{34}$$

$$T = \lambda \gamma t \tag{35}$$

$$\alpha = u/(L\lambda\gamma) \tag{36}$$

$$\beta = \gamma - 1/\gamma \tag{37}$$

$$\gamma = (R_\infty/\phi_f)^{1/2} \tag{38}$$

For numerical solution, the overall time domain is divided into a series of time periods separated at discrete times, $T^k:k=0,1,2,\ldots,m,(m+1),\ldots,$ as sketched in Fig. 3. Next Eq.(30) is rearranged as

$$\alpha \frac{\partial S}{\partial X} + \beta S + \frac{\partial S}{\partial T} - \int_{T^n}^{T} S dT = Q \quad : \quad 0<X\leq1 \ , \quad T^m<T<T^{m+1} \tag{39}$$

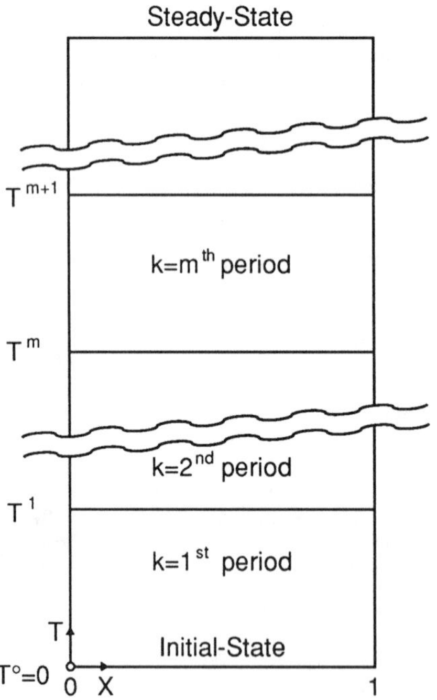

FIGURE 3. Periods considered for numerical solution of the Buckley-Leverett problem

in which

$$Q = \sum_{k=1}^{m} \int_{T^{k-1}}^{T^k} S dT \tag{40}$$

The solution at T^m time obtained from the previous period is used as the initial condition

$$S = S^m \quad ; \quad 0 \leq X \leq 1 \ , \ T = T^m \tag{41}$$

The boundary condition is given by

$$S = e^{T/\gamma} \quad ; \quad X=0 \ , \ T^m < T < T^{m+1} \tag{42}$$

Eqs. (39)-(42) are solved for consecutive time periods until the solution reaches a steady profile. Then the solution of the original problem, Eqs. (27)-(29), is obtained using Eqs. (33)-(35). In the following, the numerical solutions by the quadrature and cubature methods are presented.

Quadrature Formulation

Eq.(39) can be written as

$$L_x S + L_t S = Q \tag{43}$$

in which the linear operators L_x and L_t are defined, respectively, by

$$L_x = \alpha \partial / \partial X + \beta \tag{44}$$

and

$$L_t = \partial / \partial T - \int_{T^n}^{T} dT \tag{45}$$

Let a_{ij} and b_{ij} denote the quadrature weighting coefficients determined from Eq.(2) for the linear operators, L_x and L_t, respectively. Then, applying the method of quadrature according to Civan and Sliepcevich[5] for the grid system shown in Fig. 4 Eq.(43) is represented by the following algebraic equations:

$$\sum_{k=1}^{N^x} b_{ik} S_{kj} + \sum_{k=1}^{N^t} a_{jk} S_{ik} = Q_i ; \ i=2,3,\ldots,N^x \ \text{and} \ j=2,3,\ldots,N^t \tag{46}$$

Separating the initial and boundary point values Eq.(46) can be written as, for convenience

$$\sum_{p=2}^{N^x} \sum_{q=2}^{N^t} \{\delta_{jq} b_{ip} + \delta_{ip} a_{jq}\} S_{pq} = Q_i - (b_{i1} S_{1j} + a_{j1} S_{i1}) \ ;$$

$$i=2,3,\ldots,N^x \quad \text{and} \quad j=2,3,\ldots,N^t \tag{47}$$

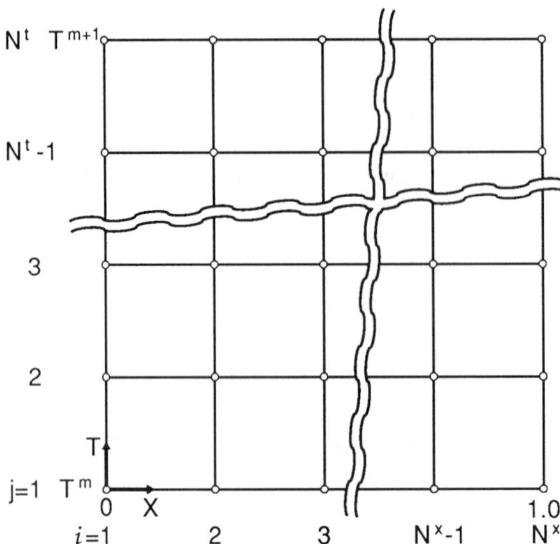

FIGURE 4. The grid system over a typical
period used for the quadrature solution

δ_{ij} is the Kronecker delta which is equal to 1 for i=j and 0 otherwise.
In Eq.(47), Q_i is given by approximating Eq.(40) by the quadrature as

$$Q_i = \sum_{k=1}^{m} \left\{ \sum_{r=1}^{N^t} c_{N,r}^t \; S_{ir} \right\}^k \tag{48}$$

in which $c_{N,r}^t$ denotes the quadrature weighting coefficients determined
by Eq.(2) for the linear operator $L = \int_{T^{k-1}}^{T^k} dT$ appearing in Eq. (40).

The initial conditions are

$$S_{i1} = S_{i1}^m \quad ; \quad i=1,2,\ldots,N^x \quad , \quad j=1 \tag{49}$$

The boundary values are given by

$$S_{1j} = e^{T/\gamma} \quad ; \quad i=1 \quad , \quad j=1,2,\ldots,N^t \tag{50}$$

Eq.(47) represents $(N^x-1)(N^t-1)$ set of linear algebraic equations which
can be solved by an appropriate numerical method. In the present study
a transpose elimination method according to Wassyng[10] has been used.
This procedure is repeated over the consecutive time periods.

For numerical calculations the grid system for $N^x=11$ and $N^t=11$ was
considered. The dimensionless time period was taken equal to $\gamma=8.944$.
Using $N^t=11$, this corresponds to a dimensionless time step of $\Delta T=0.8944$
or an actual time step of $\Delta t=1$ day according to Eq.(35). Since $N^x=11$
grid points were used, the dimensionless grid spacing is $\Delta X=0.1$. This
corresponds to an actual grid spacing of $\Delta x=100$ feet according to
Eq.(33). In their finite difference solution, Kazemi et al.[9] have
used $\Delta t=0.005$ days, $N^x=50$ grid points, and $\Delta x=20.4$ feet. In

comparison, the quadrature solution used 200 times larger time steps and 4.9 times larger grid spacing than Kazemi et al.[9]. In Fig. 5 the numerical solution obtained by the quadrature method is compared with the finite difference solution and the analytical solution by Kazemi et al.[9]. As can be seen in Fig. 5 the quadrature solution matches the analytic solution better than the finite difference solution. In spite of the very small steps used, the finite difference solution suffers from numerical dispersion. The quadrature solution is accurate and rapid and allows for use of much larger time steps and grid spacing.

<u>Cubature Formulation</u>

For numerical solution, the computational molecule shown in Fig. 6 and having n=61 nodes was selected. Thus, the order of polynomial to be fitted is k-1=10. Since the complete number of monomials is k(k+1)/2=66, k(k+1)/2-n=66-61=5 monomials of the 10th order need to be deleted (see Table 1 for two variable functions.) However, any arbitrary selection of five monomials does not always yield a successful unique solution. In the present case, a unique solution has been obtained by deleting x^{10}, x^9y, x^8y^2, x^7y^3, x^6y^4.

Eq.(39) can be written as

$$LS = Q \tag{51}$$

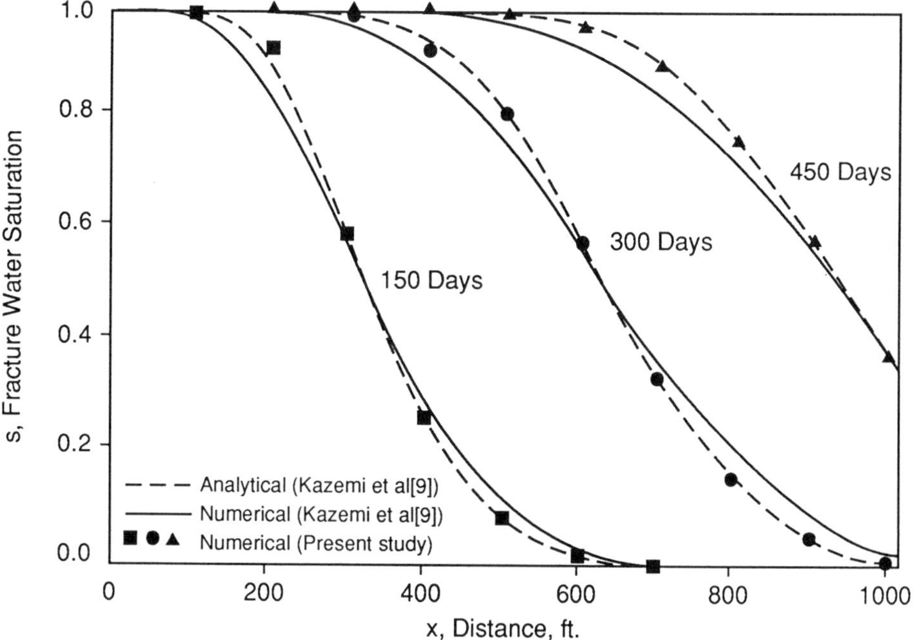

FIGURE 5. Solution of the Buckley-Leverett problem by the quadrature method.

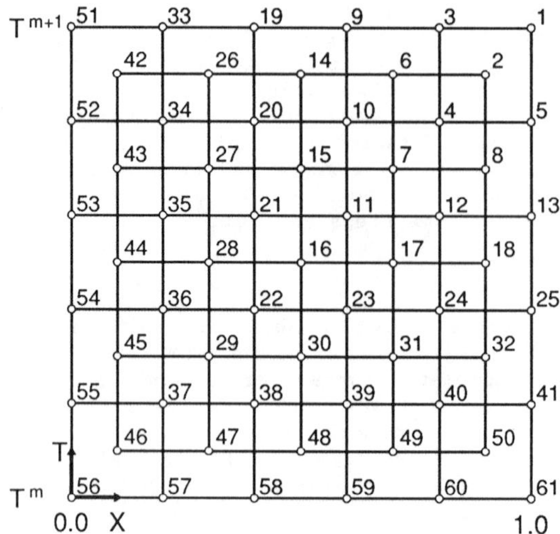

FIGURE 6. The grid system over a typical
period used for the cubature solution.

in which the linear operator is given by

$$L = \alpha \partial/\partial X + \beta + \partial/\partial T - \int_{T^n}^{T} dT \tag{52}$$

Let a_{ij} denote the cubature weighting coefficients obtained by solving Eq.(5). Hence, Eq.(51) can be represented by the following algebraic equation

$$\sum_{j=1}^{61} a_{ij} S_j = Q_i \quad ; \quad i=1,2,\ldots,50 \tag{53}$$

The initial conditions given by Eq.(41) are

$$S_i = S_i^m \quad ; \quad i=56,57,\ldots 61 \tag{54}$$

and the boundary conditions given by Eq.(42) are

$$S_i = e^{T/\gamma} \quad ; \quad i=51,52,\ldots,55 \tag{55}$$

Thus, Eq.(53) can be rearranged as

$$\sum_{j=1}^{50} a_{ij} S_j = Q_i - \sum_{j=51}^{61} a_{ij} S_j \quad ; \quad i=1,2,\ldots,50 \tag{56}$$

In Eq.(56)

$$Q_i = \sum_{k=1}^{m} \left\{ \sum_{j=1}^{61} b_{ij} S_j \right\}^k \tag{57}$$

in which b_{ij} denote the cubature weighting coefficients determined by Eq.(5) for the linear operator $L = \int_{T^{k-1}}^{T^k} dT$ appearing in Eq.(40).

The numerical solutions of the algebraic equations represented by Eq.(56) are obtained using the transpose elimination method according to Wassyng[10]. These solutions are repeated over the consecutive time periods.

In Fig. 7, the numerical solution obtained via the cubature method is compared with the analytic and the finite difference solutions presented by Kazemi et al.[9]. The cubature solution used a dimensionless period of γ=4.472. Based on the grid system shown in Fig. 6, this corresponds to a dimensionless time step of ΔT=0.8944 or an actual time step of Δt=1 day due to Eq.(35). The dimensionless grid spacing is ΔX=0.2 and thus the actual grid spacing is Δx=200 feet which is twice as large as the grid spacing used for the quadrature solution. For each period, the quadrature solution has used 121 grid points and the cubature method 61 grid points for comparable accuracy (see Figs. 5 and 7). Since the cubature method used half the number of grid points the cubature method is about twice as fast as the quadrature method. As can be seen from Fig. 7, the cubature solution fits better to the analytic solution than the finite difference solution of Kazemi et al.[9] who used Δt=0.005 day time steps and Δx=20.4 feet grid spacing. Thus, the cubature method has used 200 times larger time steps and 9.8 times larger grid

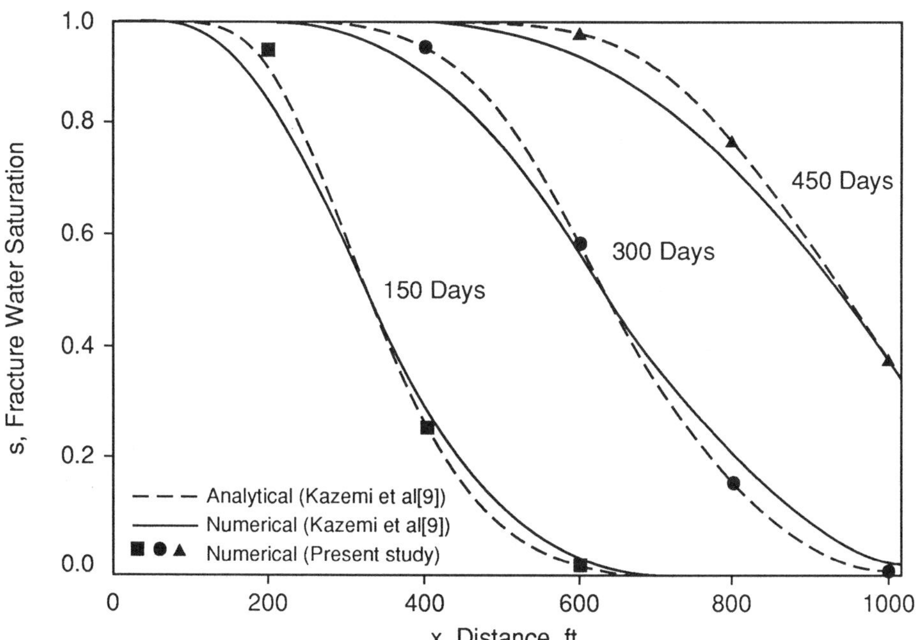

FIGURE 7. Solution of the Buckley-Leverett problem
by the cubature method.

spacing as compared to Kazemi et al.[9]. Hence, the cubature solution
should be proportionally faster than the finite difference solution for
comparable accuracy.

CONCLUSIONS

The comparisons of the numerical solutions generated by the quadrature
and the cubature methods with the exact analytic solution given by
Kazemi et al.[9] indicated that both methods yield rapid and accurate
solutions. For comparable accuracy these methods allow for use of much
larger time steps and grid spacing than the finite difference method.

ACKNOWLEDGMENT

The support from the School of Petroleum and Geological Engineering and
the College of Engineering computing facilities at the University of
Oklahoma is appreciated.

REFERENCES

1. Engels, H., Numerical Quadrature and Cubature, Academic Press, New
York, 1980.

2. Bellman, R., Kashef, B.G., and Casti, J., "Differential Quadrature:
A Technique for the rapid solution of nonlinear partial differential
equations," J. Comp. Phys., Vol. 10, 1972, pp. 40-52.

3. Civan, F. and Sliepcevich, C.M., "Differential Quadrature for Multi-
dimensional Problems," J. Math. Anal. Appl., Vol. 101, No. 2, 1984, pp.
423-443.

4. Civan, F., Differential Cubature for Multi-Dimensional Problems,
Modeling and Simulation, Proceedings of the Twentieth Annual Pittsburgh
Conference, May 4-5, 1989, Vogt, W.G. and Mickle, M.H., Eds., Vol. 20,
Part 5, pp. 1843-1847, Instrument Society of America, North Carolina,
1989.

5. Civan, F. and Sliepcevich, C.M., "Solving Integro-Differential
Equations by the Quadrature Method," published in Payne, F.R. et al.
(Eds.), Integral Methods in Science and Engineering, Hemisphere Publ.
Co., New York, 1986, pp. 106-113.

6. Hamming, R.W., Numerical Methods for Scientists and Engineers, 2nd
ed., McGraw-Hill, New York, 1973.

7. Bjork, A. and Pereyra, V., "Solution of Vandermonde Systems of
Equations," Math. Comput., Vol. 24, 1970, pp. 893-903.

8. deSwaan, A., Theory of Waterflooding in Fractured Reservoirs, Soc.
Pet. Eng. J., pp. 117-122, April 1978.

9.   Kazemi, H., Gilman, J.R., and El-Sharkaway, A.M., Analytical and
Numerical Solution of Oil Recovery From Fractured Reservoirs Using
Empirical Transfer Functions, SPE #19849 paper, Proceedings of the 64th
Annual Technical Conference and Exhibition of the Soc. Pet. Eng. held in
San Antonio, TX, October 8-11, 1989, pp. 827-842.

10. Wassyng, A., Solving Ax=b: A method with reduced storage
requirements, SIAM J. Numer. Anal., 19(1), 1982, pp. 197-204.

11. Abramowitz, M. and Stegun, I.A., eds., Handbook of Mathematical
Functions with Formulas, Graphs, and Mathematical Tables, p. 71, Ninth
Printing, National Bureau of Standards, Applied Mathematics Series-55,
U.S. Government Printing Office, Washington, D.C., November 1970.

Numerical Solution of Second Order Differential Equations

WILLIAM SQUIRE
Department of Mechanical and Aerospace Engineering
West Virginia University
Morgantown, West Virginia 26506, USA

ABSTRACT

A modification of Devogelaere's method for second order differential equations in which the first derivative does not appear is presented. Equations with the derivative present are treated as a special case of two simultaneous equations. The Runge-Kutta Nystrom method is used to start the integration.

INTRODUCTION

Computer programs for the numerical solution of ordinary differential equations subject to initial conditions are almost always written for a set of first order equations. Such programs are general since any explicit system of differential equations can be put in the required form by a straightforward procedure described in almost all texts on differential equations. However, it would often be convenient to have programs available implementing special methods for forms which arise frequently. The extensive use of microcomputers now makes it possible for small research groups and even individuals to maintain libraries of such programs.

This paper considers some methods for the numerical integration of second order equation, a type which often arises in physical applications. The presentation and examples are for a single second order equation but the extension to a system is straightforward.

BACKGROUND

The Runge-Kutta method was extended to handle second order equations directly by Nystrom (1925) and to higher order equations by Zurmuhl [1940, 1955]. This work is summarized by Collatz [1966]. More recently there has been some interest, summarized by Brankin [1989], in very high order methods for second order equations, but this paper will only consider fourth order methods for second order differential equations.

Devogelaere [1955] developed a fourth order method for the second order form:

$$X'' = F(X, t) \tag{1}$$

which while fourth order (in accuracy) required only two evaluations of F per step compared to the three required by Nystrom's method, Scraton [1964] has shown the relation of Devogelaere's method to Radeau quadrature and how higher order procedures can be generated. However the decrease in number of function evaluations compared to order of the error is not as favorable.

Runge-Kutta-Nystrom Methods

As this method is not widely known and we will be using it as a starting method we present it for convenient reference. The equation

$$X''=F(X,V,t) \quad V=X'$$ (2)

is advanced from t to t+h by

$$K1 = h^2 \ F(X,V,t)/2$$ (3a)

$$K2 = h^2 \ F(x+h/2V+K1/4,V+K1/h,t+h/2)/2$$ (3b)

$$K3 = h^2 F(x+h/2V+K1/4,v+K2/h,t+h/2)/2$$ (3c)

$$K4 = h^2 F(x+hV+K3,V+2K3/h,t+h)/2$$ (3d)

$$X(t+h) = X + hV +(K1+K2+K3)/3$$ (3e)

$$V(t+h) = V +(K1+2K2+2K3+K4)/(3h)$$ (3f)

It is easily seen that if F does not depend on V then $K2 = K3$ so that only three rather than four function evaluations are needed. A listing of a FORTRAN subroutine implementing this procedure is given in Table 1.

TABLE 1

```
C     RUNGE-KUTTA-NYSTROM FOR X''=A(X,V,T)
         SUBROUTINE RKN(FAXVT,H,N,T,X,V)
         DIMENSION T(0:N) ,X(0:N),V(0:N)
         H2=H/2.
         H3=H*3.
         HH=H*H2
         DO 10 I=0,N-1
             XI=X(I)
             TI=T(I)
             VI=V(I)
             HK1=HH*FAXVT(XI,VI,TI)
             HK2=HH*FAXVT(XI+H2*VI+HK1/4.,VI+HK1/H,TI+H2)
             HK3=HH*FAXVT(XI+H2*VI+HK1/4.,VI+HK2/H,TI+H2)
             HK4=HH*FAXVT(XI+H*VI+HK3,VI+HK3/H2,TI+H)
             T(I+1)=TI+H
             X(I+1)=XI+H*VI+(HK1+HK2+HK3)/3
             V(i+1=VI+(HK1+2.*(HK2+HK3)+HK4)/H3
10 CONTINUE
         RETURN
         END
```

Devogelaere's Method

While this method is not self starting, the starting procedure is
relatively simple and it is easy to change the step size during the
integration. An intermediate point is used during the calculation at
which the value of X'', which we will denote by A is computed but X',
denoted by V is not. The value of A is saved for the next step.

The procedure starts with

$$X(t-h/2) = Xa = Xo - h/2 \ Vo \ + h^2/ \ 8 \ Ao \tag{4a}$$

$$Aa \ = F(Xa \ , t-h/2)$$

We then proceed by

$$Xa \ = Xo + h/2 \ Vo \ + h^2/ \ 6 \ (Ao - {}^1/4 \ Aa \ )+ 0(h^4) \tag{5a}$$

$$X_1 = Xo + h \ Vo + h^2/ \ 6 \ (Ao +2Aa) + 0(h^4) \tag{5b}$$

$$A_1 = F(X_1, t \ + \ h) \tag{5c}$$

$$V_1 = Vo + h/6 \ (Ao + 4Aa + A_1) + 0(h^4) \tag{5d}$$

We then update by $t = t + h$, $Xo = X_1$ $Vo = V_1$ and $Ao = A_1$ and repeat the
procedure.   It sould be noted that (5d) is Simpson's rule which is a
Radeau quadrature rule.

NEW METHODS

We first present a slight modification of Devogelaere's method in which
the auxilary point is treated as an ordinary point. This requires a
more elaborate starting procedure as an accurate value of V must be
obtained at the first point and more data is retained for the next step.
The modified method uses only one new function evaluation per step but
this is counterbalanced to a considerable extent by an effectively
larger step size.

We then consider solving the more general form

$$X'' = F \ (X, V, t), \ V = X' \tag{6}$$

by using the additional equation

$$X''' = V'' \ = G(X, V, t) \tag{7}$$

This is then handled as a system of two equations with no derivatives
but with some simplifications compared to the general case so that after
the starting step only one evaluation of G is required for each step.

Modified DeVogelaere Method

Essentially Eq. (5) with X_1 replaced by X_2 and Xa replaced by X_1 and h
replaced by 2h is used. After computing X, V and A at $t + h$ which are

designated by the subscript 1 by some procedure such as Runge-Kutta MYSTROM the integration continues by

$$t = t + h \tag{8a}$$

$$X_2 = X_0 + 2h\,V_0 + h^2/3\,(Ao + 2A_1) \tag{8b}$$

$$A_2 = F(X_2, t) \tag{8c}$$

$$V_2 = V_0 + h/3\,(Ao + 4A_1 + A_2) \tag{8d}$$

update by Xo = X1, X1 = X2 etc and proceeds to the next step. A FORTRAN program implementing the procedure is shown in Table 2.

Equation with First Derivative

The given equation

$$X'' = F(X,\ V,\ t) = A \tag{9}$$

is supplemented by

$$V'' = X''' = G(X,\ V,\ t) = B \tag{10}$$

where

$$G(X,\ V,\ t) = \frac{\partial A}{\partial t} + \frac{\partial A}{\partial X} + A\,\frac{\partial A}{\partial V} \tag{11}$$

TABLE 2

```
C     MODIFIED DEVOGELAERE METHOD FOR X''=A(X,T)
         SUBROUTINE DVND(FAXT,H,N,T,X,V)
         DIMENSION T(O:N),X(O:N),V(O:N)
         H3=H/3
         HT=H*2
C     STARTING PROCEDURE
         AO=FAXT(X(O),T((O)))
         CALL RKND(FAXT,H,1,T,X,V)
         A1+FAXT(X(1),T(1)
C     INTEGRATION CYCLE
         DO 10 I = 2,N
            T(I)=T(I-1)+H
            X(I)=X(I-2)+HT*(V(I-2)+H3*(2*A1+AO))
            A2=FAXT(X(I),T(I))
            V(I)=V(I-2)+H3*(4*A1+A2+AO)
            AO=A1
            A1+A2
10    CONTINUE
      RETURN
      END
```

The integration is started by doing the first step say by the Runge-Kutta-Nystrom method and computing the value of A an B.
It is then continued by

$$X_2 = X_1 + hV_1 + h^2/2 \, A_1 + h^3/24 \, (5B_1 - B_0) \tag{12}$$

$$V_2 = V_0 + 2 \, h \, A_0 + 2/3 \, h^2 \, (2B_1 + B_0) \tag{12b}$$

$$B_2 = G \, (X_2, \, V_2, \, t) \tag{12c}$$

$$A_2 = A_0 + h/3 \, (B_0 + 4B_1 + B_2) \tag{12d}$$

$$V_2 + V_1 + h/2 \, (A_1 + A_2) + h^2/12 \, (B_1 - B_2) \tag{12e}$$

and then updating and proceding to the next step.
Eq/ (12b) is a Taylor series as can be seen by writing

$$X \, (t + h) = X \, (t) + h \, X' \, (t) + h^2/2 \, X'' \, (t) + h^3/6X'' \, (t)$$

$$+ \, h^4/24 \, y'''' \, (t) \tag{13}$$

and approximating $X''''(t)$ by $(X'''(t) - X'''(t-h))/h$. Eq (12c) which is the analog of Eq. (5b) is obtained by a similar argument while (12c) is again Simpson's rule. Eq. (12e) used to improve V_2 is a "quadrature by differentiation" formula given by Lanczos (1956) which has an order $0(h^5)$. A computer routine implementing this procedure is shown in Table 3.

It should be noted that there is more flexibility in this case and that many variations are possible which should be explored more fully as the performance of the procedure depends on the weakest step. Variations such as replacing (12b) with an integration from t to t + h instead of from t - h to t + h were tried but did not improve the accuracy.

NUMERICAL RESULTS

A fourth order method should integrate a differential equation having the solution

$$X = t^4 \tag{14}$$

to the accuracy of the computer arithmetic. This was checked out for

$$X'' = 12 \, t^2 \tag{15}$$

subject to

$$X(o) = X'(o) = 0 \tag{16}$$

TABLE 3

```
C     MODIFIED DEVOGELAERE FOR X''=A(X,V,T)
         SUBROUTINE DVMWD(FAXVT,FBXVT,H,N,T,X,V)
         DIMENSION T(0:N),X(0:N),V(0:N)
         H2=H/2
         H3=H/3
         H6=H/6
         H12=H/12
         HT=2*H
C     STARTING PROCEDURE
               A0=FAXVT(X(0),V(0),T(0))
               B0=FBXVT(X(0),V(0),T(0))
               CALL RKN(FAXVT,H,1,T,X,V)
               A1=FAXVT(X(1),V(1),T(1)
               B1=FBXVT(X(1),V(1),T(1))
C     MAIN LOOP
         DO 10 I=2,N
               T(I)=T(I-1)+H
               X(I)=X(I-1)+H*(V(I-1)+H2*(A1+H12*(5*B1-B0)))
               V(I)=V(I-2)+HT*(A0+h3*(2*B1+B0))
               V(I)=(I-1)+H2*((A2+A1)+H6*(B1-B2))
               A0=A1
               A1=A2
               B0=B1
               B1=B2
10    CONTINUE
      RETURN
      END
```

The original DeVogelaere method failed because the starting step has an
error $O(h^3)$. The three programs presented in this paper worked for both
eq. (15) and for

$$X'' = 12 \sqrt{X} \tag{17}$$

Scraton (1964) compared the original Devoglaere method with
Runge-Kutta-Nystrom

$$X'' = -tX \tag{18}$$

$X(o) = 1$ and $X'(o) = 0$

going from 0 to 3 with h = 0.5. Table 4 shows his results and those
obtained by our modified methods with a step size of 0.25. They are
slightly superior with the same number of evaluations as the original
Devogelaere method and $^2/3$ the number of the Runge-Kutta-Nystrom
method.

TABLE 4

t	Correct Solution	Runge - Kutta	DeVogelaere	DVND h=.25	DVMWD h=.25
0.0	1.000000	1.000000	1.000000	1.00000	1.00000
0.5	0.979253	0.979167 (-86)	0.979219 (-34)	.979221 (-32)	.979231 (-22)
1.0	0.838812	0.838609 (-203)	0.838704 (-108)	.838699 (-113)	.838670 (-142)
1.5	0.497890	0.497757 (-133)	0.497830 (-60)	.497733 (-167)	.497596 (-294)
2.0	-0.014979	-0.014487 (+492)	-0.014571 (+408)	-.015000 (-21)	-.015292 (-313)
2.5	-0.509797	-0.508159 (+1638)	-0.508499 (+1298)	-.509474 (+323)	-.509865 (+68)
3.0	-0.694729	-0.692671 (+2058)	-0.693099 (+1630)	-.694170 (+559)	-.694393 (+336)

The results with a step size of 0.5 had a much larger error which is difficult to understand because we integrated the differential equations for cos (X) and exp (X) for these step size obtaining the results shown in tables 5 and 6. While the modified Devogelaere method with its fewer function evaluations is somewhat inferior to Runge-Kutta-Nystrom they are obviously of comparable order and there is no major deterioration at the larger step size.

TABLE 5		Comparison for X" = -X			
		10^7 (Cos(t) - solution)			
		R K N		D V M W D	
t	Cos(t)	h=.25	h=.50	h=.25	h=.5
0.25	.968912	- 4		- 4	
0.50	.877583	- 22	- 216	- 55	- 216
0.75	.731689	- 53		- 176	
1.00	.540302	- 93	- 1300	- 352	- 3,420
1.25	.315322	- 137		- 573	
1.50	.070737	- 179	- 2777	- 817	- 10,680
1.75	- .178246	- 214		- 1075	
2.00	- .416147	- 234	- 3896	- 1329	- 17,196
2.25	- .628174	- 235		- 1567	
2.50	- .801144	- 215	- 3922	- 1777	- 23,638
2.75	- .924302	- 169		- 1952	
3.00	- .989992	- 101	- 2415	- 2086	- 27,337

TABLE 6   Comparison for X" = X

$$10^6 \text{ (Exp (t) - solution)}$$

t	Exp(t)	R K N h=.25	R K N h=.50	D V M W D h=.25	D V M W D h=.5
.25	1.28403	8		8	
.50	1.64872	19	284	43	284
.75	2.11700	33		88	
1.00	2.71828	52	766	147	1680
1.25	3.49034	78		224	
1.50	4.48169	114	1652	324	3985
1.75	5.75460	163		455	
2.00	7.38906	230	3302	626	8214
2.25	9.48774	321		851	
2.50	12.1825	444	6400	1144	15671
2.75	15.6426	613		1527	
3.00	20.0855	839	11772	2027	29081

DISCUSSION

The generalization of Devogelaere's method to handle the general second order equations make it a promising approach since second order equation arise so frequently in physical problems. While Runge-Kutta method are the method most widely used by engineers, numerical analysts are divided with many favoring more efficient methods i.e. methods requiring fewer function evaluations even though they require a starting procedure. Obviously it will be difficult to introduce a significant change in computing practice. The failure of engineers who deal so often with second order equations to use the Nystrom method which handles such problems directly instead of rewriting the problem as a set of first order equations illustrates the difficulty.

The logical steps in the development of a successful computer package are:

(1) to consider the possible variants of the general method and test them on a wide variety of problems
(2) try to develop a simple starting procedure which can be integrated in the subroutine
(3) write the program to handle a number of equations simultaneously

The last step should not be very difficult once the details of the method are settled. In some languages such as MATLAB^{TM} or PASCAL SC^{TM} which provide for functions of a matrix argument there would be very little difference. In a FORTRAN implementation it would be necessary to use loops and use a subroutine rather than a function to compute F and G.

REFERENCES

1. Brankin, B. et al, Algorithm 670: A Runge-Kutta-Nystrom Code, ACM
    Trans. Math. Softw., vol. 15, no. 1, 1989.

2. Collatz, L., The Numerical Treatment of Differential Equations, 3rd
    ed. pp. 61-68, 537, Springer Verlag, New York, 1966.

3. Devogelaere, R., A Method for the Numerical Integratim of Second
    Order without Explicit First Derivatives, J. Res. MBS, vol. 54, pp.
    119- , 1955.

4. Nysrom, R.j., On the Numerical Integration of Differential
    Equations, Acta. Soc. Fenn., vol. 50, no. 13, 1925.

5. Zurmuhl, R., Numerical Integration of Ordinary Differential
    Equation of Second and Higher Order, Z. Angew. Mat. Mech., vol. 20, pp.
    104-116, 1940.

6. Zurmuhl, R., Runge-Kutta Methods for Integrating the Nth Order
    Differential Equations Z. Angew. Mat. Mech., vol. 35, pp. 100-110, 1957.

Progressive Aitken Midpoint Quadrature

WILLIAM SQUIRE
Department of Mechanical and Aerospace Engineering
West Virginia University
Morgantown, West Virginia 26506, USA

ABSTRACT

It is shown that applying Aitken extrapolation to a sequence of midpoint rule evaluations each using three times as many modes as the previous one is an effective procedure for evaluating integrals. Mildly singular integrands for which the Romberg method requires modification can be handled directly by the proposed method.

INTRODUCTION

Squire (1990) compared the performance of Romberg (Richardson) extrapolation and Aitken extrapolation of midpoint and trapoizoid rule evaluations of several integrals. It was found that the midpoint rule generally gave more accurate results than the trapezoid rule and that Aitken extrapolation was superior to the Romberg procedure. In particular it could handle integrals such as $\int_0^1 \ln x \, dx$ or $\int_0^1 \sqrt{x} \, dx$ without any modification such as that developed by Fox and Hayes (1970) for Romberg quadrature.

This paper presents a procedure, implemented in a FORTRAN function which computes a sequence of midpoint rule evaluations each using three times as many modes as the previous evaluations. This reuses all the previous values. Such a sequence is referred to in the literature as being progressive.

THEORY OF THE METHOD

Writing the large n point midpoint rule approximation for $\int_a^b f(x) \, dx$ as

$$M_n = h \, S_n \tag{1}$$

where

$$Sn = \sum_{k=1}^{n} f\left[\, a+(k - {}^{1}\!/_{2})h \,\right] \qquad\qquad (2)$$

and

$$h=(b-a)/n \qquad\qquad (3)$$

allows a sequence of midpoint rule values to be computed by

$$Tn = \sum_{k=1}^{n} f[a+(6k-5)h/6] + f[b-(6k-5)h/6] \qquad\qquad (4a)$$

$$M3n = (Mn + hTn)/3 \qquad\qquad (4b)$$

In this way a sequence of midpoint rules values is obtained with all
function values being reused. However, it is easily seen that each new
evaluation requires twice as many function evaluations as all the
previous integral evaluations added together. This makes it essential
to get an accurate result from as few integral evaluations as possible.

A method originated by Aitken (1926) has found many applications some
of which are discussed by Joyce (1971), is used. The relation

$$V_a = \frac{(V_3 - V_2)^2}{(V_3 - V_2) - (V_2 - V_1)} \qquad\qquad (5)$$

where V_1, V_2 and V_3 are three successive estimates is used. This is exact
for a geometric progression. For example it will evaluate

$$1/3 = \frac{1}{1 + 2} = 1 - 2 + 4 - 8 + \ldots \qquad\qquad (6)$$

exactly from any three consecutive partial sums! However, it will fail
for an arithmetic progression since the denominator vanishes.

Squire (1975) has used extrapolation for the evaluation of integrals
over a semi infinite range. For example:

$$\int_0^\infty \sin x \, dx/x = \pi/2 \qquad\qquad (7)$$

can be evaluated accurately by splitting the integral at the zeroes of
the integrand, evaluating number of these subintegrals and extrapolating
the partial sums. The number of subintegrals required can also be
reduced by applying Aitken extrapolation to the sequence obtained from
the original partial sums. In the present application we do not
subdivide the range of integration but apply the extrapolation to the

sequence of values obtained by applying the midpoint rule with a finer and finer subdivision.

We could, of course, apply the Romberg procedure to this sequence just as easily. Apparently, Aitken extrapolation generally converges better but there is one disadvantage. For the Romberg procedure there is a general theorem that convergence along the diagonal is faster than for either a row or a column so there is a good stopping criterion. There is no corresponding theorem for the Aitken procedure. We have stopped when two elements along in a row differ by less than a specified tolerance and have found this to generally be quite conservative.

```fortran
C FUNCTION FOR PROGRESSIVE AITKEN MIDPOINT QUADRATURE
      FUNCTION PAMFUND (GRAND,XLO,XHI,ATOL)
      DIMENSION A (9, 5)
      IMPLICIT REAL *8 (A-H), REAL*8(O-Z)
      COMMON/PAMFUND/A,NFE
C***********INTEGRATION FOLLOWS***********************
      SPAN = XHI - XLO
      SUM = GRAND(.5*(XLO+XHI))
      A(1,1) = SPAN * SUM
      H = SPAN/3
      SUM = SUM + GRAND(XLO+h/2.) + GRAND(XHI-H/2.)
      A(2,1) = H * SUM
      DO 50 I = 3, 9
         H6 = H
         H = H/3.
         H2 = H/2.
         I1 = I - 1
         I2 = I 2
         DO 30 X = H2, SPAN,H6
            SUM = SUM + GRAND(XLO + X) + GRAND(XHI - X)
30    CONTINUE    ! X LOOP
      A(i,1) = H * SUM    ! ADDS NEW MIDPOINT RULE VALUE
      JUP = (I+1)/2        ! SETS UPPER J FOR AITKEN EXTRAPOLATION
      DO 40 J = 2 , JUP
         J1 = J - 1
         D1 = A(I,J1) - A(I1,J1)
         D2 = A(I1,J1) - A(I2,J1)
         IF (D1 .NE. D2) THEN
            D = D1*D1/(D1-D2)
         ELSE
            D = 0
         ENDIF
         A(I,J) = A(I,J1) - D
         IF (ABS(D)  .LT ATOL) THEN
            PAMFUND = A(I,J)
            NFE = 3**(I-1)
            GO TO 70
         ENDIF
40    CONTINUE
50    CONTINUE
      PAMFUND=A(9,2)
      NFE = 3**8
70    RETURN
      END
```

COMPUTER PROGRAM AND RESULTS

Table 1 shows the listing of a double precision FORTRAN function subprogram implementing the algorithm described in the previous section. The integrand is supplied by an external issue defined function or by an intrinsic library function. It should be emphasized that variations are possible some of which will be discussed in the concluding section.

The program begins with one point and three point midpoint rule evaluations of the integral. Then it enters a loop which permits up to seven more evaluations by adding more modes for a total of 56 function evaluations. Each time a new midpoint rule value is obtained the row is filled out with Aitken extrapolation. The value obtained by the Aitken extrapolation is returned if it differs from the previous value in the row by less than the specified tolerance. Our experience has shown that this is generally a conservative estimate of the error. We have also, on the basis of our limited experience, accepted the occurrence of a vanishing denominator as an indication that convergence had been reached. We only encountered it when the values were close to the limit of the computer arithmetic and the values were differing because of roundoff. By not proceeding further on the row, the opportunity to get a more accurate result by the small cost of another Aitken extrapolation is sacrificed but the possibility of an appreciable error is avoided. One weakness of Aitken extrapolation is that the effect of poor initial values carries through longer than in the Romberg procedure and shows up in the outer elements of the row. Therefore the values do not necessarily improve across the row.

If the tolerance is not met in the ninth row, a statement to that effect is printed out and the elements of the last row are printed out. To keep from terminating the execution the next to the last element of this row is returned. Care must be observed in setting the tolerance. The approximate value of this integral must be taken into account or the tolerance may be so small that satisfying it is beyond the capability of the computer arithmetic or so large that a very poor approximation will meet it. With double precision (16 digit arithmetic) it probably should not be less than 10^{-14} x $|\text{Integral}|$.

Numerical Results

The first problem we considered was

$$Q = \int_0^1 (n+1) \, x^n dx = 1 \tag{8a}$$

the integral being normalized so that the value of the integral is 1.0 which was evaluated for a number of values of n in order to see how the algorithm works as n is increased. By keeping the value of the integral as 1.0 the tolerance is directly related to the number of correct figures wanted. Obviously for a sufficiently large n the computation will fail by underflow at a small node, but we are interested in the difficulty caused by the increasingly sharp peak as n is increased. Table 2 shows the results for n = 4,8,16 and 32 for tolerances of 10^{-4} to 10^{-4}. The effect of increasing n is evident but not very great. Also, the effect of round-off in lending the accuracy is clearly shown.

TABLE 2) $(m+1) \int_0^1 x^m \, dx$

x^4	NFE	x^8	NFE
1.00000005029927	81	1.00000376813339	81
1.00000005029927	81	1.00000000659232	243
1.00000000006621	243	1.00000000000907	729
1.00000000000009	729	0.999999999997526	729
0.999999999999999	729	0.999999999999964	2187
0.999999999999985	2187	1.00000000000011	6561

x^{16}	NFE	x^{32}	Tol
1.00000030455560	243	1.00001525158542	10^{-4}
1.00000000048509	729	1.00000002224212	10^{-6}
1.00000000013896	729	1.00000002017609	10^{-8}
0.999999999999825	2187	1.00000000000021	10^{-10}
0.999999999999825	2187	1.00000000000046	10^{-12}
1.00000000000023	6561	1.00000000000046	10^{-14}

The second problem considered is

$$Q = \frac{n + 1}{k^{(n+1)}} \int_0^k X^n \, dx \tag{8b}$$

which tests the effect of increasing the range of integration.   Table 3
shows the result for some values of k and n.   The accuracy remains high
when the value of the unnormalized integral becomes astronomical.

   The effect of round off is clearly shown in the cases where the
occuracy decreased as the number of nodes increased.   The effect of
stopping when the tolerance is met and not combining across the row is
shown when the number of function evaluations does not change when the
tolerance is increased.   We have the strange result for X^{32} that
decreasing the tolerance decreased the accuracy!

Table 3)

$$\frac{m+1}{k^{m+1}} \int_0^k x^m \, dx$$

$$\int_0^3 x^{10} \, dx = 16104.2727272727 \qquad \int_0^9 x^{10} \, dx = 285282360.81818$$

Normalized Integral	NFE	Normalized Integral	NFE
1.00001232201557	81	1.00001232201556	81
1.00000002385729	243	1.00000002385729	243
1.00000000003374	729	1.00000000003373	729
0.999999999987501	729	0.999999999987494	729
0.999999999999994	2187	1.00000000000000	2187

$$\int_0^{10} x^{49} \, dx = 0.2 \cdot 10^{49} \qquad \int_0^{30} x^{74} \, dx = .811022383617811 \times 10^{10^9}$$

Normalized Integral	NFE	Normalized Integral	NFE
1.00000023783398	729	1.00000267922707	729
1.00000031311744	729	1.00000000371081	2187
1.00000000016699	2187	1.00000000285084	2187
0.999999999998746	6561	0.999999999998823	6561
0.999999999998746	6561	0.999999999998708	6561

We then consider three mildly singular integrals

$$\int_0^1 \sqrt{x} \, dx = \frac{2}{3} \tag{9a}$$

$$\int_0^1 \ln x \, dx = -1 \tag{9b}$$

$$\int_0^1 dx/\sqrt{x} = 2 \tag{9c}$$

In (9a) the integrand has an infinite slope at the lower limit, while in (9b) and (9c) it becomes infinite. As noted, Romberg integration requires modification in such cases but Table 4 shows that the present method encounters no difficulties.

Table 4   Mildly Singular Integrands

Integral	NFE	Tol

$$\int_0^1 \sqrt{x}\ dx = 2/3$$

0.666663642950920	81	10^{-4}
0.666666647781942	243	10^{-6}
0.666666667235211	729	10^{-8}
0.666666666669442	2187	10^{-10}
0.666666666666575	6561	10^{-12}

$$\int_0^1 \ln x\ dx = -1$$

-1.00000197115050	81	10^{-4}
-1.0000000850575	729	10^{-6}
-0.999999999934605	2187	10^{-8}
-0.999999999999609	6561	10^{-10}
-1.00000000000003	6561	10^{-12}

$$\int_0^1 \frac{dx}{\sqrt{x}} = 2.0$$

2.00000024676599	243	10^{-4}
2.00000007718496	729	10^{-6}
1.99999999983902	2187	10^{-8}
1.00000000000611	6561	10^{-10}
2.00000000000575	6561	10^{-12}

We conclude with two examples of highly oscillatory integrands for which we do not have an exact value.

$$Q_1 = \int_0^{10} \cos(\pi/2\ x^2)\ dx \tag{10a}$$

$$Q_2 = \int_0^{100\pi} \sin x\ [(100\pi)^2 - x^2]^{1/2} dx \tag{10b}$$

Besides being highly oscillatory Eq. (10b) has an infinite slope at the upper limit. Table 5a shows that (10a) is relatively easy to handle but Table 5b indicates that (10b) is considerably more difficult.

TABLE 5 Oscillatory Integrands

$$\int_0^{10} \cos\pi\, x^2/2 \; dx \qquad\qquad \int_0^{100\pi} \sin x\, [(100\pi)^2-x^2]^{1/2}\, dx$$

Integral	NFE	TOL	Integral	NFE
0.499897908158802	243	10^{-4}	299.386900040499	729
0.499897908158802	243	10^{-6}	298.424234305151	2187
0.499898817895015	729	10^{-8}	298.424630616956	2187
0.499898817899268	729	10^{-10}	298.435512459688	6561
0.499898826551577	2187	10^{-12}	298.435512459688	6561
0.499898826551577	2187	10^{-14}		

Only first 7 digits
are reliable

Intractable Problem

Equations over a semi-infinite range can be transformed into equations
over a finite range by a simple change of variable. For example by
letting
$z = 1/x$

$$\int_1^{\infty} \sin(x)\, dx/x = \int_0^1 \sin\,(1/z)\, dz/z \qquad\qquad (11)$$

Our method fails for this integral probably because the integrand is not
of bounded variation near the origin. Also the original integral is
what is called conditionally convergence. $\int^{\infty}\sin x^2\, dx$ exhibits similar
behavior.[1]

DISCUSSION

The Aitken extrapolation procedure appears to be a simple and powerful
approach to evaluating integrals though not as efficient as specialized
methods for special forms. For example a Gaussian rule of order n which
will be exact for powers up to 2n-1 would be much more efficient for the
first two examples. However, many cases arising in practice do not fall
into such special classes and there is the desire for general purpose
programs which will not require the user to make choices.

As noted, some variation in our procedure are possible. On the basis of
the experience gained in this investigation I would suggest that
instead of entering the tolerance the program be modified so that the
user enters the number of significant figures desired and how many modes

to use in the first midpoint rule evaluation. Then the tolerance would be set by taking the value obtained after tripling the number of nodes and multiplying by 0.1 raised to the number of significant figures.

It should also be noted that a given accuracy for an integral over a largerange can be obtained with somewhat fewer function evaluations by splitting the range of integration which naturally leads to the question of incorporating the basic idea into an adaptive procedure. There appears no reason why an adaptive procedure cannot be based on Aitken extrapolation in the same way that the widely used CADRE program uses Romberg extrapolation.

REFERENCES

Aitken, A.C., On Bernoulli's Numerical Solution of Algebraic Equations, Proc. Roy. Soc Edinburgh, vol 46A, pp. 289-305, 1926

Fox, L. and Hayes, L., On the Definite Integration of Singular Integrands, SIAM Rev., vol 12 no. 3, pp. 449-457, 1970

Joyce, D.C., Survey of Extrapolation Processes in Numerical Analysis, SIAM Rev., vol. 13 no. 4, pp. 435-490, 1971.

Squire, W., Partition-Extrapolation Methods for Numerical Quadratures, Intern J. Computer Maths., vol. 5B no.1, pp. 81-91, 1975.

Squire, W. ACCESS, Forthcoming, 1990.

TRANSPORT PROCESSES

Summary

Transport processes include a broad range of topics such as fluid dynamics, electromagnetic waves, magnetohydrodynamics, convective heat transfer, and conduction.

An invited paper by Adriana Nastase, Director of the Aerodynamics Institute, Aachen, Germany, describes various fluid dynamics problems associated with supersonic aircraft and space vehicles. A study of the optimum shapes for these vehicles to minimize the drag force is presented. The procedure described solves two three-dimensional boundary value problems and two successive enlarged variational problems.

Fang and Paraschivoiu use the Green's function theory to solve for flow over NACA 0012 airfoils. A graphical presentation of a pressure coefficient that includes the effect of discontinuity due to shock wave is given. An iterative scheme is used to solve for the nonlinear potentials. Shock integrals are used to capture sharp shock jumps.

Nair and Payne convert the differential equation for flow to a Volterra equation and then solve the integral equation iteratively by an improved Picard method. The results were compared with the finite difference and finite element methods. For two different examples, the accuracy of the results are comparable to the finite difference method but superior to the finite element method.

An analytical model is developed by Lee and Wilson to predict performance of large-scale MHD components. The boundary layer integral equation is solved and an approximate velocity profile in turbulent flow is analytically predicted. The near the wall velocity model for turbulent flow agrees well with the Spalding formula. Various flow parameters are compared with experimental data.

Leith and Maveety used a bifurcation theory to describe the growth rate of heat flux in Rayleigh-Bénard convection. The low Rayleigh number behavior that includes the hydrodynamic instability mechanisms is discussed. The results are compared with experimental data and the authors conclude that the finiteness of container size introduces further nonlinearity in the bifurcation theory.

Tzeng, Bredow, and Fung studied the scattering from a large number of scatterers in an incoherent field. The electromagnetic wave scattering is directed toward radar applications. A Monte Carlo simulation that uses a T-matrix approach was employed. The authors conclude that the scattering results from spherical models agree with experimental data and the results apply to randomly distributed scatterers of arbitrary shapes.

Chen and Fung studied the computational speed of the spectral domain method to calculate the scattered field of electromagnetic waves. Their data show the CPU time saving is substantial when compared to the moment method. The numerical studies are for three wave numbers.

An improved inverse conduction solution to calculate surface heat flux is described by Litkouhi and Naraghi. Higher order influence functions in the Duhamel integral is used. The numerical data for orders of influence functions are compared. They show higher order influence functions can resolve rapid changes in the surface heat flux.

The mathematical description of a coupled initial boundary value system by Jódar is an accurate approximate solution of thermal conduction problems. The solution of the one-dimensional diffusion equation is studied. The error bound indicating the accuracy of a finite series solution is described in detail.

The surface temperature for the IBM thermal conduction module is calculated for 1 Watt of dissipated energy by Haji-Sheikh and Kinsey. The Green's function solution is used and a numerical solution of the resulting Volterra equation is presented. The boundary condition external to the integrated chip package is nonelementary; hence, the Monte Carlo method is used. The data presented are for two geometrical orientations.

A. Haji-Sheikh

Over the Design of Supersonic Aircraft and Space Vehicles of Minimum Drag in Supersonic and Hypersonic Flow

ADRIANA NASTASE
Lehrgebiet Aerodynamik des Fluges
RWTH, 5100, Aachen, Templergraben 55, Germany

Abstract: The optimum-optimorum configuration of the space vehicle is the configuration for which the shapes of its surface and <u>also</u> of its planprojection are <u>simultaneously</u> determined in such a manner that its drag attains its minimum at a given cruising Mach number M_∞. The problem of the determination of the optimum-optimorum configuration of a <u>space vehicle of variable geometry</u> which presents a <u>minimum drag at two cruising Mach numbers</u> M_∞ and M_∞^* are here also considered.

1. INTRODUCTION

The author proposes the global optimization techniques for the design of optimal shapes of the space vehicles and supersonic aircraft of the future. The global optimization technique is a scientifical optimization strategy which gives the possibility to determine the best values of <u>all</u> geometrical characteristics of the shapes of space vehicle or supersonic aircraft i.e. camber, twist, thickness distributions and <u>also</u> the similarity parameters of their planprojections in order to obtain a minimum for a chosen functional which is, generally, the drag functional.

Several strategies for the optimization can be used. The most known are the variational and the evolution methods.

Global or local conditions deduced from experience, obtained from formerly design or missions and some new one introduced by the designer, must be incorporated in the optimization strategies. These auxiliary conditions can be of different natures as, for example, of aerodynamical, geometrical, thermal or structural origins.

The equations of surfaces of the space vehicles (and of the supersonic aircraft) are supposed to be expressed in the form of two-dimensional expansions which can be polynomes, double Fourier series, splines etc.. The planforms of the space vehicles (and of the supersonic aircraft) can be characterised by their similarity parameters.

A class of admissible space vehicles (or supersonic aircraft) is defined through some given common properties.

The coefficients of the expansions of the surfaces and the similarity parameters of the planforms are considered as free and must be determined after the performing of the global optimization process inside of this class.

Great decisions must be made in the choice of flight regime at which the optimization process is performed and the choice of the starting software (i.e. Full-Linearized, Full-Potential, Euler, Thin Navier-Stokes, Full Navier-Stokes or Zonal Codes (Potential/Navier-Stokes Potential/Boundary Layer etc.) for the determination of the velocity field over the class of space vehicles or supersonic aircraft.

Some results of the author concerning the design of fully-optimized shapes of wing alone (Fig. 1a), of integrated wing-fuselage configuration (Fig. 1b) and of integrated wing-fuselage-flaps configuration of variable geometry (Fig. 1b,c) are here presented.

2. THE OPTIMUM-OPTIMORUM THEORY

The optimum-optimorum theory introduced by the author in [1] , [2] , [3] and [4] was used for the determination of the optimum-optimorum shape of the wing alone at a given cruising Mach number M_∞ as in [1] - [5] , [8] , [16].

More recently, the variational problem concerning the determination of the optimum-optimorum shape of the integrated wing-fuselage configuration (at a cruising Mach number M_∞) was considered by the author in [11] - [15]. For the integrated wing-fuselage configuration all its geometrical parameters i.e. the distributions of cambers, twists, thicknesses and also the similarity parameters of the planprojections of the wing and of the fuselage are optimized in order to obtain a minimum drag (at a given cruising Mach number M_∞). This is a proposal of the author for the design of the optimal shapes of the both vehicles of a two stage configuration (like Sänger and Horus).

The numerical analysis performed by the author on about hundred optimized wings leads to the conclusion that the dimensionsless span ℓ_{opt} shows a strong dependence on cruising Mach number M_∞ as it can be seen in the optimal hyperbola (Fig. 5b) (Hereby is $B = \sqrt{M_\infty^2 - 1}$).

Therefore, the further step in the optimization of the entire configuration of the space vehicle is to determine the optimum-optimorum shape of the integrated wing-fuselage configuration of variable geometry. This variable geometry can be realized with the help of movable leading edge flaps as formerly proposed by the author in [1], [9], [10], [16] for the delta wing alone. The integrated wing-fuselage configuration with movable flaps can be optimized at two supersonic cruising Mach numbers M_∞^* and M_∞ ($M_\infty^* < M_\infty$). At the higher supersonic Mach number M_∞ the integrated configuration of the space vehicle is flying with the flaps in retracted position. At the lower supersonic Mach number M_∞^* the space vehicle is flying with the flaps in open position. This is the authors proposal for the design of the shape of a single-stage vehicle (like Hotol and Hermes).

The determination of the optimum-optimorum configuration of the space vehicle with variable geometry leads to the solving of
- two three-dimensional boundary-value problems for the axial disturbance velocity u on the space vehicle with flaps (in retracted and open positions) and of
- two successive enlarged variational problems (with free boundary) for the space vehicle with flaps (in retracted and in open positions).

The wing-fuselage configuration with flaps in retracted position is here considered as a wing alone, for which the surface is discontinuous along the junction lines between the wing and the fuselage. The wing-fuselage configuration with flaps in open position is considered also as a wing alone for which the surface is now discontinuous along the junction lines between the wing and the fuselage and between the wing and the leading edge flaps.

The author has proposed in [1], [2], [3], [4], [7] a method for the design of fully-optimized shape of the aircraft configurations and space vehicles which she called it optimum-optimorum theory. This theory allows the simultaneous determination of the optimal shapes of the surface and of the plan-projection of the space vehicle in order to obtain a minimum drag. The determination of the shape of the optimum-optimorum space vehicle leads to an extended variational problem for the drag functional C_d, i.e.

$$C_d \equiv \int_{S(x_1,x_2)} F\left[x_1, x_2, Z(x_1,x_2) \right] dx_1 \, dx_2 = \min. \tag{1}$$

Here the function $Z(x_1,x_2)$ and also the boundary $S(x_1,x_2)$ of the integral are a priori unknown and are determined by the solving of this extended variational problem. The optimum-optimorum space vehicle is chosen among a set of space vehicles, which are defined through some common properties. In the frame of the optimum-optimorum theory of the author, two space vehicles belong to the same set if:
- their surfaces can be piecewise approximated through a superposition of

homogeneous polynomes of the same degree;

- their planprojections are polygons which can be related through affine transformations, and

- the shapes of the space vehicles of the set fulfill the same auxiliary conditions (of geometrical or aerodynamical nature).

The parameters of the optimization are the coefficients Z_{ij} of homogeneous polynomes of the equations of the surfaces and the similarity parameters (ν_1, ν_2,..., ν_n) of the planprojections of the space vehicles of the set. In order to solve this enlarged variational problem for the determination of the extremum of the drag functional $C_d{}^{(t)}$ with free boundary the author uses her hybrid, numerical-analytical method.

This method starts with the remark, that the dependence of the drag functional $C_d{}^{(t)}$ versus the coefficients Z_{ij} of the polynomes, which piecewise approximate the surfaces of the space vehicles, is a quadratic form, while the dependence versus the similarity parameters of the planform are nonlinear and very complicated. The method presents two steps.

- In the first step the set of similarity parameters of the planform (ν_1, ν_2,..., ν_n) are considered as given. The boundary of the drag functional $C_d{}^{(t)}$ is now a priori known. The optimal value of the coefficients of polynomial expansions of the surface of the space vehicle are obtained by solving a linear, algebraic system. These optimal coefficients determine uniquely the value of the drag functional $(C_d{}^{(t)})_{opt}$, for the prescribed set of similarity parameters of the planform. This value of $(C_d{}^{(t)})_{opt}$ represents a "point" of what is called here lower limit hypersurface of the drag functional $C_d{}^{(t)}$ i.e.

$$(C_d^{(t)})_{opt} = f(\nu_1, \nu_2,..., \nu_n) \tag{2}$$

Each of these points can be analytically determined.

- In the second step, through systematical variation of the set of similarity parameters the "position" of the minimum of this hypersurface is numerically (or graphically) determined and gives the best set of similarity parameters (ν_1, ν_2,..., ν_n) of the planform, as presented in (Fig. 2), for two similarity parameters. The optimal set of similarity parameters together with a chosen area S_0 of the plan-projection determine the shapes of the planform and of the surface of the optimum-optimorum space vehicle of a given set of space vehicles. The optimum-optimorum space vehicle is exactly the optimal space vehicle corresponding to this optimal set of similarity parameters. The minimum value of the "ordinate" of the hypersurface represents the drag coefficient of the optimum-optimorum space vehicle of the set. The above theory was successfully used by the author for the effective design of the shape of optimum-optimorum delta wing Adela [1] – [7], and of the optimum-optimorum shape of the integrated wing-fuselage configuration (at cruising Mach number M_∞ = 2) as in [11] - [15] , [25] . The shape of delta wing Adela is given in (Fig. 3) and the modification in the shape of the wing Adela, due to the fuselage integration (in the section $\tilde{x}_1$ = 0,6) is given in (Fig. 3). A further new application of the optimum-optimorum theory taken here into consideration is the determination of the shape of the entire space vehicle which is an integrated wing-fuselage-flaps configuration of minimum drag at two cruising Mach numbers M_∞ and M_∞^* . Two variational problems in cascade are here occurring

- one for the determination of the optimum-optimorum shape of the space vehicle at the higher cruising Mach number M_∞ with the flaps in retracted position and

- the second for the determination of the optimum-optimorum shapes of the flap-surface and its planprojection in such a manner, that the entire space vehicle (with the flaps in open position) is of minimum drag at the supersonic cruising Mach number M_∞^* .

3. DETERMINATION OF THE AXIAL DISTURBANCE VELOCITIES

Let us refer the integrated thick, lifting delta wing to a three-orthogonal system of axes $Ox_1x_2x_3$ having the apex O of the wing as origin. The plane Ox_1x_2 is the plane of symmetry of the integrated wing and the axis Ox_1 is the bisectrix of the angle of the integrated wing, in the plane Ox_1x_3, at its apex (the shock-free entry direction). The integrated thick, lifting delta wing surface is supposed to be flattened in the plane Ox_1x_2 (Fig. 4a) and is considered in a parallel stream with the undisturbed velocity $\vec{V}_\infty$ at a moderate angle of attack α (measured between the Ox_1 - axis and $\vec{V}_\infty$).

In the framework of linearised theory for flattened integrated thick, lifting delta wings at moderate angle of attack α , in the boundary value problem concerning the determination of the axial disturbance velocity u the effect of lift can be separated from the effect of thickness. Further the following two delta wing components will be separately considered. The <u>thin integrated delta wing</u> which is the skeleton surface of the thick, lifting integrated delta wing and is considered at the same angle of attack α and <u>the thick-symmetrical integrated delta wing</u> which has the same thickness distribution as the thick, lifting integrated delta wing but its skeleton surface is a plane. This component is considered at zero angle of attack.

The skeleton surface $Z(x_1,x_2)$ of the integrated delta wing is supposed to be continuous but, for the sake of generality, the thickness distributions $Z^*(x_1,x_2)$ on the lateral sides OA_1C_1 and OA_2C_2 (corresponding to the wing) and $Z'^*(x_1,x_2)$ on the central part OC_1C_2 (corresponding to the fuselage) are supposed to be different. Further this wing will be called initial integrated delta wing. The author introduced, as in [1] - [15] a well-suited affine transformation in order to obtain dimensionless coordinates

$$\tilde{x}_1 = \frac{x_1}{h_1} \;,\; \tilde{x}_2 = \frac{x_2}{\ell_1} \;,\; \tilde{x}_3 = \frac{x_3}{h_1} \;,\; B = \sqrt{M_\infty^2 - 1} \tag{3}$$

$$(\; y = \frac{y}{\ell} \;,\; \ell = \frac{\ell_1}{h_1} \;,\; \nu = B\ell \;,\; \bar{\nu} = Bc' \;,\; \bar{k} = \frac{\bar{c}'}{\ell} \;)$$

A transformed integrated delta wing is obtained, which has the maximal depth 1 and the half-span 1 (Fig. 4b). The traces $\tilde{C}_1$ and $\tilde{C}_2$ of the junction lines $\tilde{O}\tilde{C}_1$ and $\tilde{O}\tilde{C}_2$ (between the wing and the fuselage) have the following positions on the axis $\tilde{C}\tilde{y}$ (parallel to axis Ox_2): $y_c = \pm \bar{k}$. The transformed integrated delta wing is placed in a supersonic flow with the cruising Mach number $\tilde{M}_\infty = \sqrt{1 + \nu^2}$. Between the dimensionless axial disturbance velocities u, u^* and $\tilde{u}$, $\tilde{u}^*$ and the dimensionless downwashes w, w^*, w'^* and $\tilde{w}$, $\tilde{w}^*$, $\bar{w}^*$ of the initial and transformed integrated delta wing components there are the following relations:

$$u = \ell\,\tilde{u} \;,\; w = \tilde{w} \;,\; u^* = \ell\,\tilde{u}^* \;,\; w^* = \tilde{w}^* \;,\; w'^* = \bar{w}^* \tag{4}$$

Further the assumption is made, that the downwashes $\tilde{w}$, $\tilde{w}^*$ and $\bar{w}^*$ are expressed in form of superpositions of homogeneous polynomes in $\tilde{x}_1$ and $\tilde{x}_2$ i.e. - on the thin component of the transformed integrated wing

$$\tilde{w} = \sum_{m=1}^{N} \tilde{x}_1^{m-1} \sum_{k=0}^{m-1} \tilde{w}_{m-k-1,k} \, |\tilde{y}|^k \tag{5}$$

and on the thick-symmetrical component of the transformed integrated wing

$$\tilde{w}^* = \sum_{m=1}^{N} \tilde{x}_1^{m-1} \sum_{k=0}^{m-1} \tilde{w}^*_{m-k-1,k} \, |\tilde{y}|^k \tag{6a}$$

if $\bar{k} < \tilde{y} < 1$ (here $\bar{k} = \bar{\nu}/\nu$ is supposed constant) and

$$\bar{w}^* = \sum_{m=1}^{N} \tilde{x}_1^{m-1} \sum_{k=0}^{m-1} \bar{w}^*_{m-k-1,k} \, |\tilde{y}|^k \;, \tag{6b}$$

if $\tilde{y} < \bar{k}$. The coefficients $\tilde{w}_{ij}$, $\tilde{w}^*_{ij}$ and $\bar{w}^*_{ij}$ and the similarity parameter ν are unknown and will be determined through the fully-optimization process. The axial disturbance velocity $\tilde{u}$ on the thin component of the transformed integrated thick-lifting delta wing with subsonic leading edges and a central ridge is of the form

$$\tilde{u} = \sum_{n=1}^{N} \tilde{x}_1^{n-1} \left\{ \sum_{q=0}^{E(\frac{n}{2})} \frac{\tilde{A}_{n,2q}\,\tilde{y}^{2q}}{\sqrt{1 - \tilde{y}^2}} + \sum_{q=1}^{E(\frac{n-1}{2})} \tilde{C}_{n,2q}\,\tilde{y}^{2q} \cosh^{-1}\frac{1}{\sqrt{\tilde{y}^2}} \right\} \tag{7}$$

If the integrated thin delta wing has supersonic leading edges it results in:

$$\tilde{u} = \sum_{n=1}^{N} \tilde{x}_1^{n-1} \left\{ \sum_{q=0}^{n-1} \tilde{H}_{nq}\, \tilde{y}^q \left[\cos^{-1}M_1 + (-1)^q \cos^{-1}M_2 \right] + \sum_{q=0}^{E(\frac{n}{2})} \frac{\tilde{P}_{n,2q}\, \tilde{y}^{2q}}{\sqrt{1 - \nu^2\tilde{y}^2}} + \right.$$

$$\left. + \sum_{q=1}^{E(\frac{n-1}{2})} \tilde{C}_{n,2q}\, \tilde{y}^{2q} \cosh^{-1}\frac{1}{\sqrt{\nu^2\,\tilde{y}^2}} \right\} \tag{8}$$

The axial disturbance velocity $\tilde{u}^*$ of the thick-symmetrical integrated delta wing component of the transformed integrated thick, lifting delta wing with subsonic leading edges is according to [7] - [9]

$$\tilde{u}^* = \sum_{n=1}^{N} \tilde{x}_1^{n-1} \left\{ \sum_{q=0}^{E(\frac{n}{2})} \frac{\tilde{P}^*_{n,2q}\, \tilde{y}^{2q}}{\sqrt{1 - \nu^2\tilde{y}^2}} + \sum_{q=0}^{n-1} \tilde{H}^*_{nq}\, \tilde{y}^q \left[\cosh^{-1}M_1 + (-1)^q \cosh^{-1}M_2 \right] + \right.$$

$$\left. + \sum_{q=1}^{E(\frac{n-1}{2})} \tilde{C}^*_{n,2q}\, \tilde{y}^{2q} \cosh^{-1}\frac{1}{\sqrt{\nu^2\,\tilde{y}^2}} + \sum_{q=0}^{n-1} \tilde{G}^*_{nq}\, \tilde{y}^q \left[\cosh^{-1}R_1 + (-1)^q \cosh^{-1}R_2 \right] \right\} \tag{9}$$

If the thick-symmetrical integrated delta wing has supersonic leading edges in the formula (9) the terms $\cosh^{-1}M_1$ and $\cosh^{-1}M_2$ are to be replaced with $\cos^{-1}M_1$ and $\cos^{-1}M_2$. Here the following notations will be made

$$R_1 = \sqrt{\frac{(1+\bar{\nu})\,(1-\nu\tilde{y})}{2\,(\bar{\nu}-\nu\tilde{y})}} \quad , \qquad R_2 = \sqrt{\frac{(1+\bar{\nu})\,(1+\nu\tilde{y})}{2\,(\bar{\nu}+\nu\tilde{y})}} \tag{10a}$$

$$M_1 = \sqrt{\frac{(1+\nu)\,(1-\nu\tilde{y})}{2\,\nu(1-\tilde{y})}} \quad , \qquad M_2 = \sqrt{\frac{(1+\nu)\,(1+\nu\tilde{y})}{2\,\nu(1+\tilde{y})}} \tag{10b}$$

The coefficients of $\tilde{u}$ for the thin transformed integrated delta wing are related to the coefficients of the downwash $\tilde{w}$ and the coefficients of $\tilde{u}^*$ for the thick-symmetrical transformed integrated delta wing are related to the coefficients of the downwashes $\tilde{w}^*$ and $\bar{w}^*$ through the following linear and homogeneous relations

$$\tilde{A}_{n,2q} = \sum_{j=0}^{n-1} \tilde{a}^{(n)}_{2q,j}\, \tilde{w}_{n-j-1,j} \tag{11}$$

$$\tilde{P}^*_{n,2q} = \sum_{j=0}^{n-1} \left(\tilde{p}^{*(n)}_{2q,j}\, \tilde{w}^*_{n-j-1,j} + \bar{p}^{*(n)}_{2q,j}\, \bar{w}^*_{n-j-1,j} \right) \tag{12}$$

The constants $\tilde{a}^{(n)}_{2q,j}$, $\tilde{p}^{*(n)}_{2q,j}$, $\bar{p}^{*(n)}_{2q,j}$ etc. are functions only of the similarity parameter ν.

Let us now consider the <u>open integrated delta wing</u> (with flaps in open positions) (Fig. 1c), (Fig. 4e) at cruising Mach number M^*_∞ $(B^* = \sqrt{M^{*2}_\infty - 1}\,)$. The downwashes on the wing and fuselage are unchanged and are given for the transformed configuration in the formulas (3), (4) and (5) and for the initial configuration as in the formulas (2). The downwashes $\tilde{\tilde{w}}$ and $\tilde{\tilde{w}}^*$ on the thin- and thick-symmetrical components on the flaps of the transformed open integrated wing (Fig. 4d) are supposed to be expressed in the form of superposition of homogeneous polynomes i.e.

$$\tilde{\tilde{w}} = \sum_{n=1}^{N} \tilde{x}_1^{n-1} \sum_{k=0}^{n-1} \tilde{\tilde{w}}_{n-k-1,k} \, |\tilde{y}|^k \tag{13}$$

on the thin transformed flaps component and

$$\tilde{\tilde{w}}^* = \sum_{n=1}^{N} \tilde{x}_1^{n-1} \sum_{k=0}^{n-1} \tilde{\tilde{w}}^*_{n-k-1,k} \, |\tilde{y}|^k \tag{14}$$

on the thick-symmetrical transformed flaps component. Between the downwashes and axial disturbance velocities on the initial and transformed flaps there are the following relations

$$\tilde{\tilde{w}}' = \tilde{\tilde{w}} \, , \quad \tilde{\tilde{w}}'^* = \tilde{\tilde{w}}^* \, , \quad \tilde{\tilde{u}}' = \ell\, \tilde{\tilde{u}} \, , \quad \tilde{\tilde{u}}'^* = \ell\, \tilde{\tilde{u}}^* \tag{15}$$

The following notations are further made:

$$\bar{\nu}^* = B^* c' \, , \quad \tilde{\nu}^* = B^* \ell \, , \quad \nu^* = B^* L \, , \quad k^* = \frac{L}{\ell} \tag{16}$$

The transformed open integrated delta wing is placed in a supersonic flow with the cruising Mach number $M_\infty^* = \sqrt{1 + \nu^{*2}}$. The axial disturbance velocities $\tilde{\tilde{u}}$ and $\tilde{\tilde{u}}^*$ on the thin and thick-symmetrical transformed open integrated delta wing with subsonic leading edges at the cruising Mach number M_∞^* are obtained by the author here under the following form

$$\tilde{\tilde{u}} = \sum_{n=1}^{N} \tilde{x}_1^{n-1} \left\{ \sum_{q=0}^{n-1} \tilde{\tilde{K}}_{nq} \tilde{y}^q \left[\cosh^{-1} N_1 + (-1)^q \cosh^{-1} N_2 \right] + \sum_{q=0}^{E(\frac{n}{2})} \frac{\tilde{\tilde{A}}_{n,2q}\, \tilde{y}^{2q}}{\sqrt{k^{*2} - \tilde{y}^2}} \right. +$$

$$\left. + \sum_{q=1}^{E(\frac{n-1}{2})} \tilde{\tilde{C}}_{n,2q}\, \tilde{y}^{2q} \cosh^{-1} \sqrt{\frac{k^{*2}}{\tilde{y}^2}} \right\} \tag{17}$$

and

$$\tilde{\tilde{u}}^* = \sum_{n=1}^{N} \tilde{x}_1^{n-1} \left\{ \sum_{q=0}^{n-1} \tilde{\tilde{K}}^*_{nq}\, \tilde{y}^q \left[\cosh^{-1} N_1^* + (-1)^q \cosh^{-1} N_2^* \right] + \sum_{q=0}^{E(\frac{n}{2})} \frac{\tilde{\tilde{P}}_{n,2q}\, \tilde{y}^{2q}}{\sqrt{1-\nu^{*2}\tilde{y}^2}} \right.$$

$$\left. + \sum_{q=1}^{E(\frac{n-1}{2})} \tilde{\tilde{C}}^*_{n,2q}\, \tilde{y}^{2q} \cosh^{-1} \sqrt{\frac{1}{\nu^{*2}\tilde{y}^2}} + \sum_{q=0}^{n-1} \tilde{\tilde{G}}^*_{nq}\, \tilde{y}^q \left[\cosh^{-1} R_1^* + (-1)^q \cosh^{-1} R_2^* \right] \right.$$

$$\left. + \sum_{q=0}^{n-1} \tilde{\tilde{H}}^*_{nq}\, \tilde{y}^q \left[\cosh^{-1} M_1^* + (-1)^q \cosh^{-1} M_2^* \right] \right\} \tag{18}$$

In the formulas (17) and (18) the following notations have been made:

$$N_1 = \sqrt{\frac{(1+k^*)(k^*-\tilde{y})}{2k^*(1-\tilde{y})}} \quad , \quad N_2 = \sqrt{\frac{(1+k^*)(k^*+\tilde{y})}{2k^*(1+\tilde{y})}} \tag{19a}$$

$$R_1^* = \sqrt{\frac{(1+\bar{\tilde{\nu}}^*)(1-\bar{\tilde{\nu}}^*\tilde{y})}{2\bar{\tilde{\nu}}^*(\bar{k}-\tilde{y})}} \quad , \quad R_2^* = \sqrt{\frac{(1+\bar{\tilde{\nu}}^*)(1+\bar{\tilde{\nu}}^*\tilde{y})}{2\bar{\tilde{\nu}}^*(\bar{k}+\tilde{y})}} \tag{19b}$$

$$M_1^* = \sqrt{\frac{(1+\tilde{\nu}^*)(1-\tilde{\nu}^*\tilde{y})}{2\tilde{\nu}^*(1-\tilde{y})}} \quad , \quad M_2^* = \sqrt{\frac{(1+\tilde{\nu}^*)(1+\tilde{\nu}^*\tilde{y})}{2\tilde{\nu}^*(1+\tilde{y})}} \tag{19c}$$

$$N_1^* = \sqrt{\frac{(1+\nu^*)(1-\tilde{\nu}^*\tilde{y})}{2\nu^*(k^*-\tilde{y})}} \quad , \quad N_2^* = \sqrt{\frac{(1+\nu^*)(1+\tilde{\nu}^*\tilde{y})}{2\nu^*(k^*+\tilde{y})}} \tag{19d}$$

These new formulas are obtained by the author by using the results of high conical flow theory of Germain [17], the hydrodynamic analogy of Carafoli [18], [19] and the principle of minimum singularities [10], [21].

The coefficients of the axial velocities $\tilde{u}$ and $\tilde{u}^*$ are related to the coefficients of the downwashes $\tilde{w}$, $\bar{\tilde{w}}$ and $\bar{w}^*$, $\tilde{w}^*$ and $\bar{\tilde{w}}^*$ through linear and homogeneous relations of the form:

$$\tilde{\bar{A}}_{n,2q} = \sum_{j=0}^{n-1} (\tilde{a}_{2q,j}^{(n)} \, \tilde{w}_{n-j-1,j} + \bar{\tilde{a}}_{2q,j}^{(n)} \, \bar{\tilde{w}}_{n-j-1,j}) \tag{20}$$

$$\tilde{\bar{P}}_{n,2q}^* = \sum_{j=0}^{n-1} (\tilde{p}_{2q,j}^{*(n)} \, \tilde{w}_{n-j-1,j}^* + \bar{p}_{2q,j}^{*(n)} \, \bar{w}_{n-j-1,j}^* + \bar{\tilde{p}}_{2q,j}^{*(n)} \, \bar{\tilde{w}}_{n-j-1,j}^*) \tag{21}$$

The coefficients $\tilde{a}_{2q,j}^{(n)}$, $\bar{\tilde{a}}_{2q,j}^{(n)}$, $\bar{p}_{2q,j}^{(n)}$, $\tilde{p}_{2q,j}^{(n)}$, $\bar{\tilde{p}}_{2q,j}^{*(n)}$ etc. are functions only on the similarity parameters $\bar{\tilde{\nu}}^*$, $\tilde{\nu}^*$ and ν^*.

The theoretical determined pressure coefficient C_p according to the present theory (i.e. by using the formulas (7) and (9) with $G_{nq}^* = 0$) are in good agreement with experimental results for a large range of Mach numbers ($M_\infty = 1,25 - 2,2$) and angles of attack α ($|\alpha| < 10°$) as it can be seen in (Fig. 6a,b) for the longitudinal central section and (Fig. 7a,b) for the transversal section $\tilde{x}_1 = 0,599$ of the upper side of the optimum-optimorum delta wing Adela and for the angles of attack $\alpha = -8°, +8°$.

This agreement between theory and experiment is due to the accuracy of the solutions of the boundary value problems for the axial disturbance velocities $\tilde{u}$ and $\tilde{u}^*$ given in formulas (7) and (8). These solutions for $\tilde{u}$ and $\tilde{u}^*$ present the following advantages in comparison with the ones obtained in the frame of slender body theory [26] - [29] :
- they fulfil the full-linearised partial differential equation, which is hyperbolic, includes the influence of Mach number M_∞ and does not need any restrictions con-

cerning the magnitude of span;

- the boundary conditions along the characteristic surface, i.e. the Mach cone of the apex of the integrated wing, and at the infinity (forward) are satisfied;

- according to the hydrodynamic analogy of Carafoli [18], [19] the singularities in these solutions of u and u^* are <u>located only along the singular lines</u> (i.e. along the leading edges of the wing, along the junction lines of the wing-fuselage configuration etc.) and therefore are easier to be applied as the solutions for axial disturbance velocities given in [28], [29], which are obtained by using singularities located on the whole wing surface;

- these singularities are chosen according to the principle of minimum singularities [14], [15] and therefore the potential solutions for u and u^* given here are <u>matched with a boundary layer solution</u> and are zonal solutions similar as in [31];

- the solutions (8) and (9) for u and u^* can be also used for the calculation of pressure distribution and of aerodynamic characteristics of space vehicle which shape is given in discrete form. The surface of the integrated wing can be piecewise approximated in form of polynomial expansions which are obtained by using the two- dimensional minimal quadratic error similar as in [7].

4. OPTIMIZATION OF THE AIRCRAFT WITH RETRACTED FLAPS

The optimization of the shape of the thin and thick-symmetrical integrated wings, components of the thick, lifting integrated delta wing are further treated. The variational problem for the thin integrated delta wing component (for a given value of the similarity parameter ν) leads to the determination of the coefficients $\tilde{w}_{ij}$ of the downwashes $\tilde{w}$ in such a manner, that the drag coefficient

$$C_d \equiv \ell \sum_{n=1}^{N} \sum_{m=1}^{N} \sum_{k=0}^{m-1} \sum_{j=0}^{n-1} \tilde{\Omega}_{nmkj} \cdot \tilde{w}_{n-j-1,j} \tilde{w}_{m-k-1,k} = \min. \tag{22}$$

In addition the following auxiliary conditions must be fulfilled. The lift coefficient $(C_\ell = \ell \tilde{C}_\ell)$ is given

$$\tilde{C}_\ell \equiv \sum_{n=1}^{N} \sum_{j=0}^{n-1} \tilde{\Lambda}_{nj} \, \tilde{w}_{n-j-1,j} = \frac{C_{\ell_0}}{\ell} \tag{23}$$

The pitching moment coefficient $(C_m = \ell \tilde{C}_m)$ is also given

$$\tilde{C}_m \equiv \sum_{n=1}^{N} \sum_{j=0}^{n-1} \tilde{\Gamma}_{nj} \, \tilde{w}_{n-j-1,j} = \frac{C_{m_0}}{\ell} \tag{24}$$

The axial disturbance velocity u vanishes along the leading edge

$$\tilde{F}_t \equiv \sum_{j=0}^{E(\frac{t}{2})} \tilde{\Psi}_{tj} \, \tilde{w}_{t-j-1,j} = 0 \, , \quad (t = 1,....,N) \tag{25}$$

The coefficients $\tilde{\Omega}_{nmkj}$, $\tilde{\Lambda}_{nj}$, $\tilde{\Gamma}_{nj}$ and $\tilde{\Psi}_{tj}$ are only functions of the similarity parameter ν. The corresponding Hamilton's operator H is

$$H \equiv \ell\,\tilde{H} = \ell\left[\tilde{C}_d + \lambda^{(1)}\tilde{C}_\ell + \lambda^{(2)}\tilde{C}_m + \sum_{t=1}^{N}\lambda_t\tilde{F}_t\right] \tag{26}$$

In this formula the Lagrange's multipliers $\lambda^{(1)}$, $\lambda^{(2)}$ and λ_t are functions of the similarity parameter ν. By cancellation of the coefficients of each independent variation $\delta\tilde{w}_{\theta\sigma}$ entering in the first variation of H the following equations are obtained

$$\sum_{n=1}^{N}\sum_{j=0}^{n-1}\left[\tilde{\Omega}_{n,\theta+\sigma+1,\sigma,j} + \tilde{\Omega}_{\theta+\sigma+1,n,j,\sigma}\right]\tilde{w}_{n-j-1,j} + \lambda^{(1)}\tilde{\Lambda}_{\theta+\sigma+1,\sigma} + \lambda^{(2)}\tilde{\Gamma}_{\theta+\sigma+1,\sigma} +$$

$$+ \lambda_{\theta+\sigma+1}\,\tilde{\Psi}_{\theta+\sigma+1,\sigma} = 0 \tag{27}$$

$$(1 \leq \theta+\sigma+1 \leq N\,,\quad \theta = 0,1,..,(N-1)\,)$$

These equations together with the auxiliary conditions (23), (24) and (25) form a linear algebraic system of equations which determines uniquely the optimum values of the coefficients $\tilde{w}_{ij}$ as well as the Lagrange's multipliers $\lambda^{(1)}$, $\lambda^{(2)}$ and λ_t for a given value of the similarity parameter ν. Similarly, the optimization of the shape of thick-symmetrical integrated delta wing component leads to the determination of the values of the $\tilde{w}^*_{ij}$ and $\bar{w}^*_{ij}$ of the downwashes $\tilde{w}^*$ and $\bar{w}^*$ (on the wing and on the fuselage) in such a manner, that the drag coefficient

$$C_d^* \equiv \ell\,\tilde{C}_d^* = \ell\sum_{n=1}^{N}\sum_{m=1}^{N}\sum_{k=0}^{n-1}\sum_{j=0}^{n-1}\left\{\left[\tilde{\Omega}^*_{nmkj}\,\tilde{w}^*_{n-j-1,j} + \tilde{\Omega}'^*_{nmkj}\,\bar{w}^*_{n-j-1,j}\right]\tilde{w}^*_{m-k-1,k} + \right.$$

$$\left. + \left[\bar{\Omega}^*_{nmkj}\,\tilde{w}^*_{n-j-1,j} + \bar{\Omega}'^*_{nmkj}\,\bar{w}^*_{n-j-1,j}\right]\bar{w}^*_{m-k-1,k}\right\} = \min. \tag{28}$$

Additionally, the following auxiliary conditions must be fulfilled:
- the cancellation of the thickness of the wing along its leading edges

$$\bar{F}^*_t \equiv \sum_{m=t+1}^{N}\sum_{k=0}^{m-1}\tilde{d}^{*(t)}_{mk}\,\tilde{w}^*_{m-k-1,k} = 0 \tag{29}$$

- the continuity of class C_1 of the surface along the junction line between the wing and the fuselage

$$\tilde{E}^*_t \equiv \sum_{m=t+1}^{N}\sum_{k=0}^{m-1}\tilde{c}^{*(t)}_{mk}\,(\tilde{w}^*_{m-k-1,k} - \bar{w}^*_{m-k-1,k}) = 0 \tag{30}$$

$$\tilde{G}^*_t \equiv \sum_{m=t+1}^{N}\sum_{k=0}^{m-1}\tilde{g}^{*(t)}_{mk}\,(\tilde{w}^*_{m-k-1,k} - \bar{w}^*_{m-k-1,k}) = 0 \tag{31}$$

$$\tilde{L}_t^* \equiv \sum_{m=t+1}^{N} \sum_{k=0}^{m-1} \tilde{\ell}_{mk}^{*(t)} \, (\tilde{w}_{m-k-1,k}^* - \bar{w}_{m-k-1,k}^*) = 0 \tag{32}$$

$$(t = 0,1,...,(N-1))$$

- the given relative volume of the wing

$$\tilde{\tau} \equiv \sum_{m=1}^{N} \sum_{k=0}^{m-1} \tilde{\tau}_{mk}^* \, \tilde{w}_{m-k-1,k}^* = \tilde{\tau}_0 \sqrt{\ell} \tag{33}$$

- the given relative volume of the fuselage

$$\bar{\tau} \equiv \sum_{m=1}^{N} \sum_{k=0}^{m-1} \bar{\tau}_{mk}^* \, \bar{w}_{m-k-1,k}^* = \bar{\tau}_0 \sqrt{\ell} \tag{34}$$

The corresponding Hamilton's operator H^* of this variational problem is

$$H^* \equiv \ell \, \tilde{H}^* = \ell \left[\tilde{C}_d^* + \mu^{(1)}\tilde{\tau}^* + \mu^{(2)}\bar{\tau}^* + \sum_{t=1}^{N} (\mu_t \tilde{F}_t + \right.$$

$$\left. + \bar{\mu}_t \bar{E}_t + \eta_t \tilde{G}_t + \bar{\eta}_t \tilde{L}_t) \right] \tag{35}$$

Here $\mu^{(1)}$, $\mu^{(2)}$, μ_t, $\bar{\mu}_t$, η_t, $\bar{\eta}_t$ are Lagrange's multipliers. If the first variation of H^* is cancelled, the following equations are obtained

$$\sum_{n=1}^{N} \sum_{j=0}^{n-1} \left\{ \left[\tilde{\Omega}_{n,\theta+\sigma+1,\sigma,j}^* + \tilde{\Omega}_{\theta+\sigma+1,n,j,\sigma}^* \right] \tilde{w}_{n-j-1,j}^* + \left[\tilde{\Omega'}_{n,\theta+\sigma+1,\sigma,j}^* + \right. \right.$$

$$\left. \left. + \bar{\Omega}_{\theta+\sigma+1,n,j,\sigma}^* \right] \bar{w}_{n-j-1,j}^* \right\} + \mu^{(1)}\tilde{\tau}_{\theta+\sigma+1,\sigma}^* + \sum_{t=1}^{N} \left[\mu_t \, \tilde{d}_{\theta+\sigma+1,\sigma}^{*(t)} + \right.$$

$$\left. + \bar{\mu}_t \, \tilde{c}_{\theta+\sigma+1,\sigma}^{*(t)} + \eta_t \, \tilde{g}_{\theta+\sigma+1,\sigma}^{*(t)} + \bar{\eta}_t \, \tilde{\ell}_{\theta+\sigma+1,\sigma}^{*(t)} \right] = 0 \tag{36}$$

and

$$\sum_{n=1}^{N} \sum_{j=0}^{n-1} \left\{ \left[\bar{\Omega}_{n,\theta+\sigma+1,\sigma,j}^* + \tilde{\Omega'}_{\theta+\sigma+1,n,j,\sigma}^* \right] \tilde{w}_{n-j-1,j}^* + \left[\bar{\Omega'}_{n,\theta+\sigma+1,\sigma,j}^* + \right. \right.$$

$$\left. \left. + \bar{\Omega'}_{\theta+\sigma+1,n,j,\sigma}^* \right] \bar{w}_{n-j-1,j}^* \right\} + \mu^{(2)}\bar{\tau}_{\theta+\sigma+1,\sigma}^* - \sum_{t=1}^{N} \left[\bar{\mu}_t \, \tilde{c}_{\theta+\sigma+1,\sigma}^{*(t)} + \right.$$

$$\left. + \eta_t \, \tilde{g}_{\theta+\sigma+1,\sigma}^{*(t)} + \bar{\eta}_t \, \tilde{\ell}_{\theta+\sigma+1,\sigma}^{*(t)} \right] = 0 \tag{37}$$

$$(1 \leq \theta+\sigma+1 \leq N, \quad \theta = 0,1,..,(N-1))$$

These equations together with the auxiliary conditions (29) – (34) form a linear algebraic system of equations which determines uniquely the optimum values of the coefficients $\tilde{w}_{\theta\sigma}^*$ and $\bar{w}_{\theta\sigma}^*$ as well as the values of Lagrange's multipliers for a given value of ν. By using the hybrid analytical-numerical method of the author [1] - [4] the optimum-optimorum shape of the thick, lifting integrated delta wing is determined. The limit line I of the drag functional $C_d^{(t)}$ of the thick, lifting integrated delta wing is introduced in (Fig. 9) i.e.

$$(C_d^{(t)})_{opt} = f(\nu) \ , \qquad (C_d^{(t)} = C_d + C_d^*) \tag{38}$$

Each point of the limit line is analytically determined by solving a classical variational problem for a given value of parameter ν. The value $\nu = \nu_{opt}$, for which the limit line I attains its minimum is obtained numerically by systematical variation of the similarity parameter ν as in [4] , [14] , [15] . The fitting of an integrated fuselage produces an important modification of the optimal thickness distribution as it can be seen in (Fig. 1a,b), for the section $\tilde{x}_1 = 0,6$.

<u>5. OPTIMIZATION OF THE FLAPS SHAPE</u>

The optimization of the shape of the thin and thick-symmetrical open integrated wing (i.e. the wing-fuselage configuration with flaps in open position) at the second, lower Mach number M_∞^* is here considered (Fig. 1c). The shape of the integrated wing-fuselage configuration (with flaps in retracted position) is determined by the solution of the precedent variational problem and remains unchanged in this second variational problem, which consists in the determination of the shape of the flap in such a manner, that the entire space vehicle (with flaps in open position) is of minimum drag at the second cruising Mach number M_∞^* . The variational problem of the thin open integrated delta wing is firstly considered. The downwashes $\tilde{w}$ (on the transformed thin integrated wing-fuselage configuration) and $\tilde{\tilde{w}}$ on the flap are given as in formulas (3) and (12). The coefficients $\tilde{w}_{ij}$ of w are previously determined by the precedent variational problem and are here supposed known and constant. The optimization of the thin flap component (for a given value of ν) leads to the determination of the coefficient $\tilde{\tilde{w}}_{ij}$ of the downwash $\tilde{\tilde{w}}$ in such a manner, that the drag coefficient C_d' at the cruising Mach number M_∞^* attains its minimum i.e.

$$C_d' = \ell \sum_{n=1}^{N} \sum_{m=1}^{N} \sum_{k=0}^{m-1} \sum_{j=0}^{n-1} \left\{ \left[\tilde{\Omega}_{nmkj} \, \tilde{w}_{n-j-1,j} + \tilde{\tilde{\Omega}}_{nmkj} \, \tilde{\tilde{w}}_{n-j-1,j} \right] \tilde{w}_{m-k-1,k} + \right.$$

$$\left. + \left[\tilde{\Omega}'_{nmkj} \, \tilde{w}_{n-j-1,j} + \tilde{\tilde{\Omega}}'_{nmkj} \, \tilde{\tilde{w}}_{n-j-1,j} \right] \tilde{\tilde{w}}_{m-k-1,k} \right\} = \min. \tag{39}$$

with the following auxiliary condition (at the cruising Mach number M_∞^*).

  - The lift coefficient C_ℓ' is given:

$$\tilde{\tilde{C}}_\ell \equiv \sum_{n=1}^{N} \sum_{j=0}^{n-1} (\tilde{\Lambda}'_{nj}\, \tilde{w}_{n-j-1,j} + \tilde{\tilde{\Lambda}}_{nj}\, \tilde{\tilde{w}}_{n-j-1,j}) = \frac{C'_{\ell_0}}{\ell} \tag{40}$$

- The pitching moment coefficient C'_m is given

$$\tilde{\tilde{C}}_m \equiv \sum_{n=1}^{N} \sum_{j=0}^{n-1} (\tilde{\Gamma}'_{nj}\, \tilde{w}_{n-j-1,j} + \tilde{\tilde{\Gamma}}_{nj}\, \tilde{\tilde{w}}_{n-j-1,j}) = \frac{C'_{m_0}}{\ell} \tag{41}$$

- The axial disturbance velocity $\tilde{\tilde{u}}$ vanishes along the leading edge

$$\tilde{\tilde{F}}_t \equiv \sum_{j=0}^{E(\frac{t}{2})} \tilde{\tilde{\Psi}}_{tj}\, \tilde{\tilde{w}}_{t-j-1,j} = 0 \quad , \quad (t = 1,...,N) \tag{42}$$

- The wing and the flap surface are continuous of class C_1 along the junction line between the wing and the flap

$$\tilde{\tilde{E}}_t \equiv \sum_{t=1}^{N} \sum_{k=0}^{m-1} \tilde{\tilde{c}}^{(t)}_{mk} (\tilde{\tilde{w}}_{m-k-1,k} - \tilde{w}_{m-k-1,k}) = 0 \tag{43}$$

$$\tilde{\tilde{G}}_t \equiv \sum_{t=1}^{N} \sum_{k=0}^{m-1} \tilde{\tilde{g}}^{(t)}_{mk} (\tilde{\tilde{w}}_{m-k-1,k} - \tilde{w}_{m-k-1,k}) = 0 \tag{44}$$

$$\tilde{\tilde{L}}_t \equiv \sum_{t=1}^{N} \sum_{k=0}^{m-1} \tilde{\tilde{\ell}}^{(t)}_{mk} (\tilde{\tilde{w}}_{m-k-1,k} - \tilde{w}_{m-k-1,k}) = 0 \tag{45}$$

All the coefficients $\tilde{\Lambda}'_{nj}$, $\tilde{\tilde{\Lambda}}_{nj}$ etc. are depending only on the similarity parameters ν^* and $\tilde{\nu}^*$ ($\tilde{\nu}^*$ is here constant). The corresponding Hamilton's operator of this variational problem is

$$\tilde{\tilde{H}} \equiv \ell\, \tilde{\tilde{H}}' = \ell \left[\tilde{\tilde{C}}_d + \tilde{\tilde{\lambda}}^{(1)}\tilde{\tilde{C}}_\ell + \tilde{\tilde{\lambda}}^{(2)}\tilde{\tilde{C}}_m + \sum_{t=1}^{N} (\bar{\tilde{\lambda}}_t \tilde{\tilde{F}}_t + \tilde{\mu}_t \tilde{\tilde{E}}_t + \bar{\mu}_t \tilde{\tilde{G}}_t + \tilde{\tilde{\mu}}_t \tilde{\tilde{L}}_t) \right] \tag{46}$$

<u>Remark:</u> The Lagrange's multipliers $\tilde{\tilde{\lambda}}^{(1)}$, $\tilde{\tilde{\lambda}}^{(2)}$, $\bar{\tilde{\lambda}}_t$, $\tilde{\mu}_t$, $\bar{\mu}_t$ and $\tilde{\tilde{\mu}}_t$ are only functions of the similarity parameter ν^*. If the first variation of the Hamilton's operator $\tilde{\tilde{H}}$ is cancelled ($\delta\tilde{\tilde{H}} = 0$) the following equations are obtained:

$$\sum_{n=1}^{N} \sum_{j=0}^{n-1} \left\{ \left[\tilde{\tilde{\Omega}}'_{n,\theta+\sigma+1,\sigma,j} + \tilde{\tilde{\Omega}}'_{\theta+\sigma+1,n,j,\sigma} \right] \tilde{w}_{n-j-1,j} + \left[\tilde{\tilde{\Omega}}'_{n,\theta+\sigma+1,\sigma,j} + \right. \right.$$

$$\left. \left. + \tilde{\tilde{\Omega}}_{\theta+\sigma+1,n,j,\sigma} \right] \tilde{w}_{n-j-1,j} \right\} + \tilde{\tilde{\lambda}}^{(1)} \tilde{\tilde{\Lambda}}_{\theta+\sigma+1,\sigma} + \tilde{\tilde{\lambda}}^{(2)}\tilde{\tilde{\Gamma}}_{\theta+\sigma+1,\sigma} + \bar{\tilde{\lambda}}_{\theta+\sigma+1,\sigma}\tilde{\tilde{\Psi}}_{\theta+\sigma+1,\sigma} +$$

$$+ \sum_{t=1}^{N} \left[\tilde{\mu}_t\, \tilde{\tilde{c}}^{(t)}_{\theta\sigma} + \bar{\mu}_t\, \tilde{\tilde{g}}^{(t)}_{\theta\sigma} + \tilde{\tilde{\mu}}_t\, \tilde{\tilde{\ell}}^{(t)}_{\theta\sigma} \right] = 0 \tag{47}$$

$$(1 \leq \theta+\sigma+1 \leq N , \quad \theta = 0,1,..,(N-1))$$

This algebraic system, together with the auxiliary conditions (26)-(31) determine uniquely the coefficients $\tilde{\bar{w}}_{n-j-1,j}$ of the downwashes and the Lagrange's multipliers $\tilde{\bar{\lambda}}^{(1)}$, $\tilde{\bar{\lambda}}^{(2)}$, $\bar{\lambda}_t$, $\tilde{\mu}_t$, $\bar{\mu}_t$ and $\tilde{\bar{\mu}}_t$ (as functions of the similarity parameter ν^*).

Let us consider now the second variational problem concerning the optimization of the thick-symmetrical component of the flap by cruising Mach number M_∞^* . The downwashes coefficients $\tilde{\bar{w}}_{ij}^*$ are determined in such a manner that the drag coefficient $\tilde{\bar{C}}_d{}'^*$ of the thick-symmetrical component of the open integrated wing attains its minimum (at cruising Mach number M_∞^*) i.e.

$$C_d'^* \equiv \ell \sum_{n=1}^{N} \sum_{m=1}^{N} \sum_{k=0}^{m-1} \sum_{j=0}^{n-1} \left\{ \left[\tilde{\Omega}_{nmkj}^{**} \, \tilde{w}_{n-j-1,j}^* + \bar{\Omega}_{nmkj}^{**} \, \bar{w}_{n-j-1,j}^* + \right. \right.$$

$$+ \left. \tilde{\bar{\Omega}}_{nmkj}^{**} \, \tilde{\bar{w}}_{n-j-1,j}^* \right] \tilde{w}_{m-k-1,k}^* + \left[\tilde{\Omega}'^{**}_{nmkj} \, \tilde{w}_{n-j-1,j}^* + \bar{\Omega}'^{**}_{nmkj} \, \bar{w}_{n-j-1,j}^* + \right.$$

$$+ \left. \tilde{\bar{\Omega}}'^{**}_{nmkj} \, \tilde{\bar{w}}_{n-j-1,j}^* \right] \bar{w}_{m-k-1,k}^* + \left[\tilde{\Omega}_{nmkj}^{(**)} \, \tilde{w}_{n-j-1,j}^* + \right.$$

$$+ \left. \bar{\Omega}_{nmkj}^{(**)} \, \bar{w}_{n-j-1,j}^* + \tilde{\bar{\Omega}}_{nmkj}^{(**)} \, \tilde{\bar{w}}_{n-j-1,j}^* \right] \tilde{\bar{w}}_{m-k-1,k}^* \right\} = \min. \tag{48}$$

Additionally auxiliary conditions are considered:

- the flap is of null-thickness along its leading edges

$$\tilde{\bar{F}}_t^* = \sum_{m=t+1}^{N} \tilde{\bar{d}}_{mk}^{*(t)} \, \tilde{\bar{w}}_{m-k-1,k}^* = 0 \tag{49}$$

- the thick-symmetrical flap is integrated i.e. along the junction line between the wing and the flap the surface must be continuous of class C_1 i.e.

$$\tilde{\bar{E}}_t'^* \equiv \sum_{m=t+1}^{N} \sum_{k=0}^{m-1} \tilde{\bar{c}}_{mk}^{*(t)} \, (\tilde{\bar{w}}_{m-k-1,k}^* - \tilde{w}_{m-k-1,k}^*) = 0 \tag{50}$$

$$\tilde{\bar{G}}_t^* \equiv \sum_{m=t+1}^{N} \sum_{k=0}^{m-1} \tilde{\bar{g}}_{mk}^{*(t)} \, (\tilde{\bar{w}}_{m-k-1,k}^* - \tilde{w}_{m-k-1,k}^*) = 0 \tag{51}$$

$$\tilde{\bar{L}}_t^* \equiv \sum_{m=t+1}^{N} \sum_{k=0}^{m-1} \tilde{\bar{\ell}}_{mk}^{*(t)} \, (\tilde{\bar{w}}_{m-k-1,k}^* - \tilde{w}_{m-k-1,k}^*) = 0 \tag{52}$$

$$(\ t = 0,1,..,(N-1) \ )$$

- the relative volume τ' of the flap is given:

$$\tilde{\bar{\tau}}^* \equiv \sum_{m=1}^{N} \sum_{k=0}^{m-1} \tilde{\bar{\tau}}_{mk}^* \, \tilde{\bar{w}}_{m-k-1,k}^* = \tau_0' \, \sqrt{\ell} \tag{53}$$

The Hamilton's operator $\tilde{\bar{H}}$ of this variational problem is

$$\tilde{\bar{H}}^* = \ell \, \tilde{\bar{H}}'^* = \ell \left[\tilde{\bar{C}}_d^* + \mu^* \tilde{\bar{\tau}}^* + \sum_{t=1}^{N} (\tilde{\mu}_t^* \tilde{\bar{F}}_t^* + \bar{\mu}_t^* \tilde{\bar{E}}_t + \tilde{\eta}_t^* \tilde{\bar{G}}_t^* + \bar{\eta}_t^* \tilde{\bar{L}}_t^*) \right] \tag{54}$$

Here μ^*, $\tilde{\mu}_t^*$, $\bar{\mu}_t^*$, $\tilde{\eta}_t^*$ and $\bar{\eta}_t^*$ are the Lagrange's multipliers. If the first variation of $\tilde{\bar{H}}^*$ is cancelled the following equations are obtained

$$\sum_{n=1}^{N} \sum_{j=0}^{n-1} \left\{ \left[\tilde{\bar{\Omega}}^{(**)}_{n,\theta+\sigma+1,\sigma,j} + \tilde{\bar{\Omega}}^{(**)}_{\theta+\sigma+1,n,j,\sigma} \right] \tilde{\bar{w}}^*_{n-j-1,j} + \left[\tilde{\Omega}^{(**)}_{n,\theta+\sigma+1,\sigma,j} \right. \right.$$

$$\left. + \tilde{\bar{\Omega}}^{**}_{\theta+\sigma+1,n,j,\sigma} \right] \tilde{w}^*_{n-j-1,j} + \left[\bar{\Omega}^{(**)}_{n,\theta+\sigma+1,\sigma,j} + \tilde{\bar{\Omega}}'^{**}_{\theta+\sigma+1,n,j,\sigma} \right] \bar{w}^*_{n-j-1,j} \right\} +$$

$$+ \mu^* \tilde{\bar{\tau}}^*_{\theta+\sigma+1,\sigma} + \sum_{t=1}^{N} \left[\tilde{\mu}_t^* \tilde{\bar{d}}^{*(t)}_{\theta+\sigma+1,\sigma} + \bar{\mu}_t^* \tilde{\bar{c}}^{*(t)}_{\theta+\sigma+1,\sigma} \right.$$

$$\left. + \tilde{\eta}_t^* \tilde{\bar{g}}^{*(t)}_{\theta+\sigma+1,\sigma} + \bar{\eta}_t^* \tilde{\bar{\ell}}^{*(t)}_{\theta+\sigma+1,\sigma} \right] = 0 \tag{55}$$

$$(1 \leq \theta+\sigma+1 \leq N, \quad \theta = 0,1,..,(N\text{-}1))$$

These equations, together with the auxiliary conditions (52)-(56) form a linear algebraic system which determines uniquely the values of the coefficients $\tilde{\bar{w}}^*_{\theta\sigma}$ of the downwash $\tilde{\bar{w}}^*$ and the values of the Lagrange's multipliers μ^*, $\tilde{\mu}_t^*$, $\bar{\mu}_t^*$, $\tilde{\eta}_t^*$ and $\bar{\eta}_t^*$ as function of the similarity parameter ν^*. The best value of the similarity parameter ν^* for the thick-lifting flap (at cruising Mach number M_∞^*) can be also determined by using the hybrid numerical-analytical method of the author as in [1] - [4]. The optimal value of ν ($\nu = \nu_{\text{opt}}$) is the position of the minimum of the lower limit line of the drag functional $(\tilde{\bar{C}}_d^{(t)})_{\text{opt}}$ (Fig. 7)

$$(\tilde{\bar{C}}_d^{(t)})_{\text{opt}} = f(\nu) \qquad (\tilde{\bar{C}}_d^{(t)} = \tilde{\bar{C}}_d + \tilde{\bar{C}}_d^*) \tag{56}$$

6. AGREEMENT WITH EXPERIMENTAL RESULTS

The aerodynamic characteristics of the optimum-optimorum wing Model Adela (Fig. 3) were measured in the frame work of a DFG research contract, by the author and collaborators, in trisonic wind tunnel (section 60 x 60 cm^2) of the DFVLR-Köln. The theoretically predicted values of the lift and pitching moment coefficients C_ℓ and C_m according to the above theory are in very good agreement with the experimental results for the all range of Mach numbers ($M_\infty = 1,25 - 2,2$) and angles of attack α ($|\alpha| < 14°$) taken here into consideration as in [7] [23]. The dependence of lift- and pitching moment coefficients C_ℓ and C_m versus the angle of attack α is linear in supersonic flow also at higher angle of attack α as in (Fig. 8a,b) and (Fig. 9a,b), but non-linear with respect to the Mach number M_∞ as in (Fig. 10a,b) and (Fig. 11a,b). Recently [35], [37] the measurements of C_ℓ and C_m on wedged delta wing model (Fig. 12), performed at angles of attack $|\alpha| \leq 16°$ and higher supersonic Mach numbers ($M_\infty = 2,4 \div 4,0$) are in good agreement with the values of C_ℓ and C_m predicted by the present theory for supersonic leading edges (Fig. 13a,b) for all the ranges of angles of attack α (i.e. $|\alpha| < 16°$) for C_ℓ and C_m at $M_\infty = 2,4 \div 3,2$. At higher Mach numbers i.e. $M_\infty = 3,6 \div 4,0$ the range of angles of attack α for a good agreement is more reduced i.e. $|\alpha| \leq 10°$ for C_ℓ and $|\alpha| \leq 12°$ for C_m . The time of calculation of the aerodynamic characteristics of the space vehicle, by using own softwares according to this theory is less than 4 seconds on Cyber 175!

7. CONCLUSIONS

The optimum-optimorum theory of the author can be successfully applied for the global-optimization of the entire configuration of the space vehicle, which shape presents the following advantages: a) it is <u>total integrated</u> (i.e. wing-fuselage and wing-flap integration) and therefore has no drag due to corners; b) it is of minimum drag for two different supersonic/hypersonic cruising Mach numbers and therefore is useful for space vehicle; c) it is of high lift due to Kutta auxiliary conditions along the leading edges; d) it presents a reduced drag and increased lift for a large range of Mach numbers and angles of attack. The hybrid numerical-analytical method of the author which allows the effective determination of the optimum-optimorum shape of the space vehicle presents the following advantages: a) it is <u>accurate</u> because it allows the <u>simultaneous</u> optimization of all geometrical parameters of its shape; b) it is <u>flexible</u> while it can be applied to the optimization of complex shape of space vehicle and allows, to add or to suppress some auxiliary conditions and to change the cruising Mach number chosen for the optimization; c) it is <u>able to determine</u> the shape of the <u>space vehicle of variable geometry</u> in order to obtain a minimum drag at two, very different, supersonic/hypersonic cruising Mach numbers; d) <u>it is fast</u> (6 sec. computer time at Cyber 175) for the full-optimization of the space vehicle shape!

REFERENCES

1. <u>NASTASE, A.:</u> Use of Computers in the Optimization of Aerodynamic Shapes (in Rom.).Ed. Acad. Romania, 1973, 280p.

2. <u>NASTASE, A.:</u> Eine graphisch-analytische Methode zur Bestimmung der Optimum-Optimorum Form dünner Deltaflügel in Überschallströmungen, RRST-MA 1, 19, (1974) (Romania).

3. <u>NASTASE, A.:</u> Eine graphisch-analytische Methode zur Bestimmung der Optimum-Optimorum Form symmetrisch-dicker Deltaflügel in Überschallströmungen. RRST-MA 1, 19, (1974) (Romania).

4. <u>NASTASE, A.:</u> Die Theorie des Optimum-Optimorum Tragflügels im Überschall. ZAMM 57 (1977). (Germany)

5. <u>NASTASE, A.:</u> Modern Concepts for Design of Delta Wings for Supersonic Aircraft of Second Generation. ZAMM 59, (1979). (Germany)

6. NASTASE, A.: New Concepts for Design of Fully-Optimized Configurations for Future Supersonic Aircraft. ICAS- Proceedings (1980), Munich.

7. NASTASE, A.: Optimierte Tragflügelformen in Überschallströmungen. Herchen Verlag Frankfurt, 1988, 300p.

8. NASTASE, A.: Contribution a l'Etude des Formes Aerodynamiques Optimales. Faculte des Sciences de Paris en Sorbonne, 150p., 1970, These.

9. NASTASE, A.: The Thin Delta Wing with Variable Geometry, Optimum for Two Supersonic Cruising Speeds. RRST-MA 3, 14, (1969). (Romania)

10. NASTASE, A.: The Delta Wing of Symmetrical-Thickness with Variable Geometry, Optimum for Two Supersonic Cruising Speeds. RRST-MA 6, 15, (1970). (Romania)

11. NASTASE, A.: Wing Optimization and Fuselage Integration for Future Generation of Supersonic Aircraft. Israel Journal of Technology, Jerusalem (1985).

12. NASTASE, A.: Computation of Wing-Fuselage Configuration for Supersonic Aircraft. Numerical Methods in Fluid Mechanics II, (Ed. K. Oshima), Tokyo, (1987).

13. NASTASE, A.: Computation of Fully-Optimized Wing-Fuselage Configuration for Future Generation of Supersonic Aircraft. Integral Methods in Science and Engineering (Ed. F. Payne, C. Corduneanu, A. Haji-Sheikh, T. Huang), Hemisphere Corp., Washington D.C., 650p., 1986.

14. NASTASE, A.: Optimum-Optimorum Wing-Fuselage Integration in Transonic-Supersonic Flow. Proceedings of High Speed Aerodynamics, (Ed. A. Nastase), 220p, Herchen Verlag Frankfurt, 1987.

15. NASTASE, A.: Optimum-Optimorum Integrated Wing-Fuselage Configuration for Supersonic Transport Aircraft of Second Generation, ICAS-Proceedings, London (1987).

16. NASTASE, A.: Optimum Aerodynamic Shape by Means of Variational Method (in Rom.). Ed. Acad. of Romania, 1969, 240p.

17. GERMAIN, P.: La Theorie des Mouvements Homogenes et son Application au Calcul de Certaines Ailes Delta en Regime Supersonique. Rech. Aero., 7, 3-16, (1949). (France)

18. CARAFOLI, E.: About the Hydrodynamic Character of the Solutions of Conical Flow Used in the Theory of Polygonal Wings II, Com. Acad. Romania, (1952).

19. CARAFOLI, E., MATEESCU, D., NASTASE, A.: Wing Theory in Supersonic Flow. Pergamon Press, London, 1969, 500p.

20. VAN DYKE, M.: Perturbation Methods in Fluid Mechanics. Acad. Press, New York, (1964).

21. NASTASE, A.: L'Etude du Comportement Asymptotique des Vitesses Axiales de Perturbation au Voisinage des Singularites. RRST-MA 4, 17, (1972).

22. NASTASE, A.: Validity of Solution of Three-Dimensional Linearised Boundary Value Problem for Axial Disturbance Velocity, in Transonic-Supersonic Flow.

ZAMM 65, (1985).

23. NASTASE, A., SCHEICH, A.: Theoretical Prediction of Aerodynamic Characteristics of Wings in Transonic-Supersonic Flow at Higher Angles of Attack and Its Agreement with Experimental Results. ZAMM 67, 5, (1987).

24. STOLLINGS, R.L., LAMB, M.: Wing Alone Aerodynamic Characteristics for High Angles of Attack at Supersonic Speeds. NASA-TP 1889, (1981).

25. NASTASE, A.: The Optimum-Optimorum Theory and its Application to the Optimization of the Entire Supersonic Transport Aircraft. Computational Fluid Dynamics (Ed. G. de Vahl Davis, C. Fletcher), North Holland 1988.

26. KÜCHEMANN, D.: The Aerodynamic Design of Aircraft. Pergamon Press, London, 1978, Chap. 5.

27. KEUNE, F., BURG, K.: Singularitätenverfahren der Strömungslehre. Braun Verlag Karlsruhe, 1975.

28. JONES, R., COHEN, D.: Aerodynamics of Wings at High Speeds. Princeton New Jersey, 1957, (Ed. Donovan, A., and Lawrence, H.).

29. FERRARI, C.: Interaction Problems. Princeton New Jersey, 1957, (Ed. Donovan, A., and Lawrence, H.).

30. SZEMA, K.Y., SHANKAR, V., RIBA, W.L., GORSKI, J.: Full Potential Treatment of Flows over 3-D Geometries Including Multibody Configurations. Numerical Methods in Fluid Mechanics II, Tokyo, 1987 (Ed. K. Oshima).

31. HOLST, T.L., THOMAS, S.D., KAYNAK, U., GUNDY, K.L., FLORES, J., CHADERJIAN, N.: Computational Aspects of Zonal Algorithms for Solving the Compressible Navier-Stokes Equations in Three Dimensions. Numerical Methods in Fluid Mechanics II, Tokyo, 1987 (Ed. K. Oshima).

32. NASTASE, A.: The Design of Optimum-Optimorum Shape of Space Vehicle. Proceedings of the First International Conference on Hypersonic Flight in the 21st Century (Ed. M.E. Higbea and J.A. Vedda), University of North Dakota, Grand Forks (1988).

33. NASTASE, A.: The Space Vehicle of Variable Geometry, Optimum for Two Supersonic Cruising Speeds. ZAMM 69 (1989).

34. NASTASE, A., STANISAV, E.: Prediction of Pressure Distribution on Optimum-Optimorum Delta Wing at Higher Angles of Attack in Supersonic Flow and Its Agreement with Experimental Results. ZAMM 70 (1990).

35. NASTASE, A., RUDIANU, C.: Theoretical Prediction of Pressure Distribution on Wedged Delta Wing at Higher Supersonic Mach Numbers and Its Agreement with Experimental Results. ZAMM 70 (1990).

36. NASTASE, A.: The Design of Supersonic Aircraft and Space Vehicles by Using Global Optimization Techniques. Collection of Technical Papers ISCFD, Nagoya, 1989.

37. NASTASE, A., HONERMANN, A.: Theoretical Prediction of Aerodynamic Characteristics of Delta Wings with Supersonic Leading Edges, in Supersonic-Hypersonic Flow and Its Agreement with Theoretical Results. ZAMM 71 (1991).

FIGURES 1a,b,c

FIGURES 2 , 3

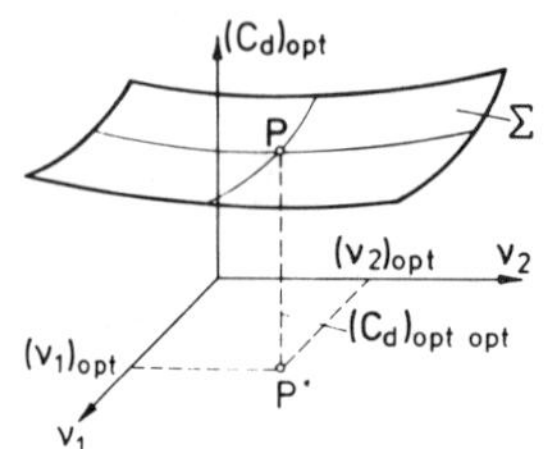

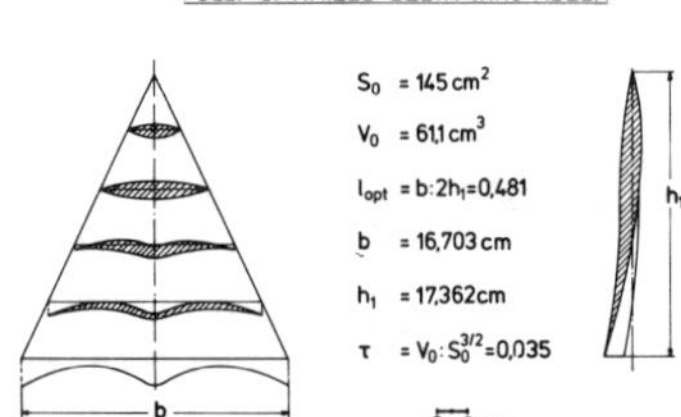

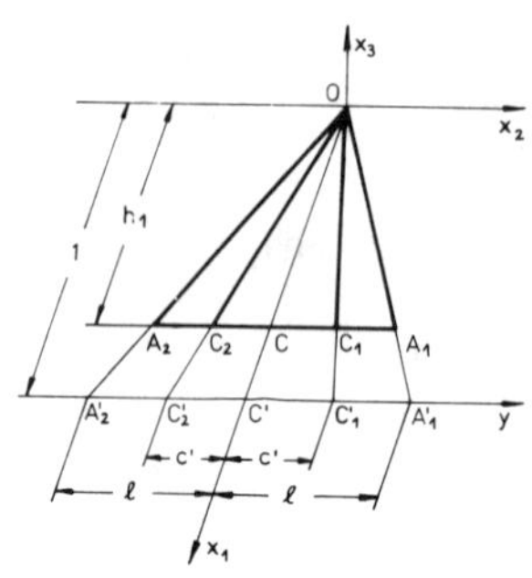

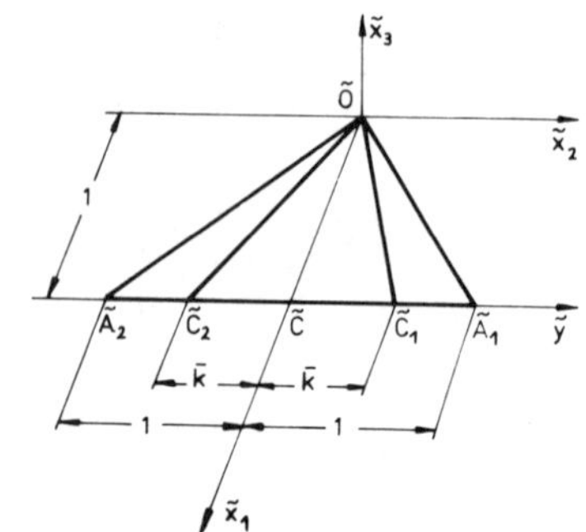

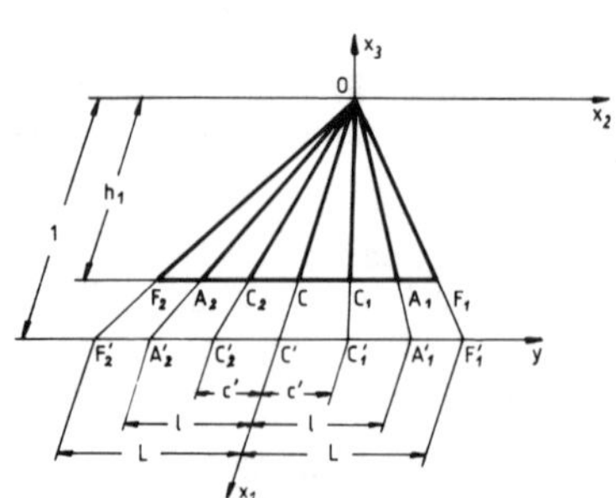

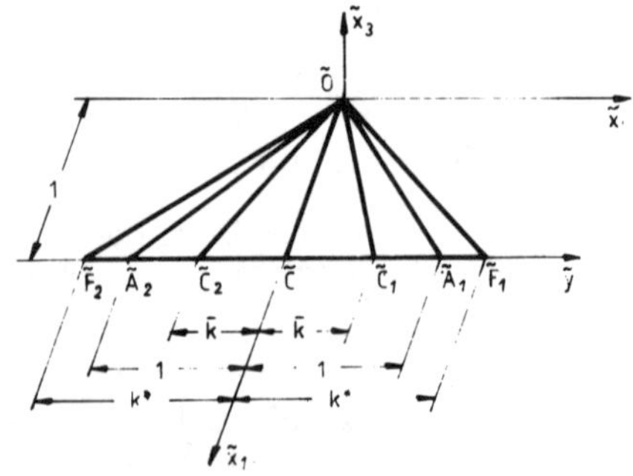

FIGURES 4a-d

FIGURES 5a,b

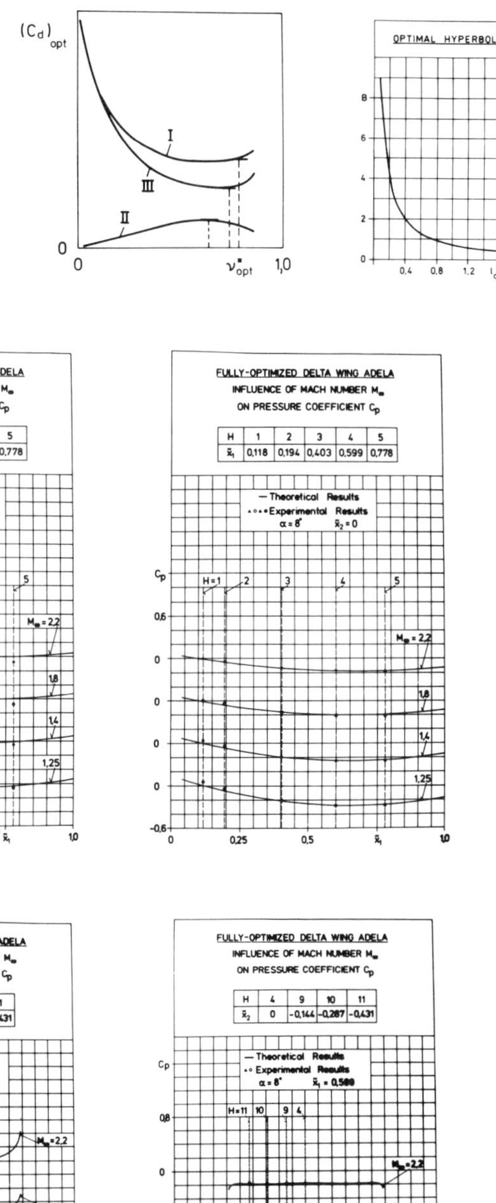

FIGURES 6a,b and 7a,b

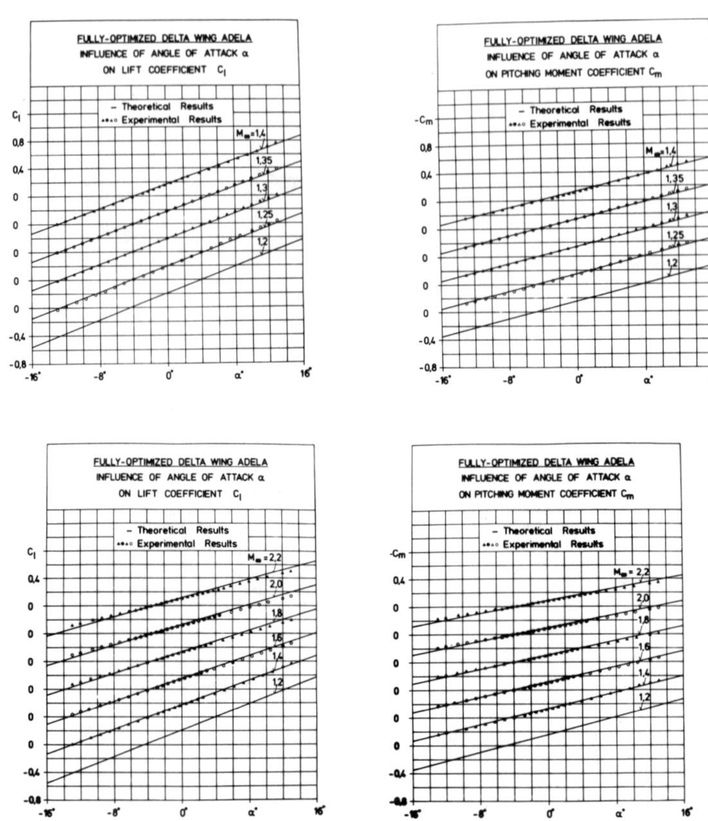

FIGURES 8a,b and 9a,b

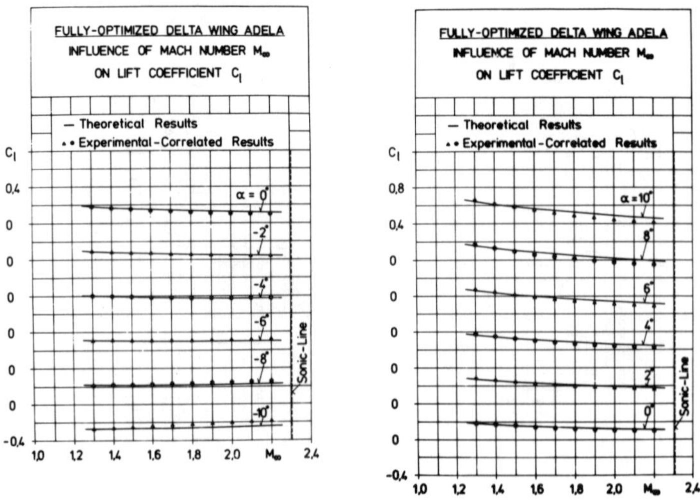

FIGURES 10a,b

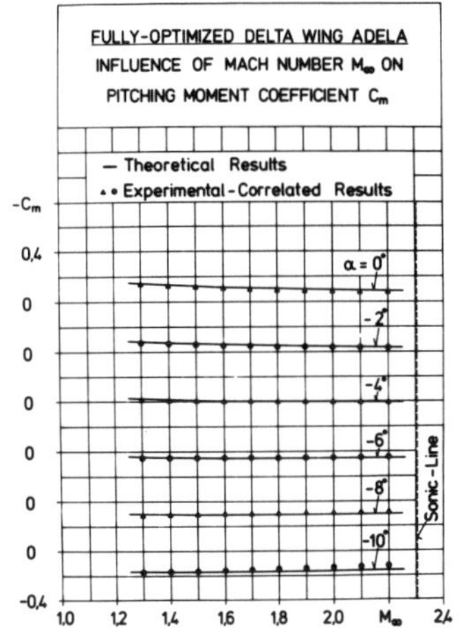

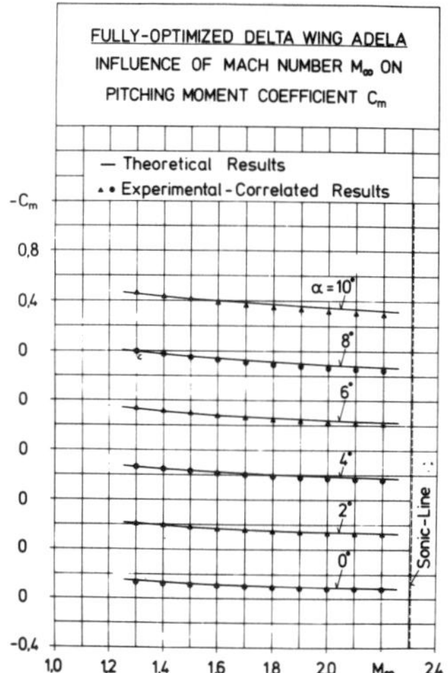

FIGURES 11a,b

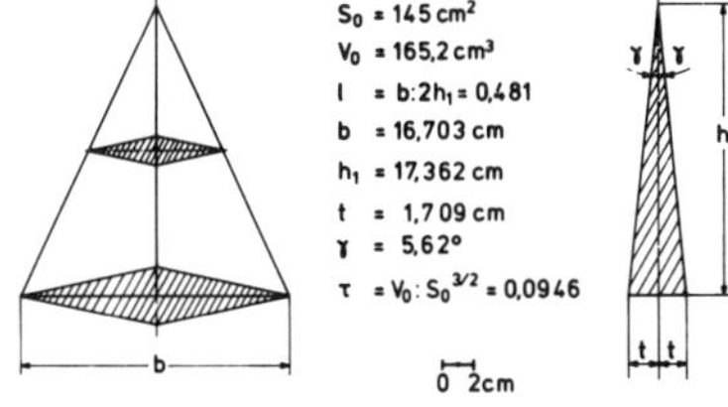

FIGURE 12

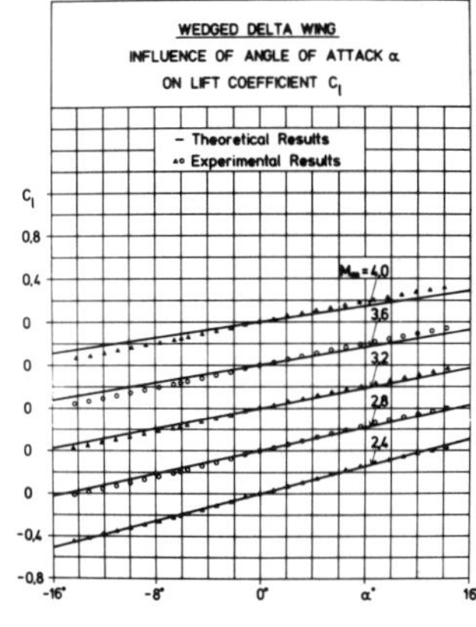

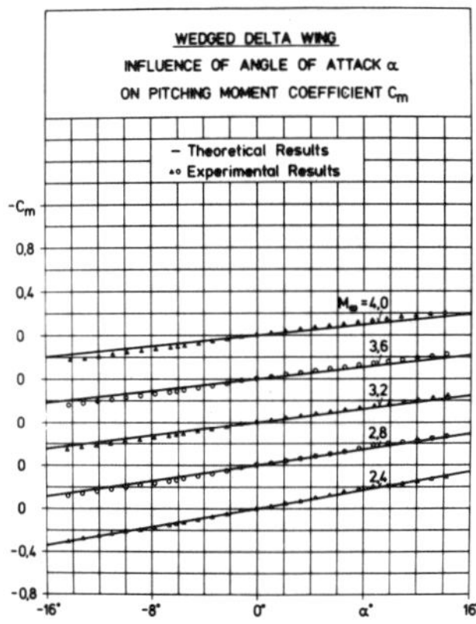

FIGURES 13a,b

A Field Integral Method for Solutions of the Full Potential Equation in Rectangular Grids

Z. FANG and I. PARASCHIVOIU
Department of Mechanical Engineering
Ecole Polytechnique de Montreal
Montreal, Quebec, Canada

A field integral method based on Green's function theory has been developed to solve the full potential equation for subsonic and transonic flows. In this integral method, potential values in the flow region are determined by potential values represented by boundary integrals and a volume integral. The boundary potential values are obtained by implementing the boundary integrals along boundary segments where a linear potential relation is assumed. The volume integral is evaluated in rectangular grids only once and can be stored in computer memory for further usage.

INTRODUCTION

The integral method for solving partial differential equations is a very useful tool in computational fluid dynamics. This method provides an alternative approach to the finite difference and finite element methods for solving transonic potential flows. One of the advantages with integral method is that it can be easily applied to solve flow problems over complex configurations. A major technical obstacle involved with other methods seems to be the difficulty in generating suitable grids for flows with complex configurations[1]. In the integral method, computational grids include two part: boundary grids and field grids. Boundary grids are used for implementation of the boundary integrals while field grids are used for evaluation of certain volume integrals instead of the conventional purpose of replacing partial differential equations. Therefore, many kinds of grids can be adopted for field grids required by integral method.

Another advantage of the integral method is that solutions can be obtained by only using surface panels and solutions everywhere else can be represented by the solutions on surface panels. For nonlinear equations only the discretized simultaneous equations for surface points are solved in iterative solution procedures. Generally the number of grid points on surfaces is much smaller than the number of grid points in the field. Thus, the integral method could take less computer CPU time.

Computational schemes based on integral formulation have been developed for the linear Prandtl–Glauert equation[2–3]. More recently, the full potential equation has been solved using integral method[4–12]. Applying integral method to solve the unsteady potential equation has been studied[13–14]. The full potential equation has also been solved using integral method for flows over multi–element airfoils[15] and over the entire aircraft[16–17]. For transonic flows, some methods combine the integral method with finite volume method[5] and some other methods use the integral method incorporated with finite difference method[7]. It is not clear that how they treat the shock integrals resulting from the Green function transformation.

In this work, the field integral method is applied to solve the full potential equation for compressible flows. The boundary linear integral method[18] is used in this work. The coefficient matrix is formulated on body grid points only. The nonlinear full potential equation is solved by Newton–Raphson iteration procedure. In each iteration the influence matrix does not change and needs to be decomposed only once. Only the source term in the volume integral needs to be evaluated in each iteration. After potential values on the body have been solved, potential values anywhere else can be calculated. For transonic flows, shock integrals are calculated and contributions are added to the right hand side of the discretized simultaneous equations. Our results show that shock integrals are necessary for capturing sharp shock jumps.

In this work, two dimensional flows over NACA0012 airfoil are calculated. Numerical results are compared with experimental data[19]. Good agreement between numerical results and experimental data shows that the boundary linear integral method provides an alternative tool for analytical and design purposes in aeronautical engineering.

GOVERNING EQUATION AND BOUNDARY CONDITIONS

In the Cartesian coordinates, the conservative form of the full potential equation is

$$\nabla \cdot (\rho \nabla \Phi) = 0 \tag{1}$$

where Φ is the full potential, ρ is the density, and ∇ is defined by

$$\nabla = \frac{\partial}{\partial x}\, \hat{i} + \frac{\partial}{\partial y}\, \hat{j} + \frac{\partial}{\partial z}\, \hat{k}. \tag{2}$$

The full potential equation given by Eq. (1) is not well suitable for integral representation because the left hand side of this equation is not a Laplacian operator. However, the full potential equation can be written in another form

$$\nabla^2 \Phi = \sigma. \tag{3}$$

The source term σ is expressed by

$$\sigma = M^2 \Phi_{ss} \tag{4}$$

where M is the Mach number and Φ_{ss} is the rate of velocity change along the stream line direction.

Boundary condition of Eq. (1) at the infinity is given by

$$\Phi_\infty = V_\infty\, (x\, cos\alpha + y\, sin\alpha). \tag{5}$$

The surface boundary conditions for inviscid flows is the tangential flow conditions given by

$$\mathbf{n} \cdot \nabla \Phi = 0 \tag{6}$$

where $\mathbf{n}$ is the normal vector on the body surface. The Kutta condition is also implemented, which requires that the flow leaves the trailing edge smoothly.

Using Green's theorem, Φ can be represented by the sum of boundary integrals and a volume integral

$$\Phi(x_p, y_p) = \oint \left[(\mathbf{n} \cdot \nabla \Phi)\Phi_L - \Phi(\mathbf{n} \cdot \nabla \Phi_L) \right] ds + \int_V \sigma \, \Phi_L \, dV \tag{7}$$

where the surface integral is on boundaries of a single connected region and the volume integral is inside the region. $\Phi(x_p, y_p)$ is the potential value at an arbitrary location P inside the region or on the boundary. For two dimensional problems Φ_L is given by

$$\Phi_L = \frac{1}{2\pi} \ln \left[(x - x_p)^2 + (y - y_p)^2 \right]^{0.5}. \tag{8}$$

In Eq. (7), boundary integrals include those at the infinity. When a numerical procedure is employed, it is inconvenient to perform the integrals at infinity directly. However, integrals at the infinity can be transformed into a simple solution called undisturbed potential by using Green's function theory. The undisturbed potential is defined by

$$\Phi_f = V_\infty \, (x \cos\alpha + y \sin\alpha). \tag{9}$$

The undisturbed potential satisfies the Laplace equation $\nabla^2 \Phi = 0$ everywhere and also satisfies the boundary condition at the infinity. Considering a region surrounded by a boundary at the infinity and applying Green's function theory to the undisturbed potential, we have

$$\Phi_f(x_p, y_p) = \oint_\infty \left[(\mathbf{n} \cdot \nabla \Phi_f)\Phi_L - \Phi_f(\mathbf{n} \cdot \nabla \Phi_L) \right] ds. \tag{10}$$

Thus, integrals at the infinity in Eq. (7) can be represented by the undisturbed potential Φ_f. By substituting the undisturbed potential into Eq. (7) solutions can be expressed by

$$\Phi(x_p, y_p) = \Phi_f + \int_{B+C+SH} \left[(\mathbf{n} \cdot \nabla \Phi)\Phi_L - \Phi(\mathbf{n} \cdot \nabla \Phi_L) \right] ds + \int_V \sigma \, \Phi_L \, dV \tag{11}$$

where B represents body surfaces, C denotes wakes or cuts between region boundaries, and SH is along shock waves. On body surfaces and wakes, we have

$$\int_{B+C} (\mathbf{n} \cdot \nabla \Phi)\Phi_L \, ds = 0 \tag{12}$$

because of the tangential boundary conditions. Across shocks, we have

$$\int_{SH} \Phi(\mathbf{n} \cdot \nabla \Phi_L) \, ds = 0 \tag{13}$$

since the tangential velocities are assumed to be equal across shock waves. Thus, Eqs. (7) becomes

$$\Phi(x_p, y_p) = \Phi_f - \int_B \Phi(\mathbf{n} \cdot \nabla \Phi_L) \, ds - \Gamma \int_C (\mathbf{n} \cdot \nabla \Phi_L) \, ds$$

$$+ \int_{SH} (\mathbf{n} \cdot \nabla \Phi) \Phi_L \, ds + \int_V \sigma \, \Phi_L \, dV \qquad\qquad (14)$$

where circulation is

$$\Gamma = \Phi_+ - \Phi_- \qquad\qquad (15)$$

which is determined from the solutions.

NUMERICAL PROCEDURE

In Eq. (14) potential values are represented by boundary integrals plus an extra volume integral. For incompressible flows, the source term σ is zero and there is no shock. The incompressible potential depends only on boundary integrals on the body surface and the wake. If the potential values on body surface are determined, potential solutions everywhere can be calculated. For compressible flows, the source term has to be evaluated and if supersonic flow appears, shock integrals have to be performed.

To determine the potential solution on the body surface, Eq. (14) is applied to points on the body surface. Therefore, the body surface needs to be cut into many pieces of small panels. Applying the surface integrals to each panel will result a system of simultaneous equations. The potential solution on the body surface are obtained from these simultaneous equations. Within a boundary segment, a linear potential formulation is assumed

$$\Phi = \Phi_i + \frac{\xi}{l}(\Phi_{i+1} - \Phi_i) \qquad\qquad (16)$$

where l is the segment length and ξ varies from zero to l. Applying integral equations on the body surface and introducing this linear potential relation into these boundary integrals, a set of linear simultaneous equations results

$$\left[A_{i,j} \right]\left[\Phi_i \right] = \left[F_i \right] \qquad\qquad (17)$$

where $i = 1, \cdots \cdots m.$ and $j = 1, \cdots \cdots m.$

with m the total number of nodes on boundaries. Because the elements of the matrix depend on geomitry information only, the matrix [A] is identical in each iteration. Thus, this matrix needs to be decomposed only once.

For compressible flows, the source term must be evaluated throughout the volume. For transonic flows, the integral for shocks is needed because the normal derivative of potential across shock wave is discontinuous. In this work, potential values across shock wave are assumed to be continuous because the tangential velocities across shock are equal.

The volume integral in Eq. (14) is evaluated in a field grid obtained by finite element discretization. In each element volume, the source term inside the volume integral is assumed to be a constant. Thus,

$$\int_V \sigma \, \Phi_L \, dV = \sum_1^M \sigma_i \int_{V_i} \Phi_L \, dV. \qquad\qquad (18)$$

The integral $\int_{V_i} \Phi_L \, dV$ can be calculated analytically and Formulations are given in the Appendix at the end of this paper.

RESULTS AND DISCUSSION

The solution procedure discussed in above sections was applied to compressible flows over the NACA 0012 airfoil for a range of Mach numbers from subsonic to transonic flows. Numerical results are compared with experimental data[19].Numerical solutions are obtained in a rectangular grid with 1219 node points. Boundary grid consists of 30 panels on the upper surface and 30 panels on the lower surface with most panels near the leading and trailing edges. A wake network consisting of two panels in the boundary grid is appended to the trailing edge of the airfoil. The wake is aligned with the chord bisector. The wake panels are divided into 11 smaller grid segments when building up the field grid. The boundary of the field grid is located at a distance of six chord length around the airfoil. Beyond that the source distributions are small thus, can be neglected.

The nonlinear potential are solved by an iterative procedure. First, the linear Laplace equation for incompressible flows is solved. Thus, the source distribution can be evaluated. The iterative solution procedure is performed by using the first evaluated source distribution. In each iteration, a system of simultaneous equations on the boundary grid is solved and then the source distribution is calculated inside the field grid. If supersonic flows appears, shock integrals are calculated. The contributions of the volume and shock integrals are added in the right hand side of the system equations. Because the matrix of the system equations does not change in the iteration loops, the matrix is decomposed in advance only once to speed up the solution. The iteration number for subsonic flows is between 7 to 10 for the system residual to reach the order of 10.0E-07. For transonic flows, more iterations are needed to reach the same system residual order. For the flows with a Mach number of 0.8 at the infinity and a zero degree angle of attack, the number of iterations is about 35.

In Figs. 1 and 2, pressure coefficients for compressible flows over NACA 0012 airfoil with Mach numbers of 0.5 and 0.7 at the infinity and a zero degree angle of attack are presented. Good agreement between numerical results and experimental data in these cases are shown. In Fig. 3, pressure coefficients for flows over NACA 0012 airfoil with a Mach number of 0.5 at the infinity and an angle of attack of 2.0° is presented. Numerical results predicted lower pressure coefficients along the upper surface of the airfoil. Pressure coefficients for flows over NACA 0012 with Mach numbers of 0.8 and 0.83 at the infinity and a zero angle of attack are shown in Fig. 4 and Fig. 5 respectively. Numerical solutions are fairly close to experimental data in these cases. Pressure coefficients for Mach number of 0.75 at the infinity and angles of attack of 1.0° and 2.0° are shown in Fig. 6 and Fig. 7 respectively. In the lifting case, numerical solutions predict lower pressure coefficients on the upper surface of the airfoil.

CONCLUDING REMARKS

Numerical study in this work shows that the boundary linear integral method provides an alternative approach for solving transonic potential flow problems. With this method, the full potential equation is solved only on the body surface. The potential solutions in the entire flow field are represented by the solutions on the body surface and are used for calculating the source distribution. Since the number of grid points on the body surface is much smaller than that in the entire flow field, the requirement for computer CPU time would be reduced.

With the boundary linear integral method, the grid system is devided into two parts: the surface grids and the field grids. The surface grids are used to obtain the potential solution on the surface and the filed grids are used to calculate the source distribution and the field integral. Since there is no specific requirement for the field grids, many kinds of grid can be adopted by the boundary integral method. Thus, this method precludes the difficulty in grid generation encountered in other field methods.

In this work, the shock integrals generated by Green's function theory are performed explicitly when the flow is transonic. Our results show that the shock integrals are necessary to capture sharp shock jumps. In this approach, the contribution of the shock integrals is added to the right hand side of the discretized simultaneous equations. This approach is a shock–capturing method bacause there is no requirement for the pre–known shock locations. Numerical results are fairly good compared with experimental data.

APPENDIX

The volume integral given by Eq. (18) is calculated analytically and the results are presented.

Introducing $\bar{x} = x - x_p$ and $\bar{y} = y - y_p$, the element volume integral becomes

$$IV = \int_{V_i} \Phi_L \, dV = \frac{1}{2\pi} \int_{\bar{x}_A}^{\bar{x}_B} \left[\int_{\bar{y}_L}^{\bar{y}_U} \ln(\bar{x}^2 + \bar{y}^2) \, d\bar{y} \right] d\bar{x} \tag{19}$$

where

$$\bar{x}_A = x_A - x_p \tag{20}$$

$$\bar{x}_B = x_B - x_p \tag{21}$$

$$\bar{y}_U = y_U - y_p \tag{22}$$

$$\bar{y}_L = y_L - y_p \tag{23}$$

where the subscript U represents the upper boundaries and L denotes the lower boundaries and A is the left boundary and B is the right boundary as shown in Fig.8. Carrying out the integration, we can obtain

$$IV = F(\bar{x}_B, \bar{y}_U) - F(\bar{x}_B, \bar{y}_L) - F(\bar{x}_A, \bar{y}_U) + F(\bar{x}_A, \bar{y}_L) \tag{24}$$

where the limit function F is

$$F(x,y) = x\,y\,\ln(x^2+y^2) - 3\,x\,y + y^2 \arctan\frac{x}{y} + x^2 \arctan\frac{y}{x} \tag{25}$$

If the singularity location (x, y) is on the node point of the finite integral volume, $x=x_p$ and $y=y_p$. Thus, the limit function value becomes zero.

REFERENCES

1. Holst, T. L., Slooff, J. W., Yoshihara, H., and Ballhaus, W. F. Jr., "Applied Computational Transonic Aerodynamics," edited by B. M. Spee and H. Yoshihara, AGARD-AG-266, pp. 86, 1982.

2. Tseng, K., Morino, L., "Nonlinear Green's Function Methods for Unsteady Transonic Flows," *Transonic Aerodynamics*, edited by D. Nixon, AIAA, New York, pp. 565-603, 1982.

3. Piers, W. J., Slooff, J. W., "Calculation of Transonic Flow by Means of A Shock-Capturing Field Panel Method," Proceedings of AIAA Computational Fluid Dynamics Conference, paper No. 79-1459, pp. 147-156, 1979.

4. Johnson, F. T., James, R. M., Bussoletti, J. E., Woo, A. C., "A transonic Rectangular Grid Embedded Panel Method," AIAA paper No. 82-0953, presented at AIAA/ASME 3rd Joint Thermophysics, Fluids, Plasma and Heat Transfer Conference, St. Louis, Missouri, June 7-11, 1982.

5. Oskam, B., "Transonic Panel Method for the Full Potential Equation Applied to Multicomponent Airfoils," *AIAA Journal,* Vol. 23, No. 9, pp. 1327-1334, 1985.

6. Crown, J. C., "Calculation of Transonic Flow Over Thick Airfoils by Integral Methods," *AIAA Journal*, Vol. 6, No. 3, pp. 413-423, 1968.

7. Sinclair, P. M., "An Exact Integral (Field Panel) Method for the Calculation of Two-Dimensional Transonic Potential Flow Around Complex Configurations," *Aeronautical Journal,* June/July, pp. 227-236, 1986.

8. Kandil, O. A., Hu, H., "Full-Potential Integral Solution for Transonic Flows with and without Embedded Euler Domains," *AIAA Journal*, Vol. 26, No. 9, pp. 1079-1086, 1988.

9. Masson, C., "Panel Method for Transonic Flows," MSc. Thesis, Ecole Polytechnique of Montreal, August, 1989.

10. Erickson, L. L., Strande, S. M., "A Theoretical Basis for Extending Surface-Paneling Methods to Transonic Flow," *AIAA Journal,* Vol. 23, No. 12, pp. 1860-1867, 1985.

11. Chu, L., Yates, E., Kandil, O., "Integral Equation Solution of the Full Potential Equation for Transonic Flows," AIAA paper No. 89-0563, presented at AIAA 27th Aerospace Sciences Meeting, Reno, Nevada, January 9-12, 1989.

12. Wilson, D. E., "A New Singular Integral Method for Compressible Potential Flow," AIAA paper No. 85-0481, presented at AIAA 23rd Aerospace Sciences Meeting, Reno, Nevada, January 14-17, 1985.

13. Hounjet, M. H. L., "Transonic Panel Method to Determine Loads on Oscillating Airfoils with Shocks," *AIAA Journal,* Vol. 19, No. 5, pp. 559-566, 1981.

Unsteady Transonic Flow," *AIAA Journal,* Vol. 23, No. 4, pp. 537-545, 1985.

15.   Halsey, N. D., "Calculation of Compressible Potential Flow About Multielement Airfoils Using a Source Field-Panel Approach," AIAA paper No. 85-0038, presented at AIAA 23rd Aerospace Sciences Meeting, Reno, Nevada, January 14-17, 1985.

16.   Erickson, L. L., Madson, M. D., Woo, A. C., "Application of the TranAir Full-Potential Code to the F-16A," *Journal of Aircraft,* Vol. 24, No.8, pp. 540-545, 1987,

17.   Sinclair, P. M., "A Three-Dimensional Field-Integral Method for the Calculation of Transonic Flow on Complex Configurations-Theory and Preliminary Results," *Aeronautical Journal,* pp. 235-243, June/July, 1988.

18.   Moran, J., *Theoretical and Computational Aerodynamics,* John Wiley & Sons, 1984.

19.   Thibert, J. J., Grandjacques, M., Ohman, L. H., "NACA-0012 Airfoil, An Experimental Data Base for Computer Program Assessment," AGARD-AR-138, pp. A1.1-A1.10, May, 1979.

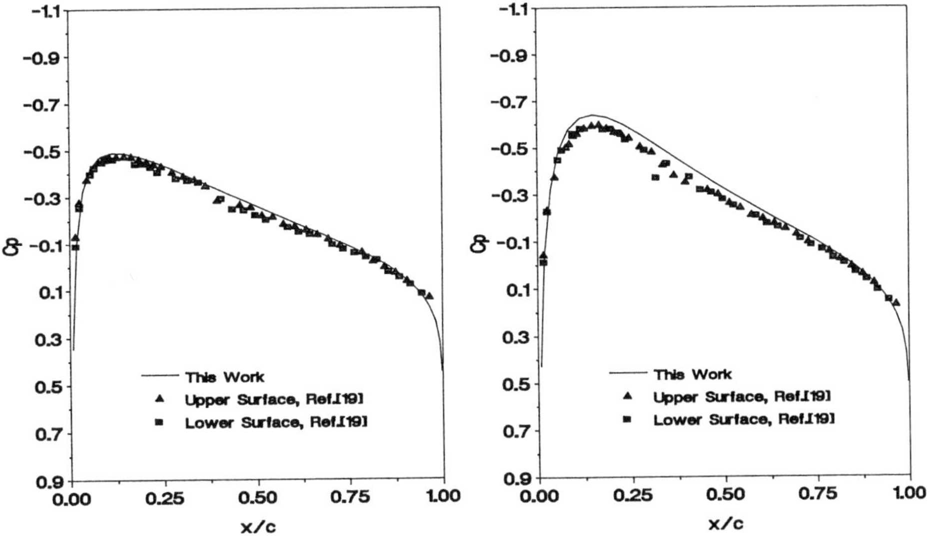

Figure 1. Pressure coefficient
M_∞ = 0.5, α = 0.0

Figure 2. Pressure coefficient
M_∞ = 0.7, α = 0.0

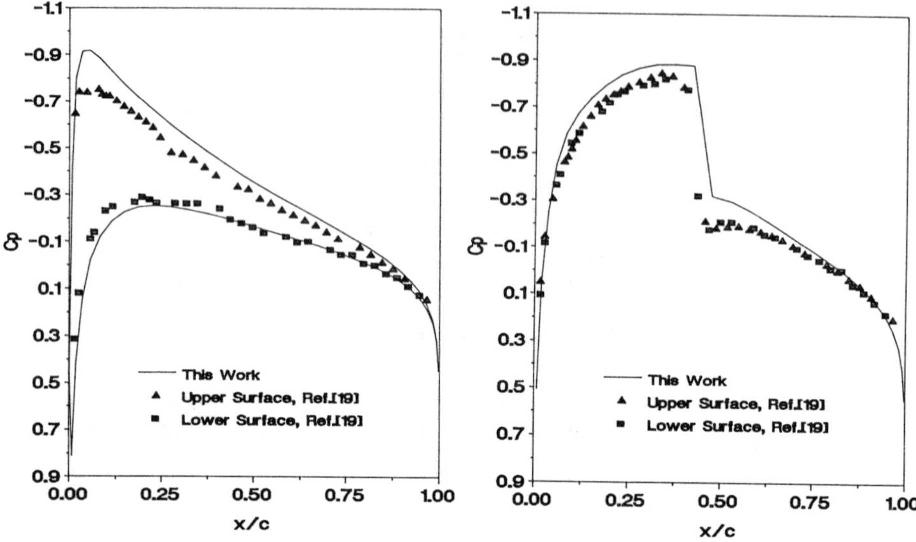

Figure 3. Pressure coefficient
$M_\infty = 0.5$, $\alpha = 2.0$

Figure 4. Pressure coefficient
$M_\infty = 0.8$, $\alpha = 0.0$

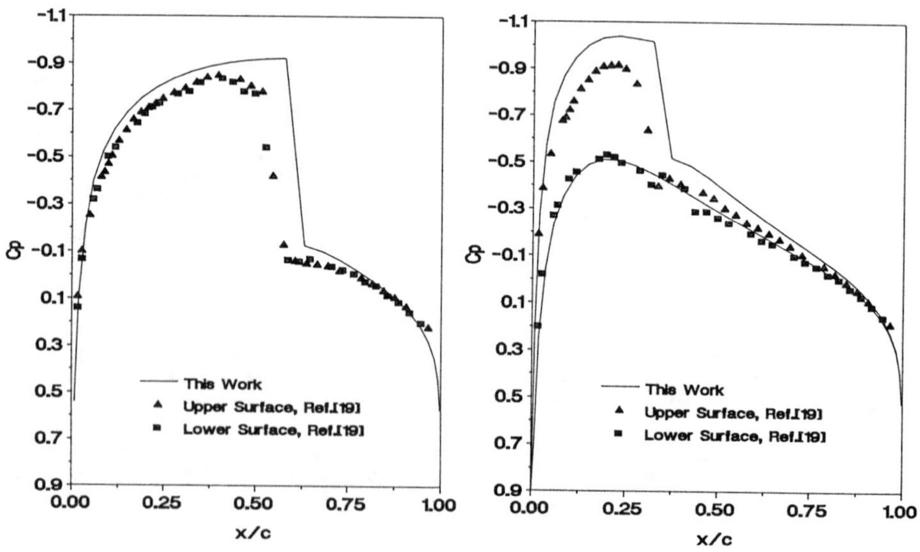

Figure 5. Pressure coefficient
$M_\infty = 0.83$, $\alpha = 0.0$

Figure 6. Pressure coefficient
$M_\infty = 0.75$, $\alpha = 1.0$

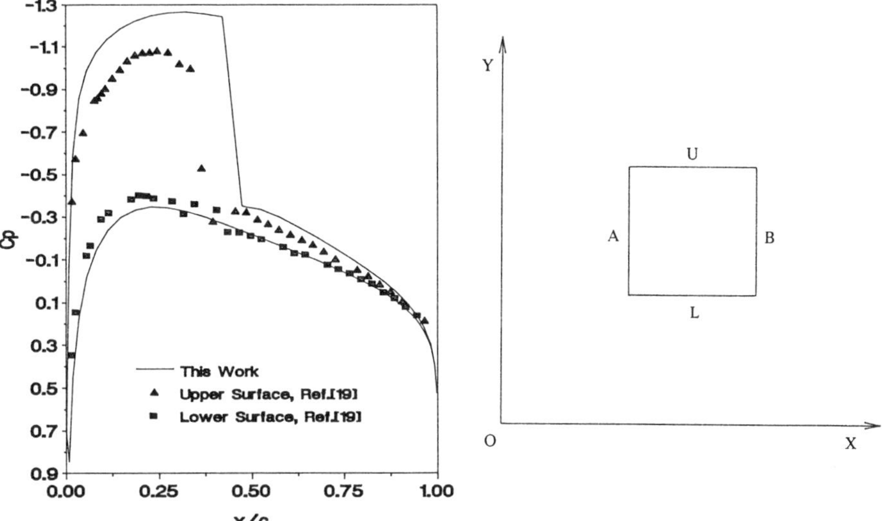

Figure 7. Pressure coefficient
$M_\infty = 0.75$, $\alpha = 2.0$

Figure 8. A finite integral volume

A Triad of Solutions for 2-D Navier-Stokes: Global, Semi-Local and Local

MURALIDHARAN NAIR
Department of Aerospace Engineering
Emory-Riddle University
Melbourne, Florida, USA

FRED R. PAYNE
Aerospace Engineering Department
The University of Texas at Arlington
Arlington, Texas 76019-0018, USA

ABSTRACT

Two Newtonian fluid applications, parallel channel and driven-cavity flows, were computed by three distinct philosophies: 1) "local" via finite difference methods (Ghia, etal 1977; Raithby and Torrance, 1974; Abdallah, 1987 were major sources); 2) "semi-local" via finite elements (Taylor and Hughes, 1981); and 3) a relatively new (Payne, 1980) conversion of DE to Volterra IE/IDE which are solved iteratively by an improved Picard scheme. The latter method is termed "Direct, Formal Integration (DFI) and seems well-suited to nonlinear systems.

The specific implementations chosen show that FDM and DFI generally exceed FEM in performance. FDM and DFI are clearly competitive; one method or the other excels in speed or accuracy, depending upon the flow case and parametric values. Of course, any final conclusions are risky because another user can be clever in code optimization.

DFI appears rather easier to implement. A recent DFI extension to turbulent flows (3-D, 8 coupled nonlinear PDE containing 28 distinct derivatives) had good success. A serendipity is DFI's "cluster property" whereby like physical causes and effects are amalgamated into single terms in the Volterra equations; this should promote the modelling of "real world" problems.

I.   INTRODUCTION

By "global", "semi-local", and "local" are meant solutions obtained, respectively, via integral, finite element and finite difference methods. "Local", in the mathematical analysis sense, can, of course, be attained upon finite machine only in some limiting sense. All computer arithmetic must be done on a finite interval and therefore "semi-local". The three numerics used fall roughly within the cited classes.

Work supported, in part, by NASA Ames Grants NSG-2077, 1980, and NCC 2-386, 1985-88; ASEE-NASA Ames Faculty Fellowships, 1988-1989; and Organized Research Fund of the Graduate Dean, UTA, 1986-88. Taken, in part, from M. Nair's Dissertation, 1990, UTA.

The primary thrust of this work was extension of a new integral method (Payne, 1980) to two-dimensional, incompressible Navier-Stokes. Prior applications include boundary-layers (Payne and Ko, 1984; Mokkapati and Payne, 1986; Mokkapati, 1988), population dynamics (Payne and Nair, 1988), nonlinear heat conduction (Mokkapati and Payne, 1988), turbulence (Ahn, 1989), microbursts (Wan, 1988), elliptics (Payne and Nair, 1989), field theory and various model equations in turbulence and heat transfer. The FEM and FDM codes were based upon current literature and run as comparisons.

II. DIRECT, FORMAL INTEGRATION (DFI), WHAT IS IT?

Second author's experience in turbulence, basically global phenomena, and search for alternative solvers of the fluid mechanical equations lead, in 1980, to formal integration of DE systems and thereby converting them to Volterra integral/integro-differential equations. He and Ko (1984) found improvements to the classic Picard iteration scheme, much used in analysis for existence and uniqueness proofs, so that "DFI" ran faster, on identical step-sizes, than all fourth-order Runge-Kutta codes tested. DFI is so simple in concept and execution that seniors prefer it and it appears ideally suited for nonlinear systems which need not be linearized, either implicity or explicitly. Granted, DFI's usual trapezoidal quadrature involves linearization but Richardson extrapolation to Romberg quadratures obviates this.

"Direct, Formal Integration (DFI)" consists of three stages:

1) Formally integrate any DE system, from any chosen fixed point, along any chosen trajectory.
2) Study the resulting Volterra forms for new insights into the mathematics, the physics and the numerics of the problem (some will arise)
3) Solve the Volterra system of integral or integro-differential equations via an improved ("micro-") Picard iteration.

Under DFI an nth order ODE may be integrated k-times, k < n, yielding an IDE; more efficient is n-times integration which removes all derivatives and yields an IE. An nth order PDE under n integrations in one variable remains an IDE; the remaining derivatives must be approximated by FDM unless "NAD" is used (see below). Consider a first order, non-linear equation, namely a form of Riccati's with solution u = tanh (y):

$$u'(y) = 1 - u^2 \text{ subject to } u(0) = 0$$

Stage I:

$$u(y) = u(0) + y - \int_0^y u^2(s)\, ds$$

Stage II:

a) IC is incorporated [u(0) = 0]
b) u is ~ linear for small y
c) u' is maximum at y = 0
d) If u is bounded as y $\to \infty$ then u $\to$ constant

Stage III:

1. Predictor: $u_0(y) = u(0) + u'(0)\, y$
$$= y \quad \text{[1st order Predictor]}$$

2. Corrector (Picard): $u_1(y) = u_0 - y{**}3/3$

$$u_2(y) = u_1 + 2y{**}5/15 - y{**}7/63 \quad \text{[Taylor series]}$$

This procedure is optimal for polynomial non-linearities such as in fluids (quadratic). Speed of this procedure is about twice that of Runge-Kutta solvers and 2-3 times as fast as FDM tried for Laplace in the plane (Nair, 1990). Several hundred ODE/PDE systems have been solved, 1980-89, many in senior-graduate courses. The most complex to date was a non-linear set of eight PDE with 28 distinct derivatives for 3-D turbulence in a straight channel (Payne, 1990). The single application of "Natural Anti-Derivative (NAD)" was 2-D, incompressible Euler flow; "NAD" extends DFI by eliminating all derivatives. This elliptic problem was solved exactly in a single pass without sweeping; granted, the velocity field was linear and thus the trapezoid quadrature was without inherent error (machine error was also zero).

III. NAVIER-STOKES UNDER DFI:

The steady-state, two-dimensional, incompressible fluid continuity and Navier-Stokes equations become under one and two, respectively, integrations in y:

$$\nabla \cdot \underline{u} = 0 \rightarrow v(x, y) = v(x,0) - \frac{\partial}{\partial x} \int_0^y u(x, s)\, ds \tag{1}$$

$$\underline{u} \cdot \nabla u = -\frac{1}{\rho} \frac{\partial P}{\partial x} + \nu\, \nabla^2 u :$$

$$\nu \frac{\partial u}{\partial y}(x, y) = \frac{\tau_w(x)}{\rho} - \frac{\partial}{\partial x} \int_0^y \left[\frac{P}{\rho} + u^2(x, s)\right] ds + uv \Big|_{(x,0)}^{(x,y)} - \nu \frac{\partial^2}{\partial x^2} \int_0^y u(x, s)\, ds \tag{2}$$

Second Formal Integration:

$$\nu\, u(x, y) = y\left[\frac{\tau_w}{\rho} - u\,v(x, 0)\right] + \frac{\partial}{\partial x} \int_0^y (y-s)\left(\frac{P}{\rho} + u^2\right) ds$$

$$- \nu \frac{\partial^2}{\partial x^2} \int_0^y (y-s)\, u(x, s)\, ds) + \int_0^y u\, v\, ds \tag{3}$$

and similar equations for $v(x,y)$. u, v are velocity components; P is pressure; τ_w is the Newtonian stress at the lower wall and ν is the molecular viscosity.

The DFI algorithm for these Volterra equations is for (3):

$$\nu\, u\,(x, y) = \nu\, u\,(x, y - \Delta y) + G\,(x, y - \Delta y)$$

$$+ \frac{\partial}{\partial x} \int_{y-\Delta y}^{y} (y - s)\left(\frac{P}{\rho} + u^2\right) ds + \int_{y-\Delta y}^{y} u\, v\, ds - \nu \frac{\partial^2}{\partial x^2} \int_{y-\Delta y}^{y} (y - s)\, u\,(x, s)\, ds \qquad (4)$$

G is known from the prior y-step; "micro-Picard" iteration is done on each interval.

Note: 3 implicit u(x,y) occurrences,
              2 are <u>non</u>linear
              LHS is <u>linear</u>   (if viscosity = constant)

The channel flow was transformed to the unit square and had zero vertical velocity on all boundaries. The horizontal velocity, due to the "no-slip" condition of viscous flow, vanished on the top and bottom walls. Due to incompressibility the flow dynamics of this quite simple flow are linearly elliptic.

The driven cavity flow was also transformed to the unit square and had zero vertical velocity on all boundaries. The horizontal velocity vanished on three boundaries; on the top boundary the velocity was set to unity which corresponds to a uniformly constant free stream above the cavity. Dynamics here are governed by full Navier-Stokes and hence this problem is one of nonlinear elliptics. For decades this relatively simple flow has served well as a CFD "bench mark" for algorithmic testing. A multitude of workers have used various techniques (Mills, 1965; Burgraff, 1966; Pan and Acrivos, 1967; Bozeman and Dalton, 1973 is a small sample).

This pair of flows is a good test of any methodology since they incorporate separately linear and nonlinear elliptic behaviors.

IV.   RESULTS AND COMPARISONS

The velocity components in the driven cavity are shown in Figures 1 and 2; vorticity is depicted in Figure 3. To the scale shown all three methods produced the same results. Differences occur in timing and accuracy.

Table 1 gives the timing and accuracy comparisons of the methods for the channel flow at three distinct Reynolds numbers (center-line velocity times channel width divided by molecular viscosity), namely, 1, 10, 100. The higher CPU times for FEM are likely not indicative of inherent efficiency. A clever programmer can reduce considerably a code's execution time; no claim is made here that any of the codes are optimal. All three were executed by the same worker upon an IBM 4381.

Table 1 indicates DFI and FDM are comparable; DFI is slightly faster in two of three cases but FDM is generally more accurate. Oddly, FEM is more accurate for the highest Reynolds number but poorer than the other two at lower Reynolds numbers.

Table 2 compares the three methods for the same Reynolds numbers in the cavity flow. FEM is again slower and generally less accurate. FDM is now faster than DFI in two of three cases but DFI is more accurate, by an order of magnitude at Reynolds number of 100.

V. CONCLUSIONS AND RECOMMENDATIONS

DFI was successful in solving both linear and non-linear elliptic problems. For linear elliptics, Laplace, Poisson and Helmholtz, DFI is remarkably efficient (Nair, 1990); a multitude of differing sweep patterns is available by simple "do-loop" indexing. At low Reynolds numbers, 1, 10, 100, no stability problems were encountered by DFI in the channel or the cavity flow. For higher Reynolds numbers strong under-relaxation was required initially; this factor rapidly approached unity as the calculation progressed.

All methods, DFI, FDM, FEM, were executed on the same IBM 4381 and produced acceptable results. A far greater effort would be required to make conclusive judgments as to which method is faster or more accurate or easier to implement. Even then, such conclusions would be tentative until a more clever programmer implemented one or more of the methods.

DFI does offer a number of advantages; among these are:

1) Treatment of nonlinear systems without any linearization
2) Superior computer compatibility due to avoidance or reduction of divisions and subtractions
3) Ease of coding
4) Arbitrary order predictor-corrector (need not be limited to order of the DE system; corrector is explicity controllable by convergence criteria of Picard loop and any "shooting" loop for boundary value problems)
5) Ease of Romberg incorporation for enhanced accuracy (controllable)
6) "Over-write" of old data and conserving RAM-storage
7) "Cluster" property which amalgamates similar physics into a single term; insights and modelling are enhanced

Recommendations include, for DFI, more study of Romberg quadrature for elliptics and application to more technologically interesting geometries such as finite wings and bodies. DFI would be interesting to apply to the classic vorticity-stream function procedure.

A speculation on the lack of improvement under Romberg for Laplace and other elliptic operators lies in the approximation of the remaining derivatives. Second author feels that the derivatives must also be "purified" by, say, Richardson extrapolation; central difference formulas will be quite similar to those for integrals.

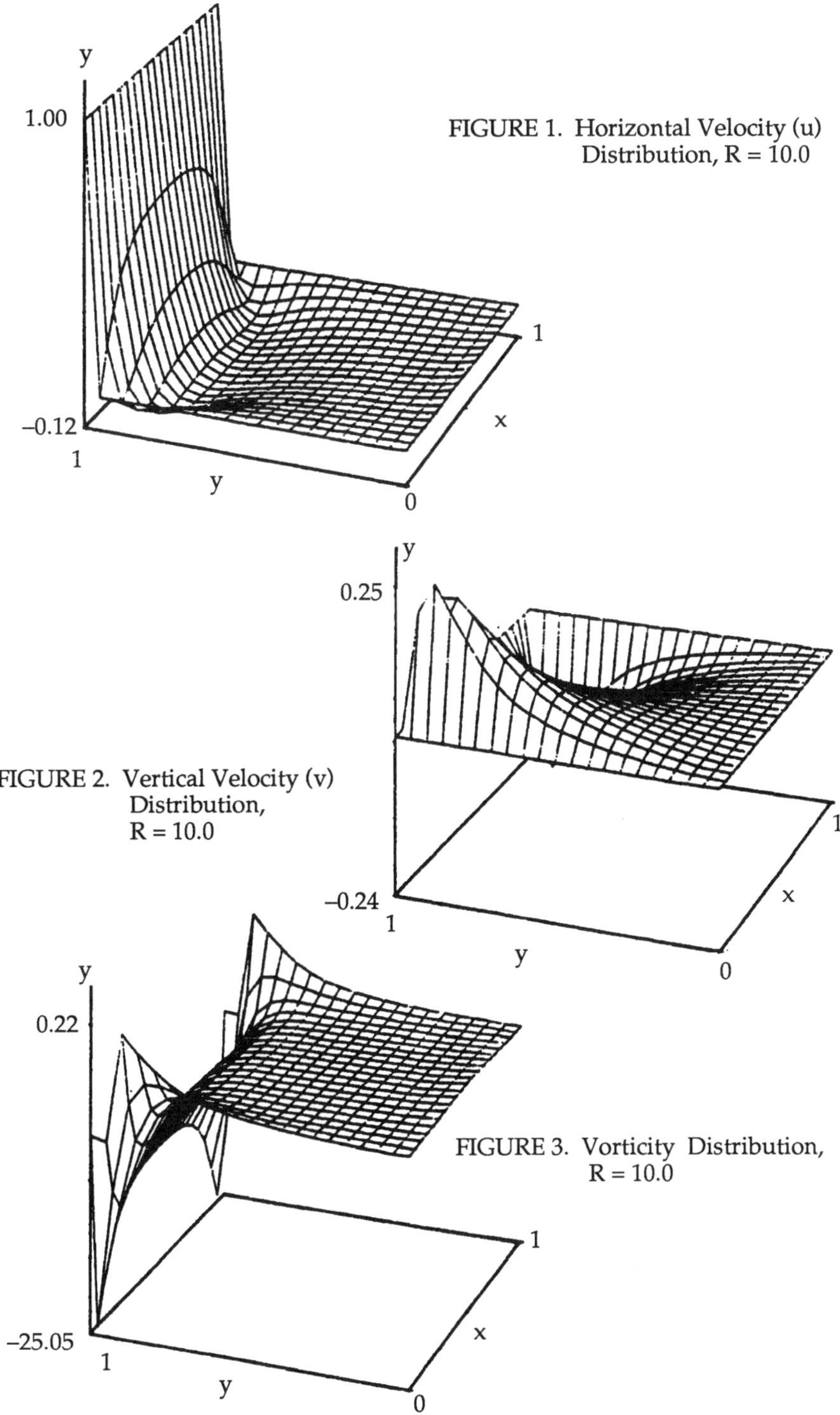

FIGURE 1. Horizontal Velocity (u) Distribution, R = 10.0

FIGURE 2. Vertical Velocity (v) Distribution, R = 10.0

FIGURE 3. Vorticity Distribution, R = 10.0

Table 1. Comparison of the Three Methods - Channel Flow Problem

	R	DFI	FDM	FEM
	1.0			
CPU(s)		29.63	37.59	174.3
Max Div		.10E-14	.10E-15	.10E-07
Vel. Err		.10E-09	.10E-16	.10E-10
Pre. Conv		.35E-14	.10E-16	.10E-05
	10.0			
CPU(s)		46.23	38.84	237.3
Max Div		.10E-06	.10E-15	.10E-05
Vel. Err		.10E-05	.10E-16	.10E-07
Pre. Conv		.50E-07	.10E-16	.10E-04
	100.00			
CPU(s)		104.68	112.56	386.1
Max Div		.42E-01	.10E-03	.10E-03
Vel. Err		.10E-02	.10E-02	.10E-03
Pre. Conv		.30E-03	.53E-02	.10E-02

Table 2. Comparison of the Three Methods - Driven Cavity Problem

	R	DFI	FDM	FEM
	1.0			
CPU(s)		82.31	86.12	129.9
Max Div		.10E-05	.10E-05	.10E-04
Pre. Conv		.50E-03	.10E-04	.10E-04
	10.0			
CPU(s)		261.15	237.21	293.56
Max Div		.62E-02	.70E-02	.10E-03
Pre. Conv		.10E-03	.65E-03	.10E-03
	100.0			
CPU(s)		53.10	36.77	278.07
Max Div		.83E-02	.56E-01	.10E-02
Pre. Conv		.10E-03	.31E-02	.10E-01

ACKNOWLEDGEMENT

In addition to NASA, ASEE, and the UTA Graduate Dean financial supports, the Department of Aerospace Engineering and three Deans of Engineering at UTA have supported, in various forms, this work over the past decade. The greatest debt is owed John L. Lumley, now at Cornell, who introduced the second author to Fredholm integral equations in several applications at the Pennsylvania State University, 1962-66.

REFERENCES

Abdallah, S. (1987)   J. COMP. PHY. no. 70, p. 182
Ahn, C-S (1989)   PhD Dissertation, UTA
Bozeman, J. D. and Dalton, C. (1973) J. COMP. PHY., no. 12, p. 348
Burgraff, O. R. (1966)   J. FLUID MECH., v. 24, 1, p. 113.
Ghia, K. N., Hankey, W. L. and Hodge, J. K. (1977) AIAA paper 77-648,
                              Albuquerque
Ko, F-T (1982)   MSAE Thesis, UTA
Mills, R. D. (1965)   J. ROY. AERO. SOC., v. 69, p. 714
Mokkapati, R. (1988)   PhD Dissertation, UTA
Mokkapati, R. and Payne, F. (1986) INTEG. METH. SCI. ENGR., p. 249,
                              Hemisphere
Mokkapati, R. and Payne, F. (1988) MIDWEST-SOUTHEAST MATH. CONF.,
                              Vanderbilt
Nair, M. (1990)   PhD Dissertation, UTA
Pan, F. and Acrivos, A. (1967) J. FLUID MECH., v. 28, 4, p. 643
Payne, F. (1980)   AE 5305 Notes (Unpublished)
Payne, F. (1990)   ASEE-NASA Report (Unpublished)
Payne, F. and Ko, F-T (1984) TRENDS NONLINEAR ANALYSIS, p. 467, N.
                              Holland
Payne, F. and Nair, M. (1988) MIDWEST-SOUTHEAST MATH. CONF.,
                              Vanderbilt
Payne, F. and Nair, M. (1989) APPL. MATH. LETTERS, v. 2, no. 1, p. 45
Raithby, G. D. and Torrance, K. E. (1974) COMP AND FLUIDS, v. 2, p. 191
Taylor, C. and Hughes, T. G. (1981) FINITE ELEMENT PROGRAMMING OF
                              THE NAVIER-STOKES EQUATIONS, Pineridge
                              Press
Wan, T. (1988)   PhD Dissertation, UTA

[Some 20 additional "DFI" references, 1981-90, available upon request.]

A New Integral Method for Magnetohydrodynamic (MHD) Flow Analysis

YING-MING LEE
Component Development and Integration Facility
MSE, Inc., Butte, Montana 59702, USA

DONALD R. WILSON, Ph.D.
University of Texas at Arlington
Aerospace Engineering Department
Arlington, Texas 76019, USA

ABSTRACT

A new magnetohydrodynamic (MHD) analysis and simulation model is developed to provide better analysis of current MHD testing data and simulation of the future higher interaction prototypical hardware at the Department of Energy's Component Development and Integration Facility (CDIF). The model will also be useful for design and performance predictions of larger scale MHD components being proposed for retrofitting of existing power plants and for various higher power density applications.

An analytical approximation of turbulent boundary layer velocity profiles has been used to describe attached and near-separated MHD flow over the entire computational domain. A new analytical description of the temperature as a function of velocity, through an MHD turbulent boundary layer for a constant but nonunity Prandtl number with MHD terms, has been derived in this study to provide a better representation of the MHD enthalpy layer profiles. The model without MHD terms has been tested against non-MHD boundary layer cases and shows good agreement. The integral equations for the MHD turbulent boundary layers based on these new models are derived in this study. Modification of a quasi-three-dimensional, coupled core flow-integral boundary layer computer code is underway by using these new models to yield an enhanced MHD turbulent flow code, which can provide reasonably accurate solutions in a short time with limited computer resources.[*]

INTRODUCTION

Magnetohydrodynamic (MHD) power generation is a method for converting thermal energy directly to electric power. A typical MHD power generation system with a bottoming steam plant would convert high temperature heat into electricity with potential efficiencies in the 50 to 60 percent range. A wide range of fuels can be burned with acceptable environmental impact. Alkali metal seed, injected for conductivity enhancement, reacts with the sulfur in coal to reduce sulfur dioxide (SO_2) emissions. Nitrogen oxides (NO_x) are reduced by adjusting the temperature distribution throughout the system. Commercially available electrostatic precipitators or baghouses can be used to reduce particulates.

This work is partially supported by the U.S. Department of Energy under contract DE-AC07-88ID12735.

The Component Development and Integration Facility (CDIF) is a Department of Energy test facility operated by MSE, Inc. MSE personnel are responsible for operating the topping cycle components for the national coal-fired MHD development program. Testing at the CDIF utilizes a TRW-designed 50-MW$_t$, slag-rejecting coal-fired combustor and coal-fired precombustor coupled to an Avco-designed supersonic MHD generator.

The MHD process is complex including electrical and gas dynamic phenomena that are coupled by the MHD interaction. This paper presents a new MHD analysis and simulation model, whose purpose is to provide enhanced insight into current MHD test data and improved simulation of future higher interaction prototypical hardware at the CDIF. The model will also be useful for design and performance predictions of larger scale MHD components being proposed for retrofitting existing power plants and for various applications.

In general, there are two approaches to solving MHD flow problems. The finite difference method can provide accurate calculations of unique MHD flow characteristics but requires significant computer resources, whereas the integral method, by using generic velocity and temperature profiles, can provide an adequate overall solution with less computer resources. The overshoot of the temperature profile in the electrode boundary layers and velocity profile in the insulator boundary layers causes special effects that are absent in non-MHD boundary layers. This paper describes an integral method with enhanced velocity and temperature profiles that provides detailed boundary layer properties while retaining computational efficiency. An analytical approximation of the turbulent boundary layer velocity profile has been used to describe attached and near-separated MHD flow over the entire computational domain. A new analytical description of the temperature as a function of velocity through an MHD turbulent boundary layer for a constant but nonunity Prandtl number has also been derived in this study to provide a better representation of the MHD enthalpy layer profile. The model has been tested against incompressible and compressible non-MHD boundary layer cases and shows good agreement. The electrode walls and sidewalls are treated differently in the modeling process. The modification of an existing MHD flow code (Wilson and Simmons, 1980) is underway, using these new models to yield an MHD turbulent flow calculation capable of providing reasonably accurate solutions with short execution times and limited computer resources.

GENERAL MAGNETOHYDRODYNAMIC EQUATIONS

The three-dimensional, steady-state equations for a plasma flowing through a magnetic field are provided by the conservation equation for mass, momentum, and energy:

Mass Continuity Equation

$$div(\rho \vec{u}) = 0 \tag{1}$$

Momentum Equation

$$\rho(\vec{u}\nabla)\cdot\vec{u} = -\nabla p - div\overline{\overline{\tau}} + \vec{j}\times\vec{B} \tag{2}$$

Energy Equation

$$\rho\vec{u}\nabla\left(h + \frac{\vec{u}^2}{2}\right) = -div(\overline{\overline{\tau}}\cdot\vec{u}) - div\vec{q} + \vec{j}\cdot\vec{E} \tag{3}$$

where ρ is mass density, p is static pressure, h is static enthalpy, $\vec{u}$ is velocity vector, $\vec{j}$ is electrical current density vector, $\vec{E}$ is electrical field vector, $\vec{B}$ is magnetic field vector, $\vec{q}$ is heat flux vector, and $\overline{\overline{\tau}}$ is stress tensor. The $\vec{j} \times \vec{B}$ (Lorentz force) and $\vec{j} \cdot \vec{E}$ (electric power per unit volume of the flow) characterize the MHD flows.

CORE FLOW AND BOUNDARY LAYER MODEL

A set of first order differential equations is derived from generalized MHD flow equations. They are inviscid core flow equations and integral boundary layer equations.

Inviscid Core Flow Equations

The steady, one-dimensional, frictionless core flow equations including MHD effects are simplified from equations (1) through (3)

$$\dot{m}_c = \rho_c u_c A_c \tag{4}$$

$$\rho_c u_c \frac{du_c}{dx} = -\frac{dp_c}{dx} + j_{yc} B_z \tag{5}$$

$$\rho_c u_c \frac{dH_c}{dx} = \rho_c u_c \left(\frac{dh_c}{dx} + u_c \frac{du_c}{dx} \right) = j_{xc} E_{xc} + j_{yc} E_{yc} - q_w \frac{P^*}{A^*} \tag{6}$$

where q_w includes convective and radiative heat losses and subscript C represents core flow variables. Also,

$$\rho_c = \rho_c (p_c, T_c) \tag{7}$$

These equations are transformed to a more convenient form for numerical integration by solving for explicit relations of two independent core flow derivatives, depending on the type of solution desired. The channel flow program has the capability of running five specific design options: specified core area, pressure, temperature, velocity, or γM^2 (refer to Wilson and Simmons, 1980, for further details).

Integral Boundary Layer Equations

The integral boundary layer equations are derived by formally integrating each boundary layer equation through or across the boundary layer at each axial location. The equations are:

Momentum Equation

$$\frac{d\theta}{dx} + \theta \left(\frac{\delta^*/\theta + 2}{u_c} \frac{du_c}{dx} + \frac{1}{\rho_c} \frac{d\rho_c}{dx} \right) = \frac{C_f}{2} + \frac{j_{yc} B_z}{\rho_c u_c^2} \delta_{JB}^* \tag{8}$$

Kinetic Energy Equation

$$\frac{d\theta^*}{dx} + \theta^* \left(\frac{2\delta^{**}/\theta^* + 3}{u_c} \frac{du_c}{dx} + \frac{1}{\rho_c} \frac{d\rho_c}{dx} \right) = c_f D + \frac{2 j_{yc} B_z}{\rho_c u_c^2} \delta_{JB}^{**} \tag{9}$$

Integral Energy Equation

$$\frac{d\phi}{dx} + \phi\left\{\frac{1}{u_c}\frac{du_c}{dx} + \frac{1}{\rho_c}\frac{d\rho_c}{dx} + \frac{1}{H_c - H_w}\frac{d(H_c - H_w)}{dx}\right\} = C_H + \frac{j_{xc}E_{xc} + j_{yc}E_{yc}}{\rho_c u_c(H_c - H_w)}(\delta^*_{JE} - \delta^*),$$ (10)

where

$$\theta = \int_0^\delta \left(1 - \frac{u}{u_c}\right)\frac{\rho u}{\rho_c u_c}\,dz$$ (11)

$$\delta^* = \int_0^\delta \left(1 - \frac{\rho u}{\rho_c u_c}\right)dz$$ (12)

$$\theta^* = \int_0^\delta \frac{\rho u}{\rho_c u_c}\left(1 - \frac{u^2}{u_c^2}\right)dz$$ (13)

$$\delta^{**} = \int_0^\delta \frac{u}{u_c}\left(1 - \frac{\rho}{\rho_c}\right)dz$$ (14)

$$D = \int_0^\delta \frac{\tau}{\tau_w}\frac{\partial(u/u_c)}{\partial z}\,dz$$ (15)

$$\phi = \int_0^\delta \left(1 - \frac{H - H_w}{H_c - H_w}\right)\frac{\rho u}{\rho_c u_c}\,dz$$ (16)

$$\delta^*_{JB} = \int_0^\delta \left(1 - \frac{j_y}{j_{yc}}\right)dz$$ (17)

$$\delta^{**}_{JB} = \int_0^\delta \frac{u}{u_c}\left(1 - \frac{j_y}{j_{yc}}\right)dz$$ (18)

$$\delta^*_{JE} = \int_0^\delta \left\{1 - \frac{j_x E_{xc} + j_y E_{yc}}{j_{xc}E_{xc} + j_{yc}E_{yc}}\right\}dz$$ (19)

$$C_f = \frac{\tau_w}{\frac{1}{2}\rho_c u_c^2}$$ (20)

and

$$C_H = -\frac{q_w}{\rho_c u_c(H_c - H_w)}$$ (21)

Core Flow and Boundary Layer Coupling

The core flow equations (7 through 9) and the boundary layer equations (8 through 10) are coupled by defining the effective core area as

$$A_c = (W - 2\delta^*_z)(Y - 2\delta^*_Y)$$ (22)

An iterative solution of the core flow equations and boundary layer equations is needed to solve the problem.

TURBULENT BOUNDARY LAYER ANALYSIS

Incompressible Non-MHD Turbulent Boundary Layer and New Velocity Profile

A commonly used physical model for the turbulent boundary-layer velocity distribution is the three-layer concept (White, 1974), where in the

> Inner layer: viscous shear dominates.
> Outer layer: turbulent shear dominates.
> Overlap layer: both types of shear are important.

The law of the wall describes the velocity distribution of the inner layer very near the wall, while the logarithmic velocity distribution is valid in the overlap layer region. Many expressions for the inner layer have been described in the literature, and a single expression for the inner layer and logarithmic region was developed by Spalding (1961). Coles (1956) developed the law of the wake for the outer portion of a turbulent boundary layer. Most of these expressions cannot be written such that either the velocity, u, or the distance from the wall, y, is an explicit function of the other. A new velocity profile was developed by Whitfield (1977), which gives the velocity explicitly as a function of distance from the wall, and can be calculated with direct properties of the boundary layer.

The composite expression for the new profile is in the inner variable y^+ and the outer variable y/θ. An analytical solution of the turbulent boundary-layer equations (including a turbulent kinetic energy equation) for the region near the wall (inner layer plus overlap layer) is used for the inner solution, and an empirical expression is derived for the outer region of the boundary layer. The following sections give a brief description of the model. Please refer to Whitfield's (1977) report for detailed derivations.

Analytical Function for Turbulent Flow in the Inner Region

This section discusses the velocity profile in the so-called viscous sublayer and buffer layers (inner region) that are normally described to extend over y^+ values of around 0 to 5 and 5 to 35, respectively. The inner variables are $u^+ = \bar{u}/v^*$ and $y^+ = yv^*/v$ while v^* is the wall-friction velocity. The following equation is used to described the velocity profile for $y^+ < 100$.

$$y^+ = \frac{1}{0.09}\tan(0.09u^+) \tag{23}$$

The velocity distribution according to equation (23) is compared with the equation introduced by Spalding (1961) in Figure 1. The two expressions provide almost identical results in the $0 \le y^+ \le 100$ region.

Analytical Function for Turbulent Flow in the Outer Region

The goal of this section is to combine the outer expression with the inner expression from the last section to make a composite expression for the complete turbulent

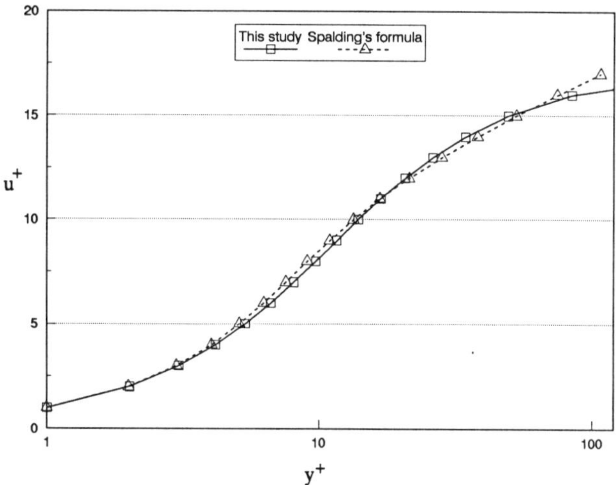

Figure 1. Velocity Distribution Near the Wall

boundary layer velocity distribution. This complete expression should have the following properties: recover the solution developed in last section for the inner region, i.e., $y^+ < 100$; when y approaches δ (edge of the boundary layer), the value u^+ should approach u_e^+ (or u/u_e approaches 1); and recover the velocity profiles similar to correlated experimental data away from the wall in an outer region variable y/θ. The composite solution u^+ is assumed as

$$u^+ = u_i^+ + u_o^+ \tag{24}$$

where u_i^+ is the inner solution and u_o^+ is the outer solution.

The inner solution was developed in the previous section. The complete velocity distribution is

$$u^+ = \frac{1}{0.09} \tan^{-1}(0.09 y^+) + \left[\left(\frac{2}{c_f} \right)^{\frac{1}{2}} - \frac{\pi}{0.18} \right] \tanh^{\frac{1}{2}} \left[a \left(\frac{y}{\theta} \right)^b \right] \tag{25}$$

where

$$\left(\frac{2}{c_f} \right)^{\frac{1}{2}} = u_e^+, \quad y^+ = \frac{Re_\theta\, y}{u_e^+\, \theta}, \quad and \quad \frac{u}{ue} = \frac{u^+}{u_e^+}$$

The a and b parameters are obtained by matching the distribution u/u_e at two points, $y/\theta = 2$ and $y/\theta = 5$. The point $y/\theta = 2$ was selected because it locates relatively close to the wall and still is outside the inner region, and $y/\theta = 5$ was used because it represents the outer region and is a point in the boundary layer where the velocity ratio, u/u_e is relatively constant for all shape factor H. The following expressions are used by Swafford (1983) to account for the flow near the separation point.

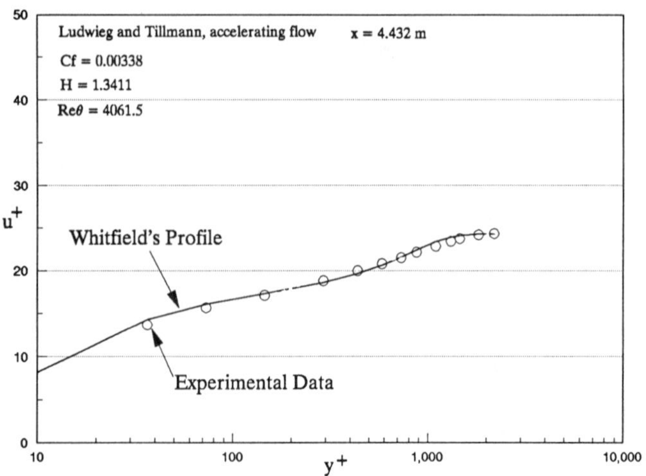

Figure 2. Present Calculations and Experimental Data of Flow 1300

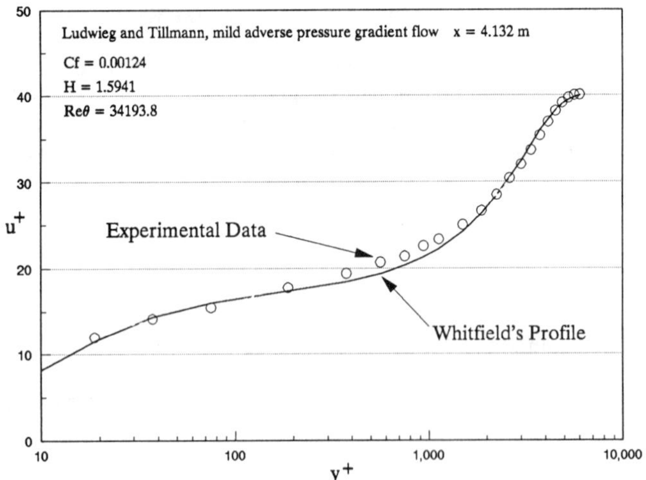

Figure 3. Present Calculations and Experimental Data of Flow 1100

For $y/\theta=2$:

$$\frac{u}{u_e} = \frac{1}{1.95}\left[\tanh^{-1}\left(\frac{8.5-H}{7.5}\right)-0.364\right] \tag{26}$$

and

For $y/\theta=5$:

$$\frac{u}{u_e} = 0.155 + 0.795\,sech[0.51(H-1.95)] \tag{27}$$

This study uses the latter set of equations to close the turbulent boundary layer equations. The experimental data from the Stanford Conference Proceedings (Coles and Hirst, 1968) have been used to verify the accuracy of the new velocity distribution. The

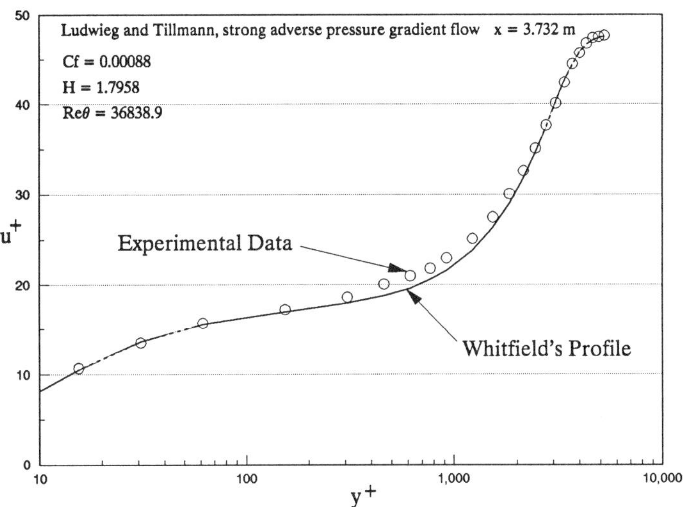

Figure 4. Present Calculations and Experimental Data of Flow 1200

comparisons include accelerating flow (1300 flow) (Figure 2), mild adverse pressure gradient flow (1100 flow) (Figure 3), and strong adverse pressure gradient flow (1200 flow) (Figure 4). The agreement between the analytical functions and experimental data are considered good.

Verification of Incompressible Turbulent Boundary Layers with New Velocity Profile

The integral techniques have been widely used in engineering calculations for turbulent flows. The 1968 AFOSR-IFP Stanford Conference demonstrated that many practical situations can be computed with these techniques. The 1980-81 AFOSR-HTTM Stanford Conference (Kline et al., 1981) showed the integral methods have been improved along with other methods during the intervening years. This study uses an integral method to provide a solution that combines good accuracy and reasonable computational speed. This section develops an integral solution for incompressible turbulent boundary layers with the new velocity profile and compares the results with different cases documented in the 1968 AFOSR-IFP Stanford Conference Proceedings. The effect of flow history has been introduced into the model to provide a better solution. The results show this combination produces a method equal or better than the first third of the methods in the proceedings. The method will be expanded into compressible flow in the following sections.

Integral Equations

The von Karman and mean energy integral equations are used in this study. The von Karman equation is the momentum equation (13) without MHD terms. It is

$$C_f = \frac{d\theta}{dx} + \theta\frac{H+2}{U_e}\frac{dU_e}{dx} \tag{28}$$

With a given distribution of U_e, the closure relationships for skin friction and the shape parameter are needed. The modified White's skin friction correlation by Swafford (1983) is used

$$c_f = \frac{0.3 e^{-1.33H}}{(\log_{10} Re_\theta)^{1.74+0.31H}} + (1.1 \times 10^{-4}) \left[\tanh\left(4 - \frac{H}{0.875}\right) - 1 \right] \qquad (29)$$

One more auxiliary equation is needed to close the set. While many people use the entrainment equation, this study prefers the mean kinetic energy integral equation, since the boundary layer thickness δ is ill defined in the entrainment equation. The mean kinetic energy integral equation (9) without MHD terms is

$$\frac{d}{dx}(U_e^3 \theta^*) = 2D^* \qquad (30)$$

where

$$D^* = \int_0^\delta \left(-\overline{uv} + v\frac{\partial U}{\partial y} \right) \frac{\partial U}{\partial y} \, dy \qquad (31)$$

Although the von Karman equation and mean energy equation are derived from the same momentum equation, they are independent because the integration process leads to different information loss. The equations (28), (29), and (30) are sufficient to solve the incompressible turbulent boundary layer problems.

Method of Solution

Two more parameters are used to solve equations (28) and (29). They are as follows:

Momentum shape factor

$$H = \frac{\delta^*}{\theta} \qquad (32)$$

Dissipation shape factor

$$H^{**} = \frac{\theta^*}{\theta} \qquad (33)$$

The velocity profile described earlier was used to carry out the relationship between H and H^{**}. This can be done in two ways: one can integrate the profile at each x location, or a correlation formula can be obtained in advance since Re_θ is very insensitive to the relationship. The correlation is approximated as

$$H^{**} = 0.3505 + 0.2182H + \frac{1.5295}{H} \qquad (34)$$

and provides two equations (28) and (30) for two unknowns H and θ. The method used to solve the equations (28) and (30) is the adaptive step-size controlled Runge-Kutta method (Press et al., 1986). This method is fifth order Runge-Kutta with monitoring of local truncation errors to ensure accuracy and adjust step size. Initial values of θ and H must be input as well as a reference to the Reynolds number, and U_e distribution as function of x.

It has been argued that the flow near the wall adjusts rapidly to changes of some external conditions, e.g., pressure gradient, while the outer layer does not respond as fast as the inner layer to the external variations due to large inertia from the large eddies. The lag equations for calculating the dissipation integral have been introduced by some researchers to compensate for this problem. Rotta (1968) suggested that the dissipation coefficient, as calculated from equation (31) at position x is considered to be the effective value of D^* at the downstream position $x + \Delta x$. Since D^* varies with $Re\theta$, a small correction must be applied to obtain agreement with the results calculated. The use of a lag equation introduces problems in specifying initial conditions if a calculation is started in a zone deviating considerably from equilibrium. Nevertheless, in real world practice, this situation is very rare.

This study uses the lag equation to calculate the flow only when the pressure parameter β is close to zero at the initial location. The pressure parameter is

$$\beta = -\frac{\delta^*}{\tau_w}\frac{dp_e}{dx} \tag{35}$$

and lag equation is

$$D^*(x + \Delta x) = D^*(x)\left(\frac{\ln(Re\theta(x))}{\ln(Re\theta(x + \Delta x))}\right)^{1.74 + 0.31 H(x + \Delta x)} \tag{36}$$

The delay Δx is assumed as

$$\Delta x = n\delta \tag{37}$$

where the total boundary thickness, δ, at position x is determined by

$$\frac{u(\delta)}{U_e} = 0.99 \tag{38}$$

and n is 5 in this study with $n = 4$ included in the results for comparison.

Results

The title of flow 1300 is "Ludwieg and Tillmann, Accelerating flow". It is a near-equilibrium boundary layer in a moderate negative pressure gradient that is similar to the flow in an MHD channel. Since it is near-equilibrium, the history effect does not influence the flow much. Figures 5 and 6 compare theory and experiment for C_f, H,

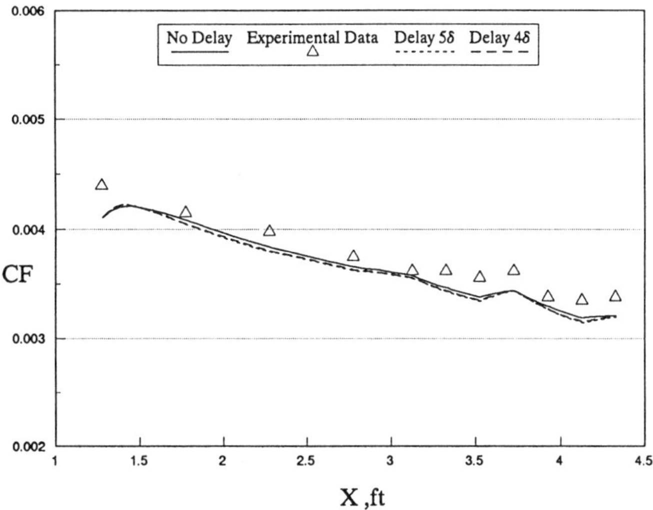

Figure 5. Skin-Friction Coefficient Comparison for Flow 1300

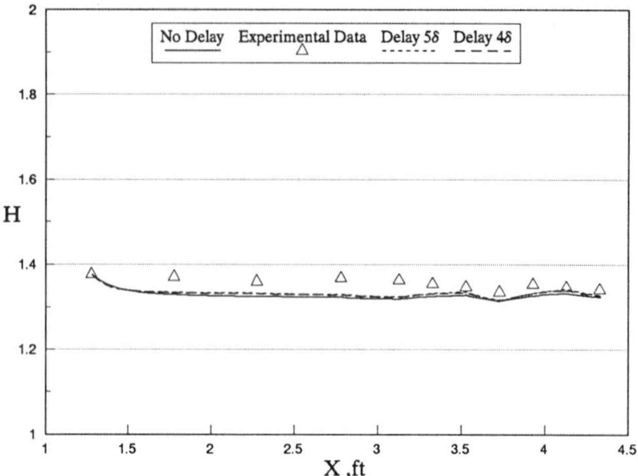

Figure 6. Shape Factor Comparison for Flow 1300

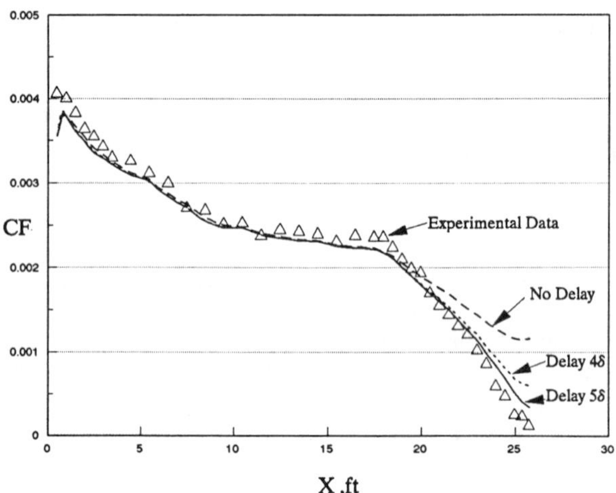

Figure 7. Skin-Friction Coefficient Comparison for Flow 2100

and delay numbers equal to 0, 4, and 5. The experiment of Schubauer and Klebanoff (flow 2100) was selected because it is a realistic airfoil-type flow commonly encountered and started with pressure parameter β equal to zero. The pressure parameter increases slowly at x = 5.5 m from 0.401 to 89.8 at the end of the run. It provides a good test for the history effect theory added into the integral technique used in this study. Figures 7 and 8 compare theory and experiment for C_f, H, and delay numbers equal to 0, 4, and 5. The history effect theory indeed provides a better solution to the problem.

Compressible Non-MHD Turbulent Boundary Layer

The new velocity profile is basically in incompressible form, where we are primary interested in compressible turbulent boundary layers. The Cole's law of corresponding stations (Coles, 1962) is used to transform the incompressible variables into compressible variables in this study. The law says

370

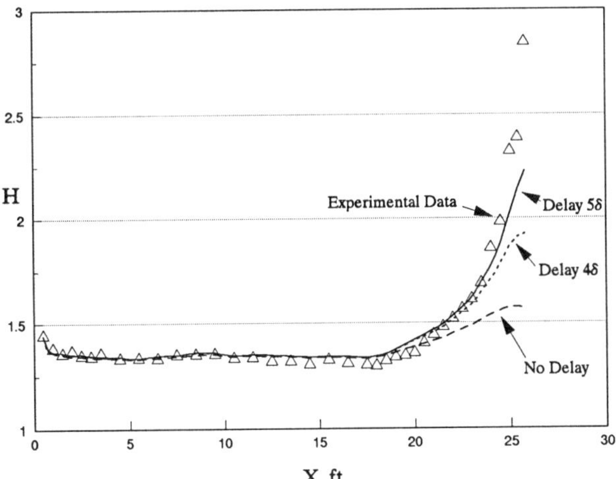

Figure 8. Shape Factor Comparison for Flow 1200

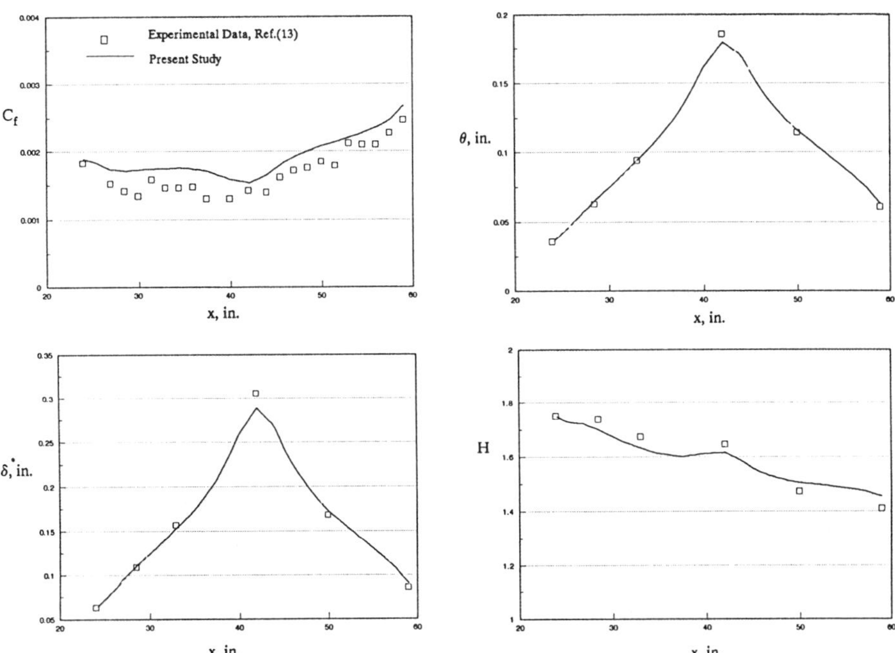

Figure 9. Boundary Layer Parameters Comparison

$$c_f Re_\theta = \frac{\rho_e \mu_e}{\rho_w \mu_w} \overline{c_f} \, \overline{Re_\theta} \tag{39}$$

where the bar ($^-$) denotes low speed flow. With the assumption, $\mu \sim T$, Lewis et al.,

(1970) used the above relation to correlate high-speed ($M_e < 8.18$) turbulent boundary layer velocity profiles successfully. The $\mu \sim T$ relation reduces equation (39) to

$$c_f Re_\theta = \overline{c_f}\,\overline{Re_\theta} \tag{40}$$

This relation provides part of the solution to the overall transformation, while the work of Winter and Gaudet (1970) provides the other relation

$$\frac{\overline{c_f}}{c_f} = F_c = \left(1 + \frac{M_e^2}{5} \right)^{\frac{1}{2}} \tag{41}$$

for adiabatic walls, $dp/dx = 0$, and air. It is assumed that equation (40) is acceptable for $dp/dx \neq 0$ if a good expression is used for $\overline{c_f}$ for $dp/dx \neq 0$. This study uses above approach of calculating the compressible turbulent boundary layer in the non-MHD and MHD flows with a modified C_f equation (29).

Verification of Compressible Turbulent Boundary Layers with New Velocity Profile

The experimental data from Winter et al. (1968) were used to verify the calculation procedures for the compressible turbulent flow described in the last section. The data set being considered in this study is the waisted body of revolution in the subsonic flow, $M_\infty = 0.597$, $Re_\infty, l = 9.98 \times 10^6$. The boundary layer properties of C_f, θ, θ^*, and H are compared in Figure 9. The overall agreement is considered good.

MHD FLOW CALCULATION PROCEDURE

The MHD channel boundary layers are divided into three distinct regions: electrode wall region, insulator wall region, and corner region. For electrode walls, current continuity provides $\delta^*_{jB} = 0$, so the MHD momentum equation has the same form as for non-MHD flows. The MHD effects reflect mainly through the enthalpy equation. For the insulator wall boundary layers, $\delta^*_{jB} \neq 0$, and the MHD effects reflect mainly through the momentum equation. The enthalpy profile remains practically the same as a non-MHD flow and can be modeled by power law or another simple profile. The corner regions are modeled by a "blending function" technique. The two-dimensional turbulent boundary layer velocity profiles derived by Whitfield and Swafford are used to represent the electrode wall velocity profiles. The electrode wall enthalpy profile is calculated by an argument that temperature is a function of velocity only, $T = T(U)$. The momentum and energy equations are combined for a constant mixed Prandtl number, Pr_m to form a second order differential equation in enthalpy. The equation provides a better solution than the common Crocco-Busemann relation. The insulator boundary layer is modeled after a simple model derived by Gertz et al. (1979), which uses a power law to represent the insulator enthalpy profile while defining a relationship to describe the difference between the actual and classical non-MHD velocity profiles. This paper discusses the procedures used in the process.

Velocity-Temperature Relations and Electrode Wall Boundary Layer Modeling

A new analytical description of temperature as a function of velocity through an MHD turbulent boundary layer for a constant but nonunity Prandtl number has been derived in this study to provide a better representation of the MHD enthalpy layer profiles. This approach used the equation resulting from combining the boundary layer momentum and energy equations and modeled the local shear stress by expressing it as a function of the local turbulent kinetic energy in the boundary layer. This approach was used by Whitfield (1978) for the non-MHD flow and showed good agreement with the experimental data. The model shows that the local total temperature must exceed the free-stream total temperature near the edge of the boundary layer for an adiabatic flat plate flow with a nonunity Prandtl number, which provides reasonable results.

This study extends Whitfield's model into MHD flow. A second order, nonlinear, ordinary differential equation is obtained and solved by a numerical method. The first assumption is that temperature is a function of velocity only at any axial location with known MHD variables. With the definition of the constant Prandtl number, Pr_m, the boundary layer momentum and energy equations are combined to yield

$$\frac{d^2\overline{h}}{d\overline{u}^2} + \left[(1 - pr_m) \cdot \frac{\partial \tau}{\partial \overline{u}} - pr_m \cdot \frac{\partial \overline{y}}{\partial \overline{u}} \cdot \overline{J} \times \overline{B} \right] \cdot \frac{1}{\tau}\frac{d\overline{h}}{d\overline{u}} + pr_m \cdot (\gamma - 1) M_\infty^2 \left(1 + \frac{1}{u_\infty \tau}\frac{\partial \overline{y}}{\partial \overline{u}} \cdot \overline{J} \cdot \overline{E} \right) = 0 \qquad (42)$$

where

$$Pr_m = c_p \left(\frac{\mu + \mu_\tau}{k + k_\tau} \right) \qquad (43)$$

The equation (42) is valid for $dp/dx = 0$, or for $Pr_m = 1$ and adiabatic wall if $dp/dx \neq 0$. Equation (42) without MHD terms is

$$\frac{d^2\overline{h}}{d\overline{u}^2} + \left[(1 - pr_m) \cdot \frac{\partial \tau}{\partial \overline{u}} \cdot \frac{1}{\tau} \right]\frac{d\overline{h}}{d\overline{u}} + pr_m \cdot (\gamma - 1) M_\infty^2 = 0 \qquad (44)$$

which is the same equation Whitfield (1978) described in his report. The expression

$$\tau = \frac{\rho}{\rho_w} \tau_w \exp(-c\eta^{2.5}) \qquad (45)$$

is used to describe the distribution of the shear stress in the boundary layer for non-MHD flows where c is a constant equal to 4 and $\eta = y/\delta$. The same expression is used to describe the MHD flows since sufficient MHD flow data does not exist.

Figure 10 shows the enthalpy versus velocity profile in a typical MHD channel with current density from 0 to 2 A/cm^2. The free stream conditions were $T_\infty = 2600$ K, $T_w = 1800$K, $M_\infty = 1.0$, $B = 3$ T, $C_f = 0.003$, and $\sigma_\infty = 6$ S/m. The curve fit equation is used to calculate the electrical conductivity distribution for a typical coal-fired MHD channel.

$$\frac{\sigma}{\sigma_\infty} = \left(\frac{P}{P_\infty} \right)^{-0.65} \left(\frac{T}{T_\infty} \right)^9 \qquad (46)$$

The result shows enthalpy profiles get fuller for increasing values of current which is as expected.

Insulator Wall Boundary Layer Modeling

The momentum equation of the MHD boundary layer presents the infulence of the body force term. For nearly constant free stream velocity terms and assuming the current to be low, Gertz et al. (1979) argue that

$$\tau = \tau_w - JBy \qquad y \ll \delta \qquad (47)$$

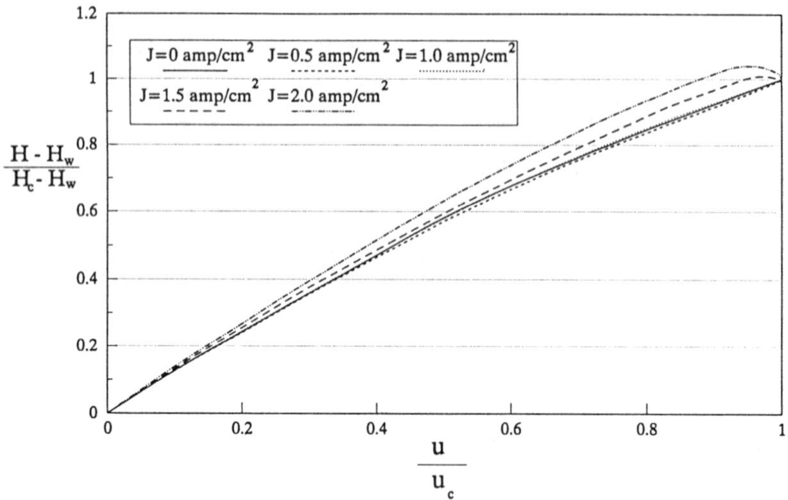

Figure 10. Enthalpy vs. Velocity Profiles in a Typical MHD Channel

If ΔU is the difference between the actual profile, and a classical profile then an excellent approximation for ΔU is

$$\Delta U = -1.53 U^*(y/d)^{1.08} \qquad (48)$$

where $d = \tau_w / JB$. This equation is used to describe the velocity profile in the insulator wall boundary layers.

SUMMARY

A new integral method for MHD flow analysis is described in this paper. An analytical approximation of turbulent boundary layer velocity profiles has been compared against incompressible and compressible non-MHD boundary layer flows and shown to provide reasonably good agreement. A new analytical description of the temperature as a function of velocity through an MHD turbulent boundary layer for a constant but nonunity Prandtl number has also been derived in this study to provide a better representation of the MHD enthalpy layer profile. The calculation result for a typical MHD channel flow shows the enthalpy model produces an expected enthalpy profile overshoot with current flow.

REFERENCES

Coles, D., "The Law of the Wake in the Turbulent Boundary Layer, "*Journal of Fluid Mechanics,*" vol. 1, pp 191-226, 1956.

Coles, D. E., "The Turbulent Boundary Later in a Compressible Fluid," R-403-PR, Rand Corp., Santa Monica, CA, September 1962.

Coles, D. E. and Hirst, E. A., Editors, "Proceedings Computation of Turbulent Boundary Layers - 1968 AFOSR-IFP Stanford Conference," vol. II, Standford University Press, Stanford, CA, 1968.

Gertz, J. et al., "Modeling of MHD Channel Boundary Layers Using an Integral Approach," 18th Symposium on Engineering Aspects of MHD, Butte, MT, June 1979.

Kline, S. J., Cantwell, B. J., and Lilley, G. M., Editors, "1980-81 AFOSR-HTTM Standford Conference on Complex Turbulent Flows," vol. I, Standford University Press, Standord, CA, 1981.

Lewis, J. E., Kubota, T. and Webb, W. H. "Transformation Theory for the Adiabatic Compressible Turbulent Boundary Layer with Pressure Gradient," *AIAA Journal*, vol. 8, no. 9, pp. 1644-1650, September 1970.

Press, W. H., Flannery, B. P., Teukolsky, S. A., and Vetterline, W. T., *Numerical Recipes (The Art of Scientific Computing)*, Cambridge University Press, Cambridge, USA, 1986.

Rotta, J. C., "Turbulent Boundary Layer Calculations with the Integral Dissipation Method," *Proceedings Computation of Turbulent Boundary Layers, 1968 AFOSR-IFP* Stanford Conference, vol. 1, pp. 177-181, Standford, USA, 1968.

Spalding, D. B., "A Single Formula for the 'Law of the Wall'," *Journal of Applied Mechanics*, Transactions of the ASME, Series E., vol. 28, pp. 455-458, September 1961.

Swafford, T. W. "Analytical Approximation of Two-Dimensional Separated Turbulent Boundary-Layer Velocity Profiles," *AIAA Journal*, vol. 21, no. 6, pp. 923-926, June 1983.

White, F. M., *Viscous Fluid Flow*, McGraw-Hill Book Company, Inc., New York, NY, 1974.

Whitfield, D. L., "Analytical Description of the Complete Two-Dimensional Turbulent Boundary-Layer Velocity Profile," Arnold Air Force Station, TN, AEDC-TR-77-79(ADA045033), September 1977; see also, AIAA Paper 78-1158, July 1978.

Wilson, D. R., and Simmons, G. A., "MHD Generator Off-design Performance and NO_x Kinetics Analysis," NASA Grant NSG 3255, Final Report, the University of Texas at Arlington, June 1980. (Also published as NASA CR-16587)

Winter, K. G. and Gaudet, L. "Turbulent Boundary-Layer Studies at High Reynolds Numbers at Mach Numbers Between 0.2, and 2.8," Aeronautical Research Council, London, R&M no. 3712, December 1970.

Winter, K. G., Rotta, J. C., and Smith, K. G. "Studies of the Turbulent Boundary Layer on a Waisted Body of Revolution in Subsonic and Supersonic Flow," Aeronatutical Research Council 30935, Reports and Memoranda No. 3633, August 1968.

Bifurcation in Rayleigh-Bénard Convection with Gases in Rectangular Containers

J. R. LEITH and J. G. MAVEETY
Department of Mechanical Engineering
University of New Mexico
Albuquerque, New Mexico 87131, USA

ABSTRACT

Bifurcation theory, in the form of a one-dimensional Ginzburg-Landau model, is applied to describe the growth rate of heat flux in Rayleigh-Bénard convection with air in rectangular containers of moderate size. The basic features of the low Rayleigh number behavior are discussed, including the relevant hydrodynamic instability mechanisms. Departure from the wide layer behavior is discussed, based upon new experimental results. Finally, we examine the issues involved in application of the "simple" nonlinear hydrodynamic modeling principles to a physical system which exhibits a size-dependent secondary bifurcation structure.

INTRODUCTION

The use of bifurcation theory provides an effective means for modeling the nonlinear dynamical behavior of physical systems, in which spatial patterns evolve, in both space and time, in a self-organized manner. Such nonlinear dynamical system models describe the slow space and time evolution of a macroscopic order parameter, in terms of a control parameter. In this paper, we utilize the one-dimensional Ginzburg-Landau equation, equivalent here to the Newell-Whitehead-Segel amplitude equation, to analyze the nonlinear growth of heat flux with increased buoyancy in finite Rayleigh-Bénard convection. Rayleigh-Bénard convection exhibits an evolution of organized flow structure, so that consideration of an amplitude envelope equation allows an initial examination of the secondary bifurcation structure.

Rayleigh-Bénard convection concerns the stability of fluid layers heated below and cooled above. At a sufficiently large buoyant force, an initially quiescent state loses stability to a convective state. In closed rectangular containers, the onset of convection is characterized by longitudinal rolls aligned parallel to the shorter horizontal side. With increase of the buoyant force in containers of gases, a secondary bifurcation structure exists (following the primary bifurcation from conductive to convective states) and is comprised of a successive loss of convection rolls. The loss of rolls with heating and subsequent gain of rolls with cooling are attributed to skewed varicose instability, which induces spatially periodic contraction and dilatation of the rolls. Dislocations in the roll pattern, which nucleate at pinches in the rolls and which execute both climb and glide (motions parallel and transverse to the roll axis), produce the bifurcation in pattern.

Many secondary bifurcations are evident in experiments with gases in wide layers [1, 2], and these result in production of turbulence at low Rayleigh number. Since the convection rolls

are essentially square at onset (width and depth = spacing of horizontal surfaces) and stable
patterns exist only for roll sizes ranging to about twice their initial width, the small to moderate size container regime is distinct from the moderate to large size container regime [3].
Whereas several instability mechanisms are manifested at increasing buoyant force in fluids
of moderate to large Prandtl number (water, oil, fluorocarbons, etc.), only skewed varicose
and oscillatory instabilities are predominant in fluids with $Pr \sim \mathcal{O}(1)$ (gases and cryogens).
As noted above, skewed varicose instability modulates the size of convection rolls and consequently does not produce a new form of convection. The onset of oscillatory convection is
characterized by a coherent transverse oscillation of the roll pattern, and nonlinear coupling
of skewed varicose and oscillatory instabilities is evident in experiments [3, 4].

ANALYSIS

Haken [5], Platten and Legros [6], Wesfreid et al.[7], and Newell [8] have shown that the
simplest differential equation of Ginzburg-Landau type which can be used to represent the
amplitude growth rate for a one-dimensional pattern in Rayleigh-Bénard convection is given
by

$$\frac{\partial A}{\partial t} = \alpha' A - \beta' A |A|^2 + \gamma' \frac{\partial^2 A}{\partial x^2} \tag{1}$$

in which A represents the slowly varying amplitude envelope of a principal measure of the
convection (*e.g.*, convective velocity or heat flux), and $\alpha', \beta',$ and γ' are parameters or scale
factors. The time and single space coordinates in (1) are denoted by t and x. The space
coordinate x is taken in a direction transverse to the roll axis. In one-dimensional systems, the
amplitude is a real function and the parameters are real. The amplitude(s) and parameters
in two-dimensional Ginzburg-Landau equations are generally complex [8]. In its most general
form, the Ginzburg-Landau equation governs generic bifurcation and growth rate of diffusive
systems.

A perturbation approach in analysis of Navier-Stokes and energy equations which govern
Rayleigh-Bénard convection yields the amplitude envelope equation in two-dimensional form,
derived as a solvability condition which preserves the uniformity of an asymptotic expansion
of the vertical velocity component [8-10]. Newell and Whitehead [9] and Segel [10] showed
that the major nonlinear contribution in the growth of convective velocity could be discerned
if the y (roll axis direction) - variation is neglected. This is accomplished by setting [10]

$$\mathcal{A}(\tau, X, Y) = f(Y) A(\tau, X) \tag{2}$$

so that the boundary conditions on Y are posed as conditions on f. Solutions $f(Y)$ are, of
course, non-trivial. In (2), X and Y are scaled from spatial coordinates x and y as

$$X = \epsilon x \qquad Y = \epsilon^{1/2} y \tag{3}$$

where $\epsilon = (Ra/Ra_c - 1)^{1/2}$, $Ra = (g\beta/\nu\alpha) d^3 \Delta T$ is the Rayleigh number and τ is the slow
time scale

$$\tau = 2 \left(a^2 + 1\right) \pi^2 \epsilon^2 Pt. \tag{4}$$

In (4), a is the dimensionless wavenumber ($a = 2\pi d/\lambda$, d = fluid layer depth, λ = wavelength of a roll pair (= $2d$ at onset of convection)), and P is a function of Prandtl number, $P = Pr/(2(1+Pr))$. In the definition of Rayleigh number, g, β, ν and α are the gravitational acceleration, coefficient of thermal expansion, kinematic viscosity and thermal diffusivity. The temperature difference is denoted by ΔT. With (2) - (4), the amplitude envelope equation simplifies to its one-dimensional form [10]

$$\frac{\partial A}{\partial \tau} = A + \delta \frac{\partial^2 A}{\partial x^2} - bA^3 \tag{5}$$

which has the same form as (1). Equation (5) is often called the Newell-Whitehead-Segel amplitude equation.

Wesfreid et al.[7] examined the one-dimensional Ginzburg-Landau equation, in terms of its similarities to Landau-Hopf bifurcation theory for phase transitions. As analyzed in [7], equation (1) with physically relevant length, time, and velocity scales can be posed for the amplitude of convective velocity as

$$\tau_o \frac{\partial V}{\partial t} = \epsilon V + \xi_o^2 \frac{\partial^2 V}{\partial x^2} - \frac{V^3}{V_o^2} \tag{6}$$

in which τ_o is a diffusion time scale related to the growth rate near onset ($1/\tau = \epsilon/\tau_o$), ξ_o is an influence length scale defined by

$$\frac{1}{\tau} = \frac{\epsilon - \xi_o^2 (a - a_c)^2}{\tau_o} \tag{7}$$

and V_o is a velocity scale. In (7), $a_c = \pi$ is the wavenumber at onset. Near the onset of convection, $\partial^2 V/\partial x^2 \sim 0$, so that in the steady state

$$V = V_o \epsilon^{1/2} \tag{8}$$

which is in excellent agreement with experimental results [11]. The characteristic (or diffusion) time exhibits the classic critical slowing down familiar from theory of second-order phase transitions, as $\tau \sim 1/\epsilon$. Experimental findings from [7] give

$$\begin{aligned}
\xi_o^2 &= c_1 d^2 \\
\tau_o &= c_2 \frac{d^2}{\alpha} \\
V_o &= c_3 \frac{\alpha}{d}
\end{aligned} \tag{9}$$

The constants c_1, c_2, and c_3 are empirical.

Previous experimental results for wide layers with air give a correlation for dimensionless heat transfer as [12]

$$Nu = 1 + 1.44 \left(1 - \frac{Ra_c}{Ra}\right), \qquad Ra < 5830 \tag{10}$$

in which $Nu = q_{conv}/q_{cond}$ is the Nusselt number and $q_{cond} = kA\Delta T/d$ is the conductive heat transfer. Since $\epsilon^2 = (Ra - Ra_c)/Ra_c$, (10) can be expressed as

$$(Nu - 1)\frac{Ra}{Ra_c} = 1.44\epsilon^2 \tag{11}$$

with ϵ as defined above. The relation between the heat flux and the convective velocity (equations (8) and (11)), is given by

$$(Nu - 1)\frac{Ra}{Ra_c} \sim \left(\frac{V}{V_o}\right)^4 \tag{12}$$

with a proportionality constant as a function of Prandtl number and the geometry.

We expect to be able to determine a geometric effect in our Rayleigh-Bénard experiments with air in moderate size containers, based upon the available theoretical and experimental results discussed above. The one-dimensional Ginzburg-Landau equation consequently suggests that the near-onset character of heat flux measurement should give

$$(Nu - 1)\frac{Ra}{Ra_c} = c\left(\frac{Ra}{Ra_c} - 1\right) \tag{13}$$

with c as an empirical constant.

Two issues are of importance in applying the bifurcation model results, equation (13), to experiments in finite containers. The first of these concerns the effects of container size on the value of the constant ,c, in (13) for description of the low Rayleigh number behavior following the onset of convection. This issue can be resolved in a simple manner, since c is the initial slope of the heat transfer curve. The remaining issue concerns the behavior of the heat transfer following secondary bifurcation (bifurcation from a pattern with n convection rolls to one with $n - 1$ rolls in experiments with increasing Rayleigh number and from $n - 1$ rolls to n rolls in experiments with decreasing Rayleigh number). We noted earlier that there are fewer secondary bifurcations of the convection roll pattern in moderate size containers than in large containers, because the reduction in wavenumber with increasing Rayleigh number is (essentially) fixed at about a factor of two. Consequently, the bifurcation of roll patterns in moderate size containers comprises a significant change in the flow structure. Our experimental results suggest that there are also significant changes in the heat flux-Rayleigh number behavior following secondary bifurcation, and we will examine in what follows the extent to which these results can be modeled by the one-dimensional Ginzburg-Landau equation.

EXPERIMENTS

Experiments were conducted with the apparatus described in [13, 14]. A schematic of the arrangement of the experiment and instrumentation used are illustrated in Figure 1. Two independent programmable temperature baths were used to circulate water through tube passages in the lower and upper horizontal heat transfer surfaces. Each heat transfer surface was constructed of a three-layer laminate, comprised of a paraffin-filled foamed aluminum slab with embedded and brazed aluminum tubes, heat flux gauges, and a chrome-plated aluminum sheet of $2.5mm$ thickness which faced the air layer.

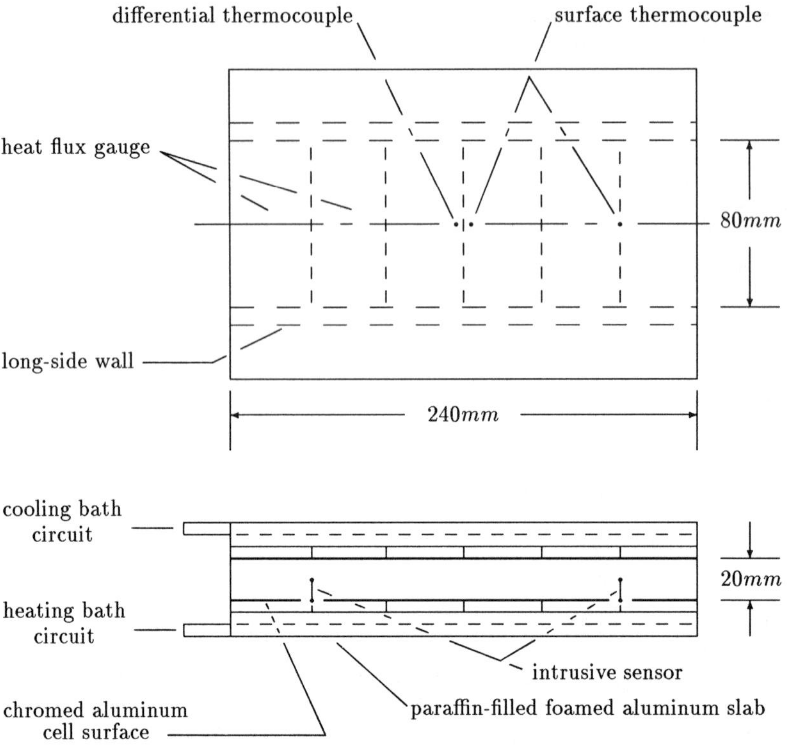

FIGURE 1: Test apparatus and instrumentation.

An air layer depth of $d = 20mm$ was used in the experiments. Dimensions of the heat transfer surfaces were $240mm \times 160mm$, which produced a container with a relative size of $12 \times 8 \times 1$. Dimensions of the heat flux gauges were $40mm \times 80mm \times 2.5mm$ thickness, so that six gauges were attached to both lower and upper heat transfer surfaces. Test geometries of smaller size were constructed by including additional short-side wall spacers and by moving the long-side walls. In the present work, results are presented for Rayleigh-Bénard experiments with $12 \times 4 \times 1$ and $10 \times 4 \times 1$ relative sizes. Short-side walls were constructed of polished acrylic, and long-side walls were constructed of either polished acrylic or clear fused quartz. Flow visualization experiments were accomplished by injecting cigar smoke through small ports in the heat transfer surfaces, with viewing of convection roll patterns through the long-side walls.

Two 36-gauge copper-constantan thermocouples were attached to each aluminum sheet for surface temperature measurement. The thermocouples were used to determine the temperature difference across the air layer, and an additional temperature difference measurement was obtained from a differential thermocouple, also attached to the aluminum sheets and located at the horizontal center of the apparatus. The heat flux gauges were constructed of copper and nickel foils, laminated to a balsa wood substrate and silver-soldered to form thermojunctions at either horizontal surface of the balsa wood [13]. Thermojunctions of the gauges were co-parallel to the long-side central vertical plane of the apparatus, so that the surface heat flux measurement was adjacent to the cross-sections of convection rolls. The design basis for the heat flux gauges was given by Hager [15].

The experimental procedure consisted of operating the two temperature baths in linear heating and cooling programs at equal rates from an initial isothermal test condition to a terminal steady state, so that the experiments were conducted at constant mean temperature and with linear increase of the temperature difference. We have consequently been able to produce extremely well-controlled experiments, characterized by a linear increase of the Rayleigh number. Installation of the two intrusive differential thermocouples in the air layer (Figure 1), as discussed in [3, 4, 13, 14], allowed an inference of bifurcation in the convection experiments. As discussed below, our previous technique in interpretation of signatures produced by the intrusive sensors is highly compatible with the heat flux measurements reported here.

RESULTS AND DISCUSSION

The experimental procedure discussed above produces a heating experiment which commences with a conduction base state, and which is swept through the onset of convection and transitions of flow structure (loss of convection rolls) initiated by the nucleation of dislocations. Cooling experiments are simply operated in reverse, commencing with an initial convective state. The onset of convection and the flow structure transitions are respectively referred to as primary and secondary bifurcations. As discussed in [7-11,19], the one-dimensional Ginzburg-Landau equation (or alternately the Newell-Whitehead-Segel amplitude equation) models convective onset as analogous to a second-order phase transition and the initial growth rate of the heat flux as a quadratic function of the Rayleigh number.

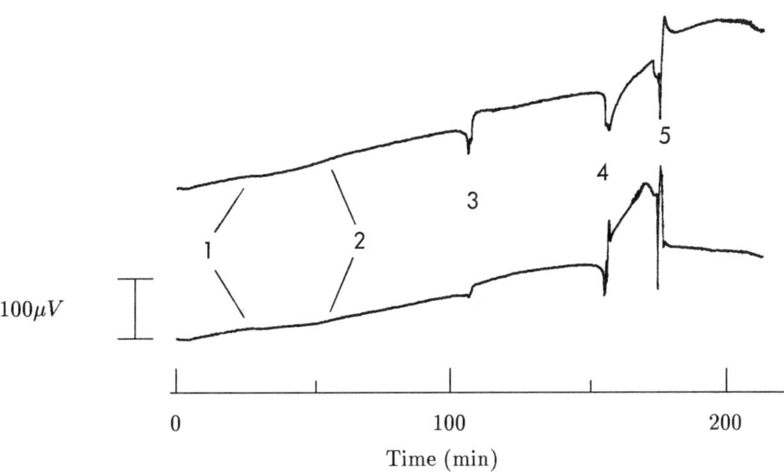

FIGURE 2: Differential thermocouple signatures for linear heating to $Ra = 10,800$ in $12 \times 4 \times 1$ container. 1: onset of convection, 2: bifurcation to $11r$, 3: bifurcation to $10r$, 4: bifurcation to $9r$, 5: brief $9r - 10r/10r - 9r$ limit cycle. $Ra_c = 1800$ [3].

Signatures from the two intrusive differential thermocouples are shown in Figure 2 for a heating experiment in a container of $12\times4\times1$ relative size. As illustrated in Figure 2, the onset of convection is denoted by the first significant change of slope in the voltage-time histories. Onset is characterized by twelve longitudinal rolls aligned parallel to the shorter horizontal side, as confirmed in independent smoke flow visualization experiments. The first of successive secondary bifurcations, a transition from twelve to eleven rolls, is evident in Figure 2 as a mild change of slope (called a "soft" transition [11] and characterized by a gradual contraction of a wall roll created by a climbing dislocation). The second of the secondary bifurcations evident in the experiment produces a dramatic and sudden discontinuity in the signatures, Figure 2, and is comprised of a rapid annihilation of a wall roll at the same end of the container as in the $12r - 11r$ transition. Dislocation climb is also the eventual mechanism by which the roll is annihilated, but propagation speed in the $11r - 10r$ transition is much more rapid than that in the $12r - 11r$ transition. For the reason apparent in Figure 2, the second of the secondary bifurcations is termed a "hard" transition [3]. An independent smoke flow visualization experiment, conducted at the same heating rate, showed a stable eleven-roll flow structure at $Ra = 2870 - 5340$. The third secondary bifurcation illustrated in Figure 2, a transition from ten to nine rolls, is also a hard transition. Previous experiments with the $12 \times 4 \times 1$ container have produced limit cycle behavior in steady-states with $9500 < Ra < 11,000$, comprised of successive annihilation of a wall roll ($10r - 9r$ transition) and nucleation and growth of a wall roll ($9r - 10r$ transition) at opposite ends of the container [3, 13]. One full cycle of this behavior is evident in Figure 2.

The average heat flux determined in the $12\times4\times1$ container experiment is illustrated in Figure 3, and the data can be interpreted with the aid of the differential thermocouple signatures in Figure 2. Heat transfer measurements illustrated in Figure 3 are represented in normalized form as

$$N = (Nu - 1)\frac{Ra}{Ra_c} \tag{14}$$

in which $Ra_c = 1800$ is the average Rayleigh number at the onset of convection, as determined from a total of 62 experiments [3, 4, 13, 14]. Data acquisition for the heat flux measurements was from a Hewlett-Packard 3456A voltmeter, and controlled by an HP-9816 computer. A sampling period of 20 sec was utilized, and voltage measurement resolution was $0.1\mu V$. Sampling period for the differential thermocouples (intrusive sensors) was $10sec$. As noted earlier, the Nusselt number is calculated as $Nu = q_{conv}/q_{cond}$. Since each experiment includes a conduction base state between the initial isothermal condition and the onset of convection, q_{cond} at $Ra > Ra_c$ is assumed from the extrapolated slope of the voltage-temperature relation for each heat flux gauge [13].

The initial slope of the heat transport curve in Figure 3 is given by

$$\left.\frac{dN}{d(Ra/Ra_c)}\right|_{Ra=Ra_c} = 0.40 \pm 0.03 \tag{15}$$

in which the error bounds are given for a 95% confidence interval, based upon the standard error of estimate and tabulated upper percentage points of the t-distribution [13]. Results which have been established for Rayleigh-Bénard convection in wide layers of air are discussed by Hollands et al. [12]. From equation (11),

$$\left.\frac{dN}{d(Ra/Ra_c)}\right|_{Ra=Ra_c} = 1.44, \qquad \frac{d}{L} \ll 0.1 \tag{16}$$

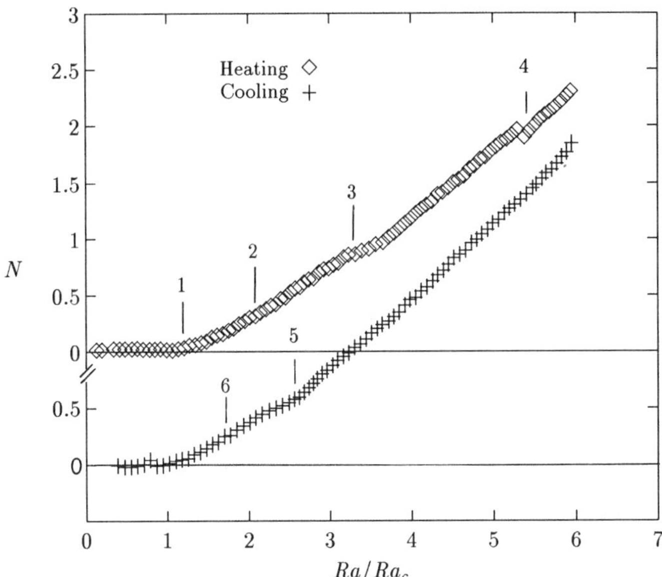

FIGURE 3: Heat flux in linear heating to, and cooling from $Ra = 10,800$ in $12 \times 4 \times 1$ Container Experiment [13]. 1: onset of convection, 2: bifurcation to $11r$, 3: bifurcation to $10r$, 4: bifurcation to $9r$, 5: bifurcation to $11r$ (cooling), 6: bifurcation to $12r$ (cooling).

The experimental result for wide layers, equation (16), compares favorably with the theoretical results of Schlüter et al. [16]. Only sparse results are available for containers of air of moderate size [12, 17, 18], but all results show a reduction in slope of the heat transport curve (compared to the results for wide layers).

We have briefly examined the effect of horizontal extent of the air layer in experiments [13]. In Figures 4 and 5, the differential thermocouple signatures and heat flux measurements are shown for an experiment with air in a container of $10 \times 4 \times 1$ relative size ($d = 20mm$, as in the $12 \times 4 \times 1$ container experiment discussed above). The average Rayleigh number at onset for this geometry is $Ra_c = 1850$, as determined in a series of 31 experiments [3, 14]. As illustrated in Figures 4 and 5, the first of successive secondary bifurcations, a transition from ten to nine rolls, is "soft" (it is in fact transparent in the differential thermocouple signatures of Figure 4). The second secondary bifurcation, a transition from nine to eight rolls, is a "hard" transition. Through the second secondary bifurcation, nonlinear behavior in the $10 \times 4 \times 1$ container is much the same as in the $12 \times 4 \times 1$ container. Limit cycle behavior, comprised of sequential forward bifurcation ($8r - 7r$ transition) and backward bifurcation ($7r - 8r$ transition) at opposite container ends, commences immediately upon reaching steady-state. This nonlinear behavior is analogous to the behavior in the $12 \times 4 \times 1$ container experiment discussed above, but the forward- and backward- transitions are more closely-spaced in time in the $10 \times 4 \times 1$ container experiment.

The bifurcation model for the growth rate of heat flux posed above (the Ginzburg-Landau equation, or Newell-Whitehead-Segel amplitude equation) predicts an initial growth rate which should be linear as in equation (13). A close examination of Figures 3 and 5 indicates

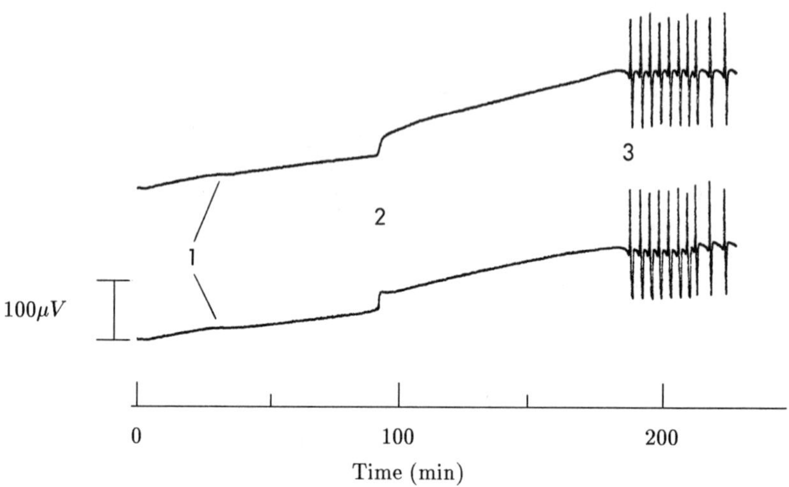

FIGURE 4: Differential thermocouple signatures for linear heating to $Ra = 10,500$ in $10 \times 4 \times 1$ container. 1: onset of convection, 2: bifurcation to $8r$, 3: onset of $8r - 7r/7r - 8r$ limit cycle. $Ra_c = 1850$ [3]

that the normalized heat flux following the onset of convection appears to follow a quadratic behavior. To first order, however, the bifurcation model prediction represents a reasonable approximation of heat flux growth with Rayleigh number in the initial pattern formation of twelve convection rolls in the $12 \times 4 \times 1$ container experiment (Figure 3) and ten convection rolls in the $10 \times 4 \times 1$ container experiment (Figure 5).

The first of successive secondary bifurcations, a "soft" transition from twelve to eleven rolls in the $12 \times 4 \times 1$ container experiment and from ten to nine rolls in the $10 \times 4 \times 1$ container experiment, can apparently be modeled by the amplitude envelope equation in its present form, if the mild nonlinearity near onset (Figures 3 and 5) can be included as a second-order effect. Ahlers [19] noted a similar behavior in experiments with a cylindrical container of liquid helium ($Pr = 0.78$), so that the initial curvature in the heat transport near onset could be expressed as

$$\left. \frac{dN}{d\left(Ra/Ra_c\right)} \right|_{Ra=Ra_c} = 0.84 + 1.28 \left(\frac{Ra}{Ra_c} - 1 \right).$$

(17)

Using the result in [19], a logical extension would be to model the growth rate near onset as

$$N = C_o + C_1 Ra + C_2 Ra^2$$

(18)

which expresses a quadratic nonlinearity. While it is apparent that the present results can be curve-fitted to a variety of nonlinear forms, the preferred modeling would be to include a nonlinearity of perturbation form in the amplitude equation which would generally describe behavior in finite containers. The important modeling consideration appears to be that the growth rate near onset in moderate size containers deviates from the linear behavior found

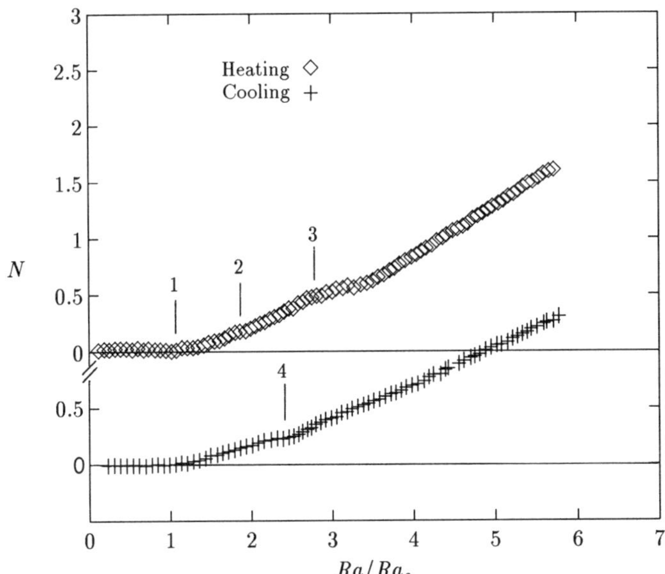

FIGURE 5: Heat flux in linear heating to, and cooling from $Ra = 10,500$, in $10 \times 4 \times 1$ Container Experiment [13]. 1: onset of convection, 2: bifurcation to $9r$, 3: bifurcation to $8r$, 4: bifurcation to $9r$ (cooling).

in wide layers and predicted by the Ginzburg-Landau equation (or Newell-Whitehead-Segel amplitude equation).

The second of successive secondary bifurcations, a "hard" transition from eleven to ten rolls in the $12 \times 4 \times 1$ container experiment and from nine to eight rolls in the $10 \times 4 \times 1$ container experiment, exhibits hysteresis: the heating transition (forward bifurcation) occurs at larger Rayleigh number than the cooling transition (backward bifurcation), so that the heat flux with increasing Rayleigh number is different from that with decreasing Rayleigh number. This type of nonlinear behavior is generally referred to as a subcritical bifurcation of inverted type [3, 20], and it cannot be described directly from the amplitude envelope equation in its present form. An amplitude equation which can model the inverted bifurcation needs an additional nonlinearity as [20]

$$\frac{\partial A}{\partial t} = (R - R_1) A + A^3 - \zeta A^5 \tag{19}$$

in which $\zeta > 0$ is a real parameter, R_1 is the Rayleigh number at forward bifurcation (roll loss), and R_2 is the Rayleigh number at backward bifurcation (roll gain). The relation between R_1, R_2 and ζ is

$$R_2 = R_1 - \frac{1}{4\zeta}, \qquad R_2 < R_1 \tag{20}$$

Additional bifurcation in pattern and heat flux could be constructed in an amplitude equation by the addition of nonlinear terms (*cf.*, equations (5) and (19)), but this type of analytical approach to model building appears to lose the generic character of bifurcation for which the Ginzburg-Landau equation is utilized in applications.

The third of successive secondary bifurcations in the present experiments is also a "hard" transition, but the bifurcation yields limit cycle behavior in steady states, which exhibits a dynamical character. Modeling of the third secondary bifurcation in the containers discussed here is difficult, and it is compounded by the near proximity of the onset of oscillatory convection in the experiments. We have not yet decided upon a satisfactory nonlinear dynamical model to accommodate this type of behavior.

CONCLUSIONS

Nonlinear dynamical behavior in finite Rayleigh-Bénard convection can be described qualitatively, and the macrostates (heat flux growth) can be described quantitatively, by application of bifurcation theory in the form of the Ginzburg-Landau equation. Stated as a "solvability condition" which predicts the amplitude envelope of an order parameter (considered here as the heat flux), the Ginzburg-Landau equation (Newell-Whitehead-Segel amplitude equation) predicts (1) the convective onset as a supercritical bifurcation, in analogy with a second-order phase transition, and (2) the initial growth rate of heat flux at low Rayleigh number. These results are in excellent agreement with previous experimental work in wide layers with gases and cryogens. The experimental results reported here for moderate size containers appear to depart from the wide layer results in a manner which can be expressed as a further nonlinearity in the bifurcation model. Such a modification will be sufficient in description of our experiments through the first of successive secondary bifurcations. Thereafter (at higher Rayleigh number), a higher-order amplitude equation and/or application of sequential amplitude equations will be necessary to model the behavior observed in experiments.

ACKNOWLEDGMENTS

This work was supported by the National Science Foundation under Grant No. MEA 83-07500 and by the Department of Mechanical Engineering, University of New Mexico.

References

1. Willis, G.E., Deardorff, J.W. and Sommerville, R.C.J. Roll-Diameter Dependence in Rayleigh Convection and its Effect upon the Heat Flux. *J. Fluid Mech.*, 54:351–367, 1972.

2. Busse, F.H. and Clever, R.M. Instabilities of Convection Rolls in a Fluid of Moderate Prandtl Number. *J. Fluid Mech.*, 91:319–335, 1979.

3. Leith, J.R. Flow Structure Transition Mechanisms in Thermal Convection of Air in Rectangular Containers. *Physica D*, 37:334–340, 1989.

4. Leith, J.R. Suppression of Flow Structure Dislocations in Thermal Convection Experiments. *Exp. Fluids*, 7:427–428, 1989.

5. Haken, H. *Advanced Synergetics. Instability Hierarchies of Self-Organizing Systems and Devices.* Springer-Verlag, Berlin, 1983.

6. Platten, J.K. Legros, J.C. *Convection in Liquids.* Springer-Verlag, Berlin, 1984.

7. Wesfreid, J., Pomeau, Y., Dubois, M., Normand, C. and Bergé P. Critical Effects in Rayleigh-Bénard Convection. *J. Physique Lett.*, 39:725–731, 1978.

8. Newell, A.C. The Dynamics of Patterns: A Survey. In J.E. Wesfreid, H.R. Brand, P. Manneville, G. Albinet, and N. Boccara, editors, *Propagation in Systems Far From Equilibrium.* Springer-Verlag, Berlin, 1988.

9. Newell, A.C. and Whitehead, J.A. Finite Bandwith, Finite Amplitude Convection. *J. Fluid Mech.*, 38:279–303, 1969.

10. Segel, L.A. Distant Side-Walls Cause Slow Amplitude Modulation of Cellular Convection. *J. Fluid Mech.*, 38:203–224, 1969.

11. Bergé, P. Rayleigh-Bénard Instabilities: Experimental Findings Obtained by Light Scattering and Optical Methods. In T. Riste, editor, *Fluctuations, Instabilities, and Phase Transitions.* Plenum Press, New York, 1975.

12. Hollands, K.G.T, Raithby, G.D. and Konicek, L. Correlation Equations for Free Convection Heat Transfer in Horizontal Layers of Air and Water. *Int. J. Heat Mass Transfer*, 18:879–884, 1975.

13. Maveety, J.G. Convective Heat Flux Measurements of Air in Rectangular Rayleigh-Bénard Cells, 1990. MS Thesis, University of New Mexico.

14. Leith, J.R. Successive Transition of Steady States in Moderate Size Containers of Air Heated Below and Cooled Above. *ASME HTD-Vol.94/AMD-Vol.89*, pages 91–98, 1987. Bifurcation Phenomena in Thermal Processes and Convection, H. Bau, L. Bertram, and S. Korpela, eds.

15. Hager, N.E. Thin Foil Heat Meter. *Rev. Sci. Inst.*, 36(11):1564–1570, 1965.

16. Schlüter, A., Lortz, D. and Busse, F. On the Stability of Steady Finite Amplitude Convection. *J. Fluid Mech.*, 23:129–144, 1965.

17. Walden, R.W., Kolodner, P., Passner, A. and Surko, C.M. Heat Transport by Parallel-Roll Convection in a Rectangular Container. *J. Fluid Mech.*, 182:205–234, 1987.

18. Motsay, R.W., Anderson, K.E. and Behringer, R.P. The Onset of Convection and Turbulence in Rectangular Layers of Normal Liquids and ^{4}He. *J. Fluid Mech.*, 189:263–286, 1988.

19. Ahlers, G. Onset of Convection and Turbulence in a Cylindrical Container. In L. Garrido, editor, *Systems Far From Equilibrium.* Springer-Verlag, Berlin, 1980.

20. Normand, C., Pomeau, Y, and Velarde, M.G. Convective Instability: A Physicist's Approach. *Rev. Mod. Phys.*, 49:581–624, 1977.

Simulation of EM Wave Scattering from a Dense Distribution of Spherical Scatterers

Y. C. TZENG, J. BREDOW, and A. K. FUNG
Electrical Engineering Department, Box 19016
University of Texas at Arlington
Arllington, Texas 76019, USA

ABSTRACT

The purpose of this study is to examine the scattering and attenuation properties of a densely populated medium by Monte Carlo simulation using the T-matrix approach. The T-matrix formulation of electromagnetic scattering for an arbitrary number of scatterers given previously by Peterson is used to solve the scattering problem of a dense distribution of spherical scatterers. In the past only the coherent scattered field from n scatterers or the incoherent field from two scatterers have been calculated. In this study a much larger number of scatterers are used in the incoherent field calculation so that the effects on scattering and attenuation due to volume fraction can be observed and compared with measurements. In particular, backscattering cross section, forward scattering cross section, and attenuation for volume fraction varying from 5 % to 50 % are calculated.

1. INTRODUCTION

The scattering of electromagnetic waves from a homogeneous medium densely populated with randomly positioned spheres of finite size has been studies by many investigators [2-9,12-15]. However, due to the complexity of the problem only the coherent field calculation or incoherent field calculation for two scatterers have been carried out. It is shown in the appendix that to perform incoherent scattered field calculations for many scatterers requires additional relations than those given by Peterson and Strom. Otherwise a symmetry will be lost and correct results cannot be obtained.

In this article only monochromatic waves are considered, and a time factor exp(-jωt) is understood and suppressed. Under this condition the electric and magnetic fields satisfy the vector Helmholtz's equation.

$$\nabla \times \nabla \times \vec{\psi} - k^2 \vec{\psi} = 0$$

The total field $\vec{E}$ is the sum of the scattered field $\vec{E}^{scat}$ from the scatterer and the incident field $\vec{E}^{inc}$. Let S denote the closed surface of the scatterer, the vector Huygen's principle states that [7].

$$\left.\begin{array}{c}\vec{E}(\vec{r}) \\ 0\end{array}\right\} = \vec{E}^{inc} + \nabla \times \int_S dS' \, k\left[\hat{n} \times \vec{E}_+(\vec{r}')\right] g(kR) +$$

$$\nabla \times \nabla \times \int_S dS' \, i\left[\hat{n} \times \vec{H}_+(\vec{r}')\right] \times g(kR) \begin{cases} \vec{r} \text{ outside S} \\ \vec{r} \text{ inside S} \end{cases}$$

where $\hat{n} \times \vec{E}_+$ and $\hat{n} \times \vec{H}_+$ are tangential fields on S, $k = \frac{\omega}{c}$, $R = |\vec{r} - \vec{r}'|$, and $g(kR) = \frac{\exp(ikR)}{4\pi kR}$ is the free space Green's function.

By solving the above equation, the scattered field can be determined for a scatterer with arbitrary shape. T-matrix descriptions of acoustic and electromagnetic wave scattering from a single scatterer have been given by Waterman[16]. To simply the problem, we assume the scatterer is a sphere. We shall derive the T-matrix for a dielectric and a conducting sphere in Section 2 and the representation of the incident wave in spherical vector wave functions in Section 3. In Section 4 we determine the T-matrix for arbitrary number of scatterers. Then the expressions for the radar cross section and attenuation are given in Section 5 and 6 followed by the numerical results and comparisons with measurements in Section 7.

2. T-MATRIX FOR A SPHERE

To derive the T-matrix, the incident and scattered fields are expanded in terms of some basis functions. The T-matrix is the relation between the expansion coefficients of the incident and the scattered fields. It can be obtained by matching the boundary conditions.

For a three-dimensional problem, the incident field and scattered field at $\vec{r}(r, \theta, \phi)$ can be represented in terms of the vector wave functions, $\vec{\psi}_{1\sigma mn}$ and $\vec{\psi}_{2\sigma mn}$, as

$$\vec{E}^{inc} = \sum_{\sigma mn} a_{1\sigma mn} \, \text{Re} \, \vec{\psi}_{1\sigma mn}(k_0\vec{r}) + a_{2\sigma mn} \, \text{Re} \, \vec{\psi}_{2\sigma mn}(k_0\vec{r}) \tag{1}$$

$$\vec{E}^{scat} = \sum_{\sigma mn} f_{1\sigma mn} \, \vec{\psi}_{1\sigma mn}(k_0\vec{r}) + f_{2\sigma mn} \, \vec{\psi}_{2\sigma mn}(k_0\vec{r}) \tag{2}$$

where

$$\left.\begin{array}{c}\vec{\psi}_{1emn}(k\vec{r}) \\ \vec{\psi}_{1omn}(k\vec{r})\end{array}\right\} = \sqrt{\gamma_{mn}} \, h_n(kr)\left[\hat{\theta} \, \frac{m}{\sin\theta} \, P_n^m(\cos\theta)\right] \begin{cases} -\sin m\phi \\ \cos m\phi \end{cases} -$$

$$\hat{\phi}\,\frac{1}{2}\left[(n-m+1)(n+m)\,P_n^{m-1}(\cos\theta) - P_n^{m+1}(\cos\theta)\right]\begin{Bmatrix}\cos m\phi\\ \sin m\phi\end{Bmatrix}\right],$$

$$\begin{Bmatrix}\vec{\psi}_{2emn}(k\vec{r})\\ \vec{\psi}_{2omn}(k\vec{r})\end{Bmatrix} = \sqrt{\gamma_{mn}}\,\frac{1}{kr}\left[\hat{r}\,n(n+1)\,h_n(kr)\,P_n^m(\cos\theta)\begin{Bmatrix}\cos m\phi\\ \sin m\phi\end{Bmatrix} + \right.$$

$$\hat{\theta}\,\frac{1}{2}\left[h_n(kr)+r\,h_n'(kr)\right]\left[(n-m+1)(n+m)\,P_n^{m-1}(\cos\theta) - P_n^{m+1}(\cos\theta)\right]\begin{Bmatrix}\cos m\phi\\ \sin m\phi\end{Bmatrix} + $$

$$\left.\hat{\phi}\left[h_n(kr)+r\,h_n'(kr)\right]\frac{m}{\sin\theta}\,P_n^m(\cos\theta)\begin{Bmatrix}-\sin m\phi\\ \cos m\phi\end{Bmatrix}\right],$$

$$\gamma_{mn} = \frac{\varepsilon_m(2n+1)(n-m)!}{4\pi\,n(n+1)(n+m)!}, \quad\text{and}\quad \varepsilon_m = \begin{cases}1, & m=0\\ 2, & m\neq 0\end{cases}$$

In the above relations $P_n^m(\cos\theta)$ is the associated Legendre function of the first kind and $h_n(kr)$ is the spherical Hankel function and its derivative is denoted by $h_n'(kr)$.

The relationship between the expansion coefficients, $a_{\tau\sigma mn}$ and $f_{\tau\sigma mn}$, defines the T-matrix for a scatterer as follows

$$f_{\tau\sigma mn} = \sum_{\tau'\sigma'm'n'} T_{\tau\sigma mn,\,\tau'\sigma'm'n'}\,a_{\tau'\sigma'm'n'} \tag{3}$$

The range of the subscripts are $\tau = 1, 2$, $\sigma = e, o$, $n = 1, 2, .., \infty$, and $m = 0, 1, .., n$. For a finite expansion, $n = 1, 2, .., L$, the total number of the elements are $2L^2 + 6L$.

For an object, the internal field can be written as

$$\vec{E}^{int} = \sum_{\sigma mn} b_{1\sigma mn}\,\text{Re}\,\vec{\psi}_{1\sigma mn}(k_1\vec{r}) + b_{2\sigma mn}\,\text{Re}\,\vec{\psi}_{2\sigma mn}(k_1\vec{r}) \tag{4}$$

and the total field outside the scatterer as

$$\vec{E}^{ext} = \vec{E}^{inc} + \vec{E}^{scat}. \tag{5}$$

Since $\quad \vec{H} = \frac{1}{i\omega\mu}\nabla\times\vec{E}$, $\ \vec{\psi}_{1\sigma mn} = \frac{1}{k}\nabla\times\vec{\psi}_{2\sigma mn}$, and $\ \vec{\psi}_{2\sigma mn} = \frac{1}{k}\nabla\times\vec{\psi}_{1\sigma mn}$, the corresponding magnetic fields are

$$\vec{H}^{inc} = \frac{1}{i\eta_0}\sum_{\sigma mn} a_{1\sigma mn}\,\text{Re}\,\vec{\psi}_{2\sigma mn}(k_0\vec{r}) + a_{2\sigma mn}\,\text{Re}\,\vec{\psi}_{1\sigma mn}(k_0\vec{r}) \tag{6}$$

$$\vec{H}^{scat} = \frac{1}{i\,\eta_0} \sum_{\sigma mn} f_{1\sigma mn}\,\vec{\psi}_{2\sigma mn}\left(k_0\vec{r}\right) + f_{2\sigma mn}\,\vec{\psi}_{1\sigma mn}\left(k_0\vec{r}\right) \tag{7}$$

$$\vec{H}^{int} = \frac{1}{i\,\eta_1} \sum_{\sigma mn} b_{1\sigma mn}\,\mathrm{Re}\,\vec{\psi}_{2\sigma mn}\left(k_1\vec{r}\right) + b_{2\sigma mn}\,\mathrm{Re}\,\vec{\psi}_{1\sigma mn}\left(k_1\vec{r}\right) \tag{8}$$

$$\vec{H}^{ext} = \vec{H}^{inc} + \vec{H}^{scat} \tag{9}$$

The tangential fields of $\vec{E}$ and $\vec{H}$ must be continuous at the boundary. The boundary conditions, for a sphere with radius b, are

$$\hat{n} \times \left(\vec{E}^{inc} + \vec{E}^{scat}\right)\Big|_{r=b} = \hat{n} \times \vec{E}^{int}\Big|_{r=b}$$

$$\hat{n} \times \left(\vec{H}^{inc} + \vec{H}^{scat}\right)\Big|_{r=b} = \hat{n} \times \vec{H}^{int}\Big|_{r=b}$$

The unit vector normal to a sphere is in the radial direction, $\hat{n} = \hat{r}$. By applying the boundary conditions, we have

$$a_{1\sigma mn}\,j_n\left(k_0 b\right) + f_{1\sigma mn}\,h_n\left(k_0 b\right) = b_{1\sigma mn}\,j_n\left(k_1 b\right)$$

$$a_{1\sigma mn}\left[j_n\left(k_0 b\right) + b\,j_n'\left(k_0 b\right)\right] + f_{1\sigma mn}\left[h_n\left(k_0 b\right) + b\,h_n'\left(k_0 b\right)\right] = b_{1\sigma mn}\left[j_n\left(k_1 b\right) + b\,j_n'\left(k_1 b\right)\right]$$

$$a_{2\sigma mn}\left[j_n\left(k_0 b\right) + b\,j_n'\left(k_0 b\right)\right] + f_{2\sigma mn}\left[h_n\left(k_0 b\right) + b\,h_n'\left(k_0 b\right)\right] = \frac{1}{\sqrt{\varepsilon_r}}\,b_{2\sigma mn}\left[j_n\left(k_1 b\right) + b\,j_n'\left(k_1 b\right)\right]$$

$$a_{2\sigma mn}\,j_n\left(k_0 b\right) + f_{2\sigma mn}\,h_n\left(k_0 b\right) = \sqrt{\varepsilon_r}\,b_{2\sigma mn}\,j_n\left(k_1 b\right)$$

where $j_n(kb)$ is the spherical Bessel function.

By eliminating $b_{1\sigma mn}$ and $b_{2\sigma mn}$ from the above equations, we find the T-matrices for a sphere with radius b as

$$T_{1\sigma mn,\,1\sigma'm'n'} = \frac{j_n'\left(k_1 b\right) j_n\left(k_0 b\right) - j_n\left(k_1 b\right) j_n'\left(k_0 b\right)}{j_n\left(k_1 b\right) h_n'\left(k_0 b\right) - j_n'\left(k_1 b\right) h_n\left(k_0 b\right)}\,\delta_{\sigma\sigma'}\,\delta_{mm'}\,\delta_{nn'}$$

$$T_{2\sigma mn,\,2\sigma'm'n'} = \frac{\left(\frac{1}{\varepsilon_r} - 1\right) j_n\left(k_1 b\right) j_n\left(k_0 b\right) + b\left[\frac{1}{\varepsilon_r} j_n'\left(k_1 b\right) j_n\left(k_0 b\right) - j_n\left(k_1 b\right) j_n'\left(k_0 b\right)\right]}{\left(1 - \frac{1}{\varepsilon_r}\right) j_n\left(k_1 b\right) h_n\left(k_0 b\right) + b\left[j_n\left(k_1 b\right) h_n'\left(k_0 b\right) - \frac{1}{\varepsilon_r} j_n'\left(k_1 b\right) h_n\left(k_0 b\right)\right]}\,\delta_{\sigma\sigma'}\,\delta_{mm'}\,\delta_{nn'}$$

$$T_{2\sigma mn,\,1\sigma'm'n'} = T_{1\sigma mn,\,2\sigma'm'n'} = 0 \tag{10}$$

where the $\delta_{pp'}$'s are the Kronecker delta functions.

For a perfectly conducting sphere with radius b, the boundary conditions become

$$\hat{n} \times \left(\vec{E}^{inc} + \vec{E}^{scat} \right) \Big|_{r=b} = 0$$

Let $\hat{n} = \hat{r}$ and apply the boundary conditions, we have

$$a_{1\sigma mn} \, j_n(k_0 b) + f_{1\sigma mn} \, h_n(k_0 b) = 0$$

$$a_{2\sigma mn} \left[j_n(k_0 b) + b \, j_n'(k_0 b) \right] + f_{2\sigma mn} \left[h_n(k_0 b) + b \, h_n'(k_0 b) \right] = 0$$

Therefore, the T-matrices become

$$T_{1\sigma mn, \, 1\sigma' m'n'} = \frac{- j_n(k_0 b)}{h_n(k_0 b)} \, \delta_{\sigma\sigma'} \delta_{mm'} \delta_{nn'}$$

$$T_{2\sigma mn, \, 2\sigma' m'n'} = \frac{- \left[j_n(k_0 b) + b \, j_n'(k_0 b) \right]}{h_n(k_0 b) + b \, h_n'(k_0 b)} \, \delta_{\sigma\sigma'} \delta_{mm'} \delta_{nn'}$$

$$T_{2\sigma mn, \, 1\sigma' m'n'} = T_{1\sigma mn, \, 2\sigma' m'n'} = 0 \tag{11}$$

3. WAVE EXPANSION FOR INCIDENT WAVE

Since the incident wave is a known function, a series representation for it must have known coefficients. In this section we will find the expressions for these coefficients. Assume a unit incident wave is a plane wave traveling in the direction specified by θ_i and ϕ_i with respect to a given coordinate system, then we have

For TM wave (v-polarization) :

$$\vec{E}^{inc} = \left[\hat{r} \sin \theta \cos \theta_i \cos (\phi - \phi_i) - \cos \theta \sin \theta_i + \right.$$

$$\left. \hat{\theta} \cos \theta \cos \theta_i \cos (\phi - \phi_i) + \sin \theta \sin \theta_i - \hat{\phi} \cos \theta_i \sin (\phi - \phi_i) \right]$$

$$\exp\!\left(i k_0 r \left[\sin \theta \sin \theta_i \cos (\phi - \phi_i) + \cos \theta \cos \theta_i \right] \right) \tag{12}$$

$$\vec{H}^{inc} = \frac{i k_0 r}{i \omega \mu r} \left[\hat{r} \sin \theta \sin (\phi - \phi_i) + \hat{\theta} \cos \theta \sin (\phi - \phi_i) + \hat{\phi} \cos \theta_i \cos (\phi - \phi_i) \right]$$

$$\exp\!\left(i k_0 r \left[\sin \theta \sin \theta_i \cos (\phi - \phi_i) + \cos \theta \cos \theta_i \right] \right) \tag{13}$$

For TE wave (h-polarization) :

$$\vec{E}^{inc} = \left[\hat{r}\,\sin\theta\,\sin(\phi-\phi_i) + \hat{\theta}\,\cos\theta\,\sin(\phi-\phi_i) + \hat{\phi}\,\cos\theta_i\,\cos(\phi-\phi_i)\right]$$

$$\exp\!\left(ik_0 r\left[\sin\theta\,\sin\theta_i\,\cos(\phi-\phi_i) + \cos\theta\,\cos\theta_i\right]\right) \tag{14}$$

$$\vec{H}^{inc} = \frac{-ik_0 r}{i\omega\mu r}\Big[\hat{r}\,\sin\theta\,\cos\theta_i\,\cos(\phi-\phi_i) - \cos\theta\,\sin\theta_i +$$

$$\hat{\theta}\,\cos\theta\,\cos\theta_i\,\cos(\phi-\phi_i) + \sin\theta\,\sin\theta_i - \hat{\phi}\,\cos\theta_i\,\sin(\phi-\phi_i)\Big]$$

$$\exp\!\left(ik_0 r\left[\sin\theta\,\sin\theta_i\,\cos(\phi-\phi_i) + \cos\theta\,\cos\theta_i\right]\right) \tag{15}$$

From equation (1) and (6), the incident wave can be represented in spherical vector wave functions as

$$\bar{E}^{inc} = \sum_{\sigma mn} a_{1\sigma mn}\,\mathrm{Re}\,\psi_{1\sigma mn}(k_0\vec{r}) + a_{2\sigma mn}\,\mathrm{Re}\,\psi_{2\sigma mn}(k_0\vec{r}) \tag{16}$$

$$\bar{H}^{inc} = \frac{k_0}{i\omega\mu}\sum_{\sigma mn} a_{1\sigma mn}\,\mathrm{Re}\,\psi_{2\sigma mn}(k_0\vec{r}) + a_{2\sigma mn}\,\mathrm{Re}\,\psi_{1\sigma mn}(k_0\vec{r}) \tag{17}$$

To find the coefficients $a_{1\sigma mn}$ and $a_{2\sigma mn}$, we only need to equate a vector component in (16) and (17) with the corresponding component in either (12) and (13) or (14) and (15). Let us consider the r-component . Then, from (16) and (17)

$$E_{r\sigma} = \sum_{n=0}^{\infty}\sum_{m=0}^{n} a_{2\sigma mn}\sqrt{\gamma_{mn}}\,\frac{1}{k_0 r}\,n(n+1)\,j_n(k_0 r)\,P_n^m(\cos\theta)\left\{\begin{array}{c}\cos m\phi\\ \sin m\phi\end{array}\right\} \tag{18}$$

$$H_{r\sigma} = \frac{k_0}{i\omega\mu}\sum_{n=0}^{\infty}\sum_{m=0}^{n} a_{1\sigma mn}\sqrt{\gamma_{mn}}\,\frac{1}{k_0 r}\,n(n+1)\,j_n(k_0 r)\,P_n^m(\cos\theta)\left\{\begin{array}{c}\cos m\phi\\ \sin m\phi\end{array}\right\} \tag{19}$$

To expand (12) through (15) in terms of the associated Legendre functions, note that the common exponential factor in these equations has the following expansion [11]

$$\exp\!\left(ik_0 r\left[\sin\theta\,\sin\theta_i\,\cos(\phi-\phi_i) + \cos\theta\,\cos\theta_i\right]\right) =$$

$$\sum_{n=0}^{\infty}\sum_{m=0}^{n} i^n(2n+1)\,j_n(k_0 r)\,\varepsilon_m\,\frac{(n-m)!}{(n+m)!}\,P_n^m(\cos\theta)\,P_n^m(\cos\theta_i)\,\cos m(\phi-\phi_i)$$

Hence, for TM wave, we have

$$E_{r\sigma} = \left[\sin\theta\,\cos\theta_i\,\cos(\phi-\phi_i) - \cos\theta\,\sin\theta_i\right]$$

$$\exp\!\left(ik_0r\left[\sin\theta\,\sin\theta_i\,\cos\left(\phi-\phi_i\right)+\cos\theta\,\cos\theta_i\right]\right)$$

$$=\frac{1}{ik_0r}\sum_{n=0}^{\infty}\sum_{m=0}^{n}i^n\left(2n+1\right)j_n\left(k_0r\right)\varepsilon_m\frac{\left(n-m\right)!}{\left(n+m\right)!}P_n^m\left(\cos\theta\right)\cos m\left(\phi-\phi_i\right)$$

$$\frac{-1}{\sin\theta_i}\left[\left(n+1\right)\cos\theta_i\,P_n^m\left(\cos\theta_i\right)-\left(n-m+1\right)P_{n+1}^m\left(\cos\theta_i\right)\right] \tag{20}$$

Upon comparing (18) and (20), we obtain

$$a_{2\sigma mn}=i^{n+1}\sqrt{\eta_{mn}}\,\frac{1}{\sin\theta_i}\left[\left(n+1\right)\cos\theta_i\,P_n^m\left(\cos\theta_i\right)-\left(n-m+1\right)P_{n+1}^m\left(\cos\theta_i\right)\right]\begin{Bmatrix}\cos m\phi_i\\[4pt]\sin m\phi_i\end{Bmatrix}, \tag{21}$$

where $\eta_{mn}=\dfrac{4\pi\,\varepsilon_m\left(2n+1\right)\left(n-m\right)!}{n\left(n+1\right)\left(n+m\right)!}$

Similarly, following the same procedure, we find

$$H_{r\sigma}=\frac{1}{i\omega\mu r}\left[ik_0r\,\sin\theta\,\sin\left(\phi-\phi_i\right)\right]\exp\!\left(ik_0r\left[\sin\theta\,\sin\theta_i\,\cos\left(\phi-\phi_i\right)+\cos\theta\,\cos\theta_i\right]\right)$$

$$=\frac{1}{i\omega\mu r}\frac{1}{\sin\theta_i}\sum_{n=0}^{\infty}\sum_{m=0}^{n}i^n\left(2n+1\right)j_n\left(k_0r\right)\varepsilon_m\frac{\left(n-m\right)!}{\left(n+m\right)!}P_n^m\left(\cos\theta\right)P_n^m\left(\cos\theta_i\right)m\sin m\left(\phi-\phi_i\right) \tag{22}$$

Therefore, we have

$$a_{1\sigma mn}=i^n\sqrt{\eta_{mn}}\,\frac{m}{\sin\theta_i}P_n^m\left(\cos\theta_i\right)\begin{Bmatrix}-\sin m\phi_i\\[4pt]\cos m\phi_i\end{Bmatrix} \tag{23}$$

For TE wave, we can get

$$E_{r\sigma}=\sin\theta\,\sin\left(\phi-\phi_i\right)\exp\!\left(ik_0r\left[\sin\theta\,\sin\theta_i\,\cos\left(\phi-\phi_i\right)+\cos\theta\,\cos\theta_i\right]\right)$$

$$=\frac{1}{ik_0r}\frac{1}{\sin\theta_i}\sum_{n=0}^{\infty}\sum_{m=0}^{n}i^n\left(2n+1\right)j_n\left(k_0r\right)\varepsilon_m\frac{\left(n-m\right)!}{\left(n+m\right)!}P_n^m\left(\cos\theta\right)P_n^m\left(\cos\theta_i\right)m\sin m\left(\phi-\phi_i\right) \tag{24}$$

Thus, the corresponding coefficient for TE wave becomes

$$a_{2\sigma mn} = i^{n-1}\,\sqrt{\eta_{mn}}\,\frac{m}{\sin\theta_i}\,P_n^m(\cos\theta_i)\begin{Bmatrix}-\sin m\phi_i\\ \cos m\phi_i\end{Bmatrix} \qquad (25)$$

Again, following the same procedure, we can get

$$H_{r\sigma} = \frac{-1}{i\omega\mu r}\left\{ik_0 r\left[\sin\theta\cos\theta_i\cos(\phi-\phi_i)-\cos\theta\sin\theta_i\right]\right\}$$

$$\exp\!\left(ik_0 r\left[\sin\theta\sin\theta_i\cos(\phi-\phi_i)+\cos\theta\cos\theta_i\right]\right)$$

$$= \frac{-1}{i\omega\mu r}\sum_{n=0}^{\infty}\sum_{m=0}^{n} i^n(2n+1)\,j_n(k_0 r)\,\varepsilon_m\frac{(n-m)!}{(n+m)!}\,P_n^m(\cos\theta)\cos m(\phi-\phi_i)$$

$$\frac{-1}{\sin\theta_i}\left[(n+1)\cos\theta_i\,P_n^m(\cos\theta_i)-(n-m+1)\,P_{n+1}^m(\cos\theta_i)\right] \qquad (26)$$

This leads to the other coefficient for TE wave to be

$$a_{1\sigma mn} = i^n\,\sqrt{\eta_{mn}}\,\frac{1}{\sin\theta_i}\left[(n+1)\cos\theta_i\,P_n^m(\cos\theta_i)-(n-m+1)\,P_{n+1}^m(\cos\theta_i)\right]\begin{Bmatrix}\cos m\phi_i\\ \sin m\phi_i\end{Bmatrix} \qquad (27)$$

These coefficients allow the incident wave to be completely determined in terms of the spherical vector wave functions. They are in the proper form to be used in satisfying boundary conditions along with the scattered fields. Also, whenever the T-matrix of a scattering problem is determined, the coefficients of the scattered fields can be evaluated in terms the incident field coefficients and the T-matrix.

4. T-MATRIX FOR ARBITRARY NUMBER OF SPHERES

In this section, we will calculate the T-matrix for N scatterers. The geometrical restriction for the problem is that the scatterers do not overlap in space. The external field for N scatterers can be represented as

$$\vec{E}^{\,ext} = \vec{E}^{\,inc} + \sum_{i=1}^{N}\vec{E}_i^{\,scat} \qquad (28)$$

where $\vec{E}_i^{\,scat}$ is the scattered field from the i-th scatterers

If we expand the above equation and represent it in matrix form, it becomes

$$\vec{E}^{\,ext} = \underset{\sim}{a}^t \, \mathrm{Re} \, \vec{\psi}(r) + \sum_{i=1}^{N} \underset{\sim}{f}_i^t \, \vec{\psi}\left(\left|\vec{r}\text{-}\vec{r}_i\right|\right) \tag{29}$$

where the superscript t denotes transposition; $\vec{\underset{\sim}{\psi}}(r)$, $\underset{\sim}{a}$, and $\underset{\sim}{f}_i$ are column matrices.

The total number of elements of $\vec{\underset{\sim}{\psi}}(r)$, $\underset{\sim}{a}$, and $\underset{\sim}{f}_i$ are denoted by their subscripts as shown in (1) and (2). As given in section 1, the range of the subscripts are $\tau = 1, 2,\ \sigma = e, o\ ,\ n = 1,$ 2, .., ∞, and m = 0, 1, .., n. For a finite expansion, n = 1, 2, .., L, the total number of the elements are $2L^2 + 6L$. In (29) the reference point for the first term is the origin and for the second term it is $\bar{r}_i$. By applying translation matrices, it is possible to re-reference both terms to the k-th scatterer and define the T-matrix for the k-th scatterer as follows:

$$\vec{E}^{\,ext} = \underset{\sim}{a}^t \, \underset{\sim}{R}(0,k) \, \mathrm{Re} \, \vec{\underset{\sim}{\psi}}\left(\left|\vec{r}\text{-}\vec{r}_k\right|\right) + \underset{\sim}{f}_k^t \, \vec{\underset{\sim}{\psi}}\left(\left|\vec{r}\text{-}\vec{r}_k\right|\right) + \sum_{\substack{i=1 \\ i \neq k}}^{N} \underset{\sim}{f}_i^t \, \underset{\sim}{\sigma}(i,k) \, \mathrm{Re} \, \vec{\underset{\sim}{\psi}}\left(\left|\vec{r}\text{-}\vec{r}_k\right|\right) \tag{30}$$

where $\underset{\sim}{R}(i,k) = \underset{\sim}{R}(-k_0\vec{a}_i)\,\underset{\sim}{R}(k_0\vec{a}_k),\ \ \underset{\sim}{\sigma}(i,k) = \underset{\sim}{\sigma}(-k_0\vec{a}_i)\,\underset{\sim}{R}(k_0\vec{a}_k)$ are the translation matrices defined and discussed in Appendix A.

By grouping together the total incident wave to the k-th scatterer, we can define the scattered coefficient matrices for the k-th scatterer, $\underset{\sim}{f}_k$, by

$$\underset{\sim}{f}_k = \underset{\sim}{T}^1(k) \left[\underset{\sim}{a}^t \, \underset{\sim}{R}(0,k) + \sum_{\substack{i=1 \\ i \neq k}}^{N} \underset{\sim}{f}_i^t \, \underset{\sim}{\sigma}(i,k) \right]^t$$

$$= \underset{\sim}{T}^1(k) \left[\underset{\sim}{R}^t(0,k) \, \underset{\sim}{a} + \sum_{\substack{i=1 \\ i \neq k}}^{N} \underset{\sim}{\sigma}^t(i,k) \, \underset{\sim}{f}_i \right]$$

$$= \underset{\sim}{T}^1(k) \left[\underset{\sim}{R}(k,0) \, \underset{\sim}{a} + \sum_{\substack{i=1 \\ i \neq k}}^{N} \underset{\sim}{\sigma}(k,i) \, \underset{\sim}{f}_i \right] \tag{31}$$

In (31) $\underset{\sim}{T}^1(k)$ represent single scattering from the k-th scatterer.

If the medium has N scatterers, k= 1,2,..N. Thus, there are N linear equations and N unknown scattered field coefficient matrices. Upon solving the N equations, we can get the unknown coefficient matrices, $\underset{\sim}{f}_k$, for the k-th scatterer. It can be written in the following form

$$\underline{f}_k = \underline{\underline{T}}^N(k)\,\underline{a} \tag{32}$$

where $\underline{\underline{T}}^N(k)$ is the T-matrix of the k-th scatterer which includes all possible multiple scattering effects due the presence of N scatterers. Clearly, the scattering coefficient matrices for every scatterer has the same form as given in (32).

If we translate all the scattered fields to the origin, the external field can be represented as

$$\vec{E}^{ext} = \underline{a}^t\,\mathrm{Re}\,\vec{\underline{\psi}}(r) + \sum_{k=1}^{N} \underline{f}_k^t\,\mathrm{Re}\,\underline{R}(k,0)\,\vec{\underline{\psi}}(r) \tag{33}$$

Then, the coefficient matrix of the total scattered field $\underline{f}$ can be written as

$$\underline{f}^t = \sum_{k=1}^{N} \underline{f}_k^t\,\underline{R}(k,0) = \sum_{k=1}^{N} \left[\underline{\underline{T}}^N(k)\,\underline{a}\right]^t \underline{R}(k,0)$$

$$\underline{f} = \sum_{k=1}^{N} \underline{R}^t(k,0)\,\underline{\underline{T}}^N(k)\,\underline{a} = \sum_{k=1}^{N} \underline{R}(0,k)\,\underline{\underline{T}}^N(k)\,\underline{a}$$

Therefore, the total T-matrix for N scatterers can be written as

$$\underline{\underline{T}}^N = \sum_{k=1}^{N} \underline{R}(0,k)\,\underline{\underline{T}}^N(k) \tag{34}$$

This work has been done by Peterson and Strom[7] and the formulae for even and odd number of scatterers are given below

For even number of scatterers :

$$\underline{\underline{T}}^N = \sum_{k=1}^{N} \underline{R}(0,k)\,\underline{\underline{T}}^1(k)\left[\underline{I} - \sum_{\substack{i=1 \\ i \neq k}}^{N} \underline{\sigma}(k,i)\,\underline{\underline{T}}^1(i)\,\underline{\sigma}(i,k)\,\underline{\underline{T}}^1(k)\right]^{-1}$$

$$\left[\underline{I} + \sum_{\substack{i=1 \\ i \neq k}}^{N} \underline{\sigma}(k,i)\,\underline{\underline{T}}^1(i)\,\underline{R}(i,k)\right]\underline{R}(k,0) \tag{35}$$

For odd number of scatterers :

$$\underline{T}^N = \sum_{\substack{k=1}}^{N} \underline{R}(0,k)\,\underline{T}^1(k)\left\{ \underline{I} - \sum_{\substack{i=1 \\ i\neq k}}^{N}\left[\underline{\sigma}(k,i) + \sum_{\substack{l=1 \\ l\neq i,\,k}}^{N} \underline{\sigma}(k,l)\,\underline{T}^1(l)\,\underline{\sigma}(l,i) \right] \right.$$

$$\left. \left[\underline{I} - \sum_{\substack{l=1 \\ l\neq i,\,k}}^{N} \underline{T}^1(i)\,\underline{\sigma}(i,l)\,\underline{T}^1(l)\,\underline{\sigma}(l,i) \right]^{-1} \underline{T}^1(i)\,\underline{\sigma}(i,k)\,\underline{T}^1(k) \right\}^{-1}$$

$$\left\{ \underline{I} + \sum_{\substack{i=1 \\ i\neq k}}^{N}\left[\underline{\sigma}(k,i) + \sum_{\substack{l=1 \\ l\neq i,\,k}}^{N} \underline{\sigma}(k,l)\,\underline{T}^1(l)\,\underline{\sigma}(l,i) \right] \right.$$

$$\left. \left[\underline{I} - \sum_{\substack{l=1 \\ l\neq i,\,k}}^{N} \underline{T}^1(i)\,\underline{\sigma}(i,l)\,\underline{T}^1(l)\,\underline{\sigma}(l,i) \right]^{-1} \underline{T}^1(i)\,\underline{R}(i,k) \right\} \underline{R}(k,0) \tag{36}$$

5. RCS FOR N SPHERES

The radar cross section (RCS) for a three-dimensional scattering problem is defined as

$$\mathrm{RCS} = \lim_{r\to\infty} 4\pi r^2 \left| \frac{\overline{E}^{scat}}{\overline{E}^{inc}} \right|^2$$

$$= \lim_{r\to\infty} 4\pi r^2 \left| \sum_{\sigma mn} \sum_{\tau'\sigma'm'n'} T^N_{1\sigma mn,\tau'\sigma'm'n'}\, a_{\tau'\sigma'm'n'}\, \Psi_{1\sigma mn} + T^N_{2\sigma mn,\tau'\sigma'm'n'}\, a_{\tau'\sigma'm'n'}\, \Psi_{2\sigma mn} \right|^2 \tag{37}$$

Under the far zone approximation, as $r\to\infty$, $h_n(k_0 r) \approx i^{-(n+1)}\dfrac{\exp(i\,k_0 r)}{k_0 r}$ and $h_n'(k_0 r) \approx i^{-n}\dfrac{\exp(i\,k_0 r)}{r}$. Thus, we have

$$\vec{\Psi}_{1\sigma mn} = \sqrt{\gamma_{mn}}\; i^{-(n+1)}\frac{\exp(i k_0 r)}{k_0 r}\left[\hat{\theta}\,\frac{m}{\sin\theta}\,P_n^m(\cos\theta) \begin{Bmatrix} -\sin m\phi \\ \cos m\phi \end{Bmatrix} - \right.$$

$$\left. \hat{\phi}\,\frac{1}{2}\left[(n-m+1)(n+m)\,P_n^{m-1}(\cos\theta) - P_n^{m+1}(\cos\theta) \right] \begin{Bmatrix} \cos m\phi \\ \sin m\phi \end{Bmatrix} \right] \tag{38}$$

$$\vec{\psi}_{2\sigma mn} = \sqrt{\gamma_{mn}}\ i^{-n}\frac{\exp(ik_0r)}{k_0r}\left[\hat{\theta}\ \frac{1}{2}\left[(n-m+1)(n+m)\ P_n^{m-1}(\cos\theta) - P_n^{m+1}(\cos\theta)\right]\left\{\begin{array}{c}\cos m\phi\\ \sin m\phi\end{array}\right\} + \right.$$

$$\left. \hat{\phi}\ \frac{m}{\sin\theta}\ P_n^m(\cos\theta)\left\{\begin{array}{c}-\sin m\phi\\ \cos m\phi\end{array}\right\}\right] \tag{39}$$

For a unit magnitude incident plane wave, the RCS for N scatterers becomes

$$\mathrm{RCS} = \frac{4\pi}{k_0^2}\left|\sum_{mn}\sqrt{\gamma_{mn}}\sum_{\tau\sigma'm'n'}\hat{\theta}\left\{i^{-(n+1)}\ T_{1emn,\tau\sigma'm'n'}\ a_{\tau\sigma'm'n'}\ \frac{m}{\sin\theta_s}\ P_n^m(\cos\theta_s)(-\sin m\phi_s) + \right.\right.$$

$$i^{-(n+1)}\ T_{1omn,\tau\sigma'm'n'}\ a_{\tau\sigma'm'n'}\ \frac{m}{\sin\theta_s}\ P_n^m(\cos\theta_s)(\cos m\phi_s) +$$

$$i^{-n}\ T_{2emn,\tau\sigma'm'n'}\ a_{\tau\sigma'm'n'}\ \frac{1}{2}\left[(n-m+1)(n+m)\ P_n^{m-1}(\cos\theta_s) - P_n^{m+1}(\cos\theta_s)\right]\cos m\phi_s +$$

$$i^{-n}\ T_{2omn,\tau\sigma'm'n'}\ a_{\tau\sigma'm'n'}\ \frac{1}{2}\left[(n-m+1)(n+m)\ P_n^{m-1}(\cos\theta_s) - P_n^{m+1}(\cos\theta_s)\right]\sin m\phi_s\right\} +$$

$$\hat{\phi}\left\{-i^{-(n+1)}\ T_{1emn,\tau\sigma'm'n'}\ a_{\tau\sigma'm'n'}\ \frac{1}{2}\left[(n-m+1)(n+m)\ P_n^{m-1}(\cos\theta_s) - P_n^{m+1}(\cos\theta_s)\right]\cos m\phi_s +\right.$$

$$-i^{-(n+1)}\ T_{1omn,\tau\sigma'm'n'}\ a_{\tau\sigma'm'n'}\ \frac{1}{2}\left[(n-m+1)(n+m)\ P_n^{m-1}(\cos\theta_s) - P_n^{m+1}(\cos\theta_s)\right]\sin m\phi_s +$$

$$i^{-n}\ T_{2emn,\tau\sigma'm'n'}\ a_{\tau\sigma'm'n'}\ \frac{m}{\sin\theta_s}\ P_n^m(\cos\theta_s)(-\sin m\phi_s) +$$

$$\left.\left. i^{-n}\ T_{2omn,\tau\sigma'm'n'}\ a_{\tau\sigma'm'n'}\ \frac{m}{\sin\theta_s}\ P_n^m(\cos\theta_s)(\cos m\phi_s)\right\}\right|^2 \tag{40}$$

To obtain the special case of backscattering, we set $\theta_s = \pi - \theta_i$, and $\phi_s = \pi + \phi_i$ in (40).

6. ATTENUATION CONSTANT FOR N SPHERES

The scattered field of N spheres can be written as

$$\vec{E}^{scat} = \frac{\exp(ik_0r)}{r}\ \vec{f}(\hat{i},\ \hat{i}) \tag{41}$$

where $\vec{f}(\hat{i},\ \hat{i})$ is the scattering amplitude of N spheres in the forward direction.

Following the optical theorem, the attenuation constant can be written as

$$\alpha_0 = \frac{1}{V}\left\{\frac{4\pi}{k_0}\,\text{Im}\!\left[\vec{f}(\hat{i},\hat{i})\right]\cdot\hat{e}_i\right\} \tag{42}$$

where $\hat{e}_i$ is the unit vector in the direction of polarization of the incident wave, and V is the volume which contains the scatterers.

For a unit volume, $V = 1$ and the attenuation constants for the following two cases become:

(a) For TM wave in θ-direction, we obtain

$$\alpha_0 = \frac{4\pi}{k_0^2}\,\text{Im}\Bigg[\sum_{mn}\sqrt{\gamma_{mn}}\sum_{\tau\dot{\sigma}\dot{m}\dot{n}}\Bigg\{ i^{-(n+1)}\,T_{1emn,\tau\dot{\sigma}\dot{m}\dot{n}}\,a_{\tau\dot{\sigma}\dot{m}\dot{n}}\,\frac{m}{\sin\theta_i}\,P_n^m(\cos\theta_i)(-\sin m\phi_i) +$$

$$i^{-(n+1)}\,T_{1omn,\tau\dot{\sigma}\dot{m}\dot{n}}\,a_{\tau\dot{\sigma}\dot{m}\dot{n}}\,\frac{m}{\sin\theta_i}\,P_n^m(\cos\theta_i)(\cos m\phi_i) +$$

$$i^{-n}\,T_{2emn,\tau\dot{\sigma}\dot{m}\dot{n}}\,a_{\tau\dot{\sigma}\dot{m}\dot{n}}\,\frac{1}{2}\left[(n-m+1)(n+m)\,P_n^{m-1}(\cos\theta_i) - P_n^{m+1}(\cos\theta_i)\right]\cos m\phi_i +$$

$$i^{-n}\,T_{2omn,\tau\dot{\sigma}\dot{m}\dot{n}}\,a_{\tau\dot{\sigma}\dot{m}\dot{n}}\,\frac{1}{2}\left[(n-m+1)(n+m)\,P_n^{m-1}(\cos\theta_s) - P_n^{m+1}(\cos\theta_s)\right]\sin m\phi_s\Bigg\}\Bigg] \tag{43}$$

(b) For TE wave in ϕ-direction, we can get

$$\alpha_0 = \frac{4\pi}{k_0^2}\,\text{Im}\Bigg[\sum_{mn}\sqrt{\gamma_{mn}}\sum_{\tau\dot{\sigma}\dot{m}\dot{n}}\Bigg\{ i^{-n}\,T_{2emn,\tau\dot{\sigma}\dot{m}\dot{n}}\,a_{\tau\dot{\sigma}\dot{m}\dot{n}}\,\frac{m}{\sin\theta_i}\,P_n^m(\cos\theta_i)(-\sin m\phi_i) +$$

$$i^{-n}\,T_{2omn,\tau\dot{\sigma}\dot{m}\dot{n}}\,a_{\tau\dot{\sigma}\dot{m}\dot{n}}\,\frac{m}{\sin\theta_i}\,P_n^m(\cos\theta_i)(\cos m\phi_i) +$$

$$-i^{-(n+1)}\,T_{1emn,\tau\dot{\sigma}\dot{m}\dot{n}}\,a_{\tau\dot{\sigma}\dot{m}\dot{n}}\,\frac{1}{2}\left[(n-m+1)(n+m)\,P_n^{m-1}(\cos\theta_i) - P_n^{m+1}(\cos\theta_i)\right]\cos m\phi_i +$$

$$-i^{-(n+1)}\,T_{1omn,\tau\dot{\sigma}\dot{m}\dot{n}}\,a_{\tau\dot{\sigma}\dot{m}\dot{n}}\,\frac{1}{2}\left[(n-m+1)(n+m)\,P_n^{m-1}(\cos\theta_i) - P_n^{m+1}(\cos\theta_i)\right]\sin m\phi_i\Bigg\}\Bigg] \tag{44}$$

7. NUMERICAL RESULTS

The radar cross section for a sphere with radius a obtained in section 5 has been normalized by πa^2. In Fig. 1. the normalized bistatic scattering coefficient for different ka and dielectric constants are shown. The results for a sphere with relative dielectric constant $\varepsilon_r = 4.431$ is shown in the left column and the results for perfectly conducting sphere is in the right column. The parameters ka and ε_r are chosen so that we can compare with the results given in Barrick [1]. Upon making the comparison very good agreements are obtained in both level and null locations between theoretical calculations in Barrick and the

T-matrix calculations.

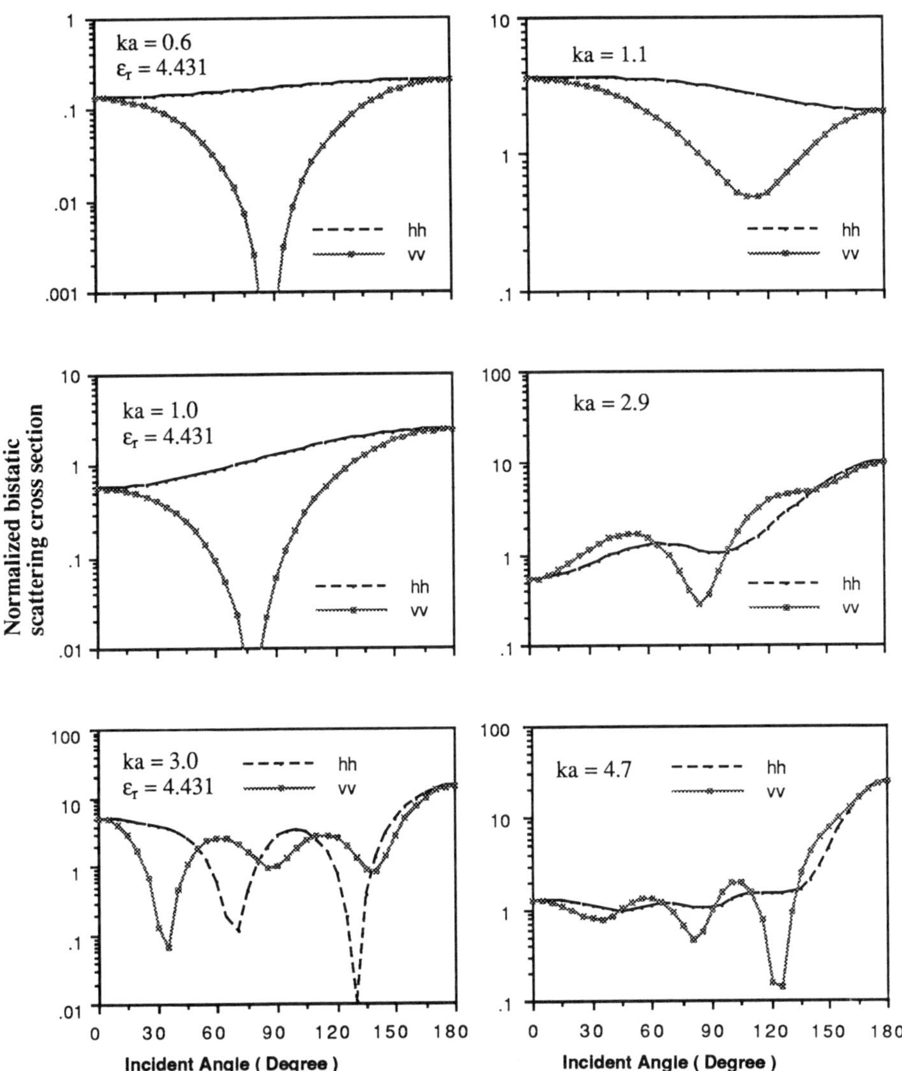

Fig. 1. Normalized bistatic scattering cross section for single sphere with different ka and ε_r

For two spheres with radius a, located at d and -d on the z-axis, the normalized backscattering cross section and the normalized bistatic scattering cross section for ka = 2 and kd = 2.25 are shown in Fig. 2. The result of the normalized backscattering cross section for hh-polarization agrees well with the result given in Peterson [7]

401

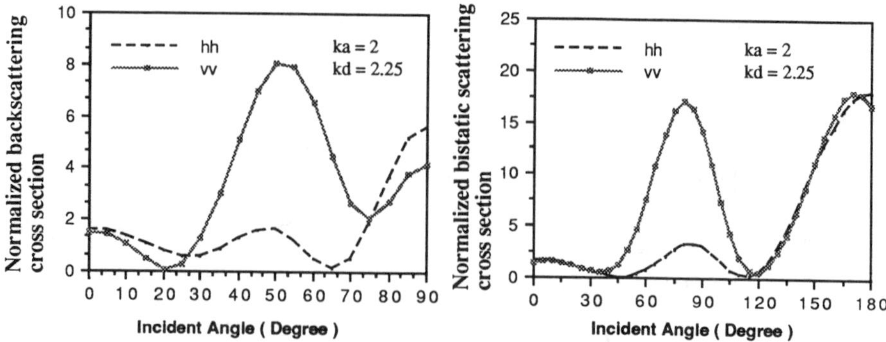

Fig.2. Normalized back scattering cross section and normalized bistatic scattering cross section for two spheres.

To test the T-matrix approach for a scattering volume involving many scatterers, we consider several volumes with length, L, width, W, and height, H. The number of spheres included in accordance with a change in the desired volume fraction. Both VV and HH polarizations are considered and calculations are done for forward and backward scattering [Fig. 3] and the associated attenuation[Fig.4]. Results shown in Fig. 3 indicate that for ka equal to 0.1 the backward scattering is generally higher than forward. As volume fraction increases both forward and backward scattering increase and the increase in forward scattering is slightly faster. Fig.4 shows that the attenuation characteristics of VV and HH polarizations are similar. Generally they both increase with an increase in the volume fraction.

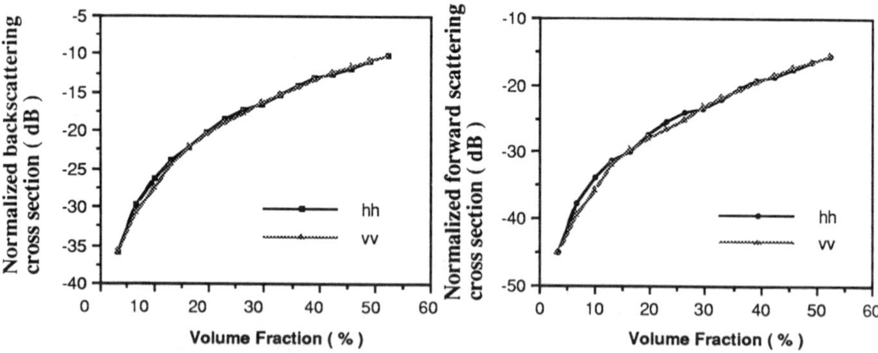

Fig. 3. Normalized backscattering radar section and normalized forward cross section for different volume fraction at ka = 0.1

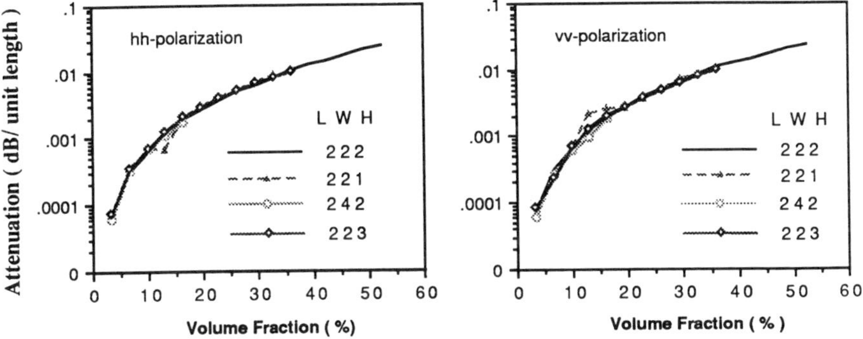

Fig. 4. Attenuation for different volume fraction at ka = 0.1

In Fig.5 a comparison is shown between the attenuation calculations using the T-matrix approach and a set of attenuation measurements at 5.3 GHz. The measurements were performed using glass marbles embedded in foam slabs. The permittivity of the marble is given by the manufacturer as 6 + j 0.036 and its radius is 0.72 cm. The general trend of the calculated attenuation value agrees with the experimental results but the rate of increase in attenuation with volume fraction is different especially at small volume fraction values.

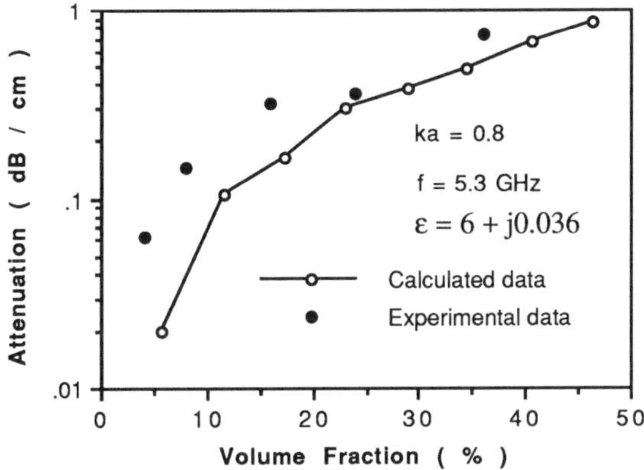

Fig. 5. Comparison between experimental data and results calculated by T-matrix approach

8. CONCLUSION AND DISCUSSION

The T-matrix approach has been generalized to handle many scatterers. This is of importance for application to volume scattering from sea ice and snow layers. It is expected that the algorithm developed can be used to simulate volume scattering from a layer of randomly distributed scatterers of arbitrary shape since it has been demonstrated experimentally that nonspherical scatterers act like spheres when they are randomly oriented.This method can be extended to non-spherical scatterers, if in addition to translation we also apply rotation theorem [10,13].

9. ACKNOWLEDGMENT

This study was sponsored by the Texas Advance Research program 1861.

10. REFERENCE

1. Barrick D. E., in *Radar Cross Section Handbook Vol. 1*, ed. G. T. Ruck, pp. 141-204, Plenum Press, New York, 1970.

2. Chew W. C., Friedrich J. A., and Geiger R., A Multiple Scattering Solution For the Effective Permittivity of a Sphere Mixture, *IEEE transactions on Geoscience and Remote Sensing,* vol. 28, no. 2, pp. 207-214, March 1990.

3. Ishimaru A., and Kuga Y., Attenuation constant of a coherent field in a dense distribution of particles, *Journal of the Optical Society of America,* vol. 72, no. 10, pp. 1317-1320, October 1982.

4. Ishimaru A., Kuga Y., Cheung R. L.-T., and Shimizu K., Scattering and diffusion of a beam wave in randomly distributed scatterers, *Journal of the Optical Society of America,* vol. 73, no. 2, pp. 131-136, February 1982.

5. Leader J. C., Polarization dependence in EM scattering from Rayleigh scatterers embedded in a dielectric slab. I. Theory, *Journal of Applied Physics,* vol. 46, no. 10, pp. 4371-4385, October 1975.

6. Leader J. C., and Dalton W. A. J., Polarization dependence in EM scattering from Rayleigh scatterers embedded in a dielectric slab. II. Experiment, *Journal of Applied Physics,* vol. 46, no. 10, pp. 4386-4391, October 1975.

7. Peterson B., and Ström S., T Matrix for Electromagnetic Scattering from an Arbitrary Number of Scatterers and Representations of E(3), *Physical Review D,* vol. 8, no. 10, pp. 3661-3678, November 1973.

8. Peterson B., and Ström S., Matrix formulation of acoustic scattering from an arbitrary number of scatterers, *Journal of Acoustical Society of America,*vol. 56, no. 3, pp. 771-780, September 1974.

9. Peterson B., and Ström S., Matrix formulation of acoustic scattering from multilayered scatterers, *Journal of Acoustical Society of America,*vol. 57, no. 1, pp. 2-13, January 1975.

10. Stein S., Addition theorems for spherical wave functions, *Quarterly of Applied Mathematics,* vol. 19, pp. 15-24, 1961.

11. Stratton J. A., *Electromagnetic Theory*, pp. 392-423, McGraw-Hill Book Co., Inc., New York, 1941.

12. Tsang L., and Kong J. A., Effective propagation constants for coherent electromagnetic wave propagation in media embedded with dielectric scatters, *Journal of Applied Physics,* vol. 53, no. 11, pp. 7162-7173, November 1982.

13. Tsang L., Scattering of electromagnetic waves from a half space of nOnspherical particles, *JRadio Science,* vol. 19, no. 6, pp. 1450-1460, November 1984.

14. Varadan V. K., Bringi V. N., Varadan V. V., and Ishimaru A., Multiple scattering theory for waves in discrete random media and comparison with experiments, *Radio Science,* vol. 8, no. 3, pp. 321-327, May-June 1983.

15. Varadan V. K., Bringi V. N., Varadan V. V., and Ma Y., Coherent attenuation of acoustic waves by pair-correlated random distribution of scatterers with uniform and Gaussian size distributions, *Journal of Acoustical Society of America,* vol. 73, no. 6, pp. 1941-1947, June 1983.

16. Waterman P. C., Symmetry, Unitarity, and Geometry in Electromagnetic Scattering, *Physical Review D,* vol. 3, no. 4, pp. 825-839, February 1971.

APPENDIX A : TRANSLATION MATRICES FOR SPHERICAL WAVE FUNCTION

Translation operation allow the shifting of one spherical coordinates to another spherical coordinates. For the vectors $\vec{r} = \vec{a} + \vec{R}$, where $\vec{r} = (r, \theta, \phi)$, $\vec{a} = (a, \vartheta, \varphi)$, and $\vec{R} = (R, \Theta, \Phi)$, the translation matrices of incident wave have the following relationship

$$\mathrm{Re}\ \vec{\psi}_{\tau\sigma mn}(k\vec{r}) =$$

$$\sum_{n'=0}^{\infty} \sum_{m'=0}^{n'} R_{\tau\sigma mn,1em'n'}(k\vec{a})\ \mathrm{Re}\ \vec{\psi}_{1em'n'}(k\vec{R}) + R_{\tau\sigma mn,1om'n'}(k\vec{a})\ \mathrm{Re}\ \vec{\psi}_{1om'n'}(k\vec{R}) +$$

$$R_{\tau\sigma mn,2em'n'}(k\vec{a})\ \mathrm{Re}\ \vec{\psi}_{2em'n'}(k\vec{R}) + R_{\tau\sigma mn,2om'n'}(k\vec{a})\ \mathrm{Re}\ \vec{\psi}_{2om'n'}(k\vec{R}) \qquad (A.1)$$

Thus, we can get

$$\mathrm{Re}\ \vec{\psi}_{\tau emn}(k\vec{r}) + i\ \mathrm{Re}\ \vec{\psi}_{\tau omn}(k\vec{r}) =$$

$$\sum_{n'=0}^{\infty} \sum_{m'=0}^{n'} [R_{\tau emn,1em'n'}(k\vec{a}) + i\ R_{\tau omn,1em'n'}(k\vec{a})]\ \mathrm{Re}\ \vec{\psi}_{1em'n'}(k\vec{R}) +$$

$$[R_{\tau omn,1om'n'}(k\vec{a}) - i\ R_{\tau emn,1om'n'}(k\vec{a})]\ i\ \mathrm{Re}\ \vec{\psi}_{1om'n'}(k\vec{R}) +$$

$$[R_{\tau emn,2em'n'}(k\vec{a}) + i\ R_{\tau omn,2em'n'}(k\vec{a})]\ \mathrm{Re}\ \vec{\psi}_{2em'n'}(k\vec{R}) +$$

$$\left[R_{\tau omn,2om\dot{n}}(\vec{ka}) - i\, R_{\tau emn,2om\dot{n}}(\vec{ka}) \right] i\, \mathrm{Re}\, \vec{\psi}_{2om\dot{n}}(\vec{kR}) \qquad (A.2)$$

By definition, the wave function can be written as

$$\vec{\psi}_{\tau omn}(\vec{kr}) = \sqrt{\gamma_{mn}} \left(\frac{1}{k} \nabla \times \right)^{\tau} \vec{kr}\, h_n(kr)\, P_n^m(\cos\theta) \left\{ \begin{array}{c} \cos m\phi \\ \sin m\phi \end{array} \right.$$

With $P_n^{-m}(\cos\theta) = (-1)^m \dfrac{(n-m)!}{(n+m)!} P_n^m(\cos\theta)$, we have

$$\vec{\psi}_{\tau o\text{-}mn}(\vec{kr}) = = (-1)^m \sqrt{\gamma_{mn}} \left(\frac{1}{k} \nabla \times \right)^{\tau} \vec{kr}\, h_n(kr)\, P_n^m(\cos\theta) \left\{ \begin{array}{c} \cos m\phi \\ -\sin m\phi \end{array} \right.$$

Therefore, we can conclude that $\vec{\psi}_{\tau e\text{-}mn} = (-1)^m \vec{\psi}_{\tau emn}$ and $\vec{\psi}_{\tau o\text{-}mn} = (-1)^{m+1} \vec{\psi}_{\tau omn}$. If we define the following functions [7]

$$\vec{f}_{nm}(\vec{kr};h_n) = i\,(-1)^m \frac{1}{\sqrt{\varepsilon_m}} \left(\vec{\psi}_{1emn} + i\, \vec{\psi}_{1omn} \right)$$

$$\vec{g}_{nm}(\vec{kr};h_n) = i\,(-1)^m \frac{1}{\sqrt{\varepsilon_m}} \left(\vec{\psi}_{2emn} + i\, \vec{\psi}_{2omn} \right) \qquad (A.3)$$

Then we can get

$$\vec{f}_{nm}(\vec{kr};j_n) = i\,(-1)^m \frac{1}{\sqrt{\varepsilon_m}} \left(\mathrm{Re}\, \vec{\psi}_{1emn} + i\, \mathrm{Re}\, \vec{\psi}_{1omn} \right)$$

$$\vec{g}_{nm}(\vec{kr};j_n) = i\,(-1)^m \frac{1}{\sqrt{\varepsilon_m}} \left(\mathrm{Re}\, \vec{\psi}_{2emn} + i\, \mathrm{Re}\, \vec{\psi}_{2omn} \right)$$

$$\vec{f}_{n,-m}(\vec{kr};j_n) = i\, \frac{1}{\sqrt{\varepsilon_m}} \left(\mathrm{Re}\, \vec{\psi}_{1emn} - i\, \mathrm{Re}\, \vec{\psi}_{1omn} \right)$$

$$\vec{g}_{n,-m}(\vec{kr};j_n) = i\, \frac{1}{\sqrt{\varepsilon_m}} \left(\mathrm{Re}\, \vec{\psi}_{2emn} - i\, \mathrm{Re}\, \vec{\psi}_{2omn} \right)$$

From the addition theorem, the functions $\vec{f}_{nm}$ and $\vec{g}_{nm}$ can be written as

$$\vec{f}_{nm}(\vec{kr};j_n) = \sum_{n'=0}^{\infty} \sum_{m'=\dot{n}'}^{n'} e^{i\,(m-m')\varphi} \left[C_{mn,m\dot{n}'}(\vec{ka})\, \vec{f}_{n'm'}(\vec{kR};j_n) + D_{mn,m\dot{n}'}(\vec{ka})\, \vec{g}_{n'm'}(\vec{kR};j_n) \right]$$

$$\vec{g}_{nm}(\vec{kr};j_n) = \sum_{n'=0}^{\infty} \sum_{m'=\dot{n}'}^{n'} e^{i\,(m-m')\varphi} \left[C_{mn,m\dot{n}'}(\vec{ka})\, \vec{g}_{n'm'}(\vec{kR};j_n) + D_{mn,m\dot{n}'}(\vec{ka})\, \vec{f}_{n'm'}(\vec{kR};j_n) \right]$$

$$(A.4)$$

where

$$C_{mn,m'n'}(k\vec{a}) = \frac{(-1)^{m'}}{2} \sum_{\lambda=|n-n'|}^{n+n'} i^{(n'-n+\lambda)}(2\lambda+1) \sqrt{\frac{(2n+1)(2n'+1)(\lambda-m+m')!}{n(n+1)n'(n'+1)(\lambda+m-m')!}}$$

$$\begin{pmatrix} n & n' & \lambda \\ 0 & 0 & 0 \end{pmatrix} \begin{pmatrix} n & n' & \lambda \\ m & -m' & -m+m' \end{pmatrix}$$

$$[n(n+1)+n'(n'+1)-\lambda(\lambda+1)] \, j_\lambda(ka) \, P_\lambda^{m-m'}(\cos\vartheta)$$

$$D_{mn,m'n'}(k\vec{a}) = \frac{(-1)^{m'}}{2} \sum_{\lambda=|n-n'|+1}^{n+n'} i^{(n'-n+\lambda)}(2\lambda+1) \sqrt{\frac{(2n+1)(2n'+1)(\lambda-m+m')!}{n(n+1)n'(n'+1)(\lambda+m-m')!}}$$

$$\begin{pmatrix} n & n' & \lambda-1 \\ 0 & 0 & 0 \end{pmatrix} \begin{pmatrix} n & n' & \lambda \\ m & -m' & -m+m' \end{pmatrix}$$

$$\sqrt{[\lambda^2-(n-n')^2][(n+n'+1)^2-\lambda^2]} \, j_\lambda(ka) \, P_\lambda^{m-m'}(\cos\vartheta)$$

Rewrite the above equation we have

$$\vec{f}_{nm}(k\vec{r};j_n) = \sum_{n'=0}^{\infty} \sum_{m'=0}^{n'} \frac{i\sqrt{\varepsilon_{m'}}}{2}$$

$$\left(\left\{\left[(-1)^{m'}\cos(m-m')\varphi \, C_{mn,m'n'}(k\vec{a}) + \cos(m+m')\varphi \, C_{mn,-m'n'}(k\vec{a})\right] + \right.\right.$$

$$i\left[(-1)^{m'}\sin(m-m')\varphi \, C_{mn,m'n'}(k\vec{a}) + \sin(m+m')\varphi \, C_{mn,-m'n'}(k\vec{a})\right]\right\} \text{Re} \, \vec{\psi}_{1em'n'}(k\vec{R}) +$$

$$\left\{\left[(-1)^{m'}\cos(m-m')\varphi \, C_{mn,m'n'}(k\vec{a}) - \cos(m+m')\varphi \, C_{mn,-m'n'}(k\vec{a})\right] - \right.$$

$$i\left[(-1)^{m'+1}\sin(m-m')\varphi \, C_{mn,m'n'}(k\vec{a}) + \sin(m+m')\varphi \, C_{mn,-m'n'}(k\vec{a})\right]\right\} i \, \text{Re} \, \vec{\psi}_{1om'n'}(k\vec{R}) +$$

$$i\left\{\left[(-1)^{m'}\sin(m-m')\varphi \, D_{mn,m'n'}(k\vec{a}) + \sin(m+m')\varphi \, D_{mn,-m'n'}(k\vec{a})\right] + \right.$$

$$i\left[(-1)^{m'+1}\cos(m-m')\varphi \, D_{mn,m'n'}(k\vec{a}) - \cos(m+m')\varphi \, D_{mn,-m'n'}(k\vec{a})\right]\right\} \text{Re} \, \vec{\psi}_{2em'n'}(k\vec{R}) +$$

$$i\left\{\left[(-1)^{m'}\sin(m-m')\varphi \, D_{mn,m'n'}(k\vec{a}) - \sin(m+m')\varphi \, D_{mn,-m'n'}(k\vec{a})\right] - \right.$$

$$\left.\left.i\left[(-1)^{m'}\cos(m-m')\varphi \, D_{mn,m'n'}(k\vec{a}) - \cos(m+m')\varphi \, D_{mn,-m'n'}(k\vec{a})\right]\right\} i \, \text{Re} \, \vec{\psi}_{2om'n'}(k\vec{R})\right)$$

$$(\text{A.5})$$

Therefore, the translation matrices become

$$R_{1emn,1emn'} = (-1)^m \frac{\sqrt{\varepsilon_m \varepsilon_m'}}{2} \left[(-1)^{m'} \cos(m-m')\,\varphi\, C_{mn,mn'}(k\vec{a}) + \cos(m+m')\,\varphi\, C_{mn,-mn'}(k\vec{a}) \right]$$

$$R_{1emn,1omn'} = (-1)^m \frac{\sqrt{\varepsilon_m \varepsilon_m'}}{2} \left[(-1)^{m'+1} \sin(m-m')\,\varphi\, C_{mn,mn'}(k\vec{a}) + \sin(m+m')\,\varphi\, C_{mn,-mn'}(k\vec{a}) \right]$$

$$R_{1omn,1emn'} = (-1)^m \frac{\sqrt{\varepsilon_m \varepsilon_m'}}{2} \left[(-1)^{m'} \sin(m-m')\,\varphi\, C_{mn,mn'}(k\vec{a}) + \sin(m+m')\,\varphi\, C_{mn,-mn'}(k\vec{a}) \right]$$

$$R_{1omn,1omn'} = (-1)^m \frac{\sqrt{\varepsilon_m \varepsilon_m'}}{2} \left[(-1)^{m'} \cos(m-m')\,\varphi\, C_{mn,mn'}(k\vec{a}) - \cos(m+m')\,\varphi\, C_{mn,-mn'}(k\vec{a}) \right]$$

$$R_{1emn,2emn'} = i \frac{(-1)^m}{2} \sqrt{\varepsilon_m \varepsilon_m'} \left[(-1)^{m'} \sin(m-m')\,\varphi\, D_{mn,mn'}(k\vec{a}) + \sin(m+m')\,\varphi\, D_{mn,-mn'}(k\vec{a}) \right]$$

$$R_{1emn,2omn'} = i \frac{(-1)^m}{2} \sqrt{\varepsilon_m \varepsilon_m'} \left[(-1)^{m'} \cos(m-m')\,\varphi\, D_{mn,mn'}(k\vec{a}) - \cos(m+m')\,\varphi\, D_{mn,-mn'}(k\vec{a}) \right]$$

$$R_{1omn,2emn'} = i \frac{(-1)^m}{2} \sqrt{\varepsilon_m \varepsilon_m'} \left[(-1)^{m'+1} \cos(m-m')\,\varphi\, D_{mn,mn'}(k\vec{a}) - \cos(m+m')\,\varphi\, D_{mn,-mn'}(k\vec{a}) \right]$$

$$R_{1omn,2omn'} = i \frac{(-1)^m}{2} \sqrt{\varepsilon_m \varepsilon_m'} \left[(-1)^{m'} \sin(m-m')\,\varphi\, D_{mn,mn'}(k\vec{a}) - \sin(m+m')\,\varphi\, D_{mn,-mn'}(k\vec{a}) \right]$$

$$(A.6)$$

Since $f_{n0} = f_{n,-0}$ and $C_{mn,0n'} = C_{mn,-0n'}$ and $D_{mn,0n'} = D_{mn,-0n'}$, a little modifications must be done for $m' = 0$. The equations (A.2) and (A.4) can be rewritten as

$$\mathrm{Re}\, \vec{\psi}_{\tau emn}(k\vec{r}) + i\,\mathrm{Re}\, \vec{\psi}_{\tau omn}(k\vec{r}) =$$

$$\sum_{n'=0}^{\infty} \sum_{m'=1}^{n'} \left[R_{\tau emn,1emn'}(k\vec{a}) + i\, R_{\tau omn,1emn'}(k\vec{a}) \right] \mathrm{Re}\, \vec{\psi}_{1emn'}(k\vec{R}) +$$

$$\left[R_{\tau omn,1omn'}(k\vec{a}) - i\, R_{\tau emn,1omn'}(k\vec{a}) \right] i\,\mathrm{Re}\, \vec{\psi}_{1omn'}(k\vec{R}) +$$

$$\left[R_{\tau emn,2emn'}(k\vec{a}) + i\, R_{\tau omn,2emn'}(k\vec{a}) \right] \mathrm{Re}\, \vec{\psi}_{2emn'}(k\vec{R}) +$$

$$\left[R_{\tau omn,2omn'}(k\vec{a}) - i\, R_{\tau emn,2omn'}(k\vec{a}) \right] i\,\mathrm{Re}\, \vec{\psi}_{2omn'}(k\vec{R}) +$$

$$\sum_{n'=0}^{\infty} \left[R_{\tau emn,1e0n'}(k\vec{a}) + i\, R_{\tau omn,1e0n'}(k\vec{a}) \right] \mathrm{Re}\, \vec{\psi}_{1e0n'}(k\vec{R}) +$$

$$\left[R_{\tau omn,1o0n'}(k\vec{a}) - i\, R_{\tau emn,1o0n'}(k\vec{a}) \right] i\,\mathrm{Re}\, \vec{\psi}_{1o0n'}(k\vec{R}) +$$

$$\left[R_{\tau emn,2e0n'}(k\vec{a}) + i\, R_{\tau omn,2e0n'}(k\vec{a}) \right] \mathrm{Re}\, \vec{\psi}_{2e0n'}(k\vec{R}) +$$

$$[R_{\tau omn,2o0n'}(k\vec{a}) - i\, R_{\tau emn,2o0n'}(k\vec{a})]\, i\, \mathrm{Re}\, \vec{\psi}_{2o0n'}(k\vec{R})$$

$$\vec{f}_{nm}(k\vec{r};j_n) = \sum_{n'=0}^{\infty} \sum_{m'=1}^{n'} \frac{\varepsilon_{m'}}{2}\left\{ e^{i(m-m')\varphi}\left[C_{mn,m'n'}(k\vec{a})\,\vec{f}_{n'm'}(k\vec{R};j_n) + D_{mn,m'n'}(k\vec{a})\,\vec{g}_{n'm'}(k\vec{R};j_n)\right] + \right.$$

$$\left. e^{i(m+m')\varphi}\left[C_{mn,-m'n'}(k\vec{a})\,\vec{f}_{n'-m'}(k\vec{R};j_n) + D_{mn,-m'n'}(k\vec{a})\,\vec{g}_{n'-m'}(k\vec{R};j_n)\right]\right\} +$$

$$\sum_{n'=0}^{\infty} e^{i\,m\varphi}\left[C_{mn,0n'}(k\vec{a})\,\vec{f}_{n'0}(k\vec{R};j_n) + D_{mn,0n'}(k\vec{a})\,\vec{g}_{n'0}(k\vec{R};j_n)\right]$$

Equate the last term of the above equations, we can find

$$\sum_{n'=0}^{\infty} e^{i\,m\varphi}\left[C_{mn,0n'}(k\vec{a})\,\vec{f}_{n'0}(k\vec{R};j_n) + D_{mn,0n'}(k\vec{a})\,\vec{g}_{n'0}(k\vec{R};j_n)\right]$$

$$= \sum_{n'=0}^{\infty} i\left\{ [\cos m\varphi\, C_{mn,0n'}(k\vec{a}) + i\sin m\varphi\, C_{mn,0n'}(k\vec{a})]\, \mathrm{Re}\, \vec{\psi}_{1e0n'}(k\vec{R}) + \right.$$

$$[\cos m\varphi\, C_{mn,0n'}(k\vec{a}) + i\sin m\varphi\, C_{mn,0n'}(k\vec{a})]\, i\, \mathrm{Re}\, \vec{\psi}_{1o0n'}(k\vec{R}) +$$

$$i\left[\sin m\varphi\, D_{mn,0n'}(k\vec{a}) - i\cos m\varphi\, D_{mn,0n'}(k\vec{a})\right]\, \mathrm{Re}\, \vec{\psi}_{2e0n'}(k\vec{R}) +$$

$$\left. i\left[\sin m\varphi\, D_{mn,0n'}(k\vec{a}) - i\cos m\varphi\, D_{mn,0n'}(k\vec{a})\right]\, i\, \mathrm{Re}\, \vec{\psi}_{2o0n'}(k\vec{R})\right\}$$

Therefore, we can get

$$R_{1emn,1e0n'} = (-1)^m \sqrt{\varepsilon_m}\, \cos m\varphi\, C_{mn,0n'}$$

$$R_{1emn,1o0n'} = (-1)^{m+1} \sqrt{\varepsilon_m}\, \sin m\varphi\, C_{mn,0n'}$$

$$R_{1omn,1e0n'} = (-1)^m \sqrt{\varepsilon_m}\, \sin m\varphi\, C_{mn,0n'}$$

$$R_{1omn,1o0n'} = (-1)^m \sqrt{\varepsilon_m}\, \cos m\varphi\, C_{mn,0n'}$$

$$R_{1emn,2e0n'} = i(-1)^m \sqrt{\varepsilon_m}\, \sin m\varphi\, D_{mn,0n'}$$

$$R_{1emn,2o0n'} = i(-1)^m \sqrt{\varepsilon_m}\, \cos m\varphi\, D_{mn,0n'}$$

$$R_{1omn,2e0n'} = i(-1)^{m+1} \sqrt{\varepsilon_m}\, \cos m\varphi\, D_{mn,0n'}$$

$$R_{1omn,2o0n'} = i\,(-1)^m\,\sqrt{\varepsilon_m}\,\sin m\varphi\;D_{mn,0n'} \tag{A.7}$$

Follow the same procedures for g_{nm}, we can obtain

$$R_{2\sigma mn,\tau\sigma'm'n'} = R_{1\sigma mn,\tau\sigma'm'n'};\quad \text{for } \tau \neq \tau' \tag{A.8}$$

By substituting j_n in $C_{mn,m'n'}$ and $D_{mn,m'n'}$ to h_n, we can get the translation matrices of scattering waves $\sigma(k\vec{a})$ easily.

A Comparison Between a Frequency Domain Method and the Standard Moment Method in Surface Scattering

K. S. CHEN and A. K. FUNG
Electrical Engineering Department, Box 19016
University of Texas at Arlington
Arligton, Texas 76019, USA

ABSTRACT

The electromagnetic boundary condition is expressed in the form of an integral equation involving the incident and the scattered fields. It is shown that by using a spectral representation for the scattered field the integral equation can be solved for the unknown scattered field amplitude at a significant saving in computation as compared to the use of the standard moment method. The computational time for obtaining the scattering coefficient from 10 random surface patches of 20-wavelength each using 12 points per wavelength is 3.2 minutes with this method on VAX 8700 and is 1 hour using the moment method.

1. INTRODUCTION

In numerical study of electromagnetic wave scattering from random surfaces, the use of moment method is a common practice among investigators. Three major steps are involved in this method: (1) convert the integral equation for the surface current into a matrix equation; (2) solve for the unknown surface current by matrix inversion and (3) calculate the scattered field in terms of the surface current and subsequently the scattering coefficient. Axline and Fung [1] were among the first to apply this method to compute the scattered fields from a computer generated random surface. The numerical simulation by moment method was also applied to study the ranges of validity of theoretical scattering models [2, 3, 4, 5]. Moreover, the moment method has been used to study the backscattering enhancement phenomenon from very rough random surfaces[6,7]. All the afore mentioned studies were restricted to two-dimensional problems. The reason is that in extending from two to three dimensioal problems, the size of the matrix to be inverted for the surface current increases from n^2 to n^4, where n is the number of unknown surface current points. Thus, the demand on computer memory space becomes excessive for a, say, 10 wavelengths wide surface. In an effort to find a more efficient approach to simulate electromagnetic wave scattering from a randomly rough surface, we have developed a frequency domain method. The idea is to use a spectral representation for the scattered field in the boundary condition which becomes an integral equation relating the unknown scattered field amplitude to the incident field. For simplicity numerical calculations have been performed for two-dimensional scattering problems to assess the amount of computational time relative to that required by a standard moment method [1]. It is found that what took an hour on a VAX 8700 computer using the moment method was reduced to 3.2 minutes. Efforts are currently underway to develop an algorithm for three-dimensional problems.

2. THE SPECTRAL DOMAIN METHOD

The theoretical basis for the frequency domain technique in a three-dimensional scattering problem is discussed in this section. With the time factor, $\exp(j\omega t)$, understood, we use a spectral representation for the scattered field in the boundary condition relating the incident to the scattered field. Then, we perform an integration in the spatial domain to remove the edge effects of a surface patch and to transform the boundary condition into an integral equation in spectral form. The property of this spectral form is such that the unknown amplitude function is easily determined. The specific details are given below.

2.1 Determination of the Unknown Field Amplitude

It is well known that the form of the scattered field is

$$E^s = \int_{-\infty}^{\infty} dk_x \int_{-\infty}^{\infty} dk_y \ \alpha(k_x, k_y) \ e^{jk_x x + jk_y y - j k_z \cdot z}$$

$$+ \int_{-\infty}^{\infty} dk_x \int_{-\infty}^{\infty} dk_y \ \beta(k_x, k_y) \ e^{jk_x x + jk_y y + j k_z \cdot z}$$

For surfaces with rms slope less than 20 degrees and incident and scattered angles 30 degrees from grazing, multiple scattering effects should be small and the second term in the scattered field expression may be ignored. Thus,

$$E^s = \int_{-\infty}^{\infty} dk_x \int_{-\infty}^{\infty} dk_y \ \alpha(k_x, k_y) \ e^{jk_x x + jk_y y - j k_z \cdot z}$$

The incident and the scattered fields should satisfy the tangential boundary condition, $\hat{n} \times \overline{E} = 0$, on a perfectly conducting surface. To illustrate the method consider a horizontally polarized field (or y-polarized) incident in the xz-plane. The x -component of the tangential field is

$$E_y^i + E_y^s + Z_y\, E_z^s = 0$$

Since surface depolarization is usually more than an order of magnitude smaller than the like and the rms surface slope is usually less than unity, we can drop the third term in the above equation for like polarization calculations and write

$$e^{-jk_{xo}x\,+jk_{zo}z(x,\,y)} = -\int\!\!\int \alpha\left(k_x,\,k_y\right) e^{jk_x x + jk_y y\, -\,j k_z\cdot z(x,\,y)}\, dk_x dk_y \tag{1}$$

In (1) the expression on the left hand side is the assumed form of the incident field and α is the unknown amplitude function to be determined. To do so we multiply both sides of (1) by the auxiliary function, $e^{-jk_{xo}x - jk_{yo}y + jk_{zo}z(x,\,y)}$, and the pattern function, $G(x,y)$, and integrate over the illuminated area yielding

$$\int\!\!\int e^{-j2k_{xo}x - jk_{yo}y + j2k_{zo}z(x,\,y)} G(x,\,y)\,dxdy$$

$$= -\int \alpha\left(k_x,\,k_y\right) e^{j\left(k_x - k_{xo}\right)x\, +\, j\left(k_y - k_{yo}\right)y\, -\, j\left(k_z - k_{zo}\right)z(x,\,y)} G(x,\,y)\,dxdy\,dk_x dk_y$$

$$\tag{2}$$

After the integration over the x,y variables, (2) is an integral equation in spectral variables. Let

$$I\left(k_x,\,k_y\right) = \int\!\!\int e^{j\left(k_x - k_{xo}\right)x\, +\, j\left(k_y - k_{yo}\right)y\, -\, j\left(k_z - k_{zo}\right)z(x,\,y)} G(x,\,y)\,dxdy$$

Note that $I\left(k_x,\,k_y\right)$ is a function which has a sharp peak around k_{xo} and k_{yo}. Since $\alpha\left(k_x,\,k_y\right)$ is a slow varying function relative to $I\left(k_x,\,k_y\right)$, we can approximate the right hand side of (2) as

$$\int\!\!\int \alpha\left(k_x,\,k_y\right) I\left(k_x,\,k_y\right) dk_x\, dk_y \cong \alpha\left(k_{xo},\,k_{yo}\right) \int\!\!\int I\left(k_x,\,k_y\right) dk_x dk_y \tag{3}$$

and

$$\alpha\left(k_{xo},\,k_{yo}\right) = -\frac{1}{I_o}\int\int e^{-j2k_{xo}x-jk_{yo}y+j2k_{zo}z(x,y)}G(x,y)\,dx\,dy \qquad (4)$$

where

$$I_o \triangleq \int\int I\left(k_x,\,k_y\right)dk_x\,dk_y$$

2.2 Relating $\alpha\left(k_x,\,k_y\right)$ to the Scattering Coefficient

Next, we need to find the relationship between the scattering coefficient and $\alpha(k_x,\,k_y)$. To do so, let us consider the ensemble average of the following product

$$\left\langle\alpha\left(k_x,\,k_y\right)\alpha\left(k'_x,\,k'_y\right)^*\right\rangle = \frac{1}{|I_o|^2}\int\left\langle e^{-j2k_xx-jk_yy+j2k_zz(x,y)}G(x,y)\,dx\,dy\right.$$

$$e^{j2k'_xx'+jk'_yy'-j2k'_zz(x',y')}G\left(x',y'\right)dx'\,dy'\Big\rangle$$

$$=\frac{1}{|I_o|^2}\int e^{-j2k_x(x-x')-j2\left(k_x-k'_x\right)x'-jk_y(y-y')-j\left(k_y-k'_y\right)y'}$$

$$e^{-4\sigma^2\left(k_z^2+\,k_z'^2-2k_zk'_z\,\rho\right)/2}G\left(x,y\right)G\left(x',y'\right)dx\,dx'\,dy\,dy'$$

$$=\frac{1}{|I_o|^2}\int e^{-j2k_x\,\xi-j2\,\Delta\,k_x\,x'-jk_y\,\eta-j2\,\Delta k_y\,y'}$$

$$e^{-2\sigma^2\left(k_z^2+\,k_z'^2-2k_zk'_z\,\rho(\xi,\,\eta)\right)}G\left(\xi+x',\,\eta+y'\right)G\left(x',y'\right)dx'\,dy'\,d\xi\,d\eta \qquad (5)$$

where ρ is the surface correlation function and σ is the surface rms height.

If $k_x = k'_x$, $k_y = k'_y$, $k_z = k'_z$, we obtain another expression,

$$\left\langle\alpha\left(k_x,\,k_y\right)\alpha^*\left(k_x,\,k_y\right)\right\rangle = \left\langle\left|\alpha\left(k_x,\,k_y\right)\right|^2\right\rangle$$

$$= \frac{1}{|I_0|^2} \int e^{-j2k_x \, (x - x') - jk_y \, (y - y') - (2k_z \, \sigma)^2 \, (1 - \rho)} G \, (x, y) \, G \, (x', y') \, dx \, dx' \, dy \, dy'$$

$$= \frac{1}{|I_0|^2} \int e^{-j2k_x \, \xi - j \, k_y \, \eta - (2k_z \, \sigma)^2 \, (1 - \rho(\xi, \eta))}$$

$$G \, (\xi + x', \eta + y') \, G \, (x', y') \, d\xi \, d\eta \, dx' \, dy' \tag{6}$$

Let us assume that the pattern function is Gaussian,

$$G \, (x, y) = e^{- \, (x^2 + y^2) / g^2}$$

Then, the following integrals given below by (7) and (8) can be evaluated

$$\int \int G \, (x + \xi, y + \eta) \, G \, (x, y) \, dx \, dy = \int \int \exp\{ -\frac{(x + \xi)^2}{g^2} - \frac{x^2}{g^2} - \frac{(y + \eta)^2}{g^2} - \frac{y^2}{g^2} \} \, dx \, dy$$

$$= \exp\{ -\frac{\xi^2}{2g^2} \} \int e^{-2 \left(x + \frac{\xi}{2} \right)^2 / g^2} dx \, \exp\{ -\frac{\eta^2}{2g^2} \} \int e^{-2 \left(y + \frac{\eta}{2} \right)^2 / g^2} dy$$

$$\cong \exp\{ -\frac{\xi^2 + \eta^2}{2g^2} \} \left(g \sqrt{\frac{\pi}{2}} \right)^2 = \frac{g^2 \pi}{2} \exp\{ -\frac{\xi^2 + \eta^2}{2g^2} \} \tag{7}$$

$$\int \int e^{-j2\Delta k_x x - j2\Delta k_y y} G \, (\xi + x, \eta + y) \, G \, (x, y) \, dx \, dy$$

$$\cong \frac{g^2 \pi}{2} e^{\left[-\frac{\xi^2}{2g^2} + j \Delta k_x \xi - \frac{g^2 \Delta k_x^2}{2} \right]} e^{\left[-\frac{\eta^2}{2g^2} + j \Delta k_y \eta - \frac{g^2 \Delta k_y^2}{2} \right]} \tag{8}$$

Using (7) and (8) we can rewrite (5) and (6) as (9) and (10).

$$\langle \alpha(k_x, k_y) \, \alpha(k'_x, k'_y)^* \rangle = \left(\frac{1}{I_0} \right)^2 g^2 \frac{2\pi}{2} \int \int e^{-j2k_x \xi - j k_y \eta - 2 \, \sigma^2 (k_z^2 + k_z'^2 - 2 k_z k_z' \rho(\xi, \eta))}$$

$$\exp\{-\frac{\xi^2}{2g^2}+j\,\Delta\,k_x\xi-\frac{g^2\Delta\,k_x^2}{2}-\frac{\eta^2}{2g^2}+j\,\Delta\,k_y\eta-\frac{g^2\Delta\,k_y^2}{2}\}d\xi d\eta \tag{9}$$

$$\langle|\alpha(k_x,k_y)|^2\rangle=\left(\frac{1}{I_0}\right)^2 g^2\,\frac{\pi}{2}\int\int e^{-j2k_x\xi-jk_y\eta-(2k_z\sigma)^2(1-\rho(\xi,\eta))}\exp\{-\frac{\xi^2+\eta^2}{2g^2}\}d\xi\,d\eta \tag{10}$$

Now we are ready to calculate the ensemble average of the scattered field with the aim of expressing it in terms of (10)as follows

$$\langle|E^s|^2\rangle=\int\langle\alpha(k_x,k_y)\,\alpha(k'_x,k'_y)^*\rangle\,e^{j(k_x-k'_x)x+j(k_y-k'_y)y-j(k_z-k'_z)z}dk_x dk'_x dk_y dk'_y \tag{11}$$

Substitute (9) into (11) and integrate $k_x{'}$ and $k_y{'}$ out using the saddle point method ,we have

$$\langle|E^s|^2\rangle=\left(\frac{1}{I_0}\right)^2 g^2\frac{2\pi}{2}\cdot\left(\frac{2\pi}{g^2}\right)\int e^{-j2k_x\xi-jk_y\eta-(2k_z\sigma)^2(1-\rho(\xi,\eta))}$$
$$\exp\,[-(\xi^2+\eta^2)/\,2g^2]\,d\xi\,d\eta\,dk_x\,dk_y \tag{12}$$

Substituting (10) into (12) we have

$$\langle|E^s|^2\rangle=\frac{2\pi}{g^2}\int\langle|\alpha(k_x,k_y)|^2\rangle\,dk_x\,dk_y \tag{13}$$

From Ulaby et al (1982) Chapter 12 , the scattering coefficient is related to the integrand of (13) as follows

$$\sigma^o=4\pi k_s^2\cos^2\theta_s\,\frac{2\pi}{g^2}\langle|\alpha(k_x,k_y)|^2\rangle \tag{14}$$

Since α is given by (4) we can evaluate the scattering coefficient using (14).

3. SPECIAL CASE

It is a straightforward to reduce the method given in Section 2 to a two dimensional problem. Here, the scattered field is of the form

$$E^s = \int_{-\infty}^{\infty} dk_x \, \alpha(k_x) \, e^{jk_x x - jk_z \cdot z}$$

The boundary condition for a y-polarized incident field is

$$E_y^i + E_y^s = 0$$

The unknown coefficient can be shown to be

$$\alpha(k_{xo}) = -\frac{1}{I_o} \int e^{-j2k_{xo}x + j2k_{zo}z(x)} G(x) dx \qquad (15)$$

where $I_o \triangleq \int I(k_x) \, dk_x$,

$$I(k_x) = \int e^{j(k_x - k_{xo})x - j(k_z - k_{zo})z(x)} G(x) \, dx$$

If a Gaussian pattern $G(x) = e^{-x^2/g^2}$ is assumed, then we obtain an expression similar to (10) as

$$\langle |\alpha(k_x)|^2 \rangle = \frac{1}{|I_o|^2} g \sqrt{\frac{\pi}{2}} \int e^{-j2k_x \xi - (2k_z\sigma)^2(1 - \rho(\xi))} \exp\{-\frac{\xi^2}{2g^2}\} d\xi \qquad (16)$$

The ensemble average of the scattered field is

$$\langle |E^s|^2 \rangle = \frac{\sqrt{2\pi}}{g} \int \langle |\alpha(k_x)|^2 \rangle \, dk_x \qquad (17)$$

Following the steps given in [Ulaby, et al., 1982, Ch. 12] we can relate the integrand in (17) to the backscattering coefficient as follows:

Let

$$\langle |E^s|^2 \rangle = \int f(k_x)\, dk_x$$

Then

$$\Delta \langle |E^s|^2 \rangle \approx f(k_x)\, \Delta k_x = f(k_x)\, k_S \cos \theta_S \Delta \theta_S$$

The average intensity P at a receiver located at a distance R away from the illuminated length L is

$$P = \frac{\Delta \langle |E^s|^2 \rangle}{\Delta \theta_S} \frac{L \cos \theta_S}{R}$$

The scattering coefficient is defined in terms of P as

$$\sigma^\circ = \frac{2\pi R}{L} P = 2\pi k_S \cos^2 \theta_S\, f(k_x) \tag{18}$$

By substituting the integrand in (17) into (18), the scattering coefficient is obtained as

$$\sigma^0 = 2\pi k_s \cos^2 \theta_s\, \frac{\sqrt{2\pi}}{g} \langle |\alpha(k_x)|^2 \rangle \tag{19}$$

4. SIMULATION RESULTS

For simplicity only calculations for the two-dimensional scattering problem are given here to justify the validity of the method and the approximations used. This is done by comparing numerical simulation results of the Frequency Domain Method (FDM) with those of the Moment Method (MM). In each case 50 field samples have been averaged to obtain a backscattering angular curve. Gaussian random surfaces with different correlation lengths and rms heights are studied at three different wavelengths. To study the effect of the surface correlation length we show in Figure 1 (a) to (c) the values of the correlation length normalized by the wavenumber, kL, at 1.57, 3.14, and 6.28, respectively while the rms surface height normalized by the wavenumber,kσ, is kept at 0.3 . The agreement between MM and FDM is within a dB for all the computed incidence angles. Note that the agreement is better at smaller incident angles than at larger incident angles. This may be due to multiple scattering effects neglected in our formulation. Figure 2 (a)-(d) show another set of comparisons with different rms heights. The values of kσ range from 0.5 to 2.0 units while keeping KL at 6.28 units. This causes the rms slopes to change from 0.11 to 0.45. It is seen that the agreement between

MM and FDM is very good and it deteriorates as the surface rms slope increases
and also as the incident angle increases. This trend is consistent with the neglect
of multiple scattering effects. Figure 3 (a)-(c) illustrates the frequency
dependence of the backscattering coefficient with kσ varying from 0.2 to 0.8 and
kL changing from 1.57 to 6.28 all with the same rms slope of 0.18. Here again the
agreement is satisfactory. Thus, for small rms slopes and angles of incidence less
than 60 degrees, the validity of the FDM is confirmed.

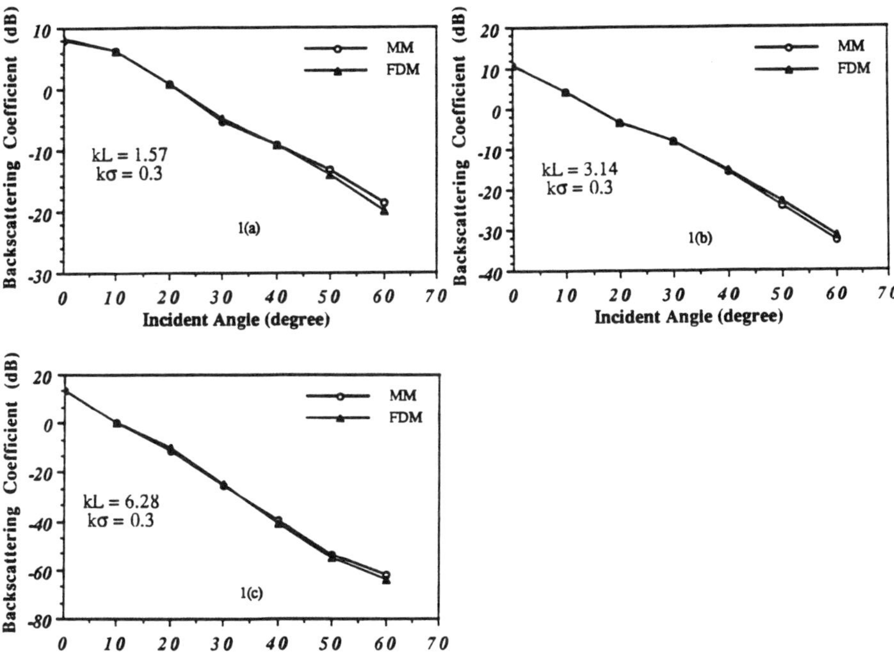

FIGURE 1. Comparison of the backscattering coefficient calculated from the
Frequency Domain Method (FDM) and the Moment Method (MM) for Gaussin
random surface with correlation length L=4.0, kσ=0.3: (a) kL=1.57, (b) kL=3.14,
(c) kL=6.28.

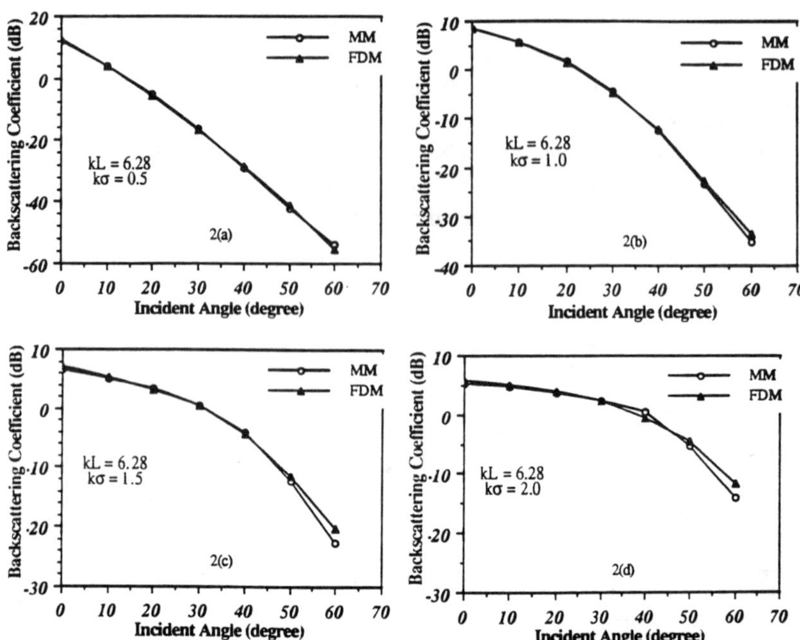

FIGURE 2. Comparison of the backscattering coefficient calculated from the Frequency Domain Method (FDM) and the Moment Method (MM) with correlation length L=4.0, kL=6.28: (a) ks=0.5, (b) kσ=1.0, (c) kσ=1.5, (d) kσ=2.0.

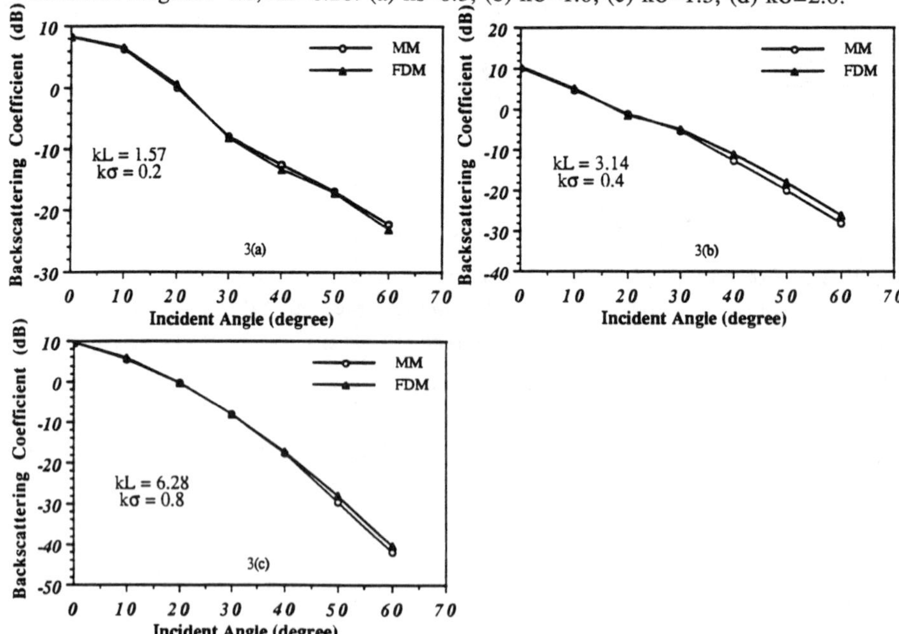

FIGURE 3. Comparison of the backscattering coefficient calculated from the Frequency Domain Method (FDM) and the Moment Method (MM) with correlation length L-4.0, standard deviation σ=0.3: (a) λ=16.0, (b) λ=8.0, (c) λ=4.0.

5. CONCLUSIONS

This paper presents a frequency domain method for the simulation of wave scattering from randomly rough surfaces. The validity of the present method has been demonstrated by comparing its backscattering coefficient calculations with those obtained from the standard moment method. It requires much less computer memory and CPU time. Its numerical implementation is simple and straightforward. Although the algorithm of this method for a 3-D problem will involve a 4-fold integral, it is not more complicated conceptually . It appears that the new method can provide a useful alternative to simulation of wave scattering from three-dimensional random surfaces where the moment method is not yet a practical approach with today's computer.

ACKNOWLEDGEMENT

This study was supported by the Texas Advanced Research Program and the CHPC of the University of Texas System.

REFERENCES

1. Axline. R. M. and Fung, A. K.,"Numerical computation of scattering from a perfectly conducting random surface," *IEEE Transactions on. Antennas and Propagation*, Vol. 26, pp. 482-488, 1978.

2. Fung, A. K. and Chen, M. F.,"Numerical simulation of scattering from simple and composite random surfaces," *Journal of the Optical Society of America.*, Vol. 2, pp. 2274-2284, 1985.

3. Chen, M. F. and Fung, A. K., "A numerical study of the regions of the validity of the Kirchhoff and small-perturbation rough scattering models," *Radio Science*, Vol. 23, No. 2, pp. 163-170, 1988.

4. Thorsos, E. I., "The validity of the Kirchhoff approximation for rough surface scattering using a gaussian roughness spectrum," *Journal of the. Acoustic Society of America*, Vol. 83, No. 1, pp. 78-83, 1988.

5. Thorsos, E. I., and Jackson, D. R., "The validity of the perturbation approximation for rough surface scattering using a gaussian roughness spectrum," *Journal of the Acoustic Society of America* , Vol. 86, No. 1, pp. 261-277, 1989.

6. Nieto-Vesperinas, M. and Soto-Crespo, J. M., "Monte Carlo simulations for scattering of electromagnetic waves from perfectly conductive random rough

surfaces," *Optics Letters*, Vol. 12, No. 12, pp. 979-981, 1987.

7. Soto-Crespo, J. M. and Nieto-Vesperinas, M., "Electromagnetic scattering from very rough random surfaces and deep reflection gratings," *Journal of the Optical Society of America*, Vol. 6, No. 3, pp. 367-384, 1989.

8. Ulaby, F. T., Moore, R. K., and Fung, A. K., *Microwave Remote Sensing: Active and Passive*, Addison-Wesley, Reading, Massachusetts, Vol. 2, 1982.

Use of First and Higher Order Influence Functions in the Duhamel's Integral Solution of the IHCP

B. LITKOUHI and M. H. N. NARAGHI
Mechanical Engineering Department
Manhattan College
Riverdale, New York 10471, USA

Abstract

In this paper, the use of higher-order influence functions for the Duhamel's integral solution of the linear inverse heat conduction problem (IHCP) is investigated and discussed. Formulations are provided for estimation of surface heat flux using the first- and second-order influence functions. The method of least squares is employed in the estimation procedure in the presence of extra data due to use of future temperature measurements or due to temperature measurements at different interior locations. An example problem for a semi-infinite geometry is solved using the first-order influence functions and the results are compared with those given in an earlier work by J. V. Beck using the zero-order influence functions.

Introduction

The determination of the boundary conditions from the interior temperature measurements known as the inverse heat conduction problem (IHCP) is of fundamental importance in engineering practice and accordingly has received considerable attention over the last four decades. There are many applications for which direct measurement of the surface conditions (temperature or heat flux) is not feasible. This is mostly due to the difficulties involved in mounting a measuring probe on the surface boundary. Some example problems are those involving heat transfer in combustion engines, gun barrels, gas turbines, rocket nozzles, quenching processes, and various nuclear reactor applications. For such problems, it becomes necessary to determine the surface conditions from the interior temperature measurements.

Many methods have been proposed and numerous papers have been written for the solution of the IHCP. A detailed discussion of different approaches and a comprehensive list of various papers can be found in an excellent book written by Beck, Blackwell, and St. Clair [1]. Overall, the approaches used for the solution of the IHCP may be divided into two categories: those which employ integral methods [2-7], and those which utilize the difference methods [8-15]. In general, difference methods provide greater flexibility due to their ability to handle nonlinear (temperature dependent thermal properties) as well as linear (temperature independent thermal properties) problems. Use of nonlinear algorithms for linear problems, however, does not provide an efficient approach since it requires excessive computations due to unnecessary iterations. For linear problems, it is easier and faster to use integral methods.

Slotz [2] was one of the earliest investigators who obtained a linear solution by numerical inversion of Duhamel's integral equation. In his solution, the surface heat flux was obtained at different time steps in a sequential manner. His solution was found to be unstable for small time steps. The use of a small time step is very important in the solution of the IHCP. This is due to the fact that by using small time steps, one can obtain the maximum amount of information about the surface condition. This becomes particularly important in the solution of problems with rapid transient changes of the surface conditions. Beck [3] considerably improved the Slotz's approach by utilizing a number of future data and the least squares technique to generate solutions for a much smaller time step than the Slotz's method. In his solution, he employed the influence functions or the **"fundamental"** solutions which were due to a uniform approximation of the surface heat flux over a number of equally-spaced time intervals. Other powerful integral methods are those presented by Hills and Mulholland [6] and Murio [7].

This paper investigates an extension of the Beck's method [3] by utilizing the first and higher order influence functions which are due to linear and higher order approximations of the surface heat flux. The method of least squares is employed in the estimation procedure in the presence of extra data due to the use of future temperature measurements and different interior locations. An example problem is solved using the first order influence functions. The results show that the use of the first order influence function can provide a smaller stability limit than the use of the zero order influence function presented in reference [3].

Statement of the IHCP

Consider the boundary value problem of heat conduction in a semi-infinite solid exposed to an arbitrary time-variable surface heat flux as shown in Fig. 1. It is assumed that the thermal properties do not change with temperature. In the inverse heat conduction, it is desired to estimate the surface heat flux, q(t), from the temperature measurements obtained at discrete times in an interior location x.

One way to estimate the surface heat flux history is to divide the time region 0 to t into a number of small time intervals, Δt, and to approximate the heat flux over each time interval by a simple function. Since the problem is homogeneous and linear, the temperature response at location x and at time t can be obtained by simply superimposing the temperature responses at x and time t due to the approximated heat fluxes over the prior time intervals. That is,

$$T(x,t) = \sum_{j=1}^{M} T^{(j)}(x,t) \tag{1}$$

where M represents the number of time intervals and $T^{(j)}(x,t)$ is the temperature response at location x and time t due to the approximated $q(t)$ over the jth time interval.

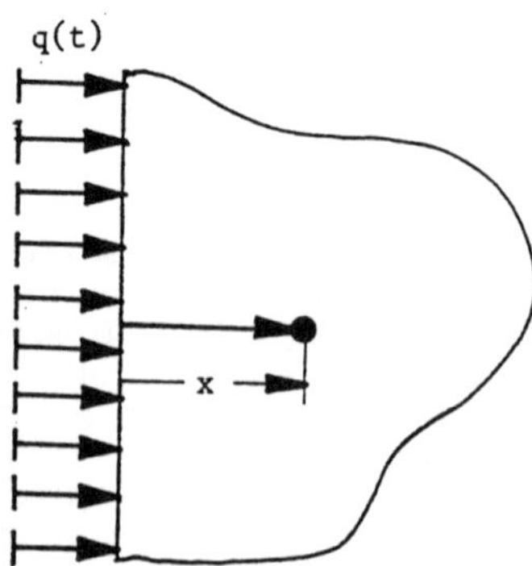

Figure 1.   Geometry of semi-infinite body heated by a time-varying surface heat flux.

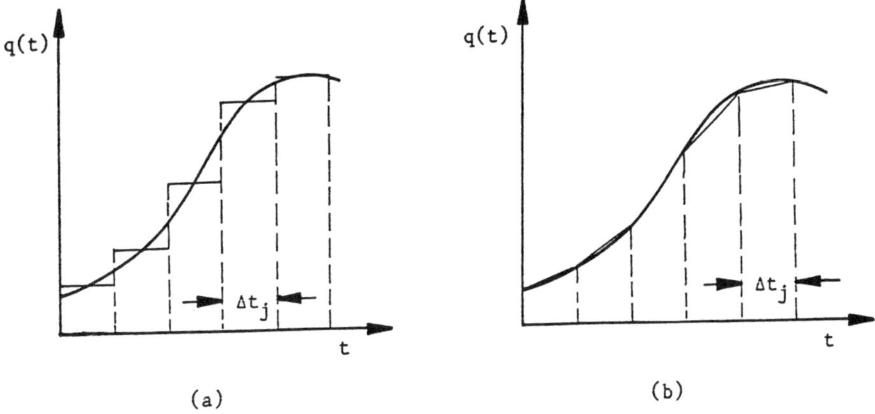

(a)   (b)

Figure 2.   Uniform and linear heat flux approximations over each time interval.

In the simplest form of approximation the heat flux histories are assumed to have a constant value in each time interval [3] as shown in Fig. 2a. Another type of approximation for $q(t)$ is a linear approximation (see Fig. 2b) which can represent $q(t)$ more accurately than the uniform approximations shown in Fig. 2a.

Higher order approximations such as parabolic or cubic approximations are also possible; however, it should be noted that for any type of approximation chosen, the exact solution to the corresponding direct problem should be available. These exact solutions which are called influence functions, are explained in more detail in the following section.

Influence Functions

The simplest form of an influence function (here termed the influence function of zero-order and denoted by $\phi^{(0)}(x,t)$) is defined as the temperature solution at location x and time t due to a constant surface heat flux input q=1 for t>0. In the case of a semi-infinite solid with zero initial temperature, this solution is available and is given by [16],

$$\phi^{(0)}(x,t) = \frac{2\sqrt{\alpha t}}{k} ierfc\left(\frac{x}{2\sqrt{\alpha t}}\right) \tag{2}$$

The temperature response at location x and time t due to a uniform heat flux q_j during the jth time interval (see Fig. 3a) can be written in terms of $\phi^{(0)}$ as,

$$T^{(j)}(x,t) = q_j \phi^{(0)}(x,t-t_{j-1}) \qquad for \;\; t_{j-1} \leq t < t_j \tag{3a}$$

$$= q_j[\phi^{(0)}(x,t-t_{j-1}) - \phi^{(0)}(x,t-t_j)] \;\; for \;\; t \geq t_j \tag{3b}$$

The first order influence function for the semi-infinite geometry is define as the temperature response at location x and time t due to a linear time dependent surface heat flux given by,

$$q(t) = t \;\; for \;\; t > 0$$

$$= 0 \;\; for \;\; t < 0 \tag{4}$$

(Note that the slope $\frac{dq}{dt} = 1$).

425

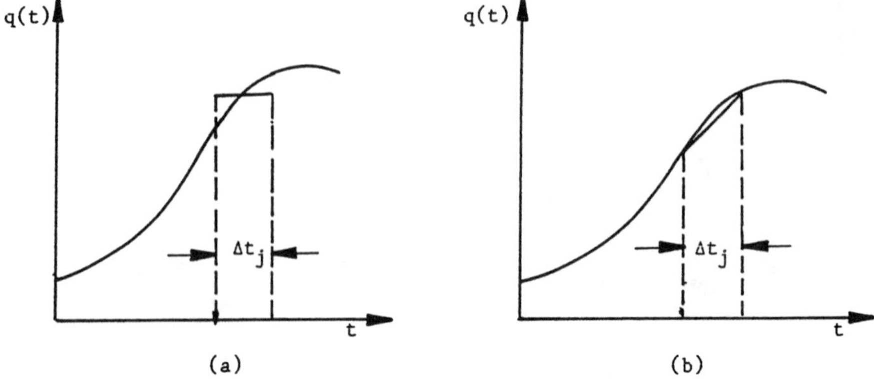

(a)   (b)

Figure 3.   Uniform and linear heat flux approximations over the jth time interval.

This influence function is also available and is given by,

$$\phi^{(1)}(x,t) = \frac{8\sqrt{\alpha}}{k} t^{3/2} i^3 erfc\left(\frac{x}{2\sqrt{\alpha t}}\right) \tag{5}$$

Then, one can show that the temperature response at location x and time t due to a linear time dependent surface heat flux during the jth time interval (see Fig. 3b) can be written as,

$$T^{(j)}(x,t) = q_{j-1}\phi^{(0)}(x,t-t_{j-1}) + \left[\frac{dq}{dt}\right]_{\Delta t_j} \phi^{(1)}(x,t-t_{j-1}) \quad for \quad t_{j-1} \le t < t_j \tag{6a}$$

$$= q_{j-1}\phi^{(0)}(x,t-t_{j-1}) - q_j\phi^{(0)}(x,t-t_j) + \left[\frac{dq}{dt}\right]_{\Delta t_j} [\phi^{(1)}(x,t-t_{j-1}) - \phi^{(1)}(x,t-t_j)]$$

$$for \quad t_{j-1} \le t < t_j \tag{6b}$$

Similarly, the second order influence function gives the temperature response at location x and time t due to a 2nd order time dependent surface heat flux given by,

$$q(t) = t^2 \quad for \quad t > 0$$

$$= 0 \quad for \quad t < 0 \tag{7}$$

(Note that in this case $\frac{1}{2!}\frac{d^2 q}{dt^2} = 1$).

Solution to this influence function for the semi-infinite geometry is given by,

$$\phi^{(2)}(x,t) = \frac{32\Gamma(3)\sqrt{\alpha}}{k} t^{5/2} i^5 erfc\left(\frac{x}{2\sqrt{\alpha t}}\right) \tag{8}$$

Then, it can be shown that the temperature response at location x and time t due to a 2nd order time dependent surface heat flux during the jth time interval, is

426

$$T^{(j)}(x,t) = q_{j-1}\phi^{(0)}(x,t-t_{j-1}) + \left[\frac{dq}{dt}\right]_{j-1}\phi^{(1)}(x,t-t_{j-1}) + \frac{1}{2}\frac{d^2q}{dt^2}\phi^{(2)}(x,t-t_{j-1})$$

$$\text{for}\quad t_{j-1}\leq t < t_j \qquad (9a)$$

$$= q_{j-1}\phi^{(0)}(x,t-t_{j-1}) - q_j\phi^{(0)}(x,t-t_j) + \left[\frac{dq}{dt}\right]_{j-1}\phi^{(1)}(x,t-t_{j-1}) - \left[\frac{dq}{dt}\right]_j\phi^{(1)}(x,t-t_j)$$

$$+ \frac{1}{2}\left[\frac{d^2q}{dt^2}\right]_{\Delta t_j}[\phi^{(2)}(x,t-t_{j-1}) - \phi^{(2)}(x,t-t_j)] \qquad \text{for}\quad t\geq t_j \qquad (9b)$$

In general, for an nth order time dependent surface heat flux during the jth time interval, one can write

$$T^{(j)}(x,t) = q_{j-1}\phi^{(0)}(x,t-t_{j-1}) + \left[\frac{dq}{dt}\right]_{j-1}\phi^{(1)}(x,t-t_{j-1}) + \frac{1}{2}\left[\frac{d^2q}{dt^2}\right]_{j-1}\phi^{(2)}(x,t-t_{j-1}) + \ldots$$

$$+ \frac{1}{n!}\left[\frac{d^nq}{dt^n}\right]_{\Delta t_j}\phi^{(n)}(x,t-t_{j-1}) \qquad \text{for}\quad t_{j-1}\leq t < t_j \qquad (10a)$$

and

$$T^{(j)}(x,t) = q_{j-1}\phi^{(0)}(x,t-t_{j-1}) - q_j\phi^{(0)}(x,t-t_j) + \left[\frac{dq}{dt}\right]_{j-1}\phi^{(1)}(x,t-t_{j-1}) - \left[\frac{dq}{dt}\right]_j\phi^{(1)}(x,t-t_j)$$

$$+ \frac{1}{2!}\left[\frac{d^2q}{dt^2}\right]_{j-1}\phi^{(2)}(x,t-t_{j-1}) - \frac{1}{2!}\left[\frac{d^2q}{dt^2}\right]_j\phi^{(2)}(x,t-t_j) + \ldots$$

$$+ \frac{1}{n!}\left[\frac{d^nq}{dt^n}\right]_{\Delta t_j}[\phi^{(n)}(x,t-t_{j-1}) - \phi^{(n)}(x,t-t_j)] \qquad \text{for}\quad t\geq t_j \qquad (10b)$$

where the corresponding nth order influence function is,

$$\phi^{(n)}(x,t) = \frac{2^{(2n+1)}\Gamma(n+1)\sqrt{\alpha}}{k} t^{(2n+1)/2} i^{(2n+1)} erfc\left(\frac{x}{2\sqrt{\alpha t}}\right) \qquad (10c)$$

This influence function is the solution to a semi-infinite solid initially at zero temperature, exposed to a time dependent surface heat flux given by

$$q(t) = t^n \qquad \text{for}\quad t > 0$$

$$= 0 \qquad \text{for}\quad t < 0 \qquad (11)$$

Eqs. (10a) and (10b) are rather general expressions and can be applied to any surface heat flux with an nth order polynomial approximation over the jth time interval. However, the use of this equation for n > 2 involves tedious and time consuming analytical work. In the following sections the zero, first, and second order influence functions are employed to find the solution to a simple inverse heat conduction problem [3].

Simple IHCP

By substituting Eqs. (3b), (6b), and (9b) into Eq. (1) and evaluating T(x,t) at time $t_M = M\Delta t$, one can write

$$T_M = T_0 + \sum_{j=1}^{M-1} q_j\Delta\phi^{(0)}_{M-j} + q_M\Delta\phi^{(0)}_0 \qquad (12)$$

(For constant heat flux approximation)

$$T_M = T_0 + q_0\phi^{(0)}_M + \sum_{j=1}^{M}\left[\frac{dq}{dt}\right]_{\Delta t_j}\Delta\phi^{(1)}_{M-j} \qquad (13)$$

(For linear heat flux approximation)

$$T_M = T_0 + q_0\phi_M^{(0)} + \left[\frac{dq}{dt}\right]_{t_0}\phi_M^{(1)} + \frac{1}{2}\sum_{j=1}^{M}\left[\frac{d^2q}{dt^2}\right]_{\Delta t_j}\Delta\phi_{M-j}^{(2)}, \tag{14}$$

(For 2nd order heat flux approximation)

where $T_M = T(x, t_M) = T(x, M\Delta t)$ and

$$\phi_n^{(\cdot)} = \phi^{(\cdot)}(x, n\Delta t) \quad ; \quad (\phi_0^{(\cdot)} = 0) \tag{15a}$$

$$\Delta\phi_n^{(\cdot)} = \phi_{n+1}^{(\cdot)} - \phi_n^{(\cdot)} \tag{15b}$$

If the temperature at location x in the body at each time step is known (measured), then Eq. (12) can be employed to estimate the surface heat flux at each time step,

$$q_M = \frac{\hat{T}_M - T_0 - \sum_{j=1}^{M-1} q_j\Delta\phi_{M-j}^{(0)}}{\Delta\phi_0^{(0)}} \tag{16}$$

where $\hat{T}_M$ is the measured temperature at time t_M. In the form given by Eq. (16), the heat flux histories (q_M's) can be determined at different time steps one after another by marching forward in time for M=1, 2, 3, ... While calculating each new component, the fluxes at previous times are known.

The approach presented by Eq. (16) was introduced by Slotz [2] and Beck [3] and is referred to as "exact matching" to distinguish it from approximate matching obtained by using least squares. The application of this approach to the cases with linear and second order heat flux approximations is not as straight forward as the case with uniform heat flux approximation since the first and second derivatives of heat flux appearing in Eqs. (13) and (14) should be approximated in an appropriate way such that these equations can be explicitly solved for the unknown q_M. In the following, it is shown how this approach can be extended to the cases with the linear and 2nd order heat flux approximations.

For the case with the linear heat flux approximation, given by Eq. (13), the heat flux during the jth time interval can be presented by,

$$q(t) = q_{j-1} + \frac{q_j - q_{j-1}}{t_j - t_{j-1}}(t - t_{j-1}) \quad for \quad t_{j-1} \leq t \leq t_j \tag{17}$$

Taking the derivative of q(t), gives

$$q'(t) = \frac{q_j - q_{j-1}}{\Delta t} \tag{18}$$

Now by substituting Eq. (18) into Eq. (13) and simplifying the results, one can show that,

$$T_M = T_0 + q_0\left(\phi_M^{(0)} - \frac{\Delta\phi_{M-1}^{(1)}}{\Delta t}\right) + \frac{1}{\Delta t}\sum_{j=1}^{M-1} q_j\Delta^2\phi_{M-(j+1)}^{(1)} + q_M\frac{\Delta\phi_0^{(1)}}{\Delta t} \tag{19}$$

Solving for q_M gives,

$$q_M = \frac{\Delta t(\hat{T}_M - T_0) + q_0(\Delta\phi_{M-1}^{(1)} - \Delta t\phi_M^{(0)}) - \sum_{j=1}^{M-1} q_j\Delta^2\phi_{M-(j+1)}^{(1)}}{\Delta\phi_0^{(1)}} \tag{20}$$

where

$$\Delta^2\phi_n^{(\cdot)} = \phi_n^{(\cdot)} - 2\phi_{n+1}^{(\cdot)} + \phi_{n+2}^{(\cdot)} \quad ; \quad \Delta\phi_0^{(\cdot)} = \phi_1^{(\cdot)} - \phi_0^{(\cdot)} = \phi_1^{(\cdot)} \quad since \quad \phi_0^{(\cdot)} = 0 \tag{21}$$

For the case with the second order heat flux approximation, given by Eq. (14), one possible expression for the heat flux during the jth time interval is,

$$q(t) = q_{j-1} + q'_{j-1}(t - t_{j-1}) + \frac{(q'_j - q'_{j-1})}{2\Delta t}(t - t_{j-1})^2 \qquad for \quad t_{j-1} \leq t \leq t_j \tag{22a}$$

with the first and second derivatives given by

$$q'(t) = q'_{j-1} + \frac{q'_j - q'_{j-1}}{\Delta t}(t - t_{j-1}) \quad ; \quad q''(t) = \frac{q'_j - q'_{j-1}}{\Delta t} \tag{22b,c}$$

Another possible expression is,

$$q(t) = q_{j-1} + q'_{j-1}(t - t_{j-1}) + \frac{q_j - q_{j-1} - q'_{j-1}\Delta t}{\Delta t^2}(t - t_{j-1})^2 \qquad for \quad t_{j-1} \leq t \leq t_j \tag{23a}$$

with the first and second derivatives given by,

$$q'(t) = q'_{j-1} + \frac{2(q_j - q_{j-1} - q'_{j-1}\Delta t)}{\Delta t^2}(t - t_{j-1}) \quad ; \quad q''(t) = \frac{2(q_j - q_{j-1} - q'_{j-1}\Delta t)}{\Delta t} \tag{23b,c}$$

Equations (22a) and (23a) are equivalent. If Eq. (22a) is used, at each time step, the slope of the heat flux will be estimated first and then the heat flux, while if Eq. (23a) is used, the heat flux will be determined first and then its slope. To show the difference, both equations are utilized here.

By substituting Eqs. (22a,b,c) into Eq. (14) and simplifying the results, one can show that,

$$q'_0 = q'_1 = \frac{(\hat{T}_1 - T_0) - q_0 \phi_1^{(0)}}{\phi_1^{(1)}} \tag{24a}$$

$$q_1 = q_0 + q'_1 \Delta t \tag{24b}$$

$$q'_M = \frac{2\Delta t(\hat{T}_M - T_0) - 2\Delta t q_0 \phi_M^{(0)} - q'_0(2\Delta t \phi_M^{(1)} - \Delta\phi_{M-1}^{(2)}) + \sum_{j=1}^{M-1} q'_j \Delta^2 \phi_{M-(j+1)}^{(2)}}{\phi_1^{(2)}} \tag{24c}$$

$$q_M = q_{M-1} + \frac{q'_{M-1} + q'_M}{2}\Delta t \qquad for \, M = 2, 3, 4, ... \tag{24d}$$

Similarly, the substitution of Eqs. (23a,b,c) into Eq. (14) gives,

$$q_1 = \frac{(\hat{T}_1 - T_0)\Delta t + q_0(\phi_1^{(1)} - \Delta t \phi_1^{(0)})}{\phi_1^{(2)}} \tag{25a}$$

$$q'_0 = q'_1 = \frac{q_1 - q_0}{\Delta t} \tag{25b}$$

$$q_M = \frac{(\hat{T}_M - T_0)\Delta t - q_0(\Delta t \phi_M^{(0)} - \Delta\phi_{M-1}^{(2)}) - \Delta t q'_0(\phi_M^{(1)} - \Delta\phi_{M-1}^{(2)}) + \sum_{j=1}^{M-1}[q_j\Delta^2\phi_{M-(j+1)}^{(2)} - \Delta t q'_j\Delta\phi_{M-(j+1)}^{(2)}]}{\phi_1^{(2)}} \tag{25c}$$

$$q'_M = \frac{2(q_M - q_{M-1})}{\Delta t} - q'_{M-1} \qquad for \quad M = 2, 3, 4, ... \tag{25d}$$

In the derivation of Eqs. (24) and (25), it was assumed that no prior information with regard to the value of q'_0 is available and therefore, the linear heat flux approximation has been employed during the first time interval. However, if the value of q'_0 is known, Eqs. (24c,d) and (25c,d) can also be used for M=1.

Use of Least Squares

In the IHCP, because of the damping effect, there is a time delay in the internal temperature response which can cause noticeable error in the estimation of the surface condition in the case of the exact matching of the data. In other words, the interior temperature response at a particular time t is due to a surface heat flux prior to that time. Therefore, in order to have a good estimation of q at time t_M, it is important to use temperature measurements at times greater than t_M (here, called as future times $t_{M+1}, t_{M+2}, \ldots$). The use of temperature measurements obtained from more than one interior location (multiple sensors), can also help to estimate the surface heat flux more accurately; however, it is less effective than the use of the future temperatures [1].

In the case where extra data (due to temperature measurements at the future times and at more than one location) are available, the surface heat flux can be estimated by using the method of least squares. The least squares procedure has been applied earlier by Beck [3] for the case with constant heat flux approximation. It can also be applied to the cases with linear and higher order approximations. Here, for simplicity, it is applied only to the case with the linear surface heat flux approximation.

Suppose that it is desired to use i future temperature measurements. It is assumed that the heat flux after time t_M temporarily varies linearly with time with the slope being equal to that of the Mth time interval as shown in Fig. 4. Then, one can show that

$$\dot{q}_{M+i} = (i+1)q_M - iq_{M-1} \tag{26}$$

From Eq. (19), the temperature solution for time t_{M+i} is given by

$$T_{M+i} = T_0 + q_0\left(\phi_{M+i}^{(0)} - \frac{\Delta\phi_{M+i-1}^{(0)}}{\Delta t}\right) + \frac{1}{\Delta t}\sum_{j=1}^{M-1} q_j\Delta^2\phi_{M+i-(j+1)}^{(1)} + \frac{1}{\Delta t}\sum_{j=M}^{M+i-1} q_j\Delta^2\phi_{M+i-(j+1)}^{(1)} + q_{M+i}\frac{\Delta\phi_0^{(1)}}{\Delta t} \tag{27}$$

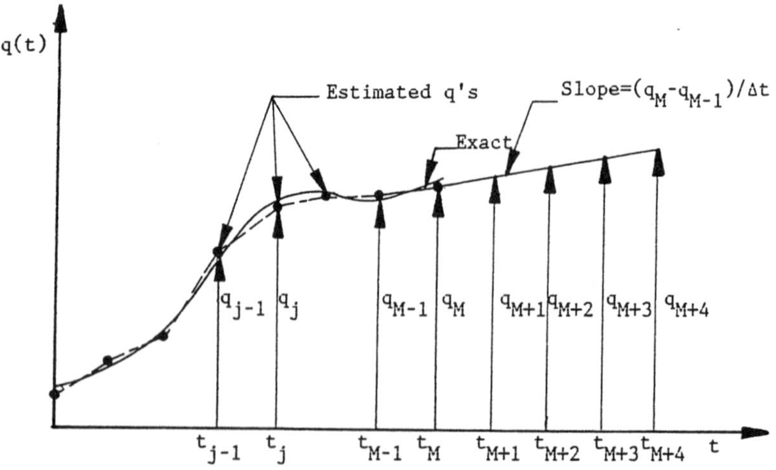

Figure 4.   Estimated heat fluxes at various time steps. Note the temporary assumption of linear heat flux for times $t \geq t_M$.

Substituting Eq. (26) into Eq. (27) and simplifying the second summation on the right hand side of the later equation, yields

$$T_{M+i} = T_0 + q_0 \left(\phi^{(0)}_{M+i} - \frac{1}{\Delta t} \Delta \phi^{(1)}_{M+i-1} \right) + \frac{1}{\Delta t} \sum_{j=1}^{M-1} q_j \Delta^2 \phi^{(1)}_{M+i-(j+1)} - q_{M-1} \frac{\phi^{(1)}_i}{\Delta t} + q_M \frac{\phi^{(1)}_{i+1}}{\Delta t} \tag{28}$$

which gives the computed temperature at time $t_{M+i} = (M+i)\Delta t$.

Now if the temperature measurements are available at R future times from N interior locations, one can minimize the least squares error between the computed and measured temperatures given by,

$$S = \sum_{k=1}^{N} \sum_{i=0}^{R} (\hat{T}_{k,M+i} - T_{k,M+i})^2 \tag{29}$$

with respect to q_M. Note that $\hat{T}$ denotes the measured temperature and subscript k represents the location of the sensor. Differentiating S with respect to q_M and set it equal to zero, gives

$$2 \sum_{k=1}^{N} \sum_{i=0}^{R} (\hat{T}_{k,M+i} - T_{k,M+i}) \frac{\partial T_{k,M+i}}{\partial q_M} = 0 \tag{30}$$

Substituting for $\frac{\partial T_{k,M+i}}{\partial q_M}$ from Eq. (28) into Eq. (30) and solving for q_M gives,

$$q_M = \frac{\displaystyle\sum_{k=1}^{N} \sum_{i=0}^{R} [\Delta t(\hat{T}_{k,M+i} - T_0) - q_0(\Delta t \phi^{(0)}_{k,M+i} - \Delta \phi^{(1)}_{k,M+i-1}) + q_{M-1} \phi^{(1)}_{k,i}] \phi^{(1)}_{k,i+1} - \displaystyle\sum_{j=1}^{M-1} q_j \Phi_{MjNR}^{(1)}}{\displaystyle\sum_{k=1}^{N} \sum_{i=0}^{R} [\phi^{(1)}_{k,i+1}]^2} \tag{31}$$

where

$$\Phi^{(1)}_{MjNR} = \sum_{k=1}^{N} \sum_{i=0}^{R} \Delta^2 \phi^{(1)}_{k,M+i-(j+1)} \phi^{(1)}_{k,i+1} \tag{32}$$

Note that in the above equations, the future temperatures represent those for times greater than t_M. If there is no future temperature (R=0) and one temperature sensor (N=1), then Eq. (31) reduces to Eq. (20) which is for the exact matching of the data. To demonstrate the validity of the proposed new approach, an example is solved using linear flux approximation and first order influence function, $\phi^{(1)}(\cdot)$.

<u>Example</u>

Consider a semi-infinite solid, initially at zero temperature, heated with a constant surface heat flux, q_s. The physical dimensions and property values are shown in Fig. 5. From the exact direct solution of this problem, one can write,

$$T(x_1 = L, \theta) = 2\sqrt{\theta}\, ierfc \left(\frac{1}{2\sqrt{\theta}} \right) \tag{33}$$

$$T(x_2 = 2L, \theta) = 2\sqrt{\theta}\, ierfc \left(\frac{1}{\sqrt{\theta}} \right) \tag{34}$$

and

$$q_1(\theta) = q_s erfc \frac{i}{2\sqrt{\theta}}) \tag{35}$$

where q_1 refers to q at location x_1 and Θ is the dimensionless time defined by,

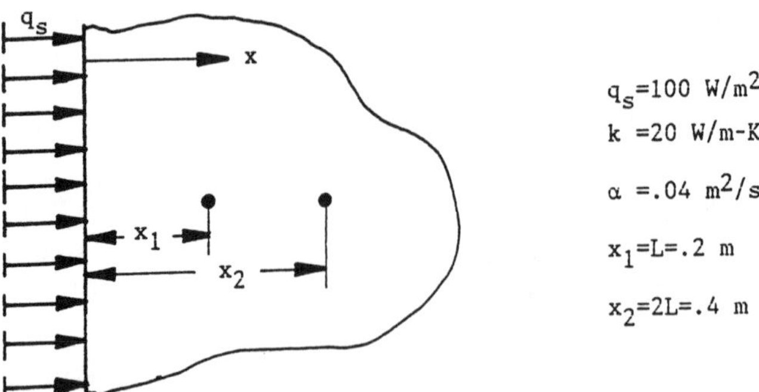

Figure 5.   Geometry and property values for the example problem.

$$\theta = \frac{\alpha t}{L^2} \tag{36}$$

To generate **data**, the exact solutions given by Eqs. (33) and (34) were employed to obtain temperatures at locations x_1 and x_2, up to 6 significant figures. The generated temperatures were then used in Eq. (31) with N=1 (one sensor) and R=1 (one future temperature) to solve two different inverse heat conduction problems; one was to estimate the constant surface heat flux, q_s, using the generated temperatures at $x_1 = L$, and the other was to estimate the time-varying flux $q_1(\theta)$ (at point 1) using the generated temperatures at $x_2 = 2L$. Later, the estimated fluxes were employed to reproduce the input temperatures using Eq. (19). The results are displayed in Figures 6, 7, and 8.

Fig. 6 shows the normalized estimated surface heat flux versus dimensionless time, Θ for various time steps. It can be seen that for $\Delta\theta$ = .05 oscillation exists which grows with time and becomes unbounded (unstable). The results for $\Delta\theta$ = 0.08, 0.09, 0.1, and 0.125 are also oscillatory; however, these oscillations decrease with time and die out. As $\Delta\theta$ becomes smaller, the oscillation becomes stronger which results in taking a longer time for estimated q_s to reach the exact value. For example, for $\Delta\theta$ = 0.08 which is close to .075 (the stability limit), the time required for the normalized surface heat flux to approach 1, is about 1.4. For $\Delta\theta$ = 0.075 (the stability limit), the estimated q_s also converges toward the exact value, but it takes a much longer time.

Fig. 7 shows the ratio of the reproduced input temperatures to the original input temperatures for x = L versus Θ. Except for the first few time steps, the two temperatures are very close and their ratio approaches 1. Fig. 8 shows $q_1(\theta)$, the exact time-varying heat flux at $x_1 = L$ obtained from Eq. (35) and its estimated values plotted versus time. For this case, it can also be seen that, the stability limit is about .075. For all time steps greater than .075, the results are stable, while for all computations with $\Delta\theta <$ 0.075, the results are unstable.

The use of small time steps is very important in the solution of the IHCP. This is due to the fact that by using small time steps, one can obtain the maximum amount of information about the surface condition. However, as shown in the above results, unlike the direct problem, small time steps cause difficulties in the solution of the IHCP.

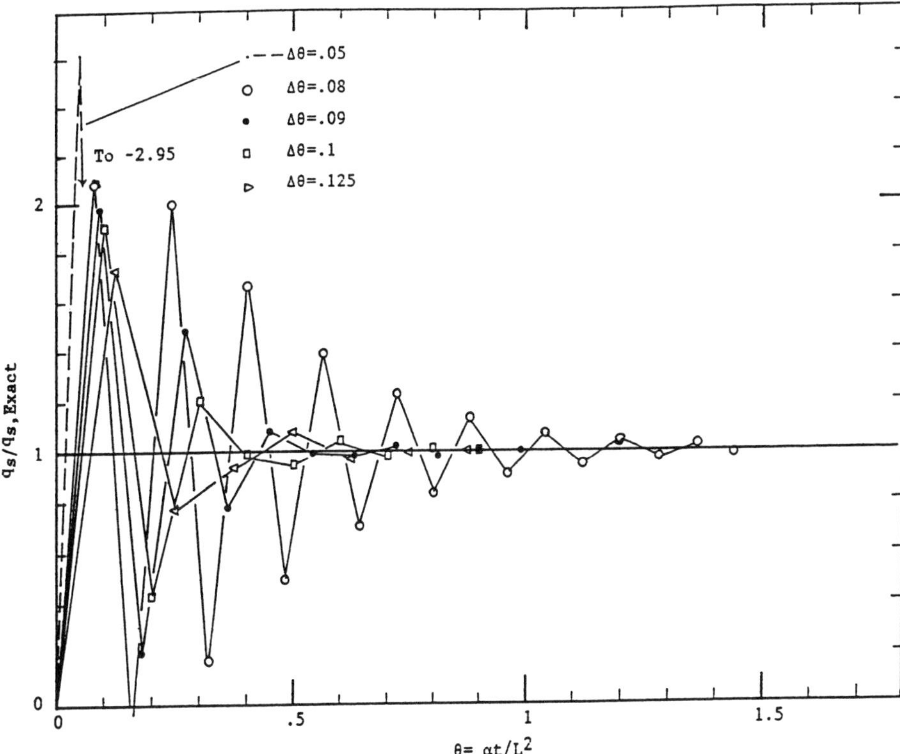

Figure 6.   Normalized estimated surface heat flux versus dimensionless time.

The example problem considered here has also been considered and solved by Beck [3] in a similar manner using the zero-order influence functions, $\phi^{(0)}(\cdot)$'s. In his solution, for the case with linear heat flux approximation and R=1 (use of one future temperature), the stability limit was .31 which is more than four times greater than the limit presented in this paper using the first-order influence functions, $\phi^{(1)}(\cdot)$'s. By lowering the stability limit, smaller time steps can be used resulting in a better estimation of the surface heat flux.

Conclusion

An improved method for the solution of the linear IHCP has been presented. Use of the first and higher order fundamental solutions (influence functions) were investigated and discussed. Although, the formulations described were for the estimation of the surface heat flux in a semi-infinite geometry, they are also applicable for other geometries as long as the corresponding influence functions are available. To demonstrate the validity of the method, an example problem has been solved using the linear flux approximation and the first order influence functions. The results show that the method produces accurate surface heat flux and temperatures. It is also shown that the use of first order influence functions can result in a reduction of the time step (the stability limit) by as much as a factor of 4 over the method previously presented [3] using the zero order influence functions.

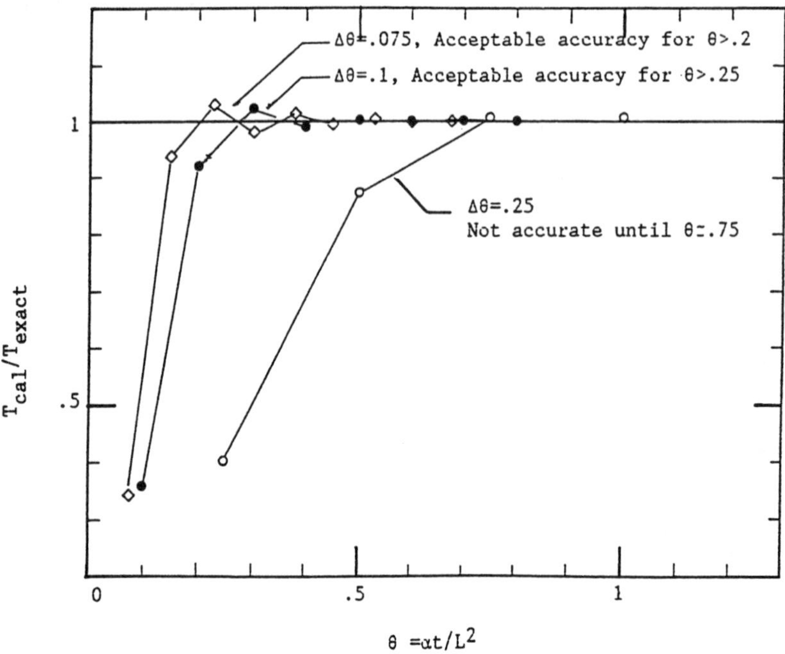

Figure 7.   Normalized reproduced temperature at $x = L$ using the estimated surface heat flux.

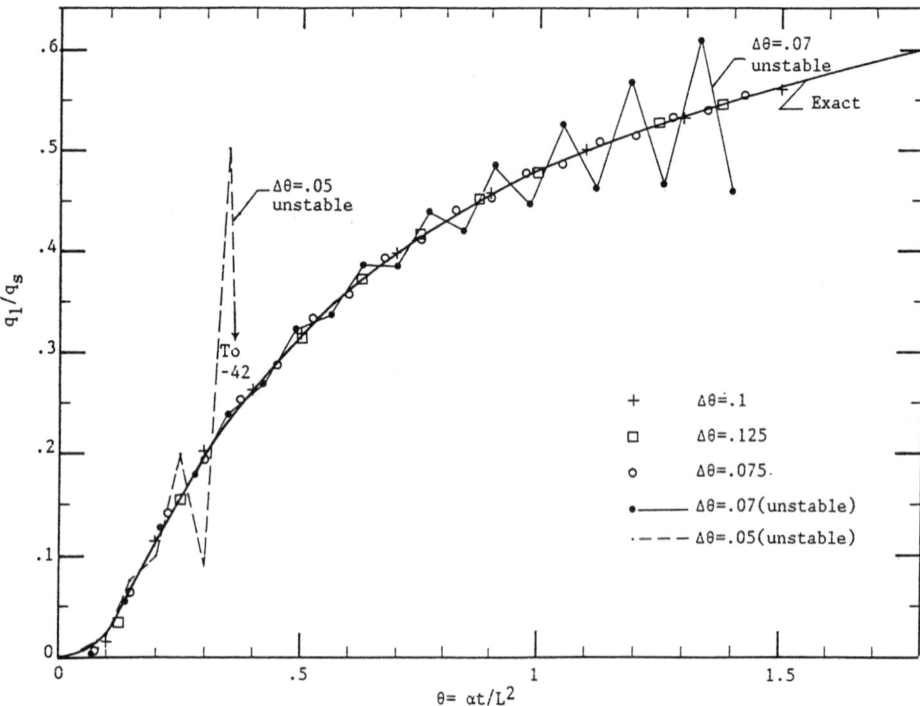

Figure 8.   Normalized estimated heat flux at $x = L$ versus dimensionless time.

Nomenclature

$erfc(\cdot)$	Complementary error function
$i^n erfc(\cdot)$	Repeated integral of the error function
k	Thermal conductivity
L	Dimension defined in Fig. 5
M	Time index
N	Number of sensors
q	Heat flux
q_0	Heat flux at time zero
q_s	Constant surface heat flux
q_M	Heat flux at time $t_M = M\Delta t$
R	Number of future temperatures
S	sum of the squares function
t	Time
T	Temperature
T_0	Initial temperature
$\hat{T}$	Measured temperature
x	Cartesian coordinate
α	Thermal diffusivity
$\Gamma(\cdot)$	Gamma function
Θ	Dimensionless time defined by Eq. (36)
$\phi^{(n)}(\cdot)$	nth order influence function
$\Phi^{(1)}_{MjNR}(\cdot)$	Defined by Eq. (32)

References

1. Beck, J.V., Blackwell, B., and St. Clair, Jr., C.R., "Inverse Heat Conduction", Wiley, 1985.

2. Slotz, Jr., G., "Numerical Solutions to an Inverse Problem of Heat Conduction for Simple Shapes", **J. Heat Transfer**, Tras. ASME, Ser. C, Vol. 82, pp. 20-26, 1960.

3. Beck, J.V., "Surface Heat Flux Determination Using an Integral Method", **Nucl. Eng. Des.**, vol. 7, pp. 170-178, 1968.

4. Sparrow, E.M., Haji-Sheikh, A., and Lundgren, T.S., "The Inverse Problem in Transient Heat Conduction", **J. Appl. Mech.**, vol. 31, Ser. E, pp. 369-375, 1964.

5. Arledge, R.G., and Haji-Sheikh, A., "An Iterative Approach to the Solution of Inverse Heat Conduction Problems", **J. of Numerical Heat Transfer**, Vol. 1, pp. 365-376, 1978.

6. Hills, R.G., and Mullholland, G.P., "The Accuracy and Resolving Power of One-dimensional Transient Heat Conduction Theory as Applied to Discrete and Inaccurate Measurements", **Int. J. Heat Mass Transfer**, vol. 22, pp. 1221-1229, 1979.

7. Murio, D.A., "The Mollification Method and the Numerical Solution of an Inverse Heat Conduction Problem", **SIAM J. Sci. Stat. Comput.**, vol. 2, no. 1, pp. 17-34, March 1981.

8. Beck, J.V., and Wolf, H., "The Nonlinear Inverse Heat Conduction Problem", ASME paper 65-HT-40, ASME-AIChE Heat Transfer Conf., Los Angeles, California, 1965.

9. Beck, J.V., Litkouhi, B., and St. Clair, Jr., C.R., "Efficient Sequential Solution of the Nonlinear Inverse heat Conduction Problem", **Numerical Heat transfer**, Vol. 5, pp. 275-286, 1982.

10. Williams, S.D., and Curry, D.M., "An Analytical and Experimental Study for Surface Heat Flux Determination", **J. Space cr. Rockets**, vol. 14, pp. 632-637, 1977.

11. Beck, J.V., "Comments on an Analytical and Experimental Study for Surface Heat Flux Determinations", **J. Space cr. Rockets**, vol. 15, pp. 381-382, 1978.

12. Garifo, L., Shrock, V.E., Spedicato, E., "On the Solution of Inverse Heat Conduction Problem by Finite Differences", **Eng. Nucl. (Milan)**, vol. 22, pp. 452-464, 1975.

13. Muzzy, R.J., Avila, J.H., and Root, R.E., "Topical Report: Determination of Transient Heat Transfer Coefficients and the Resultant Surface Heat Flux from Internal Temperature Measurements", General Motors Rept., GEAP-20731, 1975, San Jose, Calif.; available from Technical Information Center, P.O. Box 62, Oak Ridge, Tenn. 37830.

14. Krutz, G.W., Shoenhals, R.J., and Hore, P.S., "Application of the Finite-Element Method to the Inverse Heat Conduction Problem", **Numer. Heat Transfer**, vol. 1, pp. 489, 1978.

15. Bass, B.R., "Application of the Finite Element Method to the Nonlinear Inverse Heat Conduction Problem Using Beck's Second Method", **Trans. ASME J. Eng. Ind.**, vol. 102, pp. 168-176, 1982.

16. Carslaw, H.S. and Jaeger, O.C., "Conduction of Heat in Solids, 2nd Edition, Oxford University Press, London, 1959.

A Method for Solving Coupled Initial Boundary Value Systems of Second Order Partial Differential Equations: Exact and Accurate Approximate Solutions

LUCAS JÓDAR

Departamento de Matemática Aplicada
Universidad Politécnica de Valencia
46071 Valencia, Spain

ABSTRACT

In this paper a series solution for initial boundary value problems related to systems of second order partial differential equations is given. Approximate solutions obtained by truncation of the infinite series and substituting certain matrix exponentials by Padé approximants are constructed and error bounds of them in terms of the data are given.

INTRODUCTION

Many physical systems can not be described by a single partial differential equations but, in fact, are modeled by a coupled system of interlocking equations. So, in the study of propagation of signals in a system of electrical cables led to the investigation of a system of coupled linear partial differential equations. Some results related to these systems have been obtained in [3],[7],[9],[13], and others. Also, systems of linear partial differential equations appear in the study of temperature distribution in a composite heat conductor [4]. Numerical methods for solving these systems of partial differential equations may be found in [8],[12] and [16]. Methods based on the transformation of the original system into a new one with independent equations may be found in [3],[7],[9] and [18]. Explicit solutions of initial boundary value problems for systems of first order partial differential equations are given in [17]. The aim of this paper is to find explicit solutions for coupled systems of linear second order partial differential equations of the type

$$AU_{xx}(x,t)-BU_t(x,t)=0, \quad 0 \leqq x \leqq p, \ t>0, \ p>0 \tag{1}$$

$$U(0,t)=0, \ U(p,t)=0, \ t>0 \tag{2}$$

$$U(x,0)=f(x), \quad 0 \leqq x \leqq p, \tag{3}$$

where the unknown $U(x,t)$ and $f(x)$ take values in $\mathbb{R}^m$ and A,B are mxm real matrices, elements of $\mathbb{R}_{mxm}$, such that

A and B are symmetric positive definite $\hfill (4)$

Explicit solutions are of considerable interest both in explaining the physical phenomena and for checking results obtained by numerical methods. This paper is organized as follows. Section 2 provides a matrix separation of variables method for solving the initial-boundary value problem (1)-(3) when the matrices A and B satisfy the condition (4). Thus an infinite vector series solution of the problem is obtained. In section 3 we study the error of the approximate solutions when we truncate the series solution obtained in section 2. In section 4 computable approximate solutions are presented and the accuracy of the method is studied.

If C is a matrix in $\mathbb{R}_{mxm}$, we denote by $\|C\|$ the norm defined by the square root of the maximum eigenvalue of $C^T C$, where C^T is the transpose matrix of C, [15],p.41. We denote by $\|C\|_\infty$ the norm defined by

$$\|C\|_\infty = \max_i \left\{ \sum_{j=1}^{m} |c_{ij}| \right\} \hfill (5)$$

The set of all eigenvalues of C is denoted by $\sigma(C)$. If I denotes the identity matrix and z is an eigenvalue of C and $N(z,n)=\{x; (C-zI)^n x=0\}$, then the index $\nu(z)$ is the least integer ν such that $N(z,\nu)=N(z,\nu+1)$.

A MATRIX SEPARATION OF VARIABLES METHOD

As B is a positive definite matrix in $\mathbb{R}_{mxm}$, from [2], p.59, there exist an unitary matrix M such that

$$B=MCM^{-1} \hfill (6)$$

where $C=[\text{diag}(c_i); 1\leq i\leq m]$ and c_i lies in the positive real line. Now if we consider the change

$$u=M^{-1}U \hfill (7)$$

and we denote by P the matrix

$$P=M^{-1}B^{-1}AM \hfill (8)$$

then, the problem (1)-(3) takes the form

$$Pu_{xx}(x,t)-u_t(x,t)=0, \hfill (9)$$

$$u(0,t)=0, \quad u(p,t)=0, \quad t>0 \hfill (10)$$

$$u(x,0)=M^{-1}f(x)=g(x), \quad 0\leq x\leq p \hfill (11)$$

Let us suppose we are looking for solutions $u(x,t)$ of the homogeneous problem (9)-(10), of the form

$$u(x,t)=T(t)X(x) \hfill (12)$$

where $T(t)$ lies in $\mathbb{R}_{mxm}$ and $X(x)$ lies in $\mathbb{R}^m$.

Let λ be a complex number and let us consider the ordinary differential matrix equations

$$T'(t)-\lambda PT(t)=0, \tag{13}$$

$$X''(x)-\lambda X(x)=0 \tag{14}$$

Note that if $T(t),X(x)$, satisfy (13) and (14), respectively, then $u(x,t)$ defined by (12) satisfies

$$Pu_{xx}(x,t)-u_t(x,t)=PT(t)X''(x)-T'(t)X(x)=PT(t)\lambda X(x)-\lambda PT(t)X(x)=0$$

Furthermore if $X(x)$ satisfies the boundary value conditions

$$X(0)=0, \quad X(p)=0 \tag{15}$$

then $u(x,t)$ defined by (12) satisfies the boundary value conditions (10). Let us take $\lambda=-\rho^2$, where ρ is a positive number. The general solution of (14) is

$$X(x)=\cos(\rho Ix)D_1+\sin(\rho Ix)D_2$$

where D_1,D_2 are arbitrary vectors in $\mathbb{C}^m$. By imposing the conditions (15) it follows that

$$X(0)=D_1=0 \quad \text{and} \quad X(p)=\sin(\rho Ip)D_2=0$$

Thus, in order to find nontrivial solutions of the problem (14)–(15), the complex number ρ must verify the condition $\sin(\rho p)=0$. This means that a set of eigenvalues and eigenfunctions of the boundary value problem (14)–(15) is given by

$$\lambda_n=-(n\pi/p)^2, \quad X_n(x)=\sin((n\pi xI/p))D, \quad n{\geq}1; \quad D{\in}\,\mathbb{C}^m \tag{16}$$

Considering equation (13) with $\lambda=\lambda_n$, we have the matrix differential equation

$$T'(t)+P(n\pi/p)^2T(t)=0 \tag{17}$$

whose general solution takes the form

$$T_n(t)=\exp(-(n\pi/p)^2tP)Q \tag{18}$$

where Q is an arbitrary matrix in $\mathbb{C}_{mxm}$. Thus a family of solutions $u_n(x,t)$ of the homogeneous boundary value problem (9)–(10), is given by

$$u_n(x,t)=\exp(-(n\pi/p)^2tP)\sin((n\pi xI/p))D_n; \quad D_n{\in}\,\mathbb{C}^m \tag{19}$$

Now we are looking for a solution of the initial-boundary value problem (9)(10),(11), of the form

$$u(x,t)=\sum_{n{\geq}1}\exp(-(n\pi/p)^2tP)\sin((n\pi xI/p))D_n \tag{20}$$

where D_n is a vector in $\mathbb{C}^m$.

The vectors D_n appearing in (20) must be chosen so that $u(x,0)=g(x)$ for all x in $[0,p]$. If we impose in (20) that $u(x,0)=g(x)$, it follows that

$$g(x)= \sum_{n\geq 1} \sin((n\pi xI/p))D_n \tag{21}$$

Note that as $\sin((n\pi xI/p))$ is a diagonal matrix, the condition (21) will be verified if each component of $g(x)$ satisfies any sufficient condition such that its Fourier sine series converges to the value of $g(x)$ at each point x and we take

$$D_n=(2/p)\int_0^p \sin((n\pi xI/p))g(x)dx, \quad n\geq 1 \tag{22}$$

Now we are going to prove that the formal series defined by (20),(22), is a solution of problem (9)-(11). Let $t_o>0$ and let δ be such that $0<\delta<t_o$, then if we denote by $B(t_o,\delta)$ the rectangle $[0,p]\times[t_o-\delta,t_o+\delta]$, we will demonstrate:

(i) The series (20),(22), is uniformly convergent and defines a continuous function in $B(t_o,\delta)$.

(ii) The function $u(x,t)$ defined by (20),(22), admits partial derivatives $u_t(x,t),u_{xx}(x,t)$, for $(x,t)\in B(t_o,\delta)$, and these derivatives may be obtained taking termwise partial derivatives in the series (20).

Note that as A is a real positive definite matrix and M is an unitary real matrix satisfying (6), the matrix $M^{-1}AM$ is positive definite. On the other hand, as $C^{-1}=[\operatorname{diag}(c_i^{-1}); 1\leq i\leq m]$ is a positive definite diagonal matrix, from $[10]$,p.304-308, the matrix P defined by (8) is a positive definite matrix. Let us suppose that the distinct eigenvalues λ_i, $1\leq i\leq s$, of the matrix P are ordered such that

$$0<\lambda_1<\lambda_2<\ldots<\lambda_s \tag{23}$$

and let m_i be the index of the eigenvalue λ_i, $[6]$,p.556.

Let us consider the complex function $w(z)=\exp(-(n\pi/p)^2 tz)$, where $t>0$ is considered as a fixed parameter. Then from $[6]$,p.559, it follows that

$$\exp(-(n\pi/p)^2 tP)= \sum_{k=1}^{s} \sum_{j=0}^{m_k-1} (P-\lambda_k I)^j E(\lambda_k)w_{kj}/j! \tag{24}$$

where $E(\lambda_k)$ is the spectral projection of P associated to the eigenvalue λ_k and

$$w_{kj}=w^{(j)}(\lambda_k)=(n\pi/p)^{2j}(-1)^j\exp(-(n\pi/p)^2 t\lambda_k)t^j \tag{25}$$

From (20),(22) and (24), the vector series arising in the right hand side of (20) is convergent if for $1\leq k\leq s$, $0\leq j\leq m_k-1$, the series

$$W_{k,j}(x,t)= \sum_{n\geq 1} (P-\lambda_k I)^j E(\lambda_k)\sin((n\pi xI/p))w_{kj}/j! \tag{26}$$

is norm convergent. From the Riemann-Lebesgue [19] ,p.45, there exists a positive constant $\bar{M}$ such that the coefficients D_n defined by (22) satisfy

$$\|D_n\| \leq \bar{M}, \quad n \geq 1 \tag{27}$$

Let P_{kj} be defined by

$$P_{kj} = \|(P- \lambda_k I)^j E(\lambda_k)\| /j! \ , \quad 1 \leq k \leq s, \quad 0 \leq j \leq m_k - 1 \tag{28}$$

Then, from (23),(25),for $(x,t) \varepsilon B(t_o, \delta)$, it follows that

$$\|(P- \lambda_k I)^j E(\lambda_k) \sin((n\pi xI/p)) w_{kj} D_n/j!\,\|$$

$$\leq \bar{M} P_{kj} (n\pi/p)^{2j} (t_o + \delta)^j \exp(-(n\pi/p)^2 (t_o -\delta)\lambda_k) \tag{29}$$

Since $\lambda_k > 0$, it is clear that the numerical series with general term defined by by (29) is convergent. Now, from the Weierstrass criterion [1],it follows that the series defined by (26) is absolutely and uniformly convergent in $B(t_o, \delta)$. Thus, the series (20),(22), defines a continuous function at any point (x,t) with $t>0$, provided the continuity of $f(x)$.

To prove the existence of the partial derivatives $u_t(x,t)$ and $u_{xx}(x,t)$, we will use the derivation theorem for functional series, see theorem 9.14 of [1] . Note that the partial derivative with respect to the variable t of the vector functions appearing in the general term of the series (20) takes the form

$$-P(n\pi/p)^2 \exp(-(n\pi/p)^2 tP) \sin(n\pi xI/p) \ D_n \tag{30}$$

Reasoning in an analogous way to the proof of the uniform convergence of the series (20),(22) in the rectangle $B(t_o, \delta)$, one gets the uniform convergence in $B(t_o, \delta)$, one gets the uniform convergence in $B(t_o,\delta)$ of the series with general term given by 30). Hence and from theorem 9.14 of [1] ,we have

$$u_t(x,t)=-P \sum_{n\geq 1} (n\pi/p)^2 \exp(-(n\pi/p)^2 tP)\sin(n\pi xI/p)D_n = \sum_{n\geq 1} (\partial/\partial t)(u_n(x,t))$$

where $u_n(x,t)$ is defined by (19).

If we apply twice the theorem 9.14 of [1], it follows that

$$u_{xx}(x,t)=- \sum_{n\geq 1} (n\pi/p)^2 \exp(-(n\pi/p)^2 tP) \sin(n\pi xI/p)D_n = \sum_{n\geq 1} (\partial^2/\partial x^2)(u_n(x,t))$$

Thus, for $0 \leq x \leq p$ and $t>0$, it follows that

$$Pu_{xx}(x,t)-u_t(x,t)= \sum_{n\geq 1} \{P(u_n)_{xx}(x,t)-(u_n)_t(x,t)\} =0$$

where $u_n(x,t)$ is defined by (19).

We recall that if $f(x)$ is continuous in the interval $[0,p]$,then, under some of the following conditions, the Fourier sine series of $f(x)$ converges to $f(x)$:

(i) f(x) is locally of bounded variation in a neighbourhood $[x-\delta, x+\delta]$, with $\delta > 0$, see [19], p.57.

(ii) f(x) admits one-side derivatives $f'_R(x)$ and $f'_L(x)$, see corollary 1 of [5], p.95.

From the previous comments and from (7), the following result has been established:

THEOREM 1. Let f(x) be a continuous $\mathbb{C}^m$ valued function such that each component f_i of f for $1 \leq i \leq m$, satisfies some of the above properties (i) or (ii) If A,B are matrices satisfying the condition (4), then the vector series

$$U(x,t) = M \sum_{n \geq 1} \exp(-(n\pi/p)^2 tP) \sin(n\pi x I/p) D_n \qquad (31)$$

where D_n is defined by (22) and M is an unitary real matrix in $\mathbb{R}_{m \times m}$ satisfying (6).

Note that the solution $U(x,t)$ provided by theorem 1 has some drawbacks from a computational point of view. First of all $U(x,t)$ is an infinite series and secondly, the general term of (31) requires the computation of the exponentials $\exp(-(n\pi/p)^2 tP)$ and this means that the knowledge of the exact eigenvalues of P as well as the index of each eigenvalue is necessary. This motivates the work of the next section where we try to avoid the above inconveniences.

APPROXIMATE SOLUTIONS, ERROR BOUNDS AND ACCURACY

The aim of this section is to show that the solution proposed by theorem 1 may be used to obtain computable accurate approximate solutions of problem (1)-(3). Let us consider the hypotheses and the notation of the previous section and let q_o and Q be the positive constants defined by

$$q_o = \nu_1 + \nu_2 + \ldots + \nu_s \quad m; \quad Q = \max\{P_{hj}, \ 1 \leq h \leq s, \ 0 \leq j \leq \nu_h - 1\} \qquad (32)$$

where P_{hj} is defined by (28) and ν_h is the index of the eigenvalue λ_h of P. From (24),(27),(28) and (32), it follows that

$$\|\exp(-(k\pi/p)^2 tP)\| \leq q_o \bar{M} v_k(t), \qquad (33)$$

where

$$v_k(t) = (k\pi/p)^{2q_o} t^{q_o} \exp(-tb_k); \quad b_k = (k\pi/p)^2 \lambda_1 \qquad (34)$$

Let us suppose we are interested in the computation of the solution of problem (1)-(3) in the domain $[0,p] \times [t_o, t_1]$. An easy computation shows that

$$v'_k(t) = (k\pi/p)^{2q_o} t^{(q_o-1)} \exp(-tb_k)(q_o - tb_k)$$

and thus $v'_k(t)$ is negative if and only if, $q_o < tb_k$. Let k_o be the smallest positive integer k such that

$$q_o < tb_k, \quad \text{for } k \geq k_o \qquad (35)$$

Then, for every $t \varepsilon [t_o, t_1]$, $v_k(t)$ is a decreasing function of t on this interval and from (33) it follows that

$$\|\exp(-(k\pi/p)^2 tP)\| \le q_o \overline{M} v_k(t_o) \tag{36}$$

Let $S_n(x,t)$ be the n-th partial sum of the series (31) and let $t\epsilon[t_o,t_1]$, then from (20),(22),(27) and (36), for $n>k_o$ it follows that

$$\|U(x,t)-S_n(x,t)\|$$

$$\le \|M\| \sum_{k\ge n+1} \|\exp(-(k\pi/p)^2 tP)\| \ \|\sin(k\pi xI/p)D_k\| \tag{37}$$

$$\le q_o \|M\| Q\overline{M} \sum_{k\ge n+1} v_k(t_o)$$

Since M is a real unitary matrix, we have $\|M\|=1$ and from (37) we may write

$$\|U(x,t)-S_n(x,t)\| \le q_o Q\overline{M} \sum_{k\ge n+1} v_k(t_o) \tag{38}$$

Let T and T_o be the positive constants defined by

$$T=(\pi/p)^2 t_o; \quad T_o=(\pi^2 t_1/p^2)^{q_o} \tag{38}$$

From (36)-(38), for $(x,t)\epsilon[0,p]\times[t_o,t_1]$ and $k>k_o$, it follows that

$$\|U(x,t)-S_n(x,t)\| \le q_o T_o Q\overline{M} \sum_{k\ge n+1} k^{2q_o}\exp(-Tk^2) \tag{39}$$

Now we are interested in bounding the numerical series appearing in the second member of (39) in terms of the data.

Let us consider the discrete function $y(k)$ defined by

$$y(k)=(2q_o \log(k))/k - kT, \quad k\ge 1 \tag{40}$$

where T is defined by (38). An easy computation shows that $y(k)$ decreases for $k\ge 3$ and that $y(k)\to -\infty$ as $k\to\infty$. Thus the following negative number is well defined

$$a=\sup\{y(k); \ k\ge 3 \text{ and } y(k)<0\} \tag{41}$$

Note that from (40) it follows that $y(k)<0$ if and only if, $2q_o\log(k)<k^2 T$.

From (40),(41), it follows that

$$k^{2q_o}\exp(-k^2 T) \le \exp(ak), \quad k\ge 3 \tag{42}$$

If we denote by L the positive constant

$$L=q_o T_o \overline{M} Q \tag{43}$$

from (39),(42) and (43), for $n>\max(2,k_o)$ and for $(x,t)\epsilon[0,p]\times[t_o,t_1]$ we have

$$\|R_n(x,t)\| = \|U(x,t)-S_n(x,t)\| \le L \sum_{k\ge n+1} \exp(ak)=L\exp(a(n+1))/(1-\exp(a)) \tag{44}$$

If we are interested in finding the smallest value of n so that the error $R_n(x,t)$ be uniformly bounded by a prefixed number ε for $(x,t)\varepsilon[0,p]x[t_o,t_1]$ then, from (44), n must be a positive integer with $n>\max(2,k_o)$, such that

$$L \exp(a(n+1))< \varepsilon(1-\exp(a)) \tag{45}$$

Let $b=\exp(a)$, then condition (45) is equivalent to the following one

$$n+1> \{[\log(\varepsilon/L)](1-b)\}/a \tag{46}$$

Thus if n is a positive integer such that $n>\max(2,k_o)$ and satisfies the condition (46), then the truncation error $R_n(x,t)$ satisfies

$$\|R_n(x,t)\|<\varepsilon \quad , \text{ for } (x,t) \varepsilon [0,p]x[t_o,t_1] \tag{47}$$

Hence the following result has been established:

THEOREM 2. Let us consider the hypotheses and the notation of theorem 1, let ε be a prefixed positive number and let M,q_o,T_o,T and k_o defined by (27), (32),(35) and (38), respectively. If k_o satisfies (46), $n>\max(2,k_o)$,(x,t) belong to $[0,p]x[t_o,t_1]$ and $S_n(x,t)$ denotes the n-th partial sum of the series solution $U(x,t)$ defined by (31), then the truncation error $R_n=U-S_n$, satisfies $\|R_n(x,t)\|<\varepsilon$, uniformly for $(x,t)\varepsilon[0,p]x[t_o,t_1]$.

Theorem 2 avoids the computationa drawbacks of the infinite expression for the solution $U(x,t)$ given in theorem 1. No we are interested in obtaining accurate approximate solutions avoiding the computation of the matrix exponentials appearing in the finite approximation $S_n(x,t)$, given in theorem 2.

This will be done by approximating the matrix exponential of a matrix by its Padé approximant of an appropriate degree. For the sake of clarity in the presentation of the next result we recall some concepts that will be used below and that may be found in $[11]$,p.398 and $[14]$.

Let q be an integer $q>1$ and let $N_{qq}(z)$ be the polynomial

$$N_{qq}(z)= \sum_{k=o}^{q} \frac{(2p-k)!p!z^k}{(2p)!k!(p-k)!}$$

and let us introduce the Padé function

$$R_{qq}(z)=(N_{qq}(-z))^{-1}N_{qq}(z) \tag{48}$$

If A is a matrix in $\mathbb{R}_{mxm}$ and j is chosen so that $2^{j-1} \geq \|A\|_\infty$,and we define by $F_{qq}(A)$:

$$F_{qq}(A)=[R_{qq}(A/2^j)]^{2^j} , \quad q>1 \tag{49}$$

and let γ be defined by

$$\gamma=2^{3-2q} \frac{(q!)^2}{(2q)!(2q+1)!} \tag{50}$$

then, from $[11]$, p.398, it follows that

$$\|\exp(A)-F_{qq}(A)\|_\infty \leq \|A\|_\infty \, \exp(2\|A\|_\infty) \tag{51}$$

Let P be the matrix defined by (8) and let j_o be a positive integer satisfying the condition

$$2^{j_o-1} \geq (\pi k_o/p)^2 t_1 \, \|P\|_\infty \tag{52}$$

then, if we define F_{qq} by (49) taking j_o and if $t_o \leq t \leq t_1$, and we define

$$F_{qq}(k,t)=F_{qq}(-(k\pi/p)^2 tP), \quad k\geq o \tag{53}$$

from (48)-(50), it follows that

$$\|\exp(-(k\pi/p)^2 tP)-F_{qq}(k,t)\|_\infty \leq \gamma(k\pi/p)^2 t \, \|P\|_\infty \, \exp(2(k\pi/p)^2 t \|P\|_\infty) \tag{54}$$

Considering (53) for $k=1,2,\ldots,k_o$ and from (27),(54), we may write

$$\sum_{k=1}^{k_o} \|F_{qq}(k,t)-\exp(-(k\pi/p)^2 tP)\|_\infty \, \|\sin(k\pi xI/p)D_k\|_\infty$$

$$\leq \gamma \, M k_o (k_o\pi/p)^2 t_1 \, \|P\|_\infty \, \exp(2(k_o\pi/p)^2 t_1 \|P\|_\infty) \tag{55}$$

Let q large enough so that the constant γ defined by (50) satisfies

$$\gamma < \frac{\exp(-(2k_o\pi/p)^2 t_1 \|P\|_\infty) \, \varepsilon}{m^{1/2} M k_o (k_o\pi/p)^2 t_1 \, \|P\|_\infty} \tag{56}$$

If we define by $S_n(x,t,q)$ the finite series obtained by substituting in S_n, each matrix exponential $\exp(-(k\pi/p)^2 tP)$ by $F_{qq}(k,t)$, i.e.,

$$S_n(x,t,q)=M \sum_{k=1}^{n} F_{qq}(k,t) \sin(k\pi xI/p)D_k \, , \quad n > \max(2,k_o) \tag{57}$$

Thus if q satisfies (56), from the inequality

$$m^{-1/2} \|C\|_\infty \leq \|C\| \leq m^{1/2} \|C\|_\infty$$

valid for any mxm matrix C, $[11]$,p.45, and from (55) it follows that

$$\|S_{k_o}(x,t)-S_{k_o}(x,t,q)\| < \varepsilon \tag{58}$$

uniformly for $(x,t)\varepsilon [0,p] \times [t_o,t_1]$.

From the previous comments and theorems 1 and 2, the proof of the following result has been established:

THEOREM 3. Let us consider the hypotheses and the notation of theorems 1 and 2, let $0<t_o<t_1$ and let ε be a prefixed positive number and let q>1 satisfying (56). Then the finite sum $S_{k_o}(x,t,q)$ defined by (57), satisfies $\|U(x,t)-S_{k_o}(x,t,q)\| < 2\varepsilon$, uniformly for $(x,t) \varepsilon [0,p] \times [t_o,t_1]$, and $U(x,t)$ is the exact solution of problem (1)-(3).

ACKNOWLEDGEMENTS. This paper has been supported by a grant from the Conse-

lleria de Cultura,Educación y Ciencia de la Generalitat de Valencia and the
D.G.I.C.Y.T grant PS87-0064.

REFERENCES

1. Apostol,T.M.,Mathematical Analysis, Addison-Wesley, Reading MA, 1958.

2. Bellman, R., Introducción al Análisis Matricial (Spanish),Ed. Reverté,
    Barcelona, 1965.

3. Branin,JR.,F.R., Transitient Analysis of Lossless Transmission Lines,
    Proc. IEEE (Letters), vol.55, pp.2012-2013, 1967.

4. Cannon,J.R., and Klein,R.E.,On the Observability and Stability of the
    Temperature Distribution in a Composite Heat Conductor, SIAM J. Applied
    Maths., vol.24, pp.569-602, 1973.

5. Churchill,R.V., and Brown,J.W.,Fourier Series and Boundary Value Pro-
    blems, McGraw-Hill, New York, 1978.

6. Dunford,N., and Schwartz.J.,Linear Operators I, Interscience, New York,
    1957.

7. Dommel, H.W., A Method for Solving Transitient Phenomena Systems, Rep.5.8
    Proc. 2nd Power Systems Computation Conference, Stockholm, Sweden.

8. Fornberg,B.,On a Fourier Method for the Integration of Hyperbolic equa-
    tions, SIAM J. Numer. Anal., vol.12, pp.509-528, 1975.

9. Fung-Yuel, C., Transitient Analysis of Lossless Coupled Transmission Lines
    in a Non-homogeneous Dielectric Medium, IEEE Trans. Microwave Theory and
    Techniques, MIT, vol. 18, pp.616-626, 1970.

10. Gantmacher, F.R., The Theory of Matrices, vol.I, Chelsea, New York, 1959.

11. Golub, G.H., and Van Loan, C.F., Matrix Computations, Johns Hopkins Univ.
    Press, Baltimore, 1985.

12. Kreiss, H.O., and Oliger,J.,Comparison of Accurate Methods for the Inte-
    gration of Hyperbolic Equations, Tellus, vol. 24, pp. 199-215, 1972.

13. Kuznetsov, P.I., and Stratonovich,R.L.,The Propagation of Electromagnetic
    Waves in Multiconductor Transmission Lines, Macmillan, New York, 1964.

14. Moler,C.B.,and Van Loan, C.P., Nineteen Dubious Ways to Compute the Expo-
    nential of a Matrix, SIAM Review, vol. 20, pp. 801-836, 1978.

15. Ortega,J.M.,and Rheinboldt,W.C.,Iterative Solution of Nonlinear Equa-
    tions in Several Variables, Academic Press, New York, 1970.

16. Orszag, S.A., Numerical Simulation of Incompresible Flows Within Simple
    Boundaries I: Galerkin (spectral) Representations, Stud. Appl. Maths., vol.
    50, pp. 293-327, 1971.

17. Yee, K.S., Explicit Solutions of Initial Boundary Value Problems of a sys-
    tem of Lossless Transmission Lines, SIAM J. Appl. Maths., vol. 24, pp. 62-
    80.

18. Zachmanoglou, E.C., Thoe, D.W., Introduction to Partial Differential Equations, William and Wilkins, New York, 1976.

19. Zygmund, A., Trigonometric Series, 2nd ed., vols. I and II, Cambridge Univ. Press, London, 1977.

Integral Equations with Monte Carlo Boundary Conditions: Conduction in Electronic Devices

A. HAJI-SHEIKH and S. P. KINSEY
Mechanical Engineering Department
University of Texas, Arlington
Arlington, Texas 76019-0023, USA

ABSTRACT

The recent proliferation of supercomputers and parallel processors has re-energized the use of the Monte Carlo method. It is commonly believed that the Monte Carlo method is a powerful tool but it is computationally slow in comparison with numerical methods. To the contrary, it is demonstrated that for a special class of problems, the Monte Carlo method can be used to accelerate a solution and produce higher accuracy. This class of problems seeks the solution of partial differential equations, e. g., the diffusion equation, in a regular geometry physically attached to one or more irregular bodies. The solution in the regular domain is analytical, while the boundary conditions for the irregular domain are statistically determined. A numerical example illustrates the procedure.

INTRODUCTION

Monte Carlo is a method that statistically simulates physical events. Researchers have been fascinated with the method since it was introduced early in this century. It was developed to solve difficult physical problems, e. g., neutron diffusion. The list of scientists who contributed to the development of Monte Carlo includes Einstein, Smoluchowski, Lord Rayleigh, Markov, Kolmogorov, Courant, and many others. Unfortunately, in later years researchers attempted to use Monte Carlo mainly for solving simple problems. In comparison with other analytical and numerical methods the results were less than satisfactory. However, Monte Carlo stands alone for solving difficult problems. In heat transfer, there are numerous radiation problems for which Monte Carlo is the only plausible solution method. Here, a class of thermal diffusion problems are introduced which do not have exact solutions and for which numerical solutions are inadequate. These problems involve regular-shaped bodies connected to other irregular bodies, see Fig. 1. A regular-shaped body I is connected to arbitrarily shaped bodies II and III. The bodies are three-dimensional and a numerical simulation requires simultaneous solution of the temperature distribution within the primary region, I, and the contacting bodies, II, III, etc. Moreover, if the regular body, I, contains a point heat source, line heat source, or a planar heat source, the numerical solution is inadequate.

For example, energy release in a computer silicon chip is within a thin planar zone. The chips usually have simple geometries and the temperature solution can be accommodated analytically. The temperature distribution in the regular body is obtainable by classical methods. However, chips are physically connected to heat sinks, wires, and other auxiliary devices. The emphasis of a conduction analysis is

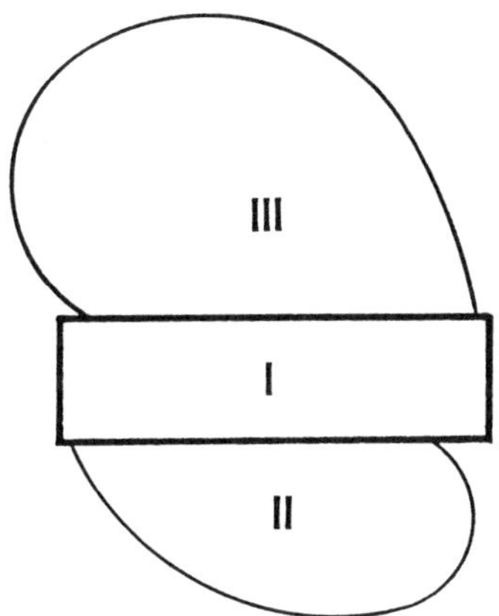

Figure 1. Arbitrary, irregular bodies connected to a regular-shaped body.

to investigate the integrity of the chips by calculating temperature distribution within the chip. A knowledge of temperature distribution in auxiliary devices often is not needed.

A simple example is selected to describe the method of solution and the auxiliary Monte Carlo procedure for the boundary condition. To demonstrate the effectiveness of the Monte Carlo method, a complex conduction problem is analyzed. The numerical example calculates the effect of geometrical eccentricity on the chip temperature for the IBM Thermal Conduction Module (TCM) (Chu et al., 1982), (Oktay and Krammerer, 1982), and (Blodgett and Barbour, 1982). The exact analytical method is used to solve for temperature in the chip and to accommodate the effect of the infinitesimal planar heat source. The Monte Carlo method is used to predict a relationship between temperature and heat flux where the chip is in contact with the heat sink while the Green's function solution yields the temperature distribution in the chip.

TEMPERATURE INSIDE THE REGULAR BODY

When the geometry of the primary region, e. g. region I, is under assumption of material homogeneity, the classical theories yield the temperature solution. In this paper, the Green's function solution method has been employed to solve the diffusion equation. The energy equation for the temperature field inside region I in Fig. 1 is

$$\nabla^2 T_I + \frac{g}{k} = \rho C_p \frac{\partial T_I}{\partial t} \tag{1}$$

The thermophysical properties are assumed to be independent of position and temperature. The solution of Eq. (1) for a parallelepiped geometry is selected; however, all other regular geometries can be accommodated in a similar manner. The Green's function solution is widely available in the literature. The Green's function for prescribed heat flux at all surfaces is (Ozisik, 1980)

$$G(x,y,z,t|x',y',z',\tau) = \sum_{l=0}^{\infty} \sum_{m=0}^{\infty} \sum_{n=0}^{\infty} \frac{f_l f_m f_n}{abc} \cos(\gamma_l x)\cos(\gamma_l x')\cos(\lambda_m y)\cos(\lambda_m y')$$

$$\cos(\mu_n z)\cos(\mu_n z')\exp[-(\gamma_l^2 + \lambda_m^2 + \mu_n^2)(t - \tau)] \tag{2}$$

The origin of the coordinate system is at the center of the parallelepiped body (region I). Region I is bounded by x=±a, y=±b, and z=±c planes. The eigenvalues γ_l, λ_m, μ_n are $l\pi/a$, $m\pi/b$, $n\pi/c$ and each of the coefficients f_l, f_m, or f_n has a value of 2 except when the corresponding l, m, or n is zero; in that case, 2 is replaced by 1. The temperature distribution within region I is

$$T_I(\text{x,y,z,t}) = T_0 + \frac{\alpha}{k} \int_0^t d\tau \int_R G(\text{x,y,z,t}|\text{x',y',z'},\tau)g(\text{x',y',z'},\tau)\text{dx'dy'dz'}$$

$$+ \alpha \int_0^t d\tau \int_R G(\text{x,y,z,t}|\text{x',y',z'},\tau)_{\text{on s}} \left(\frac{\partial T}{\partial n}\right)_{\text{on s}} \text{ds'} \tag{3}$$

The quantity $(\partial T/\partial n)_{on\ s}$ is $-q_w(\text{x',y',z'},\tau)/k$ where q_w is the surface heat flux. The initial temperature distribution is regarded as zero. For simplicity of this presentation, all boundaries of region I are considered insulated, except where region I is in contact with regions II and III. The exact representation of the boundary conditions on all surfaces of region I is not available in a simple and explicit form. If x, y, and z, in Eq. (3), are replaced by the coordinates of a point on the boundary, x_s, y_s, and z_s, Eq. (3) becomes

$$T_s(\text{x}_s,\text{y}_s,\text{z}_s,\text{t}) = T_0 + \frac{\alpha}{k} \int_0^t d\tau \int_R G(\text{x}_s,\text{y}_s,\text{z}_s,\text{t}|\text{x',y',z'},\tau)g(\text{x',y',z'},\tau)\text{dx'dy'dz'}$$

$$+ \alpha \int_0^t d\tau \int_R G(\text{x}_s,\text{y}_s,\text{z}_s,\text{t}|\text{x',y',z'},\tau)_{\text{on s}} \left(\frac{\partial T}{\partial n}\right)_{\text{on s}} \text{ds'} \tag{4}$$

Both $T_s(\text{x}_s,\text{y}_s,\text{z}_s,\text{t})$ and $q_w(\text{x',y',z'},\tau)$ in Eq. (4) are unknown. The temperature solution in regions II and III yields relations between the surface temperature and surface heat flux. For irregular geometries, the Monte Carlo method is a useful method of solution. This procedure is described in detail in Example 1.

Example 1

Region I in Fig. 2 is connected to region II. Region II is a so-called heat sink. The temperature in region II is one-dimensional. Heat transfers from region II to a reservoir at temperature T_R and to ambient air at temperature T_g. The random walk technique (Haji-Sheikh, 1988) is used to derive a relation between heat flux and temperature at the contact surface between regions I and II. The random walk method is used in region 2. When heat flux at the contact surface is prescribed, the boundary is treated as a reflective barrier. All random walks can originate from node 0, treating the contact surface as a reflective barrier. According to the following probability relation,

$$T_s = P_+ T_1 + p_g T_g + \frac{q_w \Delta\text{x}/\text{k}}{1 + 0.5\text{hP}\Delta\text{x}^2/\text{kA}} \tag{5}$$

$$P_+ = \frac{1}{1 + 0.5\text{hP}\Delta\text{x}^2/\text{kA}} \quad \text{and} \quad p_g = \frac{0.5\text{hP}\Delta\text{x}^2/\text{kA}}{1 + 0.5\text{hP}\Delta\text{x}^2/\text{kA}}$$

where h is heat transfer coefficient on the surface of region II, A is the cross-section area, and P is the perimeter of the cross section.

A random walk leaving node 0 has two options: either it goes to point 1 with probability of p_+ or it is terminated with probability of p_g. Notice that a score

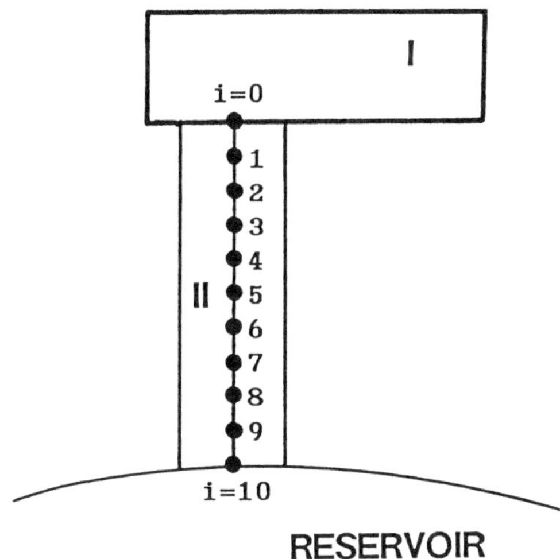

Figure 2. Geometry of Example 1.

equal to $(q_w\Delta x/k)/(1 + 0.5hP\Delta x^2/kA)$ must be recorded once a random walk leaves this reflective barrier. A random walk at any other point i within region I can move to point i+1 or i-1 with equal probabilities. The difference equation for energy balance yields the probability relation.

$$T_i = p_+ T_{i+1} + p_- T_{i-1} + p_g T_g \tag{6}$$

$$p_+ = p_- = \frac{1}{2 + hP\Delta x^2/kA} \quad \text{and} \quad p_g = \frac{hP\Delta x^2/kA}{2 + hP\Delta x^2/kA}$$

Any random walk returning at node 0 will be terminated and the corresponding score is the unknown temperature at node 0. As stated earlier, if the random walk enters air, it will be terminated and the score is the air temperature. If a random walk arrives at the reservoir, the score for that random walk is the reservoir temperature.

The elapsed time for one step in a random walk is obtained from equation

$$\frac{\alpha\Delta\tau^*}{\Delta x^2} = \frac{1}{2 + hP\Delta x^2/kA} \tag{7}$$

and the duration of a random walk, τ^* is the number of steps (before it is terminated) multiplied by the elapsed time for one step.

Table 1 is prepared using the parameters $T_g=T_R=0$, h=400 W/m^{2}K, k = 200 W/m-K, P=0.628x10^{-3} m, A=0.314x10^{-7} m^2, and Δx = 0.002 m. Region II is divided into 10 increments. The probability functions are p_+ = 0.463, p_- = 0.463, and p_g = 0.074. The data in Table 1 show the number of random walks terminated after 2 steps, 4 steps, etc. A large number of random walks in Table 1 are terminated after two steps. The average of all scores is the temperature at the point of origination of the random walk which is the temperature at the contact

Table 1. The number of random walks returned to point 0
after 2, 4, 6, . . . steps for 10,000 random walks sent.

Number of Steps	N_0	Number of Steps	N_0	Number of Steps	N_0
2	4631	18	33	34	5
4	1026	20	34	36	3
6	392	22	6	38	2
8	236	24	9	40	2
10	159	26	8	42	1
12	90	28	1	44	1
14	60	30	9	46	0
16	44	32	4	48	1

surface, T_s. The relation between temperature and heat flux at the contact surface is

$$T_s(t) = \frac{q_w(t)\Delta x/k}{1 + 0.5hP\Delta x^2/kA} + \frac{1}{N_w}\sum N_0(\tau^*)T_s(t - \tau^*) \tag{8}$$

Equations (4) and (8) produce an integral equation that must be solved numerically. The assumptions that I and II are in perfect contact and conduction in II is one-dimensional now will be relaxed. The result is a more difficult problem discussed in the next example.

Example 2

To illustrate the procedure, geometry I is selected with dimensions similar to the IBM TCM (Agonafer et al., 1987; Blodgett and Barbour, 1982; Oktay and Krammerer, 1982). In the TCM, geometry II is a spring-loaded piston which presses on the silicon chip, Fig. 3. The shape of the piston-hat combination in the TCM makes it difficult to find an exact Green's function for region II; therefore, a Monte Carlo method for the TCM (Haji-Sheikh and Lakshminarayanan, 1988) is used to statistically estimate the relationship between temperature $T_{IIo}(x,y,z,t)$ and $T_{Io}(x,y,z,t)$. The Monte Carlo method is computationally suitable for this complex problem.

The complete solution for the TCM system is carried out in three steps: (1) Developing a Monte Carlo relation between temperatures on the Io and IIo surfaces (2) finding the temperature at a set of nodal points on Io and IIo surfaces iteratively and (3) using the calculated temperature to predict the temperature distribution within the chip.

The Monte Carlo procedure allows the departure of many random walks from each required point on the IIo surface. The floating random walk method (Haji-Sheikh, 1988) is used while the random walk is in the piston. The probability that a random walk traverses the gap and goes to the chip or the hat is

$$p_g = \frac{\Delta r/R_c k_{al}}{1 + \Delta r/R_c k_{al}} \tag{9}$$

where $R_c = Z_g/k_g$, and Z_g is the gap distance between the piston and the hat or chip. Otherwise, the random walk moves a distance Δr along the normal toward the interior of the piston.

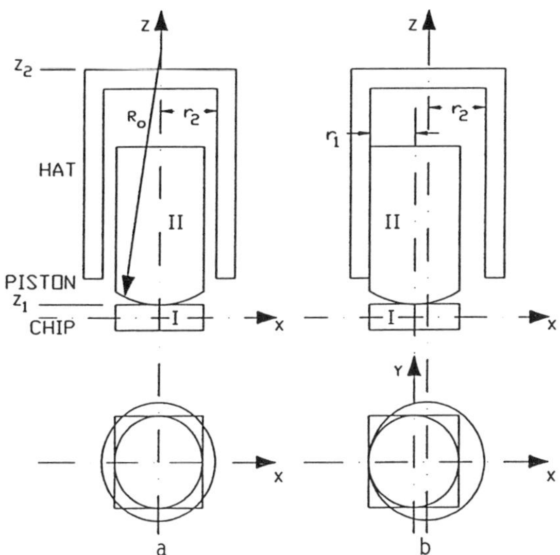

Figure 3. Schematic of piston/hat/chip arrangement; (a) eccentrically located piston and (b) concentrically located piston.

The numerical value of the probability given in Eq. (9) depends on the size of the gap and the thermal conductivity of the gas within the gap. The procedure produces a history of random walks to be stored for subsequent use. Once a history of random walks is available, there is no need to repeat the Monte Carlo procedure during the temperature calculation. The random walks wander through the piston until they either cross the gap and arrive at a point on the Io surface or terminate in the hat. For each random walk, there is a time $\Delta\tau^*$ associated with each step, and the sum of all $\Delta\tau^*$'s is the duration of that random walk, τ^*. After termination of a random walk, the duration of the walk (the travel time) and the coordinates of its termination point are recorded.

If a random walk terminates on the chip with an elapsed time of τ^*, the temperature of the chip at a time t-τ^* at the point of termination is to be tallied. The absorption from other surfaces such as the top surface of the piston is ignored because the probability that a random walk becomes absorbed at the top of the piston in the spring is negligibly small (~ 0.000005).

A computer program using the procedure (Haji-Sheikh and Lakshminarayanan, 1988) defines the coordinates and elapsed time at the point of termination of each random walk if it has arrived at the chip surface. Otherwise, the temperature of the hat or the initial temperature of the piston/hat is tallied. Initially, it is assumed that the piston/hat system is isothermal, hence the initial temperature is the same as that of the hat. It was observed that after a random walk steps on the hat's surface the probability of its return to the piston is negligibly small. In order to implement the probabilities just described, it is essential to have available sequences of random numbers uniformly distributed between 0 and 1 (Knuth, 1969).

The area on the chip covered by the piston is divided into 5 radial increments and 20 angular increments. Every nodal point on the upper surface of the chip is paired with a nodal point on the lower surface of the piston. Due to symmetry, only 56 of the 101 nodal points are used.

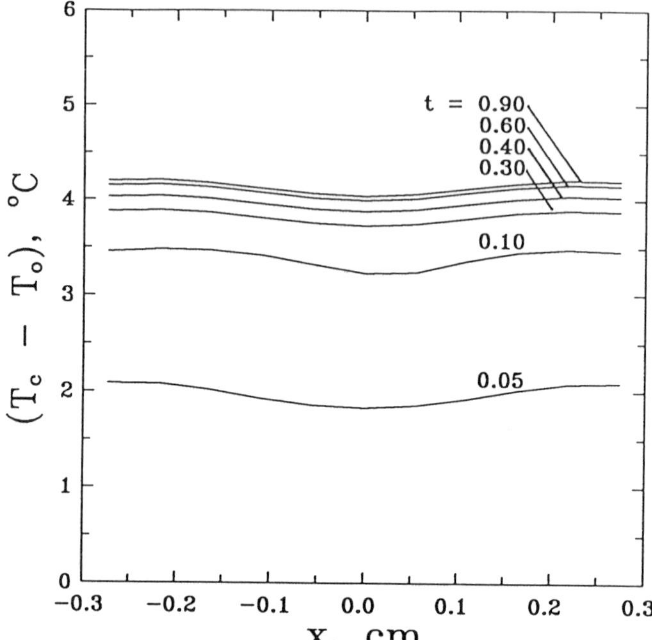

Figure 4. Temperature rise distribution on the contact surface of the chip along the x-axis for 1 Watt of energy released (eccentrically located piston).

Five hundred random walks were permitted to depart from each nodal point. The temperature at any nodal point at time t on the piston's lower surface after a release of N_w random walks is obtained from

$$T_{IIo}(x,y,z,t) = \frac{1}{N_w}\left(N_h T_h + N_o T_o + \sum_{j=1}^{N_c} T_{Io}(x_j,y_j,c,t\text{-}\tau^*)\right) \tag{10}$$

where N_h is the number of random walks that arrived at the hat within time t, N_o is the number of random walks remaining in the piston, and N_c is the number of random walks that arrived at the chip within time t. A random walk is said to remain within the piston when the time associated with that walk τ^* is larger than t. Otherwise, the random walk is terminated either at the chip site or within the hat. Since it is assumed that the module is initially at the temperature of the hat, $T_o = T_h$, then N_h and N_o can be lumped together.

Most of the 28,000 random walks (at 56 nodes) used were terminated within 0.6 second but 22 had a duration of longer than 0.6 second, seven longer than 0.7 second, three longer than 0.9 second, and two longer than 1 second. The maximum influence of the random walks with a duration of longer than 0.6 second on the temperature data in Fig 4. is estimated to be less than 0.015°C. Since the piston controls the transient duration of the module, the steady-state condition is achieved after about 0.6 second.

The computed values of the surface temperature rise along the x-axis, Fig 4., represent the expected temperatures at the nodal points for 1 Watt of energy released at the surface z = 0 for an eccentrically located piston. The maximum

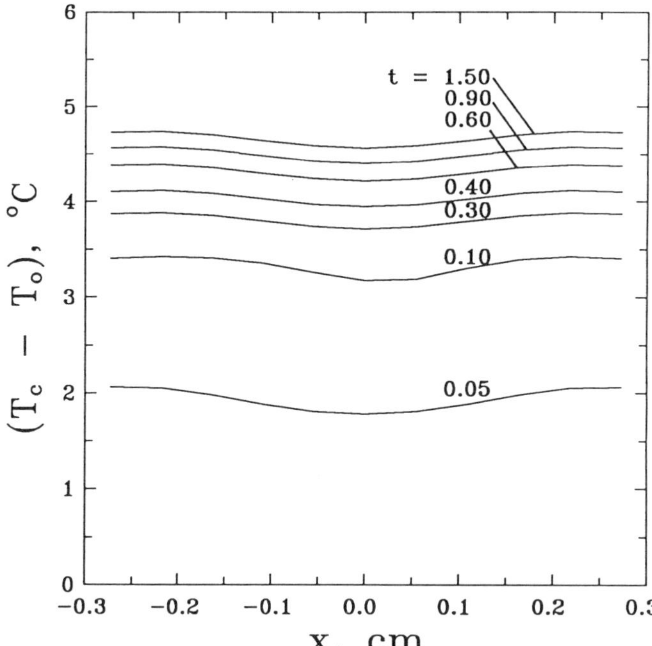

Figure 5. Temperature rise distribution on the contact surface of the chip along the x-axis for 1 Watt of energy released (concentrically located piston).

surface temperature on the $z = c$ surface appears along the diagonal, where a temperature rise of approximately 4.5°C is obtained for power dissipation of 1 Watt on the $z = 0$ surface. The maximum temperature within the chip is obviously greater and its magnitude depends on the size of the heating element.

Once the surface temperature is calculated, the maximum temperature at the heat source is 0.5° to 1°C/W higher than the surface temperature. The computed values are for the most favorable case when there is a line of contact between the piston and the hat. A tilted piston with two points of contact instead of a line of contact is expected to cause a higher chip temperature. For tilted pistons, the numerical simulation by Oktay and Kammerer (1982) indicates a slightly higher chip temperature. The results for a concentrically located piston show higher chip temperatures as indicated in Fig. 5.

The thermal conductivity and thermal diffusivity of the chip are assumed equal to that of pure silicon, 146 W/m-K and 0.883×10^{-4} m^2/s, respectively. The gas in the module is considered to be pure helium with k = of 0.165 W/m-K. For comparative purposes, the geometrical configurations closely resemble those reported by Chu et al. (1982). The dimensions of the chip used in the calculations are a=0.00273 m, b=0.00273 m, and c=0.0003 m and the heat generation inside the chip is assumed to be uniformly spread over a slab of dimensions d=0.0026 m, e=0.0026 m, and $f=0.175 \times 10^{-6}$ m, located centrally inside the chip.

The methods discussed in this section are equally applicable when the boundary conditions are nonlinear. For example, a minor modification permits inclusion of the radiation effect between the chip and the piston in the TCM.

NOMENCLATURE

a,b,c	half dimensions of chip in x, y, z directions, m
A	cross sectional area, m^2
C_p	specific heat, J/kg-K
d,e,f	half dimensions of heat generating region in x, y, z directions, m
f_l, f_m, f_n	constants in the Green's function, Eq. (2)
g	generation term, W/m^3
G	Green's function in series form, Eq. (2)
h	heat transfer coefficient, W/m^2-K
k	thermal conductivity, W/m-K
k_{al}	thermal conductivity of aluminum piston, W/m-K
k_g	thermal conductivity of fluid in gap, W/m-K
l,m,n	indices in Eq.(2)
n	normal to the surface
N_c	number of random walks arrived at chip surface
N_o	number of random walks remaining in the piston
N_w	number of random walks
P_+, P_-, P_g	probabilities of Eq.(5)
P	perimeter of cross section, m
q_w	surface heat flux
r	radial coordinate
r_1, r_2	see Fig. 3
R	region of integration in the chip
R_c	contact resistance of gap between piston and chip
s	region of integration on surface of chip
t	time, seconds
T	temperature, °C
T_h	hat temperature, °C
T_o	initial temperature of piston/hat/chip, °C
T_s	surface temperature of chip, °C
x,y,z	coordinates, m
x',y',z'	dummy variables
Z_g	gap distance, m
z_1, z_2	see Fig. 3

Greek

α	thermal diffusivity, m^2/sec
$\gamma_l, \lambda_m, \mu_n$	eigenvalues in the Green's function
τ	dummy time variable, seconds
τ^*	random walk duration
ρ	density, kg/m^3

Subscripts

c	chip
g	gas
h	hat
i,j	indices for nodal points on piston and chip
s	surface of region I
I	region I
Io	contact surface on region I
II	region II
IIo	contact surface on region II
0	node 0 in Example 1

REFERENCES

Agonafer, D., Chu, R. C., and Hwang, U. P., "Numerical Modeling of the Internal and External Thermal Resistance of a Thermal Conduction Module," in *Cooling Technology for Electronic Equipment*, Ed. W. Aung, Hemisphere Publishing Corp., Washington, D. C., 1988.

Blodgett, A. J. and Barbour, D. R., "Thermal Conduction Module: A High-Performance Multilayer Ceramic Package," IBM Journal of Research and Development, vol. 26, no. 1, pp. 30-36, 1982.

Chu, R. C., Hwang, U. P. and Simons, R. E., "Conduction Cooling for an LSI Package: A One-Dimensional Approach," IBM Journal of Research and Development, vol. 26, no. 1, pp. 45-54, 1982.

Haji-Sheikh, A. "Monte Carlo Methods," in *Handbook of Numerical Heat Transfer*, Eds. Minkowycz et al., Wiley, 1988.

Haji-Sheikh, A. and Lakshminarayanan, R. "A Multimethod Study of Thermal Conduction in Bodies in Contact," in *Cooling Technology for Electronic Equipment*, Ed. W. Aung, Hemisphere Publishing Corp., Washington, D. C., 1988.

Knuth, D. E., *The Art of Computer Programming: Seminumerical Algorithms.*, vol. 2, Addison-Wesley, Reading, Massachusetts, 1969.

Oktay, S. and Krammerer, H. C., "A Conduction Cooled Module for High-Performance LSI Devices," IBM Journal of Research and Development, vol. 26, no. 1, pp. 55-66, 1982.

Ozisik, M. N., *Heat Conduction*, Wiley, 1980.

SYSTEMS AND DYNAMICS

Summary

The dynamical systems encountered in various applied fields of science and technology are of a very diverse nature. Mathematical methods have been advanced under various assumptions, and significant progress has been achieved, both theoretically and practically.

The papers in this section deal with nonlinear feedback systems, for which the problem of predicting the autonomous behavior is considered, with analytical optimization methods (Wiener's optimum filter, Pontrjagin's maximum principle, and optimization in bounded state space, and a testing procedure for a parallel system.

The invited paper by D. P. Atherton, Professor of Control Engineering, University of Sussex, England discusses feedback systems; two basic methods (the describing function and Tsypkin's approach) are considered and briefly reviewed. Software based on these two approaches is presented. The case of chaotic motion is also dealt with, and several examples are detailed.

The paper by Chang on optimization methods analyzes a conjecture regarding two sets of conditions and the optimization operator. Several illustrations of the conjecture are presented and discussed (a generalization of Wiener's optimization in complex s-domain is one of these illustrations). It remains an open problem whether or not the conjecture is always valid.

The testing procedure considered in the third paper by Singh and Singh is thoroughly discussed as well as a modified version. Reliability of the system is investigated and a FORTRAN program with adequate characteristics is developed.

C. Corduneanu

Predicting the Autonomous Behavior of Simple Nonlinear Feedback Systems

D. P. ATHERTON
University of Sussex
Falmer, Brighton, BN1 9QT, UK

1.   INTRODUCTION

Although simple autonomous feedback systems containing a
single nonlinear element and a linear dynamic element have
been studied for many years, there are no general methods
available for accurately predicting their behaviour. Of
particular interest are the problems of stability and the
existence of limit cycles or other forms of continuous
motion. For control system type problems, where typical
nonlinear elements may occur as, or be best approximated by,
linear segmented characteristics, then methods requiring
nonlinear elements to have continuous derivatives are not
applicable. Two approaches which can be used are the approximate describing function method and Tsypkin's method, which
can be used for quantised type nonlinearities. In this paper
we briefly review both methods including the describing function approach for multiple inputs and the determination of
the orbital stability of predicted limit cycles. Software
based on these two approaches is described. The describing
function software allows iteration and higher harmonic balancing in order to try and obtain exact limit cycle solutions.
Section 4 discusses the possibility of chaotic motion in
these systems and recently proposed conjectures regarding
necessary conditions for the existence of chaos are presented. The paper concludes with a set of carefully chosen
examples which illustrate the scope and limitations of the
describing function, Tsypkin methods and simulation in determining the autonomous motion of several relay feedback
systems, which may exhibit simple limit cycles, combined mode
oscillations or chaotic motions.

2.   DESCRIBING FUNCTION APPROACH

The original describing function method was developed independently by several workers [1,2] in the 1940's who were
interested in the stability of nonlinear feedback systems
with block diagrams of the form shown in Fig. 1, where n(x)
is a nonlinear charactersistic and G(s) a linear transfer
function. It was observed that when a limit cycle existed in
such a system it was often almost sinusoidal at the input to
n(x), due to the low pass filtering provided by G(s).

It was therefore decided to replace the nonlinearity by a
complex gain equal to the ratio of the fundamental output
from the nonlinearity to the magnitude of an applied sinusoidal input. This complex gain was called the describing
function but since it is possible to extend the concept to
other inputs and also multiple inputs as discussed below it
is preferably called the sinusoidal input describing
function.

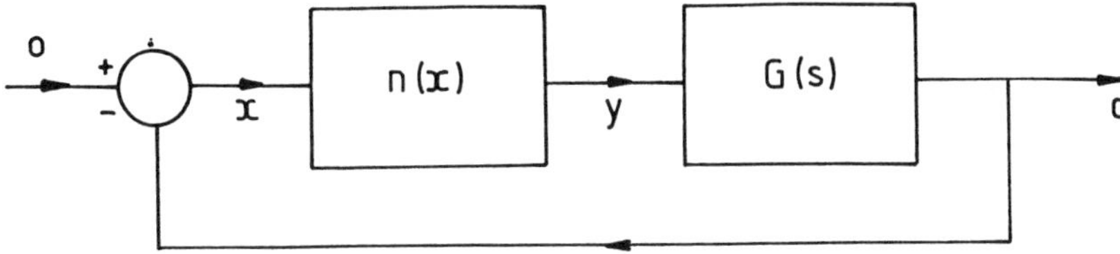

FIGURE 1   Block Diagram

2.1   Evaluation of the Describing Function

Taking the input to n(x) as $x(t) = a \cos \theta$, where $\theta = \omega t$,
then the fundamental in phase and quadrature components of
the output $y(\theta)$ are given by

$$a_1 = (1/\pi) \int_0^{2\pi} y(\theta) \cos \theta \, d\theta \qquad (1)$$

and

$$b_1 = (1/\pi) \int_0^{2\pi} y(\theta) \sin \theta \, d\theta \qquad (2)$$

The corresponding describing function, N(a) is given by

$$N(a) = (a_1 - jb_1)/a = N_p(a) + jN_q(a) \qquad (3)$$

and if n(x) is a single valued symmetrical odd nonlinearity,
N(a), becomes

$$N(a) = (4/a\pi) \int_0^{\pi/2} y(\theta) \cos \theta \, d\theta \qquad (4)$$

or, with a change of variable

$$N(a) = (4/a^2) \int_0^a x n(x) p(x) \, dx \qquad (5)$$

where $p(x) = 1/\pi (a^2 - x^2)^{1/2}$ is the amplitude probability
density function of a sinusoidal process. When the nonlinearity is a double valued symmetrical odd characteristic
such that for a sinusoidal input the characteristic $n_1(x)$ is
followed for $\dot{x} > 0$ and without discontinuity $n_2(x)$ for $\dot{x} < 0$,
then it is easily shown that:

$$a_1 = (4/a) \int_0^a x n_p(x) p(x) \, dx \qquad (6)$$

and

$$b_1 = (4/a\pi) \int_0^a n_q(x)\,dx \qquad (7)$$

where $n_p(x) = \{n_1(x) + n_2(x)\}/2$, $n_q(x) = \{n_2(x) - n_1(x)\}/2$

If we consider approximating a nonlinearity by the best
linear gain, that is the gain that minimises the mean squared
value of the error, e, between the output from the nonlinearity and the linear gain, K, then for a sinusoidal input to
a single valued nonlinearity $K = N(a)$. This is why the describing function approach is often known as quasilinearisation and this definition can, of course, be used for any
form of input. Working in terms of the nonlinear characteristic has several advantages, for example, it is easily
seen from Eqn.(7) that the quadrature component depends on
the area, or hysteresis, of the nonlinearity and it can be
easily shown that if a single valued odd symmetrical $n(x)$
lies in a sector such that $k_1 x < n(x) < k_2 x$ for $0 < x < a$
then $k_1 < N(a) < k_2$ for any a. Solutions for possible limit
cycles in the feedback system of Fig. 1. are then found from

$$1 + N(a)G(s)|_{s\,=\,j\omega} = 0 \qquad (8)$$

2.2   Multiple Input Describing Functions

Investigation of various facets of the behaviour of the
feedback loop of Fig. 1 may require that the nonlinearity
input be considered as a sum of several signals. In this
case, the quasilinearisation formulation has as many gains as
unrelated input signal components. Important two input cases
are briefly considered in this section.

<u>Sine plus bias describing function.</u>   When the input $x = \gamma + a\cos\theta$, a bias plus sinusoid then the two gains are, the
gain to the sinusoid

$$N(a,\gamma) = (2/a^2) \int_{-a}^a xn(x+\gamma)p(x)\,dx \qquad (9)$$

and the gain to the bias

$$N_\gamma(a,\gamma) = (1/\gamma) \int_{-a}^a n(x+\gamma)p(x)\,dx \qquad (10)$$

The limit of the bias gain as γ tends to zero is known as the
d.c. incremental gain and is given by

$$N_{i\gamma}(a) = \lim_{\gamma\to 0} N_\gamma(a,\gamma) = \int_{-a}^a n'(x)p(x)\,dx \qquad (11)$$

where $n'(x) = dn(x)/dx$. Integrating Eqn.(11) by parts and
using Eqn. (5) gives

$$N_{i\gamma}(a) = N(a) + (a/2)(dN(a)/da) \qquad (12)$$

Solutions for any limit cycles which are approximately
sinusoidal with bias in the feedback loop of Fig. 1 are then
obtained from the two equations

$$1 + N(a,\gamma)G(s)\big|_{s=j\omega} = 0 \tag{13}$$

and

$$1 + N_\gamma(a,\gamma)G(0) = 0 \tag{14}$$

These equations yield the three unknowns, γ, a and ω_o, the
angular frequency of the limit cycle.

<u>Unrelated sinusoids.</u>   For two sinusoids, $x_1 = a_1\cos\omega_1 t$ and
$x_2 = a_2\cos(\omega_2 t+\theta)$, with ω_1 and ω_2 incommensurate frequencies
it is easily shown that the two gains are given by

$$N_{a_1}(a_1,a_2) = (2/a_1^2)\int_{-a_2}^{a_2}\int_{-a_1}^{a_1} x_1 n(x_1+x_2)p_1(x_1)p_2(x_2)dx_1 dx_2 \tag{15}$$

and

$$N_{a_2}(a_1,a_2) = (2/a_2^2)\int_{-a_2}^{a_2}\int_{-a_1}^{a_1} x_2 n(x_1+x_2)p_1(x_1)p_2(x_2)dx_1 dx_2 \tag{16}$$

where $p_1(x_1)$ and $p_2(x_2)$ are the probability densities of x_1
and x_2.   If a feedback loop has more than one describing
function predicted limit cycle then it may exhibit a combined
mode, that is an oscillation containing two predominant
frequencies.   When this is the case the two equations

$$1 + N_{a_1}(a_1,a_2)G(j\omega_1) = 0 \tag{17}$$

and

$$1 + N_{a_2}(a_1,a_2)G(j\omega_2) = 0 \tag{18}$$

must be satisfied.

<u>Related sinusoids.</u>   The situation of x being a sum of harmon-
ically related sinusoids is obviously of considerable
interest since this will be the exact form of a simple limit
cycle.   Unfortunately, the analytical solution for the required gains becomes intractable for most nonlinear characteristics $n(x)$ even when only two harmonic components are
considered.   For the simple case of $n(x) = x^3$ and $x = a\cos$
$\omega t + b\cos(n\omega t + \theta)$ then the two gains, which are complex,
can be calculated but each one is a function of the four
parameters a, b, n and θ.   We return to this problem in
section 2.4 when we discuss a computational approach.

463

2.3 Stability of Limit Cycle Solutions

If a limit cycle is predicted by Eqn. (8) it will only exist in practice if it is stable. The limit cycle is stable if the variational equation

$$q(D)\Delta x(t) + p(D)n'\{x^*(t)\}\Delta x(t) \quad = \quad 0 \tag{19}$$

is stable, where $G(s) = p(s)/q(s)$ and D denotes the differential operator. This equation has periodically time-varying coefficients, and there are two problems in determining its stability. First, only the sinusoidal describing function approximation to the solution $x^*(t)$ is available, and secondly only a sufficient condition for instability is known. This is if the time-average equation

$$q(D)\Delta x(t) + p(D)\overline{n'\{x^*(t)\}}\Delta x(t) \quad = \quad 0 \tag{20}$$

is unstable. Using the describing function solution this equation is easily shown to be equivalent to

$$1 + N_{i\gamma}(a)G(s) \quad = \quad 0 \tag{21}$$

which corresponds to checking the stability of a small perturbation, unrelated to the limit cycle, injected into the loop.

If a small signal of the same frequency as the sinusoidal limit cycle and with relative phase β is injected into the loop at the nonlinearity input, then it can be shown that the gain through the nonlinearity to this small synchronous perturbation, δ, is

$$N_{i\delta}(a) \quad = \quad N(a) + (a/2)(dN(a)/da)(1+\exp(-2j\beta)). \tag{22}$$

Experimental results [3] indicate that within the limits of describing function accuracy a limit cycle will be stable if both Eqn.(21) and

$$1 + N_{i\delta}(a)G(s) \quad = \quad 0 \tag{23}$$

have no roots with positive real parts. These incremental describing function results are readily extendable to two unrelated sinusoids.

2.4 A computational approach

The problem of related sinusoidal inputs discussed above is relatively easily solved computationally. The nonlinearity input waveform can be represented by 2^N discrete data points and the corresponding nonlinearity output data points computed. Using FFT techniques the frequency content of this waveform can be computed and the output of G(s) found from a summation of the response to the individual harmonic terms. The choice of N is obviously a compromise between speed of computation and the avoiding of aliaising errors.

For limit cycle computation the software written [4] allows
iteration of signals around the loop at a fixed frequency and
harmonic balancing of several harmonics. Its use is
described in later examples.

3.   THE TSYPKIN METHOD

Tsypkin's method [5, 6] can be used to determine limit cycles
in the system of Fig. 1 when the nonlinearity, $n(x)$, is of
the relay type. The method has two major advantages in that
the limit cycle waveform and its stability can be determined
exactly. The starting point of the method is to assume a
periodic waveform at the relay output, which typically for a
relay with dead zone, shown in Fig. 2(a), to be considered
here, will have the form shown in Fig. 2(b). Waveforms of
this type can be considered as a summation of several
periodic pulse trains of the form of $y_i(t)$ shown in Fig.
2(c). It is straightforward to show that $y_i(t)$ can be
expressed as the Fourier series

$$y_i(t) = (h\Delta t_i/T) + (h/\pi) \sum_{n=1}^{\infty} (1/n)\{\sin n\omega\Delta t_i \cos n\omega(t-t_i)$$

$$+ (1-\cos n\omega\Delta t_i\} \sin n\omega(t-t_i)\} \tag{24}$$

If $y_i(t)$ is the input to $G(s)$ which for ease of presentation
we assume is such that $\lim_{s\to\infty} sG(s) = 0$ then the output $c_i(t)$
of $G(s)$ is given by

$$c_i(t) = (h\Delta t_i G(0)/T) + (h/\pi) \sum_{n=1}^{\infty} (g_n/n)\{\sin n\omega\Delta t_i$$

$$\cos(n\omega t - n\omega t_i + \phi_n) + (1-\cos n\omega\Delta t_i)$$

$$\sin(n\omega t - n\omega t_i + \phi_n)\} \tag{25}$$

and its derivative by

$$\dot{c}_i(t) = (\omega h/\pi) \sum_{n=1}^{\infty} g_n\{-\sin n\omega\Delta t_i \sin(n\omega t - n\omega t_i + \phi_n)$$

$$+ (1-\cos n\omega\Delta t_i)\cos(n\omega t - n\omega t_i + \phi_n)\} \tag{26}$$

where

$$G(jn\omega) = g_n \exp(j\phi_n) = U_G(n\omega) + jV_G(n\omega) \tag{27}$$

If we define an A locus of a transfer function $G(s)$ by

$$A_G(\theta,\omega) = \text{Re } A_G(\theta,\omega) + j \text{ Im } A_G(\theta,\omega) \tag{28}$$

with

$$\text{Re } A_G(\theta,\omega) \quad = \quad \sum_{n=1}^{\infty} \quad V_G(n\omega)\sin n\theta \; + \; U_G(n\omega)\cos n\theta \qquad (29)$$

and

$$\text{Im } A_G(\theta,\omega) \quad = \quad \sum_{n=1}^{\infty} \quad (1/n)\{V_G(n\omega)\cos n\theta \; - \; U_G(n\omega)\sin n\theta\} \qquad (30)$$

then $c_i(t)$ is expressible in terms of $\text{Im } A_G$ and $\dot{c}_i(t)$ in terms of $\text{Re } A_G$, where Re and Im denote real part and imaginary part, respectively.

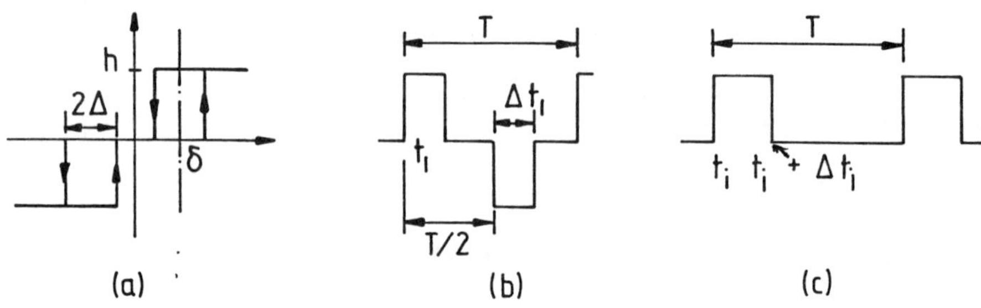

FIGURE 2 Relay with deadzone and waveforms

For the waveform of Fig. 2(b)

$$y(t) \quad = \quad \sum_{i=1}^{2} \quad y_i(t) \quad \text{with } y_2(t) \quad = \quad -y_1(t+0.5T) \qquad (31)$$

and the resultant output of $G(s)$, $c(t)$ is given by

$$c(t) \quad = \quad (2h/\pi)\{\text{Im } A_G^o(-\omega t+\omega t_1,\omega)-\text{Im } A_G^o(-\omega t+\omega t_1+\omega\Delta t_1,\omega)\} \qquad (32)$$

and

$$\dot{c}(t) \quad = \quad (2\omega h/\pi)\{\text{Re } A_G^o(-\omega t+\omega t_1,\omega)-\text{Re } A_G^o(-\omega t+\omega t_1+\omega\Delta t_1,\omega)\} \qquad (33)$$

where t_1 is the instant at which $y_1(t)$ switches positive, Δt_1 is the pulse duration, and A^o denotes the A locus with odd values of n only in Eqns. (29) and (30). Since the relay input $x(t) = -c(t)$, necessary conditions for the limit cycle to exist are

$$x(t_1) \quad = \quad -c(t_1) \quad = \quad \delta + \Delta$$

$$\dot{x}(t_1) \quad = \quad -\dot{c}(t_1) \; > \; 0 \qquad (34)$$

and

$$x(t_1 + \Delta t_1) \;=\; -c(t_1 + \Delta t_1) \;=\; \delta - \Delta$$

$$\dot{x}(t_1 + \Delta t_1) = \dot{c}(t_1 + \Delta t_1) < 0 \tag{35}$$

These conditions which can be written taking $t_1 = 0$ and $\Delta t_1 = \Delta t$ without loss of generality, become

$$A_G^o(0,\omega) \;-\; A_G^o(\omega\Delta t,\omega) \quad \text{must have}$$

$$\begin{aligned}
\text{I.P.} \;&=\; -\pi(\delta + \Delta)/2h \\
\text{R.P.} \;&<\; 0
\end{aligned} \tag{36}$$

and

$$A_G^o(0,\omega) \;-\; A_G^o(-\omega\Delta t,\omega) \quad \text{must have}$$

$$\begin{aligned}
\text{I.P.} \;&=\; \pi(\delta - \Delta)/2h \\
\text{R.P.} \;&<\; 0
\end{aligned} \tag{37}$$

where R.P. and I.P. denote real part and imaginary part, respectively.

Since closed form expressions for the A_G loci can be found for any given transfer function $G(s)$, necessary conditions for the assumed limit cycle are positive solutions for ω and Δt_1 for Eqns. (36) and (37) and satisfaction of the inequalities.

Sufficiency can be determined once a solution has been found by checking that the waveform $x(t)$ does not pass through the relay switching levels such as to give additional switchings to those assumed. This is easily done computationally since with a graphical display the solution waveform $x(t)$ equal to $-c(t)$ given by Eqn. (32) can be shown.

3.1   Stability of a Limit Cycle Solution

The stability of a limit cycle solution may be checked approximately using Eqn. (20) with the exact relay incremental gain, N_i, which can be shown to be given by

$$N_i \;=\; (h\omega/\pi)\left[\,[\dot{x}(0)]^{-1} - [\dot{x}(\Delta t)]^{-1}\,\right] \tag{38}$$

However, if the plant transfer function is written in state space form (A, B, C) then the A matrix of the periodically time-varying variational equation can be shown to be $A+Bn'\{x^*(t)\}C$, where $x^*(t)$ is the limit cycle solution [7]. This matrix is piecewise constant being equal to A except over the infinitesimal time intervals it takes the relay to switch. It is known that [8] such an equation has assymptotic orbital stability if, and only if, the eigenvalues of the matrix

$$Q = \sum_{i=1}^{m} \exp A_i \delta t_i \tag{39}$$

have magnitude less than unity, apart from one which will have unit magnitude; where A_i are the constant values of the A matrix in the time intervals δt_i and

$$\sum_{i=1}^{m} \delta t_i = T. \tag{40}$$

For the case of the relay output waveform of Fig. 2(b) the matrix, Q, because of symmetry may be taken over $T/2$, and becomes

$$Q = \exp[A(T/2-\Delta t)]\exp[hBC/|\dot{x}(\Delta t)|]\exp[A\Delta t]\exp[hBC/|\dot{x}(0)|] \tag{41}$$

3.2   Computational Procedures and Extensions

Software for evaluating limit cycles using the Tsypkin method has been written [9], and requires as input the relay characteristic, the transfer function G(s) and guesses for the limit cycle solution parameters. If the relay has no dead zone only one nonlinear algebraic equation is solved for the limit cycle frequency, in contrast to the relay with dead zone where there are two unknowns the limit cycle frequency and the pulse width. To obtain solutions for asymmetrical oscillations the number of equations and unknowns doubles. The known solutions for the A_G loci are stored for different elementary transfer functions in terms of which any given transfer function can be expressed.

If it is known that a limit cycle has several switchings per period then it is in principle possible to solve for it using the Tsypkin approach. The difficulty, however, is in providing reasonable estimates of the limit cycle parameters for the increased number of nonlinear algebraic equations. Similarly if the relay is replaced by a quantization characteristic more nonlinear algebraic equations are involved. Solutions can also be obtained if the limit cycle can be approximated by a linear segmented waveform.

4.   CHAOTIC MOTION

Although chaotic motion can exist in feedback systems of the form of Fig. 1, no exact methods are available for predicting its existence and in particular for what range of system parameters the motion will exist. In this section we discuss some results which enable both the Tsypkin and describing function methods to shed some light on this difficult problem. Ogorzalek conjectured [10] that for a system with $G(s) = -\beta_o/(s^n + \alpha_{n-1}s^{n-1} + \ldots \alpha_o)$ and nonlinearity, n(x) chaotic motion was likely to exist if:-

(i)   The Hurwitz section of the linear system with n(x) replaced by K satisfies $K_2 < K < K_1$, where $K_1 = \alpha_0/-\beta_0$ and K_2 is finite.

(ii)   The nonlinearity, n(x), intersects the line through the origin of slope K_1 at least once to ensure the existence of an equilibrium point.

(iii)   The slope of n(x) at all the intersection points with the lines of slope K_1 and K_2 should be either greater than the upper bound K_1 or less than the lower bound K_2, thus ensuring the instability of the equilibria.

More recently after some studies using the describing function method, Genesio and Tesi [11] conjectured that the system of Fig. 1 might be expected to exhibit chaos if:

(iv)   The application of the describing function method to the system gives a stable predicted limit cycle, x*(t).

(v)   There exists an equilibrium point, E, different from the one that generates the predicted limit cycle, which has one eigenvalue in the open right half plane and the others in the left half plane.

(vi)   The abscissa x_E belongs to the range of the predicted limit cycle, i.e. x*(t) = x_E for some time t.

(vii)   As the filtering properties of G(s) deteriorate the motion may be expected to change from a limit cycle to chaotic motion.

In section 5.4, an example is considered which supports these conjectures.

5.   SOME EXAMPLES

Several examples of relay feedback systems are considered in this section to illustrate difficulties in determining the behaviour of simple nonlinear feedback systems and the need for a variety of approaches for investigating these systems.

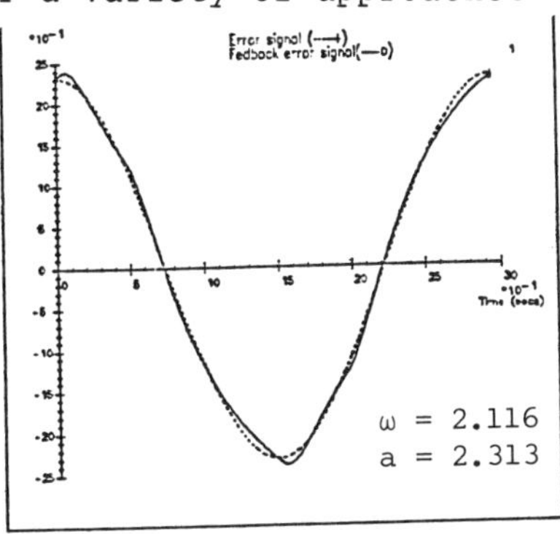

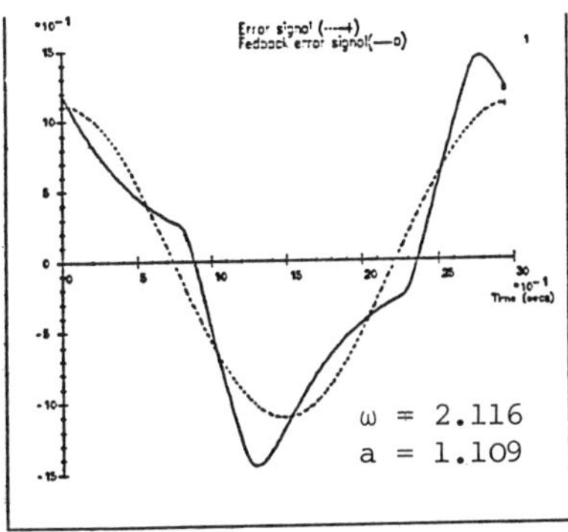

(a)   Stable solution   (b) Unstable solution

FIGURE 3 Describing function solution for Example 1

469

5.1   Example 1

The nonlinearity is taken as a relay with dead zone with h =
δ = 1 and Δ = 0 and the linear plant transfer function is 3e⁻
^{s}/(1+0.5s)(1+0.1s).   The describing function method predicts
two limit cycles, one stable with $\omega\Delta t > 90°$ and the other
unstable with $\omega\Delta t < 90°$.   The solutions obtained using the
software are shown in Fig. 3.   Iterating the stable solution
and then balancing on four harmonics results in the exact
limit cycle of Fig. 4(a) whereas convergence could not be
obtained for the unstable limit cycle, and the nearest 'fit'
is shown in Fig. 4(b).   No difficulty was encountered finding
these limit cycles by the Tsypkin method and they are shown
in Fig 5.   Unless the precise form of the unstable limit
cycle is known, it is also not normally possible to locate it
by simulation.

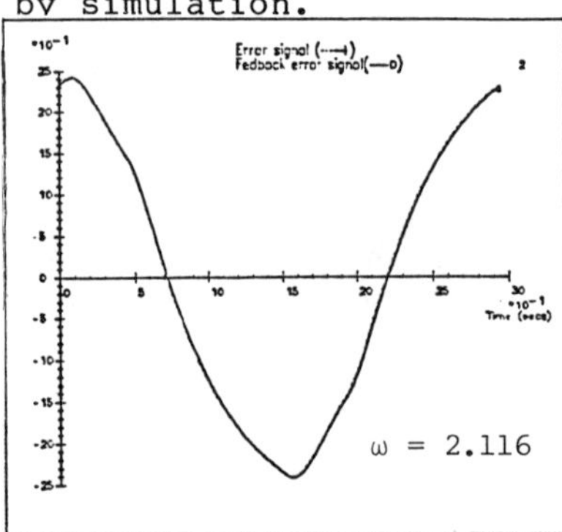

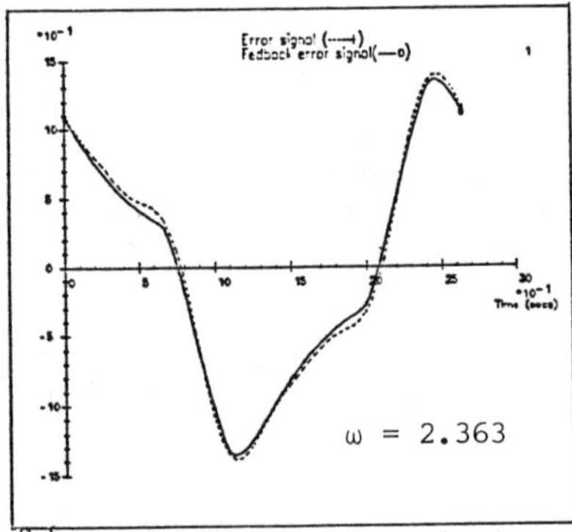

(a) Iterated stable solutions   (b) Iterated unstable solution
FIGURE 4 Improved results for Example 1

Even when exact initial conditions on the limit cycle, as
found from the Tsypkin solution, are used in a simulation,
'tracking' is usually only possible for a few cycles provided
the unstable root is not much greater than unity.

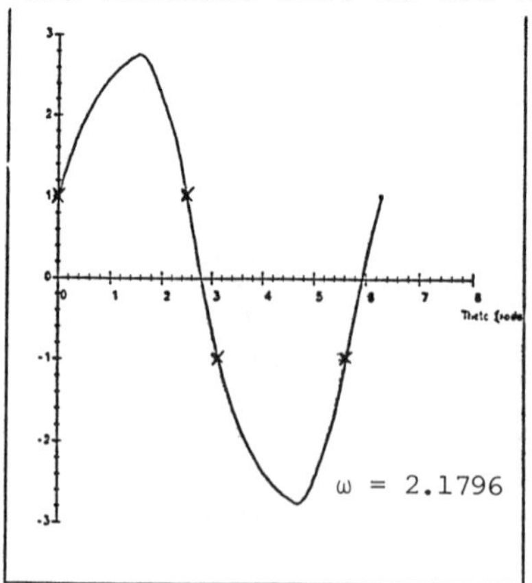

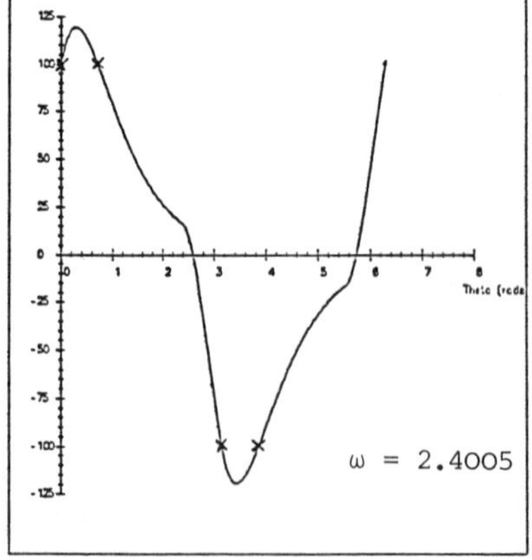

(a) Stable Limit cycle   (b) Unstable limit cycle

FIGURE 5 Tsypkin solutions for Example 1

5.2   Example 2

Sliding motion is known to occur in relay feedback systems
and is in fact used in the technique of variable structure
control system design. When sliding takes place the relay
switches rapidly between on and off, or vice versa, with the
relay input at the switching level and the relay output,
known as the equivalent control, having an average value over
the sliding regime whose magnitude can be determined
theoretically. Often the equivalent control can be approximated by one or more linear segments and since it is a
relatively straightforward extension to the Tsypkin method to
accommodate periodic pulse trains of the form of Fig. 2(c),
but with sloping rather than flat tops, a feature for
investigating limit cycles with sliding has been added to the
software [9]. Here we consider a feedback system with a
relay having h = δ = 1 and Δ = 0, and G(s) = 8(s−0.5)/(s^2−
0.5s+1). The Tsypkin program yields the invalid solution,
with ω = 0.534 shown in Fig. 6(a). Although the solution
passes through the switching levels at appropriate times it
is invalid since after the switching at A the input should
remain above the line AC. This reversal after switching
appears to be a good indicator for the possible presence of
sliding and can be used to try and determine a sliding limit
cycle, by assuming the relay will have a linear output over
an interval AB. Further iteration to approximate the
equivalent control by two linear segments gives the excellent
approximation for the sliding limit cycle, with ω = 0.386,
shown in Fig. 6(b).

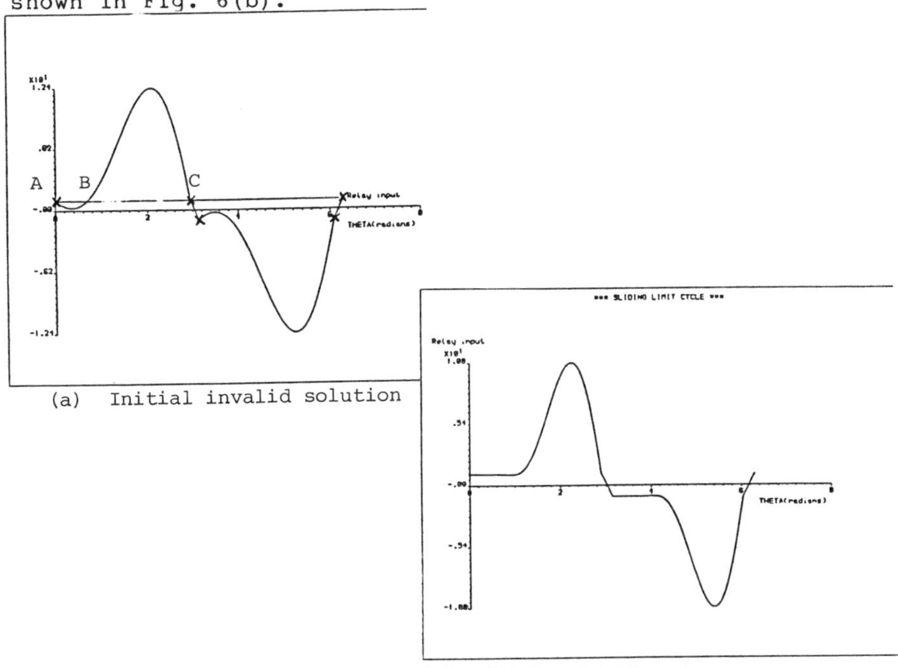

(a)   Initial invalid solution

(b)   Sliding limit cycle

FIGURE 6 Diagrams for Example 2

5.3   Example 3

Here we consider a situation where the describing function method indicates two possible limit cycles by considering a system with an ideal relay, that is h = 1 and $\delta = \Delta = 0$, and $G(s) = 12(s+1)^2/s^3(s^2 + 0.3\lambda s + \lambda^2)$. The Nyquist diagram of G(s) is shown in Fig. 7(a) for $\lambda = 4.0$.

Decreasing (increasing) λ decreases (increases) the ratio, $\alpha = |G(j\omega_1)|/|G(j\omega_2)|$ where ω_1 and ω_2 are the smaller and larger frequencies, respectively, for which $G(j\omega) = 180°$. Using the single sinusoidal describing function method and Eqn. (21) to test the stability one obtains the result that the limit cycle of frequency ω_1 is always unstable and that at ω_2 stable if $\alpha > 2$. The two sinusoidal input theory [3] shows that a combined oscillation is possible for $\alpha < 2$ for $a_1/a_2 < 0.91$. $\alpha = 2$ corresponds to $\lambda = 4.05$.

The Tsypkin method gave a stable limit cycle at ω_2 and an unstable one at ω_1 for $\lambda > 3.73$. The limit cycle at ω_2 became unstable for $\lambda < 3.73$ but remained almost sinusoidal. The unstable limit cycle at ω_1 for $\lambda = 3.73$, which has appreciable distortion, is shown in Fig. 7(b) and no solution could be found for this limit cycle when λ was further decreased. Digital simulations confirmed these results to the expected accuracy although the combined mode began to appear for λ somewhat greater than 3.73.

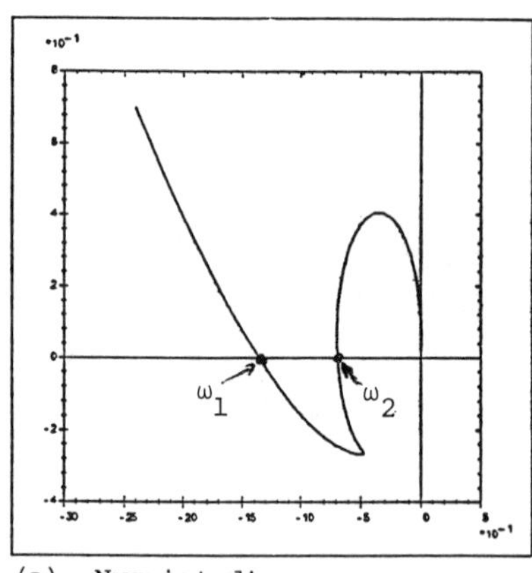

(a)   Nyquist diagram

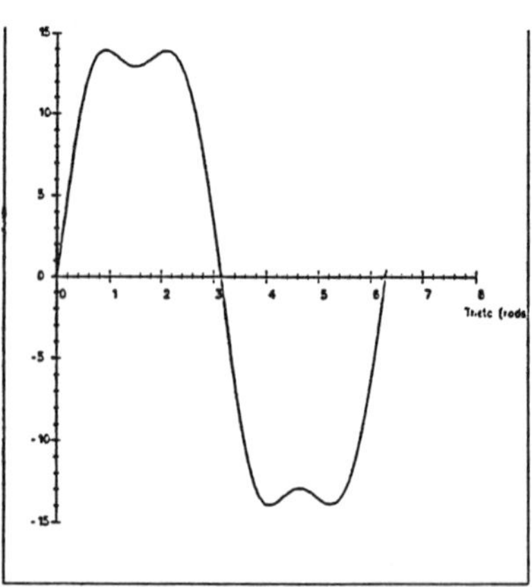

(b) Limit cycle

FIGURE 7 Diagrams for Example 3

5.4   Example 4

A feedback loop with n(x) an ideal relay and $G(s) = -1/(s^3 + \alpha_2 s^2 + \alpha_1 s + \alpha_0)$ is considered [12, 13]. The upper and lower bounds of the Hurwitz sector K_1 and K_2 are respectively α_0 and $\alpha_0 - \alpha_1\alpha_2$. Since the slope of n(x) is either zero or infinite the Ogorzalek conjecture is satisfied for

$$\alpha_0 > \alpha_0 - \alpha_1\alpha_2 > 0 \qquad\qquad\qquad [42]$$

Using the sine plus bias describing function it is also shown
in reference [13] that a limit cycle can exist about the
equilibrium points $x = \pm 1/\alpha_0$ which have two unstable
eigenvalues, that the equilibrium point at the origin has one
unstable eigenvalue, and that the limit cycle magnitude
condition (vi) in section 4 is met, again provided the conditions of Eqn. (40) are satisfied. An investigation using
the Tsypkin method [12] was undertaken for $\alpha_0 = 4$, $\alpha_1 = 1.25$
and $0 < \alpha_2 < 3.2$. With $2.8 < \alpha_2 < 3.2$ several unstable limit
cycles were found, one essentially sinusoidal with amplitude
decreasing as α_2 decreased, and others of smaller amplitude
having the form shown in Fig. 8(a), and named the spiral
type, with different numbers of oscillations per period.
Simulations of several systems with similar characteristics
revealed chaotic motion provided the amplitude of the
approximately sinusoidal limit cycle exceeded the amplitudes
of several unstable spiral limit cycles. For $\alpha_2 = 2.8$ the
spiral limit cycle found by the Tsypkin method and shown in
Fig. 8(a) became stable and for $\alpha_2 < 2.5$ a limit cycle with
two oscillations per half period became stable. The system
response goes unbounded outside the approximately sinusoidal
limit cycle and Fig. 8(b) shows a response from just inside
this limit cycle for $\alpha_2 = 3.0$ which results in chaotic
motion. It is clearly seen how this motion consists of jumps
between different unstable spiral limit cycles. Fig. 9 shows
the results of a simulation study for this system [13] for α_0
$= 4$ and varying α_2 and α_1.

It can be shown that the describing function predicted limit
cycle frequency is $\alpha_1^{1/2}$ and that G(s) has improved filtering
properties as α_1 is increased, hence the stable asymmetric
limit cycle for large α_1.

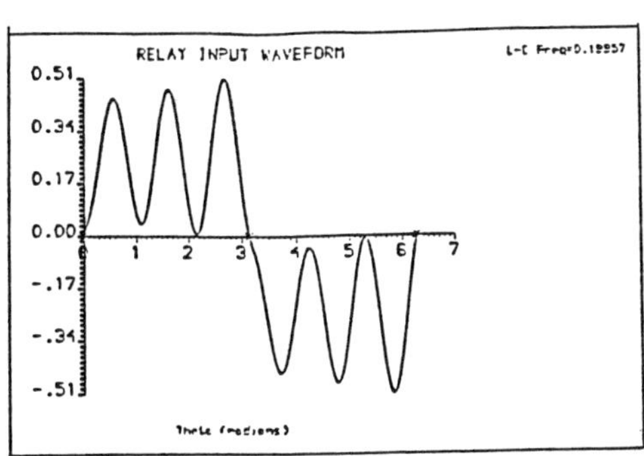

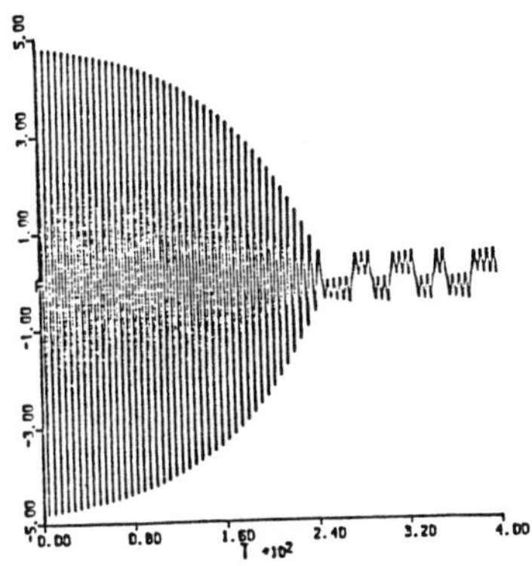

(a)   Stable limit cycle for $\alpha_2 = 2.8$ (b)   Response for $\alpha_2 = 3.0$

FIGURE 8 Responses for Example 4

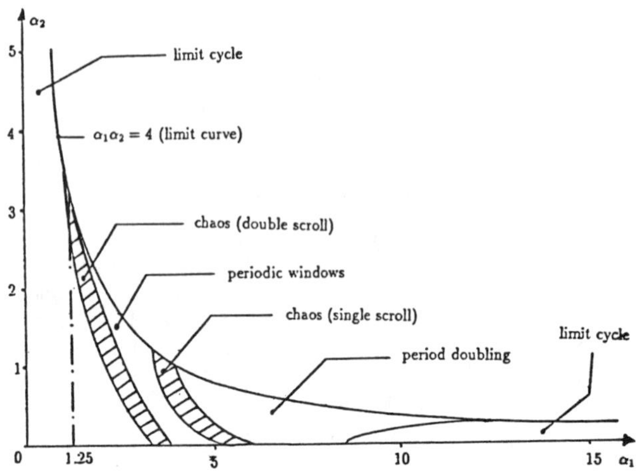

FIGURE 9 Simulation results for Example 4 [13]

6.   CONCLUSIONS

In this paper two methods have been discussed which can be
used to investigate the autonomous behaviour of simple nonlinear feedback systems. Both methods have been shown to
reveal far more information about system behaviour, than
might be expected from a cursory examination of their applicability, when implemented in software routines with
graphical output.

Many open problems still remain, for example, the stability
of the computational procedure in iterating from an
approximate describing function solution; necessary and
sufficient conditions for orbital stability; orbital
stability conditions for the Tsypkin method when the
derivative of the relay input is discontinuous at the switching instants, and the need to be able to obtain tighter
conditions for regions of chaotic motion. Although the describing function and Tsypkin methods can be extended to continuous multivariable systems and digital control systems
[14] it has not been possible to discuss these approaches
here. Needless to say, similar difficulties exist and in
many instances additional problems arise.

7.   REFERENCES

1.   Gelb, A and Vander Velde, W.E. 'Multiple-input Des-
      cribing Functions and Nonlinear System Design', pp. 49-
      211, McGraw-Hill, New York, 1968.
2.   Atherton, D. P., 'Nonlinear Control Engineering:
      Describing Function Analysis and Design', pp. 75-156,
      Van Nostrand Reinhold, London, 1975.

3.   Choudhury, S. K. and Atherton, D. P. 'Limit Cycles in
     High Order Nonlinear Systems', Proc. I.E.E., vol. 121,
     no. 7, pp. 717-724, 1974.
4.   McNamara, O.P. 'Computer Aided Design of Nonlinear Con-
     trol Systems using Describing Function Based Methods',
     D.Phil Thesis, University of Sussex, England 1987.
5.   Tsypkin, Ya. Z. 'Relay Control Systems', Cambridge
     University Press, Cambridge, England, 1984.
6.   Atherton, D. P. 'Oscillations in Relay Systems',
     Trans.Inst. M.C., vol. 3, no. 4 pp. 171-184, 1982.
7.   Balasubramanian, R., Stability of Limit Cycles in
     Feedback Systems Containing a Relay', I.E.E. Proc.D.,
     vol 128, no. 1, pp. 24-29, 1981.
8.   Willems, J.L. 'Stability Theory of Dynamical Systems'
     pp. 112-113, Nelson, London, 1970.
9.   Wadey, M.D 'Extensions of Tsypkins' Method for Computer
     Aided Control System Design', DPhil Thesis, University
     of Sussex, England, 1984.
10. Ogorzalek, M.J. 'Some Observations on Chaotic Motion in
     Single Loop Feedback Systems' Proc. 25th CDC, Athens,
     vol. 1, pp. 588-589, 1986.
11. Genesio, R. and Tesi, A., 'Chaos Prediction in Nonlinear
     Feedback Systems', Report 18/89, Dipartimento di Sistemi
     e Informatica, Universita di Firenze, 1989.
12. Amrani, D. and Atherton, D. P., 'Designing Autonomous
     Relay Systems with Chaotic Motion', Proc. 28th CDC,
     Tampa, vol. 1 pp. 512-517, 1989.
13. Genesio, R. & Tesi, A. 'Chaos Prediction in a Third
     Order Relay System' Mem. GT-01-90, Dipartimento di
     Sistemi e Informatica, Universita di Firenze, 1990.
14. Atherton, D. P., McNamara, O. P., Wadey, M.D., and
     Goucem, A., 'SUNS, The Sussex University Nonlinear
     Control System Software' Proc. 3rd IFAC/IFIP CADCE
     Symposium, Copenhagen, pp. 173-178, 1985.

A Global View of Analytical Optimization Methods

SHELDON S. L. CHANG
Department of Electrical Engineering
State University of New York
Stony Brook, New York 11794, USA

ABSTRACT

Recent developments in analytical optimization methods can be
summarized as the study of changes in the necessary condition for
optimality in response to changes in constraint. Strengthened
constraint brings about relaxed necessary condition and vice versa.
While the necessary condition and constraint taken together defines
uniquely an optimum solution in many applications, the uniqueness
can be proved only for a smaller class of problems. The following
methods are studied from this point of view: 1. Wiener's optimum
filter, 2. Pontryagin's maximum principle, 3. Optimization in bounded
state space.

This work was supported by the National Science Foundation under
Grant DMC-8702465

I. A CONJECTURE IN MATHEMATICAL OPTIMIZATION

<u>Definitions</u> (i) Let C denote a set of conditions on function space S, and f denote a function in S. If f satisfies all the conditions in C, we write

$$C(f) = 1 \tag{1}$$

otherwise

$$C(f) = 0 \tag{2}$$

(ii) Let C_a and C_b denote two sets of conditions. If, for every f in S, $C_a(f) = 1$ implies $C_b(f) = 1$, we write

$$C_a \supset C_b \tag{3}$$

If the inverse is also true, we write

$$C_a = C_b \tag{4}$$

(iii) If, for every f in S, the condition that both $C_a(f) = 1$ and $C_b(f) = 1$ implies $C_u(f) = 1$, and the inverse is also true, we write

$$C_u = C_a \cup C_b \tag{5}$$

<u>The Optimization Operator P</u>

Optimization is a selection operation. Let $P[C]$ denote the selection among all f's satisfying $C(f) = 1$, a set E of functions which maximize a value criterion defined by P. The solution to $P[C]$ is said to be unique if E has one and only one member. Mathematical optimization is to generate from $P[C]$ a set of optimizing conditions C_0 such that

$$P[C] \supseteq C \cup C_0 \tag{6}$$

C_0 is necessary if $\supseteq$ is valid, necessary and sufficient if $=$ is valid in (6).

<u>A Conjecture</u>

Let C_1 and C_2 be two sets of conditions on function space S, and P be an optimization operator on S. Let C_{01} and C_{02} denote the respective optimizing conditions:

$$P[C_1] \supseteq C_1 \cup C_{01}$$

$$P[C_2] \supseteq C_2 \cup C_{02}$$

If $C_1 \supset C_2$, then $C_{02} \supset C_{01}$

The above conjecture will be illustrated by three examples in the
subsequent sections.

II. A GENERALIZATION OF WIENER OPTIMIZATION IN COMPLEX s-DOMAIN

Wiener's original work on optimum filtering and prediction was
formulated in the time domain. [1] It was reformulated in the s-
domain by Chang. [2] In the following, the necessary conditions for
a general optimization problem in s-domain are derived for two cases
with different constraints on the optimizing function. The necessary
conditions are then shown to be also sufficient for a restricted
class of problems which include Wiener's problem.

The General Problem

$$J = \frac{1}{2\pi j} \int_{-j\infty}^{j\infty} R(F(s), F(-s),s)ds = \text{minimum} \tag{7}$$

where $R(F(s), F(-s),s)$ is a rational function in the two variables
$F(s)$ and $F(-s)$, and is real for imaginary values of s. The optimizing $F_0(s)$ is to be found under two different cases of constraint:

Case (i)   F(s) is such that

$$\lim_{s\to\infty} sR(F(s), F(-s),s) = 0 \tag{8}$$

Case (ii) F(s) is analytic in the RHP (right half of the s-plane) in
addition to (8).

Let $F_0(s)$ be the optimizing function. Any other admissible F(s) can
be expressed as

$$\tag{9}$$

$$F(s) = F_0(s)+kF_1(s)$$

where $F_1(s)$ satisfies the same conditions as F(s). For any $F_1(s)$,
(7) must hold as $K\to 0$.

$$\tag{10}$$

$$J = J_0 + kJ_1 + J_2(k^2)$$

where J_0 is the same as (7) with F(s) and F(-s) replaced by $F_0(s)$
and $F_0(-s)$ respectively.

Let F and $\bar{F}$ denote the two variables $F(s)$ and $F(-s)$ respectively in R. J_1 can be expressed as

$$J_1 = \frac{1}{2\pi j}\int_{-j\infty}^{j\infty} \frac{\partial R}{\partial F}(F, \bar{F}, s)\, F_1(s)ds + \frac{1}{2\pi j}\int_{-j\infty}^{j\infty} \frac{\partial R}{\partial \bar{F}}(F, \bar{F}, s)F_1(-s)ds \qquad (11)$$

By substituting $-s$ for s in the first integral, (11) is reduced to

$$J_1 = \frac{1}{2\pi j}\int_{-j\infty}^{j\infty} \left[\frac{\partial R}{\partial F}(F(-s), F(s),-s) + \frac{\partial R}{\partial \bar{F}}(F(s),F(-s),s)\right] F_1(-s)ds \qquad (12)$$

$J_2(k^2)$ is of the order k^2 or smaller. Since k can be either positive or negative, (7) and (10) implies

$$J_1 = 0 \qquad (13)$$

as a necessary condition. Let $B(s)$ denote the bracketed function in the integrand of (12):

$$B(s) \triangleq \frac{\partial R}{\partial F}(F(-s),F(s),-s) + \frac{\partial R}{\partial \bar{F}}(F(s),F(-s),s) \qquad (14)$$

For case (i) $J_1 = 0$ is to hold for any and all arbitrary $F_1(-s)$. The only solution is

$$B(s) = 0 \qquad (15)$$

For case (ii), $F_1(-s)$ is analytic in the LHP, and (13) is equivalent to the condition

B(s) is analytic in the LHP $\qquad (16)$

A stronger constraint in Case (ii) leads to a weaker necessary condition (16).

A Restricted Class

The function R in (7) is of the following form: $\qquad (17)$

$$R(F(s), F(-s),s) = A_0(s)F(s)F(-s) + A_1(-s)F(s) + A_1(s)F(-s) + A_2(s)$$

where $A_0(s)$, $A_1(s)$, and $A_2(s)$ are real functions of the variable s. For imaginary s, $A_0(s)$ is real and positive definite and $A_2(s)$ is real. The term $J_2(k^2)$ in (10) becomes

$$J_2(k^2) = \frac{k^2}{2\pi j} \int_{-j\infty}^{j\infty} A_0(s) \, F_1(s) F_1(-s) \, ds > 0 \tag{18}$$

The condition $J_1 = 0$ is then both necessary and sufficient for (7).

B(s) as defined in (14) becomes

$$B(s) = 2A_0(s) \, F(s) + 2A_1(s) \tag{19}$$

In the optimum filtering and reconstruction problem, the received

signal y(t) is the sum of transmitted signal x(t) and a noise n(t) not

correlated to x(t): $y(t) = x(t) + n(t)$

The processed signal f(t) * y(t) is to be closest to x(t) in the least

mean square sense:

$$\left< \left[f(t) * y(t) - x(t) \right]^2 \right> = \text{minimum} \tag{20}$$

In its s-domain formulation (20) becomes (7) with

$$R = (F(s)-1) \, (F(-s)-1) \Phi_{xx}(s) + F(s)F(-s) \Phi_{nn}(s) \tag{21}$$

where Φ_{xx} and Φ_{nn} are spectral density functions of x(t) and n(t)

respectively. In the reconstruction problem, f(t) is arbitrary and

Case (i) applies. In the optimum filtering problem, f(t) = 0 for

t<0 and case (ii) applies.

Equation (19) gives

$$B(s) = 2 \, [\Phi_{xx}(s) + \Phi_{nn}(s)] F(s) - 2\Phi_{xx}(s) \tag{22}$$

The optimizing conditions (15) and (16) lead to unique solutions of

$F_0(s)$. [2]

III. PONTRYAGIN'S MAXIMUM PRINCIPLE [3]

There is a relaxation of constraint on the control variable u(t).
In variational calculus, u(t) is required to be a continuous function.
Neighboring trajectories can be generated only by infinitesmal variations in u(t). In the maximum principle, u(t) is only required to
be measurable. Thus a finite change of u(t) for an infinitesmal
interval of time is allowable. By comparing neighboring trajectories
of the latter type, Pontryagin proved that $H(\psi, x, u, t)$ is a global
maximum in the variable u. Variational calculus requires only that
H is locally maximum in u.

The maximum principle also confirms our conjecture in its transversality conditions. A stronger condition on the terminal values of the state vector x leads to a weaker condition on the terminal values of the adjoint vector Ψ and vice versa.

IV. OPTIMUM CONTROL IN BOUNDED STATE SPACE

In many applications, the state vector is bounded. For instance, in flight control the angular rates of the control surfaces are bounded Due to structure requirements, various accelerations are also bounded. Gamkrelidze [4] and Chang [5] gave two different necessary conditions for this problem. Chang's necessary condition is simpler and also shown to be sufficient if the state equations are linear and the bounded region is convex. [5] Chang's essential result is as follows:

The additional constraint that x is bounded by a surface S brings about a relaxation in the adjoint differtential equation:

$$\frac{d\psi}{dt} = -\frac{\partial H}{\partial x} + \zeta \tag{23}$$

where ξ is a vector of arbitrary magnitude pointing in the direction of the outward normal of S if x is a boundary point on S, and ξ is zero if x is an interior point. Equation (23) includes Pontryagin's adjoint equation as a special case when S is at infinity and every x is an interior point.

V. CONCLUDING REMARKS

In the above sections, a conjecture on mathematical optimization is stated but not proven. The conjecture is valid for all known analytical results on optimization including the Lagrange multipliers.

References

[1] Wiener, N., *The Extrapolation, and Smoothing of Stationary Time Series*, John Wiley & Sons, Inc., New York 1949.

[2] Chang, Sheldon S. L., *Synthesis of Optimum Control Systems*, Mc Graw Book Company, New York, 1961.

[3] Boltjanski, V. G., Gamkredlidze, R. V., and Pontryagin, L. S., *The Theory of Optimal Processes. I. Maximum Principle*, Izv Akad. Nauk SSSR Ser. Mat. 24 (1960), No. 1. Translated by L. W. Neustadt, Space Technology Laboratories, Report No. 9810. 32-01, October 1960.

[4] Gamkredlidze, R. V., Optimal Control Processes with Restricted Phase Coordinates, Izv. Akad. Nauk SSSR Ser. Mat. 24 (1960), 315-356. Translated by L. W. Neustadt, Space Technology Laboratories Report, March 1961.

[5] Chang, Sheldon S. L., An Extension of Ascoli's Theorem and Its Applications to the Theory of Optimal Control, *Transactions of the American Mathematical Society*, Vol. 115, March 1965, pp 445-470.

A Methodology for Computation of the Exact Distribution for an Acceptance Testing Procedure for a Parallel System

ASHOK K. SINGH and ANITA SINGH
Department of Mathematics
New Mexico Institute of Mining and Technology
Socorro, New Mexico 87801, USA

ABSTRACT

There are situations in statistical testing where the moment generating function (mgf) of the test statistic is available, but it is not possible to analytically invert the mgf to obtain its distribution. The problem of computing the exact distribution for an acceptance testing procedure for a parallel system is an example. We propose and investigate a numerical method for inversion of the bilateral Laplace transform to find an approximation to the distribution of the test statistic.

1.   INTRODUCTION

The problem of acceptance testing for a parallel (1-out-of-n:G) system of components with exponentially distributed life times was considered by Yan and Mazumdar (1987). They proposed a statistical test based on the product of the total component life times on test for accepting or rejecting the system. Rajgopal and Mazumdar (1988) proposed and investigated a test statistic based on the logarithms of the total component life times on test. They consider various approximations for computing percentiles of the test statistics. The purpose of the present paper is to approximate the cumulative distribution function (cdf) of the test statistic by numerical inversion of the bilateral Laplace transform of the probability density function (pdf).

The problem of numerical inversion of the usual Laplace transform has been extensively investigated. Bellman et al (1966) have derived an explicit approximation formula based on a numerical quadrature. Barrodale (1974) has applied linear programming to compute approximate inverse in the l_∞ - norm. Stehfest (1969) has given an algorithm for the numerical inversion, and has demonstrated the superiority of their algorithm over the explicit formula of Bellman et al. This algorithm, however, is for the one-sided Laplace transform. We propose a numerical scheme involving piecewise cubics for inversion of the bilateral Laplace transform, and use it to compute the cdf of the test statistic of Rajgopal and Mazumdar. We have tested our method on the mgf of the standard normal distribution; the results look reasonable.

2. MODEL ASSUMPTIONS AND NOTATIONS

1. The system is made of n independent components in parallel.

2. The life time p.d.f. of component j is exponential with hazard rate λ_j:

$$f(x;\lambda_j) = \lambda_j e^{-\lambda_j x}, \qquad x > 0 \, (\lambda_j > 0), \; j = 1,...,n.$$

The reliability of the parallel system is given by

$$R(t) = 1 - \prod_{j-1}^{n} [1 - \overline{e}^{\lambda_j t}] \approx 1 - \prod_{j-1}^{n} \lambda_j t$$

if $\lambda_j < < 1/t$ for all j.

3. Component testing is done with replacement. The test on component j is stopped after observing a given number of failures K_j, j = 1,2,...,n. Let T_j be the total time on test for component j. The pdf is T_j is known to be gamma with parameters K_j and λ_j.

3. THE SYSTEM RELIABILITY DEMONSTRATION TEST

The acceptance test of Yan and Mazumdar rejects the system iff

$$Q = \prod_{j-1}^{n} T_j < A \tag{1}$$

where A is chosen so as to satisfy the risk requirements:

P(accept system | R ≤ R_o) ≤ α (consumer risk)

P(reject system | R ≥ R_1) ≤ β (producer risk)

where R_0, R_1, α, β are given constants.

Rajgopal and Mazumdar modified the test (1) as follows:

Reject the system iff

$$S = \sum_{j-1}^{n} \ln T_j < B \tag{2}$$

They use the fact that $\Sigma \ln(T_j \lambda_j)$ has the same pdf as $\Sigma_{j-1}^{n} \ln (X_j)$, where X_j is gamma with parameters K_j and 1, and compute percentiles of the pdf of S by several approximation methods.

This paper is an attempt to compute the cdf of S via numerical inversion of its bilateral Laplace transform.

4. THE MOMENT GENERATING FUNCTION AND ITS NUMERICAL INVERSION

The mgf of a r.v. Y is defined as $M_Y(t) = E(e^{tY})$. For the r.v. S, we have $M_S(t) = \int_{-\infty}^{\infty} e^{-ts} f_S(s)\, ds$ which is the bi-lateral Laplace transform [Andrews and Shivamoggi (1988)] of the pdf $f_S(s)$ of S, evaluated at -t.

Using independence of X_j, the mgf of S is easily computed as:

$$M_S(t) = \prod_{j=1}^{n} \frac{\Gamma(K_j - t)}{\Gamma(K_j)}, \quad K_j > t \tag{3}$$

Our problem is to solve the following Fredholm integral equation of the first kind:

$$\mathcal{L}\{f\}(t) = \int_{-\infty}^{\infty} e^{-ts} f_S(s)\, ds = \prod_{j=1}^{n} \frac{\Gamma(K_j - t)}{\Gamma(K_j)} \tag{4}$$

We follow the approach suggested by Barrodale (1974), and use an approximating function for $f(x)$ involving unknown parameters. We will estimate the parameters by minimizing the residual vector in the l_2 - norm.

We describe, step-by-step, our method for solving the integral equation (3).

<u>Step 1.</u> Use the substitution $u = 1/(1+e^s)$ and write the integral in (3) as

$$\mathcal{L}\{f\}(t) = \int_0^1 \frac{u^{t-1}}{(1-u)^{t+1}} f\left(\ln \frac{1-u}{u}\right) du \tag{5}$$

Let $0 < u_0 < u_1 < ... < u_p < 1$ be a fixed set of nodes. We approximate $f\left(\ln \dfrac{1-u}{u}\right)$ by piecewise cubics in each sub-interval:

$$f\left(\ln \frac{1-u}{u}\right) = a_j + b_j(u - u_j) + c_j(u - u_j)^2 + d_j(u - u_j)^3, \tag{6}$$
$$u_j \le u \le u_{j+1}, \quad j = 0, 1, ..., p-1$$

Using the approximation (6), we can integrate (5) to obtain

$$\mathcal{L}\{f\}(t) = \sum_{j=0}^{P-1} a_j A_j(t) + \sum_{j=0}^{P-1} b_j B_j(t) + \sum_{j=0}^{P-1} c_j C_j(t) + \sum_{j=0}^{P-1} d_j D_j(t) \tag{7}$$

where $A_j(t) = H_j(0,t)$, $B_j(t) = H_j(1,t)$,
$\quad C_j(t) = H_j(2,t)$, $D_j(t) = H_j(3,t)$, and

$$H_j(q,t) = h_j^{q+1} \int_0^1 \frac{V^q (u_j + hV)^{t-1}}{(1 - u_j - hV)^{t+1}} dV \tag{8}$$

where $h_j = u_{j+1} - u_j$.

The functions $H_j(a,t)$ were computed by using Gaussian integration formula for moments [Abramowitz and Stegun(1972)].

<u>Step 2.</u> Let t_1, t_2,...,t_m be m distinct values of t. Using (6) and (3) we get a linear model:

$$\underline{Y} = X\underline{\beta} + \underline{e} \tag{9}$$

where $\underline{Y}$ is the mx1 vector with i-th row

$$Y_i = \prod_{j=1}^{n} \frac{\Gamma(K_j - t_i)}{\Gamma(K_j)},$$

X is the m x 4P matrix with i-th row given by

$$(A_0(t_i), ..., A_{P-1}(t_i), B_0(t_i), ..., B_{P-1}(t_i),$$
$$C_0(t_i), ..., C_{P-1}(t_i), D_0(t_i), ..., D_{P-1}(t_i))$$

$\underline{\beta} = (a_0,...,a_{P-1}, b_0,...,b_{P-1}, c_0,...,c_{P-1}, d_0,...,d_{P-1})'$ is the 4P x 1 vector of unknown coefficients, and $\underline{e}$ is the vector of unobservable errors introduced by the approximation used in Steps 1 and 2.

<u>Step 3.</u> The estimate of $\underline{\beta}$ which minimizes the residual vector $\underline{e}$ in the l_2 - norm (i.e. which minimizes $\underline{e}'\underline{e}$) is given by [Vinod and Ullah (1981)]

$$\hat{\underline{\beta}} = (X'X)^{-1} X'\underline{Y}. \tag{10}$$

It is a well-known fact that the integral equation (4) is an ill-conditioned problem. This ill-condioning forces the positive definite matrix X'X of (9) to be not numerically invertible. To avoid this problem, we use the method of ridge regression, which is essentially the conditioning method proposed by Riley (1955). The ridge regression estimate of $\underline{\beta}$ is given by [Vinod and Ullah(1981)]:

$$\hat{\underline{\beta}} = (X'X + kI)^{-1} X'\underline{Y} \tag{11}$$

where k > 0 is a constant to be chosen. In practice, k should be taken to be a small positive number such that X'X + kI is not ill-conditioned.

5. RESULTS OBTAINED FOR A TEST PROBLEM

We have developed a FORTRAN program using the IMSL subroutine LSADS for solving $A\underline{x}$ = b for symmetric positive-definite matrix A for computation of $\underline{\beta}$ given by (1). For our test problem, we used the well-known standard normal pdf and its bilateral Laplace transform:

$$\mathcal{L}\left\{\frac{1}{\sqrt{2\pi}}\, e^{\frac{-x^2}{2}}\right\}(t) = e^{t^2/2} \tag{12}$$

The pdf f(x) obtained from (10) was integrated to obtain the cdf of the random variable. Figures 1 and 2 show plots for the standard normal cdf and its approximation obtained from numerically inverting its bilateral Laplace transform. We have taken 6 equally-spaced nodes (P=5) with $u_0 = 0.001$, $u_5 = 0.999$; the ridge constant k was chosen to be 5 (Figure 1) and 10 (Figure 2). Encouraged by the results obtained on the test problem, we proceded to invert (3). To our dismay, this did not yield satisfactory results.

6. A MODIFICATION OF THE PROCEDURE

The estimator of $f_X(x)$ obtained at Step 3 has two problems:

(a) $\displaystyle\int_{-\infty}^{\infty} f_X(x)\, dx \neq 1.$

(b) $f_X(x)$ may have jumps at P-1 points, since the piecewise cubic (6) has not been forced to be continuous at the intermediate nodes $u_1,..,u_{P-1}$. We add one more step to the method proposed in Section 4.

<u>Step 4.</u> We take care of (a) by forcing the integral to be equal to 1. This leads to the linear constraint

$$\sum_{j=0}^{P-1} a_j A_j(0) + \sum_{j=0}^{P-1} b_j B_j(0) + \sum_{j=0}^{P-1} c_j C_j(0) + \sum_{j=0}^{P-1} d_j D_j(0) = 1 \tag{13}$$

The continuity of the piecewise cubic approximation (6) is obtained from the conditions

$$S_j(u_{j+1}) = S_{j+1}(u_{j+1}), \quad j=1,2,...,P-1$$

which lead to the linear constraints

$$a_j - a_{j+1} + b_j\, h_j + c_j\, h_j^2 + d_j\, h_j^3 = 0, \quad j=0,1,...,P-2. \tag{14}$$

We combine (11) and (12) to obtain the following P restrictions on the parameter $\underline{\beta}$:

$$R_{Px20}\underline{\beta}_{20x1} = \underline{r}_{Px1} \tag{15}$$

where

$R = (R_{ij})$ with first four rows ($i = 1, .., 4$) given by

$$R_{ij} =$$

$$\begin{cases} 1, & j = i \\ -1, & j = i + 1 \\ h_i, & j = i + 5 \\ h_i^2, & j = i + 10 \\ h_i^3, & j = i + 15 \\ 0, & \text{otherwise} \end{cases}$$

the last row of the matrix R is

$$R_{5j} = (A_0(0)...A_{P-1}(0)...D_0(0)...D_{P-1}(0)),$$

and $\underline{r} = \begin{pmatrix} r_1 \\ ... \\ r_P \end{pmatrix}$ with $r_i = \begin{cases} 0, & i = 1,...,P-1 \\ 1, & i = P \end{cases}$.

The problem at this stage is to minimize

$$\underline{e}'\underline{e} = (\underline{Y} - X\underline{\beta})'(\underline{Y} - X\underline{\beta})$$

subject to

$$R\underline{\beta} = \underline{r}.$$

The restricted regression approach [Vinod and Ullah] yields:

$$\hat{\beta}_{rest} = \hat{\beta} - (X'X + kI)^{-1} R'[R(X'X + kI)^{-1} R']^{-1}(R\hat{\beta} - r) \tag{16}$$

where $\hat{\beta}$ is the estimator given by (11).

We have not been able to add Step 4 to our program yet. Work is in progress. We anticipate that the results will improve after these modifications.

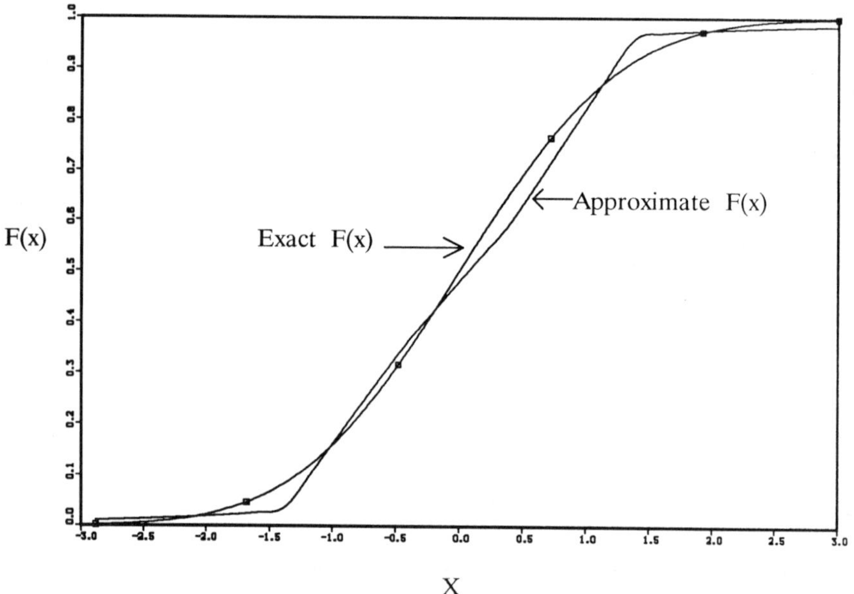

Figure 1 : Plot of the exact N(0,1) cdf and its approximation

Ridge   Parameter   K = 10

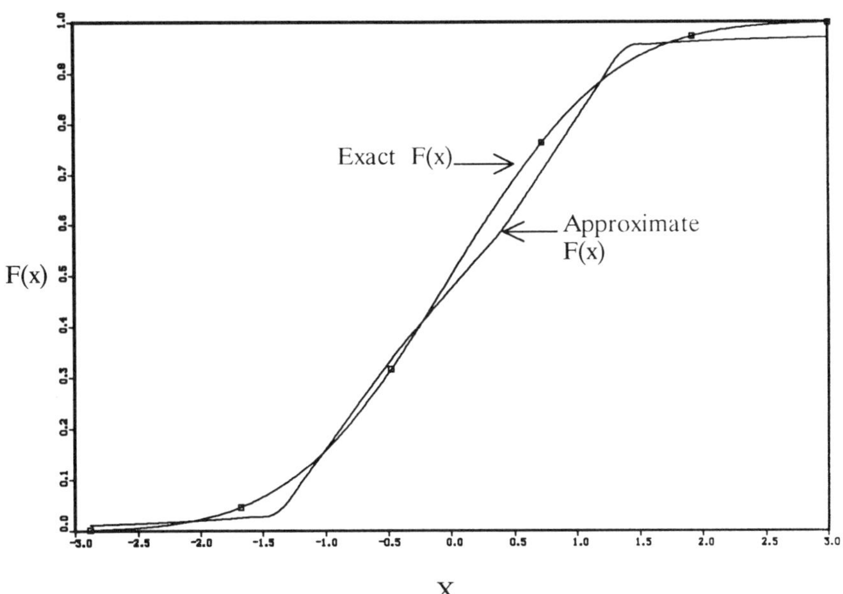

Figure 2 : Plot of the exact N(0,1) cdf and its approximation

REFERENCES

Abramowitz, M. and Stegun, I., Handbook of Mathematical Functions, pp. 921-922, Dover Publications, Inc., New York, 1972.

Andrews, L.C., and Shivamoggi, B.K., Integral Transforms for Scientists and Applied Mathematicians, pp. 214-216, MacMillan Publishing Company, New York, 1988.

Barrodale, I., Linear Programming Solutions to Integral Equations, in Numerical Solutions of Integral Equations, ed. L.M. Delves and J. Walsh, pp. 97-109, Clarendon Press, Oxford, 1974.

Rajgopal, J., and Mazumdar, M., A Type-II Censored, Log Test-Time Based, Component Testing Procedure for a Parallel System, IEEE Transactions on Reliability, Vol. R-37, pp. 406-412, 1988.

Riley, J., Solving Systems of Linear Equations with a Positive Definite, Symmetric but Possibly Ill-conditioned Matrix, Mathematical Tables and Other Aids to Computation, Vol. 9, pp. 96-101, 1955.

Stehfest, H., Algorithm 368, Collected Algorithms from CACM.

Vinod, H.D., and Ullah, A., Recent Advances in Regression Methods, pp. 61-62, Marcel Dekker, Inc., New York, 1981.

Yan, J.H., and Mazumdar, M., A Component Testing Procedure for a Parallel System with Type II Censoring, IEEE Transactions on Reliability, Vol. R-36, pp. 425-428, 1987.

SUPERCOMPUTER APPLICATIONS

Summary

Collected in this section is a group of papers unified by their use of supercomputers. Two papers relate recent attitudes about, use of and future designs of supercomputers. The remainder of the papers address the many body problem. One paper studies a classical version and the others deal with various techniques applied to the quantum many-body problem.

John W. D. Connolly, Director of the University of Kentucky Supercomputer Center, presents an invited paper entitled "The Grand Challenges, the Global Village and the Great Debate," in which he discusses the new class of problems approachable on supercomputers, the means of communicating between computers and between computer users, and the great debate as to what configurations of computers, ranging from small to large, should be made available at computer installations.

Pattnaik presents a paper describing possible computer architectures of the 1990's, primarily focusing on massively parallel, vector processors. He defines the kinds of problems for which these machines will best be suited, and relates the different approaches to communicating between the many processors in the large computers.

Next, Kelly studies the classical motion of an arch by simulating it as a collection of incompressible particles. Large sets of coupled equations of motion with various boundary conditions were solved by employing a *Cray XMP/24* supercomputer, revealing interesting physical phenomena in the response of the arch.

The first of five papers dealing with the quantum many-electron problem is by Schwartzman, who uses a many-body enhanced expression for the magnetic susceptibility of a solid to predict magnetic phase boundaries for lattice expanded transition metals. He discusses vector-parallel processing required to accomplish the numerical tasks.

Fry and his colleagues use the same techniques to study magnetic properties of the high temperature superconductor, La_2CuO_4, concluding that a Fermi liquid ground state assumed by conventional band theory is unstable toward the formation of magnetic order, and perhaps must therefore be revised before an explanation of superconductivity in these compounds is attempted.

Brener and his coauthors present a paper describing the conversion of a standard computer code for performing electronic band structure calculations of cubic solids to treat arbitrary crystal structures and large numbers of atoms per unit cell, and to include a new technique for correcting the approximate single particle energies obtained. Vector and parallel operations are evaluated to accomplish this task.

Zaider, Fry, and Orr continue the series of many-electron studies with an interesting application of a 25 year old technique for correcting single electron approximations which has only recently become practical because of improvements in supercomputers. His ambitious goal is to compute accurately the energy loss spectra for charged particles in biomolecules, since direct measurements are not possible at the microscopic level. Accurate many-body response functions provide this quantity if sufficient computer power is available.

Kaiser and Marroum present techniques for accelerating the iterative cycles in many-electron calculations used in describing solid surfaces and interfaces. Since the number of iterations to achieve self-consistency may be very large, this is still a prime consideration, even on a modern supercomputer.

Gryko and Allen also present surface studies, but their interest lies in the dynamical reconstruction of atoms occurring at a solid surface. An interesting motion picture was presented at the conference which made the process involved in reconstruction very clear. This sort of presentation, recorded directly from a supercomputer, may become more common as the ability of supercomputers to produce enormous quantities of information increases and our ability to digest cannot keep up with the output. Unfortunately the movie is not recorded in this proceedings, but the authors may be contacted directly.

The quantum many-body theme continues in the last three papers. Miller compares mean-field and path-integral approaches to describe the quantum behavior of an excess light particle (electron, positron, etc.) in fluids. The path-integral approach can be exact, in principle, but very expensive to apply to a real physical system.

Worrell and Reese use the path-integral method described by Miller to treat positronium annihilation in a Lennard-Jones fluid. The path integrals are evaluated as a function of fluid density using a Monte Carlo method. Algorithms to accomplish this on a supercomputer are discussed.

In the last paper, Kestner and Yang review a number of quantum many electron techniques and focus on the fast-Fourier-transform method as applied to molecules. The method requires many repeated Fourier transforms back and forth between r-space and k-space in order to simplify evaluation of certain quantum mechanical operators. Evidence is presented that the method can be made cost effective for a large class of systems when carefully implemented on parallel-vector processors.

John L. Fry

Scientific Supercomputing in the 90's: The Grand Challenges, the Global Village and the Great Debate

JOHN W. D. CONNOLLY
Center for Computational Sciences
University of Kentucky
Lexington, Kentucky 40506, USA

INTRODUCTION

The subtitle of this paper is shorthand for the essential issues which are currently important in the supercomputing world. The Grand Challenges (a term invented by Ken Wilson, the Nobel laureate who more than anyone is responsible for the dissemination of supercomputing imperatives) are those important problems which require the largest possible computing resource. The Global Village (a term due to Marshall McLuhan) refers to the worldwide computer network which is the most vital factor in the current supercomputer environment. The Great Debate is a reference to the continuous discussion which is found on every campus and in every research laboratory which now depends on large scale computing, namely what is the appropriate computational environment to make progress on the grand challenges and to link into the global village?

The great expansion of interest, especially in the academic and basic research communities in the use of supercomputers was sparked exactly five years ago with the establishment of the NSF National Centers. In the very short time since then, they, together with the NSFnet which links them, have become the leaders in the innovative applications of forefront computing. The grand challenges were defined in these centers, the global village has grown out of the connections between them, and the great debate is raging within their walls at this very moment. At the same time there has been a significant spin-off from the NSF centers in terms of academic regional and state centers, which now number about 20-25 around the country. This includes the recently established Supercomputing Facility and Center for Computational Sciences at the University of Kentucky.

I think it is appropriate on the fifth anniversary of the NSF Supercomputer Initiative, to reexamine the issues and to delineate the directions which we will travel along in the next 5-10 years.

(I) THE GRAND CHALLENGES:

The most important issue five years ago, and still the "bottom-line", is <u>what</u> <u>is</u> <u>a</u> <u>supercomputer</u> <u>good</u> <u>for</u>? There is so much discussion in the literature, the trade press and the popular media, about hardware issues, that we sometimes tend to forget that it is more important to measure output than input, although it might be much easier to do the latter.

The ultimate justification of supercomputing five years ago was the science produced by the researchers who used them. This remains the most important issue today. Computer technology is doubling every two years, and the various centers on the most part have been keeping ahead of this curve. In general, this means that the scientific output is more than doubling every two years as well. Another way to say this, is that the computational science done in the next two years will be more than all the computational science done up to the present time.

The "grand challenges" are those scientific and engineering problems which require a supercomputer for their solution, or if (as is often the case) no solution is possible, at least to make progress. A recent White House report (The Federal High Performance Computing Program, Sept. 1989) described a focussed national program to bring the resources of the computational research community on these problems. An appendix to this report described a list of critical areas where the various federal agencies will concentrate their supercomputing cycles.

At the top of the list is environmental modeling in which computer modeling will be used to refine the predictions on global climate change, pollution, acid rain and ozone depletion. It is extremely important to reduce the error bars in these predictions, since they are the only way we have of assessing the consequences of the atmospheric perturbations of modern society, and they will be the basis of future governmental decisions. We had better be certain of the numerics of the climate models before we start using them to make policy.

Also, near the top in priority are materials research and biomolecular modeling. The computer simulation of the behavior of semiconductors and superconductors are critical to the future of the electronics industry. The electronic behavior of materials in nanostructure devices is still not very well understood, and the nature of high temperature superconductivity in copper oxide ceramics is still a mystery. The quantum mechanical equations which describe this behavior are sufficiently complex so that their solution is at the very limit of the capacity of present day computers. Similar equations are required to understand the structure and function of molecules, but the theory reaches the limits of supercomputers for even the simplest molecules. Even the very approximate theories for predicting the behavior of biochemical systems are also consuming many gigaflop's. However, it is quite clear that today's laughably crude models and pitifully slow machines are providing the beginnings of predictive power at the subcellular level.

The impact of future computation is inevitable.

The list of grand challenges also contains many other topics from the most esoteric (astrophysics and quantum chromodynamics) to the most practical (oil recovery and airplane design). Also, there are even now a few brave social scientists and humanists discovering the value of supercomputing to their disciplines, so that it will soon be difficult to find a field in which supercomputers are not important.

The new Federal Program in High Performance Computing described by the White House Report of September 1989 outlines a $2 billion five-year plan to develop the resources necessary to   attack this daunting list of insoluble problems. This is designed to be a coordinated multiagency (e.g., NSF, NIH, DoE, NASA and DARPA) initiative focusing on four major themes:

(1)   *Hardware*: The aim will be to develop a machine at least two orders of magnitude faster than today (in the teraflops range). This will necessarily have to be a massively parallel processor for which current super-computer codes are particularly unsuited. How to achieve the teraflops capability will require a concerted effort in the second theme:

(2)   *Software*: This component of the initiative is probably the most important and the most difficult. Its goals are to build the "apparatus" on top of the hardware which will be both powerful and convenient for the assault on the grand challenges. The main question is who are the builders? Are they industrial computer scientists, numerical mathematicians, university graduate students, or some as yet undefined professional? Software design and development may become a respectable discipline of its own.

(3)   *Networking*: This is the new parameter in the research equation. From the crude beginnings of five years ago, NSFnet has blossomed into the worldwide "internet", which allows the interactive use of supercomputers everywhere. The goal of the initiative is to achieve a gigabaud capacity for all major research institutions in the U.S. If the funding holds up, speculation is that this goal will be reached ahead of schedule.

(4)   *Human Resources*: The aim here is to alleviate the shortage of trained computational professionals by augmenting the computer science schools of the country and to strengthen their currently weak ties to the grand challenges of computational

science. Perhaps one can think of eliminating the traditionally confrontational barriers between them.

(II)   THE GLOBAL VILLAGE:

Five years ago, the average researcher had no access to any kind of network. There were a few exceptions, CSnet for computer scientists, ARPAnet for some Defense contractors, MFEnet for the magnetic fusion energy community. Currently, there are two major networks (which are rapidly becoming indistinguishable) to which almost every US researcher can connect. Together they provide a new facility for the information age. If supercomputers will be the principal generators of information in the future, then the global network will provide the principal means of exchange for this information.

BITNET, which started a few short years ago, with one connection from New York to New Haven, now reaches over 3000 nodes in every state, in every industrial country in the Western world, and several in third world countries. It has been reported that several nodes in the Eastern bloc countries will soon join. It is now a functional global mail network, which is robust, inexpensive, easy to use, and ubiquitous. Even though it is slow and non-interactive, it is now a indispensable tool for communication between all researchers.

INTERNET, which is a newer, and more interactive network. Through the impetus of NSFnet, gateways have been built to every network of importance for researchers, to become the "network of networks". According to the latest count, NSFnet has more than 600 members, and connections to thousands more through its connections to other nets, regional and state included. It has now at least ten international connections to networks in Canada, Europe and the Far East. It is now possible to log in to a supercomputer in Paris as easily as to one in Lexington, Kentucky. This is done through the new satellite T1 connection between Cornell University and the University of Montpellier in France, which is the main link between NSFnet and EASInet (EASI=European Academic Supercomputer Initiative).

Despite the unpredicted (and unpredictable) success of the global network, there are several problems, which will be items of contention, and attention, for many years to come.

(1)   The weakest links in the chain are the local area networks (LAN's) which exist on every campus and in every research laboratory. Very often, only the most important and affluent researchers are connected. This will have to change soon, since it is obvious that the global

network is most useful when it (like the telephone) becomes accessible to everyone. Universities and other research institutions must make this one of the top priorities in the future development of infrastructure.

(2) Management is a critical problem. The hundreds of interlocking networks are separate managed and uncoordinated. If the network is not working, it is almost never possible to discover who is responsible. Quite clearly, no one is in charge, and more seriously, no one seems to want to be in charge. It even appears that if someone wanted to be in charge, they would not know how to go about it.

(3) A consequence of the lack of management is that each user of the network has infinite access to the network. There is no control, such as exists for every supercomputer, on how many bits or packets can be sent. Anyone, with a little ingenuity, can saturate the network. Although rapidly improving technology has so far allowed the capacity to increase more rapidly than demand, this is clearly a problem which needs to be solved soon so that the gigabaud internet will be as useful as it needs to be.

(III) THE GREAT DEBATE:

The question is often asked: "Why use a supercomputer when a large number of small computers can do the same job?" This is often used by the ill-informed as an argument to avoid investing $10-20 million in a supercomputer and to utilize the often-underutilized PC's or workstations which one already owns. The principal parameter in this debate is the hardware capacity, so that 1000 Megaflop machines are equated to a Gigaflop machine.

What is ignored in this debate is the realization that supercomputers are for solving problems of the grand challenge variety, not for demonstrating that a simple algorithm can be easily distributed around a network. In fact, the usual definition of a supercomputer implies that it should be a general purpose machine which could be used on any grand challenge problems.

In actuality there are probably no examples of a network of workstations being used to do useful science on a grand challenge problem. This is not to say that this will not happen in the future, but there is a matter of a great deal of software to be generated in the meantime.

A new factor in the debate is the massively parallel processor. There are now a large number of such machines on the market. As I pointed out in a previous section, this is the most likely future supercomputer configuration. But again there is a tremendous amount of software development to be done before such a machine can replace a supercomputer. For example, advanced time-sharing systems are needed to simultaneously provide service to hundreds of users. The common wisdom states that the software gap between the MPP and the traditional supercomputer is rapidly closing.

A sub-theme of the great debate is the matter of control. The expense of a supercomputer usually mandates that it be shared by many researchers. This means a sacrifice of the right to dominate a machine. Distributed computing does not have this defect. However, distributed computing is often the result of distributed budgets, which is a political not a technical question.

There are at least two red herrings found in the great debate: The first of these is cost effectiveness. The advocates of small machines would have us believe that there is an inverse economy of scale, so that 1000 megaflop machines are always cheaper than a gigaflop machine. Again, the fallacy here is to consider only hardware costs, and to leave out software, maintenance, peripherals and personnel costs. The second is "peak performance". Even some governmental experts who should know better insist on defining a supercomputer in terms of its maximum performance rating. The only sensible definition is one based on average performance. (The average can be defined with respect to grand challenge codes.)

The greatest complication in the great debate is that computing technology is changing so rapidly. The factor of two every two years, which is the historical average growth rate, is probably faster than any other industry. This allows one to say that one should not buy a supercomputer today since we will soon have a better one at the same cost, or equivalently, we shall soon be able to buy one with the same capacity at half the price. However, those with experience in this business know that the same argument will hold later, so that one can always justify the delay of a decision. It can always be used to rationalize inaction.

(IV)   THE OPTIMUM APPROACH: (As of 1990)

Because of the exponential technology curve, one is required to almost continuously upgrade, in order to stay at or near the leading edge of supercomputing, and therefore to remain competitive in the grand challenge arena. A new supercomputer every two years is not uncommon. The best strategy is to include a constant upgrade line item in one's computing budget. This avoids the difficulty of re-justifying the goals of supercomputing every two years.

In order to maintain supercomputing capacity on a campus or in a research laboratory, it is necessary to avoid the fallacy of the inverse economy of scale and the concomitant distribution of resources. This can only be done by centralizing control of the computing budget.

In a supercomputing environment, it is necessary to provide fast user-friendly access for every researcher whose research merits it. At the present time, the best solution is a fast local network of workstations, PC's and terminals. Whichever actually sits on a researcher's desk depends on the local taste in software.

The network connections should also be extended to the outside world, so that every researcher has access to those supercomputers and databases necessary for the rapid transfer of codes and data.

Whereas it is clear that the best environment for solving grand challenge problems is a large capacity standard supercomputer, it is probable that the optimum configuration of a will change within the next five years. In order to prepare for this cross-over, many centers are incorporating an experimental facility of massively parallel processors. This may be a prudent approach in light of the volume of software development which will be necessary before the switch occurs. The difficulty here is in the choice of architecture. No one has a clear crystal ball in this regard.

(V) THE KENTUCKY APPROACH:

In order to approach the optimum computing environment, we at the University of Kentucky have taken seriously the principles outlined in the previous section. The first of these is a centralized administration of all computing resources. The Vice-President for Information Systems has control over all computers, all networks, all libraries (soon to be incorporated into the computing world) and all printing services. This tends to avoid the disaster of uncoordinated, incompatible, distributed computing systems.

One favorable consequence of centralized control is a centralized computing budget. This allows the implementation of a "zero charging algorithm". Since the cost is centralized, it does not need to be passed on to the end user. Maximum flexibility is maintained in the use of the computing resources.

The principal computing engine at the University of Kentucky is an IBM 3090/300E supercomputer (soon to be upgraded to a 3090/600J). This is an extremely flexible machine which is friendly to most users (including administrators), and which is applicable to all grand challenge problems. Even though it is somewhat slower than the maximum supercomputer (Cray YMP) it often outperforms it on large memory jobs.

The centralized supercomputer facility is coupled through a mesh of ethernets and a somewhat slower backbone to a wide variety of workstations and terminals, which are chosen according to the personal taste of their owners. High priority is given to maintaining an open system so that all can be accommodated.

An experimental parallel processing center is planned which is anticipated to contain a variety of configurations. This will be limited in use to a few numerically intensive projects, and in no way is considered to be a general substitute for the central supercomputer.

The keystone of the whole system is the Center for Computational Sciences. This is a multidisciplinary center which has a substantial budget outside the regular computing budget. Its funds are used to support the users (typically graduate students and post-doctoral associates) who are working on the grand challenges. There are no permanent members of the Center. Everyone, including the Director, is a visitor, which allows for maximum flexibility in supporting innovative, and especially interdisciplinary projects. The Center's long term goal is to provide the best possible ambiance for computing, and to attract the best possible participants, so as to become a major participant in the national supercomputing scene.

Parallel Computing in Science

P. C. PATTNAIK
IBM Research Division
T. J. Watson Research Center
P.O. Box 218
Yorktown Heights, New York 10598, USA

ABSTRACT

Recent advances in VLSI technology have significantly altered the cost effectiveness of large scale scientific computing. Improvements in silicon IC technology have made it feasible to build large scale parallel computers. In the light of these developments, here I discuss computational needs in various areas of science and technology, examine various paradigms of computation and comment on the recent emergence of stochastic computing.

Problems and Challenges

Over the course of the last two decades, computation has provided insights in science and technology that would otherwise be unobtainable. The availability of supercomputers for scientific computations in a number of areas, such as aircraft design, VLSI chip design, semiconductor device analysis, petroleum recovery and weather prediction have yielded a significant economic impact. From the advances in VLSI technology, there is a considerable expectation that computational science will enable one to accelerated the pace of understanding of nature and help various design and analysis objectives in a timely fashion.

The potential for the greatest increase in the economic impact of computational science lies in the solution of the difficult problems that are not amenable to alternate approaches. A number of such problems have been identified in areas like electronic structure in condensed matter physics, material science, molecular biology, drug synthesis, various areas in aerospace research, weather forecasting and econometrics. A brief outline of these problems, along with the resources available in current supercomputers are shown in Fig.1. Using specific examples, I will demonstrate the role of computational methods, and the limitations of a predominantly experimental approach in a wide range of disciplines.

Medium Range weather forecast(2 to 7 days): Weather forecasting in this range is done by solving atmospheric dynamics over a grid. These grids are roughly 100 miles on a side in the horizontal plane and the entire vertical atmosphere is divided into 20 to 30 grid points. Experimental weather information is used as a starting point, and then the laws of atmospheric dynamics are solved to predict weather over two days to one week. The current supercomputers can barely keep up with the task of producing the results in this coarse grid, which ignores surface topography such as mountains and lakes within a reasonable time. A forecast with improved accuracy requiring the use of finer grid can only be done with significant improvement in computational ability. A reduction of grid spacing by half will require a sixteen-fold increase in computer resources.

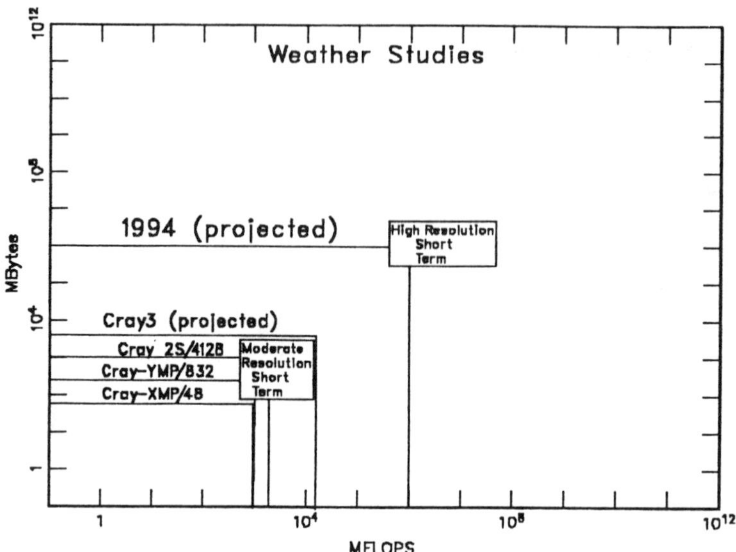

Fig. 1a, Computational needs for weather forecasting.

Semiconductor Device design and Optimization: With the constant push for miniaturization semiconductor devices have reached a point where the devices are extremely sensitive to change in processing parameters.   For example, in sub-micron MOSFETs, the threshold voltage is a sensitive function of channel length, where as in devices with channel length larger than two microns it is independent of channel length.   The same is also true for other critical parameters. Furthermore, with reduction of feature size the manufacturing of these devices becomes time consuming, thus making it less feasible to determine these dependences through controlled experiments.

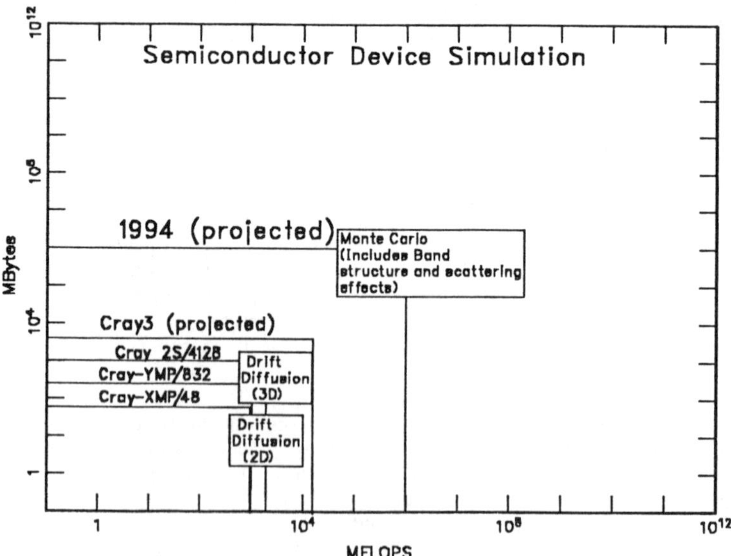

Fig. 1b, Computational needs for Semiconductor Device simulations.

Also the dynamical behavior of these devices, such as alpha particle induced failure or transient latch up, forces one, to a large extent, depend on device simulations to study and model these behaviors. With the coming of X-ray lithography, these simulations will be the prime tools for device design, augmented by a set of carefully chosen experiments. This will require orders of magnitude more computational resources than available now.

Drug Design and protein folding: Due to cost and time factors, in drug research one performs biological testing for only a very small fraction of potential drugs; only a tiny fraction of these tested compounds prove to be useful. These choices are generally made using a heuristic approach based on a very limited understanding of the underlying cause and effect relationship. A systematic study of the electronic and *in vivo* structural state of large molecules will significantly enhance the selection process and will allow one to select potential drugs more efficiently, leading to a very significant benefit for society. But this can only be achieved with massive computations, and is far beyond current computer resources.

Electronic structure studies: The study of electronic structure in condensed matter has impact in a number of areas including basic research in physics, material science, and biological science. I have already alluded to the gain in the area of drug research. The newly discovered high temperature superconductors are another example of an area of intense research.

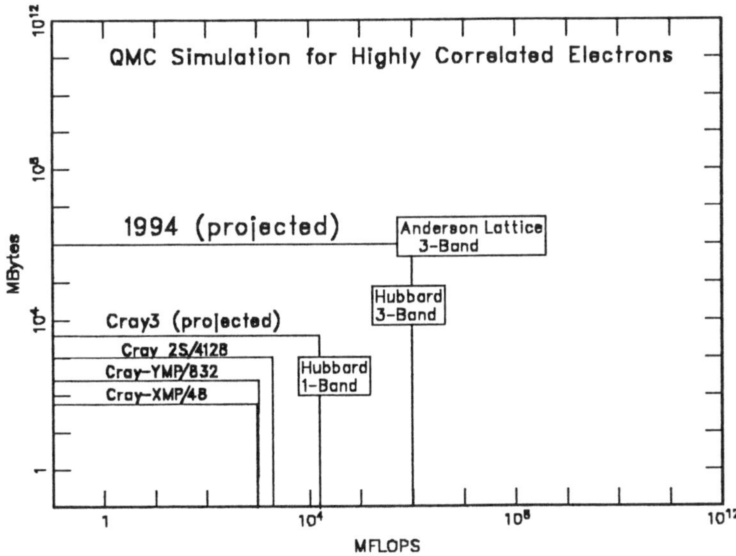

Fig. 1c, Computational needs for Quantum Monte Carlo Simulations for the study of highly correlated electronic systems (viz. High Tc superconductors).

But to gain reliable insight into the cause of the superconductivity so that along with selected experiments it can be used for material engineering, will require massive amount of computations, only possible with Tera-Flop machines.

Aerospace: In aerospace research there are a number of problems, that are solvable only with massive computation. One of them is the design of efficient hypersonic aircraft. Since the aircraft can not be tested in wind tunnels at realistic Reynolds numbers and at realistic wind speed, it is necessary to considerably overdesign for safety at the expense of efficiency and performance.

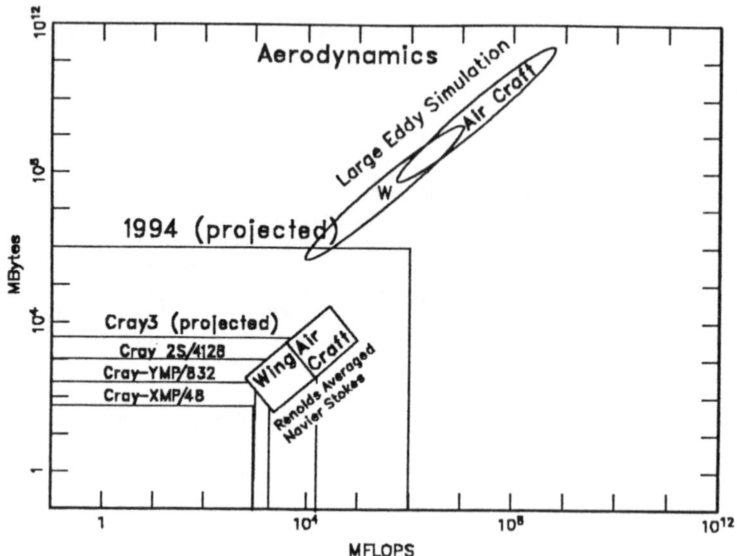

Fig.1d, Computational needs for Aerodynamic Simulations.
Useful simulation in this area again requires computers with Teraflop capacity. The same need exists for fuel design and composite material design.

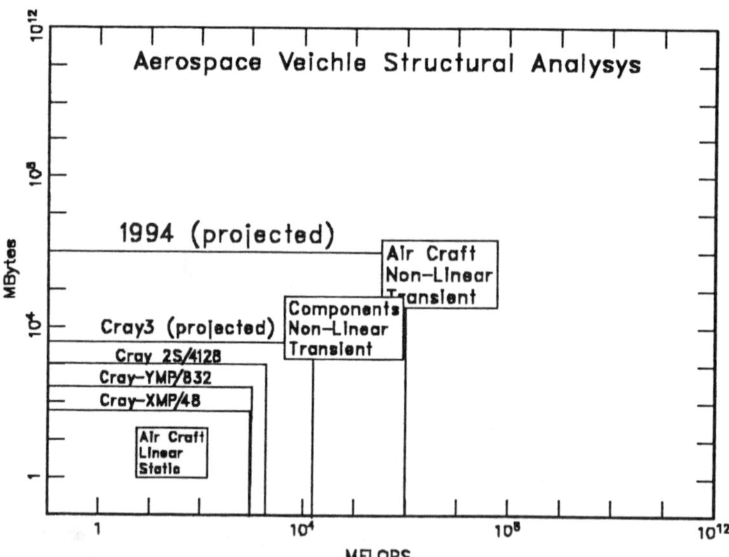

Fig.1e, Computational needs for simulations of aero space structures.

Hardware Solutions

The computing resource requirement estimates for these challenging problems leads to the conclusion that even though the approaches and algorithms used in solving these problems, with current resources, can establish the potential, the current resources are inadequate to realize this potential. Large improvements in floating point operation rate and memory availability are needed to achieve a significant fraction of the potential benefit from computation techniques.

Examining the current computer architectures, it is clear that the only feasible way to achieve the computational power required to solve these challenging problems is with the massive parallel computers, built from superscalar processors that exploit local overlap of instructions and is able to perform floating point operations at a high speed. A certain amount of debate has been taking place to identify the best parallel architecture, *viz.* SIMD *vs* MIMD etc.. Even though it is too early to settle these issues, in my opinion, in the light of design cost and the limited financial resources available to society, in this decade and the early years of the next century, MIMD architecture will be the overall choice, barring some special applications where SIMD may be prefered. My reasons for this opinion are two fold.

1) Due to the growth in personal computers and the work station market, powerful low cost superscalar processors will be readily available. The Intel i860 and IBM RS/6000 processors are typical examples. With time the use of state of the art device technologies in these processors will yield significantly more functionality and speed. This will be primarily driven by the workstation market. The wide variety of functionality currently available in Intel i860 is an example. Volume production of these processors will improve reliability while reducing the cost. Thus at any given time, for a given cost, one can built a more massive and powerful MIMD computer by connecting these superscalar processors through a network, than a SIMD machine using custom processors or boards. Furthermore due to volume production for the work station market, commercially available compilers for these single processors will be efficient and robust. Thus from market force considerations, I believe the MIMD machines in this decade have the advantage of riding on the back of the growth in the work station market.

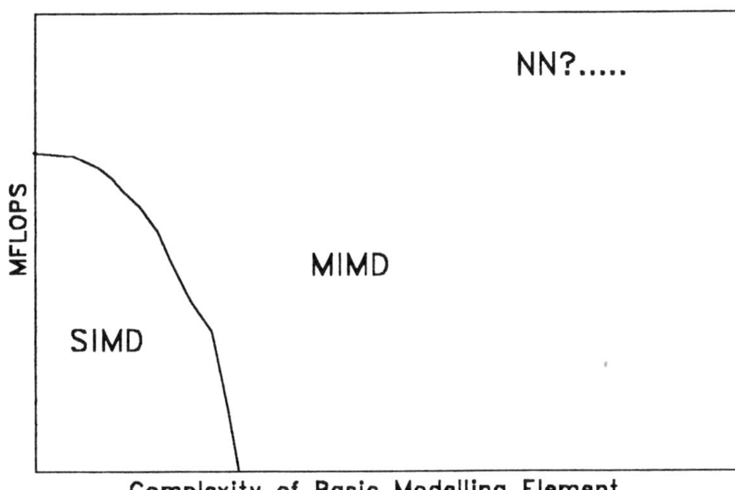

Fig 2., A conceptual divison of problem domain for efficient use of computers with SIMD and MIMD architectures.

2) The second reason for my opinion comes from the inherent nature of computing in a number of areas of science and engineering identified in figures 1(a) through 1(e). Often in science and engineering one is interested in the solution of the problem over a limited scale of energy, time or length. Thus one tries to integrates out the scales that are of little concern for the problem at hand. The only exception is when one examines the most fundamental scale of nature, as in QCD. Thus the basic element of the problem, after integrating the lower scales out, acquires a complex behavior. For example, after integrating the details of the brain and rest of the neurons, a human being "may only be represented" by a complex psychological description. After integrating the flow of electrons and holes through electron bands, the MOSFET in a circuit simulator is represented as an object with complex behavior. The same is true in other areas like device simulators, electronic structure studies, molecular dynamics etc. Examining a number of such

problems, it becomes obvious that by designing MIMD machines using available super-scalar processors, one gets local instruction overlap through pipelining etc., and exploits parallelism by allowing different processors to execution to follow different path.

In the light of these two points, the problem domains where SIMD vs MIMD is the preferred choice is shown in Fig. 2.

Software and algorithms

Besides the evolution in computer architecture from vector to massively parallel superscalar processors, the success of parallel computation, now at its infancy, depends on the development of a robust, efficient and convenient enough operating environment for experts in the application area. Thus, vital work remains to be done in the area of parallel algorithms and programming paradigms, and debugging environments. In the light of massively parallel architecture, almost all facets of these topics need a lot of rethinking and research. Fortunately, a number of people are working to achieve this objective, and research within our own laboratory and other institutions have began to yield very encouraging results.

Supercomputer Study of a Particle Arch

THERESA D. KELLY
Department of Mathematics
University of Texas at Arlington
Arlington, Texas 76019-0408, USA

ABSTRACT

The finite motions of elastic materials under initial and boundary conditions are governed classically by a system of nonlinear partial differential equations. The problems studied are concerned mostly with incompressible bodies having special geometries and applied loads, like a circular cylindrical shell under axial compression. For modelling such exceptionally difficult problems, we will develop a noncontinuum, particle approach. Our particle methodology considers all components of elastic plates in three dimensions, while others reduce their consideration to what is called the neutral or middle layer. In particular, we will be concerned with phenomena related to the bending of a finite elastic arch in three dimensions. Various loads will be applied without changing any initial or boundary conditions to determine the unstable mode of equilibrium. Snap-through inversion is then simulated and discussed. The methodology will require the solution of large systems of ordinary differential equations. In our examples, these equations are solved numerically on a Cray XMP/24.

INTRODUCTION

For a material body modelled by a continuum of points, the position of any point in the body is defined by its position vector $\vec{r}$ in some co-ordinate system. When a body is deformed, the distances between its points change. A three dimensional body cannot be deformed in such a way that parts of it move a considerable distance without the occurrence of considerable extension and compression. In fact, when a deformation occurs, the arrangement of the *molecules* is changed and the body ceases to be in its original state of equilibrium. Internal stresses result which are due to molecular forces and in this paper we will develop a three dimensional, qualitative, molecular type model of elastic behavior.

1. COMPUTATIONAL PRELIMINARIES

For the computational algorithm outlined below, the details and discussion can be found in Greenspan [10].

Consider $N=189$ particles P_i, $i=1,2, \ldots 189$. For $\Delta t>0$, let $t_k = k \times \Delta t$, $k=0,1,2, \ldots$. For each i, between 1 and 189, let M_i be the mass of P_i. At t_k let P_i be located at

Computations performed at and funded by the University of Texas Center for High Performance Computing.

$$\vec{R}_{i,k} = [x_{i,k}, y_{i,k}, z_{i,k}], \tag{1.1}$$

have velocity

$$\vec{V}_{i,k} = [v_{i,k,x}, v_{i,k,y}, v_{i,k,z}], \tag{1.2}$$

and have acceleration

$$\vec{A}_{i,k} = [a_{i,k,x}, a_{i,k,y}, a_{i,k,z}]. \tag{1.3}$$

Let the position, velocity, and acceleration be related by the recursive formulas:

$$\vec{V}_{i,\frac{1}{2}} = \vec{V}_{i,0} + \tfrac{1}{2} \times \Delta t \times \vec{A}_{i,0} \qquad\qquad \text{starter formula} \tag{1.4}$$

$$\vec{V}_{i,k+\frac{1}{2}} = \vec{V}_{i,k-\frac{1}{2}} + \Delta t \times \vec{A}_{i,k} \qquad\qquad k=1,2,3\ \ldots \tag{1.5}$$

$$\vec{R}_{i,k+1} = \vec{R}_{i,k} + \Delta t \times \vec{V}_{i,k+\frac{1}{2}} \qquad\qquad k=0,1,2,3\ \ldots \tag{1.6}$$

At t_k each particle P_i will be subjected to a force

$$\vec{F}_{i,k} = [f_{i,k,x}, f_{i,k,y}, f_{i,k,z}]. \tag{1.7}$$

This force will be determined by the interaction of P_i with surrounding particles. The relationship between force and acceleration is asssumed to be

$$\vec{F}_{i,k} = M_i \times \vec{A}_{i,k}. \tag{1.8}$$

For the force, we introduce a molecular type formula as follows. Let $\vec{R}_{ij,k}$ be the vector from particle P_i to P_j at time t_k, so that $R_{ij,k} = |\vec{R}_{i,k} - \vec{R}_{j,k}|$ is the Euclidean distance between the two particles. Then the force $\vec{F}_{ij,k}$ on P_i exerted by P_j at time t_k is assumed to be

$$\vec{F}_{ij,k} = \left\{ \left\{ -\frac{G}{R_{ij,k}{}^{p}} + \frac{H}{R_{ij,k}{}^{q}} \right\} \frac{\vec{R}_{ji,k}}{R_{ij,k}} \right\}, \tag{1.9}$$

where G is a constant of attraction, H is a constant of repulsion, p is an exponent of attraction, q is an exponent of repulsion and $G \geq 0$, $H \geq 0$, $q>p>0$. These four quantities G, H, p and q belong to the material properties of the model.

Formula (1.9) will be assumed to be valid when P_i and P_j are "neighbors", which will be defined precisely later. If P_j is <u>not</u> a neighbor of P_i, then the force on P_i exerted by P_j is taken to be

$$\vec{F}_{ij,\,k} = \vec{0}. \tag{1.10}$$

The total force $\vec{F}_{i,\,k}$ on P_i at t_k is defined to be

$$\vec{F}_{i,\,k} = \sum_{\substack{j=1 \\ i\neq j}}^{189} \vec{F}_{ij,\,k}, \tag{1.11}$$

so that,

$$\vec{A}_{i,\,k} = \frac{\vec{F}_{i,\,k}}{M_i} = \frac{1}{M_i} \times \sum_{\substack{j=1 \\ i\neq j}}^{189} \vec{F}_{ij,\,k}. \tag{1.12}$$

Once the parameters p, q, G, H are given and the definition of "neighbor" is provided, the motion of each particle can be determined by $(1.4)-(1.6)$.

Throughout the algorithm will be implemented with $\Delta t = 0.001$ and $M_i \equiv 1$ and a related FORTRAN program is given in Kelly[16].

2. THE GENERAL PLATE

Consider 189 particles P_i with respective coordinates (x_i, y_i, z_i). These positions are defined precisely as follows. Let

$$z_i = 7.0, \quad i = 1, 2, .., 189 \tag{2.1}$$

$$x_1 = -7, \ y_1 = 0 \tag{2.2}$$

$$x_{16} = -6.5, \ y_{16} = \sin 60° \tag{2.3}$$

$$x_{i+1} = x_i + 1, \ y_{i+1} = 0, \ i = 1, 2, .., 14 \tag{2.4}$$

$$x_{i+1} = x_i + 1, \ y_{i+1} = y_{16}, \ i = 16, 17, .., 28 \tag{2.5}$$

$$x_i = x_{i-29}, \ y_i = y_{i-29} + 2\sin 60°, \ i = 30, 31, .., 189 \tag{2.6}$$

For all time, a neighbor of any particle P_i in this set is defined as any point P_j whose distance to P_i is initially unity. Hence, the particles are vertices of equilateral triangles. The particles in the center region are centers of regular hexagons, and each has exactly six neighbors. As one can now see, the choice of $N=189$ allows the two row pattern to be repeated seven times describing a rectangular plate of 10×12 units.

Next, particles $P_1 - P_{189}$ were then set between the following two layers. The bottom layer will consist of particles $P_{190} - P_{351}$, each having a z-value of 6.1836. The top layer will consist of particles $P_{352} - P_{513}$, each having a z-value of 7.8164. As in the middle layer, the particles in the top and bottom layers are vertices of equilateral triangles with unit sides.

Further, the choice of location for each particle in the top layer $(P_{352} - P_{513})$ was such that it formed a tetrahedron of unit edge length with particles in the middle layer. A similar construction was implemented for the bottom layer. A neighbor of any P_i in the entire set of 513 particles is, again, any particle P_j whose distance to P_i is unity. A complete list of the neighbors of each P_i is given in Appendix I, Kelly[16].

It should be noted that not every triangle in the middle layer is used to form the tetrahedra. For the particles on the top layer (or bottom layer) to have a unit distance among themselves, we have utilized only the triangles where the position of the third vertex is on the right side of the vertical line joining the remaining two particles.

The entire three layer configuration is called a plate. We next consider a three dimensional arch. For this purpose the plate will be bent in such a manner that each of the three layers will lie on concentric cylinders.

3. FROM PLATE TO ARCH

Since particles $P_1 - P_{15}$ lie on the same straight line($y=0$, $z=7.0$), we will relocate P_1-P_{15} onto a circular arc with center $(0,0,0)$ and radius R=7. We will fix P_8 to remain at the location $(0,0,7)$. The next step is to place the particles such that the arc length (not the Euclidan distance) between two neighbors is unity. The central angle θ for each adjacent pair of particles on this arc will be 8.19°. Hence the points P_1-P_7 and P_9-P_{15} will be relocated by the formula

$$[7\sin(8.19n),\ 0,\ 7\cos(8.19n)]\quad n=\pm1,\pm2,\ ..\ \pm7. \tag{3.1}$$

The particles $P_{16}-P_{189}$ were reset in a similiar fashion on the cylinder through P_1-P_{15} which is parallel to the X-axis. The top and the bottom plates were reset similarly. The radius for the top layer of particles $P_{352}-P_{513}$ was R=7.81640. The radius for the bottom layer of particles $P_{190}-P_{351}$ was R=6.18360. A complete list of particle positions can be found in Appendix II, Kelly[16]. FIGURE 3.1 shows the middle layer $P_1 - P_{189}$ as it appears in three dimensions.

The configuration associated with FIGURE 3.1 is called an arch.

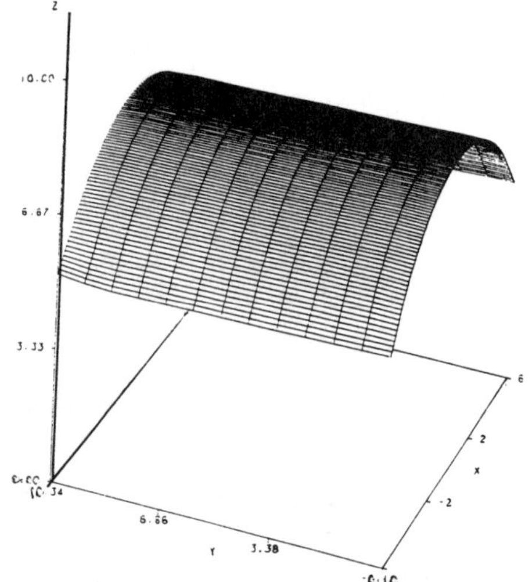

FIGURE 3.1 P_1-P_{189} in three dimensions, $\theta=8.19°$

4. COMPUTATIONAL STUDIES OF p, q, G, H

In order to determine a reasonable set of parameter values p, q, G, H, a variety of computer examples were run using the arch associated with FIGURE 3.1. For this purpose set the initial velocities of all particles P_i, i=1,2,.. 513 to (0.0, 0.0, 0.0). The particles

$$P_i,\ i=\ 1,\ 15,16,29,30,44,45,58,59,73,74,87,88,102,103,116,117,131,$$
$$132,145,146,160,161,174,175,189$$

are held fixed throughout. This is equivalent to a physical environment in which two of the edges are clamped. All particles in the bottom layer ($P_{190}-P_{351}$) and top layer ($P_{352}-P_{513}$) will be free to move. We first wish to achieve physical stability for the arch. Experiments were run using the different values for the parameters G, H, p, and q. The choice of G=1.0, H=1.0, p=2 and q=4 was made for all computations to be discussed later because these yielded the least kinetic energy after 100,000 time steps. In addition, they are computationally convenient.

Note that damping is reasonable because energy has been added to the system in the bending process. Note also that elastic behavior will be assured by not imposing an elastic limit. In the calculations, all velocities were damped using $\vec{V}_{i,k} \Rightarrow \delta \times \vec{V}_{i,k}$ where $\delta=0.98$, every 2000 time steps. In addition $\vec{V}_{i,k}$ was reset to zero every 2500 time steps during the first 100,000 time steps in order to achieve physical stability.

5. A STABLE ARCH

Consider now the arch which was generated through k=100,000 in Section 4 with p=2, q=4, G=H=1. The system was now allowed to run until k=500,000 with damping factor $\delta=0.98$ every 2000 steps from the beginning. The kinetic energy E_k for the system at k=500,000 was $E_{500,000}= 0.85273E\text{-}01$. This new set of data will be used as the starting point for the following computer experiment in which we will try to emulate snap-through inversion.

6. PUSH ALONG LENGTH OF APEX

In the following computer experiment the arch will receive a load across the length of its apex. To achieve this simulation the z-values of particles P_{358}, P_{359}, P_{372}, P_{385}, P_{386}, P_{399}, P_{412}, P_{413}, P_{426}, P_{439}, P_{440}, P_{453}, P_{466}, P_{467}, P_{480}, P_{493}, P_{494}, P_{507} will be reduced by an amount of 0.1 every 1000 time steps. We have qualitative agreement with actual physical processes in that the applied loads are not transmitted instantaneously. Since $\Delta t=0.001$, the particles in question will move downward by 0.1 every 1 unit of time. If at any point in the loading process the velocity v_z becomes positive for any apex particle, v_z will be reset to zero. To compensate for the addition of energy to the system when the load is applied, the velocities of all particles will be damped by $\delta=0.98$ every 1000 time steps. A load will be applied for a fixed number of iterations. The load will then be released allowing all particles, including the load bearing ones, to move in any direction whatsoever. If the system returns to its original stable position, then snap-through has not occurred. The procedure will then be restarted and the load will be applied for a larger number of iterations, until the system, on its own, moves in a downward direction towards a stable position which is symmetrical to the original one. An unstable modal configuration will occur in this process. Intuitively, at an unstable mode the system cannot decide immediately whether to move up or down.

Without carrying out extensive computations, various particle motions will be used over relatively short periods to estimate whether the system, in the long run, will or will not snap-through. The particles whose motions that will be observed in this computer experiment are P_8, P_{37}, P_{66}, P_{95}, P_{124}, P_{153}, P_{182} and are called *indicators*. Indicators will always reside in the middle layer.

A load was applied across the length of the apex of the arch for 40,000 iterations(t=40 units). The average change in position of the indicators was 4.06 units downward. The load was released and the system was then damped every 2000 iterations, instead of every 1000 iterations. By the time 100,000 iterations had occurred, after the release, the indicators had moved, on the average over 2 units upward; the system was returning to its original shape—snap-through did not occur. The system was again restarted with the same load applied for 50,000 iterations. The indicators had moved, on the average, 5.13 units downward. The load was released and the system was damped every 2000 iterations. This time the system continued downward on its own accord. It must be cautioned that not all particles at every iteration had downward velocities. In fact many times during the iterations a particle would move upward slightly instead of downward, indicating internal wave motions. It should also be noted that when a downward load is applied, some particles move to the right or left causing bulges in areas, which also corresponds with experimental results[32,34]. When the load was applied for 50,000 iterations and released, the indicators did not move downward initially, but in fact, moved upwards. It took the indicators 30,000 additional iterations before all had negative z-velocities.

It can now be assumed that snap-through should occur somewhere between 40,000 to 50,000 iterations from the initial state describe in Section 5. Through trial and error snap-through was found to result at approximately 44,000 iterations. The indicators had moved, on the average, 4.35 units downward from their initial positions.

An additional item to look for in snap-through is the kinetic energy E_k (see TABLE 6.1). As one can see the kinetic energy was low at the time of release, it increases dramatically during the next 6,000 iterations and then decreases in an oscillatory manner.

TABLE 6.1 Kinetic Energy E_k — Load Applied

k	E_k
44,000	0.17724E+00
50,000	0.50062E+00
60,000	0.34050E+00
70,000	0.34345E+00
80,000	0.34147E+00
90,000	0.40555E+00
100,000	0.41197E+00
110,000	0.29674E+00
200,000	0.30634E-02
300,000	0.25067E-04
400,000	0.92618E-06
500,000	0.19859E-07

The snap-through sequence of motion to relative stability is shown in FIGURES 6.1—6.3. These figures show each layer separately, starting with the top layer (P_{352}—P_{513}), middle layer (P_1—P_{189}) and finally the bottom layer (P_{190}—P_{351}). In these figures, the observer is at a rotation of 10° and a tilt of 90° from the origin. As the system moves through time, more and more of the surface will be shown, which implies that the surface is going below the eye of the observer. Hence the more visible the surface, the lower the object.   FIGURE 6.2(a)—6.2(c) shows stiffness on the two edges. This is due to the fact that the particles on these two ends are fixed (or clamped). By the time the system reaches stable equilibrium this stiffness is directed downward.

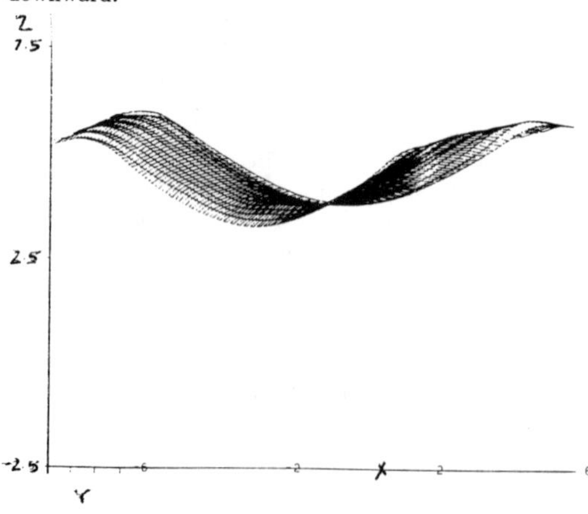

FIGURE 6.1(a) Case 1
P_{352}—P_{513}, k=50,000
(Top layer)

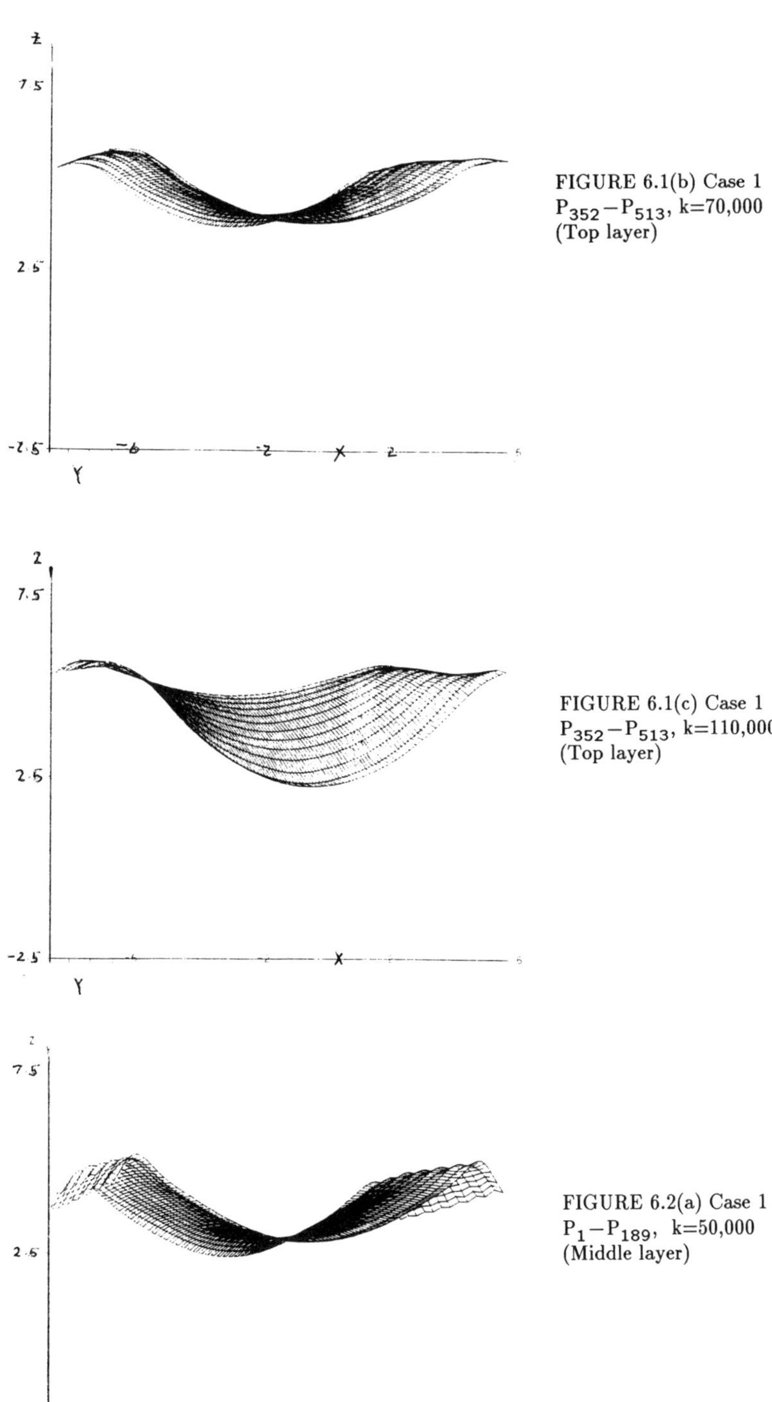

FIGURE 6.1(b) Case 1
$P_{352}-P_{513}$, k=70,000
(Top layer)

FIGURE 6.1(c) Case 1
$P_{352}-P_{513}$, k=110,000
(Top layer)

FIGURE 6.2(a) Case 1
P_1-P_{189}, k=50,000
(Middle layer)

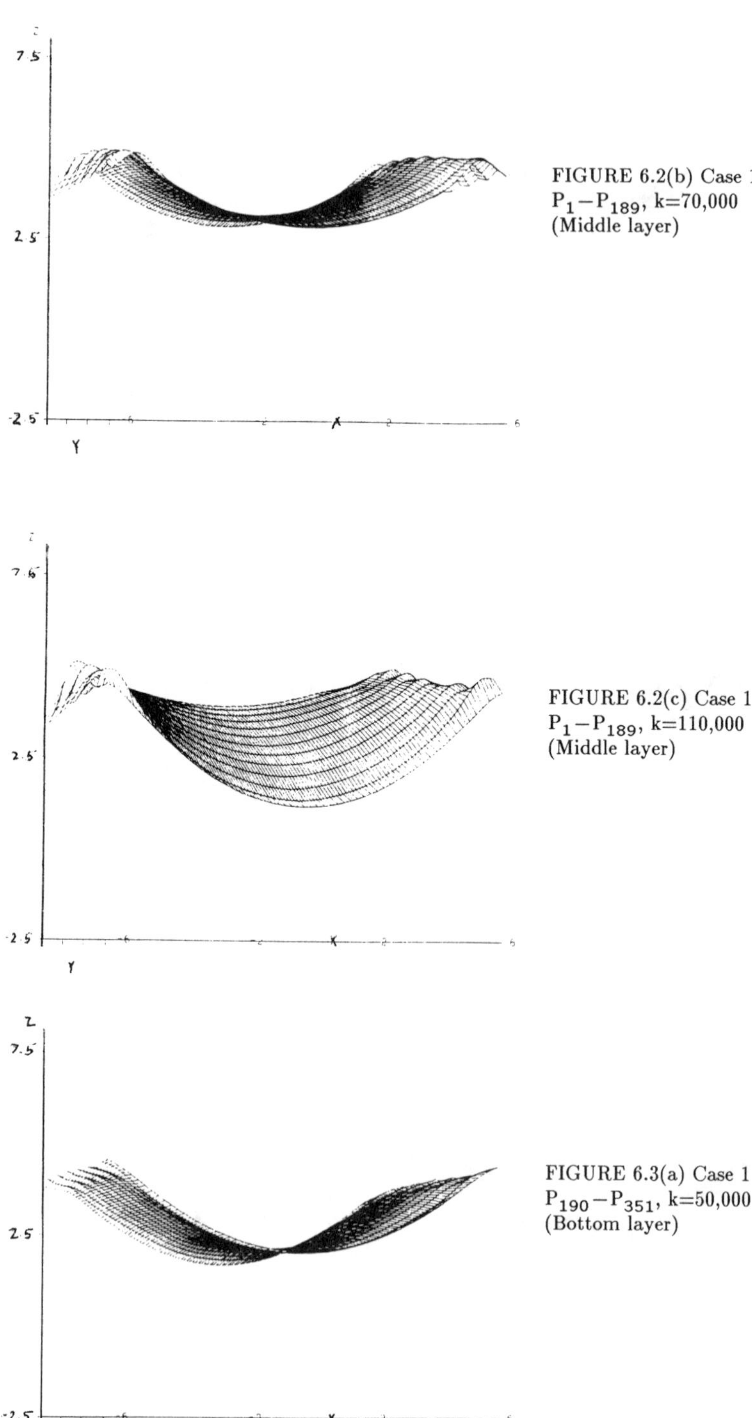

FIGURE 6.2(b) Case 1
P_1-P_{189}, k=70,000
(Middle layer)

FIGURE 6.2(c) Case 1
P_1-P_{189}, k=110,000
(Middle layer)

FIGURE 6.3(a) Case 1
$P_{190}-P_{351}$, k=50,000
(Bottom layer)

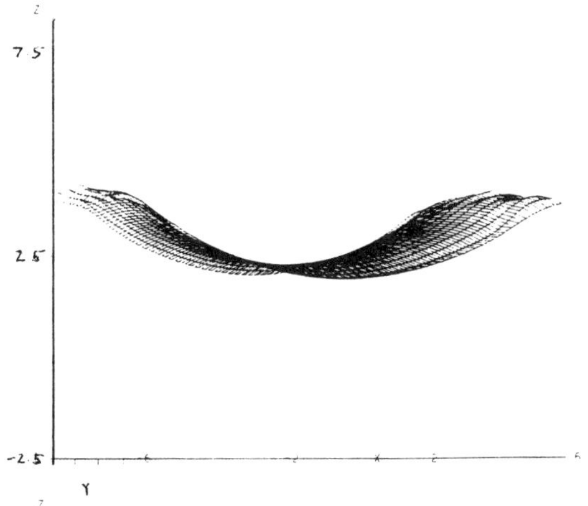

FIGURE 6.3(b) Case 1
$P_{190}-P_{351}$, k=70,000
Bottom layer)

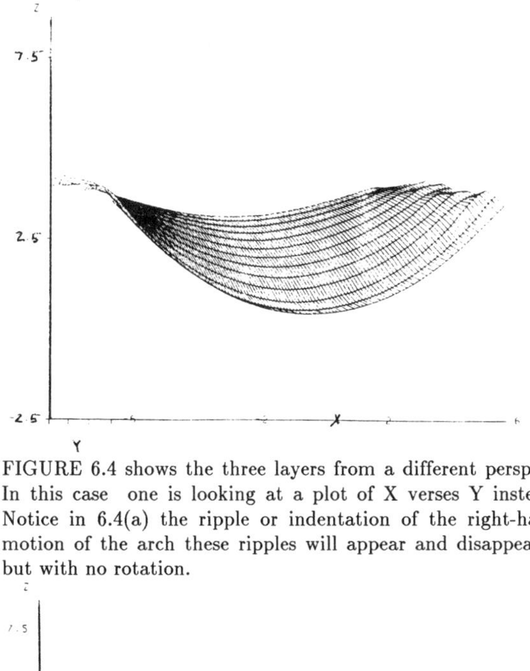

FIGURE 6.3(c) Case 1
$P_{190}-P_{351}$, k=110,000
(Bottom layer)

FIGURE 6.4 shows the three layers from a different perspective at the moment of snap-through. In this case one is looking at a plot of X verses Y instead of Y versus X, the rotation is 10°. Notice in 6.4(a) the ripple or indentation of the right-hand side of the shape; throughout the motion of the arch these ripples will appear and disappear. FIGURE 6.5 shows the same arch, but with no rotation.

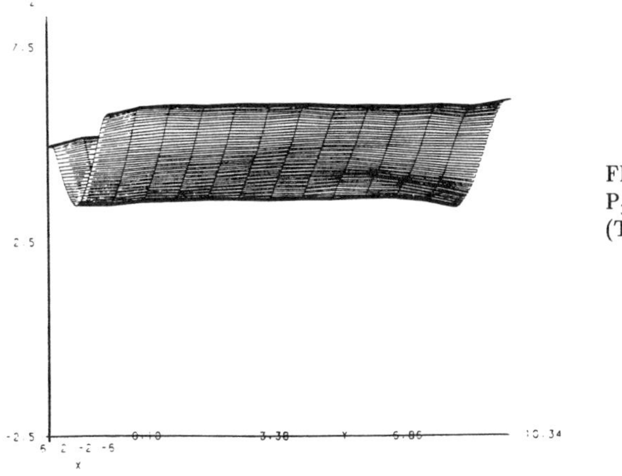

FIGURE 6.4(a) Axial View
$P_{352}-P_{513}$, k=44,000
(Top layer) load released

515

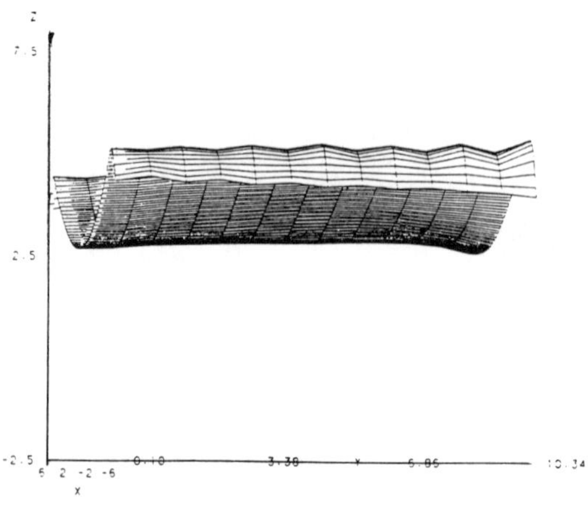

FIGURE 6.4(b) Axial View
P_1-P_{189}, k=44,000
(Middle layer) load released

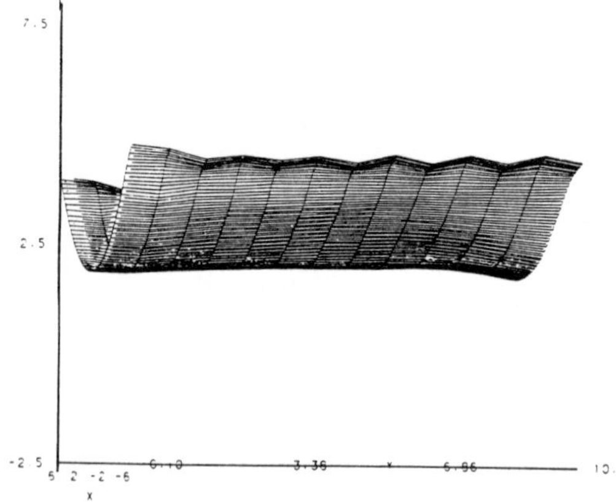

FIGURE 6.4(c) Axial view
$P_{190}-P_{351}$, k=44,000
(Bottom layer) load released

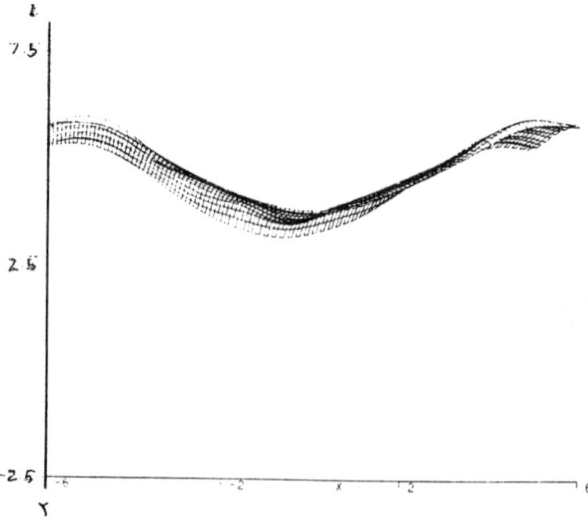

FIGURE 6.5(a) Rotation=0°
$P_{352}-P_{513}$, k=44,000
(Top layer) load released

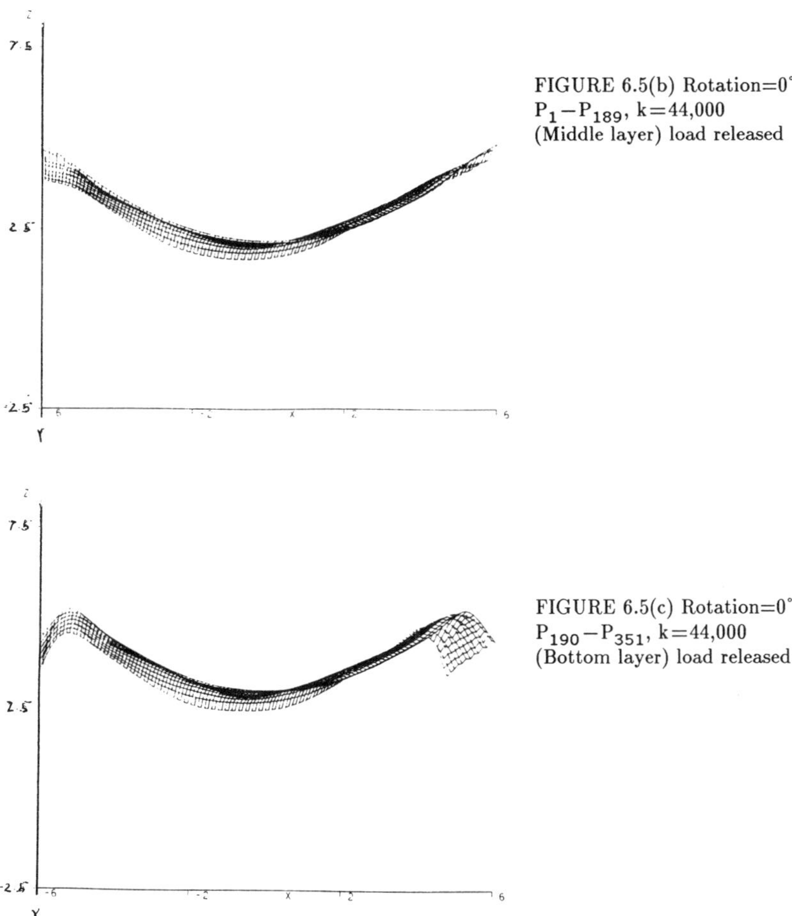

FIGURE 6.5(b) Rotation=0°
P_1-P_{189}, k=44,000
(Middle layer) load released

FIGURE 6.5(c) Rotation=0°
$P_{190}-P_{351}$, k=44,000
(Bottom layer) load released

Another item of importance to observe is the amount of the load that is distributed to each particle of the system at the moment of snap-through and how the forces between particles change in time as the arch reaches physical stability. In the following pictures, FIGURES 6.6−6.8, contour plots are shown. As one can see in FIGURE 6.6(a), the magnitude of the largest total force occurs in the top layer, concentrated about those particles where the load has been applied. In the middle layer, FIGURE 6.7(a), the maximum magnitude of the total force is again concentrated, but less so, along the center. On the bottom layer, FIGURE 6.8(a), the total force seems to be more scattered. FIGURES 6.6(b), 6.7(b), 6.8(b) shows the distribution of the magnitude of the total force at the later stages of stability. By describing this arch as consisting of three distinct, but joined, layers we are able to see on all levels the reaction of the inner and outer forces between all particles; usually one can only see what is happening on the top layer.

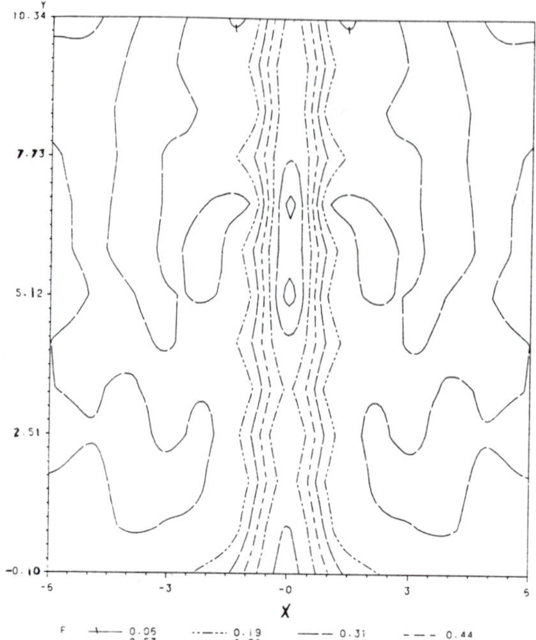

FIGURE 6.6(a) Contour plot $F_{i,k}$
$P_{352}-P_{513}$, k=44,000
(Top layer) load released

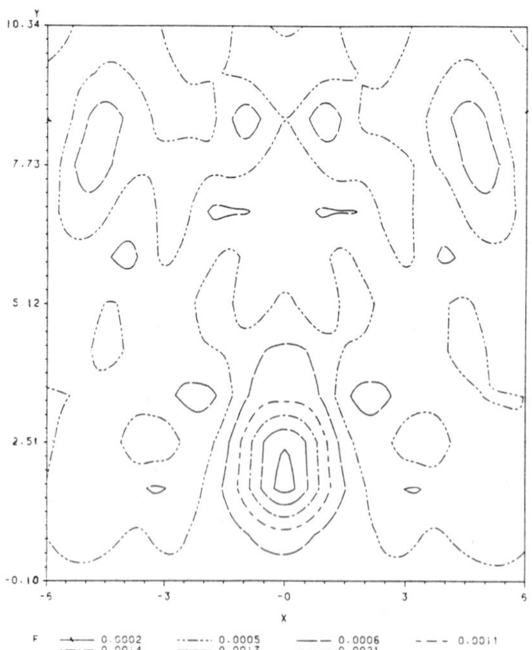

FIGURE 6.6(b) Contour plot $F_{i,k}$
$P_{352}-P_{513}$, k=300,000
(Top layer)

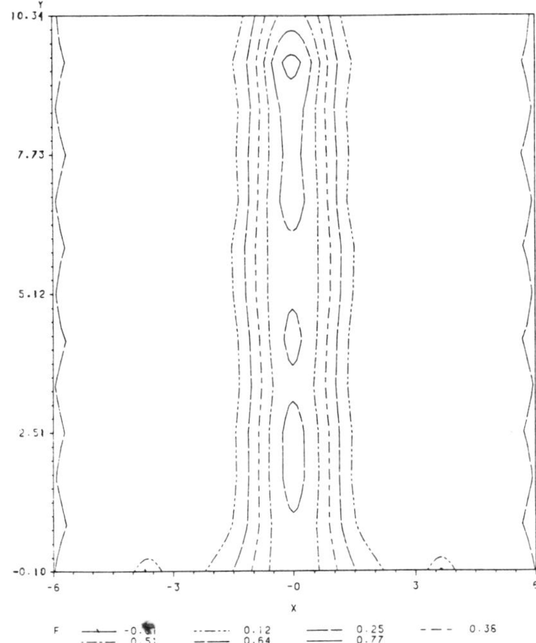

FIGURE 6.7(a) Contour plot $F_{i,k}$
$P_1 - P_{189}$, k=44,000
(Middle layer) load released

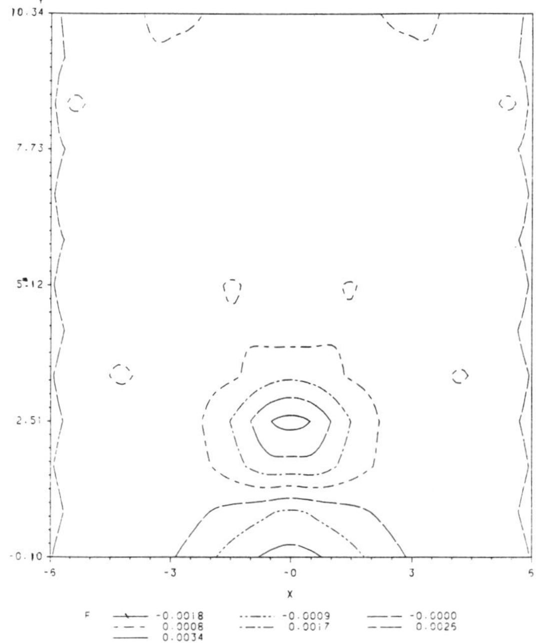

FIGURE 6.7(b) Contour plot $F_{i,k}$
$P_1 - P_{189}$, k=300,000
(Middle layer)

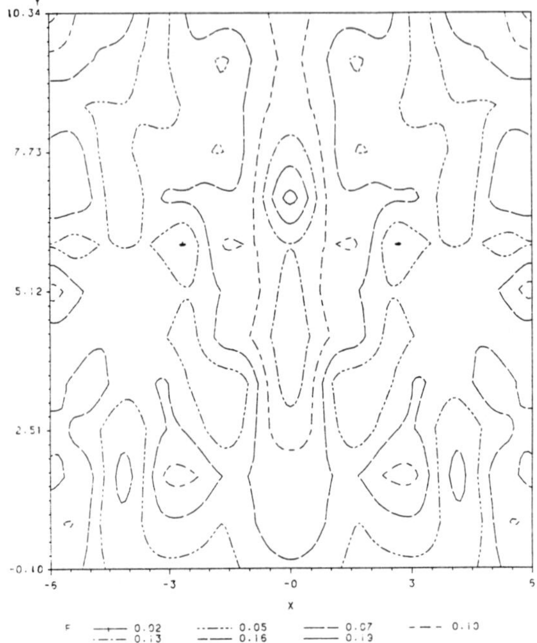

FIGURE 6.8(a) Contour plot $F_{i,k}$
$P_{190}-P_{351}$, k=44,000
(Bottom layer) load released

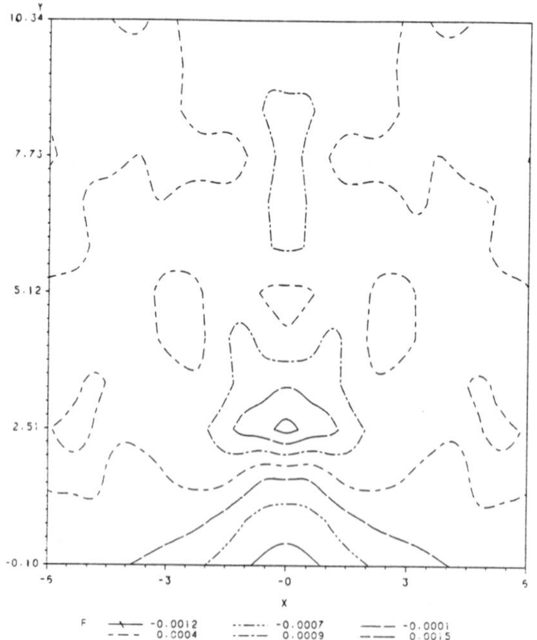

FIGURE 6.8(b) Contour plot $F_{i,k}$
$P_{190}-P_{351}$, k=300,000
(Bottom layer)

CONCLUSION

Much of the literature in elasticity theory deals with large sets of pde's. To implement these systems on a Supercomputer would require a great deal of effort. The approach in this paper was to use the molecular properties of the structure to describe its motion under applied loads which requires less effort than the traditional methods. In this computer experiment a load was applied to particles P_{358}, P_{359}, P_{372}, P_{385}, P_{386}, P_{399}, P_{412}, P_{413}, P_{426}, P_{439}, P_{440}, P_{453}, P_{466}, P_{467}, P_{480}, P_{493}, P_{494}, P_{507}. Snap-through inversion occurred after a deflection of over 4 units for the corresponding key particles P_8, P_{37}, P_{66}, P_{95}, P_{124}, P_{153}, P_{182} in the middle layer. It should be noted that the change in the kinetic energy E_k varies by only 0.0304% between 200,000 and 300,000 iterations; this says that one should not expect any extreme movements for the system of particles. In the process the kinetic energy E_k will continue to decrease in an oscillatory manner. As time progresses ripples and indentations will occur in various sections of the arch. These cannot be predicted and will effect the kinetic energy, thus causing a momentary increase in the kinetic energy. At $k=500,000(t=500$ units) the values for the velocities v_x, v_y and v_z are in the range $10^{-5} - 10^{-8}$; at any time some particles will be moving upward and others will be moving in a downward direction, but the system will always keep its characteristic concave upward appearance. This is entirely consistent with predictions from solid state physics. The study of the magnitude of the total force acting on each particle was observed through the use of contour plots. Equations (1.8), (1.9) and (1.11) may be restated in the following form (commonly found in engineering literature):

$$[\,M\,]\,\frac{d^2\,\vec{X}}{dt^2} + [\,C\,]\,\frac{d\,\vec{X}}{dt} + [\,K\,]\,\Delta\vec{X} = \vec{F}$$

where $[\,M\,]$ is a diagonal mass matrix, $[\,\underline{K}\,]$ and $[\,C\,]$ are matrices related to the parameters G, H, p and q (not necessarily constant) and $\vec{F}$ is related to the external forces.

BIBLIOGRAPHY

1. Beatty, Millard F. (1985), *Principles of Engineering Mechanics, Volume 1, Kinematics—The Geometry of Motion*, Plenum, New York.

2. Christodoulou, A.A. and A. N. Kounadis, (1986), Elastica Buckling Analysis of a Simple Frame, *Acta Mechanica*, Vol. 61, pp 153-163.

3. Ciarlet, P. G. (1988), *Mathematical Elasticity*, Studies in Mathematics & Its Applications, Vol. 20, North Holland, New York.

4. Churchill, R. V. (Editor), (1950), Elasticity, *Proceedings of the Third Symposium in Applied Mathematics of the American Mathematical Society*, Mc-Graw Hill, New York.

5. Das, T. and R. Sircar (1986), Vibration of Rectilinear Plates on Vlasov's Foundation at Large Amplitude, *Acta Mechanica*, Vol. 61, pp 217-225.

6. Dugdale, D. S. (1968), *Elements of Elasticity*, Pergamon Press, Oxford.

7. Eringen, A. Cemal and Erdoğan S. Şuhubi (1974), *Elastodynamics*, 2 Vols., Academic Press, New York.

8. Greenspan, Donald (1973), *Discrete Models*, Addison-Wesley, Reading, Mass.

9. Greenspan, Donald (1974), *Discrete Numerical Methods in Physics and Engineering*, Academic Press, New York.

10. Greenspan, Donald (1980), *Arithmetic Applied Mathematics*, Pergamon Press, Oxford.

11. Greenspan, Donald (1989), Supercomputer Simulation of Cracks and Fractures by

QuasimolecularDynamics, *J. Phys. Chem. Solids,* to be published 1989.

12. Greenspan, Donald and Vincenzo Casulli (1985), Particle Modelling of an Elastic Arch, *Applied Mathematical Modelling*, Vol. 9, June 1985, pp 215-219.

13. Greenspan, Donald and Vincenzo Casulli (1988), *Numerical Analysis for Applied Mathematics,Science, and Engineering*, Addison-Wesley, New York.

14. Gurtin, Morton E. (1981), *Topics in Finite Elasticity*, CBMS-NSF Regional Conference Series in Applied Mathematics, SIAM, Philadelphia, Pa.

15. Johnson, M. W. and E. Reissner (1958), On the Foundations of the Theory of Thin Elastic Shells, *Journal of Mathematics and Physics*, Vol. 37, pp 371-392.

16. Kelly, Theresa D. (1990), Particle Modelling of an Elastic Arch in Three Dimensions, Ph.D. Thesis, University of Texas at Arlington.

17. Komkov, Vadim (Editor), (1981), *Problems of Elastic Stability and Vibrations*, Contemporary Mathematics, Vol. 4, American Mathematical Society, Providence, Rhode Island.

18. Landau, L. D. and E. M. Lifshitz (1986), *Theory of Elasticity*, Course of Theoretical Physics, Vol. 7, 3rd Edition Revised and Enlarged, Pergamon Press, Oxford.

19. Libai, A. and J. G. Simmonds (1988), *The Nonlinear Theory of Elastic Shells: One Spatial Dimension*, Academic Press, Boston.

20. Naghdi, P. M. and L. Vongsarnpigoon (1985), Some General Results in the Kinematics of Axisymmetrical Deformation of Shells of Revolution, *Quarterly of Applied Mathematics*, Vol. 43, No.1, April 1985, pp 23-36.

21. Noor, A. K. and J. M. Housner (Editors), (1983), *Advances and Trends in Structural and Solid Mechanics*, Papers Presented at the Symposium on Advances and Trends in Structural andSolid Mechanics, Pergamon Press, New York.

22. Novozhilov, V. V. (1953), *Foundations of the Nonlinear Theory of Elasticity*, Graylock, Rochester, New York.

23. Pavlovic', M. N. (1985), A Simple Model for Thin Shell Theory-Part 2: Discretized Surface, Bending Theory and Membrane Hypothesis, *International Journal of Mechanical Engineering Education*, Vol. 13, No. 3, pp 199-222.

24. Pillasch, D. W., J. N. Majerus and A. R. Zak (1983), Dynamic Finite Element Model for Laminated Structures, *Computers and Sturctures*, Vol. 16, No. 1-4, pp 449-455.

25. Pogorelov, A. V. (1988), *Bendings of Surfaces and Stability of Shells*, Transactions of Mathematical Monographs, Vol. 72, American Mathematical Society, Providence,Rhode Island.

26. Prosser, R. T. (1966), *A New Formulation of Particle Mechanics*, Memoirs of the American Mathematical Society, Number 61, American Mathematical Society, Providence, Rhode Island.

27. Reismann, Herbert (1980), *Elasticity, Theory and Applications*, Wiley, New York.

28. Shin, Ke-Gang (1985), *On Axisymmetric Buckling Behavior of Nonlinearly Elastic Complete Spherical Shells*, University of Maryland-College Park, Maryland.

29. Smith, J. O. and O. M. Sidebottom (1969), *Elementary Mechanics of Deformable Bodies*, MacMillian, New York.

30. Sokolnikoff, I. S. (1956), *Mathematical Theory of Elasticity*, 2nd Ed., McGraw-Hill, New York.

31. Suchy, H., H. Troger, and R. Weiss (1985), A Numerical Study of Mode Jumping of Rectangular Plates, *Mathematik und Mechanik*, Vol. 65, No. 2, pp 71-78.

32. Terzopoulos, D., J. Platt, A. Barr and K. Fleishcer (1987), Elastically Deformable Models, *Computer Graphics*, Vol. 21, No. 4, July 1987, pp 205-214.

33. Terzopoulos, D. and K. Fleischer (1988), Modeling Inelastic Deformation: Viscoelasticity, Plasticity, Fracture, *Computer Graphics*, Vol. 22, No. 4, August 1988, pp 269-278.

34. Thielemann, W. F. and M. E. Esslinger (1967), On the Postbuckling Behavior of Thin-Walled, Axially Compressed Circular Cylinders of Finite Length, *Proceedings-Symposium on the Theory of Shells to honor Lloyd Hamilton Donnell*, edited by D. Muster, Univ of Houston, pp 433-477.

35. Thompson, J. M. T. (1964), The Post-buckling of a Spherical Shell by Computer Analysis, *World Conference on Shell Structures*, edited by S.J. Medwadowski, National Academy of Sciences Washington, D.C.

36. Thompson, J. M. T. and G. W. Hunt (1973), *A General Theory of Elastic Stability*, Wiley, London.

37. Thompson, J. M. T. and G. W. Hunt (1984), *Elastic Instablility Phenomena*, Wiley, New York.

38. Timoshenko, Stephen (1961), *Theory of Elastic Stablity*, 2nd Ed., McGraw-Hill, New York.

39. Timoshenko, Stephen and S. Woinowsky-Krieger (1959), *Theory of Plates and Shells*, 2nd Ed., McGraw-Hill, New York.

40. Walker, A. C. (1969), A Nonlinear Finite Element Analysis of Shallow Circular Arches, *International Journal of Solids and Structures*, Vol. 5, pp 97-107.

Instability Approach to the Magnetic Structure of Transition Metals

K. SCHWARTZMAN
Department of Physics
Indiana University of Pennsylvania
Indiana, Pennsylvania 15705, USA

ABSTRACT

The magnetic behavior of transition metal systems has been of continuing interest in condensed matter physics. A recently developed method for successfully understanding magnetism in these systems involves the analysis of instabilities of the paramagnetic phase toward magnetic order. The paramagnetic state is characterized by a wave vector dependent generalization of the Stoner susceptibility in which many-body effects are included self-consistently within the local-density approximation. The susceptibility is computed as a function of wave vector, frequency, and temperature for real systems from nearly first principles; singularities in the susceptibility are indicative of the formation of magnetically ordered structures. As a result, it is possible to determine magnetically ordered ground states, magnetic structures in the ground state, and spin-wave excitation spectra. Even though this approach is considerably more economical than total energy calculations, it does not sacrifice generality and is most conveniently studied using vector processing facilities. The method of computation, as well as the principal new physical results found, will be discussed.

1. INTRODUCTION

The physics of transition metals, despite many years of experimental and theoretical examination, remains a fertile ground for study. The vibrational structure of these systems is often highly anomalous in comparison with the spectra of 'simple' metals, and transition metals, their alloys, and compounds, form many of the known magnetically ordered systems. Among the outstanding theoretical questions are the temperature dependence of magnetic and vibrational quantities, the properties of magnetic thin films, and the vibrational behavior of shape-memory alloys, for example. Such interesting, and yet to be fully understood, behavior derives from the partially occupied tightly-bound d-states of the free-atom which lead to narrow electronic d-bands in the metallic state. The narrowness of the bands contribute to large effective masses and partial densities of states for the d-electrons. These factors, closely related to many-body effects and the multi-sheeted Fermi surfaces common to these systems, strongly influence all electronic and vibrational behavior, including ground state, equilibrium, and transport properties.

Magnetic structure is generally confined to the 3d row of the transition metals. Scandium and titanium, exhibiting hexagonal close packed (hcp) crystal structures, do not have magnetic ground states. Neither does vanadium which is body centered cubic (bcc). Bcc chromium, on the other hand, together with its (dilute) alloys with V and Mn (as well as other impurity species), is among nature's few incommensurate itinerant spin-density wave (SDW) anti-ferromagnets. The ground state magnetic structure of pure Cr (with a Néel temperature of

311K), as determined by elastic neutron scattering,[1] is an SDW with a wave vector q_{SDW}=[.95,0,0]$2\pi/_a$. Manganese is a peculiar entry in the 3d series, being the only element which does not have a conventional low temperature crystal structure. It is a complex bcc phase containing 29 atoms per unit super-cell, and exhibits anti-ferromagnetism below a Néel temperature of 96K. Manganese has simple bcc and face centered cubic (fcc) phases at temperatures above 1000K, typically above possible magnetic transition temperatures. Following Mn is the ferromagnetic series of iron (bcc), cobalt (hcp), and nickel (fcc).

Considerable technological interest has been directed towards transition metal alloys involving elements of the ferromagnetic series to produce high field magnets. More recently, a great deal of attention has been paid to growing crystal structures differing from the naturally occurring low temperature forms in thin films and layered systems. Such systems are studied for the possibility of large magnetic moments, anisotropies, and related properties. While this provides a need for theoretical predictions of magnetic properties for these novel structures, there still remain a wealth of fundamental questions regarding the naturally occurring crystal phases of the transition metals.

The motivation for the present study is to develop a theoretical and computational framework to answer these questions, as well as to provide accurate predictions for the artificially grown structures. More importantly, the computational approach which is developed here is one which can be applied to many problems in condensed matter physics involving integrations over the Brillouin zone or the Fermi surface including calculations of the electron-phonon interaction, phonon spectra and linewidths, and various transport properties such as the electrical resistivity. The overall approach utilized here is discussed in the next section and the formulation of the physical quantities that must be computed is discussed in Section 3. Methods employed in the calculation are described in Section 4 with particular attention paid to how vector-processing and parallel computing architectures may be put to use, tools which are rapidly becoming *necessary* as the scope and depth of accurate, real-system calculations increase. Finally, a summary of some of the results of the calculations is presented in Section 5.

2. INSTABILITY METHOD

A very powerful method of determining magnetic structure is derived from the response of a system to an externally applied magnetic field. In spontaneously ordering systems such as Cr, Mn, Fe, Co, and Ni, there is a *self-consistent* response to an internal field below the critical temperature. The response function most generally of interest in such systems is the paramagnetic spin susceptibility because the spin-spin correlation establishes the internal magnetic field which drives the transition to a magnetically ordered phase. The paramagnetic phase is considered not only as a matter of convenience to avoid some of the well-known difficulties inherent in a direct ferromagnetic or anti-ferromagnetic ground state calculation, but also to permit studies of excited magnetic states nearly inaccessible to other forms of calculation.

The paramagnetic state is unstable toward magnetic order if the susceptibility is singular or negative for some wave vectors in the Brillouin zone. Generally speaking, the structure of the magnetic ground state cannot be determined unambiguously if there is a range of wave vectors for which the paramagnetic susceptibility is negative. Furthermore, a series of poles in the susceptibility may be indicative of the possibility of multiple magnetic ground states. However, if the susceptibility is singular for a non-zero wave vector, and particularly if the wave vector corresponds to a maximum in the *unenhanced* susceptibility, then this wave vector may be taken to be characteristic of an SDW anti-ferromagnetic phase. Secondly, a singular or negative susceptibility at the zone center is indicative of the ferromagnetic phase. The instability approach was originally used by Penn,[2] but his calculations suffered from the lack of accurate band structures, many-body interactions, and integration techniques.

This method is very convenient for predicting magnetic ground states and their associated structures, but it also permits identification of the various mechanisms such as nesting, matrix element effects, and many-body interactions and how they conspire (or compete) to produce an magnetic instability. Secondly, the method also allows a relatively straightforward way to include temperature and frequency dependence. For example, a spin-wave spectrum can be determined from the poles of the susceptibility in $(q\text{-}\omega)$ space. Moreover, it permits the determination of magnetic transition temperatures by finding the temperature at which a singularity in the susceptibility first appears.

3 . THEORETICAL FORMULATION

The theoretical approach employed in this study is summarized here to point out its salient features, and how the approach must be extended to treat the problems of spin-wave spectra and temperature dependence. Although aspects of the theory have been discussed elsewhere,[3] these discussions have primarily been restricted to the application of the theory to ground state properties. Since spontaneous order is a consequence of particle-particle correlations the response function is singular at the critical point and, as discussed by Mattuck and Johansson,[4] the vertex in the polarization bubble will also be singular at the critical point. In the simplest application of this notion for magnetic systems, the Stoner model of the paramagnetic susceptibility is written

$$\chi(\mathbf{q},\omega) = \frac{\chi^0(\mathbf{q},\omega)}{1 - \Lambda\chi^0(\mathbf{q},\omega)} \tag{1}$$

where Λ is a constant (or perhaps $\mathbf{q}$-dependent) representation of the particle-hole contributions to the vertex in the polarization bubble, and χ^0 is the bare susceptibility. It should be noted, that Λ contains the *residual* interactions remaining after the singular part has been explicitly included in χ in the form of the denominator; secondly, the lack of frequency dependence in Λ is not as severe as it might at first seem because such variation is over significant features in the band structure, energy differences which typically contribute negligibly to the real part of the polarization bubble.

For application to real metallic systems the paramagnetic spin susceptibility can be written in a multi-band wave vector dependent generalization of the Stoner model to yield

$$\chi(\mathbf{q}) = \sum_{i,j,lm,n} I_{ij}(\mathbf{q}) \sum_{l,l'} \gamma_{ij,ll'} \left[(1\text{-}X(\mathbf{q}))^{-1}\right]_{ll',mn} I_{mn}(\mathbf{q}) \tag{2}$$

where $I_{ij}(\mathbf{q})$ is the atomic form factor which is written as a linear combinations of the radial probability densities of the free atoms, and

$$\gamma_{ij,ll'}(\mathbf{q},\omega) = -\frac{2}{N} \sum_{k,\mu,\nu} \frac{f_{k\mu}(1 - f_{k+q\nu})}{E_{k\mu} - E_{k+q\nu} - \omega} A_{i\mu}(\mathbf{k}) \, A_{j\nu}(\mathbf{k+q}) \, A_{l'\nu}(\mathbf{k+q}) \, A_{l\mu}(\mathbf{k}) \tag{3}$$

is the bare (not including many-body enhancements) *matrix* susceptibility. In Eq. (3), $E_{k\mu}$, $f_{k\mu}$, and $A_{\mu}(\mathbf{k})$ are the energy, Fermi occupation factor, and eigenvector, respectively, for an electron in the band state $(\mathbf{k},\mu)$. The integration runs over all wave vectors in the Brillouin zone (and all band combinations μ,ν) and a 9 element basis set is used (1 s, 3 p, and 5 d states as appropriate to transition metals). To good approximation, only the d-state basis functions need to be included in the eigenvector product (matrix element) in the integrand of Eq. (3). As a result, 625 sums must be performed corresponding to the indices i, j, m, and n which run over the 5 d basis states (plus an additional one sum the s and p states). However, symmetry considerations permit this number of sums to be reduced to 225. In Eq. (2), the matrix X is given in component form by

$$X_{ij,mn}(\mathbf{q}) = \sum_{l,l'} \Lambda_{ij,ll'} \, \gamma_{ll',mn}(\mathbf{q}) \tag{4}$$

where Λ contains many-body effects through the same local exchange and correlation potential used in the band structure calculation.

Studies of magnetic ground states may be performed by computation of Eqs. (2)-(4) with $\omega=0$. For non-zero temperatures, transition temperatures may be determined by including temperature variation through the Fermi factors alone[5] and finding T_c such that a pole first appears in $\chi(\mathbf{q})$; this also permits the temperature dependence of magnetic structures such as q_{SDW} to be found. However, in considering a *dynamic* response (finite ω) it is necessary to also include the orbital susceptibility in addition to the spin contribution.

4. COMPUTATIONAL CONSIDERATIONS

4.1 Overview

A feature common to many condensed matter calculations involves Brillouin zone integrals such as the one in the bare susceptibility of Eq. (3). Similar problems require Fermi surface integrations, but most of the computational considerations are analogous. An ideal approach to the Brillouin zone integration problem is provided by the analytical tetrahedron method (ATM) originally introduced by Rath and Freeman.[6] It is useful to first use the symmetry group operations of the cubic lattice to reduce the Brillouin zone to 48 irreducible elements (IE). In the ATM, the IE is further decomposed into an integral set of non-overlapping tetrahedra. An important feature of the ATM is that it yields an exact expression at zero temperature for integrals of the form

$$I_{\mu\nu}(\mathbf{q}) = \int \frac{|M_{\mu\nu}(\mathbf{k},\mathbf{q})|^2}{E_{\mathbf{k}\mu} - E_{\mathbf{k}+\mathbf{q}\nu}} f_{\mathbf{k}\mu} \, (1 - f_{\mathbf{k}+\mathbf{q}\nu}) \, d^3k \tag{5}$$

where the volume is taken over a single tetrahedron. In Eq. (5) it is assumed that the matrix elements M are constant, and that the band energies vary linearly between their values at the tetrahedron vertices. One consequence is that the intersection of the Fermi surface with a tetrahedron is a plane. While this approach was later extended[7] to allow for linear variation of the matrix elements as well, sample calculations of the susceptibility have shown that for typical numbers of tetrahedra, the neglect of matrix element variation results in at most a several percent discrepancy with the full calculation. Similar results can be expected for the calculations of other quantities if a sufficient number of tetrahedra are considered, and a significant savings in computing time is realized.

The integral in Eq. (5) as applied to the paramagnetic susceptibility produces information about the magnetic ground state. To study excited states it is necessary to generalize the ATM to non-zero temperatures and frequencies. The frequency dependence of *bare* quantities such as Eq. (5) may be correctly included simply by adding a frequency term to the denominator yielding

$$I_{\mu\nu}(\mathbf{q},\omega) = \int \frac{|M_{\mu\nu}(\mathbf{k},\mathbf{q})|^2}{E_{\mathbf{k}\mu} - E_{\mathbf{k}+\mathbf{q}\nu} - \omega} f_{\mathbf{k}\mu}(1 - f_{\mathbf{k}+\mathbf{q}\nu})\, d^3k. \tag{6}$$

The spin-wave spectrum is the curve in ($\mathbf{q}$-ω) space for which the susceptibility is singular. However, the orbital contribution to the susceptibility is important for correctly describing the spectrum and must be included. Secondly, the frequency dependence also affects the vertex corrections contributing to the many-body renormalization of the susceptibility. The first of these problems is straightforward, and the realistic modelling of the frequency dependent vertex is underway.

The inclusion of temperature dependence in the susceptibility[5] is superficially trivial: simply use the explicit form of the Fermi factors $f_{\mathbf{k}} = [\exp(\beta(E_{\mathbf{k}}-\zeta))+1]^{-1}$ where $\beta=1/k_BT$ and ζ is the chemical potential. This is appropriate in a physical sense since the band structure and atomic wave functions are essentially independent of temperature for $k_BT \ll \zeta$, as are many-body effects. Extension of the ATM to treat temperature variation can be accomplished by considering a fixed point $\mathbf{k}_0$ in a tetrahedron and expanding the numerator in Eq. (5) [or Eq. (6)] to first order about that point:

$$M_{\mu\nu}(\mathbf{k},\mathbf{q})f_{\mathbf{k}\mu}(1-f_{\mathbf{k}+\mathbf{q}\nu}) = f_{\mathbf{k}_0\mu}(1-f_{\mathbf{k}_0+\mathbf{q}\nu})\left\{\left[M_{\mu\nu}(\mathbf{k}_0,\mathbf{q}) + (\mathbf{k}-\mathbf{k}_0)\cdot\nabla_{\mathbf{k}}M_{\mu\nu}(\mathbf{k}_0,\mathbf{q})\right]\right.$$

$$\left. + \beta M_{\mu\nu}(\mathbf{k}_0,\mathbf{q})\left[f_{\mathbf{k}_0+\mathbf{q}\nu}\nabla_{\mathbf{k}}E_{\mathbf{k}_0+\mathbf{q}\nu} - (1-f_{\mathbf{k}_0\mu})\nabla_{\mathbf{k}}E_{\mathbf{k}_0\mu}\right]\cdot(\mathbf{k}-\mathbf{k}_0)\right\}. \tag{7}$$

The first term in square brackets is the zero temperature ATM with the absence of a sharp cutoff at the Fermi level. The remaining terms may be considered to be pseudo-matrix elements and computed following the standard ATM procedure. An obvious problem inherent in Eq. (7) is the rapid variation of $f_{\mathbf{k}}$ over a tetrahedron near the Fermi level, particularly for low temperatures. This can be compensated for by employing large numbers of tetrahedra in the IE: to recover the known $1/_T$ dependence of the susceptibility for the non-interacting electron gas it is necessary to use approximately 30,000 tetrahedra in the IE for T>150K. For T<150K, the variation of $f_{\mathbf{k}}$ is too rapid for even this tetrahedron density. However, much of transition metal magnetic phenomena occur above this temperature so that it is not typically necessary to consider lower temperatures (apart, of course, from T=0).

4.2 Vectorization and Parallelism

Aspects of computing the susceptibility using the ATM have been discussed elsewhere[3,5] and will not be repeated here. The algorithm may be summarized, however, by noting that only those tetrahedra in the IE which lie below or are intersected by the Fermi surface need to be considered. The occupied volume of a tetrahedron is sub-divided into smaller fully occupied tetrahedra, whose vertices are given by the set of four wave vectors $\{\mathbf{k}_i,\mu\}$ for the band μ. The integrals of Eqs. (5), (6), or (7) are then computed if the corresponding tetrahedron given by $\{\mathbf{k}_i+\mathbf{q},\nu\}$ contains at least some unoccupied states. The code for this procedure has been optimized for scalar computation and the zero temperature, static susceptibility may be calculated in between 3 and 20 minutes of CPU time on a Cray Y-MP depending on the material in question and the symmetry of $\mathbf{q}$. Because, in general, many wave vectors are needed to completely describe the magnetic ground state of a given material, between 6 and 12 CPU hours may be needed in total. This problem is compounded by frequency and temperature dependence.

A first step in accelerating the computation can be achieved by use of a parallel-processing facility in either a scalar or vector-processing cluster environment. Rotations of the wave vector $\mathbf{q}$ by the 48 symmetry group operations of the cubic lattice will generally lead to a set of degenerate wave vectors whose multiplicity depends on the symmetry of $\mathbf{q}$. Each of the *distinct* rotated wave vectors has a set of operations which can be performed in parallel including generation of the band structure for the $\{\mathbf{k}_i+\mathbf{q}\}$ and the integration over the IBZ. This generally leads to a decrease in computing time by a factor roughly equal to the number of processors running in parallel.

A generally more significant reduction in computing time can be realized by restructuring the ATM algorithm to more fully utilize vector-processing architecture. An efficient approach is to introduce three loops: the first identifies all the tetrahedra (and sub-tetrahedra) which contribute to the integral, the second tests for special conditions (such as the Fermi surface intersecting a tetrahedron along an edge) that must be accounted for, and the third performs the explicit integrations. With some care, each of these loops can be fully vectorized, and this leads to as much as a ten-fold reduction in computing time depending on the wave vector and material parameters. Analogous procedures can be applied to a wide variety of related problems, including Fermi surface integrations, which make the detailed *quantitative* description of magnetic excited states (and other phenomena) of *real* transition metal systems possible for the first time.

5. RESULTS

Calculations of the zero temperature static susceptibility have led to a detailed understanding of the anti-ferromagnetism of Cr[3] and its dilute alloys with V and Mn[8] as well as predictions of magnetic order in a variety of alternate cubic phases of the 3d transition metals.[3] Successful preliminary results for transition temperatures and the temperature dependence of the SDW in pure Cr have also been reported.[5] However, the primary new physical result of this study is the unexpected form of the magnetic ground states of the 3d *cubic* transition metals as a function of volume.

Under sufficiently large expansion of the lattice all of the transition metals will eventually become ferromagnetic in the free-atom limit in accordance with Hund's rule. This has been confirmed numerically using a total energy approach in which the ground state energies of the non-magnetic and ferromagnetic states were computed and compared.[9] Using the instability method discussed earlier, the ferromagnetic instability is predicted for lattice constants in good agreement with the total energy calculations[9] as shown in Table 1 in which the notation pm is used to refer to paramagnetic, af to antiferromagnetic, and fm to ferromagnetic. However, the new physical result is the apparent prevalence of an anti-ferromagnetic instability for a range of lattice constants less than the critical wave vector for the onset of ferromagnetism. Incommensurate itinerant anti-ferromagnetism finds its origins in both many-body effects and the nesting of sheets of the Fermi surface. The latter contribution becomes relatively less important as the volume increases which favors the tendency toward ferromagnetic order. This preliminary, but highly interesting, new result will be discussed in detail in a future publication.

6. CONCLUSIONS

The analytic tetrahedron method has been extended to treat Brillouin zone integrations for both frequency and temperature dependence. This permits a wide variety of new physical problems to be studied in non-equilibrium magnetism (and related phenomena such as transport) for *real* transition metal systems. The major difficulty of lengthy computing times has been alleviated by the development of a new algorithm for the ATM to take advantage of both parallel- and vector processing facilities.

TABLE 1. Predicted Critical Wigner-Seitz Radii (in atomic units).

System	Transition	r_{ws} (instability)	r_{ws} (total energy)
bcc Sc	pm - fm	3.37	3.65
bcc Ti	pm - af	3.16	
	af - fm	3.67	3.24
bcc V	pm - af	3.04	
	af - fm	3.19	3.17
bcc Cr	pm - af	2.67	
	af - fm	3.20	3.46
bcc Mn	pm - fm	2.43	2.53
bcc Fe	pm - af	2.10	
	af - fm	2.38	2.28

REFERENCES

1. W. C. Koehler, R. M. Moon, A. L. Trego, and A. R. Mackintosh, Physical Review **151**, 405 (1966) and references therein.

2. D. R. Penn, Physical Review **142**, 350 (1966).

3. K. Schwartzman, Y. Z. Zhao, J. L. Fry, and P. C. Pattnaik, submitted to Physical Review B.

4. R. D. Mattuck and B. Johansson, Advances in Physics **17**, 509 (1968) and references therein.

5. K. Schwartzman, E. J. Hartford, and J. L Fry, Journal of Applied Physics **67**, 4552 (1990).

6. J. Rath and A. J. Freeman, Physical Review B **11**, 2109 (1975).

7. P. C. Pattnaik, J. L. Fry, N. E. Brener, and G. Fletcher, International Journal of Quantum Chemistry Symposium **15**,499 (1981).

8. K. Schwartzman and J. L. Fry, accepted to Journal of Physics: Condensed Matter.

9. V. L. Moruzzi and P. M. Marcus, Physical Review B **38**, 1613 (1988).

Computational Approach to Determine Magnetic Order in High Temperature Superconductors

J. L. FRY, T. W. CHANG, and E. C. ETHRIDGE
Department of Physics, Box 19059
The University of Texas at Arlington
Arlington, Texas 76019, USA

P. C. PATTNAIK
IBM Thomas J. Watson Research Center
Yorktown Heights, New York 10598, USA

ABSTRACT

The paramagnetic susceptibility, $\chi(q)$, has been computed for La_2CuO_4 using a Slater-Koster band structure parameterization and a local-density, many-body enhancement theory for $\chi(q)$. Matrix elements were computed with approximate wave functions obtained from the self-consistent potentials. The random-phase approximation, χ_0, was computed with constant and k-dependent matrix elements , giving agreement with previous work using constant matrix elements, but showing dramatic reductions of $\chi_0(q)$ near the nesting vector in the [110] direction. In spite of this reduction, a pole developed in $\chi(q)$, suggesting an antiferromagnetic instability for stochiometric La_2CuO_4.

I. INTRODUCTION

After many years of slow progress in achieving superconductivity at higher critical temperatures (T_c), significant discoveries were made beginning in 1986 with the announcement[1] of the discovery of superconducting oxides with T_c exceeding 35K. The previous high had been about 23K. This discovery in doped La_2CuO_4 was followed rapidly by the discoveries of T_c around 90K in $YBa_2Cu_3O_7$ (Ref. 2) and $T_c > 120K$ in $Tl_2Ba_2CaCu_2O_8$ (Ref. 3).

These new doped compounds are ceramic materials with properties unfamiliar to most physicists and have complex structures which make interpretation of experiments and theory difficult. Because of the great technological importance of superconductivity above 79K (or better yet above room temperature), essentially every physicist in the world with any special skills to offer has attempted to contribute to the body of knowledge of this new class of materials. To date there is no generally accepted theoretical explanation of the high values of T_c achieved, nor can other properties of these compounds be explained in any consistent fashion. Theoretical understanding of the normal phase (above T_c) is still lacking. At this time few things are certain, but superconductivity in these materials is probably not caused exclusively by the electron-phonon interaction (as in all previously discovered materials), but may be dominated by some special type of electron-electron correlations. The 1989 review article by Pickett provides many references for anyone interested in the basic experimental findings and various theories proposed to explain the high values of T_c[4]

It is not the purpose of this paper to describe any new theories, but to point out the enormous computational difficulties associated with even the simplest theories for these compounds. As an example, the normal, undoped phase of La_2CuO_4 is examined here to investigate the instability of the paramagnetic phase toward the formation of some magnetically ordered phase. One of the key theoretical questions which distinguishes the various theories is whether the normal phase is a metal, behaving like a Fermi liquid, or a more complicated, strongly correlated insulator.[4] Predictions of band theory for this compound indicate that it should be a paramagnetic metal, while experiments find an insulating, antiferromagnetic material.

II.   THEORETICAL METHODS

In this paper the paramagnetic band structure computed by Pickett et al.[5] has been used to compute the density of electronic states and the wave vector dependent magnetic susceptibility of La_2CuO_4. Poles of the enhanced susceptibility are identified with magnetic instabilities of the crystal, contradicting the initial assumption of a paramagnetic ground state.

A.   Band Structure of La_2CuO_4

The electronic structure of La_2CuO_4 has been computed using a modern local-density theory to account for quantum mechanical many body effects approximately.[5] A paramagnetic ground state was assumed for the high temperature body-centered-tetragonal (bct) phase shown in Fig. 1. This band structure has been fitted[6] with an efficient Slater-Koster (SK) type interpolation scheme[7] to provide a dense sampling of energy v.s. momentum. This is needed to compute accurately the density of allowed states per unit range in energy, or other statistical quantities such as the magnetic susceptibility which is studied here. In this work the SK parameters from the fit of Ref. 6 were employed. Since the structure of the SK Hamiltonian matrix is complicated (and complex), two independent computer programs were developed, one for bct and the other for an

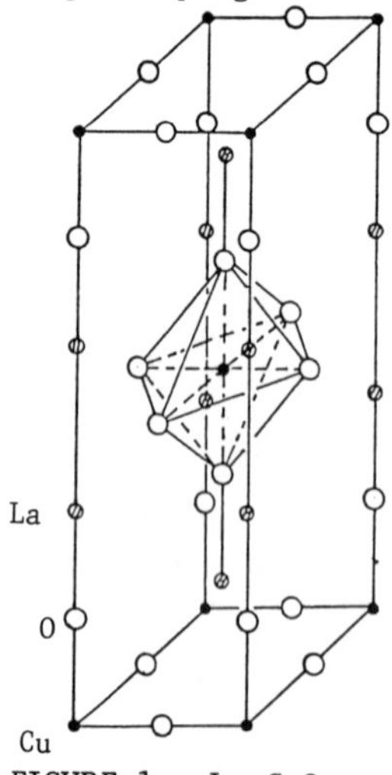

FIGURE 1. La_2CuO_4 crystal structure

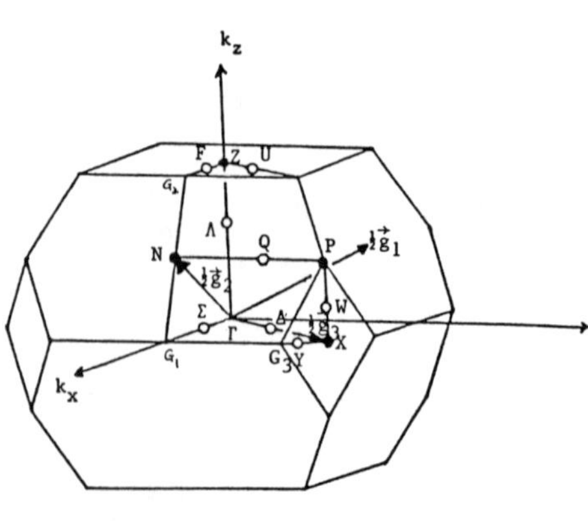

FIGURE 2.   BCT brillioun zone

arbitrary crystal structure, and both were tested against the output of computer codes of Ref. 6. After months of checking agreement between the three codes was established for the band structure of La_2CuO_4. This band structure is plotted in Ref. 6 and also Ref. 4, Fig. 4.C.2. It employed five d-orbitals for each La atom, one s and 3 p-orbitals for each Cu atom and three p-orbitals on each O atom. The two-center approximation was made in writing out the 31x31 SK Hamiltonian matrix, retaining the same number of neighboring atoms as in Ref. 6. Because it is complex, this matrix leads to great computational difficulties compared to transition metals, where a real 9x9 matrix may be employed.[8]

B. Brillouin Zone Integration

Observable quantities for a periodic solid may be expressed as integrals over a unit cell in momentum space, the first Brillouin zone (BZ). For the bct structure this zone is shown in Fig. 2. A grid of points was constructed inside the irreducible BZ (IBZ), the wedge outlined inside the BZ in Fig. 2. The IBZ was sub-divided into tetrahedrons as shown in Fig. 3, and each tetrahedron further broken into smaller tetrahedrons until a predefined maximum length of a tetrahedron was obtained. All the IBZ and hence the BZ may be completely filled with small tetrahedrons. This elemental volume shape is very convenient for numerical integration.

For example the partial density of electronic states of angular momentum type j may be written

$$N_j(\epsilon) = \frac{\Omega}{4\pi^3} \sum_n \int_{BZ} \frac{|\Psi_{nj}(k)|^2 dS_k}{|\nabla_k E_n(k)|} \tag{1}$$

where Ψ_{nj} is the jth angular momentum orbital of band n with energy $E_n(k)$, Ω is the volume of the primitive cell in r-space, k is the crystal momentum, and the integral is carried out over the surface defined by $E_n(k)=\epsilon$ in each tetrahedron. For La_2CuO_4 it was necessary to use 597 tetrahedrons in the IBZ (14,328 in the BZ) to achieve accurate values for $N_j(\epsilon)$ in Eq. 1. This required diagonalizing 180 31x31 complex matrices in the IBZ before performing the integration in Eq. 1. This process was initially done with vectorized codes on the Cray XMP-24 at the University of Texas Center for High Performance Computing. The method of integration employed was discussed at the first Integral Methods in Science and Engineering Conference.[9] Results for the total density of states as a function of energy are shown in Fig. 4. The small peak near the Fermi level was first thought to play an important role in the superconducting transition, and determining its position relative to the Fermi energy, E_F, required dense sets of integration points in k-space.

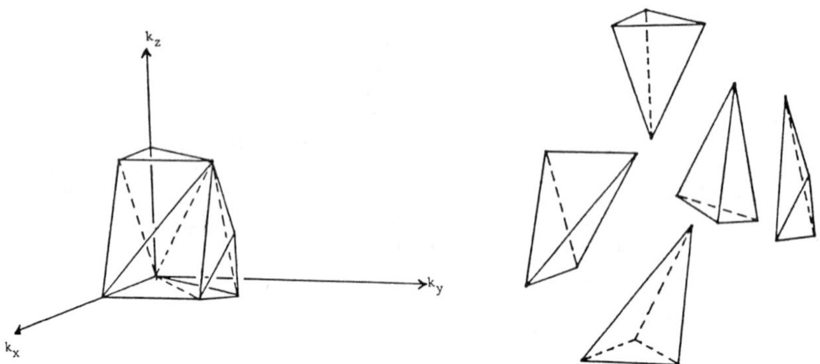

FIGURE 3.   Partitioning of BCT IBZ into tetrahedrons

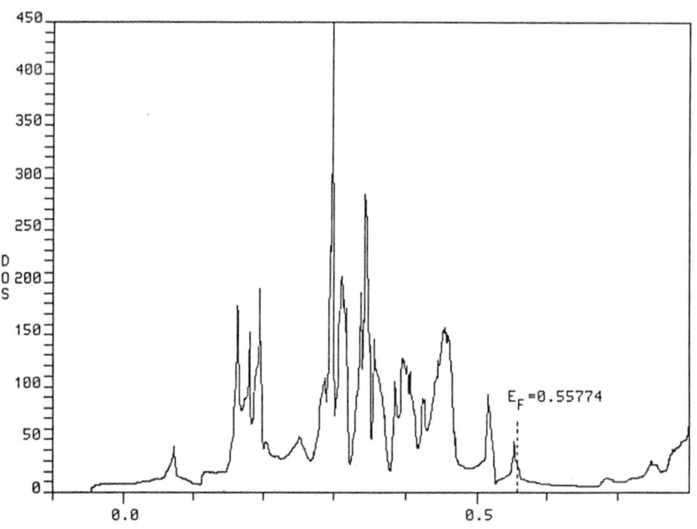

FIGURE 4.   Density of states of La_2CuO_4

C.   Magnetic Susceptibility

The magnetic susceptibility indicates a magnetic instability for the
system according to the following argument. The magnetic susceptibility
$\chi(q)$ is defined in linear response theory by

$$M(q) = \chi(q)H(q) \tag{2}$$

where $M(q)$ is the qth Fourier component of the induced magnetization and
$H(q)$ is the Fourier transform of a weak applied magnetic field. If $\chi(q)$
is predicted to be arbitrarily large, then an arbitrarily small applied
(or random fluctuation) field could generate a finite magnetization,
$M(q)$.

Thus the material could spontaneously change to a magnetically ordered
periodic structure with wavelength characterized by the q at which a

534

pole of $\chi(q)$ occurs: $\lambda=2\pi/q$. On the other hand, if $\chi(q)$ is negative for any range of q, the original assumption of a paramagnetic system is also contradicted, and a magnetic instability is also possible anywhere in that range of q.

Two expressions have been employed here for $\chi(q)$. One is the random phase approximation (RPA), which ignores correlations between electrons in the crystal:

$$\chi_0(q) = - \frac{1}{N} \sum_{\ell,n,k} \frac{[f_n(k)-f_\ell(k+q)] \; |M_{n\ell}(k,q)|^2}{E_n(k) - E_\ell(k+q) + \iota\eta} \tag{3}$$

$$M_{n\ell}(k,q) = \langle n,k| e^{-\iota q \cdot r}\sigma_+ |\ell,k+q\rangle \tag{4}$$

Here σ_+ is the spin raising operator, $f_n(f_\ell)$ is the Fermi function for band $n(\ell)$ at the point $k(k+q)$, N is the number of unit cells in the crystal and η is a positive infinitesimal quantity. The principal value of the sum is to be taken.

This RPA expression is approximate and must be corrected for many body effects in systems in which electron motions are strongly correlated. These corrections generally enhance the RPA values. The expression for the enhanced susceptibility employed here uses local density theory to provide vertex corrections,[10] and has been quite successful in explaining or predicting magnetic instabilities in transition metals.[10]

$$\chi(q) = \chi_0(q)[I-\Lambda\chi_0(q)]^{-1} \tag{5}$$

Here χ is the enhanced susceptibility, χ_0 the RPA susceptibility and Λ contains the vertex corrections. These quantities are all tensors of rank two (matrices) with reciprocal lattice vectors as indices. Poles of χ thus occur for $\det[I-\Lambda\chi_0]=0$. For simple metals, semiconductors and insulators the matrices in Eq. 5 must be of order 200 or so to obtain accurate matrices for χ upon inversion. For compounds or metals containing transition elements the order has been estimated to be well in excess of 10,000, and perhaps far in excess of that number. The time, T, to compute $\chi_0(q)$ is proportional to $N^2 xM^2 xP$, where N is the number of bands in Eq. 2, M is the number of RLV and P is the number of q points needed to map out $\chi(q)$ in order to reveal the pole structure. q is a vector in three dimensions in general. For transition metals[10] the minimum values required for accurate results are N=9, M=10,000, and P=100 along a given direction, so $T=T_o x10^{12}$, where T_o is the time to compute the sum (integral) over k in Eq. 3. Each evaluation requires a three dimensional principal value integration, and additional time is required to invert the matrix in Eq. 5 for each q. Adding to the difficulty is the problem of intermediate storage during the calculation. To address these problems requires not just larger and faster computers, but completely different techniques. The approach used here is to reduce the factor M by transforming to an "orbital" basis rather than a reciprocal lattice basis to compute χ_0.[11]

On the orbital basis the RPA susceptibility becomes

$$\gamma_{\iota j,\iota'j'}(q) = -\frac{1}{N}\sum_{\ell,n,k}\frac{[f_n(k)-f_\ell(k+q)]C_{n\iota}^{*}(k)C_{\ell j}(k+q)C_{n\iota'}(k)C_{\ell j'}^{*}(k+q)}{E_n(k) - E_\ell(k+q) + \iota\eta} \qquad (6)$$

where $C_{n\iota}(k)$ is the ith eigenvector component for band n of the Hamiltonian matrix on an orbital basis at the point k in the BZ. For a transition metal this becomes an M^2xM^2 matrix, where M is now the number of orbitals per unit cell. For a cubic transition metal M is typically 9 in a SK representation, resulting in an 81x81 matrix for χ_0. After lengthy algebra an explicit expression may be written for $\chi(q)$ on the RLV basis in terms of $\chi_0(q)$ on the much smaller orbital basis, and the matrix construction and inversion time is more feasible.

The orbital basis matrix has been computed and the matrix inversion of Eq. 5 done for a number of transition metals.[10] For La$_2$CuO4 the matrices are complex and at least 961x961. Core requirements to compute χ_0 are increased by a factor of 280 and the time to diagonalize the Hamiltonian matrices is increased by a larger factor compared to transition metals. Because of bct rather than cubic symmetry another factor of two is lost. Even on the orbital basis it was necessary to move from a Cray XMP/24 to a much larger memory Cray 2, and partition the matrix operations in Eq. 5. In spite of these efforts the results given in Section III below required many CPU hours on Cray computers.

III. MAGNETIC SUSCEPTIBILITY OF La$_2$CuO$_4$

A. RPA Susceptibility

The RPA susceptibility of La$_2$CuO$_4$ is shown in Fig. 5 along the [110] direction in k-space. The top curve was obtained by assuming that the quantum mechanical matrix elements M(k,q) in Eq. 4 are k and q independent and diagonal in other indices. The curve is very similar to results published by other authors who used this approximation to make the calculation managable.[4] Magnetic order observed experimentally in the undoped compound has a periodicity which has been associated with a q-vector near the zone boundary (1.00 in Fig. 4)[4], so these initial findings were thought to be consistent with experiment.

The lower curve in Fig. 5 was obtained by including proper k and q dependent matrix elements in calculating $\chi_0(q)$, but including only band 17 (the only one crossing the Fermi energy), while the middle curve included all 31 bands of the SK Hamiltonian and the corresponding matrix elements. Proper matrix elements dramatically reduced $\chi_0(q)$ near the zone boundary "nesting" peak. It is therefore concluded that χ_0 does not serve as a good indicator of the observed magnetic order after all.

B. Enhanced Susceptibility

The enhanced susceptibility was computed as discussed in Sec. II along the [110] direction in the BZ. Because of the enormous computational requirements it was possible to compute only the contribution from band 17. The form of Eq. 5 suggests that inclusion of more bands in χ may add more pole structure but it should not alter any structure due to band 17 only. Results are shown in Fig. 6, where strong pole structure is revealed. The peak in χ_0 near 1.00 which disappeared with the inclusion of matrix elements has reappeared in the form of a pole in χ. As seen in the figure there are poles of χ and also regions of q for which $\chi(q)$ is negative, including q=0 which corresponds to a ferromagnetic instability (infinite wavelength). Unfortunately the approach used here cannot determine which of the various possible magnetic ordering q-vectors would occur experimentally. This would require a much more complicated total energy calculations for the various magnetic orderings to find the stable (lowest energy) configuration.

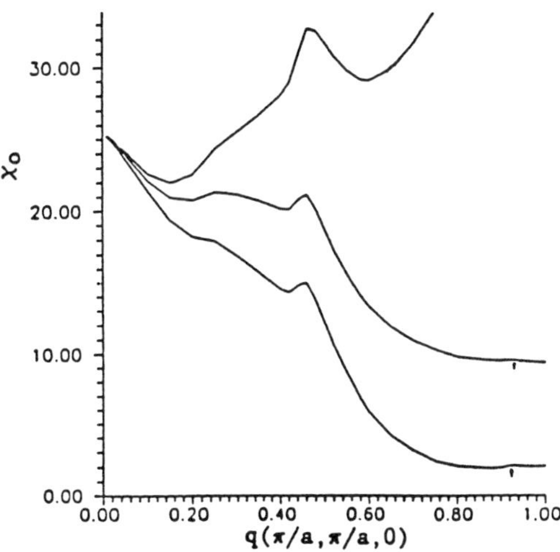

FIGURE 5.   RPA Susceptibility of La_2CuO_4 along [110] direction

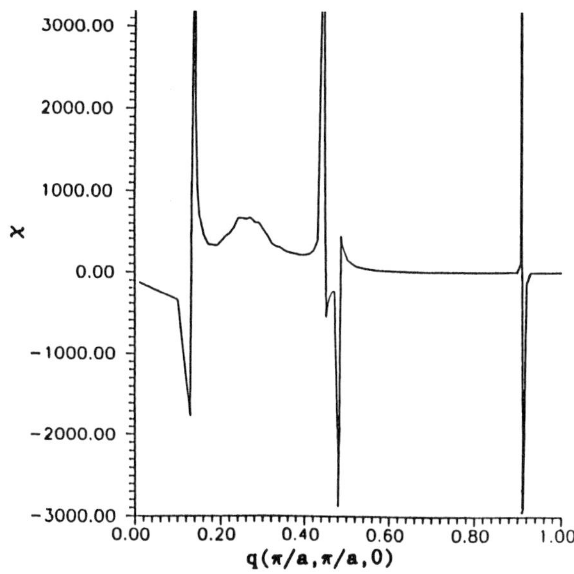

FIGURE 6.   Enhanced susceptibility of La_2CuO_4 along [110] direction

IV. CONCLUSIONS

Techniques have been developed to make tractable first principles studies of the stability of the normal phase of technologically important compounds which remain superconducting at high temperatures. The magnitude of the computational effort required the fastest and largest memory computers available and special computational procedures were needed to achieve even approximate results. Considering the complexity of the materials studied it is remarkable that such difficult problems can be addressed at all, and reflects the continued rapid growth in computational power available to scientists and engineers. The sort of problem described here could readily benefit from the use of a newly emerging technology, massively parallel computers. A project is now under way to utilize this approach in the study of complex quantum mechanical systems such as the high temperature superconductors.

In the present study of La_2CuO_4 it has has been concluded that the paramagnetic ground state, which is the basis of most band theory approximations to the electronic structure of high T_c superconductors, is unstable to a variety of magnetic transitions. Thus the usually assumed starting point to explain superconducting phase transitions already has other instabilities. It is probably not a good starting point upon which to build a reliable theory of the superconducting phase transition. Including the effects of doping could possibly change the conclusions made here about magnetic instabilities, and a Fermi liquid theory at the observed doping levels may still be appropriate. This possibility will also be the subject of future work.

ACKNOWLEDGMENTS

Computational facilities for this work were provided by UTA Academic Computing Services, The University of Texas Center for High Performance Computing and the National Magnetic Fusion Energy Computer Center. Work was supported by a U.S. Department of Energy Grand Challenge Grant and by a Grant from The Robert A. Welch Foundation.

REFERENCES

1. J. G. Bednorz and K. A. Muller, Z. Physic 64,189(1986).
2. M. K. Wu, J. R. Ashburn, C. J. Torng, P. H. Hor, R. L. Meng, L. Gao, Z. J. Huang, Y. Q. Wang and C. W. Chu, Phys. Rev. Lett. 58,908(1987).
3. Z. Z. Sheng, A. M. Herman, A. El Ali, C. Almansan, J. Estrada, T. Datta and R. J. Matson, Phys. Rev. Lett. 60,937(1988).
4. W. E. Pickett, Rev. Mod. Phys. 61,433(1989).
5. W. E. Pickett, H. Krakauer, D. A. Papaconstantopoulos and L. Boyer, Phys. Rev. B35,7252(1987).
6. D. A. Papaconstantopoulos, M. J. DeWeert, and W. E. Pickett, in **High Temperature Superconductors**, edited by M. B. Brodsky et al., MRS Symposia Proceedings No. 99 (Materials Research Society, Pittsburg, Pa, 1988).
7. J. C. Slater and G. F. Koster, Phys. Rev. 94,1498(1954).
8. K. Schwartzman, J. L. Fry and Y. Z. Zhao, Phys. Rev. B40,454(1989).
9. J. L. Fry and P. C. Pattnaik in **Integral Methods in Science and Engineering**, edited by F.R. Payne, C. C. Corduneanu, A. Haji-Sheikh and T. Huang (Hemisphere Publishing, New York, 1986)pp. 27-36.
10. N. E. Brener, G. Fuster, J. Callaway, J. L. Fry and Y. Z. Zhao, J. Appl. Phys. 63,4057(1988); Phys. Rev. B38,423(1988).
11. J. Callaway, A. K. Chatterjee, S. P. Singhal and A. Ziegler, Phys. Rev. B28, 3818 (1983).

Large Scale Scientific Computing Initiative for Solids

N. E. BRENER, J. CALLAWAY, J. M. TYLER,
and G. FUSTER
Department of Physics and Astronomy
Louisiana State University
Baton Rouge, Louisiana 70808, USA

J. L. FRY and E. C. ETHRIDGE
Department of Physics
University of Texas at Arlington
Arlington, Texas 76019-0059, USA

ABSTRACT

A project is underway to develop a general version of the program BANDPACKAGE, which will provide the capability of performing linear-combination-of-Gaussian-orbitals (LCGO) band structure calculations within the local density exchange-correlation potential approximation for complex crystals with multiple atoms per unit cell. The new codes will treat crystals with arbitrary symmetry, and will take advantage of vector and parallel processing to achieve the task. New many body corrections will also be added to the programs by computing the frequency dependent self-energy in the "GW" approximation. Results are presented here for the initial steps in using vector and parallel processing.

I. INTRODUCTION

In the decade of the 60's the availability and use of high speed digital computers became widespread in the United States. The quantum mechanical solution of many electron systems from first principles was attempted within a number of approximations for atoms, molecules and solids, and comparison with experiment lead to an understanding of the many-body approximations and the numerical methods needed to implement them.[1] While Hartree, Hartree-Fock and various statistical approximations[1] were easily applied and tested for light atoms, molecules (especially larger ones) were more difficult to treat because of computer code and speed limitations. Molecules containing a few hundred electrons could be handled, but macroscopic solids containing 10^{23} electrons had not been treated self-consistently from first principles until the latter part of the decade. One of the first techniques of solving the many-body Schrodinger equation for solids which achieved self-consistency was the linear combination of atomic orbitals (LCAO) method, which was developed for solids at Louisiana State University (LSU) and elsewhere in the late 60's.[2] The technique is now more correctly referred to as the LCGO method, since Gaussian basis sets, rather than actual atomic orbitals are usually employed. The first self-consistent results were obtained for an insulator,[3] and after some difficulty, for a metal.[4] Other computational approaches for solids also had difficulty in achieving self-consistency, but by the 70's it had been obtained by several methods.[1] The LCGO program BANDPACKAGE[2] developed at LSU has had considerable success in explaining physical properties of cubic systems, and runs very quickly on mainframe computers of the 90's. The chief advantages which the LCGO method has over other techniques are:

  (1) ability to treat potentials with arbitrary shape
  (2) ability to generate band structures rapidly at arbitrary points
      in momentum space
  (3) simple analytic expressions for electronic wavefunctions.

Because of these advantages the LCGO method is still a preferred one
where it is practical. In this paper we report progress in updating the
LCGO computer codes to include arbitrary symmetry, large numbers of
atoms per unit cell, and proven many-body corrections to the LCGO band
structure.

II. REVIEW OF THEORY

The quantum many-body problem for electrons in solids is usually
formulated in terms of an effective single particle Hamiltonian and the
corresponding time-independent Schrodinger equation

$$H_{eff} \, \Psi_n(k,r) = E_n(k)\Psi(k,r) \tag{1}$$

where in atomic units

$$H_{eff} = - \nabla^2 + V_{eff}(r). \tag{2}$$

V_{eff} includes interactions of each electron with all other electrons and
all the ions in the solid. Also included is the exchange-correlation
potential in an appropriate local density formulation.[5] Solutions of Eq.
(1) must satisfy Bloch's theorem if the solid forms an infinite,
periodic lattice:

$$\Psi_n(k,r+R_m) = \exp(ik.R_m)\Psi_n(k,r), \tag{3}$$

where R_m is any lattice translation vector for the Bravais lattice of
the crystal. In the LCAO or LCGO method this is accomplished by taking
basis functions of the form

$$\phi_i(k,r) = \surd(N)^{-1}\Sigma_m \exp(ik.R_m)u_i(r-R_m) \tag{4}$$

where u_i is a known function centered at a Bravais lattice point (or an
atomic site). Gaussian orbitals form a convenient set since they
simplify calculation of matrix elements of the effective Hamiltonian in
Eq. (3). A linear variational calculation to minimize the energy using
an expansion for the electron wave function

$$\Psi_n(k,r) = \Sigma_i \, C_{ni}(k)\phi_i(k,r) \tag{5}$$

gives a secular equation for the single particle energies $E_n(k)$ and
expansion coefficients $C_{ni}(k)$:

$$\Sigma_j \, H_{ij}(k)C_{nj}(k) = E_n(k) \, \Sigma_j \, S_{ij}(k)C_{nj}(k). \tag{6}$$

Here the Hamiltonian and overlap matrix elements are

$$H_{ij}(k) = \Sigma_m \exp(-ik.R_m)E_{ij}(R_m), \tag{7}$$

$$S_{ij}(k) = \Sigma_m \exp(-ik.R_m)S_{ij}(R_m) \tag{8}$$

with

$$E_{ij}(R_m) = \int u_i^*(r-R_m)H_{eff}\, u_j(r)d^3r, \qquad (9)$$

$$S_{ij}(r_m) = \int u_i^*(r-R_m)u_j(r)d^3r. \qquad (10)$$

Taking u_i to be a product of a Gaussian and a spherical harmonic and Fourier transforming the effective potentials in Eq. (2), analytic expressions for all the integrals may be programmed and computed. However, special efforts must be made to expedite the convergence of the Fourier series.

To solve Eq. (6), an initial effective potential must be generated, usually from a linear combination of atomic charge densities. The solutions of Eq. (6) provide the first approximation to Ψ_n in Eq. (5), which must be then be used to generate new charge densities and effective potentials. The process is iterated until the charge density, potential and/or energies satisfy pre-determined convergence criteria. The Fourier series expansions are convenient for achieving this self-consistency. Usually a reasonable guess for the starting effective potential will provide the correct high momentum Fourier coefficients, so only the first few coefficients change.

The self-consistent cycles may take considerable time, since numerical integration over the unit cell in k-space (first Brillouin zone) must be done carefully to determine the Fermi energy and iterated Fourier coefficients. For an insulator or semiconductor this is easy, but for metals and especially for magnetic metals, the iterative cycle may become unstable unless special care is taken. These problems have been successfully treated in the published codes contained in **BANDPACKAGE**.

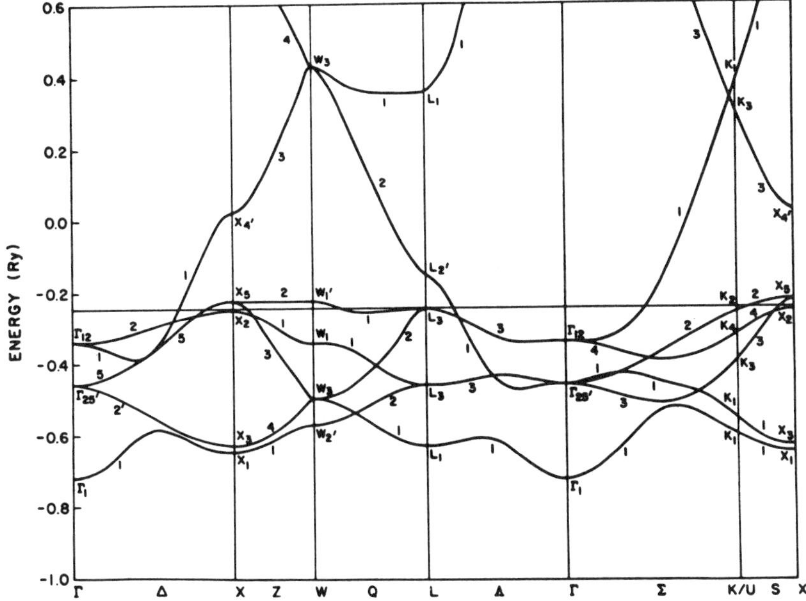

FIGURE 1.   LCGO band structure of fcc Pd.

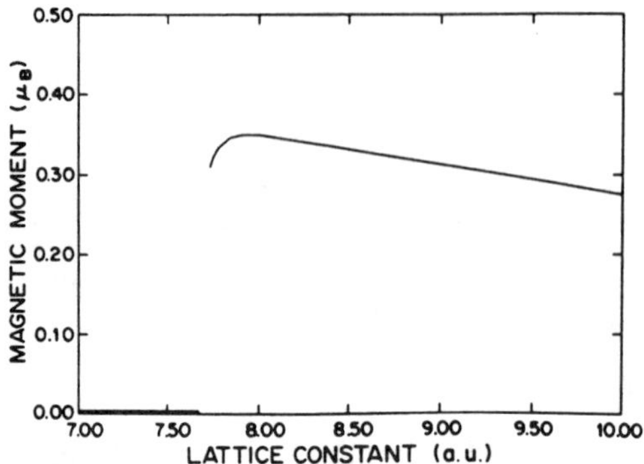

FIGURE 2. Lattice constant dependence of magnetization of fcc Pd

The band structure of a typical transition metal, face-centered-cubic (fcc) palladium is shown in Fig. 1 along symmetry axes in the first Brillouin zone (BZ).[6] Obtaining such a plot was slow in the early days of the LCGO method, but today it may be generated quite rapidly. It is possible to study properties of cubic metals as a function of cubic lattice type and as a function of the lattice constant (expanded or contracted crystals). This is demonstrated in Fig. 2, where the magnetic moment of fcc Pd is shown as a function of lattice constant. Each point on the curve represents a complete self-consistent calculation followed by computation of the magnetic moment. Note the interesting physics revealed in the abrupt transition to a magnetic state near a lattice constant of 7.75 atomic units.[6] Attempts have been made to confirm this behavior by employing techniques of molecular-beam epitaxy, with some success.[7] Calculations of the sort shown in Fig. 2 were made possible by the use of vector processors, operating in parallel, as discussed in the next section.

III. RECENT IMPROVEMENTS

To take advantage of the LCGO method for more complex crystals such as compounds with many atoms per unit cell and crystals with low symmetry it is necessary to substantially accelerate the codes in **BANDPACKAGE**. For example, the normal phase of La_2CuO_4 is body-centered tetragonal (bct) with seven atoms per unit cell and a minimal LCGO Hamiltonian matrix would be about 400x400. Compared with a cubic transition metal where the matrix is about 100x100, the matrix diagonalization time alone increases by a factor of 40. Even larger matrices are required for the newer generation of high T_c superconductors. These are yet to be understood on a theoretical basis, and have stimulated countless experiments and semi-countless explanations for their remarkable properties.[8] Reliable band structures with appropriate many-body corrections would assist in interpretation of these technologically strategic materials. This is one of the goals of the new generation **BANDPACKAGE**.

The first steps to improve the computational efficiency of the LCGO method involved utilization of vectorized codes to perform matrix diagonalizations. Parallel processing was then introduced into the self-consistent cycles. Since iterations involve identical tasks for the k points in the irreducible section of the Brillouin zone parallel processing is readily accomplished. This was done with a limited number of CPU's at first, and results for fcc chromium in Table I show the improvement in CPU and elapsed times. The transition from scalar to vector processing of matrix operations accounted for the reduction in CPU times and parallel processing further reduced the elapsed times. Overall run times were reduced by about a factor of 10 at the expense of using more CPU's. Steps are now underway to utilize _massively_ parallel computers and reduce the run times another order of magnitude or more. When this occurs the goal of making the LCGO method practical for complex crystal systems will have been achieved.

TABLE 1. Compute times for fcc Cr on IBM 3090-400

CPU's	Processing	CPU Time	Elapsed Time
1	Scalar	495.9sec	504.1sec
1	Vector	171.9	174.6
2	Vector-parallel	174.7	91.6
3	Vector-parallel	174.9	62.6
4	Vector-parallel	175.1	48.3

IV. WORK IN PROGRESS

Work in progress to improve the LCGO method still centers on the use of vector-parallel computers. Massively parallel supercomputers are expected to make the greatest reductions in elapsed time. The codes in **BANDPACKAGE** are also being modified to treat crystals with lower symmetry. This includes the hexagonal-close- packed (hcp), tetragonal, and other crystal structures of current interest. The use of group symmetry is less rewarding for low symmetry, but significant savings in computer time may be achieved in BZ or Wigner-Seitz cell integrations for functions having the full crystal symmetry. An automated way to identify symmetry is desirable for the computer codes which will be distributed to non-specialists.

Another significant effort to improve the LCGO method is the implementation of more accurate treatment of many-body corrections to the local-density band structures. A series of papers[9-10] since about 1985 has convincingly demonstrated that a theory[11] presented in the middle 60's can be quantitatively applied on mainframe computers available today. Agreement between theory and experiment is excellent for metals, insulators and semi-conductors, but the theory is expensive to implement. It is based upon a quasiparticle equation for the many-body system which resembles a typical effective Hamiltonian approach, but is exact. The equation buries the difficulty of the many-body problem in a quantity called the self-energy operator, $\Sigma(E)$, which itself depends upon the energy of the quasiparticle. The quasiparticle wave equation, sometimes referred to as the Dyson equation is, in atomic units,

$$[-\nabla^2 + V(r)]\Psi_n(r) + \int\Sigma(r,r',E)\Psi_n(r')dr' = E_n\Psi_n(r). \tag{11}$$

The difficulty in solving this equation has been finding an appropriate expression for Σ. By expanding it to lowest order in terms of a many-body screened interaction

$$W(r,r',E) = \epsilon^{-1}(r,r',E)v(r-r'), \tag{12}$$

where $v(r-r')$ is the potential of interaction between the particle at r and the particle at r' and ϵ is the dielectric response function of the system, an expression for Σ known as the "GW" approximation is obtained:[11]

$$\Sigma(E) = (i/2\pi)\int\exp(-i\delta E')G(E-E')W(E')dE' \tag{13}$$

In this equation G is the quasiparticle Green's function. A solution of the Dyson equation usually begins with a local density approximation to Σ, which is equivalent to solving Eq. (1). The band structure is then employed to compute the G and W with some approximation for the many-body response function ϵ. Published results within the "GW" method suggest that the choice of approximation for ϵ is not critical to success of the calculation, but this generally held belief is still open to question. The method of performing the integration in Eq. 13 differs from group to group, some researchers employing models and sum rules to avoid explicit evaluation of the integral, but others relying on standard expressions for ϵ analytically continued to the complex plane to perform the integration in Eq. (13) numerically. This is the most difficult part of the calculation of Σ. Once Σ has been constructed, the Dyson equation may be solved by standard techniques such as the LCGO method. The process must be iterated to achieve self-consistency. It is here that the LCGO method has advantages because of the Fourier representations of the potentials. Since the screened interaction W is most easily represented in momentum variables instead of the space variables indicated in Eq. (12), the screened interaction at each level of the self-consistent iterations is readily implemented into the band structure codes. This entire process may be coded to take advantage of vector-parallel processing as was done in the solution to Eq. (1). because of the additional integrations needed to compute G and ϵ^{-1} for each energy E in equation (13), systematic application of this many body theory must be approached with careful planning in the construction of the computer codes, even with the best, massively parallel supercomputers available in 1990.

V. CONCLUSION

Limitations on space in this article have permitted only an outline of the project being described. The well tested LCGO method for solids is being expanded in scope to treat many more systems with greater complexity. Some of the systems which will be within the scope of the next generation of computer codes are of great scientific and technological interest today. This has stimulated the investment of time on the part of researchers involved in the project and funding from the granting agencies acknowledged below. There is clear evidence that utilization of supercomputers available today will make the first part of this effort successful at a reasonable cost in manpower and CPU time. Implementation of the many-body corrections is still a great challenge, but the sort which drives the originality of numerically inclined physicists and challenges the computer industry to new efforts.

ACKNOWLEDGMENTS

Work at LSU has been supported by the National Science Foundation under Grant No. DMR-8810249. Work at UTA has been supported by The Robert A. Welch Foundation under Grant Y-707.

REFERENCES

1. J. Callaway, **Quantum Theory of the Solid State** (Academic Press, New York, 1974).
2. C. S. Wang and J. Callaway, Computer Phys. Comm. **14**, 327 (1978)
3. D. M. Drost and J. L. Fry, Phys. Rev. B5, 684 (1972).
4. J. Callaway and C. S. Wang, Phys. Rev. B7, 1096 (1973).
5. J. Callaway and N. H. March, **Solid State Physics**, edited by H. Ehrenreich and D. Turnbull (Academic Press, New York, 1984) Vol. 34, pp. 136-221.
6. H. Chen, N. E. Brener and J. Callaway, Phys. Rev. B40, 1443 (1989).
7. Z. Celinski, B. Heinrich, J. F. Cochran, W. B. Muir, A. S. Arrott, and J. Kirschner, Phys. Rev. Lett. **65**, 1156 (1990).
8. W. E. Pickett, Rev. Mod. Phys. **61**, 433 (1989).
9. M. Hybertsen and S. G. Louie, Phys. Rev. Lett. **55**, 1418 (1985).
10. Additional references are listed in the more recent paper: R. W. Godby, M. Shluter and L. J. Sham, Phys. Rev. B **37**, 10159 (1988).
11. L. Hedin, Phys. Rev. **139**, A796 (1965).
12. See the article by M. Zaider, J. L. Fry and D. E. Orr in this volume for application of the "GW" method to polymers.

Integral Methods in the Quasi-Particle Theory of Electronic Structure

MARCO ZAIDER
Center for Radiological Research
College of Physicians & Surgeons
Columbia University
630 West 168th Street
New York, New York 10032, USA

JOHN L. FRY and DAVID E. ORR
Department of Physics
University of Texas at Arlington
Arlington, Texas 76019-0059, USA

ABSTRACT

We discuss computational aspects encountered in the implementation of the GW approximation to a many-body system. To understand the features of this relatively novel technique we have selected a polymer, trans-polyacetylene, which is well characterized both theoretically and experimentally. The combination of two elements make our calculation different from previous applications of the GW method: the use of Hartree-Fock wavefunctions as starting point in the perturbation series (other calculations have used the local density approximation), together with a full calculation of the dielectric response function, $\varepsilon(\vec{q},\omega)$.

I. INTRODUCTION

Information on electron transport in condensed materials – the long-range goal of this study – can be obtained through the frequency- and wave-vector-dependent dielectric response function of the system, $\varepsilon(\vec{q},\omega)$ (1). The calculation of this quantity requires, in turn, an accurate description of the band structure of the material. Techniques currently in use for solving the Schrodinger's equation of a crystal, while providing accurate wave functions and eigenvalues for the ground state of the system, are known to overestimate [in the case of the Hartree-Fock (HF) method] or underestimate [the local density approximation (LDA)] by as much as 50-100 % the energy gaps to conduction bands. This leads to very inaccurate values for the function, $\varepsilon(\vec{q},\omega)$.

In the past, the somewhat arbitrary procedure of displacing rigidly the conduction bands – a technique known as applying the scissors operator – has been used. The application of this method requires a calibration reference, ideally some experimental data. A different approach consists of performing semi-empirical band calculations, again using experimental evidence to adjust the fitting parameters. Clearly the range of applicability of these techniques to the problem of understanding the dielectric response of an arbitrary system is severely restricted.

In a relatively recent development a number of band calculations have been performed (2-7) by taking advantage of the fact that the HF method corresponds to the first term (zero-th approximation) in a many-body perturbation series, the next term of which is the so-called

GW approximation (see below). The excitation properties of the crystals thus studied have been found to be in remarkable agreement with the experimental data, therefore raising the perspective of true ab initio self-consistent calculations.

In spite of its increasing popularity, it appears that many of the computational aspects encountered in the actual implementation of the GW method need to be systematically tested for accuracy and range of validity. As an example, the replacement in the GW approximation of the HF exchange potential with the screened Coulomb potential requires the function $\varepsilon(\vec{q},\omega)$. Obtaining the dielectric function is, under the best of circumstances, a rather demanding computational enterprise and, as a general rule, the tendency has been not to calculate $\varepsilon(\vec{q},\omega)$ directly but rather to use approximate, ad hoc models [e.g. the plasmon-pole model (5)]. The consequences of this simplification remain unclear.

In this paper we report on our approach to implementing the GW approximation. To better understand the features of this new technique we have selected a "simple" polymer, trans-polyacetylene, which (partly because of the current vogue it enjoys in material science research) is well characterized experimentally. The combination of two elements make our calculation somewhat different: the use of HF wavefunctions as starting point in the perturbation series (all other calculations have used the LDA), together with a full calculation of $\varepsilon(\vec{q},\omega)$.

The paper is organized as follows: in the next section we sketch the essentials of Green's function many-body theory; readers familiar with the main results may skip it. Next we present the GW formalism followed by explanations concerning some aspects of the calculations. We conclude with a presentation of results for trans-polyacetylene.

II. OUTLINE OF THE MANY-BODY GREEN'S FUNCTION FORMALISM.

In this section we summarize results from the theory of Green's functions which are used in this calculation. The main results of this section are Eqs.(13,20,21). For more details the reader could consult Refs.8,9. This outline follows Inkson (9).

Consider a system of N interacting particles (e.g. electrons) and let |N⟩ be the ground state of this system in the occupation number representation. Let further $\Psi(\vec{r},t)$ be a field operator acting on |N⟩ to the effect of removing from the system at time t a particle positioned at $\vec{r}$. Similarly, $\Psi^{+}(\vec{r},t)$ adds a particle to the system. In the Heisenberg picture the operators Ψ satisfy a Schrodinger-like differential equation:

$$\hat{H}(\vec{r},t)\hat{\Psi}(\vec{r},t) = ih\,\frac{\partial}{\partial t}\hat{\Psi}(\vec{r},t), \tag{1}$$

where $\hat{H}$ is the Hamiltonian of the system:

$$\hat{H}(\vec{r},t) = H_0(\vec{r},t) + v(\vec{r},\vec{r}') \tag{2}$$

$H_0(\vec{r},t)$ contains single-particle terms: the kinetic energy, the Coulomb interaction between an electron and the background charges (e.g. the fixed atoms in the lattice), and a time-dependent external potential, $\phi(\vec{r},t)$. This latter quantity represents a "probe" weekly coupled with the system, and such that

$$\phi(\vec{r},t=-\infty) = \phi(\vec{r},t=+\infty) = 0. \tag{3}$$

$v(\vec{r},\vec{r}')$ is the Coulomb interaction between electrons:

$$v(\vec{r},\vec{r}') = \frac{e^2}{|\vec{r} - \vec{r}'|}. \tag{4}$$

If we denote quantities corresponding to $\phi=0$ by $|N_0\rangle$, $\hat{\Psi}_0$, etc. then, to the extent that the probe is turned on slowly (adiabatically), one expects $|N_0\rangle$ to approach smoothly $|N\rangle$, and indeed one can show that

$$|N\rangle = \exp[\ -\frac{i}{\hbar} \int_{-\infty}^{0} \phi(t)dt\ ]|N_0\rangle\ /\ \langle N_0|N\rangle, \tag{5}$$

where

$$\phi(t) = \int d(1)\phi(1)\Psi_0^+(1)\Psi_0(1). \tag{6}$$

We use the shorthand $1 = (\vec{r}_1,t_1)$, $2 = (\vec{r}_2,t_2)$, $d(1) = d\vec{r}_1 dt_1$, etc. The single-particle Green's function, $G(1,2)$, is defined as follows:

$$G(1,2) = -i\langle N|T[\Psi(1)\Psi^+(2)]|N\rangle, \tag{7}$$

where the time-ordering operator T insures that the operator on the right acts at an earlier time than the operator on the left:

$$T[\Psi(1)\Psi^+(2)] = \Psi(1)\Psi^+(2)\Theta(t_1-t_2)\pm\Psi^+(2)\Psi(1)\Theta(t_2-t_1). \tag{8}$$

The sign $-(+)$ corresponds to fermions (bosons) and Θ is the Heaviside
unit step function.

As defined, $G(1,2)$ adds a particle to the system at $(\vec{r}_2,t_2)$ and
then measures the probability of recovering the original system, $|N\rangle$,
by removing a particle at a later time, t_1, at position $\vec{r}_1$ (we assumed
$t_1 > t_2$); it therefore measures the extent to which the system can
accommodate (and thus be described in terms of) independent particles.
Should the system be such that over large intervals $|\vec{r}_1 - \vec{r}_2|$,
$|t_1 - t_2|$, the function $G(1,2)$ remains fairly constant, one could speak
about quasiparticles of a given energy and momentum: by
Fourier-transforming a distribution $G(1,2)$ which is "flat" in
configuration space one obtains delta-like peaks in frequency and
wave-vector space – the signature of independent particles.

$G(1,2)$ satisfies an equation similar to Eq(1), namely:

$$[i h \frac{\partial}{\partial t_1} - H_0(\vec{r}_1)] G(\vec{r}_1 t_1 \vec{r}_2 t_2) +$$

$$+i \int v(\vec{r}_1,\vec{r}_3)\langle N|T[\hat{\Psi}^+(\vec{r}_3 t_1)\hat{\Psi}(\vec{r}_3 t_1)\hat{\Psi}(\vec{r}_1 t_1)\hat{\Psi}^+(\vec{r}_2 t_2)]|N\rangle d\vec{r}_3 = h\delta(1,2) \quad (9)$$

Because of the two-particle potential, $v(1,3)$, this equation involves a
new entity, the two-particle Green's function defined as:

$$G(1,2,1',2') = (i)^2 \langle N|T[\Psi(1)\Psi(2)\Psi^+(2')\Psi^+(1')]|N\rangle. \quad (10)$$

Since, by an extension of the same arguments, $G(1,2,1',2')$ satisfies an
equation involving the 3-particle Green's function, we clearly have to
deal with an infinite system of coupled equations. To actually obtain
a solution one has to approximate one of the Green's functions (say,
the n-particle one) with products of fewer-than-n-particle Green's
functions, thus reducing the system to n equations only. A special
case is the situation where $G(1,2,1',2')$ is written as products of
single-particle Green's functions: one then recovers the HF equations.

To formalize this approach the concept of <u>self-energy</u>, $\Sigma(1,2)$, is
introduced through the following definition:

$$\left[i h \frac{\partial}{\partial t_1} - H_0(\vec{r}_1) + i \int v(\vec{r}_1,\vec{r}_3) G(\vec{r}_3 t_1, \vec{r}_3 t_1^+) d\vec{r}_3 \right] G(1,2) -$$

$$\int \Sigma(1,3) G(3,2) d(3) = h\delta(1,2). \quad (11)$$

Here $t_1^+ = t_1 + \delta^+$ $(d^+ > 0)$ and, as before, H_0 includes the potential ϕ.
An equation for Σ can be obtained by comparing Eqs(9) and (11) and
using the identity:

$$-h\frac{\delta G(1,2)}{\delta\phi(3)} = G(1,2,3,3^+) - G(1,2)G(3,3^+). \tag{12}$$

The result is:

$$\Sigma(1,2) = -ih\int v(1,4)G(1,3)\frac{\delta G^{-1}(3,2)}{\delta\phi(4)} \, d(3)d(4), \tag{13}$$

where the inverse Green's function is defined through the equation:

$$\int G^{-1}(1,3)G(3,2)d(3) = \delta(1,2). \tag{14}$$

With Eqs(11,13) one can set up an iterative solution for $G(1,2)$ and $\Sigma(1,2)$. To emphasize the physical content of these equations the following quantities are introduced: Let $V(1)$ be the sum of the external potential, ϕ, and the potential of all other electrons on the electron at $(\vec{r}_1,t_1)$.

$$V(1) = \phi(1) - i\int v(1,3)G(3,3^+) \, d(3). \tag{15}$$

By definition, the (time-ordered) dielectric response function of the system is:

$$\varepsilon^{-1}(1,2) = \frac{\delta V(1)}{\delta\phi(2)}. \tag{16}$$

The screened Coulomb interaction is given by:

$$W(1,2) = \int \varepsilon^{-1}(1,3)v(3,2)d(3). \tag{17}$$

With these, Eq(13) becomes:

$$\Sigma(1,2) = -ih\int v(1,4)G(1,3)\frac{\delta G^{-1}(3,2)}{\delta V(5)}\cdot\frac{\delta V(5)}{\delta\phi(4)}d(3)d(4)d(5)$$

$$= -ih\int v(1,4)\varepsilon^{-1}(5,4)G(1,3)\frac{\delta G^{-1}(3,2)}{\delta V(5)}d(3)d(4)d(5) =$$

$$= -ih\int W(1,4)G(1,3)\frac{\delta G^{-1}(3,2)}{\delta V(4)}\,d(3)d(4). \tag{18}$$

In the _zero-th_ approximation $\Sigma^{(0)}=0$. Then:

$$[ih\frac{\partial}{\partial t_1} - H_0(\vec{r}_1) - V(1)+\phi(1)]G^{(0)}(1,2) = h\delta(1,2) \tag{19}$$

and

$$h[G^{(0)}]^{-1}(1,2) = [ih\frac{\partial}{\partial t_1} - H_0(1) - V(1)+\phi(1)]\delta(1,2)$$

$$\frac{\delta[G^{(0)}]^{-1}(1,2)}{\delta V(3)} = -\frac{1}{h}\,\delta(1,2)\delta(1,3).$$

Within the _first_ approximation:

$$\Sigma^{(1)}(1,2) = iG^{(0)}(1,2)W^{(0)}(1^+,2). \tag{20}$$

This is the so-called GW approximation (8). The corresponding Green's function is obtained from Eq(11).

At the limit of $\phi(\vec{r},t)\to 0$ the Hamiltonian is time-independent and $G(1,2)$ can be Fourier transformed from the time domain to the frequency (energy) space, $G(\vec{r}_1,\vec{r}_2,E)$. Corresponding to the Green's function equation, an eigenvector equation for the quasi-particle can be written:

$$[H_0(\vec{r}) + V(\vec{r})]\Psi_n(\vec{r}) + \int d\vec{r}'\Sigma(\vec{r},\vec{r}',E_n)\Psi_n(\vec{r}')d\vec{r}' = E_n\Psi_n(\vec{r}). \tag{21}$$

$\Psi_n(\vec{r})$ and E_n are interpreted as quasi-particle wave function and energy, respectively.

The following should be noted:
a) Since Σ is non-local and non-Hermitian the energies E_n will be generally complex quantities. The quasi-particles have a finite life-time.

b) If the dielectric screening is neglected, i.e.

$$\varepsilon^{-1}(1,2) = \delta(1,2),$$

$$(22)$$

the Hartree-Fock equations obtain, with Σ replaced by the exchange term.

III. THE IMPLEMENTATION OF THE GW METHOD.

A. <u>General equations (8)</u>.

The Fourier transform ($t \rightarrow E$) of Eq(20) is:

$$\Sigma(\vec{r},\vec{r}',E) = i\int\frac{d\omega}{2\pi}\, e^{i\delta^{+}\omega}\, G(\vec{r},\vec{r}',E+\omega)W(\vec{r},\vec{r}',\omega). \qquad (23)$$

This equation results from the convolution theorem. The factor $\exp(i\delta^{+}\omega)$, to be taken at the limit $\delta^{+}\rightarrow 0$ obtains from $t_1^{+}=t_1+\delta^{+}$ in Eq(20).

For calculating $G^{(0)}(\vec{r},\vec{r}',\omega)$ we use Hartree-Fock wave functions $|n\vec{k}\rangle=\Psi_n(\vec{k},\vec{r})$ of energy $E_n(\vec{k})$, wave vector $\vec{k}$ and band index n. A solution of Eq(11) is given by

$$G^{(0)}(\vec{r},\vec{r}',E) = \sum_{n,k}\frac{\Psi_n(\vec{k},\vec{r})\Psi_n^{*}(\vec{k},\vec{r}')}{E - [E_n(\vec{k})\pm i\delta^{+}_1]}, \qquad (24)$$

where the sign $+(-)$ is for E_n representing occupied (unoccupied) states, i.e. E_n smaller (larger) than the chemical potential μ (the Fermi level).

The screened interaction term, Eq(17), becomes

$$W(\vec{r},\vec{r}',\omega) = \sum_{q,G,G'} e^{i(\vec{q}+\vec{G})\cdot\vec{r}}\varepsilon^{-1}_{GG'}(\vec{q},\omega)v(\vec{q}+\vec{G}')e^{-i(\vec{q}+\vec{G}')\cdot\vec{r}'}, \qquad (25)$$

where $\vec{G}$ and $\vec{G}'$ are inverse lattice vectors and advantage was taken of the periodicity of the system by restricting the summation over $\vec{q}$ to the first Brillouin zone (1BZ). $v(\vec{q})$ is the Fourier transform of the Coulomb potential

$$v(\vec{q}) = \frac{4\pi e^2}{\Omega q^2}, \tag{26}$$

with $\Omega=$ the crystal volume. For the dielectric response function we use the Adler-Wiser (10,11) expression obtained in the random-phase approximation (RPA):

$$\varepsilon_{GG'}(\vec{q},\omega) = \delta_{GG'} - v(\vec{q}+\vec{G})\sum_{k,n,m} \langle n,\vec{k}|e^{-i(\vec{q}+\vec{G})\vec{r}}|m,\vec{k}+\vec{q}\rangle\langle m,\vec{k}+\vec{q}|e^{i(\vec{q}+\vec{G}')\vec{r}'}|n\vec{k}\rangle \cdot$$

$$\frac{f_m(\vec{k}+\vec{q})-f_n(\vec{k})}{E_m(\vec{k}+\vec{q})-E_n(\vec{k})-\omega+i\delta_2^+}. \tag{27}$$

Here $f_n(\vec{k})$ represents the occupation number of the state $|n\vec{k}\rangle$. To make the equations more compact we use the notation:

$$F_{q,s,s'}(\vec{k}) = \langle s\vec{k}|e^{-i\vec{q}\vec{r}}|s',\vec{k}+\vec{q}\rangle. \tag{28}$$

Eq(27) becomes:

$$\varepsilon_{GG'}(\vec{q},\omega) = \delta_{GG'} + \frac{4\pi e^2}{\Omega|\vec{q}+\vec{G}|^2}\cdot\sum_k^{1BZ}\sum_{nm}\frac{F_{q+G,nm}(\vec{k})F_{q+G',nm}(\vec{k})}{E_m(\vec{k}+\vec{q})-E_n(\vec{k})-\omega+i\delta^+}, \tag{28}$$

where n and m run over occupied and unoccupied states respectively. From Eqs(23-28) the matrix elements of Σ in the GW approximation become:

$$\langle n\vec{k}|\Sigma(\vec{r},\vec{r}',E)|n\vec{k}\rangle =$$

$$= i\sum_l\sum_{1BZ}d\vec{q}\, F_{-(q+G)nl}(\vec{k})F^*_{-(q+G')nl}(\vec{k})\frac{4\pi e^2}{|\vec{q}+\vec{G}|^2}\frac{I(E)}{(2\pi)^3}, \tag{23}$$

where

$$I(E) = \int_{-\infty}^{+\infty} \frac{d\omega}{2\pi} \, e^{i\delta^+\omega} \, \frac{\varepsilon^{-1}_{GG'}(\vec{q},\omega)}{\omega-[E_1(\vec{k}-\vec{q})\pm i\delta_1^+-E]}. \tag{30}$$

The quasi-particle energies, $E_n^{qp}(\vec{k})$, can be obtained within a good approximation by solving the equation:

$$E_n^{qp}(\vec{k}) = E_n(\vec{k}) + \langle n\vec{k}| \Sigma(E_n^{qp}(\vec{k}))-V_{ex}|n\vec{k}\rangle, \tag{31}$$

where we have used the fact that $E_n^{qp}(\vec{k})$ are obtainable by replacing in the HF equations the exchange potential, V_{ex}, with the self-energy operator. From Eq(31):

$$E_n^{qp}(\vec{k}) = \frac{E_n(\vec{k})+\langle n\vec{k}| \Sigma(0)-V_{ex}|n\vec{k}\rangle}{1-\langle n\vec{k}|d\Sigma(0)/dE|n\vec{k}\rangle}, \tag{32}$$

calculated at $\Sigma(0)=\Sigma(\vec{r},\vec{r}',E=0)$.

B. <u>The calculation of I(E).</u>

At a given value of E the integrand of Eq(30) has poles at the excitation energies of the system (contributed by ε^{-1}) in the second and fourth quadrant of the complex-ω plane, as well as the poles of the Green's function in the denominator but displaced by E. Should these poles be known (they are not) one could calculate I(E) along a semi-circle in the upper half of the complex plane (because of δ^+) by summing the residues at all poles with $\text{Im}(\omega)>0$. It follows (12,3) that only occupied states ($E_1(\vec{k})<\mu$) contribute to I(E).

Consider now the function:

$$I_1(E) = \int_{\mu-i\infty}^{\mu+i\infty} \frac{d\omega}{2\pi} \, \frac{\varepsilon_{GG'}^{-1}(\vec{q},\omega)}{\omega-[E_1(\vec{k}-\vec{q})+i\delta_1^+-E]}, \tag{33}$$

for E imaginary. The evaluation of this integral along contour B (see Fig.1) will include poles on the second and third quadrant (these latter appear when E is imaginary). It should be noticed, however, that in a small domain around E=0 we have:

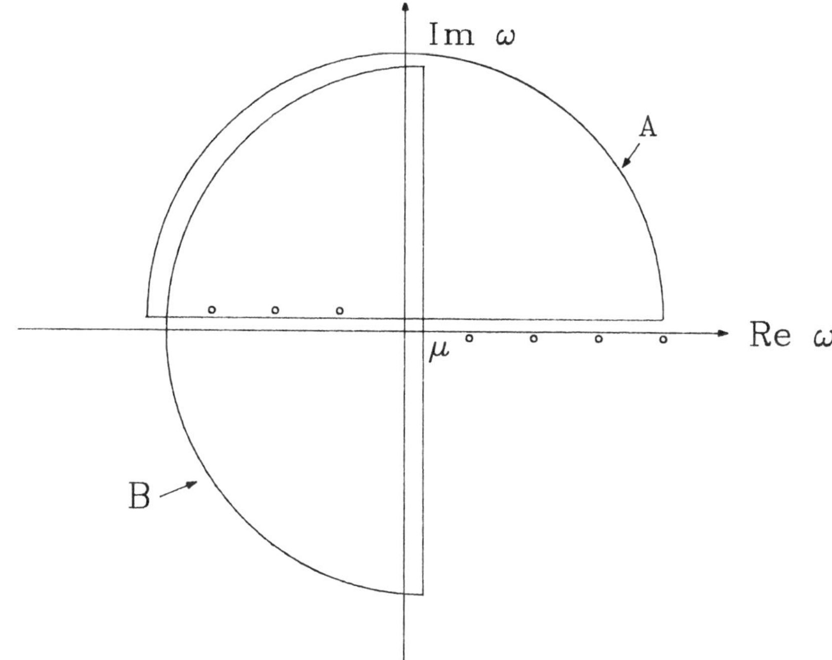

FIGURE 1. Schematic representation of the integration contours for
I(E), Eq(30) and I_1(E), Eq(33). The poles of the Green's function in
the complex ω plane are indicated with small circles.

$$I(E) = I_1(E). \tag{34}$$

We can obtain I(E), at real values of the energy E, by a) numerically
integrating I_1(E) along the imaginary line Im(ω)=μ, and b) use a Taylor
series expansion of I_1(E) around E=0 to analytically continue this
function to real energy values:

$$I(E) = I_1(0) + E\frac{dI_1(0)}{dE} +... \tag{35}$$

C. The calculation of the matrix elements, $F_{qnl}(\vec{k})$.

     In the Hartree-Fock self-consistent field method used in this
calculation one-electron crystal orbitals are expressed as linear
combinations of Bloch basis functions:

$$\Psi_1(\vec{q},\vec{r}) = \sum_{i=1}^{n} \sum_{k=1}^{\gamma_i} C_{(ik)1}(\vec{q})b_{(ik)}(\vec{q},\vec{r}). \tag{36}$$

In this equation C are q-dependent constants and the Bloch functions
are:

$$b_{(ik)}(\vec{q}) = \frac{1}{(2N+1)^{0.5}} \sum_{j=-N}^{N} e^{i\vec{q}\vec{R}_j} \Phi_k(\vec{r}-\vec{R}_j-\vec{r}_i).$$ (37)

The following notations are used in these equations: $\Phi_k(\vec{r}-\vec{R}_j-\vec{r}_i)$ is a Slater-type orbital (STO) centered at atom $\vec{r}_i$ in the cell positioned at $\vec{R}_j$; for each atom "i" there are γ_i STO's; there are $(2N+1)$ unit cells in the crystal, each one containing n atoms.

With this one obtains:

$$F_{q11'}(\vec{k}) = \sum_{i_1 i_2=1}^{n} \sum_{k_1=1}^{\gamma_{i1}} \sum_{k_2=1}^{\gamma_{i2}} C_{(i_1 k_1)1}^{*}(\vec{k}) C_{(i_2 k_2)1'}(\vec{k}+\vec{q}).$$
$$\sum_{j=-N}^{N} e^{-i\vec{k}\vec{R}_j} \langle \Phi_{k_1}(\vec{r}-\vec{R}_j-\vec{r}_{i_1}) | e^{-i\vec{q}\vec{r}} | \Phi_{k_2}(\vec{r}-\vec{r}_i) \rangle.$$ (38)

The matrix element in Eq(38) can be evaluated analytically by expressing the STO's as linear combinations of Gaussian-type orbitals (we use the STO-6G expansion).

D. <u>The evaluation of $\varepsilon(\vec{q},\omega)$.</u>

The summation over $\vec{k}$ in the expression, Eq(28), can be converted to an integral over the 1BZ with the usual conversion factor:

$$\sum_{\vec{k}} \rightarrow \frac{\Omega}{(2\pi)^3} \int d\vec{k}.$$ (39)

The integral thus obtained is of the following type (we have suppressed for clarity some of the indices):

$$J(E) = \lim_{\delta \to 0} \int dk \frac{M(k)}{E(k)-E+i\delta},$$ (40)

with obvious notations for M(k) and E(k). Only the 1-dimensional case is treated here but this approach – known as the analytic tetrahedron method (13) – can be applied to any system. Eq(40) yields:

$$ReJ(E) = P\int dk \frac{M(k)}{E(k)-E},\tag{41}$$

$$ImJ(E) = \pi\int dk\ M(k)\cdot\delta[E(k)-E].\tag{42}$$

Note that ReJ(E) and ImJ(E) are the Hilbert transforms of each other.

In a typical calculation we have data for M(k) and E(k) along a grid of k points, k_1, k_2,... The simplest approach is to linearly interpolate M(k) and E(k) between any two consecutive k points, for instance:

$$E(k) = E(k_i) + \frac{E(k_{i+1})-E(k_i)}{k_{i+1} - k_i}\ (k-k_i),\quad k_i<k<k_{i+1}\tag{43}$$

Straightforward integration yields:

$$ImJ(E) = \pi\sum_i \left|\frac{k_{i+1}-k_i}{E_{i+1}-E_i}\right|\ [M_i+(M_{i+1}-M_i)\frac{E-E_i}{E_{i+1}-E_i}\ ],\tag{44}$$

where the summation is only over those intervals $[E_i,E_{i+1}]$ which contain E; and

$$ReJ(E) = \sum_i \left|\frac{k_{i+1}-k_i}{E_{i+1}-E_i}\right|\{(M_{i+1}-M_i)\frac{|E_{i+1}-E_i|}{E_{i+1}-E_i}+[M_i+\frac{M_{i+1}-m_i}{E_{i+1}-E_i}(E-E_i)]\log\left|\frac{E_{i+1}'-E}{E_i'-E}\right|\}.\tag{45}$$

In Eq(45) the summation is over <u>all</u> intervals; also, $E_i=E(k_i)$, $M_i=M(k_i)$, $E_{i+1}'=\max(E_i,E_{i+1})$ and $E_i'=\min(E_i,E_{i+1})$.

The expressions, Eqs(44,45), while very simple to use, lead to difficulties whenever E approaches the limits of the interval $[E_i,E_{i+1}]$ because of the diverging logarithm. The source of this artifact is the first derivative of E(k) which is discontinuous at all points of the k grid because of the linear approximation. Fig.2 illustrates this point with a model calculation of ReJ(E) using $E(k)=k^2$. Also shown is the exact result obtained by direct integration. A possible solution to this problem is to replace the linear interpolation scheme with higher-order polynomials (e.g. spline fitting). We are currently studying the effects of such an approach.

IV.   APPLICATION TO TRANS-POLYACETYLENE.

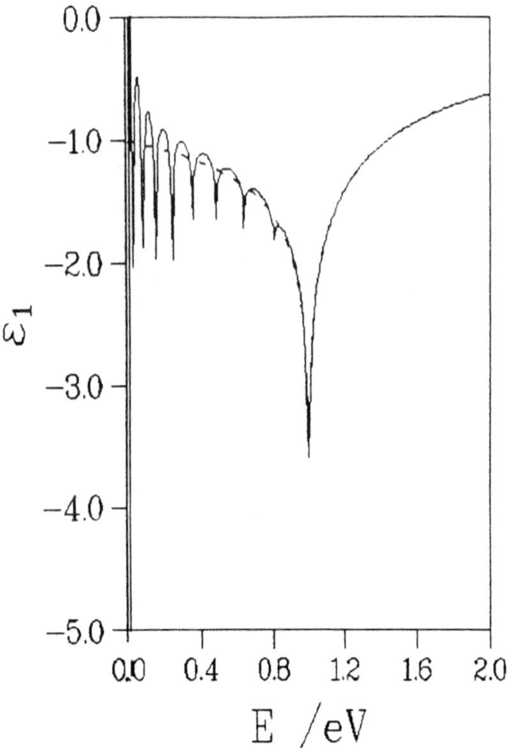

FIGURE 2. The real part, $J(E)=\varepsilon_1$, calculated exactly (dash) or using Eq(45) and a grid of 10 k points (solid line). In this example the dispersion relation $E(k)=k^2$ is used.

In this section we compare calculations using the HF method or the GW approximation in terms of energy bands and dielectric response. The polymer selected - trans-polyacetylene - has been studied quite extensively both experimentally (14,15) and theoretically (16-18). We are not aware of any previous quasiparticle calculation for this polymer. In this study we use the alternating backbone structure of polyacetylene; this is shown in Fig.3.

The HF calculation (which is also the starting point in the GW scheme) was performed using the code CRYSTAL (19) recently made available through the Quantum Chemistry Program Exchange (QCPE 577) at Indiana University. This code is based on an ab initio, self-consistent HF linear combination of atomic orbitals (LCAO) method, with crystal orbitals (CO) expressed as combinations of Bloch functions. The HF energy bands obtained are shown in Fig.4 as a function of the wave vector, k. We obtain a fundamental gap (at the zone boundary) of 5.9 eV. This is - as expected - significantly larger than measured values of 1.4 - 1.8 eV (14), a direct result of ignoring correlation effects in the crystalline system. The consequences of these difficulties in terms of electron-transport calculations are illustrated in Fig.5 where we compare our calculation for the energy-loss function, $Im[-\varepsilon(\vec{q},\omega)^{-1}]$, with results obtained from electron spectroscopy measurements by Fink and Leising (15). The

Trans–polyacetylene Model II

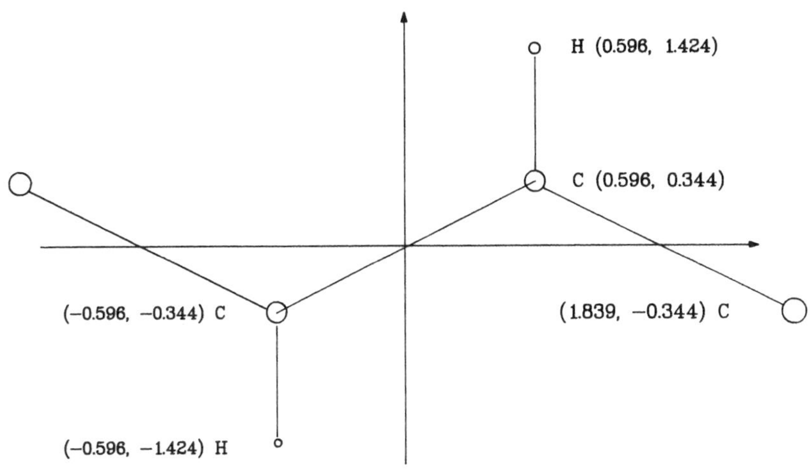

FIGURE 3. The geometry of trans-polyacetylene, Model II, used in this calculation. The numbers are in angstroms.

calculation corresponds to $q=0.1$ $(2\pi/a)$ (a=lattice constant=2.43 Å). The integrals were performed over a grid of 10 k points in the 1BZ and 9 values of G and G′ in the inverse lattice. These values appeared satisfactory although they were dictated to a large degree by limitations in computer time availability. Fig.5 also shows the effects of including (solid line) or neglecting (long dash, $G=G′=0$) local field effects in Eq(2). Within the accuracy of this calculation no significant differences were detected.

With the aid of the formalism described in the previous section we have performed a series of calculations in the GW approximation. The quasi-particle energy bands obtained are shown in Fig.6. The inclusion of the screened potential instead of the bare Coulomb exchange reduces the excitation energies; the fundamental gap is now 2.3 eV. A similar improvement is seen in the energy-loss function (Fig.7).

The results shown here, already superior to the HF calculation, do not represent necessarily a converged GW calculation. In principle one should reiterate the HF calculation by replacing the exchange integrals with the self-energy operator, obtain new quasiparticle wave functions and energies, and so on. Efficient techniques for incorporating this iterative procedure in a HF scheme have been in fact developed (20,21) and we are currently including them in the calculation. In the case of the LDA method it has been noted that the wave functions for

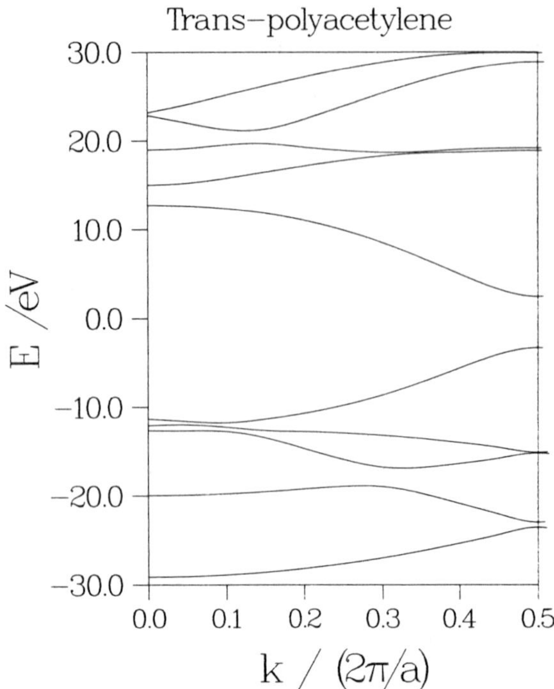

FIGURE 4. The valence (3-7) and conduction (8-12) energy bands for trans-polyacetylene obtained in the Hartree-Fock calculation.

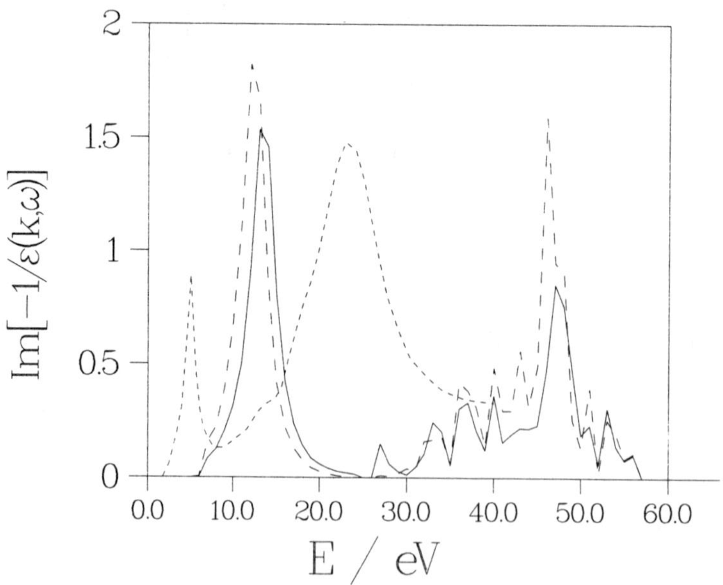

FIGURE 5. The energy-loss function for polyacetylene. Solid line: RPA calculation including local field effects, Eq(27); Long-dash: RPA calculation at G=G'=0; Short dash: the experimental results of Fink and Leising (15). The calculation is performed at q=0.05 Å.

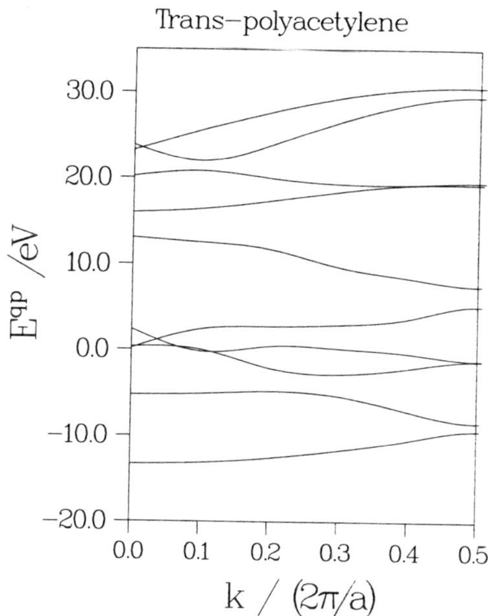

FIGURE 6. Quasiparticle energy bands for valence (3-7) and conduction (8-12) levels obtained with Eq(32) in the GW approximation.

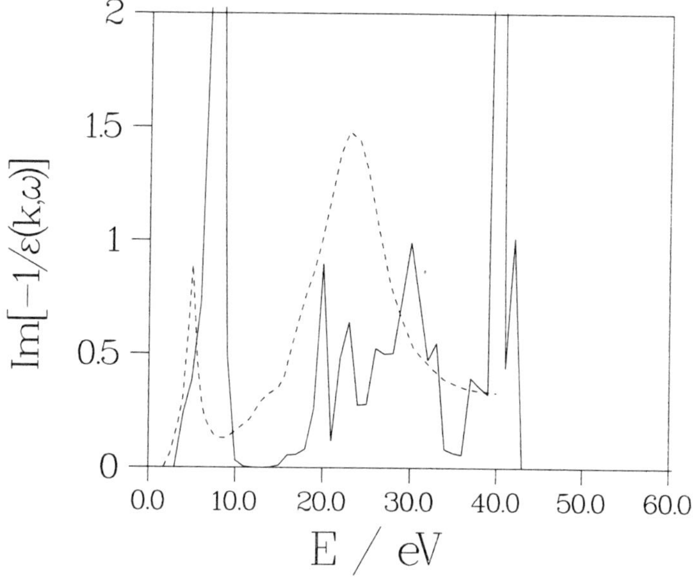

Figure 7. The energy-loss function for polyacetylene calculated in the GW approximation (solid, G=G'=0), or measured (dash, Ref.15).

quasiparticles are practically converged within the first iteration, a point that remains to be verified for the HF approach.

V. ACKNOWLEDGMENTS.

This work was performed using supercomputer time kindly made available to us by the U.S. Department of Energy under the "Grand Challenge" program; this is gratefully acknowledged. We also thank the Plasma Computer Center at the University of Texas at Austin for providing high-speed communication through its Remote Users Satellite System (RUSS) station.

This investigation was supported in part by Grant DE-FG02-88ER60631 from the Department of Energy and by Grant CA 12536 from NCI, DHHS, to the Center for Radiological Research, Columbia University, and by the Robert A. Welch Foundation at the University of Texas at Arlington.

REFERENCES.

1. Pines, D. and Nozieres, P. 1966, The theory of Quantum Liquids. W.R. Benjamin, New York.

2. Godby, R.W., Schluter, M. and Sham, L.J. 1986, Accurate exchange-correlation potential for silicon and its discontinuity on addition of an electron. Phys. Rev. Lett. 56: 2415-2418.

3. Godby, R.W., Schluter, M. and Sham, L.J. 1987, Trends in self-energy operators and their corresponding exchange-correlation potentials. Phys. Rev. B36: 6497-6500.

4. Godby, R.W., Schluter, M. and Sham, L.J. 1988, Self-energy operators and exchange potentials in semiconductors. Phys. Rev. B37: 10159-10175.

5. Hybertsen, M.S. and Louie, S.G. 1985, Electron correlation and the band gap in ionic crystals. Phys. Rev. B32: 7005-7008.

6. Hybertsen, M.S. and Louie, S.G. 1986, Electron correlation in semiconductors and insulators: band gaps and quasi-particle energy. Phys. Rev. B34: 5390-5413. 7. von der Linden, W. and Horsch, P. 1988, Precise quasiparticle energies and Hartree-Fock bands of semiconductors and insulators. Phys. Rev. B37: 8351-8362.

8. Hedin, L. and Lundqvist, S. 1969, Effects of electron-electron and electron-phonon interaction on the one-electron states of solids. In: Solid State Physics 23: 1-181.

9. Inkson, J.C., Many-body theory of solids. An Introduction. Plenum Press, New York and London, 1986. 10. Adler, S.L. 1962, Quantum theory of the dielectric constant in real solids, Phys. Rev. 126: 413-420.

11. Wiser, N. 1963, Dielectric constant with local field effects included, Phys. Rev. 129: 62-69.

12. Quinn, J.J. and Ferrell, R.A. 1958, Electron self-energy approach to correlation in degenerate electron gas. Phys. Rev. 112: 812-827.

13. Fry, J.L. and Pattnaik, P.C. 1986, Analytic approximation methods of computing multi-dimensional principal-value integrals. In: Proc. Int. Conf. Integral Meth. Scie. Eng. (Edited by Payne, F.R., Cordureanu, C.C., Haji-Sheikh, A. and Huang, T.) pp. 27-36, Hemisphere, Washington.

14. Fincher, C.R., Chen, C.E., Heeger, A.J., MacDiarmid, A.G. and Hastings, J.B. 1982, Electronic structure of polyacetylene: optical and infrared studies of undoped semiconducting $(CH)_x$ and heavily metallic $(CH)_x$, Phys. Rev. B20: 1589-1602.

15. Fink, J. and Leising, G. 1986, Momentum-dependent dielectric functions of oriented trans-polyacetylene. Phys. Rev. 34B: 5320-5328.

16. Mintmire, J.W. and White, C.T. 1983, Local-density-functional approach to all-trans-polyacetylene. Phys. Rev. B28: 3283-3289.

17. Springborg, M. 1986, Self-consistent electronic structures of polyacetylene. Phys. Rev. B33: 8475-8489.

18. Suhai, S. 1980, A priori electronic structure calculations on highly conducting polymers. I. Hartree-Fock studies on cis- and trans-polyacetylenes (polyenes). J. Chem. Phys. 73: 3843-3853.

19. Pisani, C., Dovesi, R. and Roetti, C. 1988, Hartree-Fock ab initio treatment of crystalline systems. Lecture Notes in Chemistry 48: 1-193.

20. Fry, J.L., Brener, N.E., and Bruyere, R.K. 1977, Hartree-Fock formalism for solids. II. Application to the nearly-free-electron gas. Phys. Rev. 16: 5225-5232.

21. Brener, N.E. and Fry, J.L. 1978, Hartree-Fock formalism for solids. I. Reciprocal-lattice expansion for Hartree-Fock exchange term. Phys.Rev. 17: 506-512.

Self-Consistency in Embedded Green Function Calculations of Surface Electronic Structure

J. H. KAISER and R. MARROUM
Physics Department
The University of Texas at Arlington
Arlington, Texas 76019-0059, USA

Abstract

A self-consistent full-potential method of determining surface electronic structure is described, in which the surface atomic layers are embedded onto a semi-infinite substrate. This method has advantages over more conventional slab and superlattice calculations including a description of a more realistic semi-infinite bulk and the ability to treat the termination of solid state systems with large atomic bases. This advantage is achieved through an embedding potential, derived from the substrate Green Function, which pins the calculations to the bulk Fermi energy. This pinning, however, leads to difficulties in self-consistency not encountered by the conventional calculations - the charge cannot be explicitly conserved. Our adaptation of Anderson's convergence accelerator to overcome this problem will be discussed.

Calculations of the electronic structure of surfaces provide the link between experiment and theory and assist in the explanation of chemisorption, catalytic behavior, magnetism and the structure of surfaces. There are three different geometrical approaches to these calculations: slab, supercell and the semi-infinite bulk. Slab geometry describes the surface region as a very thin film, 5 to 13 atomic layers thick surrounded on both sides by vacuum. The supercell approach is to separate an infinite number of these slabs by vacuum regions, so that the calculation has full three dimensional periodicity and is therefore essentially a bulk calculation. Both of these geometries provide good approximations to the surface region, but suffer from several drawbacks. In both of these, surfaces may interact through the small distances across the slabs and/or, in the case of supercell calculations, across the vacuum. This can lead to splitting of surface states. Additionally, with the finite number of atomic layers, the bulk continuum of electronic states cannot be reproduced. Rather this continuum is condensed into a few discrete states, which make surface densities of states somewhat unrealistic and surface states hard to identify. The embedding approach devised by Inglesfield[1], and first implemented for surfaces by Inglesfield and Benesh (as the Surface Embedded Green Function method or SEGF)[2], overcomes the difficulties of the above methods. Here, the surface region is described as a slab, one to three layers in thickness, with the vacuum to one side and the bulk, represented by an embedding potential, on the other. Thus the embedding approach describes the surface realistically as the termination of a semi-infinite solid. It provides calculations which not only represent both the surface and bulk states correctly, but which also are computationally more economical and consequently better suited to the surfaces of bulk systems with large unit cells. However, while the more conventional slab and supercell calculations have had successes in predicting surface magnetic, chemical and structural properties, the embedding method has been troubled by instabilities in the self-consistency cycle. This problem occurs in embedding because, in contradistinction to the other geometries, charge cannot be explicitly conserved since the Fermi level must be fixed in the calculation of the embedding potential. In this paper, we outline the embedding formalism, the self consistency cycle and various methods to improve convergence of the self-consistency, then discuss the nature of the difficulties and our approach to remedy them.

1. THE SURFACE EMBEDDED GREEN FUNCTION METHOD.

The embedding method is based on a variational principle in which the trial wave function $\phi(\mathbf{r})$ is defined only in the region of space of explicit interest - the surface region (region I in Figure 1). This wavefunction is extended into the substrate (region II in Figure 1) with an exact solution $\psi(\mathbf{r})$ of the bulk crystal Schrodinger equation at some energy E, which matches onto $\phi(\mathbf{r})$ in amplitude over the interface S between regions I and II. The expectation value of the Hamiltonian with ϕ in I and ψ in II is then given (in atomic units) by:

$$E = \frac{\int_I d^3r\, \phi^*(\mathbf{r})H\phi(\mathbf{r}) + \varepsilon\int_{II} d^3r\, \psi^*(\mathbf{r})\psi(\mathbf{r}) + \frac{1}{2}\int_S d^2r_s\left(\phi^*(\mathbf{r}_s)\frac{\partial\phi(\mathbf{r}_s)}{\partial n_s} - \psi^*(\mathbf{r}_s)\frac{\partial\psi(\mathbf{r}_s)}{\partial n_s}\right)}{\int_I d^3r\, \phi^*(\mathbf{r})\phi(\mathbf{r}) + \int_{II} d^3r\, \psi^*(\mathbf{r})\psi(\mathbf{r})} \tag{1}$$

The surface integrals in (1) come from the discontinuity in the normal derivatives of ϕ and ψ across interface S. The principle of embedding is that integrals through the substrate in (1) can be eliminated, provided that the Green function for the perfect crystal can be found such that, on S:

$$\frac{\partial G_0(\mathbf{r}_s,\mathbf{r}')}{\partial n_s} = 0 \tag{2}$$

The inverse of G_0 evaluated over S_0 is a generalized logarithmic derivative and relates a solution of the Schrodinger equation in region II to the amplitude[1]:

$$\frac{\partial\psi(\mathbf{r}_s)}{\partial n_s} = -2\int_S d^2r'_s\, G_0^{-1}(\mathbf{r}_s,\mathbf{r}'_s,\varepsilon)\, \phi(\mathbf{r}'_s) \tag{3}$$

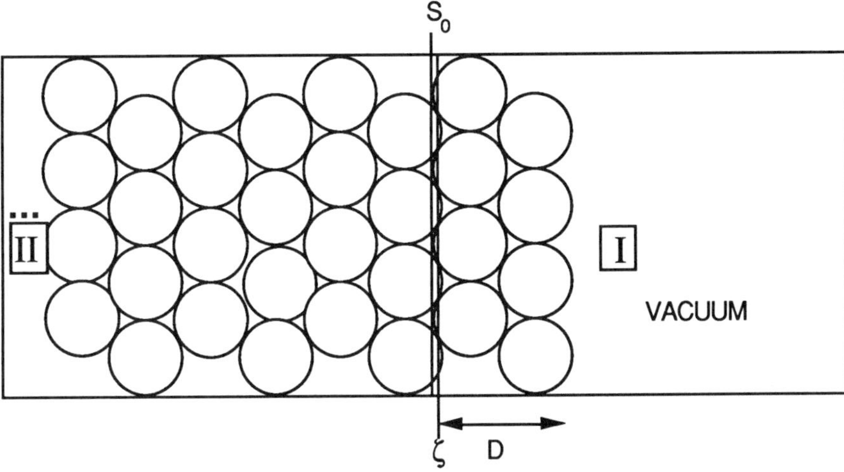

FIGURE 1. Embedding Geometry. The bulk continues indefinitely to the left. D is the thickness of the embedded slab, S_0 is the embedding plane and z denotes a plane which cuts exactly halfway between layers such that a unit cell of the embedded region will contain the total number of electrons belonging to the constituent atoms in their free state.

Substituting (3) into (1) and making use of a relationship between normalization integrals in region II and the energy derivative of the normal derivative of ψ,

$$E = \frac{\displaystyle\int_I d^3r\, \phi^* H\phi + \int_s d^2r_s \left(\phi^* \frac{\partial\phi}{\partial n_s}\right) + \iint_s d^2r_s\, d^2r'_s\, \phi^*(r'_s)\left[G_0^{-1} - \varepsilon\frac{\partial G_0^{-1}}{\partial E}\right]\phi(r_s)}{\displaystyle\int_I d^3r\, \phi^*\phi - \iint_s d^2r_s\, d^2r'_s\, \phi^*(r'_s)\left[\frac{\partial G_0^{-1}}{\partial E}\right]\phi(r_s)} \qquad (4)$$

is obtained. The variational expression for the energy depends explicitly only on ϕ in region I, with G_0^{-1} evaluated over S_0 supplying all the information about the substrate.

To minimize E, expand ϕ in terms of a set of basis functions (linearized augmented plane waves in these calculations)[1]:

$$\phi(r) = \sum_i a_i \chi_i(r) \qquad (5)$$

Substituting (5) into (4) and varying with respect to coefficient a_i^*, yields a matrix equation for the a_j :

$$\sum_j \left(H_{ij} + G_{0\,ij}^{-1} + (E-\varepsilon)\frac{\partial G_{0\,ij}^{-1}}{\partial E}\right) a_j = E \sum_j O_{ij}\, a_j \qquad (6)$$

where:

$$H_{ij} = \int_I d^3r\, \chi_i^* H\chi_j + \int_s d^2r_s\, \chi_i^* \frac{\partial\chi_j}{\partial n_s} ,$$

$$G_{0\,ij}^{-1} = \iint_s d^2r_s\, d^2r'_s\, \chi_i^*\, G_0^{-1}(r_s,r'_s)\, \chi_j$$

and

$$O_{ij} = \int_I d^3r\, \chi_i^*\, \chi_j$$

H_{ij} is the matrix element of the Hamiltonian in region I and includes an additional surface term to ensure Hermiticity. $(G_0^{-1})_{ij}$ is the matrix element of the embedding potential G_0^{-1} which provides for the joining of the surface region to the substrate. The energy derivative term in (6) gives a first order correction so that G_0^{-1} is evaluated at the correct energy, however this term will be zero since the embedding potential is always calculated precisely at E.

It is more convenient to calculate the Green function for the surface region rather than the wave functions, since the Green function allows for much easier evaluation of charge densities and surface densities of states. Expanded in terms of the basis functions χ_j , the Green function for the surface region is given by:

$$G(\mathbf{r},\mathbf{r}';E) = \sum_{ij} g_{ij}\, \chi_i(\mathbf{r})\chi_j(\mathbf{r}') \qquad (7)$$

where g_{ij} satisfies:

$$\sum_k \left(EO_{ik} - H_{ik} - G_{0\ ik}^{-1}\right) g_{kj} = \delta_{ij} \qquad (8)$$

Here, the matrix elements are defined as above and δ_{ij} is the Kronecker delta.

It is equation (8) for which solutions are sought. It is solved on a mesh of special points in the irreducible section of the first surface (2 dimensional) Brillouin zone and for complex energies in the upper half plane terminating at the Fermi energy on the real axis. G_0^{-1} is calculated (by a layer multiple scattering technique) from the reflection properties of the perfect crystal along the flat plane S_0 (see Figure 1) at those special points and energies once and for all for a given substrate[2].

2. SELF-CONSISTENCY: CONVERGENCE DAMPING AND ACCELERATION.

Once the Green function is calculated, the charge density can be obtained from its imaginary part and a new potential (Hamiltonian) can be constructed by solving Poisson's equation and using a functional dependent on the local charge density. If this new potential or some observable (such as the workfunction, total energy, etc.) differs by more than a given tolerance, the (self-consistency) cycle beginning with finding the Green function starts again. If the new (output) potential is used as input the self-consistency cycle tends to give oscillating or unstable results. Therefore, some means of damping the cycle is required.

The most naive approach to damping is linear mixing of input and output potentials:

$$V_{in}^{m+1} = (1-\alpha)V_{in}^m + \alpha V_{out}^m \qquad (9)$$

α, the mixing parameter (which, for surfaces, can take values from 0.1% to 15%), determines the relative amounts of the input and output potentials from the iteration m to form the input potential to iteration m+1. This approach, while keeping the self-consistency stable (for the right choice of α) makes convergence very slow and, consequently, calculations very costly.

A substantial improvement on the linear scheme is Anderson's method[3] which, not only provides damping, but also convergence acceleration. This method is based on finding an optimum mix of previous input (output) potentials:

$$\overline{V}_{in/out}^{m+1} = \sum_{i=0}^{N-1} \beta_i V_{in/out}^{m-i}, \qquad \sum_{i=0}^{N-1} \beta_i = 1 \qquad (10)$$

where N is the number of previous iterations to be included in the mix. This is accomplished by finding β's for the current iteration which minimize the root-mean-square of the difference between the average input and output potentials represented by equation (10):

$$D = \left(\int d^3r \, |\delta \overline{V}|^2 \right)^{\frac{1}{2}} \tag{11}$$

where

$$\delta \overline{V}^m = \sum_{i=0}^{N-1} \beta_i (V_{in}^{m-i} - V_{out}^{m-i}) \tag{12}$$

and subject to the ancilliary condition on the sum of the β's. The volume of integration in equation (11) is the entire embedded region (I). Once the β's have been optimized, a linear mix is applied to the average input potentials as in equation (9).

Anderson's method is used successfully in many self-consistent electronic structure calculations for both bulk solids and surfaces. Another more sophisticated convergence acceleration technique, Broyden's method[4,5], is also used. But in systems where there is potential for instability early in the self-consistency cycle, this method is only of use after stability has been attained. Since, as will be shown, SEGF can be such a system, we concentrate on the application and modification of the Anderson method for these calculations.

3. RESULTS FOR COPPER (001).

The convergence of our self-consistency trials for the (001) surface of one layer of copper embedded on copper are summarized in Table 1. Here, linear and Anderson (N = 2 and 3) mixing schemes are compared. The second column indicates the number of iterations, at the value of α contained in column three. Column four contains the deviation from the converged charge in the embedded region and column five indicates the work function at the final iteration. The final column is the base 10 logarithm of the distance D defined in equation (11). The conclusion is simple: linear mixing works reasonably well as does Anderson N=2, but Anderson mixing with N=3 reduces the total number of iterations by a third while providing a much greater degree of convergence.

Table 2 is the same as Table 1 but for two embedded copper layers. Again the same conclusions are reached but more iterations are required. This is because the effect of the embedding potential is diminished with increasing layer thickness. Nonetheless, there are no apparent problems with the calculations for Copper. Copper, however, has a low density of states at the Fermi level, so the pinning of the Fermi level demanded by the SEGF presents no problem. But in a system with a high density of states at the Fermi level the situation is quite different. Nickel is just such a system.

Table 1 Comparison of Convergence for one layer Cu on Cu(001)

Mixing	iter	α(%)	$\Delta Q/e$	ϕ/eV	log(D)
Linear	30	10	0.0076	5.139	-2
A2a	30	20	0.00005	5.122	-4
A3a	20	20	0.00005	5.122	-4

Table 2 Comparison of Convergence for two layers Cu on Cu(001)

Mixing	iter	α(%)	$\Delta Q/e$	ϕ/eV	log(D)
Linear	21	6	0.0007	4.99	-1
A3a	25	20	0.00005	5.07	-4

4. Results for Nickel (001)

Figure 2 illustrates the convergence of the distance parameter of D for one layer of Nickel on Nickel(001). The three curves represent the results for linear mixing (lin) and Anderson N=3 mixing, both for all potential regions (muffin-tin, interstitial and vacuum) included in the calculation of the β's (A3a) and for the muffin-tin region only (A3m). Clearly, the Anderson approach represents a very substantial improvement in convergence, both in terms of the number of iterations required and the quality of convergence (smallness of D).

In the one layer case, the embedding potential has a strong effect throughout the embedded layer, however with two layers, it loses its hold and it becomes difficult to conserve charge. Figure 3 illustrates the total charge per spin in the embedded region as the difference from the converged charge and Figure 4 presents D as in Figure 2. The curves are identified as in Figure 2. Here, although the linear mix provides a good measure of apparent convergence in charge, it is, in fact far from converged in D and the calculated workfunction of ~ 11eV is far from any physical result. The Anderson cases (A3a) for all potential regions included, however, not only do not provide the expected convergence acceleration, but are unstable regardless of the size of the mixing parameter. We discovered that this was due to the dominance of the interstitial and vacuum contributions to the calculation of D and hence β.

Since the β's determine the mixing among the previous iterations, the strong changes in the interstitial and vacuum regions can overdrive the mixing and thus the convergence acceleration. Our solution, albeit a simple one, was to only use the muffin-tin region in the calculation of the β's (but D in Figure 4 is still for the entire embedded region). The result for the one layer case (Figure 2 (A3m)) illustrates that this approximation only slows the acceleration gently. The convergence behavior for two layers is shown by the curve marked

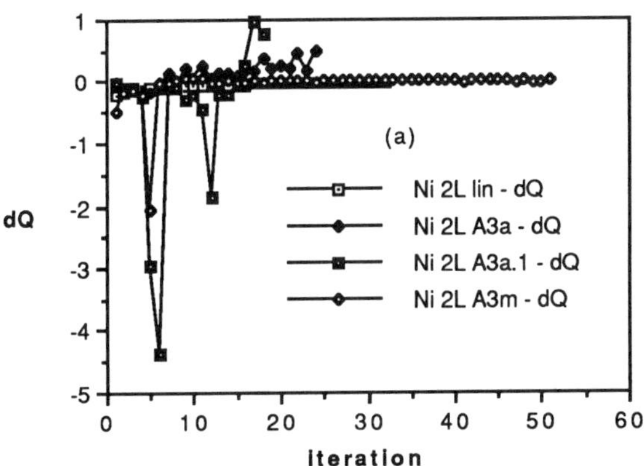

FIGURE 2. Convergence parameter D for one layer Ni on Ni(001). D in units of Hartrees/(a.u.)^3 . (1 Hartree=27.2 eV, 1 a.u.=0.0529nm). A3a indicates Anderson mixing with N=3 where β's are calculated from all potential regions, whereas A3m indicates the same but where the β's are calculated only from the muffin-tin regions. α is 2.5% for the linear case and 5% in both Anderson cases.

(A3m) in Figures 3 and 4. This calculation also gives a workfunction of 5.64eV which agrees well with published 5 (5.5eV) [6] and 7 (5.35eV) [7] layer slab and one embedded layer (5.65eV) [2] results.

5. Conclusions

This restriction of the determination of iteration mixing parameters β to the muffin-tin region now provides convergence acceleration with stability for SEGF calculations. New calculations wherein all potential terms are used either in Anderson or Broyden mixing, after stability is reached, will hopefully provide still faster convergence and so allow SEGF computations to realize their full potential.

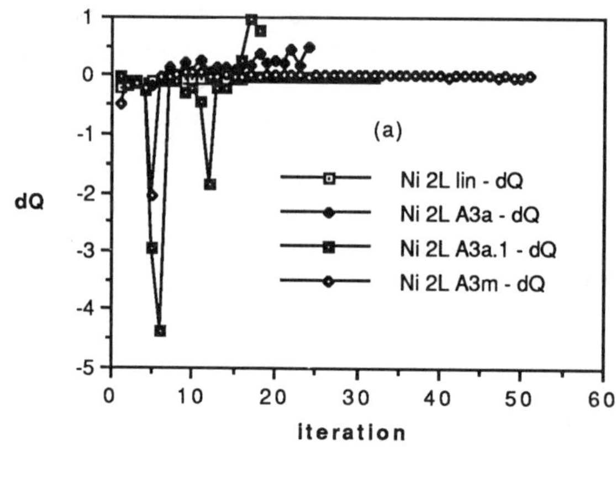

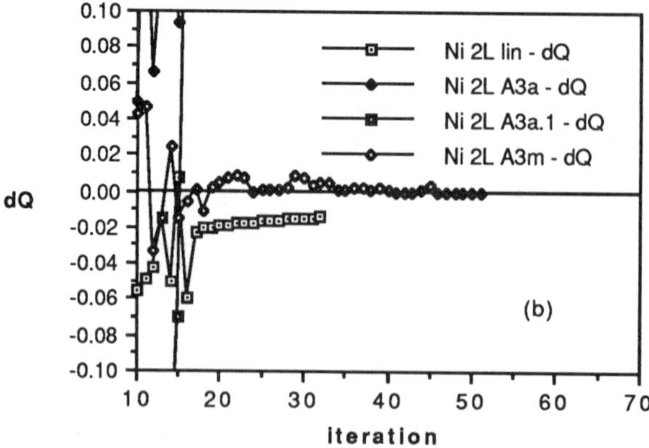

FIGURE 3. Deviation of total charge dQ (in units of e) in the embedded region from its converged value for two layers of Ni on Ni(001) (a) for all iterations and (b) for the last few iterations. Lin indicates linear mixing, the other annotations are as for Figure 2. α is 1% for the linear case, 0.5% for A3a and A3m, and 0.1% for A3a.1.

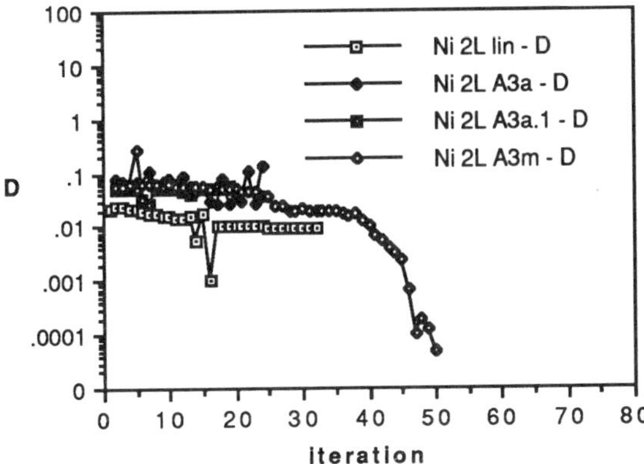

FIGURE 4. As Figure 2 but for two layers Ni on Ni(001). The legend is the same as that for Figure 3.

Acknowledgments

We would like to thank the Robert A. Welch Foundation for funding this research (grant number Y-1135), the University of Texas Center for High Performance Computing for resources on the Cray X-MP and to Mike Brand for assistance in the preparation of data for this paper.

References

1. J.E. Inglesfield, A method of embedding, J. Phys C **14**, 3785-3806 (1981).

2. J.E. Inglesfield and G.A. Benesh, Surface electronic structure: Embedded self-consistent calculations, Phys. Rev. B **37**, 6682-6700 (1988).

3. D.G. Anderson, Iterative procedures for nonlinear integral equations, J. Assoc. Comput. Mach. **12**, 547-560 (1964).

4. C.G. Broyden, A class of methods for solving nonlinear simultaneous equations, Math. Comput. **19**, 577-593(1965).

5. D.D. Johnson, Modified Broyden's method for accelerating convergence in self-consistent calculations, Phys. Rev. B **38**, 807-813 (1988).

6. H. Krakauer, A.J. Freeman and E. Wimmer, Magnetism of the Ni(110) and Ni(100) surfaces: Local-spin-density-functional calculations using the thin slab linearized augmented plane wave method, Phys. Rev. B **28**, 610-623 (1983).

7. E. Wimmer, A.J. Freeman and H. Krakauer, Magnetism at the Ni(001) surface: A high precision, all-electron local-spin-density-functional study, Phys. Rev. B **30**, 3113-3123 (1984).

Dynamics of Relaxation and Reconstruction at Semiconductor Surfaces

J. GRYKO and R. E. ALLEN
Center for Theoretical Physics
Texas A&M University
College Station, Texas 7783, USA

Abstract:
The dynamics of relaxation and reconstruction of Si(100), GaAs(110), GaAs(100), and GaAs(111) surfaces is studied using tight binding model hamiltonians and a new form of the repulsive interaction between surface atoms. It is found that relaxation or reconstruction for all four surfaces is very fast. In the case of (100) surfaces, dimerization of surface atoms is observed. The Si-Si dimers on Si(100) are initially asymmetric, but after a time ~ 1 ps they convert to symmetric form. The As-As dimers on GaAs(100) are symmetric. Since symmetric and asymmetric dimers on Si(100) are close in energy, a switching back and forth between these structures is observed on a 1 ps time scale. We conjecture that the structures thought to be symmetric dimers in scanning tunneling microscopy images may be time-averaged combinations of left-asymmetric, symmetric, and right-asymmetric dimers.

The various properties of silicon and gallium arsenide surfaces have been the subject of intense experimental and theoretical studies for many years. Although the experimental data for Si(100)[1-7] and GaAs surfaces[8-14], and theoretical calculations for Si(100)[15-21] and GaAs[22-25], are numerous, the detailed geometry of the reconstructed surfaces is still unclear. For example, the STM images of Si(100)[26,27] show that surface silicon atoms form dimers. However, the the precise geometry of these dimers is still not clear, with experimental data interpreted as supporting both symmetrical (unbuckled) and asymmetrical (buckled) arrangements of surface dimers[1-7]. For symmetric dimers, there is one basic (2x1) ordering; whereas for asymmetrical dimers there are three basic arrangements: (2x1), p(2x2) and c(2x2)[28]. Some STM images indicate that domains of symmetric and asymmetric dimers may coexist on Si(100)[27].

Theoretical calculations of the geometry of Si(100) are also inconclusive. In general, calculations based on the tight-binding (TB) approach[14] support asymmetric dimers. On the other hand, self-consistent pseudopotential calculations[17,18,19] give symmetric dimers. Although the energy difference between symmetric and asymmetric structures is found to be small, the dimerization energy obtained with different theoretical approaches varies significantly. TB methods give a dimerization energy of about 0.58 eV/dimer[29], whereas the selfconsistent pseudopotential approach with local exchange and correlation energy gives 2.06 eV/dimer. Cluster calculations using the semiempirical MINDO method give an even higher dimerization energy, namely 4.72 eV/dimer[20]. The agreement between experiment and various theories is better in the case of GaAs(110) and GaAs(111). For the non-polar (110) GaAs surface, only relaxation is observed. Both TB and self-consistent calculations[22] give similar results. The arsenic atom in the surface layer moves upward ~ 0.2Å, whereas the gallium atom moves downward ~ 0.45 Å. The combined vertical and horizontal displacements of anion and cation atoms can be described by a rotational angle. The value of this rotational angle as obtained with the use of different experimental and theoretical methods is 28±3 degrees. In the case of polar GaAs(111) (gallium-terminated), the STM images[11] shows that the surface reconstruction arises from single gallium vacancies. This was recognized earlier by Tong et al.[30], who proposed the so-called vacancy-buckling model. In this model, 25% of the Ga atoms in the surface layer are missing, and the remaining Ga atoms move downward relative to the ideal surface. The resulting reconstructed (2x2) surface has a similar configuration to the (110) surface.

In the case of the GaAs(100) (As-terminated) surface, the reconstruction arises from dimerization of As surface atoms. Although the recent STM images[9] support a (2x4) reconstruction, the theoretical calculations have been carried out for a simpler (2x1) reconstruction[8,25]. Again, the precise geometry of the As dimers is unknown, although self-consistent pseudopotential calculations[25] show that the asymmetric dimer has higher energy than the symmetric one.

Theoretical investigations of semiconductor surfaces have focused mainly on static properties. In this paper, we consider the dynamics of surface reconstruction, with the goal of understanding the energetics and time scales involved. Using molecular dynamics, with interatomic forces calculated from the Hellmann-Feynman theorem, we calculate trajectories of the surface atoms starting with ideal, unrelaxed surface positions. We use atom-atom interactions derived from TB and the electron density functional approach, rather than a model potential like that used by Weakliem et al.[31]. The TB model often describes the electronic properties of semiconductor surfaces very well. In this model the total energy of a system is expressed as a sum of two terms[32]:

$$E_{tot} = \sum_{i_{occ}} \epsilon_i + E_{repul},\tag{1}$$

where the first term is the "band energy" and the second models all other contributions. In the present work, the electron eigenvalues ϵ_i are determined by diagonalization of the TB hamiltonian for a slab of 7 or 8 Si or GaAs bilayers. We have used the sp^3s^* TB parameters of Vogl et al.[33]. For surface reconstruction calculations, TB parameters must be known over a wide range of atom-atom distances. The Vogl model uses only parameters for nearest neighbors to describe bulk properties. In our studies, the Vogl parameters were extended to next nearest neighbor distances r_{NNN} using a smooth function

$$f(r) = \frac{f_0}{r^n} \left(1 - e^{-2(r-r_{NNN})^2}\right).\tag{2}$$

where distances are in Å and f_0 and n were determined from the Harrison[34] d^{-2} scaling, by matching function and first derivative at the nearest neighbor distance. For distances greater than r_{NNN}, the TB parameters were set to zero.

The repulsive part of the total energy for Si-Si and Ga-As interactions was taken as a sum of two-body terms,

$$E_{rep} = \sum_{i>j} \phi(r_{ij}).\tag{3}$$

We have tested two functions ϕ; namely,

$$\phi_1(r) = \frac{A}{r^m} \left(1 - e^{-\alpha(r-r_{NNN})^4}\right)$$

$$\phi_2(r) = \frac{B}{r^n} e^{\frac{-\beta}{(r-r_{NNN})^4}},\tag{4}$$

where $\alpha = 0.75\text{Å}^{-4}$. As for the TB parameters, the repulsive part of the interaction was set to zero for $r > r_{NNN}$. The function $\phi_2(r)$ is similar to the repulsive part of the model potential of Sillinger and Weber[35]. The adjustable parameters A and m were determined by fitting experimental values of the lattice constants and bulk moduli of Si and GaAs. The parameters B, n, and β were determined in a similar way, with the use of experimental values of the lattice constants, bulk moduli, and cohesive energies of the bulk materials. The elastic constants of the bulk materials calculated with the use of the above interaction models are in excellent agreement with experiment[36]. Although the total energy of bulk Si as a function of lattice constant, shown in Fig. 1, is very similar for ϕ_1 and ϕ_2, the dimerization energy of two silicon atoms on Si(100) is different. For $\phi_1(r)$, the dimerization energy is 2.64 eV (for surface temperature 500K), and for $\phi_2(r)$ it is only 1.74 eV (surface temperature 300K). The dimerization energy for ϕ_2 is in good agreement with the results of pseudopotential calculations.

The As-As interaction has been determined by a different procedure. First, we carried out local density functional (LDF) calculations for a hypothetical molecule $As_2Ga_4H_{10}$. The geometry of this molecule was set up to simulate the GaAs(100) surface; i.e., the As-As and As-Ga distances and corresponding angles were the same for the above molecule and the GaAs(100) surface. The hydrogen atoms simulate bulk layers of GaAs. The total energy $E_{tot}(R)$ of this molecule was calculated as a function of the As-As distance, R. Next, we calculated the "band energy" E_{band} in the TB approximation for this molecule. The effective repulsive part of the energy was determined from[24]

$$E_{repul} = E_{tot}(R) - E_{band}(R). \tag{5}$$

In the above treatment, the attractive part of the total interaction has many-body character (i.e., it is not pairwise), whereas the repulsive part is a sum of pairwise contributions. This is similar to the model potentials of Stillinger and Weber[35] and Tersoff[37], which include n-body effects.

Using the above expressions for the total energy, we have carried out molecular dynamics simulations of the reconstruction of Si and GaAs surfaces. We started with ideal surfaces, using atomic velocities selected from the Maxwellian distribution for a given temperature. For the Si(100) surface the distance between two surface atoms is displayed in Fig. 2, and the distance between surface and subsurface atoms in Fig. 3. For both forms of the repulsive potential, the distance between two surface Si atoms decreases from 3.84 Å (ideal surface) to ~ 2.3 Å in a dimer within a very short time: ~ 0.1 - 0.25 ps. During this process, the distance between surface and subsurface atoms remains the same (Fig. 3). This means that the two surface atoms in a dimer rotate clockwise and counterclockwise respectively. This rotation is not symmetrical, however, and one atom moves much more than the other. This is illustrated in Fig. 4a, showing initially asymmetric dimerization. After about 2ps the asymmetric dimer becomes symmetric, but after an additional time of ~ 1ps it becomes asymmetric again. In addition, the direction of asymmetry is reversed. Similar behavior was seen in a molecular dynamics simulation using the Stillinger-Weber[35] potential, but the interconversion time was much shorter (about 0.2 ps). A typical approach to symmetric dimerization is shown in Fig. 4b. Again, initially there is some asymmetry, and the dimer becomes symmetric after a relatively long time, about 3 ps. The very fast first stage of dimerization, with no activation energy for this process, is due to the fact that there is no energy barrier (for either potential) for a motion of the surface atoms along a line passing through their centers. This is illustrated in Fig. 5, which shows the total energy of a 7 bilayer slab as a function of the displacement of a single surface atom.

After dimerization, the distance between two surface atoms is 2.3 Å for a surface temperature of 300K and 2.33 Å for 500K. This distance is shorter than the nearest

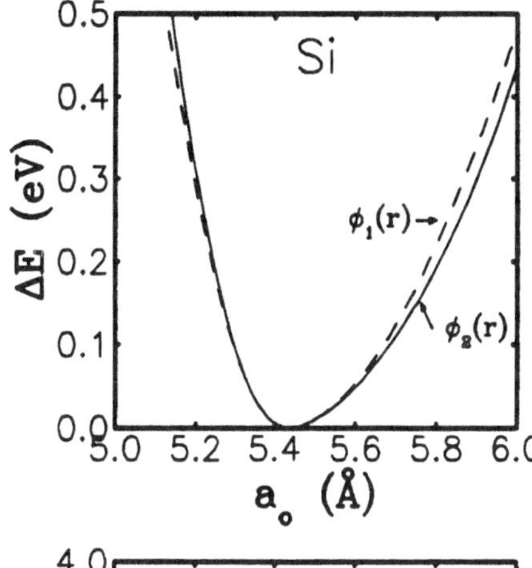

Fig. 1. Total energy of bulk silicon per atom, as a function of lattice parameter, relative to the total energy at zero presure.

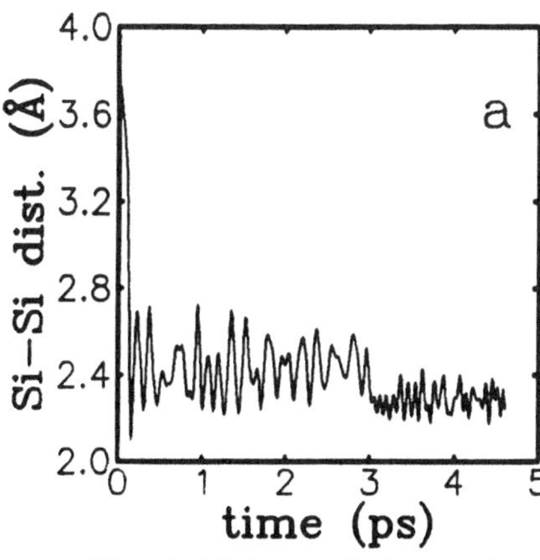

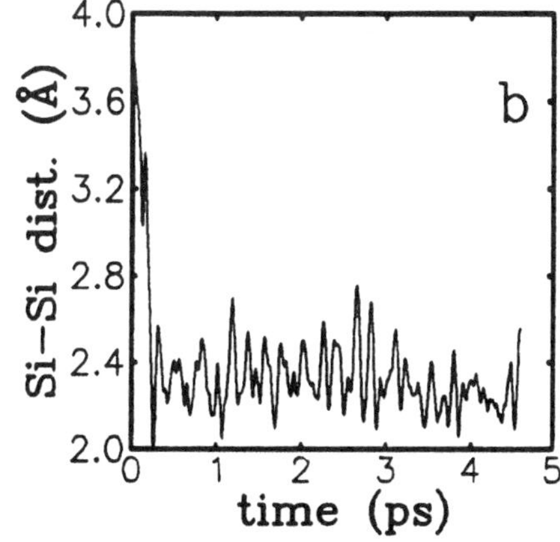

Fig. 2. Distance between two silicon atoms at Si(100) surface:
(a) repulsive potential $\phi_1(r)$, T = 300K;
(b) repulsive potential $\phi_2(r)$, T = 300K.

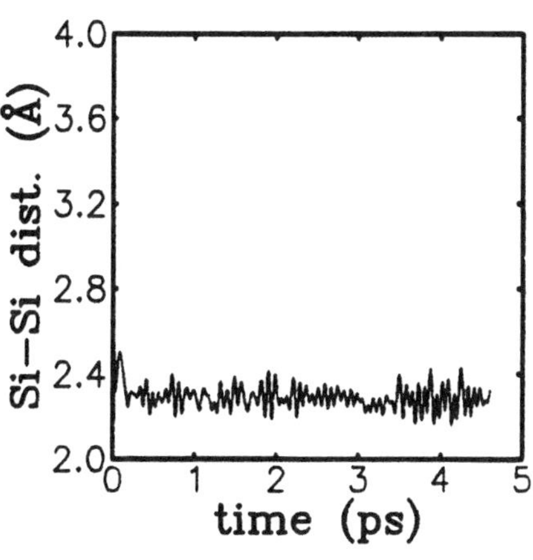

Fig. 3. Distance between surface and subsurface silicon atoms; repulsive potential $\phi_2(r)$, T = 300K.

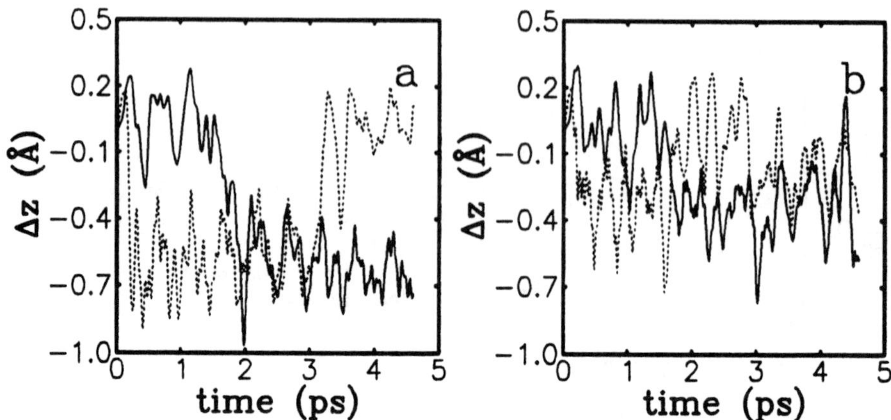

Fig. 4. Displacement of surface atoms in direction perpendicular to surface. (a) an example of asymmetric-symmetric-asymmetric interconversion (repulsive potential $\phi_1(r)$, T = 300K);(b) an example of symmetric dimerization (repulsive potential $\phi_2(r)$, T = 500K).

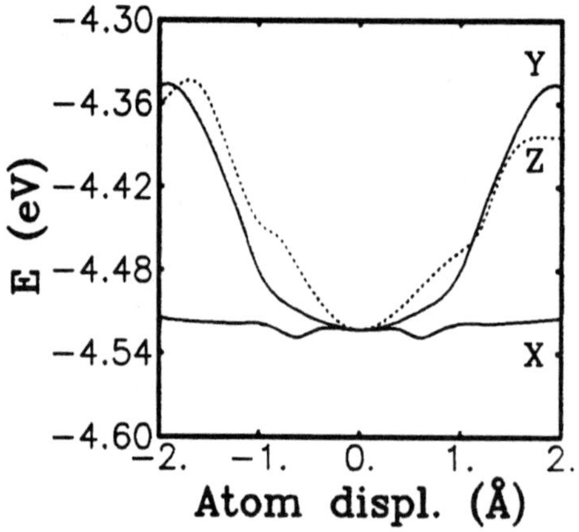

Fig. 5. Energy per atom in a slab of 7 silicon bilayers, as a function of the displacement of a single atom in various directions

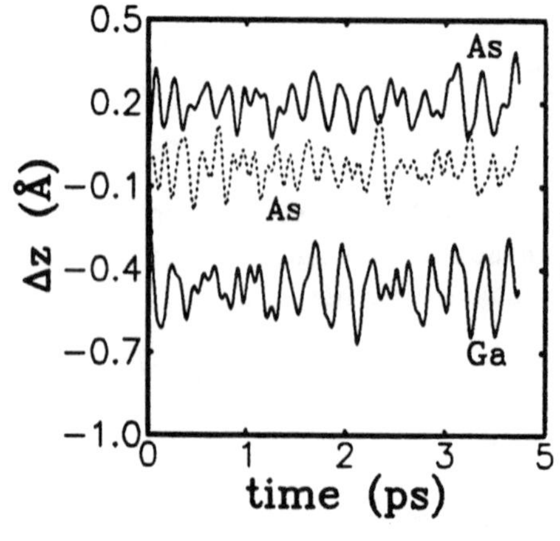

Fig. 6. Displacements of surface atoms for GaAs(110) (T = 300K).

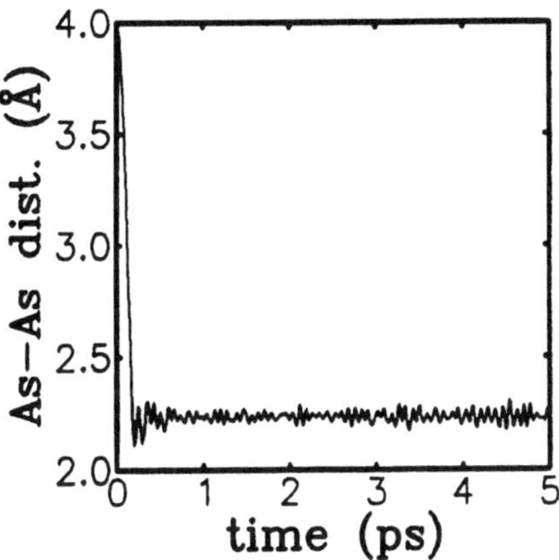

Fig. 7. As–As distance for GaAs(100) surface, T = 300K.

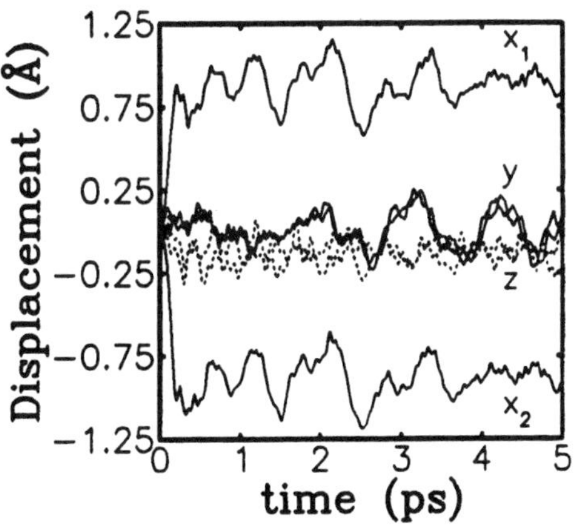

Fig. 8. Displacement of dimerizing As surface atoms on GaAs(100) surface, T = 300K.

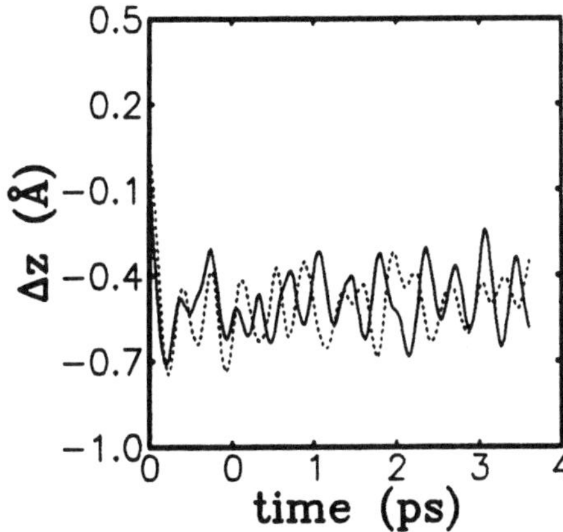

Fig. 9. Displacements of two Ga surface atoms on GaAs(111) surface, T = 300K.

neighbor distance in bulk silicon. We note that pseudopotential calculations give an even shorter bond length for a dimer, namely 2.22 Å [10] or 2.23 Å [12]. In contrast, in the model potential calculations of Weakliem *et al.* [31] the dimer bond length was 2.42 Å. This might explain their more rapid interconversion between symmetric and asymmetric dimers.

It is interesting that we obtained similar results with two different repulsive potentials. This is probably due to the fact that the dimerization energy is much higher than the average thermal energy.

The relaxation of the GaAs(110) surface is shown in Fig. 6. The surface atoms (As and Ga) move very rapidly ($\sim$ 0.1 ps) to their final positions, with the As atom moving upward and the Ga atom moving downward. Note that the equilibrium position of the subsurface As atom is not changed during relaxation. The reconstruction of the GaAs(100) surface is similar to that of the Si(100) surface. The As-As distance is shown in Fig. 7, and displacements of the As atoms in the x, y, and z directions in Fig. 8. The As-As distance after dimerization is 2.25Å, which is smaller than the Ga-As distance in bulk GaAs (2.35 Å). The dimerization of the As surface atoms is very fast, and only symmetric dimers are formed, as shown in Fig. 8. We see that the As atoms basically move along a straight line passing through their initial positions, with slight displacements toward the bulk.

The dynamics of the vacancy-buckling model for GaAs(111) is shown in Fig. 9. The initial configuration was the ideal GaAs(111) surface with one Ga atom removed per (2x2) cell. (The loss of a Ga atom is clearly a much slower process than relaxation of the resulting (111) surface.) Fig. 9 shows the displacement of two Ga surface atoms in the direction perpendicular to the surface. We note that these atoms move in a "coherent" way to their final positions, as might be expected from the vacancy-buckling structure.

In summary, we have found that tight-binding models with suitable repulsive potentials are capable of describing the dynamics of reconstruction at Si and GaAs surfaces. For Si(100), we find that both symmetric and asymmetric dimers can exist. These structures differ only slightly in energy, and in our simulations there is interconversion on a one picosecond time scale, with, e.g. left-asymmetric $\rightarrow$ symmetric $\rightarrow$ right-asymmetric. We conjecture that the STM images interpreted as a symmetric dimers may actually represent time-averaged combinations of all three types of dimers.

Acknowledgment

This work was supported by the Office of Naval Research (Grant N00014-89-J-1088). Additional support was provided by the Robert A. Welch Foundation. We wish to thank Cray Research Inc. for a research grant and for the DGAUSS program used in the local density functional calculations. Finally, we thank M. Weimer and R. Becker for a stimulating discussion of STM observations of Si(100) dimers.

REFERENCES

* Permanent address: Institute of Physical Chemistry, 01-224 Warsaw, Poland.

1. Haneman, D., Surfaces of Silicon, *Rep. Prog. Phys.*, vol. 50, pp. 1045-1086, 1987.

2. Tromp, R. M., Spectroscopy with the Scanning Tunneling Microscope: a Critical

Review, *J. Phys.: Condens. Matter*, vol. 1, pp. 10211-10228, 1988.

3. Hamers, R. J., and Köhler, Determination of the Local Electronic Structure of Atomic-Sized Defects on Si(100) by Tunneling Spectroscopy, *J. Vac. Sci. Technol. A*, vol. 7, pp. 2854-2859, 1989.

4. Martensson, P., Cricenti, A., and Hansson, G. V., Photoemission Study of the Surface States that Pin the Fermi Level at Si(100)2x1 Surfaces, *Phys. Rev. B*, vol. 33, pp. 8855-8858, 1986.

5. Holland, B. W., Duke, C. B., and Paton, A., The Atomic Geometry of Si(100)-(2x1) Revisited, *Surface Sci.*, vol. 140, pp. L269-L278, 1984.

6. Tromp, R. M., Smeenk, R. G., Saris, F. W., and Chadi, D. J., Ion Beam Crystallography of Silicon Surfaces. II. Si(100)-(2x1), *Surface Sci.*, vol. 133, pp. 137-158, 1983.

7. Sasse, A. G. B. M., Wormeester, H., Van der Hoef, M. A., and Van Silfhout, A., Transition Density of States (TDOS) of the Si(100)(2x1) Surface Derived from the $L_{2,3}$VV Auger Lineshape Compared with Cluster Calculations, *Surface Sci.*, vol. 218, pp. 553-568, 1989.

8. Larsen, P. K., van der Veen, J. F., Mazur, A., Pollmann, J., Neave, J. H., and Joyce, B. A., Surface Electronic Structure of GaAs(001)-(2x4): Angle-resolved Photoemission and Tight-Binding Calculations, *Phys. Rev. B*, vol. 26, pp. 3222-3237, 1982.

9. Pashley, M. D., Haberern, K. W., Friday, W., Woodall, J. M., and Kirchnwar, P. D., Structure of GaAs(001) (2x4)-c(2x8) Determined by Scanning Tunneling Microscopy, *Phys. Rev. Lett.*, vol. 60, pp. 2176-2179, 1988.

10. Pandey, K. C., Freeouf, J. L., and Eastman, D. E., Photoemission and Band-Structure Studies of the GaAs(110) Surface, *J. Vac. Sci. Technol.*, vol. 14, pp. 904-909, 1977.

11. Haberern, K. W., and Pashley, M. D., GaAs(111) A-(2x2) Reconstruction Studied by Scanning Tunneling Microscopy, *Phys. Rev. B*, vol. 41, pp. 3226-3229, 1990.

12. Lieske, N. P., *J. Phys. Chem. Solids*, The Electronic Structure of Semiconductor Surfaces, vol. 45, pp. 821-870, 1984.

13. Larsen, P. K., and Chadi, D. J., Surface Structure of As-stabilized GaAs(001): 2x4, c(2x8), and Domain Structures, *Phys. Rev. B*, vol. 37, pp. 8282-8288, 1988.

14. Chadi, D. J., Atomic and Electronic Structures of Reconstructed Si(100) Surfaces, *Phys. Rev. Lett.*, vol. 43, pp. 43-47, 1979.

15. Biegelsen, D. K., Bingans, R. D., Northrup, J. E., and Swartz, L.-E., Surface Reconstruction of GaAs(100) Observed by Scanning Tunneling Microscopy, *Phys. Rev. B*, vol. 41, pp. 5701-5706, 1990.

16. Yin, M. T., and Cohen, M. L., Theoretical Determination of Surface Atomic Geometry: Si(001)-(2x1), *Phys. Rev. B*, vol. 24, pp. 2303-2306, 1981.

17. Pandey, K. C., in: Proc. 17th ICPS, Eds. D.J. Chadi and W.A. Harrison (Springer, New York, 1984) pp. 55-58.

18. Pandey, K. C., Reconstruction of Semiconductor Surfaces: Buckling, Ionicity, and π-Bonded Chains, *Phys. Rev. Lett.*, vol. 49, pp. 223-226, 1982.

19. Payne, M. C., Roberts, N., Needs, R. J., Needels, M., and Joannopoulos, J. D., Total Energy and Stress of Metal and Semiconductor Surfaces, *Surface Sci.*, vol. 211/212, pp. 1-20, 1989.

20. Craig, B. I., and Smith, P. V., Slab-MINDO Calculations of the Si(100) Surface, *Surface Sci.*, vol. 218, pp. 569-579, 1989.

21. Bowen, M. A., Dow, J. D., and Allen, R. E., Si(100) Surface States: A Success for the (2x1) Asymmetric Dimer Model, *Phys. Rev. B*, vol. 26, pp. 7083-7085, 1982.

22. Ferraz, A. C., and Srivastava, G. P., Determination of the Surface Geometry of GaAs(110) by the Total Energy and Force Methods, *Surface. Sci.*, vol. 182, pp. 161-178, 1987.

23. Zhang, S. B., and Cohen, M. L., Surface States on GaAs(110), *Surface Sci.*, vol. 172, pp. 754-762, 1986.

24. Carter, J. N., Dwyer, V. M., and Holland, B. W., Empirical Tight Binding Cluster Method for Semiconductor Surfaces, Surface Sci., vol. 188, pp. L723-L728, 1978.

25. Qian, G.-X., Martin, R. M., and Chadi, D. J., First-principles Calculations of Atomic and Electronic Structure of the GaAs(110) Surface, *Phys. Rev. B*, vol. 37, pp. 1303-1307, 1988.

26. Hamers, R. J., Tromp, R. M., and Demuth, J. E., Scanning Tunneling Microscopy of Si(001), *Phys. Rev. B*, vol. 34, pp. 5343-5356, 1986.

27. Niehus, H., Köhler, U. K., Copel, M., and Demuth, J. E., A Real Space Investigation of the Dimer Structure of Si(001)-(2x8), *J. Microscopy*, vol. 152, pp. 735-742, 1988; Becker, R. S., private communication and to be published.

28. Alerhand, O. L., and Mele, E. J., Surface Reconstruction and Vibrational Excitations of Si(001), *Phys. Rev. B.*, vol. 35, pp. 5533-5546, 1987.

29. Bechstedt, F., and Reichardt, D., Total Energy Minimization for Surfaces of Covalent Semiconductors C, Si, Ge, and α-Sn. II. (100) (2x1) Surfaces, *Surface Sci.*, vol. 202, pp. 83-98, 1988.

30. Tong, S. Y., Xu, G., and Mei, W. N., Vacancy-Buckling Model for the (2x2)GaAs(111) Surface, *Phys. Rev. Lett.*, vol. 52, pp. 1693-1696, 1984.

31. Weakliem, P. C., Smith, G. W., and Carter, E. A., Subpicosecond Interconversion of Buckled and Symmetric Dimers on Si(100), *submitted to Phys. Rev. B.*

32. Chadi, D. J., (110) Surface Atomic Structures of Covalent and Ionic Semiconductors, *Phys. Rev. B*, vol. 19, pp. 2074-2082, 1979.

33. Vogl, P., Hjalmarson, H. P., and Dow, J. D., A Semi-Empirical Tight-Binding Theory of the Electronic Structure of Semiconductors, *J. Phys. Chem. Solids*, vol. 44, pp. 365-378, 1983.

34. Harrison, W. A., *Electronic Structure and the Properties of Solids* (Freeman, San Francisco, 1980).

35. Stillinger, F. H., and Weber, T. A., Computer Simulation of Local Order in Condensed Phases of Silicon, *Phys. Rev. B*, vol. 31, pp. 5262-5271, 1985.

36. Gryko, J.. and Allen, R. E., to be published.

37. Tersoff, J., Empirical Interatomic Potential for Silicon with Improved Elastic Properties, *Phys. Rev. B*, vol. 38, pp. 9902-9905, 1988.

Light Particles in Fluids: Mean Field Theory versus Path-Integration

BRUCE N. MILLER
Department of Physics
Texas Christian University
Fort Worth, Texas 76129, USA

ABSTRACT

This paper concerns the behavior of light particles (e.g. electrons, positrons, and positronium) in fluids. Two major theoretical approaches, mean-field theory versus path-integration, are compared. Applications of the theory to the calculation of positron annihilation rates and their dependence on fluid temperature and density are explored.

INTRODUCTION

This brief paper compares two different theoretical models of the interaction of a single excess light particle (e.g. an electron, positron, or positronium atom) with an atomic or molecular fluid. It emphasizes the Path-Integral representations which have been used recently to successfully model the observed behavior. A companion paper, which describes the application of Monte-Carlo method for computing physical observables in the Path-Integral representation directly follows. This work is divided into four short sections: In the first we describe the experimental phenomena associated with self-trapping. In the second we describe the earlier mean-field or density functional theory, which is differential. In the third, we outline Feynman's discretized path-integral formulation, and show how it can be adapted to the computation of the annihilation rate of a thermalized positron. In the final section we explain the relationship between the models and compare their features.

EXPERIMENTAL EVIDENCE FOR SELF-TRAPPING

When a light particle (LP) thermalizes in an atomic or molecular fluid, its mean deBroglie wavelength (about 50Å near room the atoms. In contrast with the more massive fluid particles (FPs), it can simulta-neously interact with all of the FPs within a semi-macroscopic region. While the translational degrees of freedom of most fluids approximately obey classical mechanics, any model of the LP-fluid system must account for the quantum nature of the LP, which has unique consequences in the laboratory.

Experimental studies of the behavior of a light particle in a dense fluid indicate that, in certain regions of mean fluid temperature and pressure, the LP alters the local fluid density in its vicinity. This alteration results in decreased mobility of the electron under the influence of an electric field,[1] a decrease in the decay rate of ortho-positronium (o-Ps) or an increase in the decay rate of the positron (Ps), and a change in the angular correlation of photon pairs produced in the annihilation of para-positronium (p-Ps).[2] The accepted explanation for this phenomena is that the light particle becomes localized in a semi-macroscopic volume of altered fluid density that results from the interaction of the LP and fluid atoms or molecules. This localization process is referred to as self-trapping because the presence of the LP induces the potential well in which it is stabilized.[2] Localization clearly alters the drift velocity of the electron in an applied field. It affects the decay rate of Ps or o-Ps (which has a "long" vacuum lifetime) by modifying the local density of surplus electrons available for annihilation with the positron (see Fig. 1). Localization affects the angular correlation of the decay products of the positron in p-Ps (which has a "short" vacuum lifetime) with its electron partner by increasing its average kinetic energy compared with the free positronium atom.

Experiments indicate that self-trapping is robust in the vicinity of the liquid-vapor critical point. For example, in Fig. 1 plots the o-Ps decay rate in ethane versus density are displayed on isotherms. We see that the strongest deviation from linearity, the so-called plateau region, occurs near the critical isotherm in a broad region surrounding the critical density. The small decay rate suggests that the o-Ps atom has reduced the local fluid density in its vicinity. This is consistent with the microscopic picture as the fluid properties are more easily altered here because of the large isothermal compressibility.

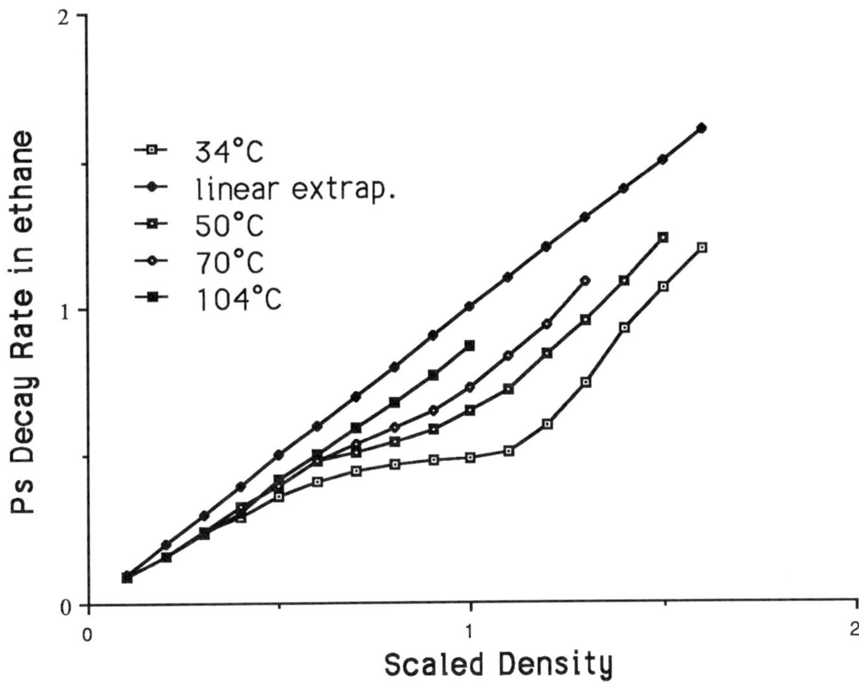

FIGURE 1. Plots of Sharma's experimental measurements of the scaled annihilation rate of ortho-positronium versus density in ethane on four isotherms. The straight line with unit slope is the linear extrapolation of the behavior at low density and represents the decay rate in the absence of localization.

MEAN-FIELD THEORY OF SELF-TRAPPING

For over two decades the central model describing self-trapping was some variant of mean field theory (MFT).[2,3] In all versions, the wavefunction, ψ, of the trapped LP satisfies the nonlinear Schrodinger equation

$$-(\hbar^2/2m)\Delta\psi(\mathbf{x}) + \psi(\mathbf{x})w_L(\mathbf{x}) = E\psi(\mathbf{x}) \qquad (1)$$

where m is the mass of the LP and w_L is the average potential

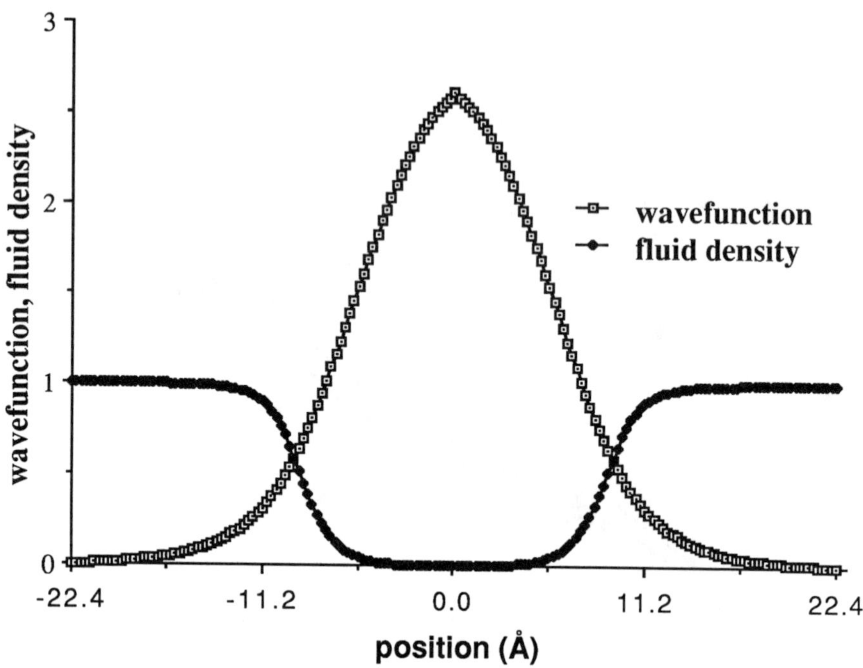

FIGURE 2. Sample scaled density and wavefunction profiles from mean field theory at the critical point of ethane.

produced by the mean local fluid density $\rho(\mathbf{x})$,
$w_L(\mathbf{x})=\int d\mathbf{x}'w(\mathbf{x}-\mathbf{x}')\rho(\mathbf{x}')$. The potential w represents the direct interaction between the Lp and FP. The equation is incomplete because we must supply a relation between ρ and ψ. The simplest is the condition of local thermodynamic equilibrium,

$$w_F(\mathbf{x}) + \mu[\rho(\mathbf{x})] = \mu_0 \qquad , \tag{2}$$

in which $w_F(\mathbf{x})=\int d\mathbf{x}'w(\mathbf{x}-\mathbf{x})\left|\psi(\mathbf{x}')\right|^2$ is the quantum averaged poten-tial experienced by an FP, $\mu[\rho(\mathbf{x})]$ is the local chemical potential of the fluid and is determined by the local temperature and density, and μ_0 is the local chemical potential at positions far from the trapped LP. (1) and (2) comprise a coupled pair which, with suitable boundary conditions, uniquely determine ρ and ψ. In principle (2) can be inverted to obtain ρ as a functional of ψ. Direct substitution in w_L results in a nonlinear Shrodinger

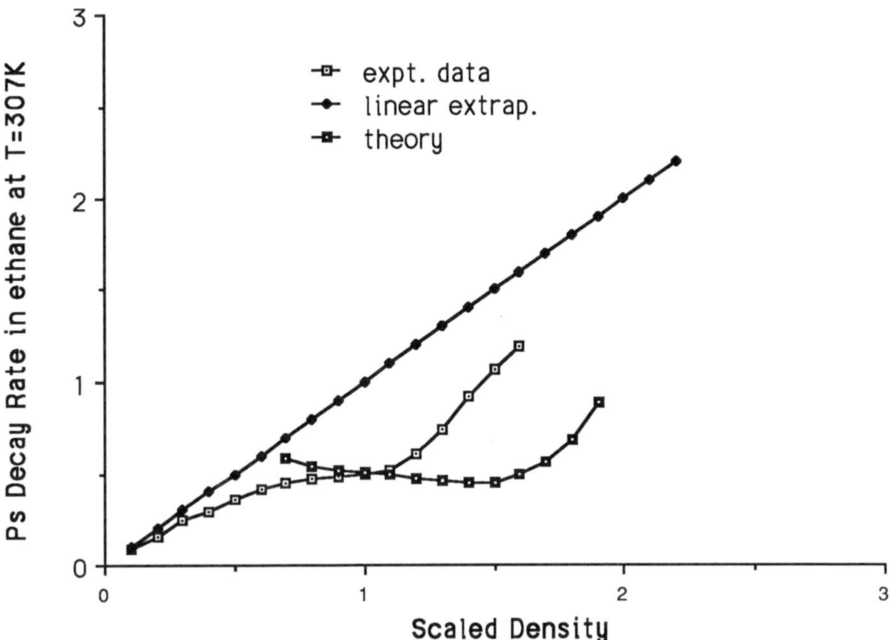

FIGURE 3. Comparison of experimental and theoretical annihilation rate of orthopositronium. The computed curve represents the contribution from the localized solution of the mean field equations.

equation for ψ. It must be supplemented by the normalization condition on ψ to represent an allowed physical state. In practice, when w is approximated by a pseudopotential ($w(r)=c\delta(r)$), spherically symmetric solutions have been obtained with numerical precision for both ground and excited states by using standard techniques[3]. A sample plot of the scaled wavefunction and local density as a function of radial distance from the LP center of mass is displayed in Fig. 2.

This is a mean-field theory because it assumes that the LP experiences the smooth potential produced by the _average_ local fluid density, i.e. it ignores fluctuations. It is the simplest mean field variant because it assumes that local thermodynamics (2) works on all length scales. With the pseudopotential representation for w, it exhibits _universality_; i.e. under a suitable scaling of

585

physical and thermodynamic quantities it takes precisely the same form for the large class fluids which obey the principle of corresponding states.[3] Other variants take into account the explicit effects of the finite range of the FP correlations.[4] All versions can be derived from a variational principle in which an effective free energy for the LP-FP system is minimized with respect to variations in ψ .[2,3,4]

MFT was used successfully to model the density dependence of the Ps annihilation rate in He at low temperatures (on the order of 5^0K).[5] As we can see from Fig. 3, which compares the predicted decay rate contributed by the ground state with experimental data for the decay rate of o-Ps in ethane, it has not fared so well at higher temperatures[6,7]. A central problem is that the existence of normalized ground state solutions turns on and off at different densities. This results in discontinuities in the predictions of physical observables, such as the decay rate, which are not observed in experiments[3] (see Fig. 1). The qualitative features of the model can be brought into closer agreement with experiment by including transitions between extended and localized states, but the results are still not satisfactory.[7]

MICROSCOPIC MODEL OF THE LP-FP INTERACTION

To improve on MFT we need to construct a microscopic model. A natural candidate for the system Hamiltonian is

$$H = \hat{p}^2/2m + \sum_i w(\mathbf{x}-\mathbf{R}_i) + \sum_{i<j} u(\mathbf{R}_i-\mathbf{R}_j) + \sum_i P_i^2/2M, \qquad (3)$$

where the first term is the kinetic energy of the LP, the second, $W(\mathbf{x},\underline{\mathbf{R}})$, is the total LP-FP interaction energy, and the third, $U(\underline{\mathbf{R}})$, and fourth are the total FP internal potential energy and kinetic energy respectively. $\mathbf{R}_i$, P_i and M are, respectively, the position, momentum and mass of an FP and u is the FP-FP interaction potential. For reasons given above, this is a hybrid representation in which the LP is treated via quantum mechanics and the FPs are treated classically. It is useful when the LP relaxation time scale is much faster than that for the FPs and is referred to as adiabatic.

In thermal equilibrium, the mean value of a physical observable, O, which may depend on both the LP and FP coordinates, but not the FP momenta, is given by

$$\langle O \rangle = (1/Z) \int d\underline{R} \, \exp[-\beta U(\underline{R})] \, \text{Tr} \, [O \exp(-\beta H')]. \tag{4}$$

The trace refers the quantum part so $H' = \hat{p}^2/2m + W(\mathbf{x}, \underline{R})$ and the classical part consists of the integral $\int dR_1 \cdots dR_N \equiv \int d\underline{R}$ weighted by the Gibb's factor $\exp(-\beta U)$. The partition function, Z, is obtained from the numerator by letting $O=1$ and, as usual, $\beta = 1/kT$.

Calculations using (4) directly are nearly impossible with present methods. For example, in a representation where H' is diagonal, it would be necessary to compute the complete set of eigenstates of H' for every realization of the collection of FP positions $\underline{R}$, and then perform the classical average. Feynman showed that the calculation could be closely approximated with an equivalent classical system by introducing the discretized path-integral (DPI).[8] Consider a ring polymer consisting of p elements with positions $\mathbf{x}_j$. Each interacts harmonically with its nearest neighbors on the chain with force constant $(pm/\hbar^2\beta^2)$ and with the FPs with potential W/p. Thus the potential energy of the polymer is

$$\Phi(\underline{\mathbf{x}}, \underline{R}) \equiv \sum_{1 \leq j \leq p} [(pm/2\beta\hbar^2)|\mathbf{x}_{j+1} - \mathbf{x}_j|^2 + (1/p)W(\mathbf{x}_j, \underline{R})] \tag{5}$$

where $\mathbf{x}_{p+1} = \mathbf{x}_1$. The mean value of the configuration dependent operator O for this system is

$$\langle O \rangle = (1/Z) \int d\underline{R} \, \exp[-\beta U(\underline{R})] \int d\underline{\mathbf{x}} \, \exp[-\beta\Phi(\underline{\mathbf{x}}, \underline{R})] O(\mathbf{x}_1, \underline{R}) \tag{6}$$

and converges to (4) in the limit $p \to \infty$. The calculation of observables from (6) using Monte-Carlo methods is the subject to the companion of the paper. Here we consider the application of (6) for the case where the observable is the positron annihilation rate, λ.

Physically the annihilation rate is proportional to the overlap of the positron and the electron density of the fluid. If $f(\mathbf{r})$ is the quantum averaged electron density, assumed frozen, of an FP fixed at the origin, then the quantum state ψ and molecule j contribute $\int |\psi(\mathbf{x})|^2 \, f(\mathbf{x}-\mathbf{R}_j)d\mathbf{x}$ to λ. Thus the best choice for the quantum mechanical decay rate operator is $\sum_{1 \leq j \leq N} f(\mathbf{x}-\mathbf{R}_j)$.

Because all of the labelled sites of the ring polymer are equivalent, we may express the mean annihilation rate as

$$\lambda = \langle \sum_{1 \leq j \leq N} f(\mathbf{x}-\mathbf{R}_j) \rangle = \langle (1/p) \sum_{1 \leq \alpha \leq p} \sum_{1 \leq j \leq N} f(\mathbf{x}_\alpha - \mathbf{R}_j) \rangle$$

$$= (1/p) \int d\mathbf{x} \int d\mathbf{R} \, f(\mathbf{x}-\mathbf{R}) \, n^{(2)}{}_{LF}(\mathbf{x}-\mathbf{R}) = \rho \int d\mathbf{r} \, f(\mathbf{r}) \, g_{LF}(\mathbf{r}), \qquad (7)$$

where, in (7), $n^{(2)}{}_{LF}(\mathbf{x}-\mathbf{R}) = \rho \, \rho_{pol} \, g_{LF}(\mathbf{x}-\mathbf{R})$ is the two point LP-FP distribution function and ρ_{pol} is the mean site density of the polymer.[9] Finally, by expressing the radial distribution in terms of the pair correlation function, $g_{LF}(r) = 1 + h_{LF}(r)$, we arrive at

$$\lambda = \rho \int d\mathbf{r} \, f(\mathbf{r}) \, (1 + h_{LF}(\mathbf{r})). \qquad (8)$$

This simple looking result supports our expectation that, in the absence of correlations, the mean decay rate is simply proportional to the average fluid density. All of the effects of localization on λ enter in the computation of the pair correlation function between polymer sites and the elements of the fluid. Because f has a short range (typically less than one Å) the experimental deviations from linearity displayed in Fig. 1 in the critical region of the fluid arise from the reduction in h_{LF} (or g_{LF}) resulting from the large compressibility. Changes in g_{LF} away from its behavior in the low fluid density regime is the most important signature of localization. Further details are explored in the next paper and are discussed at some length elsewhere.[9]

By using similar methods, it is possible to obtain expressions for the variance of λ as well as the momentum distribution of the trapped particle in terms of the distribution functions of statistical mechanics.[10] We mention in passing that the ring configuration is not ubiquitous. Observables which depend on purely quantum mechanical correlations, such as the momentum distribution, require an open (broken) chain.[10]

CONNECTION BETWEEN THE MODELS AND SUMMARY

We have described two apparently disparate theoretical models, MFT and RPI, for the same physical system. How are they related? What are their relative merits? We mentioned earlier that MFT can be derived from a density functional theory with an effective free energy, say F. Miller and Fan[9] have proved that $\exp(-\beta F)$ is a Gibbs measure in the Hilbert space for ψ. The measure can be constructed from the adiabatic model by replacing $\langle\psi|\exp(-\beta H')|\psi\rangle$ with $\exp(-\beta\langle\psi|H'|\psi\rangle)$ in the quantum trace. This approximation has a long history[11] and is notoriously problematic. However, it establishes the desired connection: MFT provides the optimal (most probable) state for a measure which roughly approximates the distribution of ψ in Hilbert space.

The chief advantage of DFT is that it is differential. The nonlinear Schrodinger equation (1) which is central to the model can be solved numerically on a modern PC. Unfortunately, with the exception of low temperatures, at present the agreement with experiment appears qualitative at best. DFT has the annoying feature that localized states turn on and off suddenly, whereas corresponding experimental measurements change smoothly. In contrast DPI is integral and requires Monte-Carlo or Molecular Dynamics for the computation and prediction of observables. These methods are time consuming and converge slowly, but provide a truer representation of the adiabatic model in that the approximation is controlled by the size of the polymer. In addition, DPI correctly accounts for fluctuations and avoids the discontinuities which are problematic in MFT. In the following paper, the Monte-Carlo method is applied to DPI to study the influence of self-trapping on the radial distribution function, g_{LF}, and the positron annihilation rate.

ACKNOWLEDGEMENTS

The author benefitted from the support of the Robert Welch Foundation of Houston, Texas through grant P-1002, the Research Foundation of Texas Christian University through grant 5-23167 and the hospitality of the Aspen Center for Physics during June, 1989. The decay rate data for ethane used to construct Fig. 1 was graciously supplied by S. C. Sharma.

REFERENCES

1. N. Gee and G. R. Freeman, Can. J. Chem. **64**, 1810 (1986).

2. For a review of both the experimental and theoretical physics of positron annihilation in fluids, including self-trapping, see I. T. Iakubov and A. G. Khrapak, Prog. Phys. **45**, 697 (1982).

3. B. N. Miller and T. Reese, Phys. Rev. A **39**, 4735 (1989).

4. J. P. Hernandez, Phys. Rev. B **11**, 1289 (1975). C. Ebner and C. Punyanita, Phys. Rev. A **19**, 856 (1979); R. M. Niemenen, M. Manninen, I. Valimaa, and P. Hautojarvi, Phys. Rev. A **21**, 1677 (1980). R. L. Moore, C. L. Cleveland and H. A. Gersch, Phys. Rev. B **18**, 1183 (1978).

5. M. J. Stott and E. Zaremba, Phys. Rev. Lett. **38**, 1493 (1977).

6. S. C. Sharma, R. H. Arganbright and M. H. Ward, Jour. Phys. B **20**, 867 (1987); S. C. Sharma, E. H. Juengurman, Phys. Lett. **144A**, 47 (1986); K. F. Canter and L. O. Roellig, Phys. Rev. A **12**, 386 (1975); T. B. Daniel and R. Stump, Phys. Rev. **115**, 1599 (1959); S. C. Sharma, A. Eftekhari and J. D. McNutt, Phys. Rev. Lett. **48**, 953 (1982).

7. T. Reese and B. N. Miller, "Self-Trapping of a Light Particle in a Dense Fluid", Physical Review A , in press.

8. R. P. Feynman, <u>Statistical</u> <u>Mechanics</u>, (Benjamin, Reading, Mass. 1972). See also D. Chandler, Y. Singh, and D. M. Richardson, J. Chem. Phys., **81**,1975 (1984); D. Thirumalai, R. W. Hall and B. J. Berne, J. Chem. Phys. **81**, 2523 (1984).

9. Y. Fan and B. N. Miller, "Localization in a Lennard-Jones Fluid: The Path-Integral Approach," J. Chem. Phys., in press.

10. B. N. Miller and Y. Fan, "Competing Theories of Localization with Applications to Positron Annihilation," Phys. Rev. A, in press.

11. M. C. Gutzwiller, Phys. Rev. **137** A, 1726 (1965); T. Ogawa, K. Kanda, and T. Matsubara, Prog. Theor. Phys. **53**, 614(1975).

Path Integral Monte Carlo Simulation of Ortho-Positronium in a 2-Dimensional Lennard-Jones Fluid

G. A. WORRELL
General Dynamics
Forth Worth, Texas 76129, USA

TERRENCE REESE
Texas Christian University
Forth Worth, Texas, USA

ABSTRACT

The decay-rate of ortho-positronium in a 2-dimensional Lennard-Jones fluid is calculated using the Path Integral Monte Carlo (PIMC) method. A series of numerical simulations on ortho-positronium in a fluid are carried out to determine the number of discretization points required to evaluate ensemble averages in the path integral representation. The decay-rate of ortho-positronium is calculated at T=200K for two fluid densities, $\rho\sigma^2$=0.3 and 0.7. Qualitative agreement with the experimentally observed anomalous density dependence of the ortho-positronium decay-rate is found.

INTRODUCTION

The phenomena associated with injecting positrons into fluids has received a great deal of attention from theorists over the past decade[1,2,3,4,5]. Perhaps the most interesting phenomenon is the effect of the fluid density on the decay-rate of ortho-positronium (Ops). To date the work has relied on various mean-field approxima-tions[6]. In this proceedings, we report the preliminary results of a path integral Monte Carlo (PIMC) simulation of Ops in a 2-dimensional Lennard-Jones (L-J) fluid. The Ops decay-rate, due to pick-off annihilation with a fluid atom electron, is calculated within the adiabatic approximation. The results are in qualitative agreement with experiment.

In the adiabatic approximation, the path integral representation of the canonical partition function of a quantum Ops atom in a classical fluid is[5,6]

$$Z = (mP/2\pi h^2\beta)^{3P/2} \int d\{R^N\} \exp(-\beta U) \int d\{r^P\} \exp(-\beta \varphi), \qquad (1)$$

$$\varphi = mP/2h^2\beta^2 \sum_{i}^{P} (r_i - r_{i+1})^2 + (1/P)\sum_{i}^{p} \sum_{j}^{N} V(|r_i - R_j|), \qquad (2)$$

$$U = \sum_{i>j} U(|R_i - R_j|) \qquad (3)$$

The P - point discretized path, or polymer representation, of the Ops atom is represented by the coordinates $\{r^P\}$. The positions of the N fluid atoms are given by $\{R^N\}$. The mass of the Ops atom is m, h is Planck's constant divided by 2π and $\beta=1/kT$. The interaction potential between the ith discrete Ops path point and the jth fluid atom is $V(|r_i - R_j|)/P$; $U(|R_i - R_j|)$ is the interaction potential between the ith and j^{th} fluid atoms.

Information about the local fluid environment around the Ops-polymer can be obtained by calculating the ops polymer-fluid pair correlation function

$$g(r) = \rho^{-1} < \sum_{i=1}^{N} 1/P \sum_{j=1}^{P} \delta(|R_i - r_j - r|)> , \qquad (4)$$

where ρ is the fluid density. The Ops decay-rate due to pick-off with a fluid atom electron is[6]

$$\lambda = \left< \sum_{i=1}^{N} 1/P \sum_{j=1}^{P} \left| \Psi(r_j - R_i) \right|^2 \right> \tag{5}$$

where Ψ is the groundstate positronium wavefunction with center of mass at r_j. In the above expressions the brackets $\langle \rangle$ denote equilibrium ensemble averages.

NUMERICAL ALGORITHM

An appealing aspect of the path integral representation of a quantum particle is that Monte Carlo methods can be used to calculate equilibrium properties; the quantum particle - polymer isomorphism provides a way to represent quantum averages as classical averages.
In practice, the number of polymer points required to achieve accurate results has to be determined by direct simulation, Fig. 1. Of course, the number of polymer points used to approximate the path integral must be large enough to justify the high temperature approximation used in the derivation of (1.). Following Feynman's argument[7] , paths that contribute the most to the integral in (1.) are those satisfying the condition

$$-Pmd^2/2h^2\beta^2 < 1 \tag{6}$$

where "d" is the maximum value of $\left| r_i - r_{i+1} \right|$. If the potential $V(r_i, \{R^N\})$ varies sufficiently slowly such that it is approximately unchanged when the path varies by more than $d \approx \lambda/\sqrt{(p)}$, where $\lambda = h(\beta/m)^{1/2}$ is the thermal wavelength, the high temperature approximation is reasonable.

From the literature, a wide range is found for the number of polymer points required to obtain accurate results. For example, argon below its triple point requires only 5 points[8], essentially a

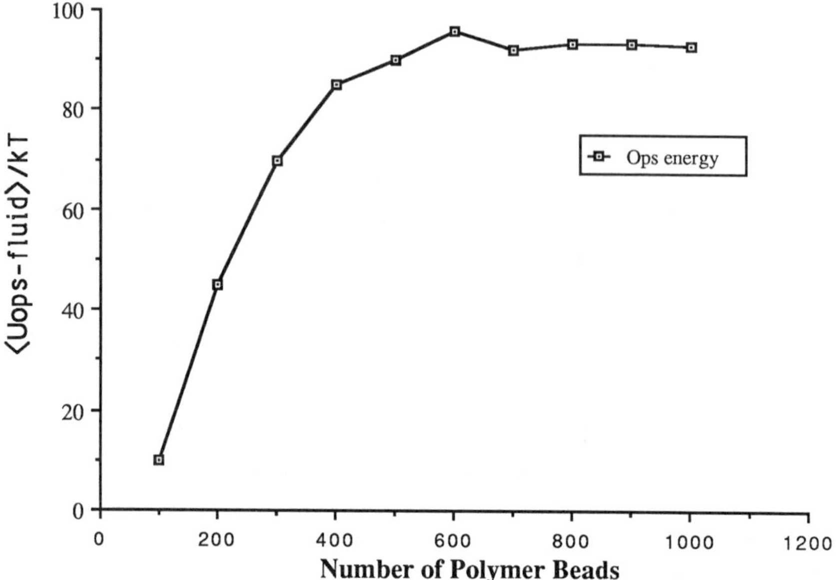

FIGURE 1. The dependence of the fluid-ops interaction energy on the number of polymer points. This is for a 2-dimensional fluid with $\rho\sigma^2=0.6$ and T=200K and is modeled as a system of 200 atoms in a periodic box of length approximately 62Å.

classical system whereas the simulation of an electron in liquid ammonia requires several thousand polymer points[9]. To evaluate the integral in (1), polymer chain configurations can be sampled according to the distribution

$$\Pi = \exp -\beta(U + \varphi) \qquad (7)$$

using the Metropolis algorithm[10]. The Metropolis algorithm can be described as follows. Let Π_s be the probability of finding the ensemble in the configuration s and let $P_{s-s'}$ be the transition probability for the configuration to change from s to s'. The Markov process describing the walk through configuration space $\{R^N, r^P\}$ is

$$\sum_s \Pi_s \, P_{s-s'} = \Pi_{s'} \tag{8}$$

It is easy to show that the transition probability

$$P_{s-s'} = P^* \, \min[1, \Pi_{s'} P^*_{s'-s}/\Pi_s P_{s-s'}], \tag{9}$$

where P^* is an arbitrary a priori transition probability and $\min[\,1, \Pi_s \cdot P^*_{s-s'}/\Pi_s P^*_{s'-s}]$ is an acceptance factor, satisfies (8) and has the required limiting distribution Π.

It is known[9], however, that such an approach is slowly convergent, except for systems that are nearly classical. The reason for the slow convergence can be understood by examining the quadratic term in the effective potential of the polymer. If a large value of "P" is necessary to use the high-temperature approximation, the force constant of the harmonic term, P/λ^2 , is large. The large force constant will constrain the polymer, energetically favoring small displacements of the polymer and result in a slow sampling of configuration space. Usually in Monte Carlo simulations the a priori transition probability P^* is chosen as a uniform distribution in the configuration space. However, as discussed above such a choice will result in an inefficient sampling of the polymer configurations.   Several approaches have been suggested to achieve a more efficient sampling of the polymer configuration space. In this work we use an approximation[9] based on a general algorithm developed by Pollock et. al.[10]. The algorithm is designed to efficiently sample the configuration space by a judicious choice of the a priori transition probability P^*. The transition probability used stems from the observation that if $P^* = \Pi$, the acceptance probability is unity; that is, all configurations generated are accepted as sample points of the ensemble. This, of course, is not a reasonable choice for P^* since Π is too complicated to sample directly. If the path integral in (1.) is rewritten in terms of conditional probabilities, ignoring the potential due to the fluid atoms, the integrand can be expressed as

$$\Pi_{s-s'} = \exp\left[\sum_{i=2}^{P} (r_i - r^*_i)^2/2\sigma_i^2\right] \qquad (10)$$

where

$$r^*_{i+1} = t_i r_i + \tau r_{P+1}/t_{i-1} \qquad (11)$$

is the mean, $t_i = \beta(1-i/P)$, $\tau = \beta/p$ and

$$\sigma_i^2 = h\tau/m \, (t_i/t_{i-1}) \qquad (12)$$

is the standard deviation.

The distribution Π, equation (10.), can now be sampled directly using Levy's[10] method for sampling conditional Brownian paths. The new points in the polymer configuration space are generated using

$$r_i = r^*_i + \sigma_i\zeta, \qquad (13)$$

where ζ is a vector of gaussian random numbers with unit variance. The above transition probability allows polymer configurations to be generated with a high acceptance. As discussed by D.F. Coker et. al.[9], the above algorithm ignores the influence of the fluid atoms on the polymer chain. In our simulations presented here, this has not been a problem. However, for highly localized polymers, this would surely be a limitation.

NUMERICAL RESULTS

In the following section we discuss the simulation of ortho-positronium in a 2-D Lennard-Jones fluid. The potential used for the fluid-fluid interaction is the Lennard-Jones 6-12 with the parameters, $\sigma_{Ar}=3.405$ and $\varepsilon_{Ar} = 117.5$. The question of a potential for the positronium-fluid interaction is difficult and has

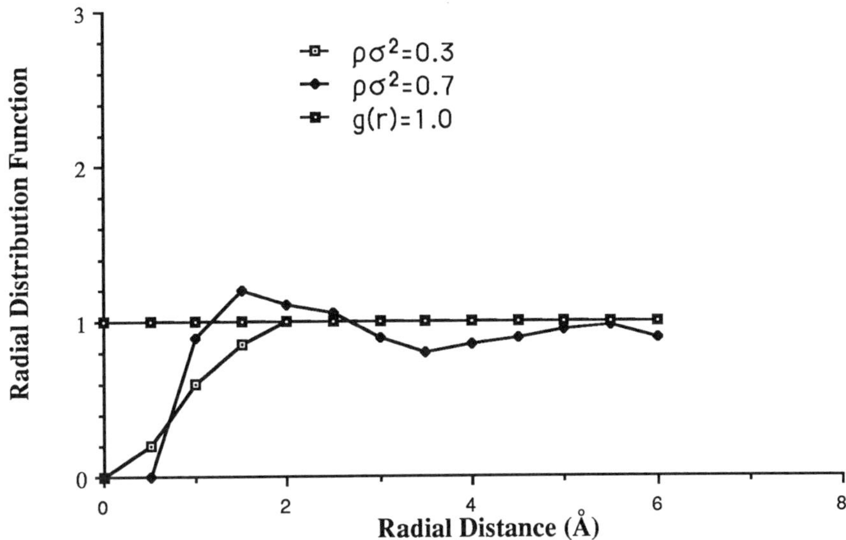

FIGURE 2. The fluid-polymer correlation function g(r) and the Ops decay rate is shown at the two densities $\rho\sigma^2$=0.3, 0.7 and T=200K.

received considerable attention in the literature[12,13]. From mean-field studies and Monte Carlo simulations of an excess electron in a fluid, it appears that the phenomenon of self-trapping is manifested even for a purely repulsive potential. For this reason and due to the lack of any definitive scattering experiments to determine a potential, we will assume a simple repulsive potential, neglecting polarization effects,

$$V(r-R) = \alpha_{Ops} (\sigma_{Ops}/|r-R|)^{12} ,$$ (14)

with α_{Ops}=4,000K and σ_{Ops}=1.7A.

To determine the number "P" of polymer points required to obtain accurate results, we have examined the P dependence of the polymer-fluid potential energy,

$$< U_{Ops-fluid}> = <\Sigma\ V(|r-R|)> .$$ (15)

In Figure 1. the average energy is shown as a function of the number of polymer points for a system at a temperature T=200K and density $\rho\sigma^2$ = 0.6. The fluid is represented using 200 atoms in a periodic box with l≈ 62A. The ratio of the thermal wavelength and box size is less than 0.5 so finite size effects should not be a problem. The figure shows that the potential energy converges for p≈ 600. The main result of this paper is shown in figure 2. For two fluid densities $\rho\sigma^2$= 0.3 and 0.6 at T=200K, the positronium decay-rate and the polymer-fluid correlation function are calculated. The simulations show that as the density of the fluid is increased, the range of the fluid-polymer correlation g(r) decreases. The decrease in the range of the fluid-polymer correlation function as the density is increased, results in a less than linear increase in the decay-rate with fluid density. Because of the decrease in the range of the correlation function, the polymer (or Ops atom) samples the fluid atoms less frequently, resulting in an increased lifetime. This appears to be a direct manifestation of the self-trapping of the Ops atom. We are in the process of simulating Ops in a 3-D Lennard-Jones fluid.

ACKNOWLEDGMENTS

The authors benefitted from discussions with Bruce Miller. The simulations were made possible with CRAY CPU time provided by the Pittsburgh Super Computing Center.

REFERENCES

1. I.T. Iakubov and A.G. Kharpak, Prog. Phys. <u>45,</u> pp 697, (1982).
2. R.M. Nieminen, I. Valimaa, M. Manninen and P. Hautojarvi, PRA <u>21</u> , pp 1677, (1980).
3. M. Tuomisaari, K. Rytsola, R.M. Nieminen and P. Hautojarvi, J. Phys. : At. Mol. Phys., <u>19</u>, pp 279, (1986).
4. B.N. Miller and T. Reese, Phys. Rev. A <u>39</u>, pp 4735 (1989).
5. J.K. Percus, <u>Positron Annihilation studies of fluids,</u> (World Scientific, Teaneck, NJ Press, 1988).
6. B.N. Miller and Yizhong Fan, this volume.

7. R.P. Feynman, <u>Statistical Mechanics</u>, (Benjamin, Reading, Mass 1972).

8. K. Singer and W. Smith, Mol. Phys. <u>57</u> , pp 761, (1986).

9. D.F. Coker, B.J. Berne and D. Thirumalai, J. Chem. Phys. <u>86</u> , pp 5689, (1987).

10. N. Metropolis, A.W. Rosenbluth, M.N. Rosenbluth, A.H. Teller and E. Teller, J. Chem. Phys. <u>21</u>, pp 1087, (1953).

11. F.B. Knight, <u>Essential of Browian Motion and Diffusion</u>, (American Mathematical Society, Providence, RI 1981).

12. J.R. Mason and R.H. Richie, Phys. Rev. Lett. <u>54</u> pp 785 (1985).

13. C.K. Au and R.J. Drachman, Phys. Rev. Lett. <u>56</u>, pp 324, (1986).

Electronic Energy Levels in Condensed Media: New Techniques for Theoreticians

NEIL R. KESTNER and JIA-AN YANG
Chemistry Department
Louisiana State University
Baton Rouge, Louisiana 70803, USA

ABSTRACT

Fast Fourier Transform methods can be used to rapidly calculate the ground and excited states of atoms and molecules. These can be coupled to Molecular Dynamics techniques to allow one to study electronic energy levels in amorphous media. Examples discussed include electron attachment to polar clusters and the spectroscopy of alkali atoms in rare gas liquids.

I. INTRODUCTION

In order to study real materials, especially impurity states of various types, it is essential to be able to handle fairly complex systems but it is also essential to introduce temperature and disorder systematically. During the past few years a number of new interrelated techniques have been developed. While these are not yet ideal in all cases, they do represent major advances and real breakthroughs not only in computational techniques but in being able to study problems which hitherto were not calculable except in some approximation schemes. These new methods are capable of taking advantage of the most advanced aspects of new super computer systems, namely both vector and parallel architectures. We will first review some of these methods and then concentrate on the one which works for a case where there is an alkali atom in an inert gas condensed phase.

First let us consider what we need in a little more detail. Let's assume that we want the electronic absorption spectrum of a molecule or atom in some condensed phase, some disordered medium. It is just as easily to study the vibrational structure but for now let us emphasize the electronic structure. Based on the Franck Condon Principle we want the excitation energy for that molecule in all of the thermally accessible configurations of the medium, but in particular for all thermally accessible configurations of the solvent around the molecule since the effect of the medium is of rather short range. If we could calculate this we would then have the inhomogeneously broadened line shape as the con-figurational average of the excitation energy for each of the possible configurations (assumed to be equal to the time average by the ergodic theorem). But how do we calculate these configurations? The most obvious method would be to use one of the standard statistical mechanical methods to find these configurations using the correct interactions between the molecules. Since the molecules are often not sypherically symmetric we will need the complete coordinates of all molecules, not only the radial distribution function around the impurity. If the medium is ordered we could use the crystal structure as have many previous workers. Alternatively, if you wanted to only have a general picture and the medium was rather simple, you could use methods like the Mean Spherical Approximation and consider the medium as a perturbation as Chandler and co-workers have done very effectively.[1] However if we want to consider cases where the wave functions are distorted by the medium and we have rather non-isotropic systems we

This work was supported by DOE Grant FG05-87ER13674.

need to consider more elaborate methods. In those cases if we want an accurate
calculation of the line shape, we would need to do a complete high level ab initio
calculation for each configuration of the medium, including not only the molecule in
question but many neighboring solvent molecules as well. Such calculations are barely
possible for even very simplistic molecules, very small solvent molecules, let alone for
large numbers of particles and higher excited states. Also the time involved to do such
calculations is many, many minutes (often hours) especially if the type of accuracy needed
is of the order of energies equivalent to better than 10 degrees Kelvin, which means that
electron correlation effects are also extremely important. With such times it would be
impossible to sample the hundreds of configurations necessary to obtain proper, reliable
statistical averages.

Therefore we need a fast way to obtain the electronic energy level and yet have the
results be accurate for the weak interactions which take place between the many molecules.
If these interactions were purely classical then one could treat the electronic part of the
calculation separately and consistently. That procedure was first used by Rossky,
Friesner, and Schnitker on the hydrated electron.[2] A more formal procedure has been
developed by Nitzan, Landman, and Barnett[3] for studies of electrons on polar clusters and
related problems. The latter group refers to the procedure as Time Dependent Hartree
Fock Theory and they are able to use the forces calculated from the electronic distribution
and the electronic energy to evaluate the way in which the electron(s) influences the
molecular configurations. This can be done self consistently to find the minimum free
energy configuration of an electron in a cluster of polar molecules or the behavior of the
electron in a polar fluid. The first people to use a similar approach were, in fact, Carr and
Parrinello[4] who evaluated systems like electrons in molten salts and most importantly,
systems like liquid semiconductors. Their work has been characterized by the use of
density functional theory for the most part since they have attempted to include a number
of electrons and do the entire calculation self consistently. All of these methods are good
if the Hartree Fock interaction is sufficient to obtain accurate results. If it is, then these
methods must be used if one desires a real self consistent result.

However, if a major part of the interaction goes beyond the Hartree Fock level, i.e.
involves electron correlation, then these methods can never yield the correct result because
the interactions are not completely specified and we are forced to try something else,
something a little less consistent, a little less appealing, but which works. One must
decouple the statistical and the electronic calculation to some extent. In addition, one must
have a very efficient method for determining the electronic energy of a system or at least
that impurity state, a method capable of yielding several excited states as well as the
ground state in tens of seconds, at most, since good statistics require that we have configurations of the order of a several hundred to a thousand.

Recently several new methods have become popular for cases involving small
numbers of quantum particles, optimally one. These are various finite difference methods
and the Fast Fourier Transform (FFT) method and the path integral methods which we
will not discuss in detail here. Let us first discuss each of these briefly, with emphasis on
the FFT methods which we have used, illustrating it with several applications.

II. SOLUTION OF ONE ELECTRON PROBLEMS

Historically, the first calculations of wave functions of an electron in a one
dimensional problem were either exact solutions or numerical finite element methods.
When Hartree first calculated Hartree Fock solutions, he and his dad use numerical integration techniques. However as we all know when the first computers were developed
Roothan realized that it was suddenly advantageous to use matrix methods and this led
him to develop basis set expansions. The success of quantum chemistry thus far is largely
a result of this switch. However today more is being demanded especially in large
systems, often at nonzero temperatures and in disordered system. The computers have
also changed and have many new unique features which implore us to explore alternatives

to the standard basis set expansions. A large number of people are beginning this search.
We might just mention Charlotte Froese who has been a real pioneer and long time
advocate of numerical methods. Her book[5] is a great reference on these issues. Also
recently Friesner[6] has advocated pseudo spectral methods. In vibrational problems there
are the many papers by Light[7] using Distributed Gaussians (DG)and Discrete Variable
Representation (DVR). These are but a few attempts in these areas.

Another major approach is the Fast Fourier Transform(FFT) method originally
developed by Feit, Fleck, and Kosloff.[8,9] It has been applied to many one electron or
quasi one electron problems by Rossky, Friesner, and Schnitker;[2] by Landman, Nitzan,
Barnett and their associates;[3] Metiu and his coworkers;[10] and recently by Berne and his
collaborators[11] to name but a few of the major groups.

Our objectives were similar to those of the Rossky and Landman group in that we
wanted to do large numbers of one electron calculations very rapidly. We were interested
primarily in the low lying excited states. Remember we were considering an effective one
electron atom imbedded in a fluid. Initially this is to be an alkali atom in an inert gas or
liquid hydrogen but ultimately the fluid could be water or other more complex molecules.
These issues ruled out the standard quantum chemistry procedures using basis sets since
our impurity would be changing position and could be altered significantly by the media as
well as by temperature. We also wanted to allow for the electron density distribution to
respond in any way, unbiased by any basis set. Our experience in these areas suggested
that we would not be able to do the calculations fast enough nor is the calculation of
excited states easy to implement in standard Hartree Fock programs. Thus we were led to
consider three alternatives: distributed gaussians(DG), finite difference or finite element
methods (FD), and Fast Fourier Transform methods (FFT). Each of these could be
implemented easily and could yield all of the data we desired. The excited states are give
most easily and most accurately by the first two methods but using successive
orthogonalization, the FFT can also achieve the lower excited states to sufficient accuracy.

We do not want to spend too much time on a comparison of the various methods
but tests were run for typical harmonic and effective one electron pseudopotentials
(Coulombic term plus some screening components). These runs were made on an IBM
3090/600E using subroutine calls to the highly vectorized ESSL (Engineering and
Scientific Subroutine Library). The distributed Gaussians were picked using the sugges-
tions of Light[7] which proved very good. They were placed at evenly spaced intervals,
which is not optimal but that would affect the accuracy more than the timing. The finite
difference results were done using a regularly spaced grid and a simple second order finite
representation of the second derivatives, again not optimal but good if large numbers of
points are used. In terms of accuracy with these approaches a minimum of eight gaussians
or eight grid points are needed per dimension, even that leads to substantial errors (10-
20%) in the few lowest eigenvalues. The time required for 7x7x7 was about 42 sec while
for the 8x8x8 it was 102 seconds so the scaling is about grid points per dimension to the
power six. To get much larger requires adapting the program to use extended core and
dynamic common. This was done for the 12x12x12 and the time was over 22 minutes.
This method was not pursued further since there is an alternative that scales a roughly n
log n, namely the Fourier Transform method. In this equation n is roughly the points per
dimension cubed. In order to get the answers in quantum chemical problems, it is
necessary to do a number of FFT's for each configuration and then propagate these in time
(or imaginary time) to achieve a converged answer. This means that we need to do
something of the order of 200 steps with about 3 FFTs and 2 Inverse FFT's at each time
step. Nevertheless this can be done sufficiently fast so that for 16x16x16 grid points, we
can get one excited state in about 12 seconds. Rossky and his colleagues[2] on the
University of Texas Cray Y-MP get this number down to about 7 seconds. The excited
states are then found as the next orthogonal member of the solutions, repeating the entire
process. Thus for two excited states we can obtain them in roughly 20-30 seconds. This

means that even for 500 or 100 configurations the three electronic states can be determined in a reasonable amount of computer time. We will later give more details on the run times of the method.

It is important to note here that in problems with lower dimensionality, such as vibrations of diatomics, the use of finite difference, distributed Gaussians or variations of collocation methods are very fast and preferable, especially if one is trying to look at very highly excited states, the area for which Light originally developed his method (specifically to study chaos in highly excited vibrational states).

In order to demonstrate why the Fast Fourier Transform method works, let's consider a typical one electron Hamiltonian, H, made up of a kinetic energy part and a potential energy part, called K and V at the moment. It is important to note that K is proportional to the momentum squared while V depends on the coordinates. In the following we will use atomic units which means simply that we will set the mass of the electron, Planck's constant divided by two time pi and the charge of the electron all equal to one, so that energies are in Hartrees (27.21 electron volts) and distances are in Bohr radii (.529 Angstroms).

Using atomic units, the time dependent wave function develops in time according to the following:

$$\Psi(x, t) = e^{-iHt} \Psi(x, 0) \tag{1}$$

where we are considering just a one dimensional case for simplicity. If this time propaga-tor is broken up into a number of pieces so we may use some other approximations, we have the following slight modification:

$$\Psi(x, N\Delta t) = e^{-iH\Delta t} e^{-iH\Delta t} \dots e^{-iH\Delta t} e^{-iH\Delta t} \Psi(x, 0) \tag{2}$$

When this continues out to large times we get a mixture of eigenstates in the result:

$$\Psi(x, N\Delta t) = \Sigma \, c_i \, e^{-iE_i N\Delta t} \psi_i(x) \tag{3}$$

But it is important to remember that $\Psi(x,0)$ can be anything, any initial choice of function.

A slight modification of this which is particularly useful is to switch to imaginary times

$$\Psi(x, Ni\Delta t) = e^{-iH\tau} e^{-iH\tau} \dots e^{-iH\tau} e^{-iH\tau} \Psi(x, 0) \tag{4}$$

$$\text{with} \quad \tau = i\Delta t$$

Now at large "times" we get a mixture of eigenstates again

$$\Psi(x, N\tau) = \Sigma \, c_j \, e^{-E_j N\tau} \psi_j(x) \tag{5}$$

but now as τ gets infinitely large

$$\Psi(x, N\tau) = {}^{-E_0 N\tau} \psi_0(x) \quad \text{times a constant} \tag{6}$$

which means that the ground state dominates the result.

However, it is easy to get the excited states as well. One has only to look for the next solution which is orthogonal to the one found and in this way one may work up to as many excited states as desired. There are, of course, practical problems involved here since each solution found has some error, some inaccuracy, and these are compounded as one considers each successively higher excited state.

Now we need to consider why a Fourier Transform method is applicable to solve this problem. Up to now any type of solution of those equations would be acceptable. However we will see that the FFT method is ideally suited to this procedure. Let us look specifically at the exponential operators involved:

$$e^{H\tau} = e^{-\tau p^2/2m \, - V(x)\tau} \tag{7}$$

It is important to note that this is not equivalent to

$$= e^{-\tau p^2/2m} \, e^{-V(x)\tau} \tag{8}$$

since the momentum and position operators do not commute. The missing terms can be expressed in terms of the commutator of position and momentum. Because of the lack of commutation, the following formula has a smaller error since to first order the two commutator expressions cancel

$$e^{H\tau} = e^{-\tau p^2/4m} \, e^{-V(x)\tau} \, e^{-\tau p^2/4m} \tag{9}$$

The above expression is often referred to in the literature as the split operator method. In Quantum Mechanics it is well known that one can work in either position or momentum space and solve the Schroedinger equation. The spatial representation is preferred when dealing with coordinates or electron density distributions in space, while the momentum representation is preferred by physicists and scattering theorists who are concerned with motions of particles. It is also well known that one can transform from one representation to another using a Fourier Transform. Also we know that the evaluation of the operator $e^{-\tau p^2/4m}$ acting on the wave function is particularly simple if we use the momentum representation since the operator is just a number, i.e. it is diagonal in the momentum. On the other hand the potential which depends on x is most easily evaluated in the coordinate representation since V and all functions of V are diagonal in the coordinates.

Specifically if

$$\varphi(k,t) = \int e^{-ikx} \, \Psi(x,t) \, dx \tag{10}$$

This is just a Fourier Transform, but when the operator above is applied to it explicitely, it yields

$$e^{-\tau k^2/4m} \, \varphi(k,t) \tag{11}$$

and the inverse Fourier Transform of this, $\chi(x,t)$, can then be multiplied by $e^{-V(x)\tau}$ and transformed again into the momentum representation, etc. Thus we can finally get $\Psi(x, t+\tau)$ for one time step, or one increment in time. This can be repeated using the above result for as many time steps as are necessary to achieve convergence.

But how do we implement this? Obviously we want to use a Fast Fourier Transform method to gain the economies of speed. We also will need to discretize our space into some reasonable number of points, partly determined by the accuracy we desire. Assuming that we have the potential in some functional form or at least, evaluated at each point in our grid, let us review the steps required.

One must first choose a grid size and time step which are appropriate. This generally requires some experimentation, although is a wide variety of problems we and others have considered a 16 point grid seems acceptable and a step size of about 3 atomic time units is also optimal. Too large a time step might lead to the wrong answer while too short a time step could cause the number of steps needed to become extremely large, and the time to be lengthened. Then the steps to be followed are:

a. Choose an initial wave function in coordinate space.
b. Transform this to the momentum representation
b. Evaluate the first kinetic energy exponential term
c. Transform that product back to the coordinate representation using an inverse transform
d. Multiply this by the exponential operator of the potential
e. Transform this back to the momentum representation
f. Evaluate the other half part of the exponential of in Kinetic energy in this momentum representation.
f. Do another inverse transform to get back to coordinate space.
g. From the imaginary time dependence of the past few steps, evaluate the estimated energy at this time step.
h. Compare the energy estimated with that of previous step.
i. If the energy has converged to an acceptable tolerance, stop. If not, use this function as input in step a and do another time step.

We have found that somewhere between 100 and 200 time steps are necessary, although the program is set to stop when the necessary criterion is reached, usually about one percent in the energy.

If an excited state is desired, one chooses a new arbitrary trial function and first orthogonalizes it to the ground state solution. In the case of finite representations of functions, such orthogonalizations are extremely fast. The results are then propagated as above, however before being sent through another time step it is reorthogonalized to make sure it remains a pure excited state.

The program we use was originally written by U. Landman and R. Barnett at the Georgia Institute of Technology for a CDC Cyber but we modified it to use IBM ESSL routines. Our older version was slightly less efficient. The speed comes from the use of the highly vectorized FFT routines, in which 256 FFTs are done at one time, and all arithmetic functions are calculated only once. This is possible since we always have the same grid. To give an example of the timings, we ran some typical cases. The 8x8x8 case took 2.4 sec, the 16 case took 25.9 sec, while the 32 grid point run took 322 seconds. These scale as roughly $n^3 \ln n^3$ while the finite difference runs were worse than $n.^6$

III. SPECIFIC APPLICATIONS

While this method has been used in many calculations, we will concentrate on only two in which we are intimately involved. The first concerns the electronic structure of the negative water dimer (a special case of our studies on electrons attached to polar clusters) and recent unpublished work on alkali atoms in inert gas solids.

In our studies of electrons attached to polar clusters and in particular to the water dimer[12] we explored several method of calculating the electron affinity, namely path integral methods which we have not discussed in this paper, Fast Fourier Transform methods in this case coupled directly to the molecular dynamics program in that the electronic density and the classical interactions of the water molecules determine all of the forces, as well as very elaborate ab initio calculations both with and without electron correlation. The electron water molecule interactions were modeled by a pseudopotential of the Gombas type using the electron density of the molecule in a density functional type of approach to electronic interactions. The water water interactions were represented by the very accurate RWK2M potential. The results of all three methods were remarkably similar both in terms of energy and electron density, justifying both the pseudopotential approach and the methodologies used. The FFT approach when coupled to the dynamics has the advantage that real time behavior can also be studied by switching from imaginary back to real times and back again. We will not discuss here the more complex issues of adiabatic surfaces which must be satisfied in this case but the reader is referred to recent papers by Landman, Nitzan, et. al.[13] as well as recent work by Rossky, Friesner, Webster[14] who consider the more general nonadiabatic surface hopping regime. For purposes of this paper it is simply important to note that this is a very viable, very accurate way to determine electronic energy states, both ground states and excited states.

Next we want to discuss the application of this method to the study of impurity levels in condensed phases, specifically we will consider alkali atoms in inert gas fluids although the method can be extended to systems of alkali atoms in other media such as liquid hydrogen. To begin we must model the Hamiltonian of an alkali in the inert gas. We have chosen to consider a simple one electron model. The alkali metal's valence electron is modeled by a simple psuedopotential composed of the Coulombic interaction with a plus one charge and an exponential repulsive term which has been fit to yield the lower excited states of the metal. In particular we use the parameterizations of Preuss,[15] although we have adjusted the parameters for lithium to yield better agreement with exper- iment. The electron inert gas potential is also represented by a pseudopotential, which for the electron xenon and electron helium was taken from work of Berne, Coker and Thirumalai,[16] although the electron helium result is very similar to the more elaborate earlier studies by Kestner, Jortner, Cohen and Rice.[17] We will report here only some of the early studies of sodium and lithium in xenon under various experimental conditions. Our procedure in this case is to use the previously fit inert gas-inert gas and inert gas-alkali atoms interactions (the latter by Baylis[18]) in a molecular dynamics program. We have been fortunate to have the MIXNVT program written by S. Gupta at Louisiana State University. After running the program long enough to establish equilibrium, we dump sample configurations at every tenth time step until we have collected 500 to 1000 repre- sentative solvent configurations around the impurity. These configurations are then put into the FFT program and three or four excited states are obtained for each configuration of the molecules. The results are then expressed in terms of a histogram of the energies using either 50 or 100 bins, typically, with the range covering the absorption peak.

The same range is used for all densities in the results shown in Figures 1 thru 4. In order to understand these results it is important that one realize that we have used a Lennard Jones 6-12 potential for the atomic interactions and thus all results are expressed relative to those parameters. In particular the relative temperature, Tr, is expressed in terms of the well depth for the inert gas reactions. Typically a liquid or solid has Tr near one. The densities are expressed relative to the zero of the potential cubed, or sigma cubed as it is usually expressed. It is a number density with low numbers being low densities. Only at low temperatures is the liquid state reached for relative densities, Dr, below about 0.7. The figures included in this paper represent only a hint at the capabilities of this method to tackle complex issues in condensed matter physics.

This method has a unique capability of then going back to find out what type of configurations contribute to what features. In early work we were able to see that one case involving three distinct peaks corresponded to three distinct sets of nearest neighbor distances. Thus this method can be used to really understand some specific spectral features in molecular terms.

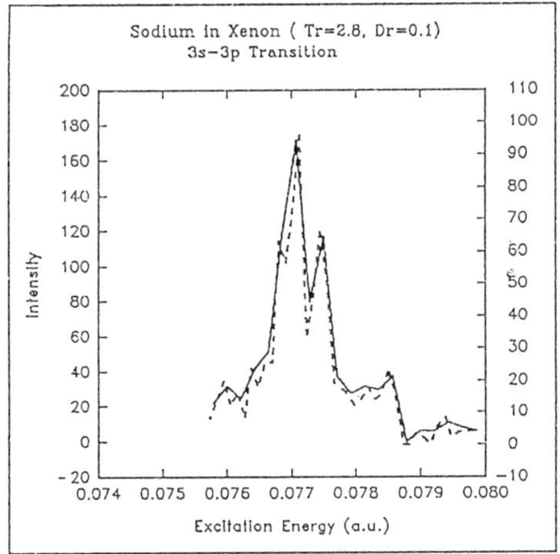

Figure 1 Sodium in Xenon, very low density, high temperature

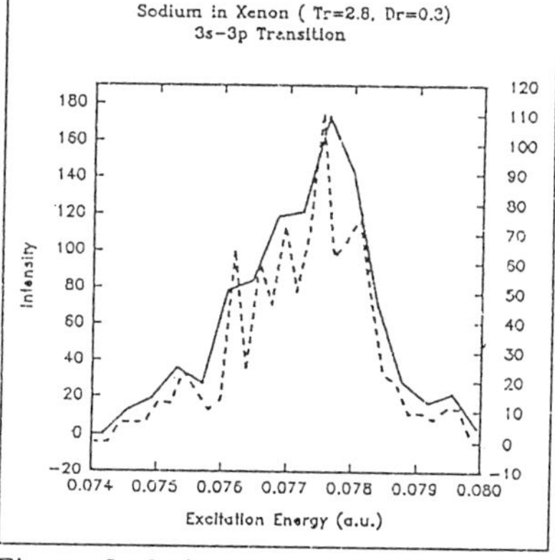

Figure 2 Sodium in Xenon, higher density, 50 and 100 bin results shown

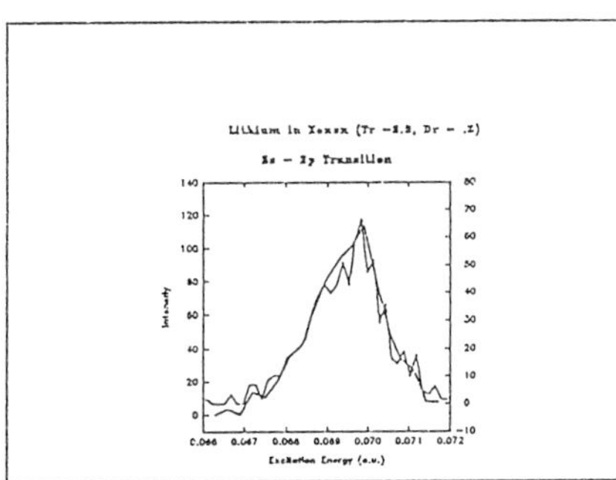

Figure 3 Lithium in Xenon medium density, high temperature

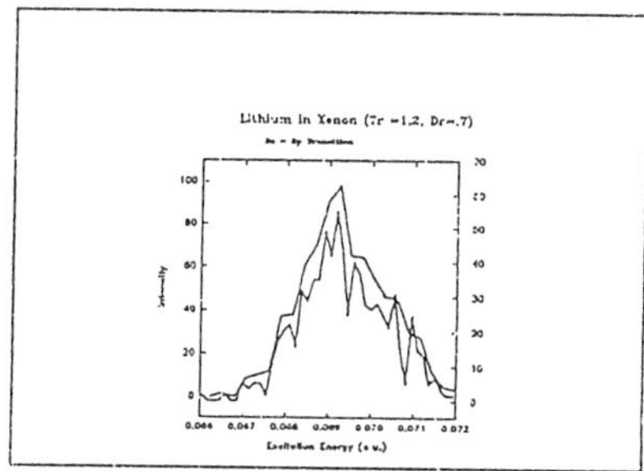

Figure 4 Lithium in Xenon high density, low temperature

IV. CONCLUSIONS

We have discussed one of the new methods which is particularly suited to new computer architectures and which allows one to study on detail, effects of temperature, density and various degrees of disorder on electronic energies. It can study a range of densities and temperatures. It can also be applied to solid state studies but there the configurations should probably be obtained from simulated annealing results rather than molecular dynamics, in order to avoid convergence bottlenecks. These new developments in computational methods finally, very slowly beginning to utilize the new computational architectures we have available.

REFERENCES

1. Schweizer, K.S.; Chandler, D., *Quantum Theory of Solvent Effects on Electronic Spectra: Prediction of the Exact Solution of the Mean Spherical Model,* J.Chem.Phys. 78,4118(1983).

2. Schnitker, J.; Motakabbir, K.; Rossky, P.J.; Friesner,R., *A Priori Calculation of Optical-Absorption Spectrum of Hydrated Electron,* Phys.Rev.Letters, 60,456(1988).
Schnitker, J.; Rossky, P.J., *Quantum Simulation Study of Hydrated Electron,* J.Chem.Phys., 86,3471(1987).

3. Barnett, R.N.; Landman U.; Cleveland C.L.; Jortner, J., Phys.Rev.Letters, 59,811(1987).
Barnett, R.N.; Landman U.; Cleveland C.L.; Jortner, J., *Electron Localization in Water Clusters. II. Surface and Internal States,* J.Chem.Phys. 88,4429(1988).

4. Car, R.; Parrinello, M., *Unified Approach for Molecular Dynamics and Density-Functional Theory,* Phys,Rev.Letters, 55,478(1985).
Fois, E.; Selloni, A.; Parrinello, M., *Approach to Metallic Behavior in Metal-Molten-Salt Solution,* Phys.Rev. B39,4812(1989).
Stich, I.; Car, R.; Parrinello, M.; Baroni, S., *Conjugate Gradient Minimization of the energy functional: A New Method for Electronic Structure Calculation,* Phys.Rev. B39,4997(1989).

5. Froese, C.F., *The Hartree Fock Method for Atoms: A Numerical Approach,* Wiley, New York, 1977.

6. Friesner, R.A., *Solution of Self-consistent Field Electronic Structure Equations by a Pseudospectral Method,* Chem.Phys.Letters, 39,116(1985).
Friesner, R.A., *Solution of the Hartree-Fock Equations by a Pseudospectral Method: Application to Diatomic Molecules,* J.Chem.Phys. 85,1462(1986).
Friesner, R.A., *Solution of the Hartree-Fock Equations for Polyatomic Molecules by a Pseudospectral Method,* J.Chem.Phys. 86,3522(1987).

7. Light, J.C.; Bacic, Z., *Adiabatic Approximation and Nonadiabatic Correction in the Discrete Variable Representation: Highly Vibrational States of Triatomic Molecules,* J.Chem.Phys. 87,4008(1987).
Hamilton, I.P.; Light, J.C., *On Distributed Gaussian Bases for Simple Model Multidimensional Vibrational Problems,* J.Chem.Phys. 84,306(1986).
Light, J.C.; Hamilton, I.P.; Lill, J.V., *Generalized Discrete Variable Approximation in Quantum Mechanics,* J.Chem.Phys. 82,1400(1985).

8. Feit, M.D.; Fleck,Jr. J.A., *Wave Packet Dynamics and Chaos in the Henon-Heiles System,* J.Chem.Phys. 80,2578(1984).
Feit, M.D.; Fleck, J.A.; Steiger, A., *Solution of the Schroedinger Equation by a Spectral Method,* J.Comput.Phys. 47,412(1982).

9. Feit, M.D.; Fleck, Jr. J.A., *Solution of the Schroedinger Equation by a Spectral Method II: Vibrational Energy Levels of Triatomic Molecules,* J.Chem.Phys. 78,301(1983).
Bisseling, R.; Kosloff R., *The Fast Hankel Transform as a Tool in the Solution of Time Dependent Schroedinger Equation,* J.Comput.Phys. 59,136(1985).
Kosloff, R., *Time-Dependent Quantum-Mechanical Methods for Molecular Dynamics,* J.Phys.Chem. 92,2087(1988).
Kosloff, D.; Kosloff, R., *A Fourier Method Solution for the Time Dependent Schroedinger Equation as a Tool in Molecular Dynamics,* J.Comput.Phys. 52,35(1983).

10.   Hellsing, B.; Nitzan, A.; Metiu, H., *A Fast Fourier Transform Method for Calculating the Equilibrium Density Matrix*, Chem.Phys.Letters, 123,523(1986).

11.   Martyna, G.J.; Berne, B.J., *Structure and Energetics of Xe_n^-*, J.Chem.Phys. 88,4516(1988).
      Coker, D.F.; Berne, B.J., *Excess Electronic States in Fluid Helium*, J.Chem.Phys. 89,2128(1988).

12.   Barnett, R.N.; Landman, U.; Dhar, S.; Kestner, N.R.; Jortner, J.; Nitzan, A., *Quantum simulations and ab initio electronic Structure Studies of $(H_2O)_2^-$*, J.Chem.Phys., 91,7797(1989).

13.   Barnett, R.B.; Landman, U.; Nitzan, A., *Relaxation Dynamics Following Transition of Solvated Electrons*, J.Chem.Phys., 90,4413(1989).
      Barnett, R.N.; Landman, U.; Nitzan, A., *Dynamics of Excess Electron Migration, Solvation and Spectra in Polar Molecular Cluster*, J.Chem.Phys., 91,5567(1989).

14.   Rossky, P.; Friesner, R.; Webster, F., *Unpublished, Reported at Boston American Chemical Society Meeting*, April, 1990.

15.   Preuss, H; Stoll, H; Wedig, U; Kruger, TH., *Combination of Pseudopotentials and Density Function*, International.J.Quant.Chem. 19,113(1981).
      Flad, J.; Stoll, H.; Preuss, H., *Calculation of Equilibrium Geometries and Ionization Energies of Sodium Clusters up to Na_8*, J.Chem.Phys. 71,3042(1979).

16.   Coker, D.F.; Berne, B.J.; Thirumalai, D., *Path Integral Monte Carlo Studies of the Behavior of Excess Electrons in Simple Fluids*, J.Chem.Phys. 86,5689(1987).

17.   Kestner, N.R.; Jortner, J.; Cohen, M.H.; Rice, S.A., *Low-Energy Elastic Scattering of Electrons and Positions for Helium Atoms*, Phys.Rev. A, 56,140(1965).

18.   Baylis, W.E., *Semiempirical, Pseudopotential Calculation of Alkali-Noble Gas Interatomic Potentials*, J.Chem.Phys. 51,2665(1969).
      Watts, R.O.; McGee, I.J., *Liquid State Chemical Physics*, John Wiley & Son, 1976.

Index

absolute p-th moment deviation,75, 76
absolute stability, 123
abstract Volterra equation, 124
acceleration schemes, 158, 160
aerodynamic simulations, 504
aerospace structures, 504
Aitken extrapolation, 307
alkali atoms, 605
amorphous media, 600
amplitude envelope equation, 377
analytic-tetrahedron method, 527
anisotropic composite, 202
antiferromagnetism, 525
approximate theorem, 93
approximate solutions, 437
a-proper mappings, 232
axisymmetric triangular
finite element, 191

back scattering, 402, 419
backward recurrence, 17
band energy, 574
BANDPACKAGE, 539, 541, 542, 543
big bang, 4
biharmonic equation, 185
BITNET, 496
Bloch basis functions, 555
bifurcation, 376
bifurcation, secondary, 379, 384
bilateral Laplace transform, 482, 483
bistatic scattering, 401
boundary conditions, Dirichlet, 249
boundary element method, 228
boundary integrals, 261
boundary layer, 362
boundary layer, compressible, 370
boundary layer, turbulent, 364
boundary value problem,
two-point, 140
boundary value systems, 437
bounded state space, 481
Brillouin parallel processing, 543
Brillouin zone, 525, 552, 567
Brillouin zone integration, 527, 533
Brillouin zone sums
vectorization of, 528
parallel computation of, 528
Buckley-Leverett equation, 289
bymoment method, 249, 250

canonical partition function, 592
censored sample, 35
ceramic materials, 531
chain relaxation dynamics, 22
chaotic motion, 469

chips, 453
composite plates, 201
compressible boundary layer, 370
compressible flow, 345
condensed media, 600
conduction, heat, 424, 448
control system, 124
coulomb interaction, 548, 606
counting process, 11
covariance operator, 54
cubature, 282
cubic metals, 542
cubic spline, 484

delta wing, 324
density function, 44
destructive processes, 25
Devogelaere'a method, 300
dielectric response theory, 546
differential equations, 270
functional, 123
ordinary 298
partial, 437
second-order, 138
diffusion, 172
Dirichlet boundary conditions, 249
discrete vortex method, 224
dislocation, 376
distributed Gaussians, 602
drag force, 319
Duhamel's integral solution, 423
dynamic universe, 3
Dyson equation, 543

eigenstates, 603
elasticity, 507
electromagnetic, 388
electromagnetic wave, 411
electron wave function, 540
electronic device, 452
electronic energy, 601
electronic energy level, 600
electronic energy states, 603, 606
electronic structure, 546, 564, 605
embedded Green's function, 564
embedded metal layers, 568, 569
enhanced susceptibility, 536
ergodic hypothesis, 8
error bounds, 442
error estimate, 67, 75, 86
estimator, 44
excitable media, 172
exterior domains, 262

feedback loop, 463

feedback systems, 460
ferromagnetism, 525
Fermi energy, 567
Fermi surface, 525
finite analytic method, 270
finite difference, 602
finite element, 260, 602
finite-element, error estimator, 195
finite element method, 249
    axisymmetric triangular, 191
    hybrid, 249
    mixed, 240
finite memory map, 92
finite part integrals 132
fixed points, nonlinear operators, 158
flap shape, 331
flow
    full potential, 343
    incompressible, 354
    hypersonic, 319
    supersonic, 319
    transonic, 342, 347
flow analysis, 361
forward recurrence, 14
fracture mechanics, 135
free-space Green's function, 260
frequency domain method, 411
floating random walk, 452
Foedo-Galerkin method, 55
Fokker-Planck equation, 111
Fredholm integral equation, 147
Fubini's theorem, 78
functional differential equation, 123
fuselage configuration, 320

Galerkin formulations, 240
Galerkin method, 202
Galerkin and related methods, 228
Ginzburg-Landau equation, 376, 377, 381
global solution, 352,
global village, 493
grand challenge, 494, 495
Green's function, 114, 117, 118, 138, 449
    embedded, 564
    free space, 260
    two-particle, 549
Green's theorem, 250
group theory, 543
GW method, 543, 552

Hamilton's operator, 329
hardware, 495
harmonic conjugates, 187
harmonic potentials, 186
Hartree-Fock theory, 601
heat conduction, inverse, 423
Helmholtz equation, 249, 260
high performance computing, 495

high-temperature superconductors, 531
Hilbert space, 53
human resources, 495
Huygen's principle, 389
hybrid finite element method, 249
hygrothermal effect, 201
hypercube, 105
hypersingular integral, 132
hypersonic flow, 319
hypersurface, 322

incompressible flow, 354
inert gas fluids, 605
infinite elements, 261
influence functions, 425
integral equation, 139
integral representation, 63, 76, 79
integral solution–numerical, 427
integration–direct formal, 354
integro-differential equations, 282
INTERNET, 496
inverse heat conduction, 423
Ito-type system, 63, 75

joint distribution function, 11

kinetic energy integral, 368
Kutta condition, 222, 225

large-scale computing, 539
lattice maps, 95
LCGO method, 540
least squares estimator, 43
least squares formulation, 243
Lennard-Jones fluid, 591
Lennard-Jones potential, 606
light particles in fluids, 581
limit cycle solutions, 464
linear transfer function, 460
Liouville-type integrals, 11
local solution, 352

Mach number, 326, 346
magnetic metals, 541
magnetic order, 531
magnetic susceptibility, 532
magnetohydrodynamics, 360
Markov process, 594
massive fluid particles, 582
maximum principle, 480
mean-field theory, 581
Metropolis algorithm, 594
mixed finite element method, 240
molecular dynamics, 22, 600, 606
molecular dynamics simulation, 574
molecular modeling, 507
monotone operators, 229
Monte Carlo method, 22, 448, 581, 587

Monte Carlo simulation, 35, 591
motion perception, 207

natural convection, 376
Navier-Stokes, 377
Navier-Stokes equations, 320, 352
networking, 495
neural networks, 206
neuropharmacology, 206
nonlinear
    boundary value problems, 228
    equations, 112
    feedback systems, 460
    integral equations, 162
    maps, 92
    operators, 158
    systems, 92
    Volterra stochastic equation, 52
nonlinearly constrained optimization, 148
numerical analysis, 307
numerical integration, 298
numerical inversion, 484
numerical quadrature, 260
numerical solution, 147
Nystrom method, 299

oil reservoir simulation, 289
one-electron problem, 602
optimization, 477
    global, 319
    with constraint, 478
    semiconductor, 502
optimum-optimorum theory, 320
ordinary differential equations, 298
ordinary differential equation,
    second-order, 138
ordinary renewal process, 14
orthogonalization, 605
ortho-positronium, 591

parallel computation of
    Brillouin zone sums, 528
parallel computing, 158, 166, 501
parallel processing, 539
paramagnetic band structure, 532
paramagnetic susceptibility, 525, 526, 527
partial differential equations, 437
particle arch, 507, 510
particle modeling 507
path integral, 591
path-integration, 581
phase space, 24
plane wave, 399
plates--theory, 184
polyacetylene, 557
polymeric material, 22
positron annihilation, 581
positronium wave function, 593

power series solution, 272
pressure, 240
pressure coefficient, 346
probability, 44
probability density function, 12
product integration, 132
projection method, 160
propagators, 117
pseudopotential approach, 572
p-th moment relative stability, 75, 81
p-th moment stability, 63

quadrature, 37, 282
quantum Monte Carlo simulation, 503
quasi-linearization formulation, 462
quasi-particle theory, 546

radar cross section, 261
random variable, 44
random walk, 450
random-phase approximation, 553
randomly rough surfaces, 412
Rayleigh-Bénard convection, 376
Rayleigh number, 377
regularization, 148, 150
renewal density functions, 15
representation theory, 92
resonance, 137
ridge regression,486
Riemann-Lebesque condition, 441
Romberg procedure, 309
Runge-Kutta methods, 299

scattered field, 416
scatterers, spherical, 388
scattering coefficient, 414
scattering, electromagnetic, 250
scattering electromagnetic wave, 261
scattering signatures, 268
scattering, surface, 411
Schrodinger equation, 540, 547
secondary bifurcation, 379, 384
self-energy, 549
self-trapping, 583
semiconductor device design, 502
semiconductor optimization, 502
semiconductor surfaces, 572
semilocal solution, 352
separation of variables, 438
shape factor, 368
shift invariant, 94
shock integrals, 344, 347
singular integral equations, 132
Skorohod theorem, 58
snap-through, 511, 512
Sobolev spaces, 230, 242
software, 495
spectral domain method, 412

spherical scatterers, 388
spin-density wave, 524
spline, 484
squared-error loss, 37
standard moment method, 411
stiffness matrix, 191
stochastic vs deterministic, 75
Stokes problem, 240
Stoner model, 526
Stoner susceptibility, 524
stress, 26, 44
stress failure, 43
stress relaxation, 27
Strouhal number, 222
successive approximations, 160
supercomputer, 493, 507
superconductors, 531
superscalar processors, 505
supersonic flow, 319
surface atomic layer, 564
surface scattering, 411
symbolic algebra, 201

TCM, 452
thermal equilibrium, 587
three-dimensional arch, 510
time-invariant operator, 126
transition metals, 525
transonic flow, 342, 347
trends in science, 3
Tsypkin method, 465

Tsypkin solution, 470
two-particle Green's function, 549

unbounded region, 250
universe, 3
unsteady flow separation, 223

variation of collocation methods, 603
variation of constants formula, 76
variation of constants method, 66
variational calculus, 480
variational principle, 586
vectorization, 528
velocity, 246
velocity, disturbance, 323
vision, 206
Volterra equation, 353, 355
Volterra series, 93
Volterra stochastic equation, 52
Volterra stochastic equation, nonlinear, 52
vortex, 174
vortex downwash, 226
vortex shedding, 222
vortical flow, 223

wave equation, 261
wave, plane, 399
wave scattering, 388
weak solution, 52
Wiener processes, 76, 80
Wiener's optimum filter, 478
wing configuration, 320